T0181626

Lecture Notes on Data Engineering and Communications Technologies

Volume 48

Series Editor

Fatos Xhafa, Technical University of Catalonia, Barcelona, Spain

The aim of the book series is to present cutting edge engineering approaches to data technologies and communications. It will publish latest advances on the engineering task of building and deploying distributed, scalable and reliable data infrastructures and communication systems.

The series will have a prominent applied focus on data technologies and communications with aim to promote the bridging from fundamental research on data science and networking to data engineering and communications that lead to industry products, business knowledge and standardisation.

** Indexing: The books of this series are submitted to SCOPUS, ISI Proceedings, MetaPress, Springerlink and DBLP **

More information about this series at http://www.springer.com/series/15362

Tamara Radivilova · Dmytro Ageyev ·
Natalia Kryvinska
Editors

Data-Centric Business and Applications

ICT Systems-Theory, Radio-Electronics, Information Technologies and Cybersecurity (Volume 5)

 Springer

Editors
Tamara Radivilova
V.V. Popovskyy department
of Infocommunication Engineering, Faculty
of Infocommunication
Kharkiv National University of Radio
Electronics
Kharkiv, Ukraine

Dmytro Ageyev
V.V. Popovskyy department
of Infocommunication Engineering, Faculty
of Infocommunication
Kharkiv National University of Radio
Electronics
Kharkiv, Ukraine

Natalia Kryvinska
Department of e-Business, Faculty
of Business, Economics and Statistics
University of Vienna
Vienna, Austria

ISSN 2367-4512   ISSN 2367-4520   (electronic)
Lecture Notes on Data Engineering and Communications Technologies
ISBN 978-3-030-43069-6   ISBN 978-3-030-43070-2   (eBook)
https://doi.org/10.1007/978-3-030-43070-2

This Springer imprint is published by the registered company Springer Nature Switzerland AG
The registered company address is: Gewerbestrasse 11, 6330 Cham, Switzerland

Preface

With this volume, we analyze challenges and opportunities for infocommunication systems usage, taking in account theory, radio-electronics, information technologies and security aspects.

Explicitly, starting with the first chapter, "Automated Subjective Assessment of Speech Intelligibility Under Diotic and Dichotic Listening" deals with the issues of the articulation tests of noised and reverberated speech were carried out under different listening modes. Developed software toolkit was used for automation of subjective assessment of speech intelligibility. The results of the articulation tests showed that the diotic representation of the speech through headphones or computer speakers leads to almost identical results provided that the distance between the listener and the computer speakers is close to critical distance value. Articulation testing for dichotic speech presentation through headphones has shown that a room can be seen as a special kind filter which can increase speech intelligibility for low SNR values, lowering it at moderate and high SNR values.

In the work called "Different Approaches to Studying the Extreme Properties of Signal Functions Synthesized with Splines," the authors concern the scope of the study of the parameters of the total energy of selective signals, which are built using quadratic spline functions, using various evaluation criteria. As a result, the use of various criteria for the study of the parameters of the total energy of selective signals makes it possible at the design stage of signals of mobile communication networks to synthesize such forms that will allow them to physically fully realize them in practice, thereby improving the technical characteristics of the radio link facilities.

The chapter authored by Volodymyr Vasylyshyn "Adaptive Complex Singular Spectrum Analysis with Application to Modern Superresolution Methods" presents adaptive variant of the Singular Spectrum Analysis (SSA) approach for complex-valued signal model. The application of the adaptive complex SSA approach as preprocessing (denoising) step to modern superresolution methods (subspace-based methods) of spectral analysis is considered in the paper. The data sequence obtained after adaptive SSA approach is used as the input data for the subspace-based methods. Simulation results confirm the performance improvement of the subspace-based methods in the conditions of low signal-to-noise ratio (SNR) when using the proposed approach. Furthermore, the performance of adaptive SSA approach depends on the quality of the noise variance estimate.

The chapter "Assortativity Properties of Barabási–Albert Networks" focuses on the problem of determining the structure of Barabási–Albert networks having an extreme assortativity coefficient is considered. Greedy algorithms for generating Barabási–Albert networks with extreme assortativity have been proposed. As it was found, an extremely disassortative Barabási–Albert network is asymptotically bipartite, while an extremely assortative one is close to be a set of complete and disconnected clusters. The estimates of boundaries for assortativity coefficient have been found. These boundaries are narrowing with increasing the network size and are significantly narrower than for networks having an arbitrary structure.

The chapter authored by Olexandr Polishchuk "Influence and Betweenness in Flow Models of Complex Network Systems" provides the analysis for functional approaches of complex network systems research. In order to study the behavior of these systems, the flow adjacency matrices were introduced. The concepts of strength, power, domain and diameter of influence of complex network nodes are analyzed for the purpose of determining their importance in the system's structure. The notions of measure, power, domain and diameter of betweenness of network nodes and edges are introduced to identify their significance in the operation process of network systems. These indicators quantitatively express the contribution of the corresponding element for the motion of flows in the system and determine the losses that are expected in the case of blocking this node or edge or targeted attack on it. Similar notions of influence and betweenness are introduced to determine the functional importance of separate subsystems of network system and the system as a whole. Examples of practical use of the obtained results in information processing and management are given.

The next chapter "Improving the Data Processing Efficiency Based on Tunable Hard- and Software Components" discusses the transfer function coefficients' influence on the characteristics of standard digital filters used in the construction of soft and hardware components. Obtained are the ratios between numerator and denominator coefficients, the cutoff frequency and the peak frequency. Using results obtained, the structure of soft and hardware devices on microprocessor basis is simplified, that, in turn, greatly simplifies the task of developing frequency-dependent components allowing both correction and adjustment of the character-istics and the system as a whole.

The chapter authored by Jan Matuszewski "Practical Application of Clustering Methods in Radar Signals Recognition System" described the system of radar signals detection, analysis and recognition. The modern electronic recognition systems should react fast and with great accuracy in the extremely complex elec-tromagnetic environment. The binary decision tree was applied at the beginning of grouping signals received from unknown sources. The paper presents some clus-tering methods to radar signal recognition based on the mathematical criteria. The concepts of this technique are described. The experiment results obtained for nearest neighbor method are presented. Clustering algorithm was tested for different methods of objects grouping and for various distance measures.

The chapter "Optimization of the Quality of Information Support for Consumers of Cooperative Surveillance Systems" discusses the place and role of cooperative airspace surveillance systems in the information support of airspace use and air traffic control systems. A brief description of the signals used in the considering systems is given. Based on the presentation of cooperative surveillance systems as two-channel data transmission systems, the statistical interpretation of consumer data transmission is considered and it is shown that the probability of information support can be an integral quality indicator of consumer's information support in the considered systems.

The authors of chapter "Adaptive Semantic Analysis of Radar Data Using Fuzzy Transform" develop of a new adaptive system of radar data semantic analysis with their nonstationarity, which is based on both numerical and logical methods of multiscanning processing of signals and methods of artificial intelligence using fuzzy transformations of the universe of signals and signal images, is proposed. The possibility of its hardware and software implementation is considered. The results of computer modeling, theoretical and experimental researches with processing of real radar signals are presented. The elements of logical analysis and algebra of finite predicates (AFP) are selected as mathematical apparatus. As experimental studies show, AFP is an appropriate tool for logical-mathematical constructions, with which it is possible to describe the radar operator actions. The basic concepts of Boolean algebra and graph theory are also used. The practical value of the work is a method for formalizing the processes of perception and transformation of signals and signal images; algorithms and software are intended for information radar systems with natural-language intellectual interface and also for support the design of information structures. Mathematical and software results can be used in the systems of automatic processing of radar information, particularly, in the intelligent radar and radio-electronic systems and complexes for monitoring of mobile air and ground objects.

The chapter "Research of Dynamic Processes in the Deterministic Chaos Oscillator Based on the Colpitts Scheme and Optimization of Its Self-oscillatory System Parameters" presents the results of researching the modern methods and devices of deterministic chaos signal generation, which are constructed on the Colpitts scheme, for the infocommunication systems are given. It is proposed to use voltage-controlled transistor capacitance equivalents on the basis of bipolar transistor structures with negative impedance as capacitive elements of oscillating systems of the Colpitts scheme-based deterministic chaos oscillators. The elements of the theory of the Colpitts scheme-based deterministic chaos oscillators with single-transistor and multi-transistor active elements are presented. The chaotic dynamics of generated electric oscillations and their statistical and informational properties are investigated.

Next chapter "Approaches to Building a Chaotic Communication System" focuses on the use of deterministic chaos in communication systems. Indeed chaotic oscillations in such systems serve as a carrier of information signals, and means of encryption, both in hardware and in software. This work is devoted to the complex analysis of the practical implementation of chaotic secure communication system. The generalized synchronization as synchronization response was selected. The circuit implementation of a modified Colpitts oscillator as a source of chaos was proposed.

The chapter called "Radiomeasuring Optical-Frequency Converters Based on Reactive Properties of Transistor Structures with Negative Differential Resistance" presented the results of the investigation of the photoreactive effect in semiconductor diodes in the dynamic mode on the basis of the solution of the transport equation under the action of optical radiation, which gave the possibility to theoretically calculate the complete resistance of the base area and obtain an analytical dependence of its components on the power of optical radiation. The authors developed a model of an optical-frequency radiomeasuring converter of optical radiation, which describes the dependence of the complete resistance of the bipolar transistor structure that lies in the base of the converter of the optical radiation power. The proposed radiomeasuring optical-frequency converters have high sensitivity in the range of low optical radiation power values, which makes it possible to reliably measure even weak optical signals.

Next chapter "Implementation of Evolutionary Methods of Solving the Travelling Salesman Problem in a Robotic Warehouse" focused on an evolutionary method for solving the traveling salesman problem in the field of pharmacy business by optimizing the work of the drug delivery device is proposed in this paper. Modifications of three methods of initialization of the initial population of the genetic algorithm are developed. The software based on the proposed evolutionary method is created. It allows changing the initial parameters of the genetic algorithm, observes the process of solving the salesman problem graphically, obtains a result in text and graphic forms.

The chapter "Unified Models of Gradation Image Correction" pointed out that the efficiency of object recognition algorithms, computer vision systems and image analysis directly depends on the quality of image preprocessing. Gradation correction is one of the stages of this preprocessing. The chapter discusses the three most common models of gradation image correction, which are able to work on any brightness scale. Also in this paper, criteria for the quality of gradation correction are formulated. Experiments have been carried out that confirm the operability and computational efficiency of the considered models.

Next chapter "Study of Approaches to the Management of the Production of Entomophages" is devoted to the study of approaches to the management of the production of entomophages, which represents a complex dynamic system and characterized by stochasticity, multidimensionality of input and output parameters, ambiguous behavior of the biological object, the presence of weakly structured processes; the processes of cultivation of entomophages from the positions of infocommunication, process and system approaches are considered; processes of production management using intelligent information processing algorithms— cognitive analysis, fuzzy logic and neural networks. Tools for creating a fuzzy cognitive map were an expansion pack Fuzzy Logic Toolbox MATLAB and a system of fuzzy conclusion Mamdani, tools for managing production using neural networks—ANFIS-editor, Fuzzy Logic Toolbox MATLAB and a system of fuzzy conclusion Sugeno, a tool for creating a structural model—Simulink MATLAB.

In work entitled "Galois Field Augmentation Model for Training of Artificial Neural Network in Dentistry," the authors consider how to label and save a large number of images that should be predicted in a single file. The technique of automatic labeling the dataset with the finite element model for training of artificial neural network in tomography is proposed. A simple transparent example of thirty-two images to be predicted in a single HDF5 file training of artificial neural network in tomography shows accuracy of 100% for training set as well for the test set. Then this technique is able to build an information model of salivary immune and periodontal status and to evaluate the correlation between salivary immunoglobulin level, inflammation in periodontal tissues and orthodontic pathology. The results showed changes in the antioxidant balance in children with atopy that were expressed in an increase in malondialdehyde level, a decrease in superoxide dismutase activity and a level of reduced glutathione. These indicators can be considered as biological markers of the development of gingivitis at the preclinical stage in children against atopic diseases.

The chapter "Deep Convolutional Neural Network for Detection of Solar Panels" describes the method of detection of roof-installed solar photovoltaic panels in low-quality satellite photos. It is important to receive the geospatial data (such as country, zip code, street and home number) of installed solar panels because they are connected directly to the local power. It will be helpful to estimate a power capacity and energy production using the satellite photos. For this purpose, a Convolutional Neural Network was used. The dataset consists of low-quality Google satellite images was used for training and testing. The experimental results show a high rate accuracy of detection with low rate incorrect classifications of the proposed approach.

The authors of chapter "Designing Network Computing Systems for Intensive Processing of Information Flows of Data" shown the systematic research of technologies and concepts used for designing and building distributed fault-tolerant web systems. The general principles of design of distributed web applications and information technologies used in the design of web systems are considered. As a result of scientific research, it became clear that data backup is a defining attribute of web systems serving a large number of customers. Thus the main role in the construction of modern web applications virgae their scaling. Scaling-up in the distributed systems apply when the performance of this or that operation demands a considerable quantity of computing resources. There are two variants of scaling, namely vertical and horizontal. Vertical scaling consists of increasing the performance of existing components in order to increase overall performance. However, horizontal scaling is used to build distributed systems. Horizontal scaling consists in the fact that the system is divided into small components and placed on various physical computers. This approach allows for adding new nodes to increase the performance of the web system as a whole. However, this imposes certain limitations on the developers of software systems, namely, providing fault tolerance on each computer node as separately and as a whole in a distributed system.

The next chapter titled "Processing Signals in the Receiving Channel for the LoRa System" discusses the features of signal processing in receiving devices using LoRa technology. A feature of the technology is the ability to build a radio network for transmitting messages over long distances under the condition of long-term battery operation. The main attention is paid to the study of the influence of interference on the work of the receiving device. It also proposes a technique for studying a receiver in a densely populated area with an estimate of the calculated and experimental data. The studies were carried out according to the "point-to-point" and "point-gateway" schemes to check the quality of processing the received information by the remote server. The test results are given.

The chapter authored by Nataliia Ivanushchak, Nataliia Kunanets and Volodymyr Pasichnyk "Information Technologies for Analysis and Modeling of Computer Network's Development" focused on the analysis the properties of computer networks of different Internet providers, develop new, improve and adapt existing methods and tools of mathematical modeling, which enable the study of their structure and parameters based on fragmentary observation data, modeling and forecasting processes for their development and structuring within the framework of the formalism of complex networks. The developed method of modeling was used for analysis, evaluation and development of processes of stability of computer networks for directed hacker attacks and distribution of computer viruses in them.

The chapter "Adaptive Space-Time and Polarisation-Time Signal Processing in Mobile Communication Systems of Next Generations" is devoted to the possibilities of improving electromagnetic compatibility in mobile communication systems of the next generations based on the application of adaptive space-time signal processing in antennas of base stations. Features of solving the problem of interference suppression in adaptive antenna arrays are considered with using recurrent adaptation algorithms realized on the basis of the minimum mean square error criterion. The results of modeling adaptation of signal processing and its analysis are presented. A promising method of interference suppression using polarization-time processing of signals using optimal stochastic control is considered.

The next chapter "Fast ReRoute Tensor Model with Quality of Service Protection Under Multiple Parameters" proposes a flow-based model of Fast ReRoute in a multiservice network with quality of service protection under multiple parameters, such as bandwidth, probability of packet loss and average end-to-end delay. In the course of solving the task within the framework of this model, a result was obtained, the use of which contributes to the optimal use of the available network resource while ensuring a given level of quality of service and quality of resilience over both the primary and backup routes in the infocommunication network. As a result of the problem solution, the primary and backup multipath were obtained, along which a given level of quality of service was ensured in terms of bandwidth, the probability of packet loss and the average end-to-end delay.

The chapter entitled "The Development of Routing Flow Model in IEEE 802.11 Multi-radio Multi-channel Mesh Networks, Shown as a Konig Graph" describes an approach to the use of Konig graphs to model routing in IEEE 802.11 standard multi-radio multi-channel mesh networks, which enabled to describe more comprehensively and in more detail all possible configurations of the mesh network. Based on the representation planar, Konig's of multi-radio multi-channel mesh network in the routing problem is solved, which is to identify those collision domains through which traffic is to be transmitted from the sender to the recipient to meet end-to-end performance requirements. The proposed model uses the optimality criterion, aiming to minimize the multi-radio multi-channel mesh network performance, i.e., intensity of the total network traffic catered with its priorities.

The next chapter titled "Processing of the Residuals of Numbers in Real and Complex Numerical Domains" discussed the procedures for the formation and use of real residuals of real numbers on a real module, as well as complex and real residues of an integer complex number on a complex module. The chapter focuses on the processing of complex and real residuals of an integer complex number by a complex module. This procedure is based on using the results of the first fundamental Gauss theorem. On the basis of the considered procedure, an algorithm was developed for determining the real deduction of an integral complex number using a complex module in accordance with which the device was synthesized for its technical implementation.

The chapter authored by Alexandr Kuznetsov, Andriy Pushkar'ov, Roman Serhiienko, Oleksii Smirnov, Vitalina Babenko and Tetiana Kuznetsova "Representation of Cascade Codes in the Frequency Domain" focused on the mathematical apparatus of the multidimensional discrete Fourier transform over finite fields. Methods for the description of linear block codes in the frequency domain are investigated. It is shown that, in contrast to iterative codes (code-products), cascade codes in the general case cannot be described in the frequency domain in terms of multidimensional spectra. Analytic expressions are obtained that establish a one-to-one functional correspondence between the spectrum of a sequence over a finite field and the spectra of the corresponding words obtained by limiting this word to a subfield. A general solution of the problem of representation of cascade codes in the frequency domain is obtained, which allows constructing in the frequency domain using computationally efficient algorithms of encoding and decoding, and the derived analytic dependences of components of multidimensional spectra.

The next chapter "The New Cryptographic Method for Software and Hardware Protection of Communication Channels in Open Environments" is devoted to presenting theoretical bases and features of practical realization the new cryptographic method for software and hardware protection of communication channels in open environments. In the theoretical part of the chapter is shown the possibility of modeling the modes of deterministic chaos in the oscillations of discrete structures formed in the form of special matrix forms of the Latin square. A mathematical model is proposed for such discrete systems, on the basis of which an analysis of the combination of conditions for discrete structures and the requirements for transformations of their evolution operators ensuring the achievement of such modes is carried out. Analyzed the principles of creating devices to implement a closed communication channel in an open data transmission environment. To solve the protection problem, it is proposed to use the AES standard cryptographic algorithm, supplemented by the author's system of dynamic generation of encryption keys. Using the modified AES cryptographic algorithm, a working model of the system for protecting a remote control of a mobile object over the radio channel has been built.

The chapter entitled "Output Feedback Encryption Mode: Periodic Features of Output Blocks Sequence" presents model of random homogeneous substitution that used for an abstract description of this formation. This property is directly related to the periodic properties of output feedback encryption mode, since it characterizes the probabilistic distribution of output blocks with certain period appearance, provided that the assumption is made that the properties of the block symmetric cipher are consistent with certain properties of the random substitution. Also in the work, specific practical tasks are solved, namely recommendations are being developed for the application of the outbound feedback on the encryption threshold, certain requirements and limitations are justified.

The next chapter "Information-Measuring System of Polygon Based on Wireless Sensor Infocommunication Network" is devoted to the development of methods for automating the optical information-measuring system of a landfill for detecting and tracking guided and unguided aviation means of destruction, anti-aircraft guided missiles, artillery and rockets during their field testing. The basic idea is to use small-sized optoelectronic stations of trajectory measurements with high-speed non-inertial drives located along the route of an ammunition flight. All optoelectronic stations are combined into a single information and measuring system. Each optoelectronic station in its area of responsibility is programmed to support the target in the predicted trajectory. The programming process is automated and carried out at the same time for everyone optoelectronic station of trajectory measurements. To determine the location of each of the optoelectronic station of trajectory measurements at the polygon, methods for measuring the distance between the nodes using wireless sensor network technology are used.

The chapter "Development and Operation Analysis of Spectrum Monitoring Subsystem 2.4–2.5 GHz Range" presents a substantiation of the effectiveness of IEEE 802.11 wireless network analysis subsystem implementation using miniature spectrum analyzers. Also, it was given an overview of firmware work scheme, development process of trial versions, monitoring system development approaches, current development stage, infrastructure for research system, reliability and scan check, our system design and hardware implementation, future work, etc. Paper also provides technical solutions on automation, optimal algorithms searching, errors correcting, organizing software according to the Model-View-Controller scheme, harmonizing data exchange protocols, storing and presenting the obtained results.

The authors of "Computer-Integrated Technologies for Fitomonitoring in the Greenhouse" show that when growing vegetables in greenhouses, it is important to take into account not only the temperature of the atmosphere in the greenhouse but also the temperature of the plant itself. The dependence of temperature of plants on illumination in the greenhouse was analyzed; the refined mathematical model of the greenhouse as an object of control, suitable for the formation of control influences, taking into account the spatial dispersion of the object, was obtained. The purpose is to develop a subsystem of phytomonitoring in the greenhouse, which will complement the traditionally existing microclimate control system (in our case tomatoes) and clarify the mathematical model of the greenhouse as a spatially dispersed control object.

The authors of the chapter "Fusion the Coordinate Data of Airborne Objects in the Networks of Surveillance Radar Observation Systems" provide a classification of surveillance radar surveillance systems of airspace, which are among the main information sources of the airspace control system and air traffic control. A brief description of the information processing process in survey radar systems for observing airspace is given and it is shown that the complexity of the processing system does not allow formalization and analysis of its robots as a whole; therefore, it is necessary to preliminarily divide the system into elements and study their functioning separately. The tasks of information processing at the stage of signal processing are considered, as well as a brief description of the primary, secondary and tertiary data processing. It is shown that the fusion of information from the same air objects can be carried out at all stages of data processing. It is shown that the transition to the assessment of the four-dimensional location (4D) of an airborne object changes the procedures for merging individual measurements carried out by various radar observation systems with different rates of data output. This is due to the fact that from the output of the primary data processing by monitoring systems, an airborne object form is issued, which includes the time to estimate the coordinates of the airborne object with the necessary accuracy.

The chapter entitled "Diakoptical Method of Inter-area Routing with Load Balancing in a Telecommunication Network" present the diacoptical method of inter-area routing with load balancing in a telecommunication network. The method allows to increase the scalability of routing solutions in comparison with the centralized approach without reducing the efficiency of the network, estimated by the maximum value of link load threshold. The method involves the decomposition of the general routing problem in a multi-area network into several routing subtasks of smaller size that can be solved for each individual area followed by combining the solutions obtained for the whole telecommunication network. The foundation of the method is a flow-based routing model based on the implementation of the concept of Traffic Engineering and focused on minimizing the maximum value of link load threshold. The results of the analysis confirmed the operability of the method on a variety of numerical examples and demonstrated the full correspondence of the efficiency of the obtained diacoptical routing solutions to the centralized approach.

The author of the chapter "Analysis of Influence of Network Architecture Nonuniformity and Traffic Self-similarity Properties to Load Balancing and Average End-to-End Delay" presents the analysis of influence of network architecture nonuniformity and traffic self-similarity properties to load balancing and average end-to-end delay. Therefore, nonuniformity of network architecture was implied that its structure could be represented by a separable graph or the one close to it. This means that the telecommunication network contained routers and links, which were simulated by articulation points and bridges, respectively. And, by nonuniformity may be implied the fact that the network could have a minimum cut, the rate of which was much less than the bandwidth of other cuts of the network. And for the analysis of influence of network architecture nonuniformity and traffic self-similarity properties to load balancing and average end-to-end delay, the mathematical model of load balancing in the telecommunication network was used, within which not only the upper threshold of traffic load of the network links in general but also certain coefficients of link utilization are minimized for maximally satisfaction of the requirements of the concept of Traffic Engineering. This made it possible to organize the load balancing process in the network more effectively and provide the best value of such an important quality of service indicator as the average end-to-end packet delay in the network.

Kharkiv, Ukraine   Tamara Radivilova
                                                             tamara.radivilova@nure.ua
Kharkiv, Ukraine   Dmytro Ageyev
                                                               dmytro.aheiev@nure.ua
Vienna, Austria   Natalia Kryvinska
                                                        natalia.kryvinska@univie.ac.at

Contents

Automated Subjective Assessment of Speech Intelligibility Under Diotic and Dichotic Listening

Arkadiy Prodeus, Vitalii Didkovskyi, Maryna Didkovska, Igor Kotvytskyi, and Daria Motorniuk

Abstract The articulation tests of noised and reverberated speech were carried out under different listening modes: (1) diotic speech presentation through headphones, (2) diotic speech presentation through computer speakers, and (3) dichotic speech presentation through headphones. Developed software toolkit was used for automation of subjective assessment of speech intelligibility. The results of the articulation tests showed that the diotic representation of the speech through headphones or computer speakers leads to almost identical results provided that the distance between the listener and the computer speakers is close to critical distance value. This result is of practical value, since it means the admissibility of using computer loudspeakers during articulation tests. Articulation testing for dichotic speech presentation through headphones has showed that a room can be seen as filter which can reduce speech intelligibility and that direct sound has a greater effect on speech intelligibility compared to earlier reflections.

Keywords Speech intelligibility · Subjective assessment · Articulation test · Additive noise · Late reverberation · Early reflections · Listening mode

1 Introduction

The subjective assessment of the intelligibility of speech distorted by noise and reverberation is important for improving simulation of the human auditory system [1, 2], calibration of instrumental (objective) speech intelligibility assessment systems

A. Prodeus (✉) · V. Didkovskyi · I. Kotvytskyi · D. Motorniuk
Department of Acoustics and Acoustoelectronics, National Technical University of Ukraine "Igor Sikorsky Kyiv Polytechnic Institute", Kyiv, Ukraine
e-mail: aprodeus@gmail.com

M. Didkovska
Department of Mathematical Methods of System Analysis, National Technical University of Ukraine "Igor Sikorsky Kyiv Polytechnic Institute", Kyiv, Ukraine

© The Editor(s) (if applicable) and The Author(s), under exclusive license to Springer Nature Switzerland AG 2021
T. Radivilova et al. (eds.), *Data-Centric Business and Applications*, Lecture Notes on Data Engineering and Communications Technologies 48,
https://doi.org/10.1007/978-3-030-43070-2_1

[3], development and certification of communication devices and channels, lecture and meeting rooms, hearing aids and cochlear implants [4–6].

Comparison of subjective quality and intelligibility estimators of speech distorted by white, pink and brown noise indicates that the results are consistent for a small range of SNR values, namely for $0 < \text{SNR} < 5$ dB [7]. Therefore, when studying the effect of listening conditions on the speech intelligibility in a wide range of SNR values, it is impractical to use speech quality as a measure of speech intelligibility.

Results of studies on the degree of influence of noise and reverberation on speech intelligibility in classrooms are presented in [5, 8–10]. It has been shown that noise is much more dangerous than reverberation, due to the closeness of noise sources (talking students), and due to the similarity of the noise and speech spectra. Typical reverberation times for audiences are close to 0.8–1.1 s, thus reverberation has little effect on speech intelligibility due to the relatively low influence of the interference of this type.

It was shown in [11] that early reflections can enlarge the SNR up to 6–9 dB because the early reflection energy can be up to 9 dB greater than that of the direct sound. For specific listening situations, e.g. when the speaker's head is turned away from the listener, early reflections can be relatively more important than the direct sound. At the same time, the possibility of SNR increasing up to 6–9 dB was questioned and results of extended studies were presented in [12]. It was shown in [12] that increasing of early reflections energy leads to improvement of speech intelligibility, but this improvement is less than in case of increasing of direct sound energy. Thus, only part of the early reflections energy is useful for speech intelligibility. It means for rooms designers that enhancement of early reflections is reasonable, but increasing the speaker's voice level is more beneficial. Perhaps, this result can be explained by the fact that not only energy, but also the phase of early reflections plays an important role. This effect can also be explained by the uneven frequency response of the room, corresponding to early reflections of sound.

It was shown in [12] that binaural listening can increase SNR up to 2–3 dB. It should be borne in mind the danger of weakening the positive effect of early reflections with the combined action of binaural listening and reverberation [1].

It was noted also in [12] that the difference in results of [11, 12] may be largely due to the experimental setups difference. First, different stimuli were used to measure speech intelligibility. Second, the simulations of sound fields were not identical: only 7 early reflections were used in [11], while 20 early reflections were used in [12]. Third, late reverberation was considered in [11], while reflections arriving later than 55 ms after the direct sound were discarded in [12]. There were also different background noise models. Ambient noise with a spectrum shape corresponding to that of an NC 40 contour was used in [11] and a diffuse stationary speech-like noise was used in [12].

Thus, the main object of this study is to present the results of testing the developed software tools under various listening modes and for a wide range of SNR values. At the same time, an experimental assessment of speech intelligibility makes it possible to evaluate the properties of rooms as a special type of filter.

Another object of research is considering the degree of influence of experimental studies organization at results of subjective assessment of speech intelligibility. Consideration of this issue will allow us to better understand the significance of such an impact.

2 Experimental Studies of the Effects of Noise and Reverberation on Speech Intelligibility

Experimental studies, conducted as part of an educational research project, of the effects of noise and reverberation on speech intelligibility have been implemented during 2017–2018 at the Department of Acoustics and Acoustoelectronics of the National Technical University of Ukraine "Igor Sikorsky Kyiv Polytechnic Institute". A special software toolkit was developed for these studies. This toolkit allowed, firstly, automating the procedure of articulation testing, and secondly, allowed students to carry out such tests at home, i.e. in the most comfortable conditions, using personal computers.

During realization of the studies, it was revealed that the development and testing of appropriate software tools is inevitably accompanied by various inaccuracies and errors. One example of such inaccuracies is insufficiently good consideration of the peculiarities of the Ukrainian phonetics when creating sets of test words [13]. Another example is the imperfection of the developed software. In addition, the results of experiments can be significantly influenced by the shortcomings of the experimental studies organization, such as, for example, a lack of training of the experiment participants or, conversely, excessive informing of participants of experiment during its realization [14].

It is natural to expect that listening modes also influence the experimental results. Speech intelligibility estimators for diotic speech presentation (the same signal is directed into different loudspeakers) through couple of acoustic monitors and through headphones were given in [15]. Dichotic listening (the different stimuli are directed into different ears) through headphones may be considered as binaural mode with all its advantages compare to monaural and diotic modes. Thus subjective assessment of speech intelligibility for dichotic listening is of special interest and these results will be presented below too.

2.1 Automated System for Subjective Assessment of the Speech Intelligibility

A software toolkit of an automated system for subjective assessment of the Ukrainian speech intelligibility was presented in [13]. The system consists of two parts: first part task is preparing of text and sound sets ("tables") and second part task is management of the articulation testing process.

Each sound table contains records of 50 monosyllables of consonant-vowel-consonant (CVC) type. The text table contains reference monosyllables which are used for comparison with the monosyllables entered by the listeners into the computer using the keyboard.

Automated articulation testing process consists of next stages: (1) modeling of distorted speech, (2) presentation of distorted speech to subjects, (3) fixing of perceived monosyllables by means of keyboard, (4) evaluation of speech intelligibility.

Preparing of Text and Sound Tables Monosyllables tables of the standard GOST R 50840-95 (Transmission of Speech Through Communication Paths. Methods for Assessing Quality, Intelligibility and Recognizability) which provides the possibility of automation of subjective articulation testing were taken as a prototype for the development of text tables. These tables were adapted to the Ukrainian in accordance with peculiarities of Ukrainian pronouncing and spelling. An important peculiarity of the standard is the use of incomplete CVC monosyllables. In this case, the subjects can add the missing letter in the CVC monosyllable by means of the computer keyboard. The main advantage of such an approach is significant reduction in the training time. However, this approach raises doubts about the reliability of the results because of its essential difference from the traditional articulation test in accordance with GOST 16600-72 (Transmission of Speech through the Radiotelephone Communications. Requirements to Intelligibility of Speech and Methods of Articulatory Measurements).

Three versions of the reference text table were created, which avoided the possibility of a double interpretation of similar sounds, such as x /h/ and г /g/, c /s/ and з /z/, etc. Moreover, the ambiguity of the entering from the keyboard of perceived monosyllables was also taken into account in this way. For example, the м'яр /myiar/ can be entered in two ways, as мйар or as м"яр. The first variant allows someone to work around the problem of apostrophe typing in the Ukrainian keyboard layout. The second version looks a bit strange, but it complies with MATLAB syntax requirements.

The sound tables were developed by recording the speaker's voice in an anechoic room. The external PRESONUS AudioBox USB sound card, the Superlux ECM 999 microphone, and the Audacity audio editor and recorder version 2.1.3 were used for recording. The recording was performed at a rate of 44,100 Hz and 16 bits accuracy. Monosyllables were embedded in the sentence "Запишіть ___ тепер" and were spoken by 7 males and 2 females. For example, the бон /bon/ monosyllable was read as "Запишіть бон тепер" ("Write down bon now").

Signal Models In studies, the following model of distorted speech signal was used:

$$y(t) = x(t) \otimes h(t) + n(t) \tag{1}$$

where h(t) is room impulse response (RIR) of the room, $\otimes$ is convolution symbol, x(t) is clear speech, n(t) is noise.

A simpler model was used instead of (1) when the reverberation influence can be neglected:

$$y(t) = x(t) + k \cdot n(t) \tag{2}$$

where $k = 10^{0.05(SNR-SNR_0)}$, $SNR = 10 \lg(D_x/D_n)$ is the "primary" signal-to-noise ratio for the clean signal x(t) with "primary" variance D_x and the noise n(t) with "primary" variance D_n, SNR_0 is the desired SNR value for the mixture (2). Note that the zones of the speech signal existence need be determined before SNR calculation. Voice activity detector (VAD) is a tool commonly used to solve this problem, and a simple VAD was developed for software under test.

If the reverberation effect cannot be neglected, it should be noted that early reflections can significantly amplify the direct sound [11] and the SNR value can be calculated considering this circumstance. Indeed, the RIR h(t) can be presented as a sum of initial and tail sections $h_e(t)$ and $h_l(t)$, respectively:

$$h(t) = h_e(t) + h_l(t), \tag{3}$$

$$h_e(t) = \begin{cases} h(t), & t \in 0 \ldots 50\,\text{ms}; \\ 0, & t \notin 0 \ldots 50\,\text{ms}, \end{cases}$$

$$h_l(t) = \begin{cases} h(t), & t > 50\,\text{ms}; \\ 0, & t \le 50\,\text{ms}, \end{cases}$$

Under this consideration, the "primary" SNR can be represented as $SNR = 10 \lg D_{x_e}/D_n$, where D_{x_e} is a variance of the amplified signal $x_e(t) = x(t) \otimes h_e(t)$. The SNR_0 and k parameters can be then calculated in accordance with (2).

Computer Programs The block diagram of developed managing software routine is shown in Fig. 1. The routine begins with a request for the mode choice—training or working. The routine also requests the subject's last name and numbers of monosyllables tables. After preparing the folders where results of speech intelligibility assessment will be saved, special program named "assistant" runs and ensure the faultless continuation of work in case of termination of the working session. Note that such periodical termination of work is necessary to avoid excessive fatigue of the listeners.

After loading of a sound table, clean speech is distorted in accordance with (1) and is presented to a listener, which fixes perceived monosyllable by means of computer keyboard. When listening process of the sound table is over, calculation of speech intelligibility is made and results of calculation are written down in special file which is placed to preliminarily prepared folders on disk. A monosyllable is considered correctly recognized if its symbolic composition fully coincides with at least one of the three referenced textual variants of articulation tables.

Fig. 1 Block diagram of
managing software routine

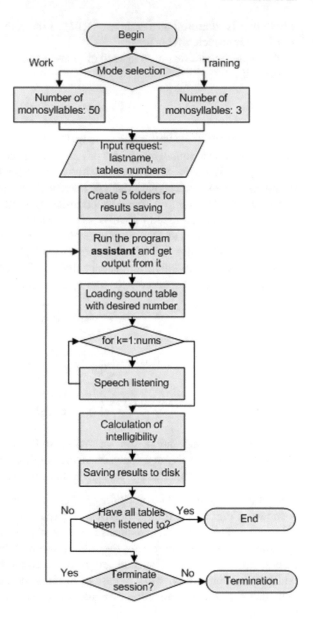

The formation of a distorted signal is provided using four programs: (1) modeling
of voice activity detector for subsequent correct calculation of the desired SNR, (2)
creation of octave bandpass comb, (3) using this comb to get, from white noise,
colored (pink and brown) noise with stepped spectra in the 90–11,000 Hz band, (4)
distortion of speech signals in accordance with (1)–(2).

Two more computer programs were used for the test results processing and
averaging over all speakers and listeners.

Thus, the developed software toolkit allows us to solve two important tasks: (1) the creation of text and sound tables, which are the basis of the articulation technique, (2) the implementation of articulation testing.

Listening Modes Articulation tests were conducted for three modes: (1) diotic speech presentation through headphones, (2) diotic speech presentation through computer speakers, (3) dichotic speech presentation through headphones.

The first case is the simplest; its use is "natural" due to the ease of a single-channel RIR measuring. The testing of the developed articulation tables and software was carried out in 4 stages: (1) with the use of undistorted speech signals, (2) with signals distorted by white and colored (pink or brown) noise at different (-10, 0 and $+10$ dB) SNR values, (3) with signals distorted by reverberation at different ($T_{20} = 0.3$; 0.6; 0.9; 1.1; 1.4; 2; 2.7 s) reverberation times, (4) with the use of signals distorted by the combined effect of pink noise (-10, 0 and $+10$ dB SNR values), and by reverberation ($T_{20} = 0.3$; 0.9; 1.4 s).

The second service mode was introduced after the results of the first listening mode were analyzed, due to the careless execution of instructions by some listeners who ignored the requirement to listen to the sound only through headphones. It should be noted that this incident turned out to be very important and instructive, since the obtained results made us pay attention to some flaws in the organization of the experiment, as well as to the special role of early reflections. Thus, articulation testing for 2nd listening mode was performed in the same 4 stages as for first listening mode.

The third listening mode case was more complicated. The six binaural RIRs from AIR database [16] belonged to lecture room of $10.8 \times 10.9 \times 3.15$ m were used for this task solving. There was parquet as floor cover, 3 glass windows, concrete walls, and wooden furniture (tables and chairs) in the room. The loudspeaker was placed at the teacher table and the microphones (artificial head) were placed at students chairs with distances d = 2.25, 4, 5.56, 7.1, 8.68, and 10.2 m between microphones and loudspeaker. The reverberation time depended little on the distance and was $RT_{60} = 0.7$–0.83 s. Early-to-late energy ratio C_{50} [8] and early reflection benefit ERB [11] values

$$C_{50} = 10 \lg \left\{ \int\limits_{0}^{0.05} h^2(t)dt / \int\limits_{0.05}^{\infty} h^2(t)dt \right\},$$

$$ERB = 10 \lg \left\{ \int\limits_{0}^{0.05} h^2(t)dt / \int\limits_{0}^{0.01} h^2(t)dt \right\},$$

are presented in Fig. 2 for all six RIRs. As can be seen from Fig. 2a, it is natural to assume that the harmful effect of late reverberation markedly reduces speech quality and intelligibility with an increase in the distance from 2.25 to 5.56 m, but the quality and intelligibility of speech remains approximately the same and

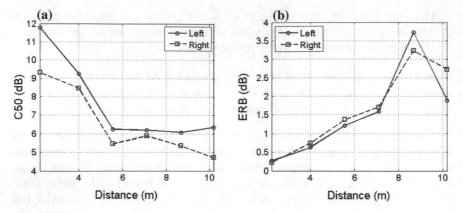

Fig. 2 Parameters C_{50} (**a**) and ERB (**b**) values

quite acceptable with a further increase in the distance (it is known that syllable intelligibility not lower than 80% is achieved under the condition of $C_{50} \geq -2$ дБ [17]). Thus, when considering only the reverberation effect on speech intelligibility, we can say that speech intelligibility is determined by the confrontation of direct sound and early reflections, on the one hand, and late reverberation, on the other hand. At the same time, early reflections should have a beneficial effect on speech intelligibility (Fig. 2b). The direct sound energy decreases with increasing distance, so the level of direct speech will be unacceptably low for large distances. This lack of direct sound energy is compensated by the addition of the energy of early reflections. Note, however, that ERB grows with increasing distance much slower than indicated in [12] and is only 3.5 dB for a distance of 8.68 m.

Another signal model was used in the third case:

$$y(t) = [x(t) + n(t)] \otimes h(t). \tag{4}$$

Model (4) is convenient when room is considered as a kind of filter. The "direct speech to noise ratio" [3]

$$SNR = 10 \lg D_x/D_n \tag{5}$$

was used for model (4).

2.2 Results of Experimental Studies

The results will be presented in the order in which they were received. This will help to better understand the causes of precisely such a development of events. The results obtained for the diotic presentation of the sound will be presented first. The results obtained for the binaural presentation will be as follows.

Diotic Speech Presentation through Headphones This articulation test was performed by 26 listeners aged 22 with normal hearing. Native language was Ukrainian for the subjects. Listening had been made through headphones in accordance with GOST R 50840-95 and GOST 16600-72 standards.

Averaged results of the monosyllables intelligibility assessment are shown in Fig. 3. The relative standard deviation was within the range of 4–15% for all estimates.

Figure 3a curves are in good agreement with the results of [18, 19] for moderate and large SNR values (0–10 dB), but the agreement isn't quite well for small SNR values (less minus 5 dB). It was shown in [19] that white noise is practically as effective as pink one at SNR < −5 dB. At the same time, it was stated in [18] that white noise is worse than pink and brown ones. Considering the results obtained, it was recommended in [13] to repeat the subjective testing for the case of SNR < − 5 dB after analyzing the possible causes of the above discrepancies.

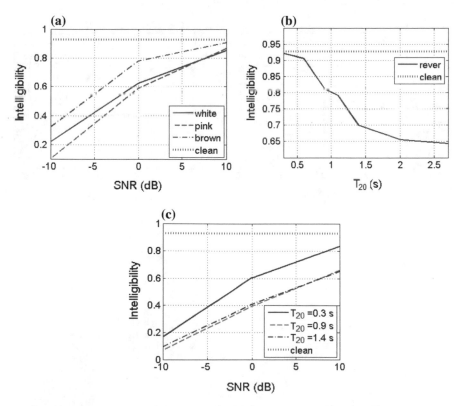

Fig. 3 Monosyllables intelligibility for signals distorted by noise (**a**), reverberation (**b**) and combined effect of noise and reverberation (**c**)

As for the curves shown in Fig. 3b, c, they are in satisfactory agreement with the well-known results of studies of the influence of noise and reverberation on speech intelligibility [5, 11].

Diotic Speech Presentation through Computer Speakers A strange phenomenon was revealed in [13] for case of noisy and reverberated signals listening. Unlike the 16 students mentioned above, three listeners have got results with a serious deviation from the averaged ones (Fig. 4). A survey of listeners showed that, firstly, they listened to signals through computer speakers, and secondly, they heavily used information about the current results of the assessment displayed on a computer monitor.

Partially obtained results can be explained by the early reflections influence. Indeed, early reflections can increase the SNR by 6–9 dB [11]. At the same time, the experimentally obtained huge increase in speech intelligibility cannot be explained only by the action of early reflections. The difficulty of increasing the SNR due to the effect of early reflections was shown in [12]. First, it was noted that when the early reflections were presented from the front, speech intelligibility was significantly

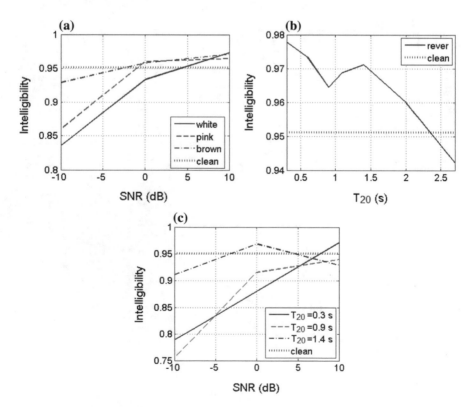

Fig. 4 Monosyllables intelligibility estimates for signals distorted by noise (**a**), reverberation (**b**) and combined effect of noise and reverberation (**c**) for case of listening through computer speakers

higher than when the early reflections were spatially distributed. Thus, only "direct" part of the energy of the early reflections was considered as useful for speech intelligibility. Second, the beneficial effect of the binaural listening mode, the advantage of which was about 2–3 dB, was noted.

As far as "direct" part of the energy of the early reflections, unfortunately, no explanations were given in [12] for how, from a physical point of view, early reflections can come from a direct sound direction in a typical situation when the speaker and the listeners look to each other. Perhaps, the mentioned effect in real rooms can be explained by a fundamentally different reason, namely, the difference in the phases and amplitudes of the early reflection signals perceived by the listener's left and right ears.

As can be seen, the mentioned in [13] huge deviation in speech intelligibility cannot be explained only by physical reasons. A thorough analysis of the most likely causes of the detected phenomenon revealed significant drawbacks of the developed software and experimental setup. Firstly, the random order of presentation of the elements of the sound tables was not implemented, which contributed to the memorization of these elements. Secondly, too detailed information about the recognition errors of perceived words was presented on the monitor screen, which also contributed to the memorization of the elements of the sound tables.

After correcting the software, the experiment was repeated and new listeners (20 students aged 20 with normal hearing) were involved in it. The same articulation tables were perceived twice by every listener: through the headphones and through the computer speakers. Results of the experiment for cases of speech distorted with noise or reverberation are shown in Figs. 5 and 6 [15].

As can be seen, intelligibility scores practically do not differ for headphones and speakers cases though light tendency of score increasing is seen for speakers. This small increase can be explained by the effect of early reflections which are weak (relatively to direct sound) due to the small distance between the listener and the sound source (about 0.6–1 m). At the same time, the speech intelligibility values of

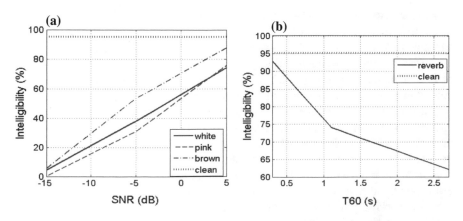

Fig. 5 Intelligibility score, headphones: noise (**a**) and reverberation (**b**) interferences

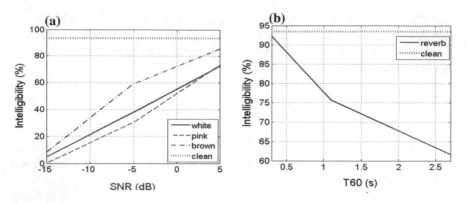

Fig. 6 Intelligibility score, PC speakers: noise (**a**) and reverberation (**b**) interferences

Figs. 5 and 6 are radically lower than those shown in Fig. 4 and are close to those shown in Fig. 3. Thus, this fact indicates a significant reliability increasing of the revised software package and the results obtained with it.

Obviously, the results obtained do not fully describe the situation when speech is presented through computer speakers or another similar sound system. Therefore, in the development of this topic, it is natural to ask the following question: how will change the intelligibility of noised speech in a room for distances which exceed the value of 0.6–1 m?

Dichotic Speech Presentation through Headphones To answer the question of room influence on noised speech intelligibility, new study was realized where binaural RIRs [16] were used as two-channel filter impulse responses for filtering of noised speech. Filtered speech was perceived by means of listening through headphones. The model (4) of the perceived signal was used in this case, and white noise was used to mask the signal.

Fifteen listeners with normal-hearing aged from 19 to 22 took part in the testing. The intelligibility score averaged over all listeners and speakers is shown in Fig. 7a. Relative standard deviation values of these estimates are shown in Fig. 7b.

As can be seen, speech intelligibility was approximately 10% higher for distances of 7–10 m than for distances of 2.25 m for SNR $= -10 \ldots -5$ dB. For SNR > 3–5 dB, speech intelligibility at distances of 4–10 m is lower than at 2.25 m distance.

An important feature of the above research was that the distance between the speaker and the subjects varied from small values to large ones. For a fixed distance, the SNR also varied from small to large. Thus, the prerequisites were created for the listeners to memorize the elements of the articulation tables as the experiment continued. To reduce the possibility of such memorization, it was necessary to modernize the conditions of experimental studies. The modernization was realised by means of varying distance value for a fixed SNR values. Proper experimental results are shown in Fig. 8. This new study involved nine listeners aged 19–22 years with normal hearing.

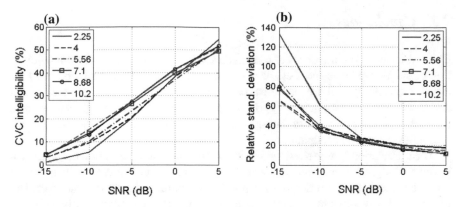

Fig. 7 Speech intelligibility estimates (**a**) and their relative standard deviations (**b**)

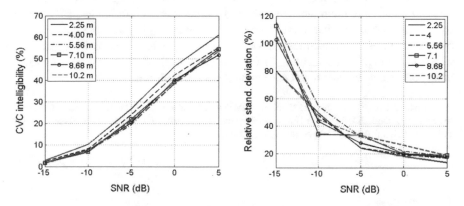

Fig. 8 Corrected speech intelligibility estimates (**a**) and relative standard deviations (**b**)

As can be seen, the results presented in Fig. 8a differ significantly from the ones of Fig. 7a. Now, distance 2.25 m provides the best speech intelligibility, distance 4 m is some worse and other distances are roughly the same and worst in terms of speech intelligibility. These results are in good concordance with results of [12] where direct sound was given a more important role compared to early reflections.

Presented in Figs. 7 and 8, the results clearly demonstrate the significant influence of the organization of experimental studies on the reliability of the results.

In further studies, it seems appropriate to evaluate the relative effects of early reflections and late reverberation on speech intelligibility. Results of [20–34] can be used in these studies.

2.3 Conclusions

CVC type monosyllables tables of Ukrainian speech and software toolkit for automation of subjective assessment of speech intelligibility were developed and used for articulation testing of noised and reverberated speech. The studies were carried out under different listening modes: (1) diotic speech presentation through headphones, (2) diotic speech presentation through computer speakers, and (3) dichotic speech presentation through headphones.

The articulation testing results showed that the diotic representation of the speech through headphones or computer speakers leads to almost identical results provided that the distance between the listener and the computer speakers is close to critical distance. This result is of practical value, since it means the admissibility of using computer loudspeakers during articulation tests.

Articulation testing for dichotic speech presentation through headphones had demonstrated, first, importance of right experimental organisation. Secondly, it was experimentally shown that direct sound plays a more important role compared to early reflections. Since the presence of early reflections leads to specific distortions of sound, it would be advisable in further studies to evaluate the relative influence of early reflections and late reverberation on speech intelligibility.

Acknowledgements The authors thank the students and teachers of the Acoustics and Acoustoelectronics Department of the National Technical University of Ukraine "Igor Sikorsky Kyiv Polytechnic Institute" who helped to implement the studies.

References

1. Blauert J (ed) (2013) The technology of binaural listening. Springer-Verlag, Berlin Heidelberg New York
2. Naida S, Pavlenko O (2018) Coupled circuits model in objective audiometry. In: Proceedings of the 2018 IEEE 38th international conference on electronics and nanotechnology (ELNANO), Kyiv, Ukraine, 24–26 Apr 2018
3. Jiang W, Schulzrinne H (2002) Speech recognition performance as an effective perceived quality predictor. In: IEEE international workshop on quality of service, Miami, Florida, May 2002
4. van Wijngaarden S, Verhave J, Steeneken H (2012) The speech transmission index after four decades of development. Acoust Aust 40(2):134–138
5. Yang W, Bradley J (2009) Effects of room acoustics on the intelligibility of speech in classrooms for young children. J Acoust Soc Am 125(2):922–933
6. Naida S, Pavlenko O (2018) Newborn hearing screening based on the formula for the middle ear norm parameter. In: Proceedings of the 2018 IEEE 38th international conference on electronics and nanotechnology (ELNANO), Kyiv, Ukraine, 24–26 Apr 2018
7. Prodeus A, Didkovskyi V, Didkovska M et al (2018) Objective and subjective assessment of the quality and intelligibility of noised speech. In: Proceedings of the 2018 IEEE 5th international scientific-practical conference "problems of infocommunications. Science and technology" (PIC S&T'2018), Kharkiv, Ukraine, 9–12 Oct 2018

8. Bradley J, Reich R, Norcross S (1999) On the combined effects of signal-to-noise ratio and room acoustics on speech intelligibility. J Acoust Soc Am 106(4):1820–1828
9. Sato H, Bradley J (2004) Evaluation of acoustical conditions for speech communication in working elementary school classrooms. J Acoust Soc Am 106(4):2064–2077
10. Bradley J, Sato H (2004) Speech intelligibility test results for grades 1, 3 and 6 children in real classrooms. In: Proceedings of the 18th international congress on acoustics (ICA), Kyoto, Japan
11. Bradley J, Sato H, Picard M (2003) On the importance of early reflections for speech in rooms. J Acoust Soc Am 113(6):3233–3244
12. Arweiler I, Buchholz J, Dau T (2009) Speech intelligibility enhancement by early reflections. In: Proceedings of 2nd international symposium on auditory and audiological research (ISAAR), Elsinore, Denmark, Aug 2009
13. Prodeus A, Bukhta K, Morozko P et al (2018) Automated system for subjective evaluation of the ukrainian speech intelligibility. In: Proceedings of IEEE 38th international conference on electronics and nanotechnology (ELNANO), Kyiv, Ukraine, 24–26 Apr 2018
14. Prodeus A, Vityk A, Dvornyk O et al (2018) Subjective evaluation of the speech intelligibility on the background of noise and reverberation. Microsyst Electron Acoust 23(2):66–73
15. Prodcus A, Bukhta K, Morozko P et al (2018) Automated subjective assessment of speech intelligibility in various listening modes. Microsyst Electron Acoust 23(3):49–57
16. Jeub M, Schäfer M, Vary P (2009) A binaural room impulse response database for the evaluation of dereverberation algorithms. In: Proceeding of international conference on digital signal processing (DSP), Santorini, Greece, 5–7 July 2009
17. Ahnert W, Schmidt W (2005) Fundamentals to perform acoustical measurements. Appendix to EASERA. http://renkusheinz-sound.ru/easera/EASERAAppendixUSPV.pdf. Accessed 30 Dec 2019
18. Prodeus A, Didkovskyi V, Didkovska M (2017) Akusticheskaya ekspertiza i korrektsiya kommugikatsionnykh kanalov (Acoustic examination and correction of communication channels: monograph). LAP LAMBERT Academic Publishing, Deutschland
19. Makarov Yu, Khorev A (2000) K otsenke effektivnosti zashchity akusticheskoy (rechevoy) informatsii (To the evaluation of the protection performance of acoustic (speech) information). http://aprodeus.narod.ru/Teaching/COBRAS/Art_2000_Makarov-Horev_K_ocenke_effect_zasch_acust.pdf. Accessed 30 Dec 2019
20. Kirichenko L, Radivilova T, Zinkevich I (2018) Comparative analysis of conversion series forecasting in e-commerce tasks. In: Shakhovska N, Stepashko V (eds) Advances in intelligent systems and computing. Springer, Cham, pp 230–242. https://doi.org/10.1007/978-3-319-70581-1_16
21. Kryvinska N (2010) Converged network service architecture: a platform for integrated services delivery and interworking. International Academic Publishers, Peter Lang Publishing Group
22. Kryvinska N (2008) An analytical approach for the modeling of real-time services over IP network. Math Comput Simul 79:980–990. https://doi.org/10.1016/j.matcom.2008.02.016
23. Kryvinska N (2004) Intelligent network analysis by closed queueing models. Telecommun Syst 27(1):85–98
24. Kryvinska N, Zinterhof P, Thanh D (2007) New-emerging service-support model for converged multi-service network and its practical validation. In: Proceedings of the IEEE first international conference on complex, intelligent and software intensive systems (CISIS-2007) in conjunction with ARES 2007, Vienna, Austria, 10–13 Apr 2007
25. Ageyev DV, Ignatenko AA, Fouad W (2013) Design of information and telecommunication systems with the usage of the multi-layer graph model. In: Proceedings of the XIIth international conference "the experience of designing and application of CAD systems in microelectronics (CADSM)", Lviv-Polyana, Ukraine, 19–23 Feb 2013
26. Ageyev DV, Ignatenko AA (2012) Describing and modeling of video-on-demand service with the usage of multi-layer graph. In: Proceedings of the XIth international conference on modern problems of radio engineering, telecommunications, and computer science (TCSET'2012), Lviv-Slavske, Ukraine, 21–24 Feb 2012

27. Abdalla H, Ageyev D (2012) Application of multi-layer graphs in the design of MPLS networks. In: Proceedings of international conference on modern problem of radio engineering, telecommunications and computer science (TCSET'2012), Lviv-Slavske, Ukraine, 21–24 Feb 2012

28. Dobrynin I, Radivilova T, Ageyev D et al (2018) Use of approaches to the methodology of factor analysis of information risks for the quantitative assessment of information risks based on the formation of cause-and-effect links. In: Proceedings of the 2018 international scientific-practical conference problems of infocommunications. Science and technology (PIC S&T), Kharkiv, Ukraine, 9–12 Oct 2018. https://doi.org/10.1109/infocommst.2018.8632022

29. Radivilova T, Kirichenko L, Ageyev D et al (2018) Decrypting SSL/TLS traffic for hidden threats detection. In: Proceedings of the 2018 IEEE 9th international conference on dependable systems, services and technologies (DESSERT), Kiev, Ukraine, 24–27 May 2018. https://doi.org/10.1109/dessert.2018.8409116

30. Ageyev D, Bondarenko O, Radivilova T et al (2018) Classification of existing virtualization methods used in telecommunication networks. In: Proceedings of the 2018 IEEE 9th international conference on dependable systems, services and technologies (DESSERT), Kiev, Ukraine, 24–27 May 2018. https://doi.org/10.1109/dessert.2018.8409104

31. Karpukhin A, Tevjashev A, Ageyev D et al (2017) Features of the use of software packages for modeling infocommunication systems. In: Proceedings of the 2017 4th international scientific-practical conference problems of infocommunications. Science and technology (PIC S&T), Kharkiv, Ukraine, 10–13 Oct 2017. https://doi.org/10.1109/infocommst.2017.8246421

32. Radivilova T, Kirichenko L, Ageyev D et al (2020) The methods to improve quality of service by accounting secure parameters. In: Hu Z, Petoukhov S, Dychka I, He M (eds) Advances in computer science for engineering and education II. ICCSEEA 2019. Advances in intelligent systems and computing, vol 938. Springer, Cham

33. Bulakh V, Kirichenko L, Radivilova T (2018) Time series classification based on fractal properties. In: Proceedings of the 2018 IEEE second international conference on data stream mining & processing (DSMP), Lviv, Ukraine, 21–25 Aug 2018. https://doi.org/10.1109/dsmp.2018.8478532

34. Soulodre G, Popplewell N, Bradley J (1989) Combined effects of early reflections and background noise on speech intelligibility. J Sound Vib 135(1):123–133

Different Approaches to Studying the Extreme Properties of Signal Functions Synthesized with Splines

Irina Strelkovskaya, Irina Solovskaya, and Anastasiya Makoganiuk

Abstract The study of parameters of the full energy of signal functions, which are built on the basis of square spline functions, with the use of various evaluation criteria, was conducted. The use of the coefficient of energy concentration (energy criterion) of signals based on quadratic splines is proposed, an analytical equation is obtained for a selective signal whose energy is concentrated near the main lobe, allowing you to choose the optimal values of the width parameter of the transition region and the shape of the spectrum signal implementations. The use of the D-criterion for signals with a limited duration and intersymbol interference is considered. It is shown that the minimization of the level of interference is possible due to the variation of the parameter of the signal spectrum in the transition region, allowing the synthesis of signals with the necessary properties during practical implementation. The use of the criterion of the size of the eye-diagrams allowed us to synthesize signals that are most resistant to shifts of the moments of sampling—jitter. The conditions under which the maximum value of the opening is reached are determined. It was established that in the research of the full energy selective signal evaluation criterion appropriate to use with differential calculus function of several variables, which allows the network in step predetermined design technology to provide the required energy signal indicators. As a result, the use of various criteria for the research of the parameters signals makes it possible at the design stage of signals of mobile communication networks to synthesize such forms that will allow them to physically fully realize them in practice, thereby improving the characteristics of the radio link equipment.

I. Strelkovskaya · I. Solovskaya (✉) · A. Makoganiuk
Educational and Research Institute of Infocommunications and Software Engineering, O.S. Popov Odessa National Academy of Telecommunications, Odessa, Ukraine
e-mail: i.solovskaya@onat.edu.ua

I. Strelkovskaya
e-mail: strelkovskaya@onat.edu.ua

A. Makoganiuk
e-mail: a.makoganyuk@onat.edu.ua

© The Editor(s) (if applicable) and The Author(s), under exclusive license to Springer Nature Switzerland AG 2021
T. Radivilova et al. (eds.), *Data-Centric Business and Applications*, Lecture Notes on Data Engineering and Communications Technologies 48,
https://doi.org/10.1007/978-3-030-43070-2_2

17

Keywords Selective signal · Total signal energy · Spline · Criterion · Indicators

1 Introduction

The rapid development of mobile technologies in the direction of 4G/5G generation networks is connected, first of all, with development in radio access technologies and an increase in the spectrum bandwidth. The use of third-generation 3G and fourth-generation 4G technologies, such as UMTS and LTE-Advanced. The active development of high-speed Wi-Fi (IEEE 802.11n/ac/ad) implies the use of MIMO-antennas and OFDM/QAM-modulation for which they have used signal functions with minimal intersymbol and inter-channel interference values. As you know, the appearance of intersymbol interference is the result of distortions and fading in the radio interface, therefore, an increase in interference often leads to a decrease in noise immunity and loss. Reducing inter-symbol interference by implementing special designs of signal functions allows you to get the right level of quality characteristics, such as the probability of errors and loss of characters. Besides, the implementation of modern mobile communication technologies is connected with problems of effective use of a frequency resource and a decrease in the level of mutual hindrances. After all, it is from a successful choice of the type of signal structures depends on the capacity of the radio channel, its noise resistance and other characteristics [1, 2].

Saving a radiofrequency resource can be achieved by using spectrally efficient modulation techniques or by using signals with a high concentration of energy in the limited frequency bandwidth. Such signals, according to [1–8], are selective signals with a finite spectrum that satisfy the first Nyquist criterion. Selective signals are quite applicable today, as they allow transmitting information without inter-symbol interference [1–8]. At the same time, studies on the possible improvement of their frequency-time structure are far from completed [1–8].

In studying the parameters of selective signals, various criteria can be used to consider them from different positions: the coefficient of dependence of the energy concentration (energy criterion) of selective signals, the D-criterion, the magnitude criterion of the eye-chart and the criterion based on the based on differential computation of functions of many variables [1–4, 9–13].

Studies of signal functions using various criteria are devoted to the work of the authors [1–4, 7, 8], in which various interpolations of spectral characteristics of signals in the frequency-time domain are considered, which best bring them closer to the best designs. According to the work [9], to select the optimal value of the width of the transition region and the shape of the signal spectrum in it, the authors proposed using the criterion of the coefficient of dependence of energy concentration (energy criterion), which allowed us to obtain the signal energy concentrated near the main lobe. For signals with a limited duration and intersymbol interference, in the work [8, 9] was used a D-criterion that allowed minimizing the level of interference due to the variation of the signal spectrum parameter in the transition region. At the stage of synthesis of selective signals, special attention in the study in the work

[9] was paid to the criterion of the magnitude of the opening of the eye-diagrams, which allows you to select the signals that are most resistant to shifts in the sampling moments—jitter.

The authors reviewed the criterion using differential computation functions of many variables in the study for the total energy extremum of the signal, which allows providing the required energy signal indicators. In such works [10–18] the authors used spline-functions that allow increasing the accuracy of interpolation while simplifying the procedure of mathematical calculations. In [19–26] it is proposed to use spline-functions to perform a polynomial approximation of signals, which is confirmed by the research results. Similar issues have been considered by the authors in [27–48].

Therefore, the search for new types of interpolation of spectral characteristics of selective signals and the study of their extreme properties of signals in the conditions of active technology development is an important task. In papers [3, 4], the authors developed a method of interpolating the spectral characteristics of the signals under consideration using cubic splines. However, it is interesting enough to study the full energy of the functions of the signal under consideration using square splines, because it will significantly reduce the number of calculations in determining the analytical type of the synthesized signal and its full energy [1–4, 9–13].

It is proposed to investigate the extremal properties of the total energy of signals constructed on the basis of quadratic splines using the methods of differential computation.

The purpose of this work is to use different approaches to the study of the total energy of selective signals, built on the basis of quadratic spline functions. The resulting relationships will allow you to calculate the energy characteristics of the signals and determine the waveform depending on the selected criterion.

2 Methods of Study of the Extreme Properties of Signal Functions Synthesized with Splines

Let's consider a signal with a finite spectrum, which meets the first Nyquist criterion [1–8], for which equality is used in the time domain:

$$g(kT) = \begin{cases} U & \text{for } k = 0, \\ 0 & \text{for } k = \pm 1, \pm 2, \ldots, \end{cases} \tag{1}$$

where T is the length of a time interval, U is the instantaneous meaning of the pulse count meaning at the time of $k = 0$.

Interpolation selective spectral density of the signal g(t), obtained in [8] is shown in Fig. 1.

As a result, the quadratic spline interpolation [1–9], received signals, the spectral density which has the form

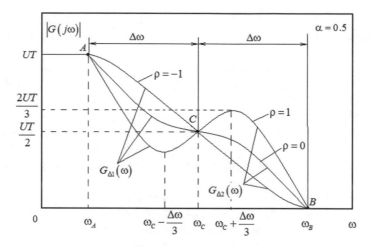

Fig. 1 Interpolation of a selective signal spectrum using a quadratic spline

$$G_{\Delta2}(\omega) = \frac{UT}{2} + \frac{\rho UT}{\Delta\omega}(\omega - \omega_C) - \frac{UT(1+2\rho)}{2(\Delta\omega)^2}(\omega - \omega_C)^2. \qquad (2)$$

The analytical expression of the selective signal g(t) obtained in [1–9] is as follows

$$g(t) = 2U\frac{\sin \omega_{c'}}{\omega_{c'}}\left[\frac{(1+\rho)\sin(\Delta\omega\, t)}{\Delta\omega\, t} - \frac{(1+2\rho)(1-\cos(\Delta\omega\, t))}{(\Delta\omega)^2 t^2}\right], \qquad (3)$$

where $U = g(0)$, $\omega A = (1 - \alpha)\omega C$; $\omega B = (1 + \alpha)\omega C$; $\omega C = \pi/T$; $\alpha = (\omega C - \omega A)/\omega C$ is a spectral rounding factor that determines the width of the transition region $[\omega_A, \omega_B]$ $(0 \le \alpha \le 1)$; $\Delta\omega = \alpha\omega_C$.

Figure 2 shows the dependence of the shape of the selective signal in the intervals between equidistant zeros on the parameter ρ.

Therefore, the signals corresponding to the first Nyquist criterion accept the value U at t = 0, and the equidistant points following the interval T have zero meanings. These signals are called selective or countdown signals [1–4, 9, 10]. Their main characteristic is the absence of inter-symbolic interference.

We will investigate the parameters of the full energy of signals construction based on quadratic spline functions using various criteria, namely: the dependence of the concentration of energy (energy criterion) of the selective signal, *D*-criterion, the criterion of the size of the eye-diagram and differential calculus of functions of many variables (method of solving the problem of extremum).

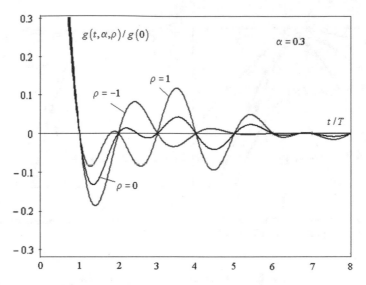

Fig. 2 Dependence of the shape of the selective signal on the parameter ρ

2.1 Dependence of Concentration of Energy of Signal Functions Synthesized by Using of Quadratic Splines

We use the energy concentration dependence coefficient to study the full energy of selective signals. Let us consider the level of concentration of energy of the signal on a given time interval depending on the parameter ρ. As a natural interval let's choose the interval $[-nT, nT]$ between zeros which are symmetrically located concerning the beginning of the coordinate system. It is important to note that when changing the parameter ρ, the position of equidistant zeros of the signal under consideration (3) does not change and inter-symbolic interference does not occur.

The full energy of the selective signal g(t) in accordance with the generalized Reyle formula (Parseval equality) is determined by the relation

$$\int\limits_{-\infty}^{\infty} g^2(t)dt = \frac{1}{2\pi} \int\limits_{-\omega_B}^{\omega_B} |G(j\omega)|^2 d\omega. \tag{4}$$

Partial energy on a segment $[-nT, nT]$ is equal

$$E_t(n, \alpha, \rho) = \int\limits_{-nT}^{nT} g^2(t)dt.$$

The full energy of selective signals g(t) using quadratic splines has the form [9, 10]

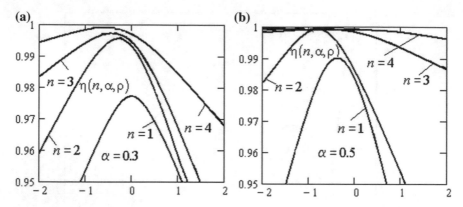

Fig. 3 Dependence of the signal energy concentration $g(t)$ on the coefficient ρ of a quadratic spline $G_{\Delta 2}(\omega_C)$ on a segment $[-nT, nT]$, $n = 1, 2 \ldots$

$$E_\omega(\alpha, \rho) = U^2 T\left[1 + \frac{\alpha}{30}(2\rho^2 - 3\rho - 12)\right],\tag{5}$$

where ρ is the ratio $G_{\Delta 2}(\omega_C) = \rho\frac{UT}{\Delta\omega}$.

For further research, we use the energy coefficient, which shows how much of the total signal energy g(t) lies on the segment, $[-nT, nT]$, $n = 1, 2 \ldots$ and allows us to investigate the dependence of the signal energy concentration g(t) on the coefficient ρ of a cubic spline $G_{\Delta 2}(\omega_C)$.

$$\eta(n, \alpha, \rho) = E_t(n, \alpha, \rho)/E_\omega(\alpha, \rho),\tag{6}$$

The results of the research are presented in Fig. 3.

The results of the study, which are shown in Fig. 3 show clearly defined extreme whose values increase with n. Optimal values of the corner coefficient are shifted as n changes and are enclosed within $-1 < \rho_{opt} < 0.5$. Thus, by setting ρ_{opt} in (3), we obtain the signal function g(t), which leads to the maximum of the function (6).

The described method—using the criterion of the energy coefficient—allows one to obtain an analytical equation for a signal function whose energy is concentrated near the main lobe. The use of such a signal function allows process simplification its practical implementation at the design stage of mobile systems.

To minimize the intersymbol interference of selective signals, we use the D-criterion. Consider the procedure for using the D-criterion in the synthesis of selective signals with a minimum level of side lobes. For this purpose, according to [2], we write the equation, which is called the D-criterion

$$D(N, \alpha, \rho) = \frac{1}{g(0)} \sum_{k=1}^{N} \left|g\left[\frac{(2k+1)T}{2}\right]\right|.\tag{7}$$

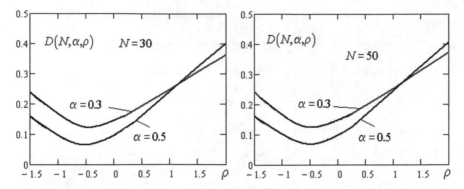

Fig. 4 Minimization of side-lobe levels using *D*-criteria

The decision of the problem is to estimate the value of ρ_{opt} that delivers the minimum of the objective function (7). Given the clear symmetry g(t), one can restrict one-sided calculation of the values of side lobes. Besides, the g(t) signal when $t \to \infty$ damping out quickly, therefore, increasing the number of components in (7) above N = 10 doesn't give a significant increase in accuracy [9, 10]. Figure 4 shows the extreme properties of selective signals using interpolation by quadratic splines relative to criterion (7).

The results of the study allow us to conclude that the use of the D-criterion in the interpolation of the parameters of selective signals allows synthesizing new signal functions, whose extreme properties will improve the characteristics of mobile systems.

2.2 Using the Eye-Diagram for the Study of Extreme Properties of Signal Functions Constructed Using Quadratic Splines

One of the criteria for the extreme properties of selective signals is the magnitude of the horizontal disclosure of the eye-diagram, which is often used to assess the to assess quality and quantity characteristics of the noise immunity of digital transmission systems. The eye-chart is a superposition of the signal segments, reduced to a common interval, and reveals the level ISI in the system, and also allows us to estimate the permissible value of random deviations of the registration moments from the optimal position. The reason for the ISI are linear distortion of the amplitude-response (AR) and the phase-response (PR) channel communication range restriction selective jamming [9, 10].

The results of constructing eye-diagram for different values of α and ρ are shown in Figs. 5, 6 and 7.

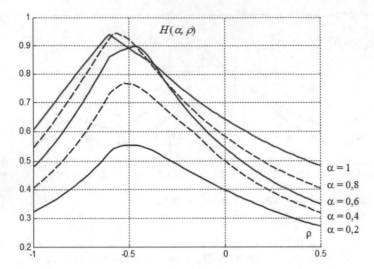

Fig. 5 The dependence of the horizontal disclosure of the eye-diagram on the parameter ρ when the parameter α is changed for a selective signal

Earlier in the works [9–13], the authors analyzed the difference in the openings of eye-charts for selective signals constructed using quadratic splines and signals based on cubic splines, which was 1%. The conditions under which the maximum value of the aperture of the chart is reached are determined. The largest opening is obtained when $\alpha = 1$, $\rho = -0.6$, with H ≈ 0.9582, which corresponds to the standard values according to the recommendation of ITU-T G.957. Where does the conclusion come from: this slight difference can be ignored, thus using a less calculating aspect, and a more basic practical implementation of the synthesized selective signals built on the quadratic splines.

2.3 Study of the Full Energy of Signal Functions Based on Quadratic Splines Using Differential Computation

Consider the evaluation criterion using the differential computation of functions of many variables, which allows providing the necessary energy indicators of the signal at the design stage of the network of a given technology.

To research the extremum of the full energy of a signal functions from parameters α and ρ, we determine the range of allowed values of parameters α and ρ, the parameter α varies from 0 to 1, and the parameter ρ from $\rho_{min} = -1$ to $\rho_{max} = 1 + \sqrt{2}$. Area of definition of D parameters α and ρ is shown in Fig. 8.

We study the dependence of the full energy $E_\omega (\alpha, \rho)$ of the signal functions g(t) on its parameters in the area of $D = \left\{ (\alpha, \rho) : \ 0 \leq \alpha \leq 1, -1 \leq \rho \leq 1 + \sqrt{2} \right\}$. Find

Fig. 6 Eye-diagram of the selective signal $g(t)$

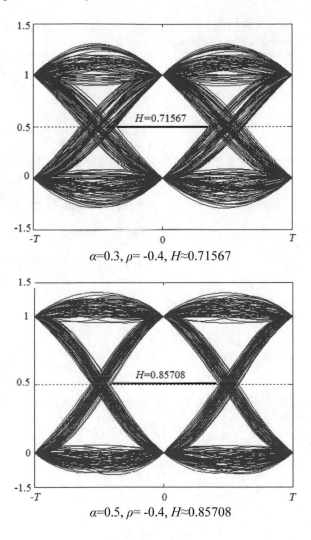

$\alpha=0.3, \rho=-0.4, H \approx 0.71567$

$\alpha=0.5, \rho=-0.4, H \approx 0.85708$

the minimum and maximum values of the function that are achieved in this area [48–50]. Consider the function.

$$f(\alpha, \rho) = E_\omega(\alpha, \rho)/U^2 T. \tag{8}$$

Using Eqs. (5) and (8), we get

$$f(\alpha, \rho) = 1 + \frac{\alpha}{30}(2\rho^2 - 3\rho - 12). \tag{9}$$

To find stationary points of extremum of a function having the form (9), we find its private derivatives.

Fig. 7 The maximum
possible horizontal opening
of the eye-diagram for a
signal

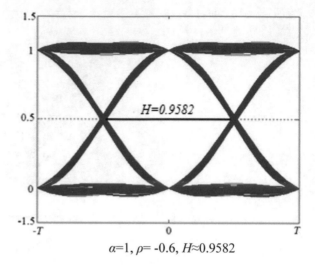

$\alpha=1, \rho=-0.6, H\approx0.9582$

Fig. 8 The area of function
$Df(\alpha, \rho)$

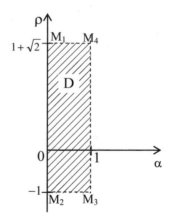

$$f'_\alpha(\alpha, \rho) = \frac{1}{30}\left(2\rho^2 - 3\rho - 12\right), \quad f'_\rho(\alpha, \rho) = \frac{\alpha}{30}(4\rho - 3),$$

that exist throughout the area D.

We will find stationary points using the conditions of the extremum of the function
of two variables [48–50] by solving the system of equations

$$\begin{cases} \frac{1}{30}(2\rho^2 - 3\rho - 12) = 0, \\ \frac{\alpha}{30}(4\rho - 3) = 0, \end{cases} \Leftrightarrow \begin{cases} 2\rho^2 - 3\rho - 12 = 0, \\ \frac{\alpha}{30}(4\rho - 3) = 0, \end{cases}$$

whence from the first equation $\rho_1 = \frac{3+\sqrt{105}}{4} \approx 3.312$, $\rho_2 = \frac{3-\sqrt{105}}{4} \approx -1.812$, from
the second equation $\alpha = 0$ and $\rho = \frac{3}{4}$. From the above, in area D no extremum
points.

We find the minimum and maximum meaning of the function $f(\alpha, \rho)$ at the boundary region D. On the segment: $M_1 M_2$ (Fig. 8) we have $\alpha = 0$, $\rho \in [\rho_{\min}; \rho_{\max}]$. Then function (9) has the form

$$f(\alpha, \rho) = f(\rho) = 1. \tag{10}$$

On the segment $M_2 M_3$, we obtain $\rho_{\min} = -1$, $\alpha \in [0, 1]$. In this case, function $f(\alpha, \rho)$ (9) takes the form

$$f(\alpha, \rho) = 1 - \frac{7\alpha}{30}. \tag{11}$$

Having defined $f'_\alpha = -\frac{7}{30}$, we have $f'_\alpha < 0$ for $\alpha \in [0, 1]$. Taking into account the fact that $f(\alpha)$ is a monotonically decreasing function, it takes the minimum value on the interval $[0, 1]$ at $\alpha = 1$ and is equal $f'(1) = \frac{23}{30} \approx 0.77$, and maximum value at $\alpha = 0$ and is equal $f(0) = 1$. In this way,

$$\underset{\alpha \in [0,1]}{f_{\max}(\alpha)} = f(0) = 1, \quad \underset{\alpha \in [0,1]}{f_{\min}(\alpha)} = f(1) = \frac{23}{30}. \tag{12}$$

On the segment $M_3 M_4$ we have $\alpha = 1$, $\rho \in [\rho_{\min}; \rho_{\max}]$. The function (9) has have the form

$$f(\alpha, \rho) = f(\rho) = 1 + \frac{1}{30}(2\rho^2 - 3\rho - 12). \tag{13}$$

Having determined $f'_\rho = \frac{1}{30}(4\rho - 3)$, we get $f'_\rho = 0$ at $\rho = \frac{3}{4}$. As $f'_\rho = \frac{4}{30} > 0$, then $\rho = \frac{3}{4}$ the minimum point for function (13). Accordingly, the minimum value of the function of the form (9) on the segment $M_3 M_4$ has have at the point $M_0\left(1; \frac{3}{4}\right)$ and is equal to

$$f\left(1; \frac{3}{4}\right) = \frac{9}{16} \approx 0.5625 \tag{14}$$

On the segment $M_1 M_4$ we have $\alpha \in [0, 1]$ and $\rho = \rho_{\max} = 1 + \sqrt{2}$. Function (9) is represented as

$$f(\alpha, \rho) = f(\alpha, \rho_{\max}) = f(\alpha) = 1 + \left(-9 + \sqrt{2}\right)\frac{\alpha}{30}. \tag{15}$$

Calculating the derivative of a function $f(\alpha)$ of the form (15), we obtain

$$f'_\alpha = -\frac{-9 + \sqrt{2}}{30} < 0.$$

Consequently, when $\alpha \in [0, 1]$ the function $f(\alpha)$ monotonously decreases and reaches the maximum value when $\alpha = 0, f_{\max}(\alpha) = 1$, and the minimum value is reached when $\alpha = 1, f_{\min}(\alpha) \approx 0.747$. Thus,

$$f_{\max}(\alpha)_{\alpha\in[0,1]} = f(0) = 1, \quad f_{\min}(\alpha)_{\alpha\in[0,1]} = f(1) = \frac{21 + \sqrt{2}}{30} \approx 0.747 \qquad (16)$$

Comparing the obtained values obtained on different segments M_1M_2 (10), M_2M_3 (12), M_3M_4 (14), M_1M_4 (16), we'll get that the maximum value of the function $f(\alpha, \rho)$ of the form (9) is reached on the segment M_1M_2 and is 1. The smallest value functions of the form (9) is achieved on the interval M_3M_4 has have at the point $M_0\left(1; \frac{3}{4}\right)$ and equals $\frac{9}{16}$. In Fig. 9 there is a view of the function $f(\alpha, \rho)$ in three-dimensional space.

In conclusion, we get

$$f_{\min}(\alpha, \rho) = f_{\min}\left(1; \frac{3}{4}\right) = \frac{9}{16},$$

$$f_{\max}(0, \rho) = 1.$$

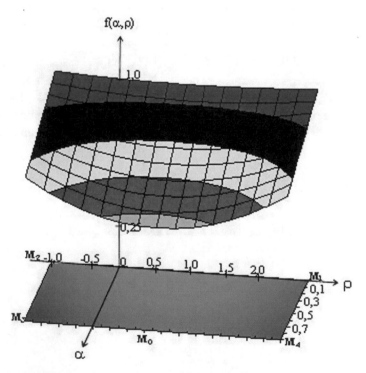

Fig. 9 Considered function $f(\alpha, \rho)$ in three-dimensional space

The conducted research allowed us to conclude that the lowest value, which reaches the full energy of a selective signal using quadratic spline, is equal to

$$E_\omega(\alpha, \rho) = E_\omega\left(1; \frac{3}{4}\right) = \frac{9}{16}U^2T,$$

And maximum

$$E_\omega(\alpha, \rho) = E_\omega(0; \rho) = U^2T.$$

Thus, when selecting an optimal selective signal waveform, infinitely small α values must be used to achieve maximum energy.

3 Conclusion

1. Parameter study performed of the full energy of signal functions using quadratic spline using the following criteria: coefficient of the dependence of energy concentration (energy criterion) of selective signals, D-criterion, opening eye-chart criterion and criterion based on the differential computation of functions of many variables.

2. The use of the coefficient of energy concentration (energy criterion) of signal functions using a quadratic spline is proposed. Retrieved parameter relationship width of the transition region and shape of the spectrum in it that when selecting the time interval $[-nT, nT]$ with $n = 1$ allow obtaining 97% of the maximum possible energy, thus for $n = 3$, the result can approach sufficiently close to 1 (Fig. 3).

3. When implementing the time characteristics of signal functions using a quadratic spline, in practice it is necessary to minimization the duration of the signals, which will inevitably lead to inter-symbolic interference. It is shown that the minimization of the level of interference is possible due to the variation of the parameter of the signal spectrum in the transition region, allowing the synthesis of signals with the necessary properties during practical implementation. With the optimal choice of the parameters of the width of the transition region and the shape of the spectrum in it, you can receive signals with the desired properties, which will lead to the possibility of practical implementation of these signals.

4. A criterion for the size of the opening of the eye-diagrams for the study of the full energy of signal functions is proposed. The conditions under which the maximum value of the aperture of the chart is reached are determined. The largest opening is obtained at $\alpha = 1$, $\rho = -0.6$, while H = 0.9582, which corresponds to the standard values according to the recommendation of ITU-T G.957. It has been substantiated that the use of the criterion of the size of the eye-diagrams allows synthesizing signals that are most resistant to shifts of the moments of sampling—jitter.

5. The parameters of the signal functions were obtained, which allow the energy indicators to be calculated and the optimal waveform to be determined in accordance with the criterion for the differential calculation of functions of many variables using the minimum and maximum values of the function that describes the signal in a given range.
6. The results obtained can be used for the synthesis of new signal functions, when the use of different criteria for the study of signal energy will improve the technical parameters of the radio tract.
7. The obtained results allow asserting that the synthesized selective signals on the basis of quadratic splines have the same properties as those based on cubic splines. In this case, the use of quadratic splines can achieve a significant reduction in the computational process.

References

1. Strelkovskaya I, Solovskaya I, Makoganiuk A (2018) Estimation of the parameters of selective signals using interpolation quadratic spline functions. In: Proceedings of the international scientific-practical conference (PICS&T-2018), Kharkiv, Ukraine, pp 327–330. https://doi.org/10.1109/Infocommst.2018.8632095
2. Strelkovskaya I, Solovskaya I, Makoganiuk A (2018) A study of the extremum of the total energy of the selective signals constructed by quadratic splines. Period Polytech Electr Eng Comput Sci 63(1):30–36. https://doi.org/10.3311/PPee.12457
3. Sukachev E, Strelkovskaya I (2004) Selective signals extremum properties in their spectra interpolation by cubic splines. Izv Vyss Uchebnykh Zaved Radioelektron 47(1):32–37. https://doi.org/10.20535/s0021347004010054
4. Strelkovskaya I (2007) Application of cubic B-splines for synthesis of selective signals. Telecommun Radio Eng 66(12):1047–1056. https://doi.org/10.1615/TelecomRadEng.v66.i12.10
5. Tomasin S, Benvenuto N (2014) Fractionally spaced non-linear equalization of faster than Nyquist signals. In: Digests of the 22nd European signal processing conference (EUSIPCO), Lisbon, Portugal, pp 1861–1865
6. Anderson JB, Rusek F, Owall V (2013) Faster-than-Nyquist signaling. Proc IEEE 101:1817–1830. https://doi.org/10.1109/jproc.2012.2233451
7. Sugiura S (2013) Frequency-domain equalization of faster-than-Nyquist signaling. IEEE Wirel 2:555–558. https://doi.org/10.1109/wcl.2013.072313.130408
8. Venkataramani R, Bresler Y (2000) Perfect reconstruction formulas and bounds on aliasing error in sub-Nyquist nonuniform sampling of multiband signals. IEEE Trans Inf Theory 46:2173–2183. https://doi.org/10.1109/18.868487
9. Strelkovskaya I, Makoganiuk A, Paskalenko S (2015) Comparative analysis of signal functions, built on the basis of quadratic and cubic splines. In: Proceedings of the second international scientific-practical conference (PICS&T-2015), Kharkiv, Ukraine, pp 173–176. https://doi.org/10.1109/INFOCOMMST.2015.7357305
10. Strelkovskaya I, Solovskaya I, Makoganiuk A (2017) Finding some QoS characteristics of self-similar traffic serviced by a mobile network. In: Proceedings of the 2nd IEEE international conference advanced information and communication technologies-2017 (AICT-2017), Lviv, Ukraine, pp 146–149. https://doi.org/10.1109/AIACT.2017.8020086
11. Strelkovskaya I, Solovskaya I (2017) Probabilistic and time characteristics of the G/M/1 QS with the Weibull distribution of arrivals. In: Proceedings of the second international conference

on information and telecommunication technologies and radio electronics (UkrMiCo'2017), Odessa, Ukraine. IEEE, pp 452–455. https://doi.org/10.1109/UkrMiCo.2017.8095416

12. Strelkovskaya I, Solovskaya I, Makoganiuk A (2019) Spline-extrapolation method in traffic forecasting in 5G networks. J Telecommun Inf Technol 3:8–16. https://doi.org/10.26636/jtit. 2019.134719
13. Strelkovskaya I, Solovskaya I (2019) Using spline-extrapolation in the research of self-similar traffic characteristics. J Electr Eng 70(4):310–316. https://doi.org/10.2478/jee-2019-0061
14. Zakharow YuV, Tozer TC, Adlard JF (2004) Polynomial spline-approximation of Clarke's model. IEEE 5:1198–1208. https://doi.org/10.1109/tsp.2004.826159
15. Sakai M, Usmani R (1983) Quadratic spline and two-point boundary value problem. Publ RIMS Kyoto Univ 19:7–13
16. De Boor C (1978) A practical guide to splines. Springer-Verlag, New York
17. Kim T, Kvasov B (2012) A shape-preserving approximation by weighted cubic splines. J Comput Appl Math 236:4383–4397. https://doi.org/10.1016/j.cam.2012.04.001
18. Bosner T, Rodina M (2011) Variable degree polynomial splines are Chebyshev splines. Adv Comput Math 383–400. https://doi.org/10.10007/s10444-011-9242-z
19. Bosner T (2005) Knot insertion algorithms for weightier splines. In: Proceedings of the conference on applied mathematics and scientific computing. Springer, pp 151–160. https://doi. org/10.1007/1-4020-3197-1
20. Mazure ML, Laurent PJ (1999) Polynomial Chebyshev splines. Comput Aided Geom Des 16:317–343
21. Zhang M, Yan W, Yuan CM, Wang DK, Gao XS (2011) Curve fitting and optimal interpolation on CNC machines based on quadratic B-splines. Sci China Inf Sci 54:1407–1418. https://doi. org/10.1007/s11432-011-4237-4
22. Zhang L, Sun R, Gao X et al (2010) An optimal solution for high-speed interpolation on consecutive micro-line segments and adaptive real-time look ahead scheme in CNC machining. Sci China Technol Sci 29:206–227
23. Zhao X, Yin Y, Yang B (2007) Dominant point detecting based non-uniform B-spline approximation for grain contour. Sci China Ser E Technol Sci 50:90–96
24. Pan Z, Chen W, Jiang Z et al (2016) Performance of global look-up table strategy in digital image correlation with cubic B-spline interpolation and bicubic interpolation. Theor Appl Mech Lett 6:126–130. https://doi.org/10.1016/j.tami
25. Viswanathan P, Chand R, Agarwal AKB (2013) Preserving convexity through rational cubic spline fractal interpolation function. J Comput Appl Math 263:262–276. https://doi.org/10. 1016/j.cam
26. Strelkovskaya I, Solovskaya I, Makoganiuk A (2017) Optimization of QoS characteristics of self-similar traffic. In: Proceedings of the 4th international scientific-practical conference problems of infocommunications. Science and technology (PICS&T-2017), Kharkiv, Ukraine. IEEE, pp 497–500. https://doi.org/10.1109/INFOCOMMST.2017.8246447
27. Strelkovskaya IV, Solovskaya IN, Makoganiuk AO (2019) Predicting self-similar traffic using cubic B-splines. In: Proceedings of the 3rd IEEE international conference advanced information and communication technologies-2019 (AICT-2019), Lviv, Ukraine. IEEE, pp 153–156. https:// doi.org/10.1109/AIACT.2019.8847761
28. Kryvinska N (2010) Converged network service architecture: a platform for integrated services delivery and interworking. In: Electronic business series, vol 2. International Academic Publishers, Peter Lang Publishing Group
29. Kryvinska N (2008) An analytical approach for the modeling of real-time services over IP network. Math Comput Simul 79(4):980–990. https://doi.org/10.1016/j.matcom.2008.02.016
30. Ageyev DV, Salah MT (2016) Parametric synthesis of overlay networks with self-similar traffic. Telecommun Radio Eng 75(14):1231–1241
31. Ageyev D, Qasim N (2015) LTE EPS network with self-similar traffic modeling for performance analysis. In: Proceedings of the 2015 second international scientific-practical conference problems of infocommunications. Science and technology (PIC S&T), Kharkov, Ukraine. IEEE, pp 275–277. https://doi.org/10.1109/INFOCOMMST.2015.7357335

32. Ageyev DV, Evlash DV (2008) Multiservice telecommunication systems design with network's incoming self-similarity flow. In: Proceedings of the international conference on modern problems of radio engineering, telecommunications and computer science (TCSET 2008), Lviv-Slavsko, pp 403–405

33. Radivilova T et al (2018) Decrypting SSL/TLS traffic for hidden threats detection. In: Proceedings of the 2018 IEEE 9th international conference on dependable systems, services and technologies (DESSERT). IEEE, pp 143–146. https://doi.org/10.1109/DESSERT.2018. 8409116

34. Ageyev D et al (2018) Method of self-similar load balancing in network intrusion detection system. In: 2018 28th international conference radioelektronika (RADIOELEKTRONIKA). IEEE, pp 1–4. https://doi.org/10.1109/RADIOELEK. 2018.8376406

35. Radivilova T, Kirichenko L, Ageiev D, Bulakh V (2020) The methods to improve quality of service by accounting secure parameters. In: Hu Z, Petoukhov S, Dychka I, He M (eds) Advances in computer science for engineering and education II. ICCSEEA 2019. Advances in intelligent systems and computing, vol 938. Springer, Cham

36. Ageyev D et al (2019) Infocommunication networks design with self-similar traffic. In: 2019 IEEE 15th international conference on the experience of designing and application of CAD systems (CADSM). IEEE, pp 24–27. https://doi.org/10.1109/CADSM.2019.8779314

37. Kirichenko L, Radivilova T, Bulakh V (2020) Binary classification of fractal time series by machine learning methods. In: Lytvynenko V, Babichev S, Wójcik W, Vynokurova O, Vyshemyrskaya S, Radetskaya S (eds) Lecture notes in computational intelligence and decision making. ISDMCI 2019. Advances in intelligent systems and computing, vol 1020. Springer, Cham

38. Kirichenko L, Radivilova T, Bulakh V (2018) Machine learning in classification time series with fractal properties. Data 4(1):5. https://doi.org/10.3390/data4010005

39. Bulakh V, Kirichenko L, Radivilova T (2018) Time series classification based on fractal properties. In: Proceedings of the 2018 IEEE second international conference on data stream mining & processing (DSMP). IEEE, pp 198–201. https://doi.org/10.1109/DSMP.2018.8478532

40. Yeremenko OS, Lebedenko TM, Vavenko TV, Semenyaka MV (2015) Investigation of queue utilization on network routers by the use of dynamic models. In: Proceedings of the IEEE second international scientific-practical conference problems of infocommunications. Science and technology (PIC S&T-2015), Kharkiv, Ukraine. IEEE, pp 46–49. https://doi.org/10.1109/ INFOCOMMST.2015.7357265

41. Yeremenko O, Tariki N, Hailan AM (2016) Fault-tolerant IP routing flow-based model. In: Modern problems of radio engineering, telecommunications and computer science (TCSET-2016): proceedings of the 13th international conference, Lviv, Ukraine, 23–26 Feb 2016. IEEE, pp 655–657. https://doi.org/10.1109/tcset.2016.7452143

42. Kryvinska N (2004) Intelligent network analysis by closed queuing models. Telecommun Syst 27:85–98. https://doi.org/10.1023/B:TELS.0000032945.92937.8f

43. Kryvinska N, Zinterhof P, van Thanh D (2007) An analytical approach to the efficient real-time events/services handling in converged network environment. In: Enokido T, Barolli L, Takizawa M (eds) Network-based information systems. NBiS 2007. Lecture notes in computer science, vol 4658. Springer, Berlin, Heidelberg

44. Kryvinska N, Zinterhof P, van Thanh D (2007) New-emerging service-support model for converged multi-service network and its practical validation. In: First international conference on complex, intelligent and software intensive systems (CISIS'07). IEEE, pp 100–110. https:// doi.org/10.1109/CISIS.2007.40

45. Strelkovskaya I, Solovskaya I, Severin N, Paskalenko S (2017) Spline approximation based restoration for self-similar traffic. East Eur J Enterp Technol 3/4(87):45–50. https://doi.org/ 10.15587/1729-4061.2017.102999

46. Lemeshko OV, Yeremenko OS, Tariki N, Hailan AM (2016) Fault-tolerance improvement for core and edge of IP network. In: Proceedings of the XIth international scientific and technical conference computer sciences and information technologies (CSIT-2016), Lviv, Ukraine. IEEE, pp 161–164. https://doi.org/10.1109/STC-CSIT.2016.758989

47. Strelkovskaya IV, Grygoryeva TI, Solovskaya IN (2018) Self-similar traffic in G/M/1 queue defined by the Weibull distribution. Radioelectron Commun Syst 61(3):173–180. https://doi.org/10.20535/S0021347018030056
48. Ahlberg JH, Nilson EN, Walsh JL (1967) Teoriya splaynov i yeye prilozheniya (The theory of splines and their applications). Moscow, Russia, Peace (in Russian)
49. Laurent PJ (1972) Approximation et optimisation. Hermann, Paris, France
50. Fikhtengolts GM (1968) Osnovyi matematicheskogo signal functions (Fundamentals of mathematical analysis). Moscow, Russia (in Russian)

Adaptive Complex Singular Spectrum Analysis with Application to Modern Superresolution Methods

V. Vasylyshyn

Abstract The adaptive variant of the Singular Spectrum Analysis (SSA) approach for complex-valued signal model is obtained. It is related with the estimation of the noise variance using the results of the random matrix theory. Application of the adaptive complex SSA approach as preprocessing (denoising) step to modern methods of spectral analysis (subspace-based methods) is considered in the paper. The data sequence obtained after adaptive SSA approach is used as the input information for the superresolution method. The results of simulation demonstrate the performance improvement of the subspace-based methods in the conditions of low signal-to-noise ratio (SNR) when using the proposed approach. The performance of the subspace-based methods without and with the use of the adaptive SSA is comparable at high SNR. Furthermore, the performance of adaptive SSA approach depends on the quality of the noise variance estimate. The application of extended data matrix with specific structure obtained from the filtered data matrix is proposed. The directions of further investigations and possible applications of presented approach in communication systems are considered.

Keywords Singular value decomposition · Adaptive singular spectrum analysis · Superresolution methods

1 Introduction

Time series analysis (analysis of series of statistical observations taken at regular intervals in time) is an important tool in various areas of science. It plays a significant role for parameter estimation, prediction, pattern classification, signal processing, signal recognition, business information processing, network traffic measurements and system identification. Time series analysis can be performed using the correlation analysis, spectral analysis, principal component analysis (PCA), nonlinear PCA,

V. Vasylyshyn (✉)
Ivan Kozhedub Kharkiv National Air Force University, Kharkiv, Ukraine
e-mail: vladvas@ukr.net

© The Editor(s) (if applicable) and The Author(s), under exclusive license to Springer Nature Switzerland AG 2021
T. Radivilova et al. (eds.), *Data-Centric Business and Applications*, Lecture Notes on Data Engineering and Communications Technologies 48,
https://doi.org/10.1007/978-3-030-43070-2_3

time-frequency analysis, singular spectrum analysis (SSA), chaotic analysis (including the correlation dimension, recurrence plots, Lyapunov exponents, surrogate data technology) [1–7]. The modern view about the possibility of description of observed data gives the embedding which combines the information theory, topology, differential dynamics, dynamical system theory and based on the Takens-Mane theorem [4].

The embedding of time series, multivariate geometry, dynamical system theory are the fundamental moments of the SSA technique [4–7]. It additionally incorporates the elements of classical time series analysis, multivariate statistics. SSA is usually used for the decomposition of input data into a sum of periodic components, noise components and trend. Moreover, it has been applied to image processing, forecasting of time series, smoothing, analysis of the time series with missing data, construction of model [5–12]. It can be used as a preprocessing step in many engineering applications including communication systems, radar systems, control systems. The application of such preprocessing step allows improving the performance of the corresponding systems. For example, SSA is employed for denoising the received signals with noise mixture.

In the case of modern communication systems (newest commercial fourth-generation long-term evolution, fifth-generation) and radar systems the performance can also be improved by using existing techniques such as multiple input- multiple output system (MIMO), orthogonal frequency division multiplexing (OFDM), MIMO-OFDM and their variants including massive MIMO, generalized frequency division multiplexing (GFDM), spectrally efficient frequency division multiplexing (SEFDM), filter bank multicarrier (FBMC), beamspace MIMO and so on [1]. The beamforming (or precoding), spatial smoothing, coding, space-time coding (Alamouti coding, space-time OFDM) and other signal processing procedures can be used [1, 2]. Additional improvement can be attained by using the properties of communication signals (cyclostationarity or periodic correlation), higher-order statistics, random matrix theory, surrogate data technology [1–4].

SSA is closely related to singular value decomposition (SVD) of the matrices with special structure including Hankel, Toeplitz, circulant, block-Hankel matrices [5–15]. However, the eigenvalue decomposition (EVD) of the covariance matrix (CM) formed from the data matrix of special structure can be used. This is one of the points of relation of SSA and subspace-based methods [2, 3]. The name of methods is caused by the fact that in order to estimate the signal parameter the signal (or noise) subspace eigenvectors are used. These methods are also known as eigenstructure methods. The methods estimate the signal parameters in the superresolution mode [2, 3]. They are widely used in many problems of communication and radar including frequency and direction of arrival estimation, time of delay estimation, collision resolution in packet radio networks, multidimensional harmonic retrieval for MIMO channel estimation (channel sounding) and so on [1–3]. Furthermore, the several adaptive subspace algorithms have been introduced in the neural network literature. They have been suggested for estimating the noise subspace eigenvector corresponding to the smallest eigenvalue of covariance matrix [2].

The list of the noise reduction methods in observation (that is methods of improving the signal-to-noise ratio) includes the SSA, wavelet transformations, spectral-subtractive algorithms, empirical mode decomposition (EMD), surrogate data technology, the total least squares (TLS) and structured TLS, spectral domain constrained estimator, time domain constrained estimator [1–18]. The part of the methods is based on the principal component analysis (PCA) or independent component analysis. In the technical literature the PCA is also known as singular system analysis [11, 12]. The singular system analysis theory with application to dynamical systems was considered in [12]. In the paper the relation of the theory and subspace-based methods is mentioned. A nonlinear generalization of PCA such as Kohonen's self-organizing maps and principal manifolds are widely used in signal and field (image) processing.

Besides the PCA (Karhunen-Loeve expansion theorem) SSA, time and spectral domain constrained estimators are based on Eckart-Young-Mirsky theorem [4, 9]. This theorem shows how to solve the mathematical problem of approximating one matrix by another matrix of lower rank. It should be noted that the lower-rank approximation is used in many areas of signal processing including adaptive filtering, spectral analysis and others [11–18]. For example, the dimensionality reduction phase in [19] is performed by low-rank approximation of a matrix composed of received signals. The narrowband interference canceller that allows suppression of the spectral leakage in the OFDM-based system was presented in [20]. The theory of optimal rank reduction was used to reduce the operational complexity of the canceller.

The attempts of investigators were bent to the search of engineering applications for SSA and to performance improvement of Basic SSA. Image (field) processing technique based on SSA was proposed in [5]. The application of the Basic SSA as a signal preprocessing scheme for modern spectral analysis methods was considered in [5, 16, 17, 21]. The joint use of SSA and surrogate data technology was considered in [17]. The influence of outliers on the performance of SSA was investigated in [6].

The overlap-SSA was proposed in [22] for segmentation, analysis and reconstruction of long-term and/or non-stationary signals. The application of the method allows improving the reconstruction and component separability in the case of the non-stationary time-series. The sliding SSA presented in [23] improves the separability of the SSA.

The application of SSA to OFDM system channel estimation was proposed in [24]. SSA allowed improving bit error rate because of improvement in SNR. Furthermore, channel estimation was done using low rank approximation of the channel correlation matrix using SSA.

The real-valued signals are usually used with SSA-based techniques. However, in the many cases is of interest to use complex presentation of signals. Complex SSA is applied to complex-valued time series by using complex-valued SVD [5, 6].

In the paper the variant of the complex SSA approach is presented. The name adaptive complex SSA is used because it is based on the estimation of the noise variance. Such an estimate is obtained using the results of the random matrix theory. However, the alternative estimate of noise variance can be used [18]. One of the differences of the paper as compared to [21] is using the improved estimate of the noise variance.

The application of the adaptive complex SSA approach as a denoising step to superresolution methods is considered in the paper. The data sequence obtained after adaptive SSA approach is used for the subspace-based methods.

The paper is organized as follows. In the next section we consider the signal model and details of the spectral presentation of the data matrix or corresponding covariance matrix. Then we consider the adaptive complex SSA approach. The application of extended data matrix with specific structure after adaptive SSA is proposed. The matrix of such form is used in the spectral analysis applications. The possible simplifications of joint using the SSA and superresolution methods are considered. Simulation results show that spectral analysis performance by superresolution method can be improved by using presented approach in the conditions of low signal-to-noise ratio.

2 Signal Model and Assumptions

2.1 Signal Model

The frequency estimation problem is a one of important problems of signal processing. It can be considered as independent problem or as part of more complicated problem such as MIMO wireless channel sounding problem [1, 25].

The problem of frequency estimation can be stated as follows: given the input data presented by model

$$y(n) = s(n) + e(n) = \sum_{v=1}^{V} x_v(n) + e(n), \tag{1}$$

to $n = 1, \ldots, N$ find the estimates of distinct parameters $\omega_v, v = 1, \ldots, V$. The form of the model is similar to the model used in the parameter estimation problem of MIMO flat-fading channels and in radar applications [25]. In Eq. (1) vth harmonic component has the form $x_v(n) = \xi_v \exp(j(\omega_v n + \varphi_v))$. Here ξ_v is the amplitude, $\omega_v = 2\pi f_v$ is the frequency, $\omega_v \in [0, \pi)$. Furthermore, φ_v is vth component phase and it is assumed that φ_v are uniformly distributed on the interval $[0, 2\pi)$. The additive white noise $e(n)$ has variance σ^2 and zero mean.

The matrix form of model (1) can be presented as follows [1, 3, 21]

$$\mathbf{y}(n) = \mathbf{B}\mathbf{x}(n) + \mathbf{e}(n) = \mathbf{g}(n) + \mathbf{e}(n), \tag{2}$$

where $n = 1, \ldots, K$, $K = N - m + 1$ and $m > V$ is the size of data vector (window size). In other words initial time series $y(n), n = 1, \ldots, N$ is divided into consecutive and overlapping segments of the smaller size

$$\mathbf{y}(n) = [y(n) \dots y(n+m-1)]^T, \tag{3}$$

where $()^T$ denotes transposing operator. The rank of covariance matrix for $\mathbf{y}(n)$ be equal (or will exceed) V due to such segmentation. This allows to provide a low-rank matrix representation of the problem. Similar manipulation is performed under spatial smoothing and other procedures of the signal processing.

In Eq. (2) $\mathbf{B} = [\mathbf{a}(\omega_1) \dots \mathbf{a}(\omega_V)]$ is the $m \times V$ Vandermonde matrix containing information about frequencies of signal harmonics, $\mathbf{x}(n)$ is the $V \times 1$ vector, $\mathbf{e}(n) = [e(n) \dots e(n+m-1)]^T$, $\mathbf{a}(\omega_v) = [1 \dots \exp(j(m-1)\omega_v)]^T$, $j = \sqrt{-1}$.

The application of segmentation to $y(n)$ results in the sequence of vectors $\mathbf{y}(n)$ in the embedding space [4, 12] that can be presented as the columns of the $m \times K$ Hankel data matrix

$$\mathbf{Y} = \begin{bmatrix} y(1) & y(2) & \cdots & y(N-m+1) \\ y(2) & y(3) & \cdots & y(N-m+2) \\ \vdots & \vdots & \cdots & \vdots \\ y(m) & y(m+1) & \cdots & y(N) \end{bmatrix} = [\mathbf{y}(1), \dots, \mathbf{y}(K)] = \mathbf{Y}_r + j\mathbf{Y}_i. \tag{4}$$

In other words $y(n)$ is mapped onto a $m \times K$ Hankel matrix consisting of m-lagged vectors (or initial one-dimensional (scalar) time series $y(n)$ is passed to a multidimensional representation $\mathbf{Y}$). It should be noted that this matrix called as a trajectory matrix can has a Toeplitz structure. The construction of the set of delayed vectors $\mathbf{y}(n)$, $n = 1, \dots, K$, caused the name of the method of such mapping as method of delays. Here $\mathbf{Y}_r$ and $\mathbf{Y}_i$ are the trajectory matrices obtained based on $y_r(n) = real(y(n))$ and $y_i(n) = imag(y(n))$.

Takens' theorem [4] describes the details of the time series embedding and reconstruction and gives sufficient conditions on choice of parameter of embedding m or suggests criteria for m. The registration of the variations is possible when size of embedding window is adequately [6]. For example, if we know that $y(n)$ has a periodic component then it is rational (with the aim of better separability of the components) to select m proportional to the period of component.

The covariance matrix (CM) for $y(n)$ will have the form [1–3, 16, 17]

$$\mathbf{R} = E\{\mathbf{y}(n)\mathbf{y}^H(n)\} = \sum_{v=1}^{V} \xi_v^2 \mathbf{a}(\omega_v)\mathbf{a}^H(\omega_v) + \sigma^2\mathbf{I} = \mathbf{BSB}^H + \sigma^2\mathbf{I}. \tag{5}$$

Here $E\{\}$ denotes mathematical expectation, $()^H$ represents the Hermitian transpose operator, $\mathbf{S} = diag(\xi_1^2, \dots, \xi_V^2)$ is the CM of signal and $\sigma^2\mathbf{I}$ is the noise covariance matrix. Here $\mathbf{I}$ is the identity matrix.

If the noise is colored then it is necessary to estimate the noise property using available data and perform prewhitening of the initial data. This transformation will change the form of noise CM to usual form.

2.2 EVD of the Covariance Matrix and SVD of the Data Matrix

The basic point of SSA and subspace-based methods is spectral decomposition of data CM or data (trajectory) matrix [2, 5].

The eigenvalue decomposition (EVD) of CM $\mathbf{R}$ has the form

$$\mathbf{R} = \mathbf{U}\Sigma\mathbf{U}^H. \tag{6}$$

The matrix $\mathbf{U} = [\mathbf{u}_1 \cdots \mathbf{u}_m]$ consists of the m orthonormal eigenvectors of $\mathbf{R}$ and Σ is a diagonal matrix containing the corresponding eigenvalues. The matrix Σ contains the ordered eigenvalues λ_i, i.e. $\lambda_1 > \ldots > \lambda_V \geq \lambda_{V+1} = \ldots = \lambda_m = \sigma^2$. It is very important point is partitioning of the eigenvectors into a signal subspace set and its orthogonal complement (i.e. noise subspace). Furthermore, it is necessary to say that such terms as spaces, norms, basis taken from functional analysis help to describe many transformations of signals in many procedures of signal processing.

Let us present matrix $\mathbf{U}$ as $\mathbf{U} = [\mathbf{U}_s \mathbf{U}_n]$ where matrix $\mathbf{U}_s = [\mathbf{u}_1 \cdots \mathbf{u}_V]$ consists of the signal-subspace eigenvectors corresponding to the V principal eigenvalues and $\mathbf{U}_n = [\mathbf{u}_{V+1} \cdots \mathbf{u}_m]$ is formed from noise-subspace eigenvectors.

In the practice only a sample CM is available i.e. an estimate of $\mathbf{R}$ based on finite number of data samples

$$\widehat{\mathbf{R}} = \frac{1}{K} \sum_{n=1}^{K} \mathbf{y}(n)\mathbf{y}^H(n) = \frac{1}{K}\mathbf{Y}\mathbf{Y}^H. \tag{7}$$

The estimates of eigenvectors and eigenvalues can be obtained due to spectral decomposition of the sample CM $\widehat{\mathbf{R}}$ which can be presented as [2]

$$\widehat{\mathbf{R}} = \widehat{\mathbf{U}}_s\widehat{\Sigma}_s\widehat{\mathbf{U}}_s^H + \widehat{\mathbf{U}}_n\widehat{\Sigma}_n\widehat{\mathbf{U}}_n^H. \tag{8}$$

Here $\widehat{\mathbf{U}}_s$ and $\widehat{\mathbf{U}}_n$ are the $m \times \widehat{V}$ and $m \times (m - \widehat{V})$ matrices of the signal-subspace and noise-subspace eigenvectors. $\widehat{V}$ is the estimate of the harmonic component number. $\widehat{\Sigma}_s$ and $\widehat{\Sigma}_n$ are the diagonal matrices containing the $\widehat{V}$ signal-subspace eigenvalues and $m - \widehat{V}$ noise-subspace eigenvalues, respectively. In the considered case of complex data matrix and complex CM we have real-valued eigenvalues and complex-valued eigenvectors. In the technical literature (meteorological literature) the eigenvectors of CM have been called empirical orthogonal functions (EOF) [5, 6, 12].

The necessity of spectral representations of data matrix or CM arises in many areas including SSA, spectral analysis methods, spectral recognition of signals, spectral recognition of modulation digital types of communication signals, image processing, spectral theory of graphs, adaptive modulation schemes in MIMO, eigen-beamforming in MIMO (MIMO-OFDM) and for systems wits antenna

Fig. 1 Eigenvalue profile in the presence of two harmonics

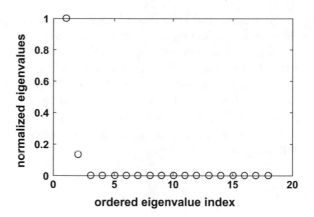

arrays, steganography and steganalysis, matched subspace detectors, nonnegative matrix/tensor factorization algorithms and in the many other applications of information processing [1–6, 16].

The profile of eigenvalues $\widehat{\mathbf{R}}$ is presented on Fig. 1.

We assume that the signal consists of two harmonic components with equal power and frequencies $f_1 = 0.3$ Hz, $f_2 = 0.311$ Hz, $N = 64$, $m = 18$ and SNR $= 2$ dB. SNR was determined as $10 \log_{10} \left(\sum_{v=1}^{V} \alpha_v^2 / \sigma^2 \right)$. Figure 1 shows that profile of observed eigenvalues (i.e. eigenvalues of sample CM) contains two signal eigenvalues and they are distinct from the noise eigenvalue distribution. It is shown in [26] that theoretical profile of noise eigenvalues can be described by approximately exponential model.

Let $m_y \leq \min\{m, K\}$ be the rank of trajectory matrix $\mathbf{Y}$. The SVD of the Hankel data matrix $\mathbf{Y}$ has the form [2, 3, 5]

$$\mathbf{Y} = \sum_{i=1}^{m_y} \mu_i \mathbf{u}_i \mathbf{v}_i^{H}, \tag{9}$$

where $\mu_i = \sqrt{\lambda_i}$ is the singular value, $\mathbf{u}_i$ and $\mathbf{v}_i$ are the left and right singular vectors of data matrix, respectively. The vectors $\mathbf{u}_i$ and $\mathbf{v}_i$ form the orthonormal basis of row space and column space of matrix $\mathbf{Y}$ respectively. Furthermore, $\mathbf{u}_i$ and $\mathbf{v}_i$ are the eigenvectors of the matrices $\mathbf{YY}^{H}$ and $\mathbf{Y}^{H}\mathbf{Y}$ respectively.

It is known that SSA is based on PCA. The principal component matrix of $\mathbf{Y}$ is given by an matrix $\mathbf{Y}_{pc} = \mathbf{U}^{H}\mathbf{Y} = \boldsymbol{\Sigma}_{sv}\mathbf{V}^{H}$. Here $\mathbf{U}$ consists of vectors $\mathbf{u}_i$ and $\mathbf{V}$ consists of $\mathbf{v}_i$. Matrix $\boldsymbol{\Sigma}_{sv}$ is a diagonal matrix with entries $\sqrt{\lambda_i}$.

The collection $(\mu_i, \mathbf{u}_i, \mathbf{v}_i)$ is called as ith eigentriple. Furthermore, $\mathbf{v}_i$ can be defined as $\mathbf{v}_i = \mathbf{Y}^{H}\mathbf{u}_i / \sqrt{\lambda_i}$. The energy contribution of the ith eigentriple to the trajectory matrix, given by the ratio $\mu_i^2 / \sum_{j=1}^{m_y} \mu_j$ is called the singular spectrum of the time series [5].

The singular and eigenvectors have a temporal structure and can be considered as time series and, for example, multirate processing can be performed on eigenvectors including modulation, filtering and decimation [2]. Furthermore, they can be used for eigen-beamforming in radar and communication systems with MIMO.

As was mentioned before the SSA approach is related with Eckart-Young-Mirsky theorem [5]. This theorem is a procedure for finding the best lower rank approximation to a given matrix. It can be obtained if for restoration of data matrix we will use only $\widehat{V}$ largest singular values and the corresponding singular vectors by the following way

$$\mathbf{Y}_{rec.} = \sum_{i=1}^{\widehat{V}} \widehat{\mu}_i \widehat{\mathbf{u}}_i \widehat{\mathbf{v}}_i^H. \tag{10}$$

The obtained matrix is the best least squares approximation of lower rank $\widehat{V}$ to the given matrix. The details of phase-space reconstruction can be found in [4]. Another class of low-rank approximation algorithms is based on the Nyström method. Furthermore, Nyström method can be used for approximation of the eigenvectors of CM. The truncated SVD is also the classical regularization method [2, 3, 5].

Moreover, SSA is related with functional analysis (theory of Hilbert-Schmidt operators, self-adjoint operators, spectral representations of operators) [5].

In the following section we describe the details of proposed adaptive complex SSA technique.

3 Adaptive Complex SSA

The SSA is a non parametric method for decomposition and classification of the time series [1–3]. It is known, that the singular values of trajectory matrix are increased by noise standard deviation due to the additive noise [9, 16]. Therefore the smallest singular value can be subtracted to compensate this effect [18, 27]. The similar fact is used to improve the performance of complex SSA approach.

The traditional estimate of noise variance can be obtained using $\widehat{\mathbf{\Lambda}}_n$ ($\widehat{\mathbf{\Lambda}}_s$) and it has a form $\widehat{\sigma}^2 = (1/(m - \widehat{V}))(\text{trace}(\widehat{\mathbf{R}}) - \text{trace}(\widehat{\mathbf{\Lambda}}_s)) = (1/(m - \widehat{V}))\text{trace}(\widehat{\mathbf{\Lambda}}_n)$ [2, 3, 16]. In this estimate the components of noise corresponding to the noise subspace are used. However, the components of noise added to signal eigenvalues are not used. It is possible to affirm that the traditional estimate of noise variance can be improved.

Several enhanced noise variance estimators were proposed [18, 26]. One of them involves the use of approximation of the profile of ordered noise eigenvalues by damped exponential function (exponential profile) [26] and the forecast of the noise values for signal subspace [18].

In the proposed noise variance estimate the results of [28–30] are used based on random matrix theory. These results were proposed for the case when the number of samples is not large compared with the dimension of the array. The results of [29, 30] were generalized and new estimate of noise variance is obtained.

In proposed adaptive complex SSA the additional denoising is performed as $\mathbf{Y}_{filt.} = \sum_{i=1}^{\widehat{V}} (\hat{\mu}_i - \hat{\sigma})\hat{\mathbf{u}}_i\hat{\mathbf{v}}_i^H$. Here we remove the noise components from the (signal + noise) subspace. The usual estimate of noise variance was used in [21]. Obviously that estimate $\hat{\sigma}$ influences on the filtration efficiency from observation noise.

The proposed noise variance estimate can be presented as $\hat{\sigma}_2^2 = \dfrac{\hat{\sigma}_1^2}{(1-\widehat{V}/K)}$. The first part of proposed estimate is based on the paper of Kritchman and Nadler [30]. Here $\hat{\sigma}_1^2 = \hat{\sigma}^2 + (1/K) \sum_{i=1}^{\widehat{V}} \dfrac{\lambda_i \hat{\sigma}^2}{(\lambda_i - \hat{\sigma}^2)}$ and this estimate can reduce the leakage from the signal subspace into noise subspace [29]. Adaptation to SNR consists in changing the level of denoising depends on SNR (i.e. reduction of denoising when SNR is high).

The proposed adaptive complex SSA includes the following steps:

Step 1. Select or estimate window length (embedding dimension) m. Construct the trajectory matrix of Hankel (or Toeplitz) form $\mathbf{Y}$ from the $\{y(n)\}_{n=1}^{N}$ [5].

Step 2. Perform the EVD of CM $\widehat{\mathbf{R}}$ or SVD of the trajectory matrix $\mathbf{Y}$.

Step 3. Estimate the harmonic component number $\widehat{V}$ [2] or the number of most significant eigenvectors (eigenvalues) for more general signal model.

Step 4. Calculate $\hat{\sigma}^2$, $\hat{\sigma}_1^2$ and $\hat{\sigma}_2^2$.

Step 5. Perform eigentriple grouping and reconstruct the denoised data matrix $\mathbf{Y}_{filt.} = \sum_{i=1}^{\widehat{V}} (\hat{\mu}_i - \hat{\sigma}_2)\hat{\mathbf{u}}_i\hat{\mathbf{v}}_i^H$.

Step 6. Perform averaging the elements of matrix $\mathbf{Y}_{filt.}$ located on cross diagonals (or diagonals for the Toeplitz matrix) and produce the filtered time series $y_{filt.}(n)$.

In the case of using EVD of CM the calculation of $\mathbf{v}_i$ can be performed as was mentioned before i.e. $\mathbf{v}_i = \mathbf{Y}^H\hat{\mathbf{u}}_i/\sqrt{\lambda_i}$.

The performed singular value spectral subtraction in some sense is similar to the spectral subtraction method [14]. However, instead of DFT of data sequence, we perform SVD of data matrix and so on. The Lanczos-bases truncated SVD can be used for efficient implementation of SSA [5]. The Prony-Lanczos method, power method and others can be used for calculation of approximations to a principal eigenvectors and eigenvalues of the CM. Hankel matrix-vector multiplication and Hankelization operator can be computed by the means of Fast Fourier Transform [5].

The averaging step can be explained in the following way. Let $\mathbf{G} = \mathbf{Y}_{filt.}$ be matrix with elements g_{ij}, $1 \leq i \leq m$, $1 \leq j \leq K$ and $w_{\min} = \min(m, K)$, $w_{\max} = \max(m, K)$. We assume that $g_{ij}^* = g_{ij}$ when $m < K$ and $g_{ij}^* = g_{ji}$ in other cases. The diagonal averaging can be performed as [5, 6]

$$
g_k = \begin{cases} \frac{1}{k} \sum\limits_{l=1}^{k} g^*_{l,k-l+1}, & for\ 1 \le k \le w_{\min} \\[3mm] \frac{1}{w_{\min}} \sum\limits_{l=1}^{w_{\min}} g^*_{l,k-l+1}, & for\ w_{\min} \le k \le w_{\max} \ . \\[3mm] \frac{1}{N-k+1} \sum\limits_{l=k-w_{\max}+1}^{N-w_{\max}+1} g^*_{l,k-l+1}, & for\ w_{\max} \le k \le N \end{cases} \tag{11}
$$

It has been shown that averaging of elements located on the cross diagonals of reduced rank matrix is equivalent to subtracting the certain information from the original signal. This information is contained in eigen-residuals corresponding to some noise-subspace eigenvalues [31]. This can be considered as a way of removing undesirable spectral information carried by the noise-subspace eigenvectors. Finite-impulse response (FIR) filter representation of truncated SVD was discussed in [32]. The FIR filters are related to the eigenvectors of CM. The embedding dimension can be estimated by false nearest neighbor approach, Grassberger-Proccacia algorithm and so on [4].

The MDL algorithm, the exponential fitting test, the bootstrap-based approaches, the approaches with using the angles between subspaces, the approaches based on surrogate data technology and approach based on Gershgorin radii can be used for the Step 3. Furthermore, the selection of most appropriate eigenvectors (singular vectors) can be performed using Pearson correlation coefficient or Kendall rank tests [5, 24].

Step 5 contains the truncation (low rank approximation). Besides of subtraction which allows reducing the perturbation of singular values conditioned by additive white noise the appropriate weighting can be used to additionally improve the denoising performance [33]. Similar weighting is used for subspace-fitting methods, MODE [2]. Moreover, Step 5 can be improved using hierarchical clustering [23].

The important term in SSA is the separability. It means orthogonality of the column and row spaces of the trajectory matrices of input sequence components [5]. Therefore it is possible to consider the separability on the level of signals, subspaces and so on. In the case of sufficiently long sequences (in the asymptotic case), SSA can separate the signals with different frequencies, the signal, and noise.

ICA analysis improves the separability of signal components. Source separation includes recovering a set of signals of which only instantaneous linear mixtures are observed. The separation problem arises in many practical applications and includes the separation of overlapping secondary surveillance radar replies (blind separation of partially overlapping data packets), separation of asynchronous OFDM signals, separation of a mixture of pseudoperiodic chaotic signals, separation of overlapping linear frequency modulated signals [1, 2].

Nowadays blind source separation (BSS) algorithms are widely used. Such algorithms include AMUSE (algorithm for multiple unknown signals extraction), EFOBI (extended fourth order blind identification) [3]. The pattern recognition approaches can be used for the separation of the signal and noise subspaces.

The measure of orthogonality of time series (or principal components) can be computed as normalized inner (scalar) product of time series segments (or principal components) or as weighted correlation. It simplifies the automatic grouping of components [5, 6, 24]. The principal angles between two subspaces can be used.

The effect of application of SSA method can be presented by using the phase portraits [4, 17]. Figure 2 presents the sample portrait of the signal.

We assume that the frequencies of two real harmonic components with equal power are $f_1 = 0.3$ Hz and $f_2 = 0.312$ Hz. The SNR was 1 dB, $N = 64$, $m = 18$. Figure 3 shows the sample phase portrait of the input sequence.

The term "sample" is used because of the portraits are changed depends on time series (from realization to realization). This can be explained by the fact that phases of harmonic components are random values.

Figure 4 corresponds to the data sequence filtered by adaptive SSA method.

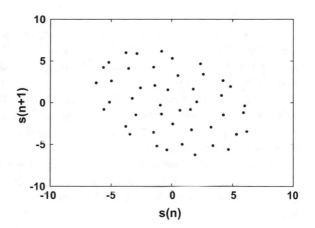

Fig. 2 Sample phase portrait of the signal

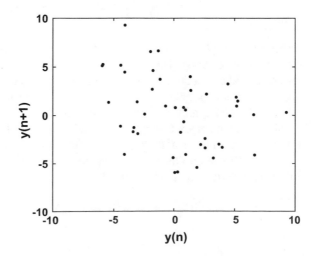

Fig. 3 Sample phase portrait of the signal with additive white noise

Fig. 4 Sample phase portrait
of data sequence cleaned by
adaptive SSA method

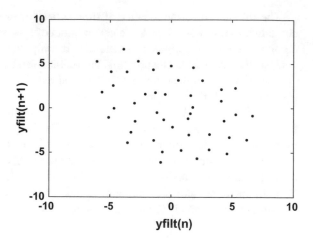

It is possible to see that structure of the phase portraits presented on Figs. 2 and 4 are rather like. Furthermore, the comparison of Figs. 2, 3, 4 allows to see that the phase portrait of data sequence cleaned by adaptive SSA method $y_{filt.}(n)$ is closer to phase portrait of signal.

The application of superresolution methods will contain the several steps. At first, the Hankel data matrix $\mathbf{Y}_{filt.H}$ is formed from $\mathbf{y}_{filt.}(n)$. Then the spectral decomposition of $\widehat{\mathbf{R}}_{filt.H} = \mathbf{Y}_{filt.H}\mathbf{Y}^H_{filt.H}$ is calculated. At this point the subspace-based methods can be calculated.

The most known signal-subspace methods are MUSIC, Min-Norm, FINE, ESPRIT [2]. Let $\widehat{\mathbf{G}}_n$ be the matrix of noise–subspace eigenvectors of $\widehat{\mathbf{R}}_{filt.H}$. Frequency estimates of Root-MUSIC can be found from the appropriately selected roots of polynomial [2, 3]:

$$P_{rm}(z) = \mathbf{a}^H(z^{-1})\widehat{\mathbf{G}}_n\widehat{\mathbf{G}}^H_n\mathbf{a}(z), \tag{12}$$

where $\mathbf{a}(z) = [1, z, \ldots, z^{M-1}]^T$, $\widehat{\mathbf{G}}_n\widehat{\mathbf{G}}^H_n = \mathbf{\Pi}^\perp$ is the spectral projector on the orthogonal subspace (noise subspace), $z = \exp(j\omega)$ [2, 34]. The paper [34] contains the sequential estimation approach which can be used together with the results of the paper. The stable polynomial approximation of a spectral projector can be used for additional simplification of proposed procedure [35].

Simplification of joint using the SSA and superresolution methods can be realized. For this aim the results of the paper of A. Gershman generalized in [36] can be used.

The additional improvement can be obtained by using the extended data matrix for the realization of superresolution methods

$$\mathbf{Y}_{ext} = [\mathbf{Y}_{filt.H}\,\widetilde{\mathbf{I}}_m\mathbf{Y}^*_{filt.H}\widetilde{\mathbf{I}}_K], \tag{13}$$

where $\widetilde{\mathbf{I}}_m$ and $\widetilde{\mathbf{I}}_K$ are exchange matrices of sizes $m \times m$ and $K \times K$, respectively, $()^*$ represents complex conjugation operator, and

$$\widetilde{\mathbf{I}}_m = \begin{bmatrix} & & 1 \\ & \cdot^{\cdot^{\cdot}} & \\ 1 & & \end{bmatrix}. \tag{14}$$

Such form of data matrix corresponds to a forward- backward averaging scheme [3]. The corresponding forward-backward CM obtained as a result of forward-backward averaging can be used instead of extended matrix. It should be noted that this extended matrix is related with forward-backward linear prediction and estimation of the persymmetric CM [2, 3].

Simulation Results

We consider the case when $N = 64$ and signal consists of two harmonic components with frequencies $f_1 = 0.3\,\text{Hz}$, $f_2 = 0.313\,\text{Hz}$ and equal power. In the considered case the frequency separation is less than Rayleigh resolution limit. $L_r = 1000$ independent simulation runs were performed for each SNR. Root-mean square error (RMSE) of frequency estimation was defined as [17]

$$RMSE = \sqrt{\frac{1}{L_r V} \sum_{l=1}^{L_r} \sum_{v=1}^{V} [(\hat{f}_{v,l} - f_v)^2]}. \tag{15}$$

Here $\hat{f}_{v,l}$ is the estimate of frequency of vth harmonic component in the lth simulation run, and f_v is the frequency of vth harmonic component.

We compared the performance of Root-MUSIC, Root-MUSIC with using SSA method (Root-MUSIC with SSA). Furthermore, Root-MUSIC with using adaptive SSA method (Root-MUSIC with adaptive SSA) and Root-MUSIC with adaptive SSA and usual estimate of noise variance (UE) were considered.

The experimental RMSE's of frequency estimation versus the SNR are presented in Fig. 5. The size of window used for realization of superresolution method was $m = 18$ and size of window for SSA was $m = 10$.

It is obvious from Fig. 5 the use of the adaptive SSA improves the performance of frequency estimation. The advantage of using the improved estimate of noise variance is observed at low SNRs.

In the next case we consider the similar case but $\Delta f = 0.011\,\text{Hz}$. The corresponding simulation results presented on Fig. 6.

The comparison of results obtained for $\Delta f = 0.013\,\text{Hz}$ (Fig. 5) and $\Delta f = 0.011\,\text{Hz}$ (Fig. 6) allows to reveal the influence of separation reduction of the signal components. The reduction of frequency separation leads to growth of the threshold SNR as compare to Fig. 5.

It is of interest to consider the dependence of RMSE of frequency estimation versus the size of segment. RMSEs of the mentioned before variants of Root MUSIC versus the size of segment are shown in Fig. 7. Here $\Delta f = 0.011\,\text{Hz}$.

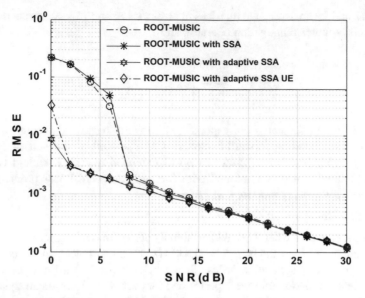

Fig. 5 Experimental RMSEs versus the SNR, $m = 18$, $\Delta f = 0.013\,\text{Hz}$

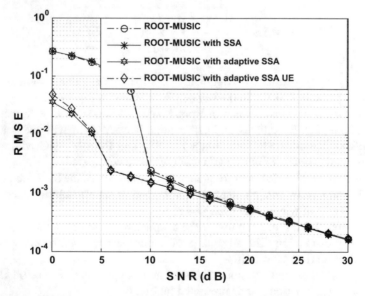

Fig. 6 Experimental RMSEs versus the SNR, $m = 18$, $10\,\Delta f = 0.011\,\text{Hz}$

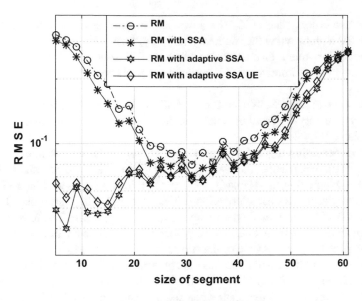

Fig. 7 Experimental RMSEs versus the size of segment, SNR = 5 dB

The similar dependences for SNR = 14 dB are presented on Fig. 8.

It is possible to see from Fig. 7 in the case of considered SNR the advantages of adaptive SSA as compared with Basic SSA are more evident when the value of

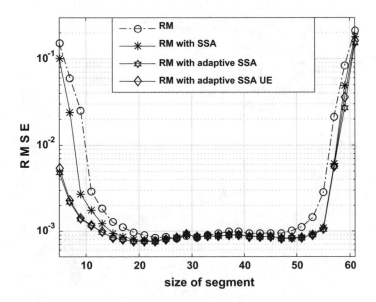

Fig. 8 Experimental RMSEs versus the segment size, SNR = 14 dB

segment size is less than $m/2$. The advantage of the improved estimate of noise estimate has a place when the segment size is less than $m/3$.

It should be noted that character of curves behavior is similar to the case of application well known spatial smoothing approaches in antenna array signal processing.

In the case of high SNR (for Fig. 8 SNR = 14 dB) the difference in performance when using the usual estimate and improved estimate of noise variance is inappreciable (Fig. 8).

Cadzow denoising algorithm consists of iterations related with matrix property mappings [10]. In particular, the rank V property mapping and the Hermitian-Toeplitz (or Hankel, block Toeplitz-Hankel) property mappings were used. The SVD representation was used for characterization of the matrices with given rank. Therefore, one iteration of Cadzow method corresponds to SSA approach. The reusing of proposed approach by analogy to Cadzow approach can improve the performance of the preprocessing and performance of the superresolution methods.

The corresponding simulation results are presented in Fig. 9. Here the abbreviation SI indicates on second iteration. In other words the data obtained after the first iteration are used as the input data when second using of adaptive SSA. The value of window used for the formation of data matrix before application superresolution method is $m = 18$ and window with size $m = 10$ is used for SSA approach. An improved estimate of noise variance was used for adaptive SSA.

As we can see the performance is improved at low SNR. However, it is necessary to say that the next reusing of adaptive SSA may not improve the performance. Such

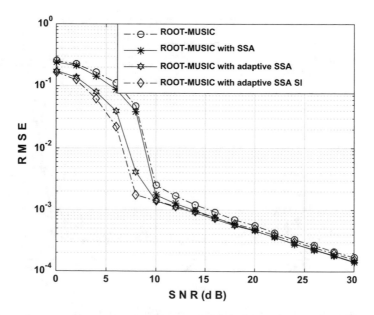

Fig. 9 Experimental RMSEs versus the SNR, $m = 18$, $\Delta f = 0.011\,\text{Hz}$

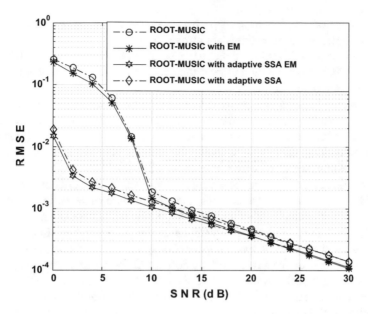

Fig. 10 Experimental RMSEs versus the SNR, SNR $= 15$ dB, $m = 18$, $\Delta f = 0.011$ Hz

as the each iteration of proposed approach requires the SVD (EVD) the small number of iteration can be recommended.

In the next case we compare the initial Root-MUSIC and Root-MUSIC with SSA for the case of usual data matrix and extended data matrix, described by Eq. (13). The abbreviation EM (Extended matrix) is used on legend for Fig. 10.

The size of window used for SSA is a half of window used for realization of superresolution method (i.e. 9 and 18 respectively). The performance improvement is evident at low and medium SNRs.

Therefore, the using of the data matrix with specific form can additionally improve the performance of adaptive SSA. In this case $\mathbf{Y}_{ext}$ is used. However it should be noted that the noise variance estimate was obtained using usual data matrix with Hankel form.

4 Conclusion

The adaptive complex SSA approach is presented. The essence of proposed technique is additional noise reduction by subtraction of noise standard deviation from the signal subspace (or signal + noise subspace) singular values of input data trajectory matrix.

Simulation results show that usage of the adaptive complex SSA method improves the performance of superresolution methods. The using of the improved noise variance estimate obtained based on the random matrix theory is recommended. The possible ways of simplification of proposed approach and joint using of adaptive SSA and superresolution methods are proposed.

The application of extended data matrix obtained from the filtered data matrix is also proposed. It additionally improves the performance of frequency estimation by superresolution methods.

The obtained results can be used as preprocessing before classification of the digital modulation type such as BPSK, QAM and so on. It is of interest to consider the problem of channel parameter estimation, the carrier frequency offset estimation for OFDM systems, estimation of multipath parameters in wireless communication, wind forecasting, structural damage detection [37–41].

Application of the bootstrap-based approaches, surrogate data theory, the elements of the blind signal separation methods, the combination of several noise-reduction approaches (for example, SSA and wavelet, SSA and surrogate data technology) probably gives the additional improvement of the performance. Moreover, it is of interest to use the results of spectral graph theory [39], spectral clustering.

The directions of the future investigations will be related with adaptation of the proposed approach for the iterative oblique SSA, sliding SSA, SSA-AMUSE, adaptive line enhancer, multidimensional harmonic retrieval problem with using higher-order SVD and tensors.

References

1. Proakis G, Salehi M (2008) Digital communications, 5th edn. McGraw-Hill, NewYork
2. Trees HLV (2002) Optimum array processing. Part IV of detection, estimation and modulation theory. Wiley-Interscience
3. Marple SL (2018) Digital spectral analysis, 2nd edn. Dover Publication Inc, Maniola, New York
4. Small M (2005) Applied nonlinear time series analysis applications in physics, physiology and finance. World Scientific Publishing Co. Pte. Ltd
5. Golyandina N, Zhigljavsky A (2013) Singular spectrum analysis for time series. Springer
6. Sanei S, Hassani H (2016) Singular spectrum analysis of biomedical signals. CRC Press, London
7. Yanai H, Takeuchi K, Takane Y (2011) Projection matrices, generalized inverse matrices, and singular value decomposition. Springer Science
8. Aivazyan SA, Buchstaber VM, Enyukov IS, Meshalkin LD (1989) Applied statistics. Classification and reduction of dimensionality. Finances and statistics. Moscow
9. Tufts DW, Kumaresan R, Kirsteins I (1982) Data adaptive signal estimation by singular value decomposition of a data matrix. Proc IEEE 70:684–685. https://doi.org/10.1109/PROC.1982.12367
10. Cadzow JA (1988) Signal enhancement—a composite property mapping algorithm. IEEE Trans ASSP 36:49–62. https://doi.org/10.1109/29.1488
11. Broomhead D, King G (1986) Extracting qualitative dynamics from experimental data. Phys D 20:217–236. https://doi.org/10.1016/0167-2789(86)90031-X
12. Broomhead D, Jones R, King G, Pike E (1987) Singular system analysis with application to dynamical systems. In: Pike ER, Lugiato LA (eds) Chaos, noise and fractals. Adam Hilger, Bristol
13. Scharf LL (1991) The SVD and reduced rank signal processing. Sig Process 25:113–133. https://doi.org/10.1016/0165-1684(91)90058-Q

14. Ephraim Y, Trees HLV (1995) A signal subspace approach for speech enhancement. IEEE Trans Speech Audio Process 3:251–266. https://doi.org/10.1109/89.397090
15. van der Veen AJ, Deprettere EF, Swindlehurst AL (1993) Subspace based signal analysis using singular value decomposition. Proc IEEE 81:1277–1308. https://doi.org/10.1109/5.237536
16. Vasylyshyn VI (2014) The signal preprocessing with using the SSA method in the spectral analysis problems. Appl Radio Electron 14(1):43–50 (in Russian)
17. Kostenko PYu, Vasylyshyn V (2015) Surrogate data generation technology using the SSA method for enhancing the effectiveness of signal spectral analysis. Radioelectron Commun Syst 58:356–361. https://doi.org/10.3103/S0735272715080038
18. Vasylyshyn V (2015) Adaptive variant of the surrogate data technology for enhancing the effectiveness of signal spectral analysis using eigenstructure methods. Radioelectron Commun Syst 58(3):116–126. https://doi.org/10.3103/S0735272715030036
19. Choi J, Evans BL, Gatherer A (2016) Space-time fronthaul compression of complex baseband uplink LTE signals. Paper presented at 2016 IEEE international conference on communications, Kuala Lumpur, 22–27 May 2016
20. Nilsson R, Sjöberg F, LeBlanc JPA (2003) Rank-reduced LMMSE canceller for narrowband interference suppression in OFDM-based systems. IEEE Trans Commun 51(12):2126–2140. https://doi.org/10.1109/TCOMM.2003.820761
21. Vasylyshyn V, Lyutov V (2018) Signal denoising using modified complex SSA method with application to frequency estimation. Paper presented at 2018 5th international scientific-practical conference problems of infocommunications. Science and technology, Kharkiv, 9–12 Oct 2018
22. Leles MCR, Sansão JPH, Mozelli LA, Guimarães HN (2018) Improving reconstruction of time-series based in singular spectrum analysis: a segmentation approach. Digit Signal Proc 77:63–76. https://doi.org/10.1016/j.dsp.2017.10.025
23. Harmouche J, Fourer D, Auger F, Borgnat P, Flandrin P (2018) The sliding singular spectrum analysis: a data-driven non-stationary signal decomposition tool. IEEE Trans Signal Process 66(1):1–13. https://doi.org/10.1109/TSP.2017.2752720
24. Krishna EH, Sivani K, Reddy K (2018) New channel estimation method using singular spectrum analysis for OFDM systems. Wireless Pers Commun 101(4):2193–2207. https://doi.org/10.1007/s11277-018-5811-5
25. Besson O, Stoica P (2003) On parameter estimation of MIMO flat-fading channels with frequency offsets. IEEE Trans Signal Process 51(3):602–613. https://doi.org/10.1109/TSP.2002.808102
26. da Costa JPCL, Haardt M, Rmer F, del Galdo G (2007) Enhanced model order estimation using higher-order arrays. Paper presented at 41st Asilomar conference on signals, systems, and computers, Pacific Grove, 4–7 Nov 2007
27. Kung SY, Arun KS, Bhaskar Rao DV (1983) State-space and singular-value decomposition methods for the harmonic retrieval problem. J Opt Soc Am 77:1799–1811. https://doi.org/10.1364/JOSA.73.001799
28. Zhidong B, Zhaoben F, Yingchang L (2014) Spectral theory of large dimensional random matrices and its applications to wireless communications and finance statistics. World Scientific Publishing Co. Pte. Ltd
29. Yazdian E, Gazor S, Bastani H (2012) Source enumeration in large arrays using moments of eigenvalues and relatively few samples. IET Signal Process 6(7):689–696. https://doi.org/10.1049/iet-spr.2011.0260
30. Kritchman S, Nadler B (2008) Determining the number of components in a factor model from limited noisy data. Chemometr Intell Lab Syst 94:19–32. https://doi.org/10.1016/j.chemolab.2008.06.002
31. Dologlou I, Carayannis G (1991) Physical interpretation of signal reconstruction from reduced rank matrices. IEEE Trans Signal Process 39(7):1681–1682. https://doi.org/10.1109/78.134407
32. Hansen PC, Jensen SH (1998) FIR filter representations of reduction-rank noise reduction. IEEE Trans Signal Process 46(6):1737–1741. https://doi.org/10.1109/78.678511

33. Van Huffel S (1993) Enhanced resolution based on minimum variance estimation and exponential data modeling. Sig Process 33:333–355. https://doi.org/10.1016/0165-1684(93)90130-3
34. Vasylyshyn V (2007) Antenna array signal processing with high-resolution by modified beamspace MUSIC algorithm. Paper presented at 6th international conference on antenna theory and techniques, Sevastopol, 17–21 Sept 2007
35. Moskvina V, Schmidt KM (2003) Approximate projectors in singular spectrum analysis. SIAM J Matrix Anal Appl 24(4):932–942. https://doi.org/10.1137/S0895479801398967
36. Vasylyshyn V (2013) Removing the outliers in root-MUSIC via pseudo-noise resampling and conventional beamformer. Sig Process 93:3423–3429. https://doi.org/10.1016/j.sigpro.2013.05.026
37. Li J, Liu G, Giannakis GB (2001) Carrier frequency offset estimation for OFDM-based WLANs. IEEE Signal Process Lett 8(3):80–82. https://doi.org/10.1109/97.905946
38. Volosyuk VK, Kravchenko VF, Kutuza BG, Pavlikov VV (2015) Review of modern algorithms for high resolution imaging with passive radar. Paper presented at the international conference on antenna theory and techniques, Kiev, 21–24 Apr 2015
39. Ortega A, Frossard P, Kovačević J, Moura JMF, Vandergheynst P (2018) Graph signal processing: overview, challenges and applications. Proc IEEE 106(5):808–828. https://doi.org/10.1109/JPROC.2018.2820126
40. Lemeshko AV, Evseeva OYu, Garkusha SV (2014) Research on tensor model of multipath routing in telecommunication network with support of service quality by greater number of indices. Telecommun Radio Eng 73(15):1339–1360. https://doi.org/10.1615/TelecomRadEng.v73.i15.30
41. Bulakh V, Kirichenko L, Radivilova T (2018) Time series classification based on fractal properties. Paper presented at the 2018 IEEE second international conference on data stream mining & processing (DSMP), Lviv, 21–25 Aug 2018
42. Ageyev DV, Salah MT (2016) Parametric synthesis of overlay networks with self-similar traffic. Telecommun Radio Eng 75(14):1231–124 https://doi.org/10.1615/TelecomRadEng.v75.i14.10

Assortativity Properties of Barabási-Albert Networks

Vadim Shergin, Serhii Udovenko, and Larysa Chala

Abstract Nodes distribution by degrees is the most important characteristic of complex networks. For SF-networks it follows a power law. While degree distribution is a first order graph metric, the assortativity is a second order one. The concept of assortativity is extensively using in network analysis. In general degree distribution forms an essential restriction both on the network structure and on assortativity coefficient boundaries. The problem of determining the structure of BA-networks having an extreme assortativity coefficient is considered. Greedy algorithms for generating BA-networks with extreme assortativity have been proposed. As it was found, an extremely disassortative BA-network is asymptotically bipartite, while an extremely assortative one is close to be a set of complete and disconnected clusters. The estimates of boundaries for assortativity coefficient have been found. These boundaries are narrowing with increasing the network size and are significantly narrower than for networks having an arbitrary structure.

Keywords Assortativity · Scale-free networks · Barabasi-Albert model

1 Introduction

The most important characteristic of complex networks is the nodes distribution by degrees, i.e. by the number of links. A network is called *Scale-Free* (SF) if this distribution follows a power law, at least asymptotically. According to the results of many studies [1–4], most of the real-world networks are scale-free. The simplest and most common model of SF-networks is the Barabasi-Albert (BA) model.

V. Shergin (✉) · L. Chala
Artificial Intelligence Department, Kharkiv National University of Radio Electronics (NURE), Kharkiv, Ukraine
e-mail: vadim.shergin@nure.ua

S. Udovenko
Informatics and Computer Engineering Department, Kharkiv National University of Economics (KhNUE), Kharkiv, Ukraine

© The Editor(s) (if applicable) and The Author(s), under exclusive license to Springer Nature Switzerland AG 2021
T. Radivilova et al. (eds.), *Data-Centric Business and Applications*, Lecture Notes on Data Engineering and Communications Technologies 48,
https://doi.org/10.1007/978-3-030-43070-2_4

Despite the importance of the degree distribution, it is not a comprehensive characteristic of networks. An important one is an *assortativity* [5, 6], that is, the tendency of nodes to connect with similar or dissimilar ones (thus networks called assortative or disassortative respectively). While degree distribution is a first order graph metric, the assortativity is a second order one. In a strict sense, the network assortativity may be applied to any of first order characteristics [5, 7] such as node weight, node betweenness, kth level node degree. In addition, assortativity may be applied to node characteristics that are not directly topology-related. But in common use assortativity is considered for the degree of the nodes in the network. It is known [1, 5, 8] that social networks have a mainly positive assortativity, while biological and technical ones are disassortative. Artificial network models (BA-model and others) are asymptotically neutral.

Despite a large number of studies of scaling and assortativity, these characteristics are studied separately which is considered as an essential disadvantage of the current state of the complex networks theory. While the structure of general-type networks with extreme assortativity is widely known [5], it is unknown what it is for SF-networks. In particular, there exists the rewiring algorithm to change the assortativity without changing the nodes distribution, but it is known nothing that the attainable limits of the assortativity coefficient are equal to.

The assortativity properties of SF-networks in general and BA-model, in particular, are studied in this work. Finding the network structure for extreme cases (extremely assortative and disassortative BA-networks) forms the first problem to deal with, while estimating the bounds of assortativity coefficient for the BA-model of SF-networks with known power exponent is another one.

2 The Effect of Network's Scaling Properties on Their Assortativity

2.1 Assortativity Properties of Networks

The term "assortative mixing" originally appeared in epidemiology and sociology, in particular for mixing by race among sexual partners [1].

In network theory, assortative mixing characterizes the tendency for vertices in networks to be connected to other vertices that are like (or unlike) them in some way. As a simplest case assortative mixing is considered according to vertex degree. An *assortativity coefficient* is defined as the correlation coefficient of the nodes by their degrees [1]. More convenient form of this coefficient is given in [5]:

$$r = \frac{S_1 N_3 - S_2^2}{S_1 S_3 - S_2^2}, \tag{1}$$

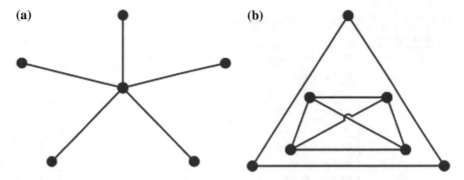

Fig. 1 Structure of extremely disassortative (**a**) and assortative (**b**) networks

where

$$S_k = \sum_{i=1}^{n} d_i^k, \quad N_3 = d^T A d = \sum_{i=1}^{n} \sum_{j=1}^{n} A_{ij} d_i d_j, \tag{2}$$

A is adjacency matrix of the network, d_i is node degrees.

In general case assortativity coefficient (1) lies in the range $-1 \leq r \leq 1; r = 0$ for random networks. In the boundary case $r = -1$ (an extremely disassortative mixing) network has *biregular* structure [5], so all vertices on the same side of the bipartition have the same degree and degrees of the different sides of the bipartition do not coincide with each other. The simplest example of such a structure is star graph (Fig. 1).

In another boundary case $r = 1$ (assortative mixing) network consists of two or more regular connected components (Fig. 1).

It was established [1, 2, 4, 6, 9–12] that the real-world networks are not neutral on its assortativity (Table 1): social networks are positively assortative while technical and biological ones are disassortative.

At a first glance to the data on Table 1 it may seems that assortativity coefficients for real-word networks are close to zero, in addition their structures are very far from extreme assortative/disassortative cases (Fig. 1), so it may seems that assortativity is not an essential property. In fact it is not so.

2.2 The Effect of Scaling on Mixing Properties

One of the key-properties of complex networks is their self-similarity or scale-freeing [2, 6, 13]. The network is called *scale-free* (SF), if the distribution of vertex degrees follows a power law, at least asymptotically, i.e. the fraction p_k of nodes, having k edges for large k is

Table 1 Assortativity values for different networks (from [2])

Kind of network	Network	Assortativity
Social	Physics co-authorship	0.363
	Mathematics co-authorship	0.120
	Company directors	0.276
Technical	Connections between autonomous systems on the internet	−0.189
	World-wide web	−0.067
	Undirected hyperlinks between Web pages in a single domain	−0.065
Biological	Neural network	−0.163
Simulated models	Experimental Erdős-Rényi (ER) graph (for sufficiently large network size)	~0
	Experimental Barabási-Albert (BA) graph (for sufficiently large network size)	~0

$$p_k \sim k^{-\gamma}, \quad 2 < \gamma \leq 3. \tag{3}$$

If fraction distribution of node degrees is power (3), then their range distribution asymptotically also follows a power-family distribution. Without loss of generality, we can assume that the nodes are ordered by a decreasing degree, so index number i of the node d_i is its range:

$$d_i \sim i^{-\beta}, \quad \beta = 1/(\gamma - 1). \tag{4}$$

One of the simplest and most popular model of SF-network is Barabási-Albert (BA) model, for which $\gamma = 3$ and thus $\beta = 1/2$. To obtain more accurate form of (4) for BA-model one should clarify and specify generative procedure for this model.

Initially net is a full graph with $m + 1$ vertices, so each vertex has m edges. Each time step one vertex with m edges is added. This edges links with some of the existing vertices according to the *preferential linking* rule [2]: for any of the existing vertices i the probability of getting a link from new vertex is proportional to the vertex degree d_i ($p_i = d_i / \sum d_i$). Thus, the vertex degree evolution equation is

$$E[d_{i,n+1}] = d_{i,n} + m \frac{d_{i,n}}{S_1(n)}, \quad S_1(n) = \sum_{i=1}^{n} d_{i,n} = 2mn - m(m+1), \tag{5}$$

with initial conditions

$$\begin{aligned} d_{i,i} &= m, \quad i = m+1, \ldots, n \\ d_{i,m+1} &= m, \quad i = 1, \ldots, m \end{aligned}, \tag{6}$$

where n is the number of vertices (regarded as time measure); $S_1(n)$ is the total number of links; $E[\ ]$ denotes expected value.

One can see that the solution of (5)–(6) asymptotically follows the power law (4). In the particular case $m = 2$ it has the form

$$d_1 = d_2 \approx \sqrt{\pi(n-1)}, \quad d_i = 2\frac{\Gamma(n-1) \cdot \Gamma\left(i - \frac{3}{2}\right)}{\Gamma\left(n - \frac{3}{2}\right) \cdot \Gamma(i-1)} \approx 2\sqrt{\frac{n-1}{i-1}}, \quad i > 2. \quad (7)$$

It follows directly from the definition (1)–(2) that network assortativity coefficient (r) depends on degree distribution d_i (4), (7), so network mixing cannot be studied apart from scaling. One can expect that the really achievable bounds of assortativity coefficient will be much narrower than ± 1.

One of the problems to be discussed is estimating the bounds of assortativity coefficient for the SF-networks with known power exponent β (4) and another one is finding the networks structure for that extreme cases.

As it follows from (2), N_3 is the only component of (1) which depends on network structure (i.e. adjacency matrix A). Thus, finding the networks structure (in extreme assortative/disassortative case) is a *binary programming* problem:

$$N_3 = d^T A d = \sum_{i=1}^{n} \sum_{j=1}^{n} A_{ij} d_i d_j \rightarrow extr \quad (8)$$

with restrictions (4) or (7) and

$$\sum_{j=1}^{n} A_{ij} = d_i, \quad A_{ij} \in \{0, 1\}, \quad A_{ij} = A_{ji}, \quad A_{ii} = 0. \quad (9)$$

3 The Structure of Extremely Assortative/Disassortative Scale-Free Networks

In general case (i.e. fixed but arbitrary degree distribution) problem (8)–(9) is *NP-complete* and its exact solution cannot be obtained in analytical form.

For small-scale networks problem (8)–(9) can be solved numerically. An adjacency matrix of extremely disassortative BA-network with $n = 32$ vertices, $\beta = 1/2$ and $m = 2$ (number of edges added with each new vertex) is shown on Fig. 2.

As one can see this structure is *not biregular* (like a star graph on Fig. 1a) but tends to be *bipartite*. Vertices can be divided into two subsets: first $k = 9$ of them are "rich" in links, other vertices are "poor", and vertices from each subsets almost never connected with their own kind.

An asymptotical solution of (8)–(9) can be found by greedy algorithm. Its pseudocode has the form:

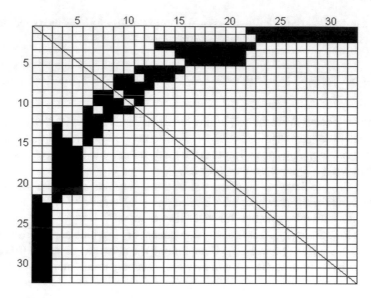

Fig. 2 An adjacency matrix of extremely disassortative BA-network with $n = 32$, $\beta = 1/2$

```
for(i=1; i<=n ; i++)      dFree[i]=d[i];
for(i=1; i<=k ;) {
    for(j=n; dFree[j] = = 0; j--) ;  // finding vertices
    //which are not connected yet, starting from poorest
    connect(i,j);
    dFree[j]--;
    dFree[i]--;
    if (dFree[i] = = 0)      i++;
}
```

The fact that this algorithm provides a minimum of (8) can be proved by contradiction. After finishing this algorithm one can change the assortativity (i.e. criterion (8)) by Degree-Preserving Rewiring (DPR) procedure [5]. DPR changes the criterion (8) as

$$\Delta N_3 = 2(d_i d_l + d_k d_j - d_i d_j - d_k d_l) = 2(d_i - d_k)(d_l - d_j) \qquad (10)$$

According to generative algorithm of disassortative network, $d_k \geq d_i$ and $d_l \geq d_j$ for any i, j, k, l. Thus $\Delta N_3 \geq 0$, i.e. assortativity coefficient cannot be decreased by DPR, hence our generative algorithm really provides the minimum of assortativity coefficient.

The general scheme of DPR and its application to disassortative network are shown on Fig. 3.

The boundary value between "rich" and "poor" subsets is determined by a balance condition: each of subsets has the half of overall network links:

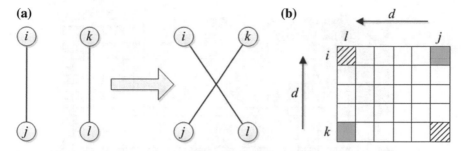

Fig. 3 The general scheme of DPR (**a**) and its application to disassortative network (**b**)

$$\sum_{i \le k} d_i = \sum_{i > k} d_i. \tag{11}$$

If vertex degree follows a power distribution (4), an approximate solution of (11) is

$$k = 1 + \lfloor n \cdot 2^{-1/(1-\beta)} \rfloor. \tag{12}$$

In general (11) has not an exact integer solution, that is the cause of exceptions from bipartiality near the bound (12), i.e. the cause of a links between vertices of the same subset (links 7–9, 8–9 and 10–11 on Fig. 2).

The structure of *extremely assortative* SF-networks also can be found in a similar way. The only difference in generative algorithms is the direction of a nested cycle (by j): to maximize the assortativity coefficient one should try to link the current vertex (having index i) with the richest among remaining, so the nested cycle is ascending.

For considered case (BA-network with $n = 32$ vertices and $\beta = 1/2$) an adjacency matrix obtained numerically is shown on Fig. 4.

One can see that the structure of an extremely assortative network tends to the "ideal" one (Fig. 1b). Such network is a set of almost isolated clusters all but the largest of which are almost regular.

4 Estimating the Boundary Values of Assortativity Coefficient for BA-Network

In disassortative case according to generative algorithm each vertex i from "rich" subset is linked with d_i of "poor" vertices which indexes (ranges) are in series (from some j_{min} up to $j_{max} = j_{min} + d_i$). Thus, according to mean value theorem for each i there exists some intermediate value $\bar{j} = f(i)$ such that the minimum value of (8) is

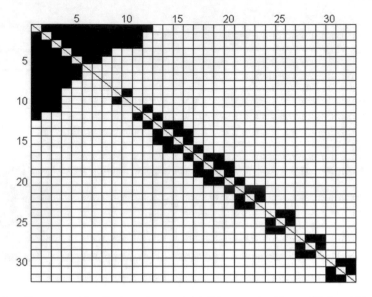

Fig. 4 An adjacency matrix of extremely assortative BA-network with $n = 32$, $\beta = 1/2$

$$N_3^{\min} = \sum_{i=1}^{k} \sum_{j \sim i} d_i d_j = \sum_{i=1}^{k} d_i \sum_{j \sim i} d_j = \sum_{i=1}^{k} d_i^2 d_{\bar{j}}. \tag{13}$$

The value of $\bar{j}$ can be estimated in assumption that vertices having indexes less than i captures all the links with vertices having indexes greater than $\bar{j}$:

$$\sum_{s < i} d_s = \sum_{s > \bar{j}} d_s. \tag{14}$$

In abstraction from integerness of all indexes and degrees, the solution of (14) for BA-model with degree distribution (7) takes the form

$$\sqrt{i} + \sqrt{\bar{j}} = \sqrt{n}. \tag{15}$$

From (14)–(15) it also follows that limit of summation in (13) is determined by the condition $k = f(k)$, thus, $k = 1 + \lfloor n/4 \rfloor$ (in accordance with (12)).

Approximating the sum (13) by the integral, in respect with (7) and (15) we obtain

$$N_3^{\min} \approx 2 \sum_{i=1}^{k} \frac{(4n)^{3/2}}{i(\sqrt{n} - \sqrt{i})} \approx 16n \cdot (\ln(n) + a), \tag{16}$$

where $a = 1 + \gamma$, $\gamma \approx 0.5772$—Euler–Mascheroni constant.

As it was mentioned, terms S_1, S_2, S_3 in (1) does not depends on mixing properties (i.e. on adjacency matrix A), but only on scaling properties, i.e. on degree distribution d_i. For BA-model having degree distribution (7) an asymptotic estimates are:

$$S_1 = 4n - 6, \quad S_2 \approx 4n(\ln(n) + c_2), \quad S_3 \approx 4c_3 \cdot n^{3/2}, \tag{17}$$

The numerical estimates of c_2, c_3 are given in Table 2.

According to (16), (17), the lower bound of assortativity coefficient (1) is estimated as

$$r_{\min} \approx \frac{4(\ln(n) + a) - (\ln(n) + c_2)^2}{c_3 \sqrt{n} - (\ln(n) + c_2)^2}, \tag{18}$$

$$\lim_{n \to \infty} r_{\min} \approx \frac{-\ln^2(n)}{c_3 \sqrt{n}}. \tag{19}$$

In the case of maximization the criterion (8) (i.e. an extremely assortative case) the first $d_1 = \sqrt{\pi(n-1)}$ vertices (i.e. the "richest" one) linked to each other, forming the largest cluster ("arrow head" on Fig. 4). The contribution of this cluster to the objective function (8) is

$$N_{31} = \sum_{i=1}^{d_1+1} \sum_{j=1}^{d_i+1} d_i d_j. \tag{20}$$

By approximating this sum by the integral in abstraction from integerness of all parameters and in respect with (7) we obtain:

$$N_{31} \approx 4\sqrt{n} \sum_{i=1}^{d_1+1} (d_i)^{3/2} \approx 4\sqrt{n} \int_0^{\sqrt{\pi n}} \left(\frac{4n}{x}\right)^{3/4} dx = 4b \cdot n^{11/8}. \tag{21}$$

To obtain an estimate of the entire objective function (8), one should add up to (21) the contributions from the rest of the clusters (i.e. from "arrow body" on Fig. 4). However, one can prove that these contributions in total are $O(n \log(n))$, so $N_3 \approx N_{31}$.

In accordance with (17), (21), the upper bound of assortativity coefficient can be estimated as

Table 2 The numerical values of parameters estimates

Parameter	a	b	c_2	c_3
Formula number	(16)	(21)	(17)	(17)
Estimated value	1.577	10.026	1.779	7.160

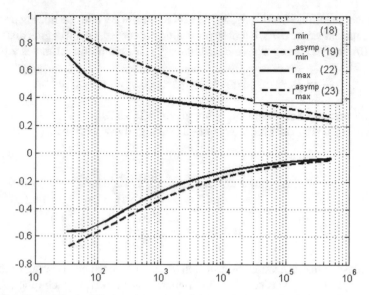

Fig. 5 Estimates of the boundaries of the assortativity coefficient for BA-network

$$r_{\max} \approx \frac{b \cdot n^{3/8} - (\ln(n) + c_2)^2}{c_3 \cdot n^{1/2} - (\ln(n) + c_2)^2},$$ (22)

$$\lim_{n \to \infty} r_{\max} \approx \frac{b}{c_3} n^{-1/8}.$$ (23)

Estimates of the lower (18)–(19) and upper (22)–(23) boundaries of the assortativity coefficient are shown on Fig. 5. As one can see, these boundaries are essentially narrowed with increasing the network size. One can also conclude that for large networks (with $n > 300$ nodes) the boundary from the positive side is much wider than from the negative one. Therefore, the fact that technical and biological networks have assortativity coefficients which are much smaller in absolute value than for social ones (Table 1), can be explained by the sign of this coefficient. Also, real-world networks not strictly follow to a BA-model, the obtained results (greedy algorithms and extremal structures) may be generalized to the general (not BA-) models of SF-networks.

5 Conclusions and Future Work

The most important characteristic of complex networks is the nodes distribution by degrees. A lot of real-world networks are found to be scale-free, i.e. the nodes degrees follow a power law. One of the models of SF-networks is BA-model.

While degree distribution is a first order graph metric, the assortativity is a second order one. The concept of assortativity is extensively using in network analysis.

Although there exists a means for changing the assortativity coefficient without changing the degree vector, but in general degree distribution forms an essential restriction both on the network structure and on coefficient bounds. Thus, network assortativity properties cannot be studied apart from scaling.

The problem of determining the structure of BA-networks having an extreme assortativity coefficient is considered. The mathematical formulation of this problem is done. Greedy algorithms for solving these problems have been developed. The structure of BA-networks with extreme assortativity has been found. An extremely disassortative BA-network is asymptotically bipartite (Fig. 2) while an extremely assortative one (Fig. 4) is asymptotically a set of clusters which are close to be complete and disconnected subgraphs.

The formulation of the problem of finding the boundary values for the assortativity coefficient of SF-networks were made. For the BA-model of SF-networks the estimates of boundaries for assortativity coefficient has been found. These boundaries are narrowing with increasing the network size and are significantly narrower than for networks having an arbitrary structure.

Generalization the obtained results to the general (not BA-) models of SF-networks is a promising direction for further research.

References

1. Newman M (2003) Mixing patterns in networks. Phys Rev E 67
2. Dorogovtsev N, Mendes J (2002) Evolution of networks. Adv Phys 51(4):1079–1187
3. Shergin VL, Chala LE, Udovenko SG (2018) Fractal dimension of infinitely growing discrete sets. In: 2018 14th international conference on advanced trends in radioelectronics, telecommunications and computer engineering (TCSET), Slavske, pp 259–263. https://doi.org/10.1109/TCSET.2018.8336198
4. Kirichenko L, Radivilova T, Bulakh V (2018) Classification of fractal time series using recurrence plots. In: 2018 international scientific-practical conference problems of infocommunications. Science and technology (PIC S&T). IEEE, pp 719–724. https://doi.org/10.1109/infocommst.2018.8632010
5. Noldus R, Van Mieghem P (2015) Assortativity in complex networks. J Complex Netw 3:507–542
6. Shergin V, Chala L (2017) The concept of elasticity of scale-free networks. In: 2017 4th international scientific-practical conference problems of infocommunications. Science and technology (PIC S&T), Kharkov, pp 257–260. https://doi.org/10.1109/INFOCOMMST.2017.8246392
7. Peel L, Delvenne JC, Lambiotte R (2017) Multiscale mixing patterns in networks. Proc Natl Acad Sci 115. https://doi.org/10.1073/pnas.1713019115
8. Muscoloni A, Cannistraci CV (2017) Rich-clubness test: how to determine whether a complex network has or doesn't have a rich-club? In: The 6th international conference on complex networks & their applications, Lyon (France), 29 Nov–01 Dec 2017, pp 291–293
9. Ageyev DV, Ignatenko AA, Wehbe F (2013) Design of information and telecommunication systems with the usage of the multi-layer graph model. In: Proceedings of the XIIth international conference on the experience of designing and application of CAD systems in microelectronics (CADSM). Lviv Polytechnic National University, Lviv-Polyana, Ukraine, pp 1–4
10. Ageyev D et al (2018) Classification of existing virtualization methods used in telecommunication networks. In: Proceedings of the 2018 IEEE 9th international conference on dependable systems, services and technologies (DESSERT), pp 83–86

11. Kryvinska N (2010) Converged network service architecture: a platform for integrated services delivery and interworking. In: Electronic business series, vol 2. International Academic Publishers, Peter Lang Publishing Group
12. Kirichenko L, Radivilova T, Zinkevich I (2018) Comparative analysis of conversion series forecasting in E-commerce tasks. In: Shakhovska N, Stepashko V (eds) Advances in intelligent systems and computing II. CSIT 2017. Advances in intelligent systems and computing, vol 689. Springer, Cham, pp 230–242. https://doi.org/10.1007/978-3-319-70581-1_16
13. Shergin V, Chala L, Udovenko S (2019) Assortativity properties of scale-free networks. In: 2019 international scientific-practical conference problems of infocommunications. Science and technology (PIC S&T). IEEE, pp 92–96

Influence and Betweenness in Flow Models of Complex Network Systems

Olexandr Polishchuk

Abstract This paper provides the analysis for functional approaches of complex network systems research. In order to study the behavior of these systems the flow adjacency matrices were introduced. The concepts of strength, power, domain and diameter of influence of complex network nodes are analyzed for the purpose of determining their importance in the systems structure. The notions of measure, power, domain and diameter of betweenness of network nodes and edges are introduced to identify their significance in the operation process of network systems. These indicators quantitatively express the contribution of the corresponding element for the motion of flows in the system and determine the losses that are expected in the case of blocking this node or edge or targeted attack on it. Similar notions of influence and betweenness are introduced to determine the functional importance of separate subsystems of network system and the system as a whole. Examples of practical use of the obtained results in information processing and management are given.

Keywords Complex Network · Network System · Complexity · Flow · Influence · Centrality · Betweenness · Stability

1 Introduction

To study any real network system (NS), whether natural or artificial, we have to form full and comprehensive representation of this system. Usually it is reached through observations, experimental and theoretical investigations and displaying the system as the models of different types [1]. When talking about network systems

O. Polishchuk (✉)
Laboratory of Modeling and Optimization of Complex Systems, Ya. S. Pidstryhach Institute for Applied Problems of Mechanics and Mathematics National Academy of Sciences of Ukraine, Lviv, Ukraine
e-mail: od_polishchuk@ukr.net

© The Editor(s) (if applicable) and The Author(s), under exclusive license to Springer Nature Switzerland AG 2021
T. Radivilova et al. (eds.), *Data-Centric Business and Applications*, Lecture Notes on Data Engineering and Communications Technologies 48,
https://doi.org/10.1007/978-3-030-43070-2_5

modeling, two main approaches may be distinguished: structural and functional. In modern NSs studies, the structural approach prevails, which is implemented in so-called theory of complex networks (TCN) [2, 3]. The subject of TCN investigation s is the creation of universal network structures models, determination of statistical features that characterize their behavior and forecasting networks behavior in case their structural properties change. Sometimes the term complex network (CN) is used to denote both structure and system [4, 5], though these are fundamentally different concepts. The laws according to which the systems operate are usually much more complicated than the features of system structure, and methods of structural studies often do not allow us to solve NS functional problems [6]. Within the scope of functional approach, system structure is analyzed in conjunction with functions implemented by components of this structure and system in general, but the function takes precedence over structure.

The theory of binary networks is completely abstracted from the functional features of the NS. Weighted networks are an attempt to "tied" the functional characteristics of the system to the elements of structure [7]. Indeed, in each particular case, the weight of CNs edges is a reflection of certain functionality of the corresponding system [8]. Network, as a structure, is considered to be dynamic if the composition of its nodes and edges changes over time. The system is a dynamic formation, even if its structure remains unchanged [9, 10]. The system forms its structure in the process of development. The structure is being developed and improved from the needs of the system and not vice versa [11]. What prompts the structure to develop, modify, or degrade? Movement of flows is one of the defining features of real NS. In some cases, providing the movement of flows is the main goal of creation and operation of such systems (transport and telecommunication systems, resource supply systems, trade and information networks, etc. [12, 13]), in others—the main process that provides their vital activity (blood and lymph flows, neuronal impulses in the human body). Stopping of flows movement leads to the termination of the NS existence.

Complexity of network structures and systems as well as of their models in general may be represented by different concepts. Network complexity is determined, in particular, by the presence of a large number of nodes and edges between them [14, 15]. Networks with relatively small number of elements are usually not considered complex. But these relatively small structures can generate unquestionably complex systems [16]. In other words, the complexity of the network structure is quantitative, and the complexity of the system is qualitative. While trying to embrace functional complexity we often have to neglect the structural complexity. Among the examples of this situations are the attempts to solve problems associated with controllability and observability of NSs. At the present stage, such problems are being solved for the simplest linear models of network systems with the number of nodes up to 100 [17, 18]. Such structures are hard to be called complex. At the same time, problems associated with controllability, observability and synchronization of large-scale systems are rather complex functional, not structural problems. This does not downplay the significance of structural approach of studies, as long as poor operation of many real systems is driven by the disadvantages of their structure [19].

This means the need to develop a conceptual apparatus and toolkit for studying the functional features of operation process of network systems components, beginning with their elements and ending with the system as a whole. Introduction and research of functional analogues of well-known structural characteristics of complex networks elements is one of the ways to solve this problem. This allows us to compare the advantages and disadvantages of functional and structural approaches to the study of NS of different types and nature, to combine them in order to create a holistic view about the state and operation process of the system, and also contribute a deeper understanding of NS behavior and solution of some practically important problems.

2 Flow Adjacency Matrices of Network Systems

The network structure is completely determined by its adjacency matrix $\mathbf{A} = \{a_{ij}\}_{i,j=1}^N$, where N is the number of CN nodes. For the most studied binary networks, the value of a_{ij} is equal to 1, if there is a connection between the nodes, and is equal to 0, if such connection is absent. Using the matrix $\mathbf{A}$ are defined the local and global characteristics of CN and studied its properties. We describe the process of system functioning on the basis of flows motion analysis by the network and introduce the following adjacency matrices of NS [20]:

(1) the matrix of the density of flows which are moving by the network edges at the current moment of time t:

$$\boldsymbol{\rho}(t, x) = \{\rho_{ij}(t, x)\}_{i,j=1}^N, \ x \in (n_i, n_j),$$

where (n_i, n_j) is the edge connected network nodes n_i and n_j, $i, j = \overline{1, N}, t > 0$;

(2) the matrix of volumes of flows that are moving by the network edges at time t:

$$\mathbf{v}(t) = \{v_{ij}(t)\}_{i,j=1}^N, \ v_{ij}(t) = \int_{(n_i, n_j)} \rho_{ij}(t, x)dl, \ t > 0,$$

(3) the integral flow adjacency matrix (IFAM) of volumes of flows passed through the network edges for the period $[t - T, T]$ to the current moment t:

$$\mathbf{V}(t) = \{V_{ij}(t)\}_{i,j=1}^N, \ V_{ij}(t) = \tilde{V}_{ij}(t)/ \max_{m,l=\overline{1,N}} \tilde{V}_{ml}(t),$$

$$\tilde{V}_{ij}(t) = \int\limits_{t-T}^{t} v_{ij}(\tau)d\tau, \ t \geq T > 0,$$

(4) the matrix of loading of network edges at time t:

$$\mathbf{u}(t) = \{u_{ij}(t)\}_{i,j=1}^{N}, \ u_{ij}(t) = v_{ij}(t)/v_{ij}^{\max}, \ t > 0,$$

where $v_{ij}^{\max}$ is bandwidth of the edge connected the network nodes n_i and $n_j, i, j = \overline{1, N}$,

(5) the integral matrix of NS loading for period $[t - T, T]$ to the moment t:

$$\mathbf{U}(t) = \{U_{ij}(t)\}_{i,j=1}^{N}, \ U_{ij}(t) = \int\limits_{t-T}^{t} u_{ij}(\tau)d\tau/T, \ t \geq T > 0.$$

The introduced above flow adjacency matrices in aggregate give a sufficiently clear quantitative picture of the system's operation process, allow us to analyze the features and predict the behavior of this process, to evaluate its effectiveness and prevent existing or potential threats [16, 19]. The matrices $\rho(t, x)$ and $\mathbf{v}(t)$ can be useful for the current analysis of network system's operation. The matrix $\mathbf{V}(t)$ enable to track the integral volumes of flows that pass through the network edges. They are especially important in predicting and/or planning the NS operation and allow us to timely respond to deploying threatening processes in the system. The matrices $\mathbf{u}(t)$ and $\mathbf{U}(t)$ enable to analyze the current and integral activity or passivity of separate system components, as well as the level of their critical loading, which can lead to crashes in the NS operation [21]. These matrices allow us to timely increase the bandwidth of network elements, build new ones or search the alternative paths of flows movement, etc. Systems of transmission, processing and analysis of information are very dynamic formations [22, 23]. Therefore, continuous monitoring of flows motion by the network is especially important in such systems. The introduced above flow adjacency matrices allow us to carry out such real-time monitoring.

During investigation of the system and forming its model we are interested in a clear identification of the NS structure. The network elements that are not involved in the system operation will be called fictitious. Examples of the existence of numerous fictitious nodes and edges can be found in many real systems, including social networks and the Internet [24]. The World Wide Web has a deep and dark web, pages of which are not indexed by any search engines [25]. Elements that are involved in the operation of particular system, but not included in its structure, will be called hidden. The identification of hidden nodes and connections plays no less important role in constructing the NS model than the search of fictitious elements. Obviously, the removal of fictitious elements contributes to overcoming the complexity problem by reducing the dimensionality of system model, and the inclusion of hidden nodes and connections—to better understanding of processes that occur in it. The flow adjacency matrices of the NS enable to identify the fictitious elements in the source network and exclude them from the system structure [16]. These matrices also allow us to carry out the search and inclusion of hidden nodes and connections in the system structure [20].

Different ways to determine both the local and global importance of the network node there are in TCN [3, 7, 26, 27]. However, the importance of a node in the structure is often not the same as the functional significance of node in the system [19].

3 Influence of Network Systems Nodes

The functional importance of the edge (n_i, n_j) in the system is determined by the value $V_{ij}(t)$, $i, j = \overline{1, N}$. We will define the functional importance of node in the following way [20]. Denote by $v_k^{out}(t, n_i, n_j)$ the volume of flows generated in node n_i and received at node n_j, which passed through the path $p_k(n_i, n_j)$ for the period $[t - T, T]$, K_{ij} is the number of all possible paths that connect nodes n_i and n_j, $k = \overline{1, K_{ij}}$, $i, j = \overline{1, N}$. Then

$$V^{out}(t, n_i, n_j) = \sum_{k=1}^{K_{ij}} v_k^{out}(t, n_i, n_j)$$

is the total volume of flows generated in node n_i and directed to accept in node n_j by all possible paths for the period $[t - T, T]$. Parameter $V^{out}(t, n_i, n_j)$ defines the strength of influence of node n_i on node n_j at the current time t, $i, j = \overline{1, N}$. Denote by $R_i^{out} = \{j_{i_1}, \ldots, j_{i_L}\}$ the set of node numbers that are the final receivers of flows generated in the node n_i (Fig. 1). Parameter

$$\xi_i^{out}(t) = \sum_{j \in R_i^{out}} V^{out}(t, n_i, n_j)/s(\mathbf{V}(t)), \quad \xi_i^{out}(t) \in [0, 1]$$

determines the strength of influence of node n_i on the system as a whole, $i = \overline{1, N}$. Here $s(\mathbf{M})$ is a sum of elements of the matrix $\mathbf{M} = \{m_{ij}\}_{i,j=1}^{M}$, i.e.

$$s(\mathbf{M}) = \sum_{i=1}^{N} \sum_{j=1}^{N} m_{ij}$$

and the value $s(\mathbf{V}(t))$ determines the total volume of flows which passed through the network for the period $[t - T, T]$.

The power of influence of node n_i on the system is determined by the parameter

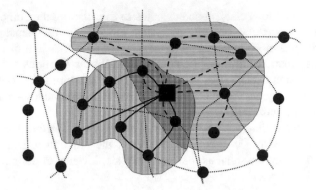

Fig. 1 Domains of input ($G_i^{in}(t)$—vertical lines) and output influence ($R_i^{out}(t)$—horizontal lines) of mapped by the square the node of network system

$$p_i^{out} = L/N, \ p_i^{out} \in [0, 1],$$

where L is the number of elements of the set $R_i^{out}(t)$ which we call the domain of influence of node n_i on the NS, $i = \overline{1, N}$. Denote by $\delta_i^{out}(t)$ the diameter of domain $R_i^{out}(t)$, as subnet of the source network, and D—diameter of CN. Parameter

$$\Delta_i^{out}(t) = \delta_i^{out}(t)/D$$

will be called diameter of influence of the node n_i on NS, $i = \overline{1, N}$. For example, the domain and diameter of influence of local government or regional media are usually limited to the relevant region of the country. At the same time, the diameter of influence of the state government and national media is equal to the "diameter" of the state as a network. The diameter of influence allows us to determine the influence of separate political parties, civic organizations, religious denominations, etc. Parameters $\xi_i^{out}(t)$, $p_i^{out}(t)$, $R_i^{out}(t)$ and $\Delta_i^{out}(t)$ will be called the output parameters of influence of the node n_i respectively, $i = \overline{1, N}$. In the simplest case, the output domain of influence of each NS's node is limited by adjacent nodes. Then the power of output influence of the node is equal to its output degree, and the diameter of influence is equal to 1. In the most complex case, the output domain of influence of all NS's nodes form a complete graph. Then the power of its output influence is equal to N, and the diameter of influence is equal to D.

So-called botnets are often presented in social online services [28]. By means of these botnets one person can create the illusion of common opinion of many people, massively distribute the disinformation, organize *DDoS*-attacks, and so on. So, in one of the most popular social networks Twitter there are huge networks of fake accounts, the number of nodes of which exceeds 350 thousand [29]. Detection of nodes-generators of such botnets and their blocking allows us to prevent many negative social and economic phenomena. Parameters $\xi_i^{out}(t)$ and p_i^{out}, $i = \overline{1, N}$,

enable to identify the botnet generators with sufficient precision, since the strength
and power of their influence on the NS are usually much higher than average.

The output parameters of influence of the node allow us to determine the tendencies of growth or decrease of the magnitude and power of this influence, as well
as the rate and direction of its spread or convolution. Indeed, if function $\frac{d\xi_i^{out}(t)}{dt}$ is
positive, then the strength of node's influence on the network over the period of time
$[t - T, T]$ increases. If this function is negative, then this strength decreases. If function $\frac{dp_i^{out}(t)}{dt}$ is positive, then the power of influence of the node n_i on NS increases.
If the values of the function $\frac{d\Delta_i^{out}(t)}{dt}$ are close to 0, then the increase of number of
nodes-final receivers of flows occurs in the domain bounded by the boundary $R_i^{out}(t)$.
If the value of the function $\frac{d\Delta_i^{out}(t)}{dt}$ is positive, the diameter of the influence of the
node n_i increases. In general, if the values of functions $\frac{d\xi_i^{out}(t)}{dt}$, $\frac{dp_i^{out}(t)}{dt}$, and $\frac{d\Delta_i^{out}(t)}{dt}$ are
positive, then such model adequately describes the process of spreading epidemics
or computer viruses that are "generated" by one source. At the same time, the greater
are the values of these functions, the faster and more threatening is this process. We
note that a sharp increase of domain, power and diameter of influence is characteristic for so-called cascading failures in the network [30]. If the function $\frac{d\xi_i^{out}(t)}{dt}$ is
negative, then the strength of node's influence on NS decreases. If functions $\frac{dp_i^{out}(t)}{dt}$
and $\frac{d\Delta_i^{out}(t)}{dt}$ are also negative then accordingly decreases the number of nodes-final
receivers of flows generated in node n_i as well as the power and diameter of influence
of this node on the network. Thus, the output influence parameters allow us to track
the dynamics of change of importance of the node n_i in NS and to simulate some
important processes in this system, $i = \overline{1, N}$.

Analysis of the behavior of derivatives of influence parameters allows us to determine the current trends in the state of system elements. However, the construction of
at least short-term forecasts for the development of such trends is no less important.
Consider the algorithm for short-term forecasting of the parameter of output strength
of NS's node for the period $[0, T]$. Let us the set

$$\{\xi_i^{out}(t_j)\}_{j=1}^J, J \geq 2,$$

determines the prehistory of values of this parameter at the moments of time

$$t_j = \frac{jT}{J} \in [0, T], j = \overline{1, J}.$$

Denote by $\boldsymbol{\Phi}(t) = \{\varphi_j(t)\}_{j=1}^J$ the system of linearly independent functions defined
on the interval $[0, T]$. Construct a function

$$\xi_i^{out}(t) = <\mathbf{a}, \boldsymbol{\Phi}(t) >_{R^J},$$

where $\mathbf{a} = \{a_j\}_{j=1}^J$ is the vector of unknown coefficients. Then the forecasted value
of parameter of the output strength of influence $\xi_i^{out}(t)$ of node n_i on the network

system at the time t_{J+l} is obtained from the ratio

$$\xi_i^{out}(t_{J+l}) = < \mathbf{a}, \mathbf{\Phi}(t_{J+l}) >_{R^J}, \; l = 1, 2, \ldots,$$

in which vector $\mathbf{a}$ is determined from the condition

$$< \mathbf{a}, \mathbf{\Phi}(t_k) >_{R^J} = \xi_i^{out}(t_k), k = \overline{1, J}.$$

The choice of the system of basic functions can be determined by the experimentally defined behavior of the parameter of strength of influence. Similarly, short-term forecasts of the behavior for other output influence parameters of the node n_i, $i = \overline{1, N}$, are carried out. For the construction of medium- and long-term forecasts of the development of system elements, other prognostic techniques are commonly used, for example, the methods of time series [31]. However, it should be borne in mind that constructing reliable long-term forecasts of many processes occurring in real systems is often practically impossible. This is confirmed by the numerous social disturbances that have taken place in Ukraine, North Africa and the Middle East over the last decades. In most cases, it was impossible to predict the appearance of such disturbances and their magnitude even several hours before they began. Long-term forecasts of financial processes, climatic phenomena and so on are often unreliable.

Denote by $v_k^{in}(t, n_j, n_i)$ the volume of flows generated in node n_j and received at node n_i, which passed through the path $p_k(n_j, n_i)$ for the period $[t-T, T]$, K_{ji} is the number of all possible paths that connect nodes n_j and n_i, $k = \overline{1, K_{ji}}$, $i, j = \overline{1, N}$. Then

$$V^{in}(t, n_j, n_i) = \sum_{k=1}^{K_{ij}} v_k^{in}(t, n_j, n_i)$$

is the total volume of flows generated in node n_j and directed to accept in node n_i by all possible paths for the period $[t - T, T]$. Parameter $V^{in}(t, n_j, n_i)$ defines the strength of influence of node n_j on node n_i at the current time t, $i, j = \overline{1, N}$. Denote by $G_i^{in} = \{j_{i_1}, \ldots, j_{i_M}\}$ the set of node numbers in which the flows are generated, which are sent for receiving in the node n_i. Parameter

$$\xi_i^{in}(t) = \sum_{j \in G_i^{in}} V^{in}(t, n_j, n_i)/s(\mathbf{V}(t)), \; \xi_i^{in}(t) \in [0, 1],$$

determines the strength of influence of NS on the node n_i, $t \geq T > 0$, $i = \overline{1, N}$.

The power of influence of the system on the node n_i is determined by the parameter

$$p_i^{in} = M/N, \; p_i^{in} \in [0, 1],$$

where M is the number of elements of the set G_i^{in} which we call the domain of influence of NS on the node n_i, $i = \overline{1, N}$. Denote by $\delta_i^{in}(t)$ the diameter of domain G_i^{in} and D—diameter of CN. Parameter

$$\Delta_i^{in}(t) = \delta_i^{in}(t)/D$$

will be called diameter of influence of NS on the node n_i. Parameters $\xi_i^{in}(t)$, $p_i^{in}(t)$, $G_i^{in}(t)$, and $\Delta_i^{in}(t)$ will be called the input parameters of influence of NS on the node n_i respectively. In the simplest case, the input domain of influence of each NS's node is limited by adjacent nodes. Then the power of input influence of the node is equal to its input degree, and the diameter of influence is equal to 1. In the most complex case, the input domain of influence of all NS's nodes form a complete graph. Then the power of its input influence is equal to N, and the diameter of influence is equal to D.

The input parameters of influence of the node allow us to determine the tendencies of growth or decrease of the magnitude and power of this influence, as well as the rate and direction of its spread or convolution. In social networks, parameters $\xi_i^{in}(t)$ and p_i^{in}, $i = \overline{1, N}$, allow us to identify users whose judgments pose the greatest attention of the Internet community, since the response to them (the strength and power of influence from the NS) is significantly higher than average.

Input and output parameters of the strength and power of influence are global dynamic characteristics of node in the NS. But determining the set of nodes-receivers of flows for a given NS's node-generator and vice versa is often an ambiguous problem. This is usually due to the type of NS and the level of flows ordering in it (for most systems with a fully ordered motion of flows—industrial, commercial, transport systems etc., the influence parameters of their nodes are sufficiently determined and predicted [19]). However, for systems with partially ordered and disordered motion of flows, the set of nodes-receivers for most or all nodes-generators and vice versa is not predetermined [16]. It should also be borne in mind that in reality the processes occured in such system and behavior of the influence parameters of the NS's nodes may be much more complicated. So a node that has directed the flow to all adjacent nodes can again become a receiver, and adjacent nodes from receivers turn into generators that direct this flow further (the spread of epidemics of infectious diseases under the so-called SIS scenario [32]). In addition, the influence parameters of NS's nodes generally are dynamic characteristics, the values of which can change significantly over time.

Special attention in TCN is given to the issue of network stability, as its ability to resist targeted external influences (hacker or terrorist attacks, etc.) [33, 34]. Attacks on the nodes with large values of input and output parameters of the strength of influence can significantly destabilize the whole system or a large part of it. These parameters allow us to define the following scenarios of attacks on the network system:

(1) a list of network nodes is being prepared in order of decreasing the values of their influence strength and the nodes from the beginning of this list are consistently

withdrawn from the structure until a predetermined level of critical losses is reached;

(2) after removing the next node, the list of nodes formed in the previous scenario is rewritten according to the same principle and the attack is carried out on the first node from the modified list.

The second scenario takes into account the need to replace blocked nodes-generators and nodes-receivers and the corresponding redistribution of flows motion through the network. Depending on the method of dealing with potential threats, the last two scenarios can be formed separately for nodes-generators (for example, search for initiators of *DDoS*-attacks), and nodes-receivers of flows (finding the most likely targets of *DDoS*-attacks).

However, there is another side of the protection problem. It consists in the timely detection and blocking of those network system nodes that present a potential or real threat and can destabilize the system operation—hacker and terrorist groups, sources of the spread of dangerous infectious diseases, and so on. The input and output influence parameters of NS's nodes allow us to identify the botnet generators with sufficient precision. Usually, the botnet generator, by sending commands to the bots created by it (information about the purpose and content of the attack), does not need and receive no feedback, that is, for such formations, inequality is performed

$$\frac{\xi_i^{in}(t)}{\xi_i^{out}(t)} << 1.$$

From these considerations it also follows that the domain and power of output influence of such nodes are sufficiently large and the domain and power of input influence are small, moreover

$$R_i^{out} \cap G_i^{in} \approx 0.$$

In real network systems there are practically no nodes that are only generators or receivers of flows. Indeed, the manufacturing of certain products requires the supply of raw materials and components, mining can not be carried out without the appropriate mining equipment, etc. Denote by RG_i the union of domains of the input and output influence of the node n_i, i.e.

$$RG_i = R_i^{out} \cup G_i^{in}.$$

The interaction strength of the node n_i with NS will be determined by the parameter

$$\xi_i(t) = (\xi_i^{in}(t) + \xi_i^{out}(t))/2,$$

and the power of this interaction—by means of the parameter p_i, which is equal to the number of elements of the set RG_i.

The other side of systems resistance is its sensitivity to small changes in the structure or operation process. Such changes can be caused by both internal and external factors, and can lead to the no less consequences than targeted attacks. In this case, the stability of structure is determined by the sensitivity to small changes in the set of its nodes and edges. The structure is unstable when such changes can lead to loss of certain network properties, such as connectivity. The stability of NS operation process is determined by its sensitivity to small changes in the volume of flows movement. For example, the systems operation may become unstable in the conditions of critical loading of part of its edges (the corresponding elements of matrices $\mathbf{u}(t)$ or $\mathbf{U}(t)$ are close to 1) or some the most important nodes in terms of strength and power of influence. Many systems are sensitive to small violations of established schedule of flows motion. Obviously, the stability of process is associated with the resistance of NS structure. If small changes (blocking some network nodes and edges) lead to loss of connectivity, this directly affects on the systems operation. If the load of certain elements of structure by flows is critical (close to their bandwidth), it also creates a threat of blocking these elements.

Node n_i, for which

$$\xi_i^{in}(t) = \xi_i^{out}(t) = 0$$

and

$$W_i^{in}(t) = W_i^{out}(t) = W_i^{tr}(t) \neq 0, \, t > T > 0, i = \overline{1, N},$$

will be called a transit node. The importance of transit node in the system is determined by the volume of flows that pass through it. Extraction from structure the transit nodes is one way to reduce the dimensionality of system model. It should be borne in mind that destabilization of important transit node operation with large value $W_i^{tr}(t)$ and high betweenness centrality can destabilize the whole system or large part of it [35].

The preferential influence $\psi_i(t)$ of node n_i for non-transit NS nodes we will determined by the ratio

$$\psi_i(t) = (\xi_i^{in}(t) - \xi_i^{out}(t))/(\xi_i^{in}(t) + \xi_i^{out}(t)), \, \psi_i \in [-1, \, 1].$$

If the value of parameter $\psi_i(t)$ is close to -1, then the preferential influence is from the node n_i on NS. If the value of parameter $\psi_i(t)$ is close to 1, then the preferential influence is from NS on the node n_i. In case $\psi_i(t) \approx 0$, $i = \overline{1, N}$, the influence is uniform on each side. The network structure (Fig. 2a) is usually much simpler than the structure of flows in it (Fig. 2b). Parameter of preferential influence allows us to determine the predominant direction of flows within the system (Fig. 2c).

Thus, passenger traffic in a country or a large city is characterized by the value of $\psi_i(t) \approx 0$, $i = \overline{1, N}$. At the same time, migration processes (refugee movement, urbanization, etc.) are characterized by a pronounced uneven distribution of the values of preferential influence.

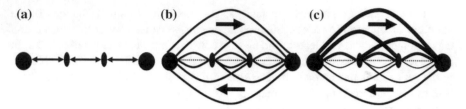

Fig. 2 Fragments: **a** Network structure; **b** structure of flows in network; **c** volumes of flows motion in network

4 Betweenness in Network Systems

One of the main concepts of TCN is the so-called centrality of the node, which allows us to determine its importance in the network: the most influential persons in social networks, key nodes on the Internet and transport networks, etc. [26, 36]. However, the notion "importance" may have different content, which led to the emergence of many definitions of the term "centrality". The most used measures of centrality in a complex network include degree centrality [37], closeness centrality [38], betweenness centrality [39], eigenvector centrality [40], percolation centrality [41], cross-clique [42], Katz [43], and PageRank centralities, harmonic [44], Freeman, and alpha centralities [37] etc. At the same time, one measure of centrality may contradict another and the centrality that is important for one problem may be insignificant for another. This phenomenon was confirmed by Krackhardt [45], who gave an example of simple network, for which the degree, betweenness, and closeness centralities took completely different values, that is, gave three different choices of the most important nodes in system structure. Hence it follows that the mentioned above definitions of centralities have a quite a relative value. This led to the introduction, along with the concepts of centrality the associated with them indicators of influence of nodes on the network structure. The main measures of the node's influence are its accessibility and expected force [27]. The accessibility of a node is determined by the number of nodes to which we can walk from it over a specified period of time. Expected force of a node's influence is determined by the number of nodes to which we can pass through two or more steps of motion (step—the transition by one edge of the network). Obviously, the above mentioned measures of centrality and influence of the node are determined solely by the properties of structure and are the characteristics of this structure, rather than systems in general.

The input and output influence parameters of a node were introduced above to determine its importance in the system. These concepts allow us to quantify the participation of separate node as a receiver or generator of flows in the process of system operation and its significance in this process. Another indicator of the importance of node interaction with NS is measure of its contribution in the transit of flows through the network. One of the most used with the degree centrality in TCN is the betweenness centrality. Perhaps the notion "betweenness" is most successful in determining the participation of NS's node in the process of joint operation and

interaction of all nodes in the network or a certain part of it. Therefore, to determine
the functional importance of a node or an edge in the system, we will use the term
"betweenness".

Denote by $P_{ij}^{K_{ij}} = \{p_{ij}^k\}_{k=1}^{K_{ij}}$ the set of paths that connect the nodes-generators
and nodes-receivers of NS flows, and contain, as an element, the edge (n_i, n_j),
$i, j = \overline{1, N}$. Let us $v_{ij}^k(t)$ is the volume of flows that have passed through path p_{ij}^k
from the node-generator to the node-receiver, and hence by the edge (n_i, n_j), for the
period $[t - T, t]$. Then the value

$$V_{ij}^{K_{ij}}(t) = \sum_{k=1}^{K_{ij}} v_{ij}^k(t)$$

defines the total volume of flows that have passed through the set of paths $P_{ij}^{K_{ij}}$, and
hence by the edge (n_i, n_j), over the same period of time. Parameter

$$\Phi_{ij}(t) = V_{ij}^{K_{ij}}(t)/s(\mathbf{V}(t)),$$

which determines the specific weight of flows passed through the edge (n_i, n_j) for
period $[t - T, t]$, will be called the betweenness measure of this edge in the process
of NS operation.

The set L_{ij} of all NS's nodes, which lie on the paths of set $P_{ij}^{K_{ij}}$, will be called the
betweenness domain, and the number η_{ij} of these nodes—the power of betweenness
of the edge (n_i, n_j) (Fig. 3). Denote by δ_{ij} the diameter of betweenness domain of
the edge (n_i, n_j). This diameter is calculated as the diameter of the set L_{ij}. Parameter

$$\Delta_{ij} = \delta_{ij}/D$$

will be called the diameter of betweenness of the edge (n_i, n_j), $i, j = \overline{1, N}$.

The parameters of measure, domain, power and diameter of betweenness of the
edge (n_i, n_j) are global characteristics of its importance in the process of NS operation, $i, j = \overline{1, N}$. They, in particular, determine how the blocking of this edge will

Fig. 3 The betweenness
domain of edge (n_i, n_j) in
the process of NS operation

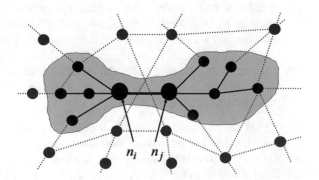

affect on the work of domain of its betweenness, the magnitude of this domain and, as a result, the whole system.

Denote by K_i the set of paths that connect nodes-generators and nodes-receivers of NS flows, and pass through a node n_i, $i = \overline{1, N}$. Let us $v_i^k(t)$ is the volume of flows passing through path p_i^k from the node-generator to node-receiver, and hence through the node n_i, for the period $[t - T, t]$. Then the parameter

$$V_i^{K_i}(t) = \sum_{k=1}^{K_i} v_i^k(t)$$

determines the total volume of flows that have gone through the set of paths $P_i^{K_i}$, and hence through node n_i, over the same period of time. Parameter

$$\Phi_i(t) = V_i^{K_i}(t)/s(\mathbf{V}(t)),$$

which determines the specific weight of flows passing through the node n_i for period $[t-T, t]$, will be called the betweenness measure of this node during the NS operation. The set M_i of all NS's nodes, which lie on the paths of set $P_i^{K_i}$, will be called the betweenness domain, and the number η_i of these nodes—the power of betweenness of the node n_i. Denote by δ_i the diameter of betweenness domain of the node n_i. Then parameter

$$\Delta_i = \delta_i/D$$

will be called the diameter of betweenness of the node n_i, $i = \overline{1, N}$.

The parameters of measure, domain, power and diameter of betweenness of the node n_i are global characteristics of its importance in the process of NS operation, $i = \overline{1, N}$. They, in particular, determine how the blocking of this node will affect on the work of domain of its betweenness, the magnitude of this domain and, as a result, the whole system.

Betweenness parameters allow us to define the following scenarios of attacks on the network system:

(1) a list of network nodes is being prepared in order of decreasing the values of their betweenness measure and the nodes from the beginning of this list are consistently withdrawn from the structure until a predetermined level of critical losses is reached;

(2) after removing the next node, the list of nodes formed in the previous scenario is rewritten according to the same principle and the attack is carried out on the first node from the modified list.

The second scenario takes into account the need to replace blocked nodes-generators and nodes-receivers of flows and the search for alternative paths of movement of transit flows that pass through blocked nodes, i.e. the corresponding redistribution of flows motion through the network. Similar scenarios of attacks are also formed for NS's edges, since in many cases the blocking of network edge is much simpler than blocking one of the nodes that it combines. The parameters of betweenness of nodes and edges allow us to estimate to what part of the NS the consequences of failures of the corresponding system element will spread and to what losses this will result in the sense of lack of supply of certain volumes of transit flows.

We have defined above the parameters of betweenness of the node, taking into account only the transit flows that pass through it. However, the importance of betweenness parameters can be significantly expanded, taking into account that the node n_i can be not only a transit, but also a generator and final receiver of flows. Then the set $P_i^{K_i}$ can be supplemented by the paths of flows that begin (generated) or end (received) in the node n_i. Denote such supplemented set by $\tilde{P}_i^{K_i}$, $i = \overline{1, N}$. Then parameter

$$\tilde{\Phi}_i(t) = (\Phi_i(t) + \xi_i^{in}(t) + \xi_i^{out}(t))/3$$

will be called a generalized measure of betweenness of the node n_i in the process of NS operation. Accordingly, the set $\tilde{M}_i$ of all NS's nodes, which lie on the paths from the set $\tilde{P}_i^{K_i}$, will be called a generalized betweenness domain, and the number $\tilde{\eta}_i$ of these nodes is the generalized betweenness power of the node n_i, $i = \overline{1, N}$. The generalized betweenness parameters take into account the interaction between all directly and indirectly connected nodes of NS (generators, receivers and transits) and allow us to form the most effective scenarios of attacks on them. Principles for creation such scenarios are described above.

5 Influence and Betweenness of Subsystems of Complex Network Systems

Denote by S the subsystem of source NS, formed on the basis of principles of ordering or subordination [46]. Let us H_S is the set of nodes that make up the structure of subsystem S, and F_S is the set of edges that combine nodes of the set H_S.

Denote by G_S^{out} the set of all nodes-generators of flows included in the set H_S, p_S^{out}—the number of elements of G_S^{out} and determine by the parameter

$$\xi_S^{out}(t) = \sum_{i \in G_S^{out}} \xi_i^{out}(t)/s(\mathbf{V}(t))$$

the strength of influence of the subsystem S on NS as a whole.

Let us

$$R_S^{out} = \bigcup_{i \in G_S^{out}} R_i^{out}$$

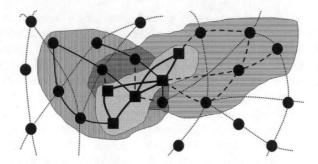

Fig. 4 Domains of input (G_S^{in}—vertical lines) and output influence (R_S^{out}—horizontal lines) of NS subsystem (subsystem nodes are reflected by squares)

is the set of numbers of nodes—final receivers of flows generated in nodes belonging to the set G_S^{out} (Fig. 4). Divide the set R_S^{out} into two subsets, namely

$$R_S^{out} = R_{S,int}^{out} \cup R_{S,ext}^{out},$$

where $R_{S,int}^{out}$ is a subset of nodes R_S^{out} belonging to H_S, and $R_{S,ext}^{out}$ is a subset of nodes R_S^{out} that belong to the supplement to H_S in the source NS. The set $R_{S,ext}^{out}$ is called the domain of the output influence of subsystem S on NS, and the number of elements $p_{S,ext}^{out}$ of this set is the power of this influence. Denote by $\delta_{S,ext}^{out}$ the diameter of the set $R_{S,ext}^{out}$ and D—the diameter of NS. Parameter

$$\Delta_{S,ext}^{out} = \delta_{S,ext}^{out}/D$$

will be called the diameter of output influence of subsystem S on NS.

The external and internal output strength of influence of the nodes-generators of flows belonging to the set G_S^{out} on subnets $R_{S,ext}^{out}$ and $R_{S,int}^{out}$ will be determined by the parameters

$$\xi(t)_{S,ext}^{out} = \sum_{i \in R_{S,ext}^{out}} \xi_i^{out}(t)/s(\mathbf{V}(t))$$

and

$$\xi(t)_{S,int}^{out} = \sum_{i \in R_{S,int}^{out}} \xi_i^{out}(t)/s(\mathbf{V}(t))$$

respectively. Then the parameter

$$\omega_S^{out}(t) = \frac{\xi_{S,ext}^{out}(t)}{\xi_{S,int}^{out}(t)}$$

determines the relative strength of influence of subsystem S on the system as a whole. Namely the smaller the value of parameter ω_S^{out}, the less the strength of influence of the subsystem S on NS. Parameters $\xi_{S,ext}^{out}$, $R_{S,ext}^{out}$, $p_{S,ext}^{out}$, $\Delta_{S,ext}^{out}$ and ω_S^{out} will be called the output influence parameters of subsystem S on NS.

Denote by R_S^{in} the set of all nodes-receivers of flows included in the set H_S (Fig. 4), p_S^{in}—the number of elements of R_S^{in} and determine by the parameter

$$\xi_S^{in}(t) = \sum_{i \in R_S^{in}} \xi_i^{in}(t)/s(\mathbf{V}(t))$$

the strength of influence of NS on subsystem S.

Let us

$$G_S^{in} = \bigcup_{i \subset R_S^{in}} G_i^{in}$$

is the set of numbers of nodes-generators from which the flows are directed to nodes belonging to the set R_S^{in}. Divide the set G_S^{in} into two subsets, namely

$$R_S^{in} = R_{S,int}^{in} \cup R_{S,ext}^{in},$$

where $G_{S,int}^{in}$ is a subset of nodes G_S^{in} belonging to H_S, and $G_{S,ext}^{in}$ is a subset of nodes G_S^{in} that belong to the supplement to H_S in the source NS. The set $G_{S,ext}^{in}$ is called the domain of the input influence of NS on the subsystem S, and the number of elements $p_{S,ext}^{in}$ of this set is the power of this influence. Denote by $\delta_{S,ext}^{in}$ the diameter of the set $G_{S,ext}^{in}$. Parameter

$$\Delta_{S,ext}^{in} = \delta_{S,ext}^{in}/D$$

will be called the diameter of input influence of NS on subsystem S.

The external and internal input strength of influence of the nodes-receivers of flows belonging to the set R_S^{in} on subnets $G_{S,ext}^{in}$ and $G_{S,int}^{in}$ will be determined by the parameters

$$\xi_{S,ext}^{in}(t) = \sum_{i \in G_{S,ext}^{out}} \xi_i^{in}(t)/s(\mathbf{V}(t))$$

and

$$\xi_{S,int}^{in}(t) = \sum_{i \in G_{S,int}^{out}} \xi_i^{in}(t)/s(\mathbf{V}(t))$$

respectively. Then the parameter

$$\omega_S^{in}(t) = \frac{\xi_{S,ext}^{in}(t)}{\xi_{S,int}^{in}(t)}$$

determines the relative strength of influence of NS on subsystem S. Namely the smaller the value of parameter ω_S^{in}, the less the strength of influence of NS on the subsystem S. Parameters $\xi_{S,ext}^{in}$, $G_{S,ext}^{in}$, $p_{S,ext}^{in}$, $\Delta_{S,ext}^{in}$ and ω_S^{in} will be called the input influence parameters of NS on subsystem S.

The behavior of derivatives of influence parameters of NS's subsystems allows us to determine the tendencies of growth or decrease of their magnitude and power, as well as the rate of distribution and growth. For a deeper study of the behavioral patterns of these parameters, it is also advisable to use the prediction methods described above.

The notion of community is important in TCN [47]. Community is a group of closely interconnected CN's nodes which are weakly interconnected with other nodes in the network. The main disadvantage of existing methods for identifying communities in the CN (methods of minimal cut, hierarchical clusterization, Girvan and Newman, modularity maximization (Louvain algorithm), methods based on clicks (Bron and Kerbosh algorithm) [48–51] etc.) along with computational complexity and resource expenditures is the lack of reliable criterion of what the group of nodes determined by any of these methods really forms the community. A pair of parameters $(\omega_S^{out}, \omega_S^{in})$ gives such criterion. Indeed, the smaller are the values of these parameters, the less is the external interaction of subsystem S with the system as a whole and the larger are intragroup interactions, which is, in essence, a community definition. Moreover, a pair of these parameters obviously allows us to determine the system-wide and internal activity or passivity of the subsystem S.

Determining the participation of subsystem S in the system operation in the sense of predominant influence allows the parameter

$$\psi_S(t) = \left(\xi_{S,ext}^{in}(t) - \xi_{S,ext}^{out}(t)\right) \big/ \left(\xi_{S,ext}^{in}(t) + \xi_{S,ext}^{out}(t)\right), \ \psi_S \in [-1, \ 1].$$

If the value of parameter $\psi_S(t)$ is close to -1, then the predominant is the influence of subsystem S on the NS, i.e. it is generally a subsystem that generates flows. If the value of parameter $\psi_S(t)$ is close to 1, then the influence of NS on subsystem S is predominant, i.e. it is generally the receiver of flows. In the case of $\psi_S(t) \approx 0$, $i = \overline{1, N}$, the influence is uniform from each side, i.e. the subsystem S is simultaneously both the generator of flows and the flows receiver. It is also non-difficult to determine the strength of predominant influence between two arbitrary subsystems of NS, the sets of nodes of which does not intersect.

Equally important for the analysis of NS operation are the parameters of betweenness of its separate subsystems, which we define as follows. Denote by $P_S^{K_S} = \{p_S^k\}_{k=1}^{K_S}$ the set of paths that combine the NS's nodes-generators and nodes-receivers of flows and pass through elements of the subsystem S. Let us $v_S^k(t)$ is the volume of flows that went through path p_S^k from the node-generator to node-receiver, and hence through the elements of subsystem S, for the period $[t - T, t]$.

Then parameter

$$V_S^{K_S}(t) = \sum_{k=1}^{K_S} v_S^k(t)$$

determines the total volume of flows that went through a set of paths $P_S^{K_s}$, and therefore through elements of the subsystem S, over the same period of time. Parameter

$$\Psi_S = V_S^{K_i}(t)/s(\mathbf{V}(t)),$$

which determines the specific weight of flows passing through elements of subsystem S for period $[t - T, t]$, will be called the betweenness measure of this subsystem during the NS operation.

The set M_S of all NS's nodes, which lie on the paths of set $P_S^{K_s}$, will be called the betweenness domain (Fig. 5), and the number η_S of these nodes—the power of betweenness of subsystem S. Denote by δ_i the diameter of betweenness domain of the node $n_i \in H_S$. Then parameter

$$\Delta_S = \max_{n_i \in H_S} \delta_i/D,$$

where D is the diameter of CN will be called the diameter of betweenness of subsystem S.

The parameters of measure, domain, power and diameter of betweenness of subsystem S are global characteristics of its importance in the process of NS operation. They, in particular, determine how the blocking of this subsystem will affect on the work of domain of its betweenness, the magnitude of this domain and, as a result, the whole system. In addition, the small values of betweenness parameters of the subsystem S may also indicate that it forms a community within the NS.

The behavior of derivatives of betweenness parameters of NS's subsystems allows us to determine the tendencies of growth or decrease of their magnitude and power, as well as the rate of distribution and growth. For a deeper study of the behavioral patterns of these parameters, it is also advisable to use the prediction methods described above.

Fig. 5 Betweenness domain of subsystem S in the process of NS operation

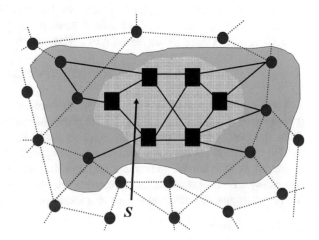

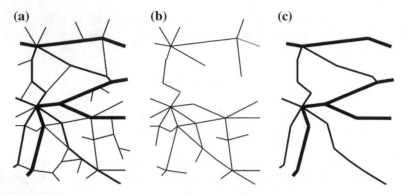

Fig. 6 Fragments: **a** Source NS; **b** 4-core of CN; **c** 0.7-core of NS

Another way to determine the most important subsystems of NS is to introduce the notion of its k-core, that is, the largest subnet of the source CN, all nodes of which have degree not less than k, and the extraction from the network structure of nodes with degree less than k [52]. Using the flow characteristics of NS allows us to introduce the concept of flow λ-core of network system, as the largest subnet of source network, for which all elements of the integral flow adjacency matrix have values not less than λ, $\lambda \in [0, 1]$, [53]. Figure 6a reflects the structure of railway transport system of the western region of Ukraine. The thickness of lines in this figure is proportional to the weight of edges—the volumes of flows passing through them. Figure 6b displays the 4-core of this network and Fig. 6c reflects the flow 0.7-core of this system.

Introduce the IFAM of λ-core by means of ratio

$$\mathbf{V}^\lambda(t) = \{V_{ij}^\lambda(t)\}_{i,j=1}^N, \quad V_{ij}^\lambda(t) = \begin{cases} V_{ij}(t), & \text{if } V_{ij}(t) \geq \lambda, \\ 0, & \text{if } V_{ij}(t) < \lambda. \end{cases}$$

We will use parameter $\sigma_\lambda(t)$ to determine the specific weight of λ-core. This parameter is equal to the ratio of volumes of flows passing by the λ-core to the volume of flows that pass through the network as a whole during the period $[t - T, t]$:

$$\sigma_\lambda(t) = s(\mathbf{V}^\lambda(t)) / s(\mathbf{V}(t)).$$

Since the main goal of the most network systems is to provide the flows motion, parameter $\sigma_\lambda(t)$ quantifies how the λ-core provides the implementation of this goal. Thus, this parameter determines the importance of subsystem, formed by λ-core, in the NS operation process as a whole. So, the spread of epidemics usually occurs on the ways of intensive movement of large masses of people, and the spread of computer viruses—on the paths of intense information traffic. The flow cores of NS with large values of λ determine the most likely paths of deploying such processes.

6 Integral Parameters of Influence and Betweenness of Complex Network Systems

The most common indicator of NS operation is the total volume of flows that pass through the network over period of time $[t - T, t]$. This indicator is determined by the value $s(\mathbf{V}(t))$, $t > T$. But it is rather relative, since it does not determine how effective the system functions compared to potential opportunities.

Let us denote

$$\mathbf{V}_{\max}(t) = \{V_{ij}^{\max}(t)\}_{i,j=1}^{N}, \ t \in [t - T, t],$$

where $V_{ij}^{\max}(t)$ is the maximum volume of flows that could pass through the edge (n_i, n_j), $i, j = \overline{1, N}$, over the same time period, taking into account the bandwidth of this edge. Parameter

$$\pi(t) = \frac{s(\mathbf{V}(t))}{s(\mathbf{V}_{\max}(t))} \in [0, 1], \ t \in [t - T, t],$$

determines how effective the NS operates compared to its potential possibilities.

Critically loaded systems are very vulnerable to increasing the volume of flows. It is difficult and sometimes impossible to find alternative paths of flows motion, since such paths that can increase the volume of flows may not exist. At the same time, the most dangerous for the stable system operation is the critical loading of its λ-core with a high specific weight. This is the negative reverse side of an attempt to maximize the efficiency of NS operation, if the bandwidth of the system nodes and edges does not increase at the same time.

Denote by R^{out} the set of all network nodes-generators and introduce parameter

$$V^{out}(t) = \sum_{i \in R^{out}} \sum_{j \in R_i^{out}} V^{out}(t, n_i, n_j), \ t \in [t - T, t].$$

Determine parameter

$$p^{out} = \mu(R^{out})/N, \ p^{out} \in [0, 1],$$

where $\mu(R^{out})$ is the power (number of elements) of subset R^{out}, which determines the specific weight of nodes-generators in the system structure. Obviously, the smaller the value p^{out}, the more vulnerable is the NS to destabilization the work of the nodes-generators of flows.

Denote by G^{in} the set of all network nodes-receivers and introduce parameter

$$V^{in}(t) = \sum_{i \in G^{in}} \sum_{j \in G_i^{in}} V^{in}(t, n_j, n_i), \ t \in [t - T, t].$$

Determine parameter

$$p^{in} = \mu(J^{in})/N, \ p^{in} \in [0, 1],$$

where $\mu(G^{in})$ is the power of subset G^{in}, which determines the specific weight of nodes-receivers in the system structure. Obviously, the smaller the value p^{in}, the more vulnerable is the NS to destabilization the work of the nodes-receivers of flows.

Any real system is open, that is, it interacts with other systems [54]. Let us that our system is the subsystem of a bigger formation—mega-system. Then, as in the previous paragraph, we can introduce the influence parameters of our system on this mega-system and vice versa, as well as the parameters of its betweenness during mega-system operation. This approach allows us to reach the level of multilayer network system interactions research.

7 Conclusions

The functional approach of network systems research is considered in this article. In order to study the process of such systems operation the flow adjacency matrices of different types were introduced. It was also analyzed, how these matrices help to investigate and forecast the peculiarities of this process, evaluate its efficiency and prevent existing and potential threats. Global dynamic influence and betweenness parameters of the network systems elements were determined. These parameters allow us to identify nodes that generate and receive flows, and transit nodes, determine the predominant direction of flows within the system, study activity, passivity, and stability of separate system components and NS in general, as well as form much more realistic scenarios of potential attacks on the system, quantify the losses from these attacks, and build the more reliable means of protecting it. The parameters of influence and betweenness of network system components defined in the article, as well as the concept of its flow cores, allow us to identify the most important subsystems for NS's operation and contribute to a better understanding of the processes that occur in them. The obtained results can be used to reduce the NS vulnerability from negative external and internal influences, to develop the modern methods for information and security systems protecting, to improve the efficiency of operation of transport and industrial networks of different types, etc.

References

1. Boccara N (2010) Modeling complex systems. Springer Science & Business Media, New York
2. Barabási A-L, Frangos J (2002) Linked: the new science of networks. Basic Books, New York
3. Boccaletti S et al (2006) Complex networks: structure and dynamics. Phys Rep 424(4):175–308. https://doi.org/10.1016/j.physrep.2005.10.009
4. Dorogovtsev SN, Mendes JFF (2013) Evolution of networks: from biological nets to the internet and www. Oxford University Press, Oxford

5. Caldarelli G, Vespignani A (2007) Large scale structure and dynamics of complex networks: from information technology to finance and natural science. World Scientific, New York
6. Northrop RB (2011) Introduction to complexity and complex systems. CRC Press, Boca Raton
7. Barrat F, Barthélemy M, Vespignani A (2007) The architecture of complex weighted networks: measurements and models. Large scale structure and dynamics of complex networks. World Scientific, London, pp 67–92
8. Newman MEJ (2004) Analysis of weighted networks. Phys Rev E 70:056131. https://doi.org/10.1103/PhysRevE.70.056131
9. Polishchuk DO, Polishchuk OD, Yadzhak MS (2016) Complex deterministic evaluation of hierarchically-network systems: IV. Interactive evaluation. Syst Res Inf Technol 1:7–16. https://doi.org/10.20535/SRIT.2308-8893.2016.1.01
10. Polishchuk O (2001) Optimization of evaluation of man's musculo-sceletal system. Comput Math 2:360–367
11. Ageyev DV, Salah MT (2016) Parametric synthesis of overlay networks with self-similar traffic. Telecommun Radio Eng 75(14):1231–1241 (English translation of Elektrosvyaz and Radiotekhnika)
12. Ageyev D et al (2019) Infocommunication networks design with self-similar traffic. In: 2019 IEEE 15th international conference on the experience of designing and application of CAD systems (CADSM). IEEE, pp 24–27. https://doi.org/10.1109/cadsm.2019.8779314
13. Daradkeh YI, Kirichenko L, Radivilova T (2018) Development of QoS methods in the information networks with fractal traffic. Int J Electron Telecommun 64(1):27–32
14. Albert R, Barabasi A-L (2002) Statistical mechanics of complex networks. Rev Mod Phys 74(1):47. https://doi.org/10.1103/RevModPhys.74.47
15. Newman MEJ (2010) Networks. An introduction. Oxford University Press, Oxford
16. Polishchuk O, Yadzhak M (2018) Network structures and systems: I. flow characteristics of complex networks. Syst Res Inf Technol 2:42–54. https://doi.org/10.20535/SRIT.2308-8893.2018.2.05
17. Lombardi A, Hörnquist M (2007) Controllability analysis of networks. Phys Rev E 75(5):056110. https://doi.org/10.1103/PhysRevE.75.056110
18. Liu Y-Y, Slotine JJ, Barabási A-L (2013) Observability of complex systems. Proc Natl Acad Sci 110(7):2460–2465. https://doi.org/10.1073/pnas.1215508110
19. Polishchuk D, Polishchuk O, Yadzhak M (2014) Complex evaluation of hierarchically-network systems. Autom Control Inf Sci 1(2):32–44. https://doi.org/10.12691/acis-2-2-1
20. Polishchuk O (2018) Flow models of complex network systems. In: Intern. Scientific-practical conf. on problems of infocommunications, science and technology, pp 317–322
21. Polishchuk OD, Tyutyunnyk MI, Yadzhak MS (2007) Quality evaluation of complex systems function on the base of parallel calculations. Inf Extr Process 26(102):121–126
22. Zurek WH (2018) Complexity, entropy and the physics of information. CRC Press, Boca Raton
23. Kryvinska N (2004) Intelligent network analysis by closed queuing models. Telecommun Syst 27:85–98. https://doi.org/10.1023/B:TELS.0000032945.92937.8f
24. Prell A (2012) Social network analysis: history, theory and methodology. SAGE, New York
25. Price G, Sherman C (2001) The invisible web: uncovering information sources search engines can't see. CyberAge Books, New York
26. Bonacich P (1987) Power and centrality: a family of measures. Am J Sociol 92(5):1170–1182. https://doi.org/10.1086/228631
27. Glenn L (2015) Understanding the influence of all nodes in a network. Sci Rep 5:8665. https://doi.org/10.1038/srep08665
28. Cao Q et al (2012) Aiding the detection of fake accounts in large scale social online services. In: 9th USENIX symposium on networked systems design and implementation, pp 197–210
29. Abokhodair N, Yoo D, McDonald DW (2015) Dissecting a social botnet: growth, content and influence in Twitter. In: 18th ACM conference on computer supported cooperative work & social computing, pp 839–851
30. Buldyrev SV et al (2010) Catastrophic cascade of failures in interdependent networks. Nature 464:1025–1028. https://doi.org/10.1038/nature08932

31. Brockwell PJ, Davis RA (2002) Introduction to time series and forecasting. Springer, Switzerland
32. Juher D, Ripoll J, Saldaña J (2013) Outbreak analysis of an SIS epidemic model with rewiring. J Math Biol 67(2):411–432. https://doi.org/10.1007/s00285-012-0555-4
33. Albert R, Jeong H, Barabási A-L (2000) Error and attack tolerance of complex networks. Nature 406:378–482. https://doi.org/10.1038/35019019
34. Holme P et al (2002) Attack vulnerability of complex networks. Phys Rev E 65:056109. https://doi.org/10.1103/PhysRevE.65.056109
35. Polishchuk O, Polishchuk D (2013) Monitoring of flow in transport networks with partially ordered motion. In: XXIII conf. Carpenko physics and mechanics institute, NASU, Lviv, pp 326–329
36. Borgatti SP (2005) Centrality and network flow. Soc Netw 27(1):55–71. https://doi.org/10.1016/j.socnet.2004.11.008
37. Freeman LC (1979) Centrality in social networks conceptual clarification. Soc Netw 1(3):215–239. https://doi.org/10.1016/0378-8733(79)90002-9
38. Bavelas A (1950) Communication patterns in task-oriented groups. J Am Acoust Soc 22(6):725–730. https://doi.org/10.1121/1.1906679
39. Freeman LC (1977) A set of measures of centrality based upon betweenness. Sociometry 40:35–41. https://doi.org/10.2307/3033543
40. Bonacich P, Lloyd P (2001) Eigenvector-like measures of centrality for asymmetric relations. Soc Netw 23(3):191–201. https://doi.org/10.1016/S0378-8733(01)00038-7
41. Piraveenan M (2013) Percolation centrality: quantifying graph-theoretic impact of nodes during percolation in networks. PLoS ONE 8(1):e53095. https://doi.org/10.1371/journal.pone.0053095
42. Faghani M, Nguyen UT (2013) A study of XSS worm propagation and detection mechanisms in online social networks. IEEE Trans Inf Forensics Secur 8(11):1815–1826. https://doi.org/10.1109/TIFS.2013.2280884
43. Katz L (1953) A new status index derived from sociometric index. Psychometrika 18(1):39–43. https://doi.org/10.1007/BF02289026
44. Marchiori M, Latora V (2000) Harmony in the small-world. Phys A: Stat Mech Its Appl 285(3–4):539–546. https://doi.org/10.1016/S0378-4371(00)00311-3
45. Krackhardt D (1990) Assessing the political landscape: structure, cognition, and power in organizations. Adm Sci Q 35(2):342–369. https://doi.org/10.2307/2393394
46. Polishchuk O, Yadzhak M (2018) Network structures and systems: III. Hierarchies and networks. Syst Res Inf Technol 4:82–95. https://doi.org/10.20535/SRIT.2308-8893.2018.4.07
47. Girvan M, Newman MEJ (2002) Community structure in social and biological networks. Proc Natl Acad Sci USA 99(12):7821–7826. https://doi.org/10.1073/pnas.122653799
48. Newman MEJ (2004) Detecting community structure in networks. Eur Phys J B 38(2):321–330. https://doi.org/10.1140/epjb/e2004-00124-y
49. Newman MEJ (2004) Fast algorithm for detecting community structure in networks. Phys Rev E 69(6):066133. https://doi.org/10.1103/PhysRevE.69.066133
50. Danon L et al (2005) Comparing community structure identification. J Stat Mech 09:P09008. https://doi.org/10.1088/1742-5468/2005/09/P09008
51. Blondel VD et al (2008) Fast unfolding of community hierarchies in large networks. J Stat Mech 10:P10008. https://doi.org/10.1088/1742-5468/2008/10/P10008
52. Dorogovtsev SN, Goltsev AV, Mendes JFF (2006) k-core organization of complex networks. Phys Rev Lett 96(4): 040601. https://doi.org/10.1103/physrevlett.96.040601
53. Polishchuk O, Yadzhak M (2018) Network structures and systems: II. Cores of networks and multiplexes. Syst Res Inf Technol 3:38–51. https://doi.org/10.20535/SRIT.2308-8893.2018.3.04
54. Scott WR (2015) Organizations and organizing: rational, natural and open systems perspectives. Routledge, London

Improving the Data Processing Efficiency Based on Tunable Hard- and Software Components

Hanna Ukhina, Valerii Sytnikov, Oleg Streltsov, Pavel Stupen, and Dmitrii Yakovlev

Abstract The article discusses the transfer function coefficients' influence on the characteristics of standard digital filters used in the construction of soft- and hardware components. Obtained are the ratios between numerator and denominator coefficients, the cut-off frequency and the peak frequency. Using results obtained, the structure of soft- and hardware devices on microprocessor basis is simplified, that, in turn, greatly simplifies the task of developing frequency-dependent components allowing both correction and adjustment of the characteristics and the system as a whole.

Keywords Technological processes · Automated control and management systems · Digital filter · Transmission coefficient · Pulsation · Characteristics correction and adjustment

1 Introduction

When data processing with complex and critical systems, such as these implementing technological processes, equipment of aviation and space engineering, nuclear

H. Ukhina (✉) · V. Sytnikov · O. Streltsov · P. Stupen · D. Yakovlev
Odessa National Polytechnic University, Odessa, Ukraine
e-mail: ukhinanna@gmail.com

V. Sytnikov
e-mail: sitnvs@gmail.com

O. Streltsov
e-mail: ovstreltsov@gmail.com

P. Stupen
e-mail: stek2000@gmail.com

D. Yakovlev
e-mail: dpyakovlev39@gmail.com

91

T. Radivilova et al. (eds.), *Data-Centric Business and Applications*, Lecture Notes on Data Engineering and Communications Technologies 48,
https://doi.org/10.1007/978-3-030-43070-2_6

power engineering, which usually include automated process control and management systems, necessary is to apply modern methods of multi-level hierarchical systems control and synthesis using high-performance multiprocessor systems [1, 2]. These complexes engagement determines the need for efficient solution of parameters correction and adjustment problem as to the software and hardware components making part to these systems [3–7].

Therefore, when building the lower level hard- and software components of the hierarchical system, there arises the task of analyzing their transfer function coefficients influence on the amplitude-frequency response (AFR) properties. When adjusting some component properties, these properties' correction can be carried out either separately or in a complex [8–10]. In this regard, the problem of rearranging the component properties is typical for adaptive and tunable devices, including filters, which have the same mathematical description [11, 12].

As the preprocessing and filtering path components, we consider commonly used standard digital filters. It is known that for ease of setup and control the high order components are built on the basis of first and second order components [1, 3, 9].

Therefore, the analysis of influence produced by the digital filter transfer function coefficients on its characteristics' properties of is carried out using the first order transfer function of the kind [13]

$$H(z) = \frac{a_0 + a_1 z^{-1}}{1 + b z^{-1}}, \tag{1}$$

where a_0, a_1, b are real coefficients of the numerator and denominator.

In this case, the low-pass filter (LPF) corresponds to the transfer function (1) that has the numerator coefficient $a_0 > 0$, and for the high-pass filter (HPF) it is $a_1 < 0$. It should be noted that the first order normalized digital filters numerator coefficients are equal, that means [14–17]:

$$a_0 = |a_1|, a_0 > 0. \tag{2}$$

Then the transfer function (1) can be written as

$$H(z) = a_0 \frac{1 \pm z^{-1}}{1 + b z^{-1}},$$

where the numerator's "+" is characteristic for LPF and "−" for HPF.

When substituting $z^{-1} = e^{-j\bar{\omega}}$ or using the Euler formula, $z^{-1} = \cos \bar{\omega} - j \sin \bar{\omega}$, where $\bar{\omega}$ is the normalized angular frequency, $\bar{\omega} = 2\pi \frac{f}{f_d}$, f, f_d are respectively, the linear frequency and the sampling frequency, we obtain a complex transmission coefficient, and on its basis we get the low and high frequencies filters' frequency response.

In accordance with the [9] AFC from the complex transfer coefficient, we obtain $H(\bar{\omega}) = \sqrt{\text{Re}^2 + \text{Im}^2}$, where Re and Im are the real and imaginary parts of the complex transfer coefficient, respectively.

2 Low Pass Filter

The frequency-dependent LPF transmission coefficient is generally described by the expression [18, 19]

$$H(\bar{\omega}) = \sqrt{\frac{(a_0 + a_1)^2 - 2a_0a_1(1 - \cos\bar{\omega})}{(1 + b)^2 - 2b(1 - \cos\bar{\omega})}} \tag{3}$$

or taking into account (2) and (3)

$$H(\bar{\omega}) = \frac{2a_0 \cos\left(\frac{\bar{\omega}}{2}\right)}{\sqrt{(1 + b)^2 - 4b \sin^2\left(\frac{\bar{\omega}}{2}\right)}} \tag{4}$$

At zero frequency $\bar{\omega} = 0$, the transmission coefficient $H(0)$ will be determined by the ratio

$$H(0) = a_0 \frac{2}{1 + b}. \tag{5}$$

Based on the fact that at zero frequency the first-order normalized filter transfer coefficient is equal to one $H(0) = 1$ [9], the numerator coefficient a_0 from (5), when filter stability condition complied $|b| < 1$ [9], is equal to

$$a_0 = \frac{1 + b}{2}, \tag{6}$$

and the normalized LPF frequency response will be

$$H(\bar{\omega}) = \frac{(1 + b) \cos\left(\frac{\bar{\omega}}{2}\right)}{\sqrt{(1 + b)^2 - 4b \sin^2\left(\frac{\bar{\omega}}{2}\right)}}.$$

The cutoff frequency $\bar{\omega}_c$ at the level c, $0 < c < 1$, is found from Eq. (4) $H(\bar{\omega}_c) = c$, Fig. 1, then [20]

$$\bar{\omega}_c = \arccos\left(-\frac{1 - 2c^2 \frac{1+b^2}{(1+b)^2}}{1 - 4c^2 \frac{b}{(1+b)^2}}\right).$$

At that the cutoff frequency $\bar{\omega}_c$ level c is set for the Butterworth filter at 0,707, and for the Chebyshev filter and the elliptical filter it is set with passband ripple ε

$$c = \frac{1}{\sqrt{1 + \varepsilon^2}}. \tag{7}$$

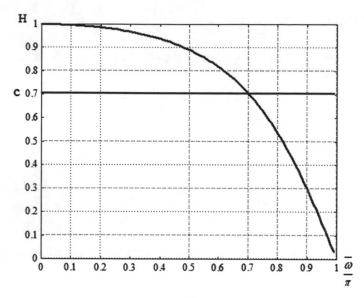

Fig. 1 AFR of the first order LPF Butterworth at $\frac{\bar{\omega}_c}{\pi} = 0,7$

and for Chebyshev's inverse filter, it is set by pulsations ε in the attenuation band

$$c = \frac{\varepsilon}{\sqrt{1 + \varepsilon^2}}. \tag{8}$$

Essential to note is that in most mathematical packages the ripple level is set in decibels: RP is the rippling index in the passband, RS is the rippling index in the attenuation band. However, in both cases

$$c = \frac{1}{\sqrt{10^{0,1RP}}} \text{ or } c = \frac{1}{\sqrt{10^{0,1RS}}} \tag{9}$$

In the presence of a gain not equal to unity, expression (4) will have the form

$$H(\bar{\omega}) = k \frac{2a_0 \cos\left(\frac{\bar{\omega}}{2}\right)}{\sqrt{(1+b)^2 - 4b \sin^2\left(\frac{\bar{\omega}}{2}\right)}}$$

In this case, the relation (5) takes the form $H(0) \cdot k = k \cdot a_0 \frac{1+b}{2}$, which after reduction will correspond to the same (5).

For the first-order low-pass filter, the following formula for determining the transfer function (3) denominator coefficient b depending on the cut-off frequency $\bar{\omega}_c$ and level c was obtained

$$b = -\left\{1 - \frac{2c^2 \sin^2\left(\frac{\bar{\omega}_c}{2}\right)}{c^2 - \cos^2\left(\frac{\bar{\omega}_c}{2}\right)} \left(1 - \frac{\cos\left(\frac{\bar{\omega}_c}{2}\right)}{\sin\left(\frac{\bar{\omega}_c}{2}\right)} \sqrt{\frac{1 - c^2}{c^2}}\right)\right\} \tag{10}$$

However this representation of the denominator coefficient b dependence on the cutoff frequency $\bar{\omega}_c$ and the level c is not quite ideal to implement. So we introduce a dummy variable ξ so that

$$c = \cos\left(\frac{\xi}{2}\right). \tag{11}$$

Or the dummy value is $\xi = 2\arccos(c)$.

Then, after substituting (11) into (10), we can obtain an expression simpler for the implementation in the system

$$b = \frac{\sin\left(\frac{\bar{\omega}_c - \xi}{2}\right)}{\sin\left(\frac{\bar{\omega}_c + \xi}{2}\right)}. \tag{12}$$

3 High Pass Filter

For normalized high-pass filters, which transfer function is given by expression (3), the frequency-dependent transmission coefficient, taking into account (2), can be generally written as

$$H(\bar{\omega}) = \frac{2a_0 \sin\left(\frac{\bar{\omega}}{2}\right)}{\sqrt{(1 + b)^2 - 4b \sin^2\left(\frac{\bar{\omega}}{2}\right)}}. \tag{13}$$

The transmission coefficient for frequency $\bar{\omega} = \pi$ is determined by the ratio

$$H(\pi) = a_0 \frac{2}{1 - b}. \tag{14}$$

Then, taking into account the normalized frequency response, the numerator coefficient a_0 from (14) is equal to

$$a_0 = \frac{1 - b}{2}, \quad |b| < 1, \tag{15}$$

and the normalized high pass filter frequency response will appear like

$$H(\bar{\omega}) = \frac{(1 - b) \sin\left(\frac{\bar{\omega}}{2}\right)}{\sqrt{(1 + b)^2 - 4b \sin^2\left(\frac{\bar{\omega}}{2}\right)}}.$$

At the level c the cutoff frequency $\bar{\omega}_c$ from (13) is equal to

$$\bar{\omega}_c = \arccos\left(\frac{1 - 2c^2\frac{1+b^2}{(1-b)^2}}{1 + 4c^2\frac{b}{(1-b)^2}}\right).$$

At that the cutoff frequency $\bar{\omega}_c$ level c is set for HPF similarly to the LPF and in accordance with Formulas (7) and (8), Fig. 2.

With the inequality of the gain 1 in the HPF, the situation is similar to the LPF.

Depending on the cut-off frequency $\bar{\omega}_c$ and level c, a relationship was found to determine the HPF transfer function (3) denominator coefficient b

$$b = 1 - \frac{2c^2\cos^2\left(\frac{\bar{\omega}_c}{2}\right)}{c^2 - \sin^2\left(\frac{\bar{\omega}_c}{2}\right)}\left(1 - \frac{\sin\left(\frac{\bar{\omega}_c}{2}\right)}{\cos\left(\frac{\bar{\omega}_c}{2}\right)}\sqrt{\frac{1 - c^2}{c^2}}\right). \qquad (16)$$

Using the relation (14) we obtain the expression for coefficient b (16) as

$$b = -\frac{\cos\left(\frac{\bar{\omega}_c + \xi}{2}\right)}{\cos\left(\frac{\bar{\omega}_c - \xi}{2}\right)}. \qquad (17)$$

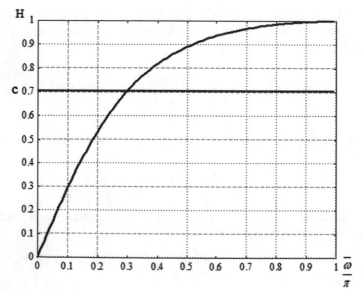

Fig. 2 AFR of the first order HPF Butterworth at $\frac{\bar{\omega}_c}{\pi} = 0,3$

4 Dummy Variable ξ

The expression obtained for the coefficients b (12) and (17) are convenient to use for hard- and software components based on microprocessor technology. It is possible to memorize the sines and cosines values in a given range of the cut-off frequency control at step preset while implementation. In these formulas, a dummy value ξ is introduced, which, based on Formula (9), can be expressed in terms of the frequency response RP and RS ripple pattern

$$\xi = 2 \arccos\left(\frac{1}{\sqrt{10^{0,1RP}}}\right), \tag{18}$$

where instead of RP, RS also can be used, Fig. 3.

From Fig. 1 it follows that when setting the ripple level of 3 dB, which corresponds to the Butterworth filter, the dummy ξ value is equal to $\pi/2$.

Then due to the trigonometric transformations we can write [21]

$$\sin\left(\frac{\overline{\omega_c}}{2} - \frac{\pi}{4}\right) = \sin\left(\frac{\overline{\omega_c}}{2}\right) \cdot \cos\left(\frac{\pi}{4}\right) - \cos\left(\frac{\overline{\omega_c}}{2}\right) \cdot \sin\left(\frac{\pi}{4}\right) = \frac{\sqrt{2}}{2}\left(\sin\left(\frac{\overline{\omega_c}}{2}\right) - \cos\left(\frac{\overline{\omega_c}}{2}\right)\right);$$

$$\sin\left(\frac{\overline{\omega_c}}{2} + \frac{\pi}{4}\right) = \sin\left(\frac{\overline{\omega_c}}{2}\right) \cdot \cos\left(\frac{\pi}{4}\right) + \cos\left(\frac{\overline{\omega_c}}{2}\right) \cdot \sin\left(\frac{\pi}{4}\right) = \frac{\sqrt{2}}{2}\left(\sin\left(\frac{\overline{\omega_c}}{2}\right) + \cos\left(\frac{\overline{\omega_c}}{2}\right)\right);$$

$$\cos\left(\frac{\overline{\omega_c}}{2} - \frac{\pi}{4}\right) = \cos\left(\frac{\overline{\omega_c}}{2}\right) \cdot \cos\left(\frac{\pi}{4}\right) + \sin\left(\frac{\overline{\omega_c}}{2}\right) \cdot \sin\left(\frac{\pi}{4}\right) = \frac{\sqrt{2}}{2}\left(\cos\left(\frac{\overline{\omega_c}}{2}\right) + \sin\left(\frac{\overline{\omega_c}}{2}\right)\right);$$

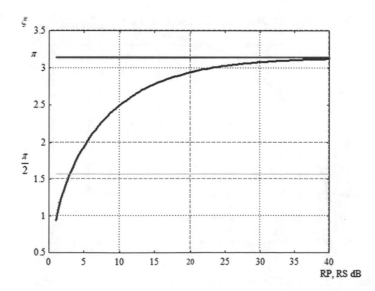

Fig. 3 Graph of fictitious value ξ dependence on the ripple pattern RP or RS, given in decibels

$$\cos\left(\frac{\overline{\omega_c}}{2} + \frac{\pi}{4}\right) = \cos\left(\frac{\overline{\omega_c}}{2}\right) \cdot \cos\left(\frac{\pi}{4}\right) - \sin\left(\frac{\overline{\omega_c}}{2}\right) \cdot \sin\left(\frac{\pi}{4}\right) = \frac{\sqrt{2}}{2}\left(\cos\left(\frac{\overline{\omega_c}}{2}\right) - \sin\left(\frac{\overline{\omega_c}}{2}\right)\right).$$

In this case, Formulas (12) and (17) will look the same.

$$b = \frac{\sin\left(\frac{\overline{\omega_c}}{2}\right) - \cos\left(\frac{\overline{\omega_c}}{2}\right)}{\sin\left(\frac{\overline{\omega_c}}{2}\right) + \cos\left(\frac{\overline{\omega_c}}{2}\right)}. \tag{19}$$

Thus, the Butterworth filter is described by the same formulas, and the difference is in determining the coefficient a_0 at Formulas (6) and (15), as well as in these low-pass filters and high-pass filters transfer function structure (3).

Usually, when using Chebyshev, Inverse Chebyshev and Elliptic filters, the frequency response RP and RS ripples are set above 3 dB. Then the dummy value ξ will be in the range $\frac{\pi}{2} < \xi < \pi$. This allows the passage to the new value θ, which would be in the range from 0 to $\pi/2$, $\theta = \xi - \frac{\pi}{2}$, $R = RP - 3dB$, Fig. 4.

Such a representation allows to simplify the formulas determining the coefficient b for other filters, excluding Butterworth filter, given that $\xi = \theta + \frac{\pi}{2}$. In this case, Formulas (12) and (17) will have the form as

$$b = \frac{\sin\left(\frac{\overline{\omega_c} - \left(\theta + \frac{\pi}{2}\right)}{2}\right)}{\sin\left(\frac{\overline{\omega_c} + \left(\theta + \frac{\pi}{2}\right)}{2}\right)}, \tag{20}$$

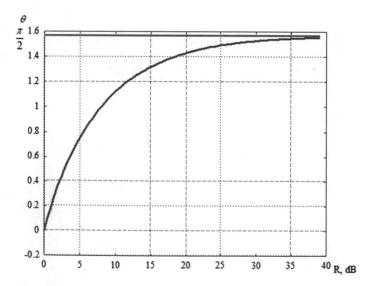

Fig. 4 Graph of value θ dependence on the ripple pattern RP or RS, given in decibels

$$b = \frac{\cos\left(\frac{\bar{\omega}_c + \left(\theta + \frac{\pi}{2}\right)}{2}\right)}{\cos\left(\frac{\bar{\omega}_c - \left(\theta + \frac{\pi}{2}\right)}{2}\right)}. \tag{21}$$

Due to trigonometric transformations, the formula for calculating the low-pass filter coefficient b will appear as

$$b = \frac{\sin\left(\frac{\bar{\omega}_c - \left(\theta + \frac{\pi}{2}\right)}{2}\right)}{\sin\left(\frac{\bar{\omega}_c + \left(\theta + \frac{\pi}{2}\right)}{2}\right)} = \frac{\sin\left(\frac{(\bar{\omega}_c - \theta) + \frac{\pi}{2}}{2}\right)}{\sin\left(\frac{(\bar{\omega}_c + \theta) + \frac{\pi}{2}}{2}\right)} = \frac{\sin\left(\frac{(\bar{\omega}_c - \theta)}{2} + \frac{\pi}{4}\right)}{\sin\left(\frac{(\omega_c + \theta)}{2} + \frac{\pi}{4}\right)} = \frac{\sin\left(\frac{(\bar{\omega}_c - \theta)}{2}\right) + \cos\left(\frac{(\bar{\omega}_c - \theta)}{2}\right)}{\sin\left(\frac{(\bar{\omega}_c + \theta)}{2}\right) + \cos\left(\frac{(\bar{\omega}_c + \theta)}{2}\right)},$$

And for high-pass filter

$$b = \frac{\cos\left(\frac{\bar{\omega}_c + \left(\theta + \frac{\pi}{2}\right)}{2}\right)}{\cos\left(\frac{\bar{\omega}_c - \left(\theta + \frac{\pi}{2}\right)}{2}\right)} = \frac{\cos\left(\frac{(\omega_c + \theta) + \frac{\pi}{2}}{2}\right)}{\cos\left(\frac{(\bar{\omega}_c - \theta) + \frac{\pi}{2}}{2}\right)} = \frac{\cos\left(\frac{(\bar{\omega}_c + \theta)}{2} + \frac{\pi}{4}\right)}{\cos\left(\frac{(\bar{\omega}_c - \theta)}{2} + \frac{\pi}{4}\right)} = \frac{\sin\left(\frac{(\bar{\omega}_c + \theta)}{2}\right) - \cos\left(\frac{(\bar{\omega}_c + \theta)}{2}\right)}{\sin\left(\frac{(\bar{\omega}_c - \theta)}{2}\right) + \cos\left(\frac{(\bar{\omega}_c - \theta)}{2}\right)}.$$

5 Bandpass Filter

On the basis of the considered first-order components and the relations obtained, it is possible to easily construct the band components and rearrange their properties. However, for better implementation, we should consider bandpass components, described by a second order equation of general form

$$H(z) = \frac{a_0 + a_1 z^{-1} + a_2 z^{-2}}{1 + b_1 z^{-1} + b_2 z^{-2}}.$$

For the bandpass component characteristic is $a_0 = -a_2, a_1 = 0$. Then the frequency response and phase response will be

Frequency response

$$H(\bar{\omega}) = \sqrt{\frac{(2a_0 \sin(\bar{\omega}))^2}{(1 - b_2)^2 + b_1^2 + 2b_1(1 + b_2)\cos(\bar{\omega}) + 4b_2 \cos(\bar{\omega})^2}}, \tag{22}$$

Phase response

$$\varphi(\bar{\omega}) = arctg\left(\frac{b_1 + (1 + b_2)\cos(\bar{\omega})}{(1 - b_2)\sin(\bar{\omega})}\right). \tag{23}$$

It should be noted that the standard normalized digital filters also belong to the bandpass components, Fig. 5.

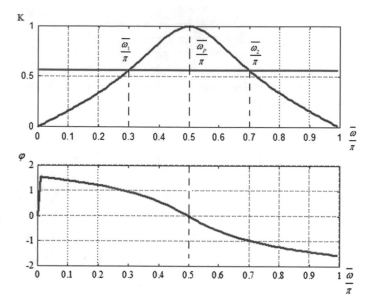

Fig. 5 The normalized digital Chebyshev bandpass filter frequency response and phase response at cut-off frequencies $\frac{\bar{\omega}_1}{\pi} = 0, 3$ and $\frac{\bar{\omega}_2}{\pi} = 0, 7$ at the ripple level RP $= 7$

To adjust the frequency response, it is necessary to determine which effect the changes in the transfer function coefficients a_0, b_1, b_2 and the ripple index in the passband produce onto amplitude and cutoff frequency of the frequency response.

From the analysis of frequency response and phase response, it follows that at the frequency response frequency peak $\bar{\omega}_p$, the phase-frequency characteristic is zero, Fig. 3. The frequency response frequency peak $\bar{\omega}_p$ does not always coincide with the middle of the passband $\bar{\omega}_0$, as shown in Fig. 3. Quite often they do not coincide, for example, refer to Fig. 6.

From Fig. 6 it follows that the peak frequency $\frac{\bar{\omega}_p}{\pi} = 0, 6$, the bandwidth $\frac{\bar{\omega}_p}{\pi}$ center frequency is 0, 53.

The frequency response analysis for bandpass filters showed that the derivative of the frequency response at $\frac{\bar{\omega}_p}{\pi}$ is zero and at the same frequency the transfer function numerator and denominator derivatives intersect, as shown in Fig. 7.

The transfer function further analysis showed that the numerator and the denominator themselves also intersect at this frequency, Fig. 8.

Then, we can equate the transfer function numerator and denominator for finding the peak frequency, however, this problem is solved much easier by equating the transfer function numerator and the denominator derivatives.

From Eq. (23) we can be write

$$H^2(\bar{\omega}) = \frac{4a_0^2 \sin^2(\bar{\omega})}{(1 - b_2)^2 + b_1^2 + 2b_1(1 + b_2)\cos(\bar{\omega}) + 4b_2 \cos^2(\bar{\omega})}.$$

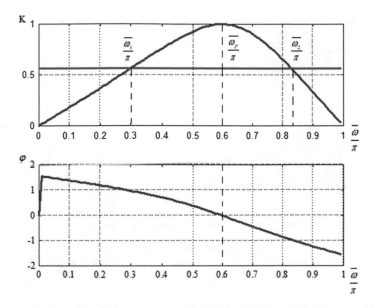

Fig. 6 Frequency response and phase response of a standardized digital Chebyshev bandpass filter at cut-off frequencies $\frac{\tilde{\omega}_1}{\pi} = 0, 3$ and $\frac{\tilde{\omega}_2}{\pi} = 0, 83$, ripple level RP $= 7$

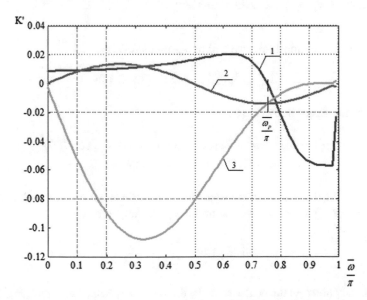

Fig. 7 Frequency response derivative (1) in general, as well as the transfer function numerator (2) and denominator (3) derivates for normalized digital Chebyshev bandpass filter at cut-off frequencies $\frac{\tilde{\omega}_1}{\pi} = 0, 5$ and $\frac{\tilde{\omega}_2}{\pi} = 0, 9$, when the ripples level is RP $= 5$, $\frac{\tilde{\omega}_p}{\pi} = 0, 7589$

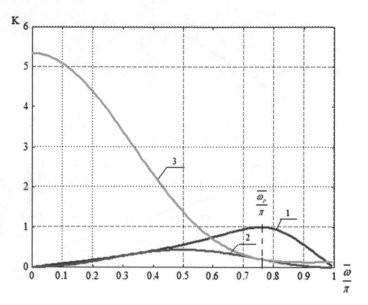

Fig. 8 Graph of frequency response (1), as well as the normalized digital Chebyshev bandpass filter transfer function numerator (2) and denominator (3) at cut-off frequencies $\frac{\bar{\omega}_1}{\pi} = 0,5$ and $\frac{\bar{\omega}_2}{\pi} = 0,9$, when the ripple level is RP $= 5$, $\frac{\bar{\omega}_p}{\pi} = 0,7589$

Then the numerator derivative will be found as follows.

$$\left(4a_0^2 \sin^2(\bar{\omega})\right)' = 8a_0^2 \sin(\bar{\omega}) \cos(\bar{\omega}), \tag{24}$$

and the denominator derivative as

$$\left((1 - b_2)^2 + b_1^2 + 2b_1(1 + b_2) \cos(\bar{\omega}) + 4b_2 \cos^2(\bar{\omega})\right)'$$
$$= -2b_1(1 + b_2) \sin(\bar{\omega}) - 8b_2 \sin(\bar{\omega}) \cos(\bar{\omega}). \tag{25}$$

Equating the derivatives with respect to the transfer function, numerator (24) and denominator (25) frequency we can find the peak frequency cosine

$$\cos(\bar{\omega}_p) = -\frac{b_1(1 + b_2)}{4(a_0^2 + b_2)}. \tag{26}$$

As already noted, at the peak frequency the frequency response is zero, Fig. 3. This made it possible to determine the peak frequency from the phase response Eq. (23). Equating to zero the frequency response numerator

$$b_1 + (1 + b_2) \cos(\bar{\omega}_p) = 0.$$

Then

$$\cos(\bar{\omega}_p) = -\frac{b_1}{(1 + b_2)}. \tag{27}$$

Equating (26) and (27) we obtain the equations of the kind

$$(1 + b_2)^2 = 4(a_0^2 + b_2).$$

Solving this equation we obtain the numerator coefficient a_0 dependence on the denominator coefficient b_2.

From the first equation we express the coefficient of the denominator, substituting from Formula (26) the value of the coefficient of the numerator.

Solving this equation, we get the following solution

$$a_0 = \frac{1 - b_2}{2}. \tag{28}$$

Thus, the frequency response amplitude is determined by the denominator coefficient b_2.

Let we consider the denominator coefficients b_1 and b_2, as well as the passband level c influence on the frequency response properties.

Then for frequency response square we can make two equations for each cutoff frequency

$$\begin{cases} \dfrac{(2a_0 \sin(\overline{\omega_1}))^2}{(1 - b_2)^2 + b_1^2 + 2b_1(1 + b_2)\cos(\overline{\omega_1}) + 4b_2\cos(\overline{\omega_1})^2} = c^2; \\ \dfrac{(2a_0 \sin(\overline{\omega_2}))^2}{(1 - b_2)^2 + b_1^2 + 2b_1(1 + b_2)\cos(\overline{\omega_2}) + 4b_2\cos(\overline{\omega_2})^2} = c^2. \end{cases} \tag{29}$$

From the first equation we express the denominator coefficient b_1, substituting from Formula (26) the value of numerator coefficient a_0.

Solving this equation, we get the following solution:

$$b_1 = -(1 + b_2)\cos(\overline{\omega_1}) \pm (1 - b_2)\sin(\overline{\omega_1})\frac{\sqrt{1 - c^2}}{c}.$$

This ratio analysis made it possible to determine that the coefficient b_1 for cutoff frequencies $\overline{\omega_1}$ is determined as follows.

$$b_1 = -(1 + b_2)\cos(\overline{\omega_1}) + (1 - b_2)\sin(\overline{\omega_1})\frac{\sqrt{1 - c^2}}{c}, \tag{30}$$

and for the cutoff frequency $\overline{\omega_2}$ as

$$b_1 = -(1 + b_2)\cos(\overline{\omega_2}) - (1 - b_2)\sin(\overline{\omega_2})\frac{\sqrt{1 - c^2}}{c}.$$

Substituting expression (30) for coefficient b_1 into the second equation from (29), we obtain a series of additional relations, whose analysis allowed us to determine the value of denominator coefficient b_2 as

$$b_2 = \frac{(\cos(\overline{\omega_2}) + \cos(\overline{\omega_1})) - 2c^2\cos(\overline{\omega_2}) - 2c\sqrt{1 - c^2}\sin(\overline{\omega_2})}{(\cos(\overline{\omega_2}) + \cos(\overline{\omega_1})) - 2c^2\cos(\overline{\omega_1}) - 2c\sqrt{1 - c^2}\sin(\overline{\omega_1})}. \tag{31}$$

Substituting the dummy value ξ from expression (11) we obtain more simplified equations for coefficients b_1 and b_2

$$b_1 = -(1 + b_2)\cos(\overline{\omega_1}) + (1 - b_2)\sin(\overline{\omega_1}) \cdot tg\left(\frac{\xi}{2}\right); \tag{32}$$

$$b_2 = \frac{\cos(\overline{\omega_1}) - \cos(\overline{\omega_2} - \xi)}{\cos(\overline{\omega_2}) - \cos(\overline{\omega_1} - \xi)}. \tag{33}$$

Considering that for Butterworth bandpass filters, $\xi = \frac{\pi}{2}$ Eqs. (32) and (33) will have the form

$$b_1 = -(1 + b_2)\cos(\overline{\omega_1}) + (1 - b_2)\sin(\overline{\omega_1}); \tag{34}$$

$$b_2 = \frac{\cos(\overline{\omega_1}) - \sin(\overline{\omega_2})}{\cos(\overline{\omega_2}) - \sin(\overline{\omega_1})}. \tag{35}$$

Taking into account that $\xi = \theta + \frac{\pi}{2}$ the following transformations can be carried out in Eqs. (32) and (33)

$$tg\left(\frac{\xi}{2}\right) = tg\left(\frac{\theta + \frac{\pi}{2}}{2}\right) = tg\left(\frac{\theta}{2} + \frac{\pi}{4}\right) = \frac{\sin\left(\frac{\theta}{2} + \frac{\pi}{4}\right)}{\cos\left(\frac{\theta}{2} + \frac{\pi}{4}\right)} = \frac{\sin\left(\frac{\theta}{2}\right) + \cos\left(\frac{\theta}{2}\right)}{\cos\left(\frac{\theta}{2}\right) - \sin\left(\frac{\theta}{2}\right)};$$

$$\cos(\overline{\omega} - \xi) = \cos\left(\overline{\omega} - (\theta) - \frac{\pi}{2}\right) = \sin(\overline{\omega} - \theta).$$

Then Eqs. (32) and (33) for the Chebyshev, inverse Chebyshev and elliptic filters will have the form

$$b_1 = -(1 + b_2)\cos(\overline{\omega_1}) + (1 - b_2)\sin(\overline{\omega_1}) \cdot \frac{\cos\left(\frac{\theta}{2}\right) + \sin\left(\frac{\theta}{2}\right)}{\cos\left(\frac{\theta}{2}\right) - \sin\left(\frac{\theta}{2}\right)}; \tag{36}$$

$$b_2 = \frac{\cos(\overline{\omega_1}) - \sin(\overline{\omega_2} - \theta)}{\cos(\overline{\omega_2}) - \sin(\overline{\omega_1} - \theta)}. \tag{37}$$

6 Conclusions

The design of first order digital frequency-dependent components is considered and described with the transfer functions characteristic to first order filters. The considered digital filters frequency response analysis shows that when designing frequency-dependent components depending on a given cut-off frequency $\bar{\omega}_c$ and level c, we can unequivocally find the numerator and denominator coefficients a_0 and b values, and for normalized filters, we can find the necessary coefficient b of the denominator.

To change the component's gain sufficient is changing the numerator coefficient a_0, leaving unchanged the denominator coefficient b. In this case, linear gain control is possible due to a change in the numerator coefficient a_0, that is characteristic to adaptive filters.

When the denominator coefficient b changed at non-standardized filters, adjusted are both the gain and the cutoff frequency. However, in order to change the cutoff frequency at a constant amplitude, the gain value correction with the transfer function numerator coefficient a_0 is necessary when a new value of the denominator coefficient b in accordance with, for example, (6) and (14).

From the analysis of formulas obtained for the order low-pass filters and high-pass filters frequency response, to denote is that:

- the numerator coefficient a_0 does not affect the frequency response shape, but only determines the frequency response gain level;
- for the same type of filters, the low-pass filter and high-pass filter denominator coefficients b are different;
- when the low-pass filter denominator coefficient $b = 0$ the LPF becomes an average in accordance with the moving average algorithm at two points [18, 22]

$$H(z) = \frac{1}{2}(1 + z^{-1}),$$

And the HPF becomes a differentiator with an accuracy reaching the gain value in accordance with the two-point difference algorithm [22, 23]

$$H(z) = \frac{1}{2}(1 - z^{-1}).$$

Based on the expressions obtained, the soft- and hardware tools implementation with microprocessors technology is greatly simplified, that makes much easier to solve the problem of constructing the specialized computer systems' frequency-dependent components whose functionality allows correcting and adjusting the characteristics of individual component or whole system when required.

References

1. Disadvantages of existing automated process control systems at nuclear power plants and their development prospects, http://tesiaes.ru/?p=13030
2. Eliseev VV (2013) Information and control systems HDM production NPP "Pulse Status, prospects". Nucl Radiat Saf 4(60): 61–64
3. Stupen PV (2000) Construction of digital filter structures for monitoring and control systems. Nucl Radiat Saf 3(4):96–100
4. Bilenko AA, Sitnikov VS (2010) Analysis construct reconfigurable computing on computer systems. Radio Electron Comput Syst 7(48):212–214
5. Stoyanov G, Uzunov I, Kawamata M (2000) Design of variable IIR digital filters using equal subfilters. In: Proceedings of IEEE international symposium on intelligent signal processing and communication systems, vol 1, pp. 141–146
6. Stoynov G, Uzunov I, Kawamata M (2001) Design and realization of variable IIR digital filters as a cascade of identical subfilers. IEICE Trans. Fundamentals E84-A(8): 1831–1838
7. Robinson A, Hardie R, Heinisch H (1997) Implementing continuously programmable digital filters with the TMS320C30/4DSP. Application Report SPRA190A. Texas Instruments
8. Bukashkin SA, Vlasov VP, Zmii BF (1984) Handbook of calculation and design of ARC-circuits. Radio and communication, Moscow
9. Sergienko AB (2006) Digital signal processing. Piter, St. Petersburg
10. Tepin VP, Tepin AV (2011) Programmable digital filters: a synthesis of the transfer function of the control laws. Proc SFU Tech Sci 1:50–57
11. Malakhov VP, Sitnikov VS, Iakovleva ID (2008) Adaptive restructuring of the digital filter in the automatic control system. Automation Automation Electr Syst Syst 1(21):158–161
12. Brus AA, Malakhov VP, Sitnikov VS (2009) The range of regulation of frequency-dependent components of computer systems car. Electr Mach Electr Equip 72:139–142
13. Khilari SS (2014) Transfer function and impulse response synthesis using classical techniques. Masters Theses 1911
14. Gupta DR (2006) Synthesis of continuous time Nyquist pulses, M.S. Thesis, University of Massachusetts Amherst
15. Benke G, Wells B (1985) Estimates for the stability of low-pass filters. IEEE Trans Acoust Speech Signal Process 33(1):98–105
16. Ćertić JD, Milić LD (2011) Investigation of computationally efficient complementary IIR filter pairs with tunable crossover frequency. AEU-Int J Electron Commun 65(5):419–428
17. Al-Radhi M (2012) Design of finite impulse response digital filters using optimal methods. Portsmouth, England
18. Brus AA, Dikusar EV, Sitnikov VS, Iatsenko TP (2010) Correction performance tunable components using a moving average algorithm. Scientific Bulletin of Chernivtsi University. Series: Computer Systems and Components 1(1): 26–30
19. Matveichuk MI, Patsar VS, Sitnikov VS (2010) Analysis of the impact factors of the transfer function of the first order nonpolynomial digital filter on the properties of amplitude-frequency characteristics. The works of the Odessa Polytechnic University 2(22): 1–5
20. Ukhina AV, Bilenko AO, Sytnikov VS (2016) Improving the efficiency software and hardware complexes in NPP APCS. Nucl Radiat Saf 3:70–76
21. Korn G, Korn T (2000) Mathematical handbook for scientists and engineers—McGraw-Hill Book Company
22. Turulin II, Bulgakova IuI (2011) Methods of synthesis of controlled digital filter based on analog prototypes. Proc SFU Tech 2:88–92
23. Petrov IS, Ukhina AV, Iatsenko TP, Sitnikov VS (2015) Reconfigurable component of the first order of specialized autonomous mobile platform computer system. SIET-2015, Odessa, pp. 24–25

Practical Application of Clustering Methods in Radar Signals Recognition System

Jan Matuszewski

Abstract The system of radar signals detection, analysis and recognition was described. The modern electronic recognition systems should react fast and with great accuracy in the extremely complex electromagnetic environment. The binary decision tree was applied at the beginning of grouping signals received from unknown sources. The paper presents some clustering methods to radar signal recognition based on the mathematical criteria. The concepts of this technique are described. The experiment results obtained for nearest neighbour method are presented. Clustering algorithm was tested for different methods of objects grouping and for various distance measures.

Keywords Electronic intelligence · Signal recognition · Clustering methods · Binary decision tree

1 Introduction

On the modern electromagnetic battlefield, where radio communications, radar and guided weapons are widely used, the signal environment is of high density. In order to ensuring the rapid system response in a dense and complex electromagnetic environment, the functions of search, interception, analysis and identification must be automated [1, 2]. The electronic intelligence system (ELINT) must be able to recognize emitters from pulse-by pulse measurement made by receiver in order to indicate the presence of known (friendly or hostile) radar, as well as to provide emitter tracking, threat assessment and platform identification. The data processing in ELINT systems, with response time constantly decreasing, presents one of the most complex and time-domain critical problem for the current technology [3–6].

The basic task to do for ELINT systems are as follows:

J. Matuszewski (✉)
Faculty of Electronics, Institute of Radioelectronics, Military University of Technology, Warsaw, Poland
e-mail: jan.matuszewski@wat.edu.pl

© The Editor(s) (if applicable) and The Author(s), under exclusive license to Springer Nature Switzerland AG 2021
T. Radivilova et al. (eds.), *Data-Centric Business and Applications*, Lecture Notes on Data Engineering and Communications Technologies 48,
https://doi.org/10.1007/978-3-030-43070-2_7

- detect electromagnetic emissions and find the direction of incoming signals,
- analyze the emitter's characteristics such as technical parameters, message traffic and content, operating role and geographic location,
- monitor any changes of situation and target parameters,
- communicate the most important data as soon as possible,
- ensure control, command and support the forces on the battlefield.

   The concept of data analysis, processing, recognition and security contains many different solutions [4, 7–13].

2   Radar Signals Recognition System

The main group of electronic threats, being respectively electronic warfare (EW) targets, are radars. Each radar has the typical parameters which may be divided into two groups, primary and secondary. The primary parameters of radar signal, measured by the receiving subsystem are: radio frequency, time of arrival, pulse width, angle of arrival and amplitude. They are passed in digital form to the pre-processor in order to sorting and dividing them into pulse chains associated with detectable radars. The main processor block derives the secondary radar parameters such as: pulse repetition interval, inter-pulse modulation (agility, stagger, dwell and switch, jitter), scan type, scan period [1, 2, 14]. By accurately determining these parameters of a radar signal it is possible to identify the type, application, platform and radar user.

   According to available bibliography concerning the subject matter, the radar (emitter) pattern ought to possess measured and non-measured features (information) such as: logical, numerical, descriptive and graphical. The measured feature is a result of measurements and calculations. The non-measured feature can be expressed by the chain of words or logical expression. Figure 1 illustrates the process of measurement and recognition of radar signals in ELINT system.

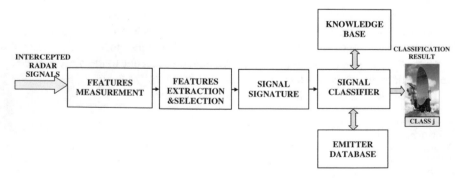

Fig. 1 Radar signals recognition system

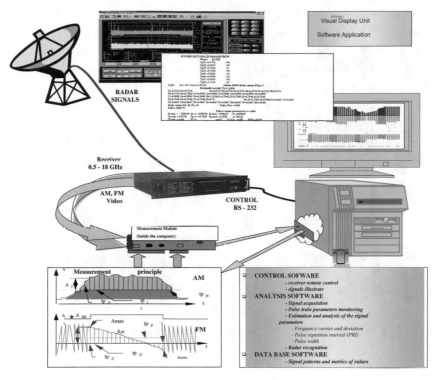

Fig. 2 A general system structure for measurement, recording, analysis and radar signals recognition

The recognition system, shown in Fig. 2, consists of the following subsystems [14, 15]:

(a) Signal receiver: provides, with very short acquisition times, long range detection of emitter radiations and measurements of their basic parameters.

(b) Signal processor: allows, in a dense electromagnetic environment, a very accurate analysis of complex radar signals, sorting the input data into pulse chains (trains) with the aim of producing one pulse chain for each detected emitter (this process is called de-interleaving), Fig. 3. This analysis allows to set up the signature of signal parameters in order to precise identification and update the emitter data base. During signal processing very often the intra-pulse analysis is done by digital recording frequency, amplitude, pulse width and AOA versus the time.

(c) Radar signal classifier: allows the feature extraction and selection/reduction the data incoming to the classifier, which successively compares the specific signal parameters (signatures) with a library of known emitter characteristics (compiled through intelligence service) to determine the possible identities of detected signals. The result of identification is presented to the operator on the display in the form of tabular or graphical option for situation monitoring,

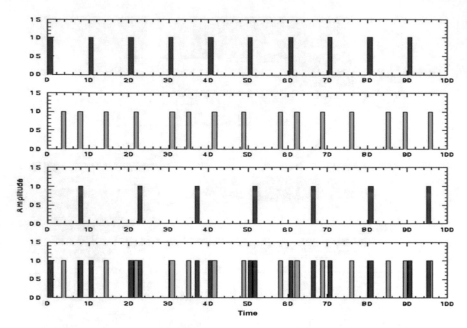

Fig. 3 Time measurement of signal pulses arrival (bottom drawing) generated by three various radars

mapping and radar interception analysis (histograms, patterns of signals, etc.) and written in the special table of detected radars which is continuously updated to produce the time-average parameters of each radar.

The radar signal consisting of N parameters will be assumed as a point in the N-dimensional Euclidean space and denoted as a vector x or in case of many samples by x_i, where index i denotes the number of the sample. These vectors create a pattern of radar emission source which is called as a radar signature. These radar signatures, formed earlier on the measured signals in the ELINT systems, are written in the emitter database. In this way, the emitter database should include the patterns of radio-electronic devices and the technical characteristics of electromagnetic sources.

In order to dividing unknown radar signals in ELINT systems into distinct groups, one group per radar or more generally per the radar's mode of operation, the clustering methods may be applied [16–19]. The task of clustering is based on dividing all samples, previously unknown radar signals into at least two groups of objects (clusters) so that the elements of one cluster are similar, and the elements from different clusters are significantly different on the set of similarity and differences.

Finally, the system matches the 'signal signature', composed by the average parameters from each group, with the characteristics of the identified emitters stored in the emitter database [14].

The main problems in radar signals recognition are the following:

- different number of measured vectors of radar signal parameters from previously unknown classes;
- lack of prior information about number and characteristics of recognized signals.

The efficient clustering algorithm must meet the following basic requirements:

- minimal computational complexity;
- capability to select the different clusters structure (shape, size, density) using minimum amount of prior information;
- specifying the number of unknown clusters;
- determining the statistical parameters of performed clusters.

Nowadays a wide variety of clustering algorithms and their modifications were developed. However, none of these approaches can create a universal algorithm for radar signals recognition. In the process of classification and recognition of unknown measured radar signals the artificial neural networks (ANNs) may be also applied [6, 20–25]. The technical and tactical parameters connected with radars' applications are very often a good source of knowledge and helpful to recognize radars on the base of intercepted signals. Majority of information useful to the knowledge-based system can be obtained from individual radar details, their platforms and relationships between them. The knowledge base employs a declarative, rule-based representation of facts about the radar parameters and its application [15].

The quality of proposed methods was examined with the use the real signals received from the electromagnetic environment. The aim of the paper is to examine if clustering methods may be used to radar signals recognition in the ELINT systems.

3 Binary Decision Tree Use for Radar Signal Recognition

In order to reduce the recognition time and simplify the classifier it is convenient at the beginning to split the feature space into unique regions (subspaces), corresponding to the classes containing radars about similar values of signal parameters, in a sequential manner. The most popular method among the nonlinearly classifiers is a binary decision tree (BDT) in which the space is split into hyperrectangles with sides parallel to the axis.

By successive sequential splitting the feature space are created the regions corresponding the respective classes. The sequence of decisions is applied to individual features, and the questions to be answered are of the form "*is parameter* $x_i \leq \alpha_s$" where α_s is the threshold value [26]. The value α_s is calculated as a mean value between the middles of two the closest classes if the distance between them is greater than earlier calculated threshold Δ_i, for example $6 \div 12$ values of root mean square RMS for this parameter.

The main idea of BDT for simplified example is demonstrated in Fig. 4 where numbers of region are assigned in a sequential way. For testing this procedure are

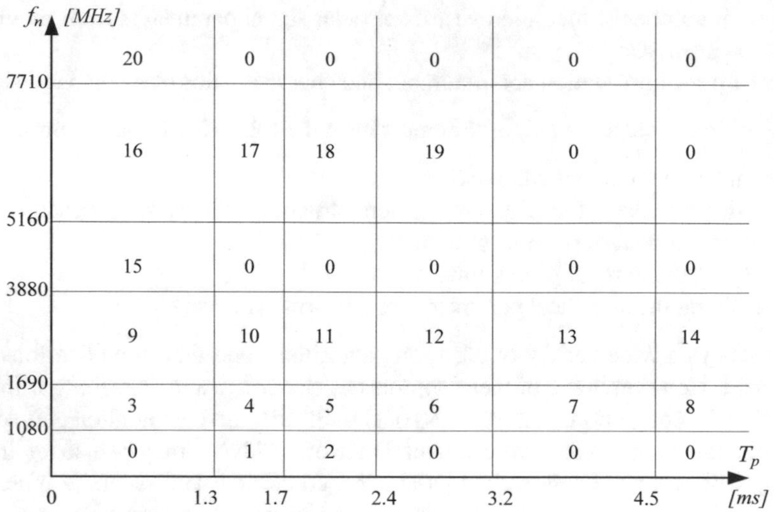

Fig. 4 The space divide on regions for two radar signal features: T_p, f_n

used the radar signals obtained from more than 120 various sources. After a preliminary analysis for two radar parameters: pulse repetition interval (T_p) and radio frequency (f_n) twenty separate regions (group of classes) containing radars with various applications were created. Number 0 means that any signal from this region was not intercepted [26].

The binary decision tree is defined as [26]

$$\text{BDT} = \left\{ (h_s, k_s, \alpha_s, i_s^l, i_s^r), \quad s = 1, 2, \ldots, K \right\} \tag{1}$$

where: h_s—number of decision node (binary divide), k_s—a feature on which the binary divide was made, ($k_s = 1$ for T_p, $k_s = 2$ for f_n), α_s—value of binary divide on the k_s feature, i_s^l—number of next decision node for the left subset of classes, i_s^r—number of next decision node for the right subset of classes, K—number of decision nodes in BDT.

Upon intercepting and measuring the feature vector, the searching of the region to which this vector will be classified, is achieved via a sequence of decisions along a path of nodes of appropriately constructed tree.

Figure 5 illustrates the respective BDT with its decision nodes and paths. The positive numbers above the ellipses mean the number of decision node, negative numbers in the rectangles–numbers of classes. Such way of class notation is connected with a way of creating a very simply algorithm for signal classification. A separate classifier was used at each node, and each of them performs a two-class decision. If in result this sequential way made decisions we get a negative number then classification process of incoming radar signal ends.

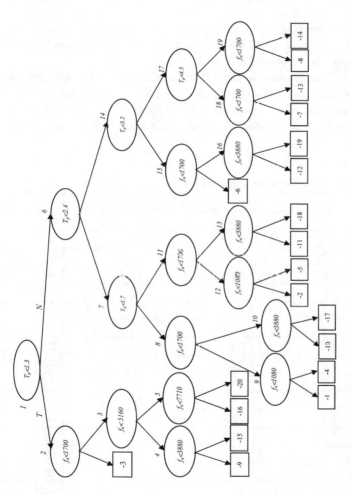

Fig. 5 A binary decision tree structure for twenty subsets of classes

The optimization of BDT contains the minimization the path length from the root node (number 1) to the end node (negative values) in this tree. Figure 3 shows that maximum 5 binary comparisons of appropriate signal parameter with earlier fixed threshold α_s is needed to classifying incoming signals to one of 20 subsets of classes in two dimensional measurement space [26].

The length of tree is equal the number of nodes in the longest created path. Note that it is possible to reach a decision without having tested all the available features. After carry out a few tests for real and simulated date it can say that total time of signal classification decreases about eight times and probability of true recognition was practically the same. This binary classification tree according with Eq. (1) can be written in the way as shown in Table 1.

This way of classification is very effective, particularly in situations when the large number of classes and radar parameters causes a great increase amount of calculations and time of signal recognition. The appropriate algorithm recognition of radar signals x_i for BDT according with (1), illustrated in Fig. 5 and in Table 1, is shown in Fig. 6, [26].

Table 1 Binary decision tree for example from Fig. 2

h_s	k_s	α_s	i_s^l	i_s^r
1	1	1.3	2	6
2	2	1700	−3	3
3	2	5160	4	5
4	2	3880	−9	−15
5	2	7710	−16	−20
6	1	2.4	7	14
7	1	1.7	8	11
8	2	1700	9	10
9	2	1080	−1	−4
10	2	3880	−10	−17
11	2	1700	12	13
12	2	1080	−2	−5
13	2	3880	−11	−18
14	1	3.2	15	17
15	2	1700	−6	16
16	2	3880	−12	−19
17	1	4.5	18	19
18	2	1700	−7	−19
19	2	1700	−8	−14

Fig. 6 Classification
algorithm with use the binary
decision tree (BDT)

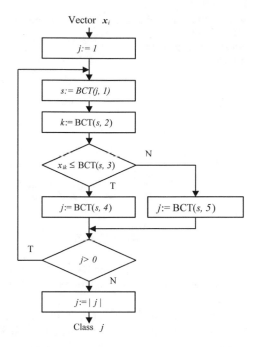

Vector x_i

$j := 1$

$s := BCT(j, 1)$

$k := BCT(s, 2)$

$x_{ik} \leq BCT(s, 3)$

N

T

$j := BCT(s, 4)$ $j := BCT(s, 5)$

T

$j > 0$

N

$j := |j|$

Class j

4 Clustering Algorithms in Recognition Radar Signals

All clustering algorithms can be divided into two groups: hierarchical and non-hierarchical Hierarchical algorithms can detect the nested clusters. For these clusters a tree is built, called distance chart with which the dendrogram can be build. Non-hierarchical algorithms calculate clusters based on the optimization of the determined criteria of quality.

Initially, it is assumed that each object creates a separate cluster, which are then merged into the most similar subsets in stages, until in the final phase only one cluster containing all of the recognized realizations will be formed. These algorithms do not require the number of groups to be specified in advance, as the number of clusters can be selected after the analysis is completed by cutting the dendrogram at the appropriate height (threshold c). All hierarchical grouping procedures (clustering methods) can be described by one general scheme—an algorithm. The differences are based on varying application of distance measures between clusters [16–19].

In practice, different names are adopted that define this sequence, such as: object image, object implementation or observation vector (parameters vector). An image of an object (pattern) can create sets of numbers, sets of logical expressions or sets of tasks that describe its structure.

The values of radar signal parameters are treated as a point in the N-dimensional Euclidean space and denoted as a vector x_i where index i denotes the number of vector sample.

Fig. 7 Illustration of points
belonging to particular
classes in the parameter
space

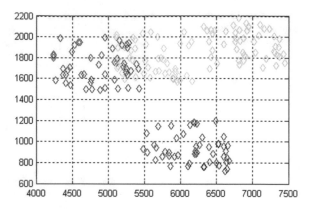

$$x_i = (x_{i1}, x_{i2}, \ldots, x_{iN})^T, \quad i = 1, 2, \ldots, n \tag{2}$$

where: x_{ij}—value of parameter j in vector i ($j = 1,2,\ldots, N$), N—number of radar
signal parameters, n—number of measured vectors x_i, K—number of formed clusters
(classes).

The coordinates of this point x_i in Cartesian coordinate system are the real numbers. The measured vectors that belong to different classes create in the observation
space the specific areas that can have different shapes, dimensions, boundaries and
structure. They can be separable or overlap in varying degree. Figure 7 shows an
example of the distribution of 4 classes in the parameter space [27, 28].

In the approach of unknown radar signals recognition the nearest neighbour (NN)
rule with the different minimum distance measures in the measurement space $\mathbf{X}$ is
used [9, 16]. The vector x_i of radar signal parameters is assigned to the class k of its
nearest neighbour. The similarity measures of unknown object $x_i \in \mathbf{X}$ are calculated
to all formed before clusters ω_k, $k \in \langle 1, K \rangle$, where K—number of cluster. These
calculated values determine the similarity measures of object x_i to respective cluster
ω_k. This algorithm allows to split the initial sample consisting from M vectors into
K-clusters (K—parameter is set by the user). The final clusters are described by
vectors of medium values of radar parameters (centroids).

In hierarchical grouping methods the following distances measures between
vectors x_i, x_j are applied [11–13]:

(a) Euclidean

$$d_E(x_i, x_j) = \left[\sum_{k=1}^{N} (x_{ik} - x_{jk})^2 \right]^{1/2} \tag{3}$$

Due to applying in calculations the square of numbers and square roots, necessary time to obtain the results of calculations is extended. The most important
one is it that values of respective vector components may be in the different

scales. The solution for this problem is applying one of normalization method for radar signal parameters.

(b) Seuclidean

$$d_S(\boldsymbol{x}_i, \boldsymbol{x}_j) = \sum_{k=1}^{N} |x_{ik} - x_{jk}|^2 \tag{4}$$

(c) Cityblock

$$d_C(\boldsymbol{x}_i, \boldsymbol{x}_j) = \sum_{k=1}^{N} |x_{ik} - x_{jk}| \tag{5}$$

(d) Minkowski of rank $s = 1, 2, \ldots$

$$d_M(\boldsymbol{x}_i, \boldsymbol{x}_j) = \left[\sum_{k=1}^{N} |x_{ik} - x_{jk}|^2 \right]^{1/s} \tag{6}$$

If $s = 2$ then the Euclidean metric is obtained. For calculation are accepted that $s = 3$.

(e) Mahalanobis

$$d_H(\boldsymbol{x}_i, \boldsymbol{x}_j) = \sqrt{\sum_{k=1}^{N} (x_{ik} - x_{jk})^t \mathbf{T}^{-1} (x_{ik} - x_{jk})} \tag{7}$$

where the inverse covariance matrix $\mathbf{T}^{-1}$ is calculated based on the set of radar signals parameters.

In many practical situations, the designer has to deal with features whose values lie within the limits of different dynamic ranges. This would mean that features with high values may have a greater impact on target functions than features with low values, although this does not necessarily reflect their individual validity in designing the classifier. This problem is solved by parameters normalization at the initial step of recognition what causes that all data will be in the similar ranges. It needs the mean value and variance for every signal parameter.

The standardization of individual radar signal parameter k ($k = 1, 2, \ldots, N$) is done in the following way:

$$\bar{x}_k = \frac{1}{N} \sum_{i=1}^{n} x_{ik}, \quad k = 1, 2, \ldots, N \tag{8}$$

$$\sigma_k^2 = \frac{1}{N-1} \sum_{i=1}^{n} (x_{ik} - \bar{x}_k)^2 \tag{9}$$

$$\bar{x}_{ik} = \frac{x_{ik} - \bar{x}_k}{\sigma_k} \qquad (10)$$

In cases where data is not evenly distributed around the center, non-linear methods are used. Then, transformations based on non-linear functions can be used to arrange data within the limits of precise intervals.

Often the so-called "softmax" scaling is used, which consists of two steps:

$$y = \frac{x_{ik} - \bar{x}_k}{r\sigma_k}, \qquad (11)$$

$$\bar{x}_k = \frac{1}{1 + \exp(-y)} \qquad (12)$$

where:

r —coefficient determined by designer

To divide the set of measurement vectors into clusters, which will correspond to a specific subset of radars determined by use the decision tree, the following grouping methods may be applied:

- Single linkage (nearest neighbour)—the distance between two clusters is determined by the distance between the two closest objects (nearest neighbors) belonging to different clusters. According to this principle, objects form clusters connecting into strings, and the resulting clusters form long "chains".
- Complete linkage method (furthest neighbor)—the distance between clusters is determined by the largest distance between any two objects belonging to different clusters (i.e. the furthest neighbors).
- Centroid—the centroid is the average point in the multidimensional space defined by these dimensions. In this method, the distance between the two clusters is defined as the difference between these centroids.
- Median—it is the same method as the centroid method, except that weighing is introduced to account for the differences between the cluster sizes.
- Group average method—the distance between clusters is defined by means of the arithmetic mean determined from all the distances of the elements forming these clusters.
- Ward method—the analysis of variance approach is used to estimate the distance between clusters. In other words, this method aims to minimize the sum of squared deviations of any two clusters that can be formed at each stage.

These grouping methods will be later applied in the algorithm for recognition unknown electromagnetic emission sources.

The main idea of these six grouping methods is shown in Fig. 8.

All hierarchical grouping procedures (clustering methods) can be described by one general scheme—an algorithm, containing with the following steps:

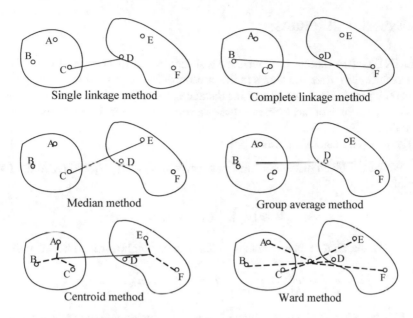

Fig. 8 Methods for grouping objects

1. Input the set of radar signal parameters measured in ELINT system.
2. Data normalization: it is responsible for transferring all loaded radar signal parameters into one dynamic range.
3. Choice of distance measure: the user selects one of five available distance measures used to calculate the distance between each pair of objects in the data set.
4. Creation of a distance matrix and its transformation into a square form: on the basis of calculations made earlier, a matrix of distances is created between all objects.
5. Selecting the grouping method: when measures of the distance between objects are calculated, you must decide how the objects should be grouped into classes. The user has six different grouping methods.
6. Creating a hierarchical connection tree: combines individual clusters from the distance matrix in accordance with the grouping rule selected earlier. After each such connection, the size of the matrix is reduced by one row. The combined clusters create a new focus. This is called an iteration that repeats until one cluster is received.
7. Creation of a matrix with grouping results: after the user defines parameters such as the grouping threshold c and the number of connection in formed dendrogram.

5   Experiment Results

In ELINT systems is often encountered a situation that for measured signals there
are not the appropriate signatures in the emitter database [14]. The starting point in
the central procedure of grouping is the distance matrix $\mathbf{D}$ between classified objects:
$x_1, x_2, \ldots, x_n$. Each of these objects makes a separate cluster. Therefore we obtain n
clusters: $G_1, G_2, \ldots, G_n$.

The grouping algorithm is as follows:

(1)   We have a given distance matrix between clusters: $G_1, G_2, \ldots, G_n$ in the form
      of

$$\mathbf{D} = [d_{ij}], \quad (i, j = 1, 2, \ldots, n) \tag{13}$$

We search for a pair of points in the nearest neighbourhood between them

$$d_{pq} = \min_{i,j}\{d_{i,j}\}, \quad (i, j = 1, 2, \ldots, n), \text{ where } p < q \tag{14}$$

(2)   Clusters G_p and G_q are combined in one new cluster with the assigned number
      p:

$$G_p := G_p \cup G_q \tag{15}$$

(3)   From the matrix $\mathbf{D}$, the line on number q is deleted and $n: = n{-}1$ is substituted.
(4)   The distance d_{pj} ($j = 1,2, \ldots, n$) is calculated between a formed cluster and all
      the other clusters, according to the selected distance measure.
(5)   Values d_{pj} are substituted into matrix $\mathbf{D}$ in the place of row p and values d_{jp} in
      the place of column p.
(6)   The steps 1 to 4 are repeated until all clusters form one cluster, that means $n =$
      1.

The results of images grouping for five separated classes (Fig. 9a) and overlapping
classes (Fig. 9b) are shown in Tables 2 and 3 [28].

Based on these calculations, it is difficult to say which of this method is better
because the obtained results are similar. The choice of distance measure is not important in this case. The values of the true classification index exceed the value of 0.8
in most cases. The bold fonts in Table 3 show the combination of these grouping
methods and metrics for which the algorithm was not able to create the appropriate
number of classes.

In further analysis, it can be noticed that in the case of not entirely separate classes,
the methods of the farthest neighbor (complete linkage) and centroid are the best.
The highest probability value of the true algorithm classification was obtained for
Euclidean and centroid distance measures methods of grouping.

One of disadvantages of the single linkage method is that if at the beginning of
recognition one object x_k is falsely assigned, then the whole of its neighbourhood

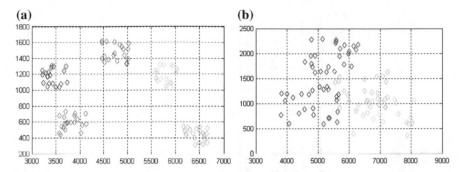

Fig. 9 Illustration of the distribution of implementation for separated (**a**) and overlapping classes (**b**)

may be falsely classified. After each hierarchical iteration, grouping procedures are obtained by dividing comparable objects into decreasing number of clusters and a modified distance matrix between clusters.

6 Images Grouping with Using Dendrogram

Using different distance measures, the set of measured radar signals will be divided into clusters which should represent the appropriate radar sources. In grouping procedures, it is assumed that each vector representing the radar signal parameters is at the beginning assign to a separate cluster. Later on, the number of clusters decreases as a result of combining them into groups with a larger number of elements. The process ends when only one group containing all elements from the analyzed set is formed [28].

In the example below (Table 4) the values of the radar signal parameters (pulse repetition interval, pulse width, antenna rotation ratio), denoted as: x_1, x_2 and x_3, are in the same scale, thus they do not require normalization.

The initial distance matrix for data from Table 4 is presented in Table 5.

The group average method was applied for merging these 8 points in separately clusters.

The common formula for merging clusters has the following form:

$$d_{pj} = \frac{n_p}{n_p + n_q} + \frac{n_q}{n_p + n_q} d_{qj} \qquad (16)$$

where: p, q is the number of combined clusters in the current step (the nearest ones).

According to this algorithm, we assume that there are eight clusters at the beginning (that means that every signal forms a cluster—as shown in Table 5) [18, 26–28].

Table 2 Results of clusters grouping for separated classes

Separate classes		The distance measures									
		Euclidean		Seuclidean		Cityblock		Mahalanobis		Minkowski	
		c	Probability of true classification	c	Probability of true classification	c	Probability of true classification	c	Probability of true classification	c	Probability of true classification
The methods of pattern grouping	Single	2.62	0.807	2.40	0.851	2.79	0.806	2.42	0.803	2.56	0.791
	Complete	3.20	0.85	3.03	0.882	3.19	0.849	3.02	0.879	3.22	0.846
	Average	3.15	0.854	2.84	0.886	3.07	0.824	2.91	0.89	3.18	0.848
	Centroid	3.28	0.854	2.91	0.884	–	–	2.83	0.886	–	–
	Median	3.29	0.853	3.04	0.882	–	–	3.07	0.885	–	–
	Ward	3.63	0.846	3.45	0.877	–	–	3.49	0.879	–	–

Table 3 Results of clusters grouping for overlapping classes

Overlapping classes		The distance measures								
		Euclidean		Seuclidean		Cityblock		Mahalanobis		Minkowski	
		c	Probability of true classification	c	Probability of true classification	c	Probability of true classification	c	Probability of true classification	c	Probability of true classification
The methods of pattern grouping	Single	–	**0.452**	–	**0.496**	–	**0.566**	2.22	0.562	–	**0.419**
	Complete	398	0.703	3.48	0.67	4.02	0.66	3.70	0.617	4.09	0.7
	Average	–	**0.714**	3.19	0.678	–	**0.715**	3.31	0.679	3.60	0.707
	Centroid	–	**0.713**	3.07	0.679	–	–	3.57	0.671	–	–
	Median	373	0.651	3.71	0.634	–	–	–	**0.631**	–	–
	Ward	437	0.695	3.90	0.666	–	–	3.40	0.645	–	–

Table 4 Values of three radar parameters (8 vectors)

$$X = \begin{bmatrix} x_1 & x_2 & x_3 \\ 2 & 2 & 3 \\ 3 & 3 & 3 \\ 4 & 3 & 4,5 \\ 6 & 5 & 6 \\ 5 & 7 & 6 \\ 6 & 6,5 & 7,5 \\ 7 & 9 & 8 \\ 8 & 9 & 10 \end{bmatrix} \quad \begin{array}{l} \text{Point 1} \\ \text{Point 2} \\ \cdot \\ \cdot \\ \cdot \\ \cdot \\ \cdot \\ \text{Point 8} \end{array}$$

Table 5 The distance matrix between 8 points (clusters)

Points	1	2	3	4	5	6	7	8
1	0	1.41	2.69	5.83	6.56	6.34	9.95	11.58
2	1.41	0	1.8	4.69	5.39	5.02	8.77	10.49
3	2.69	1.8	0	3.2	4.39	4.06	7.57	9.07
4	5.83	4.69	3.2	0	2.24	1.8	4.58	6.0
5	6.56	5.39	4.39	2.24	0	1.5	3.46	5.39
6	6.34	5.02	4.06	1.8	1.5	0	4.03	5.94
7	9.95	8.77	7.57	4.58	3.46	4.03	0	2.24
8	11.58	10.49	9.07	6.0	5.39	5.94	2.24	0

1. The minimum element in matrix **D** is $d_{12} = 1.41$ (thus $p = 1$, $q = 2$). It means that clusters G_1 and G_2 are the nearest ones.
2. Clusters G_1 and G_2 are merged in one with the assigned number 1, (because $p = 1$).

$$G_1 := G_1 \cup G_2$$

3. The row 2 is removed from the matrix because $q = 2$ (shaded elements in Table 5) and substitutes $n: = 7$ (after this iteration number of clusters will be smaller by one).
4. The new distances d_{1j} ($j = 1, 2, ...,7$) (are the same for d_{pj}) of the formed cluster G_1 with the remaining clusters are determined. Formula (14) will then take the following form

$$d_{1j} := \frac{1}{2}d_{1j} + \frac{1}{2}d_{2j} \tag{17}$$

Values on the right side of the Eq. (15) are elements from the matrix **D** before the reduction, and the values d_{1j} are put in place of the first row. In the place of first column is inserted d_{1j}, for example: (Table 6)

Table 6 The distance matrix between 7 clusters after iteration 1

Points		1, 2	3	4	5	6	7	8
	Cluster number	1	2	3	4	5	6	7
1, 2	1	0	2.245	5.260	5.975	5.68	9.36	11.04
3	2	2.245	0	3.2	4.39	4.06	7.57	9.07
4	3	5.260	3.2	0	2.24	1.8	4.58	6.0
5	4	5.975	4.39	2.24	0	1.5	3.46	5.39
6	5	5.68	4.06	1.8	1.5	0	4.03	5.94
7	6	9.36	7.57	4.58	3.46	4.03	0	2.24
8	7	11.035	9.07	6.0	5.39	5.94	2.24	0

$$d_{12} = \frac{1}{2} \times 2,69 + \frac{1}{2} \times 1,8 = 2,245; \quad d_{13} = \frac{1}{2} \times 5,83 + \frac{1}{2} \times 4,69 = 5.26$$

After iteration **1** the following clusters are formed:

$$G_1 = \{1, 2\}, \ G_2 = \{3\}, \ G_3 = \{4\}, \ G_4 = \{5\}, \ G_5 = \{6\}, \ G_6 = \{7\}, \ G_7 = \{8\},$$

In iteration **2** the smallest distance in the transformed distance matrix is between clusters $G_4 = \{4\}$ and $G_5 = \{5\}$, $(p = 4, q = 5)$. These clusters are merged, receiving $G_4 = \{5, 6\}$. After substitution $n: = 6$ the new distances are determined from the formula:

$$d_{4j} := \frac{1}{2}d_{4j} + \frac{1}{2}d_{5j}, \quad j = 1, 2, \ldots, 6$$

For example: (Table 7)

$$d_{41} = \frac{1}{2} \times 5.975 + \frac{1}{2} \times 5.68 = 5.8275$$

In iteration **3**, the clusters $G_3 = \{4\}$ and $G_4 = \{5, 6\}$, $(p = 3, q = 4)$ are merged and after substitution $n: = 5$ the new distances d_{3j} are calculated according to the formula:

Table 7 The distance matrix between 6 clusters after iteration 2

Points		1, 2	3	4	5, 6	7	8
	Cluster number	1	2	3	4	5	6
1, 2	1	0	2.245	5.26	5.8275	9.36	1.035
3	2	2.245	0	3.2	4.225	7.57	9.07
4	3	5.26	3.2	0	2.02	4.58	6.0
5, 6	4	5.8275	4.225	2.02	0	3.745	5.665
7	5	9.36	7.57	4.58	3.745	0	2.24
8	6	1.035	9.07	6.0	5.665	2.24	0

Table 8 The distance matrix between 5 clusters after iteration 3

Points		1, 2	3	4, 5, 6	7	8
	Cluster number	1	2	3	4	5
1, 2	1	0	2.245	5.273	9.36	11.035
3	2	2.245	0	3.883	7.57	9.07
4, 5 ,6	3	5.273	3.883	0	4.023	5.77
7	4	9.36	7.57	4.023	0	2.24
8	5	11.035	9.07	5.77	2.24	0

$$d_{3j} := \frac{1}{3}d_{3j} + \frac{2}{3}d_{4j}, \quad j = 1, 2, \ldots, 5$$

For example: (Table 8)

$$d_{31} = \frac{1}{3} \times 5.26 + \frac{2}{3} \times 5.28 = 5.273, \ldots \text{e.t.c.}$$

In iteration **4**, the clusters $G_4 = \{7\}$ and $G_5 = \{8\}$, $(p = 4, q = 5)$ are merged, after substitution $n: = 4$ the distance d_{4j} are calculated according to the formula: (Table 9)

$$d_{4j} := \frac{1}{2}d_{4j} + \frac{1}{2}d_{5j}, \quad j = 1, 2, 3, 4$$

In iteration **5**, the clusters $G_1 = \{1, 2\}$ and $G_2 = \{2\}$, $(p = 1, q = 2)$ are merged and after substitution $n: = 3$ the distances d_{4j} are calculated according to the formula: (Table 10)

Table 9 The distance matrix between 4 clusters after iteration 4

Points		1, 2	3	4, 5, 6	7, 8
	Cluster number	1	2	3	4
1, 2	1	0	2.245	5.273	10.198
3	2	2.245	0	3.883	8.32
4, 5, 6	3	5.273	3.883	0	4.9
7,8	4	10.198	8.32	4.9	0

Table 10 The distance matrix between 3 clusters after iteration 5

Points		1, 2, 3	4, 5, 6	7, 8
	Cluster number	1	2	3
1, 2, 3	1	0	4.81	9.572
4, 5, 6	2	4.81	0	4.9
7, 8	3	9.572	4.9	0

$$d_{4j} := \frac{2}{3}d_{1j} + \frac{1}{3}d_{2j}, \quad j = 1, 2, 3$$

In iteration **6**, the clusters $G_1 = \{1, 2, 3\}$ and $G_2 = \{4, 5, 6\}$, $(p = 1, q = 2)$ are merged and after substitution $n: = 2$ the distances d_{4j} are calculated according to the formula: (Table 11)

$$d_{4j} := \frac{3}{6}d_{1j} + \frac{3}{6}d_{2j}, \quad j = 1, 2$$

In all hierarchical grouping methods, the results of pattern grouping may be presented in a graphic form, that so-called **dendrogram** (tree of connections) [27, 28].

Dendrogram (Fig. 10) presents the structure of analysed set of points and subsequently appearing clusters. In the figure, the objects of analyzed set have been marked as well as the distances between clusters. Looking from below you can see the implementation of subsequently merged clusters.

Table 11 The distance matrix between 2 clusters after iteration 6

Points		1, 2, 3, 4, 5, 6	7, 8
	Cluster numbers	1	2
1, 2, 3, 4, 5, 6	1	0	7, 9875
7, 8	2	7, 9875	0

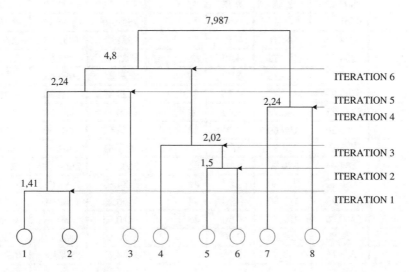

Fig. 10 The grouping dendrogram

The next example for analysis concerns the set of 50 vectors containing the parameters: radio frequency, pulse repetition interval and pulse width [18, 28], see Table 12.

For a such big set of data the grouping dendrogram is difficult to illustrate. It can be seen that data from Table 12 clearly differs after every ten elements. Thus, five groups should be formed.

Using the Euclidean distance measure, the correct result of grouping into 5 clusters is received for the threshold value c for the following methods:

- single linkage $= 2.49$
- complete linkage $= 2.50$

Table 12 The values of 3 radar parameters

Number of signal	Radio frequency	Pulse repetition interval	Pulse width	Number of signal	Radio frequency	Pulse repetition interval	Pulse width
1	8997.60	82.0	8.50	26	3245.16	1400.32	9.40
2	8998.28	81.9	9.40	27	3200.85	1410.50	9.80
3	9001.42	81.9	8.00	28	3120.20	1410.80	10.50
4	9004.31	84.1	7.50	29	2950.50	1412.00	10.90
5	9000.64	84.0	6.50	30	2950.50	1420.50	11.85
6	9003.28	84.1	8.00	31	8520.11	993.20	2.23
7	9002.99	84.1	9.50	32	8521.31	995.45	2.65
8	9003.48	84.2	8.75	33	8522.61	997.90	3.50
9	9001.03	84.1	9.00	34	8520.40	996.50	3.40
10	9004.45	84.1	7.50	35	8521.45	997.90	2.65
11	5734.73	182.72	3.40	36	8522.61	995.60	2.23
12	5869.20	265.20	3.70	37	8521.31	996.45	2.70
13	5783.83	312.79	4.30	38	8520.40	993.80	2.75
14	5734.73	182.72	3.40	39	8520.11	994.50	3.40
15	5860.20	195.60	3.70	40	8522.61	996.50	3.20
16	5783.83	312.80	4.10	41	3124.41	1762.61	8.00
17	5734.73	270.00	4.10	42	3100.20	1969.14	9.56
18	5869.20	265.20	3.40	43	3124.87	2298.08	10.00
19	5734.73	195.60	3.70	44	3100.20	1762.61	8.00
20	5860.20	182.72	4.30	45	3110.85	1860.56	8.75
21	3124.56	1400.32	9.40	46	3120.85	1969.14	9.56
22	3245.16	1412.91	12.34	47	3124.41	2298.08	10.00
23	2907.62	1413.00	15.00	48	3124.87	1860.56	9.25
24	2907.62	1413.00	15.00	49	3100.20	1762.61	9.56
25	3124.56	1412.91	12.34	50	3110.85	1860.56	8.75

- group average $= 2.48$
- Ward $= 2.52$

The centroid method is incapable of forming five groups for a selected Euclidean distance measure. It is difficult to establish which method is better, because for such a large amount of data similar results are obtained. As an example, we can compare a grouping result obtained by applying the single linkage (nearest neighbour) method for $c = 2.49$. It is the optimal threshold value, for which the Euclidean distance measure gives the correct number of five clusters. The biggest difference was obtained for Mahalanobis method, where seven clusters was formed.

7 Conclusion

All hierarchical grouping procedures require quite a large amount of computing power and thus the attempt to analyze a greater number of radar signals can be time-consuming or even unfeasible. Quite fast methods are non-hierarchical methods, however, they require giving in the initial phase the exact number of groups to which the recognized signals are to be assigned.

Nonparametric methods used in the pattern recognition algorithm are effective and their structure is not too complicated. As it results from the calculations carried out, they are especially effective in the case of well-separated clusters of any shape. The result of the classification of the large data set depends on the correct determination of the grouping threshold for a particular method of recognition. Well-used nonparametric methods can be a very good tool to reduce the volume of large data sets.

In this paper the role of different distance measures and grouping rules have been emphasized. It turns out that the final recognition effects depend on the amount of measuring data and their dispersion in the space of parameters. Those factors have important influence if an analyzed set of measuring data is very big. In a such situation, at the first stage of recognition the big set of data should be divided into separated subspaces by using for example the binary decision tree. To sum up, it must be stated that cluster methods greatly facilitate and speed up the recognition radar signals coming from unknown emitter sources for which there are no appropriate signatures in the emitter database of the electronic recognition systems.

The methods of signal recognition presented here can be adapted to the needs of any signal processing devices, such as radars, electronic recognition systems, telecommunications systems, autonomous vehicles, etc.

References

1. Adamy DL (2004) EW 102. A Second Course in Electronic Warfare. Artech House, Boston • London
2. Willey RG (2006) ELINT. The interception and analysis of radar signals. Artech House, Inc., Boston & London
3. Daradkeh YI, Kirichenko L, Radivilova T (2018) Development of QoS methods in the information networks with fractal traffic. Intl J Electron Telecommun 64(1):27–32. https://doi.org/10.24425/118142
4. Pietkiewicz T, Kawalec A, Wajszczyk B (2017) Analysis of fusion primary radar, secondary surveillance radar (IFF) and ESM sensor attribute information under Dezert-Smarandache theory. In: The 18th international radar symposium (IRS), Prague, Czech Republic, 1–10 June 2017. https://doi.org/10.23919/IRS.2017.8008199
5. Pietkiewicz T, Wajszczyk B (2018) Fusion of identification information from ELINT-ESM sensors. Proc. SPIE 10715, 2017 radioelectronic systems conference, 107150F, April (2018). https://doi.org/10.1117/12.2316489
6. Wong MD, Nandi AK (2004) Automatic digital modulation recognition using artificial neural network and genetic algorithm. Sig Process 84:351–365
7. Ageyev D, Bondarenko O, Alfroukh W, Radivilova T (2018) Provision security in SDN/NFV. In: 2018 14th international conference on advanced trends in radioelectronics, telecommunications and computer engineering (TCSET), Lviv-Slavske, pp 506–509. https://doi.org/10.1109/tcset.2018.8336252
8. Ageyev D, Kirichenko L, Radivilova T, Tawalbeh M, Baranovskyi O (2018) Method of self-similar load balancing in network intrusion detection system. In: 2018 28th international conference radioelektronika (RADIOELEKTRONIKA), Prague, Czech Republic, 1–4. https://doi.org/10.1109/radioelek.2018.8376406
9. Duda R, Hart P, Stork D (2001) Pattern classification, 2nd edn. Wiley, Inc., N.Y. (2001)
10. Pietrow D, Matuszewski J (2017) Object detection and recognition system using artificial neural networks and drones. In: 2017 signal processing symposium, SPSSympo, Jachranka, Poland, 12–14 Sept 2017, art. no. 8053689, pp 1–5. https://doi.org/10.1109/sps.2017.8053689
11. Smith YF (2010) Image processing and pattern recognition. Fundamentals and Techniques, Wiley, Inc., N.Y
12. Stąpor K (2005) Automatyczna klasyfikacja obiektów [Automatic objects classification]. Akademicka Oficyna Wydawnicza EXIT, Warsaw, Poland, (In Polish)
13. Theodoridis S, Kountroumbas K (1998) Pattern recognition. Academic Press, London
14. Matuszewski J (2012) The radar signature in recognition system database. In: 19th international conference on microwaves, radar and wireless communications MIKON 2012, Warsaw, Poland, vols 1 and 2, 21–23 May 2012, pp 617–622
15. Matuszewski J, Paradowski L (1998) The knowledge based approach for emitter identification. In: 12th international conference on microwaves and radar (MIKON). Krakow, Poland, vol 3, 20–22 May (1998), pp 810–814. https://doi.org/10.1109/mikon.1998.742832
16. Du KL (2010) Clustering: a neural network approach. Neural Netw 23:89–107
17. Klymash M, Shpur O, Peleh N, Lutsiuk I (2018) Clustering model of cloud centers for big data processing. In: 14th international conference modern problems of radio engineering, telecommunications and computer science TCSET'2018, Lviv-Slavske, Ukraine, 19–24 Feb 2018, pp 366–382
18. Matuszewski J (2018) Application of clustering methods in radar signals recognition. In: 5th international scientific and practical conference on problems and infocommunications science and technology, PIC S and T, Kharkov, 9–11 Oct 2018, pp 745–751
19. Prudyus I, Hryvachevskyi A (2016) Image segmentation based on cluster analysis of multispectral monitoring data. In: 14th international conference modern problems of radio engineering, telecommunications and computer science TCSET2018, Lviv-Slavske, Ukraine, 19–24 Feb 2016, pp 226–229. https://doi.org/10.1109/tcset.2016.7452020

20. Carter CA, Masse N (1993) Neural networks for classification of radar signals. Defence Research Establishment Ottawa, Technical Note 93–33, Ottawa, Canada
21. Matuszewski J (2018) Radar signal identification using a neural network and pattern recognition methods. In: Proceedings of the 14th international conference on modern problems of radio engineering, telecommunications and computer science, TCSET 2018, Lviv-Slavske, Ukraine, art. no. 7452040, 19–24 Feb 2018, pp 79–83. https://doi.org/10.1109/tcset.2018.8336160
22. Elmaggeed MEA, Alzubaidi AJ (2015) Neural network algorithm for radar signal recognition. Int J Eng Res Appl 5(2): 123–125 (Part 2)
23. Petrov N, Jordanov I (2013) Radar emitter signals recognition and classification with feedforward networks. In: 17th international conference in knowledge based and intelligent information and engineering systems—KES2013, procedia computer science, vol 22, pp 1192–1200. https://doi.org/10.1016/j.procs.2013.09.206
24. Rogers SK, Colombi JM, Martin CE, Gainey JC (1995) Neural networks for automatic target recognition. Neural Netw 8:1153–1184
25. Yun L, Jing-chao L (2011) Radar signal recognition algorithms based on neural network and grey relation theory. In: Proceedings of 2011 cross strait quad-regional radio science and wireless technology conference, 26–30 July 2011, Harbin, China. https://doi.org/10.1109/csqrwc.2011.6037247
26. Matuszewski J (2010) Applying the decision trees to radar targets recognition. In: 4th microwave and radar week MRW-2010—11th international radar symposium, IRS 2010, conference proceedings, art. no. 5547502, pp 301–304
27. Kołodziejska U (2009) Badanie jakości rozpoznawania obiektów przy użyciu metod nieparametrycznych. [Testing the quality of object recognition using nonparametric methods]. M.S. thesis. Dept. Electronics. Military University of Technology. Warsaw. (In Polish)
28. Małecki P (2005) Ocena przydatności metod taksonomicznych dla celów rozpoznawania źródeł emisji sygnałów radarowych. [Usability evaluation of clustering methods for goals of emitter sources signals recognition]. M.S. thesis, Dept. Electronics, Military University of Technology, Warsaw, (In Polish)

Optimization of the Quality of Information Support for Consumers of Cooperative Surveillance Systems

Ivan Obod⬤, Iryna Svyd⬤, Oleksandr Maltsev⬤, Oleksandr Vorgul⬤, Galyna Maistrenko⬤, and Ganna Zavolodko⬤

Abstract The paper discusses the place and the role of cooperative airspace surveillance systems in the information support of airspace use and air traffic control systems. A brief description of the signals used in the considering systems is given. Based on the presentation of cooperative surveillance systems as two-channel data transmission systems, the statistical interpretation of consumer data transmission is considered and it is shown that the probability of information support can be an integral quality indicator of consumers information support in the considered systems. That is defined as the product of the probability of detecting the request signals by the aircraft responder, aircraft responder availability factor, probability of detection of an air object by the requester, the probability of correct reception of on-board information and the probability of combining the flight and coordinate information. The variants for optimization each of the components of these probabilities are considered. The optimization issues of measurement parameters of signals in cooperative observation systems, which determine the probability of combining flight and coordinate information, are also considered.

Keywords Traffic control · Cooperative surveillance systems · Secondary surveillance radar · Automatic dependent surveillance · Multilateration · Wide area multilateration · Air object

I. Obod · I. Svyd (✉) · O. Maltsev · O. Vorgul · G. Maistrenko
Kharkiv National University of Radio Electronics, Nauky Ave. 14, 61166 Kharkiv, Ukraine
e-mail: iryna.svyd@nure.ua

G. Zavolodko
National Technical University "KhPI", Kyrpychova Str. 2, 61002 Kharkiv, Ukraine

133

T. Radivilova et al. (eds.), *Data-Centric Business and Applications*, Lecture Notes on Data Engineering and Communications Technologies 48,
https://doi.org/10.1007/978-3-030-43070-2_8

1   The Place and Role of Cooperative Airspace Surveillance Systems in Consumer Information Support

1.1   Classification of Cooperative Surveillance Systems (CSS)

In airspace monitoring [1–3], information on air situation is mainly provided by surveillance systems (SS) which can be divided into two main types: dependent SSs and independent ones. Independent SSs determine the location of an air object (AO) using ground-based facilities. Primary [4, 5] and secondary [6–13] SSs are samples of systems for independent observations, while in dependent SSs the AO location is determined on board and then transmitted to interested agencies. An example of dependent surveillance is the concept of automatic dependent surveillance (ADS).

Depending on whether or not the information from an AO is used to determine its position, surveillance is divided into cooperative and non-cooperative. The classification of cooperative airspace SSs [14–17] is shown in Fig. 1.

With independent cooperative observation, the location is determined on the basis of data measurement performed by local observation subsystems which communicate with the AO onboard equipment to obtain data on the AO barometric altitude, its identification index, etc.

Independent cooperative systems include the Secondary Surveillance Radar (SSR), which performs the following functions [6–13]:

- determines the AO coordinates;
- obtains flight data from the AO onboard equipment;
- transmits back data necessary for the AO control, monitoring and guidance;
- provides the AO air traffic control identification;
- provides the AO nationality radar identification.

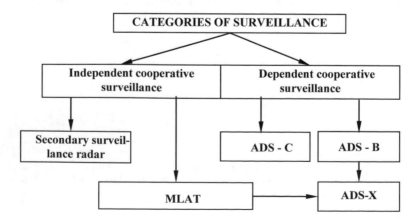

Fig. 1 Classification of cooperative airspace surveillance systems

SSRs have the following modes of interrogation: A, B, C, D, S and 1, 2, 3, 4, 5 [18].

The use of response signals (RPS) from aircraft responders enables the creation of multi-position SSs, or multilateration systems (MLAT) [19–21]. Notably, it is possible to determine the AO position both by sending request signals (RQS) from one's own transmitter and by using request signals from other transmitters.

MLAT can form the basis for creating a height accuracy control system which will arrange the AO altitude measurement and data transfer to a regional monitoring agency. By providing data on the AO position relative to its landing course and glide path, MLAT can be used to direct the AO to the runway, thus enabling the creation of a MLAT-based aircraft landing system.

There is a current trend to use MLAT in Wide Area Multilateration integrated systems [22–24].

With dependent cooperative observation, the location is determined on board the AO. This information along with other required data is transmitted to the local surveillance subsystem both by one's own means and through satellite data transfer channels.

Samples of dependent observation include the concept of automatic dependent surveillance (ADS) [25–27]. ADS is based on the availability of air-to-ground data lines through which AO-related information (location, time, etc.) is automatically transmitted to the relevant agency on the ground.

ADS is divided onto:

- ADS-A (Automatic Dependent Surveillance — Addressable);
- ADS-B (Automatic Dependent Surveillance—Broadcast);
- ADS-X (Automatic Dependent Surveillance—Extended).

ADS-A equipment, also called ADS-C (Contract), automatically sends AO onboard data on its location to the air traffic control (ATC) authority on the ground, at certain time intervals or in certain predetermined cases. The frequency of data transmission or the special cases in which a message is to be transmitted are determined by the ATC authority.

In ADS-C mode, the AO uses its onboard navigation systems to determine its location, speed, and other data. The ground system "contracts" with an AO to provide it with information at regular intervals or after certain events. This information is transmitted through the data line, which limits access to the data by other parties.

ADS-B system broadcasts data on the AO location (latitude and longitude), altitude, velocity, identification index and other information received from onboard systems. Each ADS-B location report includes an indication of the data quality, allowing users to see whether the quality of information is sufficient to support the intended function.

The AO position, speed and associated data quality indicators are usually obtained from its onboard system, which processes data received from different measuring systems, data on altitude being supplied by a barometric altimeter.

Since ADS-B messages are transmitted in broadcast mode, they can be received and processed by any suitable receiver. Therefore, the ADS-B feature supports both ground and airborne surveillance applications. Ground surveillance stations are established for aeronautical surveillance and are intended for receiving and processing ADS-B messages. For onboard use, AOs equipped with ADS-B receivers can process messages from other AOs to determine the air situation.

Mode ADS 1090 MHz (1090 ES) can be used to transmit ADS-B data. The 1090 ES facilities were developed as part of the S mode system. The standard S mode signal length is 56 bits, the +1090 MHz ES signal containing an additional 56-bit data block with ADS-B information. The duration of each ES message is 120 μs (8 μs of the preamble and 112 μs of data), and signals are transmitted at a frequency of 1090 MHz. ADS-B information is transmitted as separate messages, each of which contains a corresponding data set (AO location and barometric altitude, surface location, speed, identification index and AO type, alarm data, etc.). Data on location and speed are transmitted twice per second.

ADS-X is an integration of MLAT and ADS-B systems. Extended ADS (Extended ADS-X) integrates multilateration methods into the ADS-B monitoring infrastructure. ADS-X supports full AO fleet coverage. Since multilateration uses signals from all transponders (in Modes A/C/S), the network of ground stations does not need new avionics to be able to determine the AO position. Independent backup is required along with the verification of aircraft ADS-B performance in its self-location. This is done by activating the possibility of multilateration in the ADS-B network because each ADS-B position report is checked in real time using multilateration.

Based on the above, it can be concluded that all cooperative observation systems are request-response data transmission systems containing an RQS channel and an RPS channel. Information in these systems is encoded by primitive interval-time and positional codes, which significantly reduces the information capacity of the data channels considered.

1.2   Signals in CSSs and Their Mathematical Models

To obtain RQS and RPS, CSSs use interval-time and position codes, which are simple but have low noise immunity. Indeed, the suppression of at least one RQS pulse by interferences or the appearance of false pulses during code intervals in RQS/RPS immediately causes errors: missed signals, false alarms of the first and second kind, or RQS confusion.

It should be noted that the existing CSSs belong to a class of asynchronous data transmission systems [28] and consist of a number of transmitters and receivers that use different frequency ranges to transmit and receive signals [29–43].

Transmitters generate discrete signals $s_i(t)$, which belong to a finite set, an ensemble $\vec{S} = \{s_i(t)\};$ $i = 1, 2, \ldots, V$, and transmit them to the communication line asynchronously, independently of each other, at their own points of time. In this case, the condition $t_i \ll T_p$, is usually satisfied, where t_i is the signal duration $s_i(t)$; T_p is the RQS repetition period. Two factors impede the operation of CSSs when they are exposed to extraneous or intrasystemic interference: the use of a single RQS transmission and the underlying principle of an open queuing system with failures.

CSS signal can be written as

$$s_i(t) = U_i f_i(t) \cos(\omega_o t + \varphi),$$

where $f_i(t)$ is a binary coding sequence of 1 and 0, where $f_i(t) = 1$ corresponds to a pulse, and $f_i(t) = 0$ when there is a pause in signal $s_i(t)$ and the sequence itself; ω_o is a carrier frequency, φ is a random initial phase of high-frequency filling, which may be different for each pulse in the coding sequence; U_i is an amplitude factor that remains constant within one complete realization of this signal. The ensemble $\vec{S}$ contains V various signals (it is usually the number of CSS modes of operation), which differ from each other by the coding sequence, $f_i(t)$, and have, in the case of radio signals, the same carrying frequency ω_o. The condition of signal adequacy is that the pulse duration and the number of pulses must be the same for all signals. The number of pulses in a code is called the code value.

Let us consider the RQS noise immunity, taking into account different types of interference which CSSs may experience.

First of all, the noise immunity of RQS and, consequently, of the entire CSS is reduced due to the interference that correlates and does not correlate with the existing set of RQSs. It is worth noting that this kind of interference may just appear within a CSS or may be intentionally produced. This latter probability is created by the complete openness of request channels and is further evidence of the problems which CSSs encounter in the face of intentional interference.

When RQSs are received asynchronously, three main types of errors may occur: signal misses and false alarms, i.e., errors of the first and second kind. So, the noise immunity of these signals can be characterized by the following three error probabilities: P_p, F_1, and F_2.

It is interesting to evaluate the RQS noise immunity to uncorrelated interference. Suppose that the interference is stationary, has no aftereffect, and is statistically and structurally independent of the received signal. Let the signal consists of n elementary pulses, its function of uncertainty be R, and the threshold (logic) of the decoder equals to k.

The probability of a miss, i.e. the conditional probability that, when receiving a signal, no less than $n - k + 1$ pulses will be suppressed by interference can be derived from the following expression:

$$P_p = \sum_{i=n-k+1}^{n} C_n^i P_{10}^i (1 - P_{10})^{n-i}, \tag{1}$$

where P_{10} is the probability that at least one signal pulse will be suppressed, which depends on the specific type of interference and its intensity.

The probabilities of false alarms are somewhat more complicated. The false alarm of the first kind occurs if (1) a wrong combination with k or more interference pulses enters the decoder input (necessary condition); (2) this erroneous combination is decrypted by this decoder (sufficient condition).

The probability of such an event can be defined as

$$F_1 = \sum_{i=k}^{n} C_n^i P_{01}^i (1 - P_{01})^{n-i} P_d(i), \qquad (2)$$

where P_{01} is the probability of a false pulse emerging at the decoder input during the time interval $\delta\tau$, which depends on the type and intensity of interference; P_d is the probability of a false i-pulse combination being decrypted by the decoder.

For the false alarm of the second kind, we will consider the probability that during the time the signal passes through the decoder at least one false alarm of the second kind will occur. In this case, the inverse value will set the conditional probability that the arrival time of the received signal will be determined uniquely (understandably, with an accuracy within the pulse duration).

Then the probability of a false alarm of the second kind is

$$F_2 = 1 - \prod_{s=1}^{R} [1 - F_2(s)]^{M(s)}, \qquad (3)$$

where $F_2(s)$ is the probability of generating and decoding a false second kind combination which consists of signal pulses that give one lobe of the uncertainty function, equals to s, and i noise pulses ($k - s < i < n - s$)

$$F_2(s) = \sum_{i=k-s}^{n-s} C_{n-s}^i P_{01}^i (1 - P_{01})^{n-s-i} P_d(i + 1), \qquad (4)$$

and $M(s)$ is the number of the uncertainty function side lobes equal to s.

For given values of P_{10}, P_{01} and set upper limits of error probability P_p, F_1 and F_2, relations (1)–(3) form a system of three inequalities with three unknowns, n, k, R. From this system we may find the smallest possible n, the largest possible R and the corresponding value of k ($R < k < n$) for which error probabilities do not exceed specified upper bounds.

The system of inequalities formed by expressions (1) and (2) for n and k is solved similarly to the statistical problem of determining the minimum signal energy and the optimal threshold required to identify a single signal in the noises of given intensity by Neumann-Pearson's criterion. The number of pulses appears as the parameter equivalent to energy n, and the logic of the decoder is used as the threshold k. The difference is that, while in conventional statistical problems the probability of a false alarm depends only on the threshold value and interference intensity, in our case it is also dependent on the number of pulses in the RQS.

Chaotic impulse noise and fluctuation noise are the two types of uncorrelated interference usually dealt with in conventional problems. For CSS, the most expedient and effective type of intentional interference is chaotic impulse noise. Let us determine the probabilities of events P_{10} and P_{01} for this case.

It is common to use the Poisson flow of single pulses for modeling the flow of chaotic impulse noise. Notably, the Poisson flow of single pulses is the same idealization for impulse noise as normal noise is for fluctuation noise. For the Poisson flow, the probability of the occurrence of at least one pulse within the time interval t_0 is

$$P = 1 - \exp(-\lambda_0 t_0),$$

where λ_0 is the chaotic impulse noise intensity.

Then the probability P_{01} can be obtained from the expression

$$P_{01} = 1 - \exp(-\lambda_0 \tau_0). \tag{5}$$

The following relationship defines the probability of the suppression of a single impulse signal by interference:

$$P_{10} = 1 - \exp(-\lambda_0 \tau_p)[1 - \gamma(1 - \exp(1 - \lambda_0 \tau_0))], \tag{6}$$

where τ_p is the time of the receiver paralysis after a noise pulse has passed through it; γ is a coefficient which determines the probability that the received signal pulse will be suppressed by interference when it coincides in time with the noise pulse.

Understandably, relation (6) takes into account both inertial and interference types of suppression in systems with pulsed signals: the first type is caused by elements with a non-zero sensitivity recovery time, and the second one results from the interaction of high-frequency signal pulses and interference when they coincide in time.

The coefficient γ depends on the ratio of the amplitudes and phases of interfering oscillations and the receiver limit. If the phase difference in the interval $[0; 2\pi]$ is uniformly distributed, it is commonly assumed that γ is 0.2.

In addition, the probability of suppressing at least one of the signal pulses by Poisson's interference can be presented as:

$$P_{10p}(n) = 1 - [1 - P_{10}(1)]^n.$$

At the same time, it should be noted that, as a rule, real-life flows are made of a certain number of periodic or quasi-periodic flows, and this fact allows them to be classified as regular.

2 Statistical Interpretation of Data Transfer in CSS

As shown above, every CSS is a two-channel request and response signaling system [9–13, 44–47]. Let us consider the statistical interpretation of transmission process in CSSs representing it by points x, n, y and γ which belong to the spaces of signal parameters, interference, observations and decisions. We will use the indices 1 and 2 to indicate that the signal belongs to the request channel or response channel respectively. The converter of the responder unambiguously transforms all points of the space of responder's decisions into the space of parameters for signals transmitted over the response channel.

We will approach the problem of signal detection in CSSs as that of testing two hypotheses and try to find an optimal rule for signal detection in the case of an arbitrary value function.

Let the space C_1 contain only two points, x_{10} and x_{11}, which correspond to the absence of RQS and the presence of RQS with an amplitude equals to the detection limit value, respectively. Accordingly, other spaces also contain two points that we will denote by the same indices. The prices of the CSS decisions can then be described as a value matrix

$$\vec{C} = \left\| \begin{array}{cc} C(\gamma_{20}, x_{10}) & C(\gamma_{20}, x_{11}) \\ C(\gamma_{21}, x_{10}) & C(\gamma_{21}, x_{11}) \end{array} \right\| = \left\| \begin{array}{cc} C_{1-\alpha} & C_\beta \\ C_\alpha & C_{1-\beta} \end{array} \right\|.$$

In this case, the general expression for the average risk can be written as

$$R = \sum_{C_1} \sum_{C\Pi_1} \sum_{P_1} \sum_{C\Pi_2} \sum_{P_2} \left[\begin{array}{c} P(x_1)P(y_1|x_1)\delta(\gamma_1|y_1) \\ \times P(y_2|x_2)\delta(\gamma_2|y_2)C(\gamma_2; x_1) \end{array} \right], \tag{7}$$

where $P(x_1)$ is the a priori probability distribution of the parameter value x_1; $P(y_1|x_1)$ and $P(y_2|x_2)$ are conditional likelihood functions for implementations made by the responder and the requester; $\delta(\gamma_1|y_1)$, and $\delta(\gamma_2|y_2)$ are the decision rules that describe the algorithm of the respondent and requester's actions.

Assuming that the decision converter in the responder is ideal, we can write

$$P(y_2|x_2) = P(y_2|\gamma_1),$$

It follows that the optimal decision rule for the responder can be found by minimizing the expression (7) as a functional of $\delta(\gamma_1|y_1)$.

Since the conditional function of probabilities distribution for the decisions taken by the requester, which characterizes its performance, is

$$P(\gamma_2|x_2) = P(\gamma_2|\gamma_1) = \sum_{P_2} \delta(\gamma_2|y_2)P(y_2|\gamma_1)$$

$$= \sum_{P_2} \delta(\gamma_2|y_2)P(y_2|x_2)$$

and in order to provide $\delta(\gamma_{11}|y_1) + \delta(\gamma_{10}|y_1) = 1$, the expression (8) can be written in the following form:

$$
R = \sum_{H_1} \delta(\gamma_{11}|y_1) \left\{ \sum_{C_1} P(y_1|x_1) \sum_{P_1} P(\gamma_2|\gamma_{11})C(\gamma_2; x_1) \right.
$$

$$
\left. - \sum_{C_1} P(y_1|x_1)P(x_1) \sum_{P_2} P(\gamma_2|\gamma_{10})C(\gamma_2; x_1) \right\}
$$

$$
+ \sum_{C_1} P(x_1) \sum_{P_2} P(\gamma_2|\gamma_{10})C(\gamma_2; x_1) \tag{8}
$$

As can be seen from (8), the optimal rule that minimizes the average risk is $\delta(\gamma_{11}|y_1) = 1$, provided that

$$
\sum_{C_1} P(y_1|x_1)P(x_1) \sum_{P_2} P(\gamma_2|\gamma_{11})C(\gamma_2; x_1)
$$

$$
\leq \sum_{C_1} P(y_1|x_1)P(x_1) \sum_{P_2} P(\gamma_2|\gamma_{10})C(\gamma_2; x_1) \tag{9}
$$

Using the equation $P(\gamma_{21}|\gamma_1) + P(\gamma_{20}|\gamma_1) = 1$, it is possible to simplify (9), so the expression gets the final form

$$
\frac{P(y_1|x_{11})P(x_{11})}{P(y_1|x_{10})P(x_{10})} \geq \frac{C_\alpha - C_{1-\alpha}}{C_\beta - C_{1\ \beta}}.
$$

Thus, to be optimal in the Bayesian sense, the responder is to compare the generalized likelihood ratio with the threshold value which does not depend on either the algorithm or the requester's performance and is completely determined by the decision values given for the system as a whole.

The above expression describes a decision rule for a single RQS. Since the RQS consists of n components (code value), the complete decision rule can be written as

$$
\sum_{i=1}^{n} \frac{P(y_i|x_{i1})P(x_{i1})}{P(y_i|x_{i0})P(x_{i0})} \geq \sum_{i=1}^{n} \frac{C_\alpha - C_{1-\alpha}}{C_\beta - C_{1-\beta}}.
$$

To apply the Bayesian rule for the requester's decision-making, we introduce a modified likelihood functions for a two-link system

$$
P(y_2|x_1) = \sum_{P_1} P(y_2|\gamma_1)P(\gamma_1|x_1)
$$

$$
+ \sum_{H_1} \sum_{P_1} P(y_2|\gamma_1)\delta(\gamma_1|y_1)P(y_1|x_1).
$$

Using this expression, we can bring the Formula (8) to the following form usually applied to single-link systems:

$$R = \sum_{C_1} \sum_{H_1} \sum_{P_2} P(x_1)P(y_2|x_1)\delta(\gamma_2|y_2)C(\gamma_2; x_1). \tag{10}$$

The decision rule that minimizes the average risk (10) can be written as $\delta(\gamma_{21}|y_2) = 1$ if

$$\frac{P(y_2|x_{11})P(x_{11})}{P(y_2|x_{10})P(x_{10})} \geq \frac{C_\alpha - C_{1-\alpha}}{C_\beta - C_{1-\beta}}. \tag{11}$$

To show the algorithm for the responder's operation, we introduce a modified likelihood functions in the form

$$\begin{aligned}
\Lambda &= \frac{P(x_{11})\sum_{P_1} P(y_2|\gamma_1)P(\gamma_1|x_{11})}{P(x_{10})\sum_{P_1} P(y_2|\gamma_1)P(\gamma_1|x_{10})} \\
&= \frac{P(x_{11})[P(y_2|x_{20})(1 - P_o) + P(y_2|x_{21})P_o]}{P(x_{10})[P(y_2|x_{20})(1 - F_o) + P(y_2|x_{21})F_o]}
\end{aligned} \tag{12}$$

where $P_0 = P(\gamma_{11}|x_{11})$ is the probability that the responder will release a signal after receiving an RQS, i.e. the aircraft responder availability index; $F_0 = P(\gamma_{11}|x_{10})$ is the same probability in the absence of RQS.

So, to be optimal in the Bayesian sense, the responder has to apply statistics (12) for each implementation. In addition, the responder's performance is to be considered and compared with the threshold which is completely determined by the set decision values.

Further, when an AO is detected, this process is repeated N times, where N is the packet of RPS. The result is compared to a digital threshold, which is usually equal to half the packet.

Thus, it is the requirement to take into account the responder's performance and to consider the price functions for CSS in general that makes a signal detection process optimal according to Bayes' criterion.

3 Quality Estimation for Airspace CSS Consumer Information Support

An integral indicator of the quality of information supplied to decision makers in airspace monitoring can be the probability of information provision [9, 45], which can be obtained from the following relation:

$$P_{\text{inf}} = D_z \cdot P_0 \cdot P_{obn} \cdot P_{pri} \cdot P_{ob}, \tag{13}$$

where D_z is the probability of RQS detection; P_0 is the aircraft responder availability index; P_{obn} is the probability of AO detection by CSS; P_{pri} is the probability of the correct reception of flight information, P_{ob} is the probability of flight and coordinate data consolidation in CSS.

The quality of consumer information support can be enhanced by increasing the value of each of the components. We will consider each component of information provision in detail.

3.1 Synthesis of an RQS Optimal Detector in CSSs

Let us concentrate on the problem of synthesizing optimal devices for detecting RQS in interrogating CSS aircraft responders. Characteristically for CSSs, any useful signal at the receiver input is observed against the background of fluctuation noise and intra-system interference.

In this case, RQSs are additively summed in the medium used for transmission without any mutual synchronization. During any time interval of observation, the aircraft responder receiver is entered into by the oscillation

$$r(t) = x(t) + \mu(t) + n(t), t > t_0, \tag{14}$$

which contains useful signal $x(t)$, intra-system interference $\mu(t)$ and fluctuation interference $n(t)$, interdependent with the desired signal and intra-system interference. The fluctuation interference $n(t)$ is approximated by stationary white noise and known statistical characteristics

$$< n(t) >= 0; \; < n(t_1)n(t_2) >= 0, 5N_0\delta(t_2 - t_1); \; N_0 = const.$$

The intra-system interference is

$$\mu(t) = \sum_{j=1}^{n} \mu_j(t) = \sum_{j=1}^{n} \sum_{i=1}^{r} \mu_{ji}(t), \tag{15}$$

where $\mu_j(t)$ is the partial intra-system interference with the j-th code, $r = r(j, T_0)$ is the unknown number of interference pulses with the j-th code on the observation interval, $\mu_{ji}(t)$—is the i-th radio pulse of the j-th partial interference flux, which can be written as

$$\mu_{ji}(t) = X_{ji}(t, \tau_{ji}, A_{ji}, \varphi_{ji}) = A_{ji} f_{ji}(t - \tau_{ji}) \cos[\omega_0(t - \tau_{ji}) + \phi_{ji}], \tag{16}$$

where $f_{ji}(.)$ is the interfering radio pulse envelope, which coincides in shape with the envelope of the useful radio pulses; A_{ji} is the random amplitude of this radio pulse; τ_{ji} is the unknown time of the interfering radio pulse occurrence; $\mu_{ji}(t)$; ϕ_{ji} is the initial phase of the interfering radio pulse.

Intra-system interference (15) has a structure similar to that of the useful signal with an unknown number of interference pulses for the j-th code in the observation interval.

We assume—and this is true in practice—that the moments of appearance of any partial interfering flow pulses (15) are distributed according to Poisson's law with a flux density

$$\lambda = \lambda_i (N - 1), \tag{17}$$

where λ_i is the average number of RQSs.

We synthesize an optimal RQS detector assuming that CSS signals are received coherently; the initial phases of the useful and jamming (16) radio pulses are known (for simplicity's sake we take them to be zero); and the amplitudes of these radio pulses are constant and equal: $A_{jk} = A_{ji} = 1$.

According to Neumann-Pearson's criterion, the probability of RQS detection by the detector being synthesized is to be maximum while the probability of false alarms is fixed. The decision on the presence or absence of a useful signal in oscillation $r(t)$ is made on the basis of the a post-test probability density

$$W_{ps}(a_{mM}, \tau_{mr}) = W_{ps}[a_{mM}, \tau_{mr} | r(t)], \tag{18}$$

where $a_{mM} = a_{11}, \ldots, a_{mM}$; $\tau_{mr} = \tau_{11}, \ldots, \tau_{mr}$.

After splitting the observation interval into M equal segments, the probability density (18) can be written as

$$W_{ps}(a_{mM}, \tau_{mr}) = W_{ps}(a_{mM}, \tau_{mkM}) = \prod_{k=1}^{M} F_k, \tag{19}$$

where $\tau_{mkM} = \tau_{111}, .., \tau_{jik}.., \tau_{mkM}$ is also known as the vector of unknown moments at which interfering pulses appear during the observation interval; τ_{jik} is the time shift of the i-th radio pulse of the j-th partial interfering flow on the k-th interval; $k = k(j, k)$ is the unknown number of interfering pulses on the j-th code sequences in the k-th interval; F_k is the likelihood function on the k-th interval. The likelihood function on the k-th interval can be written as

$$F_k = \prod_{j=1}^{m} F_{jk}(a_k), \tag{20}$$

where

$$F_{ij}(a_k) = exp\left[a_k q(\tau_{jk}) - a_k \frac{E}{N_0} - 2a_k \frac{E}{N_0}R(t_{jk}) + \sum_{l_k=1}^{k} q(\tau_{jlk}) - \frac{E}{N_0}\sum_{l_k,i_k=1}^{k} R(t_{jlk})\right];$$

$$q(\tau_{jk}) = \frac{2}{N_0}\int_{T_{k-1}}^{T_k} r(t)f_{jk}(t - \tau_{jk})\cos[\omega_o(t - \tau_{jk})]dt, \tag{21}$$

is the cross-correlation function of the received signal; $q(\tau_{jlk})$ is the cross-correlation function for the interfering pulse r_{jlk} with a time shift τ_{jlk}; $E = \int_{T_{k-1}}^{T_k} r^2(t)dt = \int_{T_{k-1}}^{T_k} f_{jk}^2(t - \tau_{jk})dt$ is the radio pulse energy; $R(t_{jk}) = (1/E)\int_{T_{k-1}}^{T_k} r_{jk}(t)r_{jlk}(t)dt = \rho(t_{jk})\cos(\omega_o t_{jk})$ is the uncertainty function for useful and interfering radio pulses; $R(t_{jlk})$ is the uncertainty function for the interfering radio pulses r_{jlk} and r_{jik}; $\rho(t_{jk}) = \frac{1}{2E}\int_{T_{k-1}}^{T_k} f_{jk}(t - \tau_{jk})f_{jik}(t - \tau_{jik})dt$ is the uncertainty function envelope; $t_{jk} = \tau_{jk} - \tau_{jik}$ is the overlap interval of the interfering pulses $r_{jlk}(t)$ and $r_{jik}(t)$.

It is worth noting that, while obtaining (21), we took into account the possibility that ordinary partial interfering flows can be superposed on the useful impulse with the length τ_i by only the i-th ($i = 1, \ldots k$) impulse of the j-th partial interfering flow during the k-th interval. This assumption is acceptable since in real life the following is true for CSS:

$$\langle \rho(t_{jk})\rangle = \lambda \tau_i << 1. \tag{22}$$

The decision on the presence ($a_k = 1$) or absence ($a_k = 0$) of useful signals in the k-th interval is based on the ratios of the experimental probability densities (20)

$$\frac{W_{ps}(a_k = 1, \tau'_{m\hat{e}l})}{W_{ps}(a_k = 0, \tau'_{m\hat{e}l})} = \prod_{j=1}^{n+1} \Lambda_{jk},$$

where $\Lambda_{jk} = F_{jk}(a_k = 1)/F_{jk}(a_k = 0) = \exp[q(\tau_{jk}) - E/N_0 - (2E/N_0)R(t_{jk})]$ is the partial likelihood ratio for the signals of the j-th request code in the k-th interval; $\tau'_{m\kappa M}$ is the vector of the moments at which the impulses of noise flows appear during the k-th interval.

In this case, for each interrogative code in the k-th observation interval, it is worth comparing the random value

$$b_{jk} = [q(\tau_{jk}) - E/N_0 - (2E/N_0)R(t_{jk})]. \tag{23}$$

with a certain threshold.

If $b_{jk} > z_0$, then a decision is made about the presence of a useful signal pulse of the j-th request code in the k-th interval; if $b_{jk} < z_0$, then a decision is made about the absence of any useful signal pulses.

The structural scheme of the optimal RQS detector operating in accordance with (23) can be represented by $n + 1$ channels which perform the optimal processing of single pulses. Each processing channel is based on ordinary devices consisting of multipliers and integrators which form random variables $q(\tau_{jk})$, from which partial corrections $(E/N_0)[1 + 2R(t_{jk})]$ are subtracted, to compensate for the intra-system interference. It should be noted that the decision whether an interference pulse has been superposed or not, can be made using functional selection methods.

As seen from the above, the probability of RQS detection by an optimal device in the case of coherent signals and the orthogonality of both signals and the interference do not depend on the intensity of intra-system noise, although the latter complicates the structural scheme of the receiver. This conclusion is explained by the fact that we consider an idealized case of coherent signals with fully known parameters.

However, our optimal RQS detector for coherent signal reception is designed with due regard to intra-system interference. For this purpose, its structural scheme is supplemented with devices measuring the time intervals over which useful and interfering pulses overlap. Then $[1 + R(t_{jk})]$ is calculated, and the voltage at the outputs of these measuring instruments is used to correct the operation of threshold devices.

3.2 Assessment of the Impact of the CSS Response Channel on the Quality of Information Support

An integral quality indicator of information support for decision-makers in airspace monitoring is the information provision probability, which is defined as the product of the probability that an AO will be detected by the CSS and the probability that data from the AO responder will be adequately received. Understandably, the quality of information support can be improved by increasing the values of both probabilities in the product. We will consider the impact of the CSS request channel on the quality of AO detection.

In order to increase the probability of decision-making in the process of AO detection, different strategies of RPS processing can be used, the inter-period processing of coded signals in particular. With existing CSSs, RPSs are first decoded and then the AO detection is based on a packet of the received RPSs. If RPSs are processed in this way, logic n/n, where n is the value of the RPS code, is used in the decoder; and logic k/N, where k the is the digital threshold of AO detection, is used in the detector.

However, it is possible to change the way RPSs are processed by first performing AO detection for each RPS component, which will be followed by decoding the frame response signal.

In this regard, it may be interesting to consider the RPS detection features for these methods of RPS processing and evaluate the effect of the airborne responder's readiness factor on the AO detection.

At the output of the receiver of signals, binary quantization of signals is carried out, that is, at a fixed signal/noise ratio (q) and a selected threshold of the limit from the bottom (z_0) uniquely determined probabilities—P_{11} (probability of detecting a single pulse of the response signal) and P_{01} (probability of occurrence of noise emissions at a given time positions).

We will compare the characteristics of AO detection for both processing methods using Neumann-Pearson's criterion. Postulating a fixed level of false alarms, we will find a detection characteristic (probability of detecting a coded signal) which depends on the signal-noise ratio at the first moment of AO detection (fulfillment of the criterion for determining the beginning of an information packet).

When decoding is done with a subsequent inter-period processing of received signals (Method I), the probability that n pulse interval-time codes and false signals will pass through the decoder is defined as

$$D_d = (P_{11})^n; \quad F_d = (P_{01})^n.$$

Probabilities P_{01} and P_{11} are determined by the following relationships:

$$P_{01} = exp(-z_0^2/2); \quad P_{11} = \int_{z_0}^{\infty} x \, exp[-(x^2 + q^2)] I_0(qx) dx,$$

where $I_0(qx)$ is a modified Bessel function of the first kind of zero order.

The probabilities of detecting useful signals at the output of inter-period processing devices are calculated accordingly:

$$D_1 = \sum_{k}^{N} C_k^N D_d^N (1 - D_d)^{N-k}. \tag{24}$$

If decoding is done after the RPS inter-period processing (Method II), the probability that coded signals and false alarms will pass through the inter-period signal processing device is

$$D_{21} = \sum_{k}^{N} C_k^N P_{11}^N (1 - P_{11})^{N-k}. \tag{25}$$

Then the probability that both useful and false signals will pass through the decoder can be defined as

$$D_2 = D_1^n. \tag{26}$$

Given (25), the expression (26) can be written as

$$D_2 = \left[\sum_k^N C_k^N P_{11}^N (1 - P_{11})^{N-k} \right]^n. \tag{27}$$

Probability estimates for AO detection by CSS, resulting from (26) and (27), are shown in Figs. 1 and 2. The above calculations indicate that the quality of information service can be improved by changing the RPS processing method. Indeed, it can be seen from Fig. 1 that for a signal-to-noise ratio of 1.32 the RPS processing Method I provides the probability of AO detection equal to 0.53, while Method II raises this probability to 0.68 (Figure 3).

Figure 2 shows the probability of AO detection by CSS as a function of the aircraft responder availability index, with a fixed signal/noise ratio $q = 1, 9$ and the value of interval time codes equal to 2, for data processing Methods I and II.

These dependencies allow us to estimate the effect of the aircraft responder availability index on the quality of AO detection by CSS and clearly show that Method

Fig. 2 Air object detection characteristics

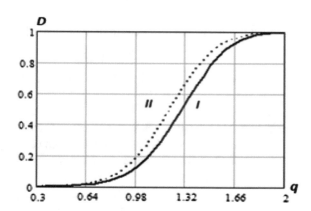

Fig. 3 Air object detection characteristics

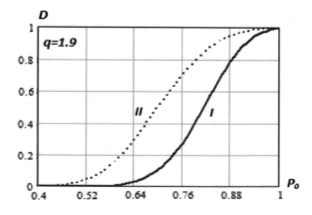

II is less sensitive to the negative factors mentioned above. Actually, if $P_0 = 0,88$, the probability of AO detection by CSS is 0.8 for the conventionally used Method I and 0.95 for Method II.

3.3 Optimization of Signal Parameter Measurement in CSSs

As shown in (13), the probability of information provision depends on the accuracy of AO coordinates estimation, which, in turn, is determined by the precision of the CSS signal parameters estimation. Therefore, we will consider the process of optimizing the measurement of signal parameters, with the CSS structural specifics accounted for.

It is due to the peculiarities of the CSS structure that the operation of its measuring devices is hampered when signals carrying some parameter data temporarily disappear. Generally, these signal drops are random and cannot be predicted.

We will investigate the nature of optimal interaction between the detector and the measuring device in a single unit for processing received signals. Simultaneously with the formation of an optimal estimate of the measured parameter, the unit is to produce a decision on the reliability of the obtained estimate, i.e. the presence of a signal in the segment for which the assessment is done.

This problem can be solved on the basis of Bayes' approach to testing statistical hypotheses with the simultaneous estimation of distribution parameters associated with these hypotheses. The method has been applied in a significant number of studies.

The peculiarity of the problem under consideration is that, when a signal is accidentally lost, the measured parameter of the vector $\vec{\alpha}$ is not lost. Despite the first-kind detection error it is possible to obtain a more or less correct estimate of the parameters $\hat{\vec{\alpha}}$ encoded in the missing signal. These figures are also valuable to the observer. It follows that, if hypothesis H_1 is accepted, regardless of whether it is true or not, the observer's attitude to the estimation errors should be characterized by the same value function $C_1(\vec{\alpha})$. If the alternative H_0 is adopted, the estimate should be considered invalid, and the value of the decision made is to be determined only by its truth, regardless of the assessment results.

Let us denote the value of signal transmission as C_β and take the value of the correct decision about signal absence to be zero.

The optimal estimate of the measured parameter against an arbitrary rule for testing hypothesis H_1 against hypothesis H_0 can generally be found from the differential equation

$$\frac{d}{d\vec{\alpha}} \int_\Omega C_1(\vec{\alpha}) p(\vec{\alpha}) \big[P p_{cn}(Y|\vec{\alpha}) + (1 - P) p_n(Y) \big] d\vec{\alpha} = 0, \qquad (28)$$

where P is the a priori probability that a signal is present; $p(\vec{\alpha})$ is the a priori distribution density of the measured parameter in space Ω; $p_n(Y)$ and $p_{cn}(Y|\vec{\alpha})$ are the distribution densities of the observed events in the absence and presence of a signal with fixed parameter values, respectively.

Considering a quadratic cost function

$$C_1(\vec{\alpha}) = C_1(\vec{\alpha} - \hat{\vec{\alpha}})^2, \tag{29}$$

the following formula for the optimal estimate can be obtained from (28):

$$\hat{\vec{\alpha}}_{onm} = P(H_1|Y)\hat{\vec{\alpha}}_{po} + P(H_0|Y)\hat{\vec{\alpha}}_{ap}, \tag{30}$$

where $P(H_1|Y)$ and $P(H_0|Y)$ are the post-test probabilities of signal presence and absence, which can be defined as

$$P(H_1|Y) = \frac{PF_1(Y)}{(1-P)F_0(Y) + PF_1(Y)} = \frac{\Lambda(Y)}{1 + \Lambda(Y)},$$

$$P(H_0|Y) = \frac{(1-P)F_0(Y)}{(1-P)F_0(Y) + PF_1(Y)} = \frac{1}{1 + \Lambda(Y)},$$

$$F_1(Y) = \int_\Omega p(\vec{\alpha})F_1(Y|\vec{\alpha})d\vec{\alpha}, \quad \Lambda(Y) = PF_1(Y)/(1-P)F_0(Y),$$

where $\hat{\vec{\alpha}}_{ap}$ and $\hat{\vec{\alpha}}_{po}$ are the a priori and post-test average values of the measured parameter, which are defined as

$$\hat{\vec{\alpha}}_{ap} = \int_\Omega \vec{\alpha}w(\vec{\alpha})d\vec{\alpha}, \quad \hat{\vec{\alpha}}_{po} = \int_\Omega \vec{\alpha}p(\vec{\alpha}|Y)d\vec{\alpha},$$

where $p(\vec{\alpha}|Y) = p(\vec{\alpha})F_1(Y|\vec{\alpha})/F_1(Y)$ is the post-test density distribution of the measured parameter.

It is evident from (30) that if the post-test probability of the signal presence is high, the optimal estimate is the well-known Bayes' estimate of a non-disappearing signal, i.e. post-test distribution gravity center. Provided the post-test probability $P(H_1|Y)$ is low, the a priori average of the measured parameter is the best estimate. In general, the optimal estimate (30) is a weighted sum of the estimates $\hat{\vec{\alpha}}_{po}$ and $\hat{\vec{\alpha}}_{ap}$, the values of the weighting factors for the given error values being completely determined by the likelihood ratio $\Lambda(Y)$, which carries information about the presence of a signal in the received segment.

Expression (30) is obtained under the assumption that, simultaneously with the evaluation of the parameters $\vec{\alpha}$ by the processing unit, a certain decision is made about the presence of a signal in the received segment. In order for this decision to be optimal, it must be made as a result of checking the inequality

$$\int_{\vec{\alpha}} C_1(\vec{\alpha})p(\vec{\alpha})\big[Pp_{cn}(Y|\vec{\alpha}) + (1 - P)p_n(Y)\big]d\vec{\alpha} = C_\beta Pp_{cn}(Y), \qquad (31)$$

where previously adopted assumptions about the cost of joint detection and evaluation errors are accounted for.

When inequality (31) holds, hypothesis H_1 should be adopted, otherwise—H_0. Using the above notation, (31) can be formulated as

$$\Lambda(Y)\left[C_\beta - \int_{\vec{\alpha}} C_1(\vec{\alpha})p(\vec{\alpha}|Y)d\vec{\alpha}\right] \geq \int_{\vec{\alpha}} C_1(\vec{\alpha})p(\vec{\alpha})d\vec{\alpha}. \qquad (32)$$

It follows that detection makes sense only if

$$C_\beta - \int_{\vec{\alpha}} C_1(\vec{\alpha})p(\vec{\alpha}|Y)d\vec{\alpha} > 0, \qquad (33)$$

otherwise the inequality (31) cannot be fulfilled and the observer has to choose hypothesis H_0, whatever the received segment is. Practically, it means that if (33) is not fulfilled, the cost of measurement errors (in comparison with the cost of signal transmission) is so high that the minimum average losses are obtained by completely abandoning measurements, i.e. making a decision about the absence of a signal regardless of the received segment. Therefore, for further consideration it can be assumed that condition (33) is fulfilled. Then, in accordance with (32), the optimal detector must adopt hypothesis H_1, i.e. confirm the accuracy of the parameter $\hat{\vec{\alpha}}$ estimate obtained by the measuring device if the likelihood ratio $\Lambda(Y)$ exceeds a certain positive threshold level

$$K(\hat{\vec{\alpha}}) = \frac{\int_{\vec{\alpha}} C_1(\vec{\alpha})p(\vec{\alpha})d\vec{\alpha}}{C_\beta - \int_{\vec{\alpha}} C_1(\vec{\alpha})p(\vec{\alpha}|Y)d\vec{\alpha}},$$

which depends on the accepted evaluation and the costs of measurement errors.

In the case of a quadratic cost function (29), the expression for an optimal detection threshold is reduced to

$$K(\hat{\vec{\alpha}}) = \frac{C_1[(\hat{\vec{\alpha}} - \bar{\vec{\alpha}}_{ap})^2 + \sigma_{ap}^2]}{C_\beta - C_1[(\hat{\vec{\alpha}} - \bar{\vec{\alpha}}_{po})^2 + \sigma_{po}^2]}, \qquad (34)$$

where $\sigma_{ap}^2 = \int_{\vec{\alpha}} (\vec{\alpha} - \bar{\vec{\alpha}}_{ap})^2 p(\vec{\alpha})d\vec{\alpha}$ and $\sigma_{po}^2 = \int_{\vec{\alpha}} (\vec{\alpha} - \bar{\vec{\alpha}}_{po})^2 p(\vec{\alpha}|Y)d\vec{\alpha}$ are the a priori and post-test distribution variances of the measured parameter.

Let us consider the detection algorithm for the case of optimal estimation done in accordance with rule (30). In this case, condition (32) can be written as

$$(1 + \Lambda(Y))^2 (C_\beta - C_1 \sigma_{po}^2) - C_1 (\bar{\bar{\alpha}}_{po} - \bar{\bar{\alpha}}_{ap})^2 > 0.$$

Based on this condition, we obtain two inequalities that have to hold:

$$\left. \begin{array}{c} C_\beta - C_1 \sigma_{po}^2 > 0 \\ \Lambda > \Lambda_{kp} = \sqrt{\frac{C_1 (\bar{\bar{\alpha}}_{po} - \bar{\bar{\alpha}}_{ap})^2}{C_\beta - C_1 \sigma_{po}^2} - 1} \end{array} \right\}, \tag{35}$$

where Λ_{kp} denotes the critical value of the optimal detection which makes it possible to fulfill the initial condition (33), in which case the problem of detection and evaluation has a physical meaning.

Given (30), (34) and (35), the optimal rule (32) can be written as

$$\Lambda^2 - (c\Delta^2 + c - 1)\Lambda - c \overset{>}{\underset{<}{}} 0,$$

where $\Delta = \frac{\bar{\bar{\alpha}}_{po} - \bar{\bar{\alpha}}_{ap}}{\sigma_{ap}}, c = \frac{C_1 \sigma_{ap}^2}{C_\beta - C_1 \sigma_{po}^2}$.

However, taking into account that $\Lambda > 0$ and $c > 0$, we obtain the following expression for the boundary separating the areas of hypotheses H_1 and H_0 in the decision space

$$\Lambda_{gr} = \frac{1}{2}(c\Delta^2 + c - 1) + \frac{1}{2}\sqrt{(c\Delta^2 + c - 1)^2 + 4c}. \tag{36}$$

In terms of practical implementation, it should be postulated that $C_1 \sigma_{ap}^2 > C_\beta$, since measurements are meaningless if errors exceed the value of the a priori parameter variance. Then $c > 1$ and (36) can be combined into

$$\Lambda_{gr} \approx c\Delta^2 + c + \frac{1 - 4c\Delta^2}{4c(\Delta^2 + 1)}.$$

Thus, when the detector and the RPS parameter evaluation device in the CSS operate together, the optimal detection algorithm is reduced to comparing the threshold Λ_{zp} with the optimal detection averaged over the measured parameter. The threshold value is derived from not only the cost of errors, but also the a priori data about the measured parameter ($\bar{\bar{\alpha}}_{ap}$ and σ_{ap}) and the results of received segment processing in the measuring device ($\bar{\bar{\alpha}}_{po}$ and σ_{po}).

It follows from the foregoing that the error signal $\left(\bar{\bar{\alpha}}_{po} - \bar{\bar{\alpha}}_{ap} \right)$, which is proportional to the mismatch between the current estimate and the value previously generated by the measuring device, can be obtained at the output of the optimal discriminator. This signal can be used by the detector to control the value of the critical threshold in accordance with rule (36). Likewise, for the optimal estimate (30) to be achieved, the likelihood ratio value must be sent from the detector to the measuring device.

References

1. Farina A, Studer F (1993) Digital processing of radar information. Radio i svyaz, Moscow
2. Obod I, Strelnitskyi O, Andrusevich V (2015) Informational network of aerospace surveillance systems. KhNURE, Kharkov
3. Ueda T, Shiomi K, Ino M, Imamiya K (1998) Passive secondary surveillance radar system for satellite airports and local ATC facilities. In: 43rd annual air traffic control association. Air Traffic Control Association, Atlantic City, pp 20–24
4. Skolnik M (2008) Radar handbook, 3rd edn. McGraw-Hill, New York
5. Lynn P (1989) Radar systems. Springer, U.S. https://doi.org/10.1007/978-1-4613-1579-7
6. Stevens M (1988) Secondary surveillance radar. Artech House, Norwood
7. Kim E, Sivits K (2015) Blended secondary surveillance radar solutions to improve air traffic surveillance. Aerosp Sci Technol 45:203–208
8. Obod I, Svyd I, Shtykh I (2014) Interference protection of questionable airspace surveillance systems: monograph. KhNURE, Kharkiv
9. Svyd I, Obod I, Zavolodko G, Maltsev O (2018) Interference immunity of aircraft responders in secondary surveillance radars. In: 2018 14th international conference on advanced trends in radioelectronics, telecommunications and computer engineering (TCSET). IEEE, pp 1174–1178. https://doi.org/10.1109/tcset.2018.8336404
10. Svyd I, Obod I, Maltsev O, Shtykh I, Zavolodko G, Maistrenko G (2019) Model and method for request signals processing of secondary surveillance radar. In: 2019 IEEE 15th international conference on the experience of designing and application of CAD systems (CADSM). IEEE, pp 1–4. https://doi.org/10.1109/cadsm.2019.8779347
11. Svyd I, Obod I, Maltsev O, Shtykh I, Maistrenko G, Zavolodko G (2019) Comparative quality analysis of the air objects detection by the secondary surveillance radar. In: 2019 IEEE 39th international conference on electronics and nanotechnology (ELNANO). IEEE, pp 724–727. https://doi.org/10.1109/elnano.2019.8783539
12. Obod I, Svyd I, Maltsev O, Maistrenko G, Zubkov O, Zavolodko G (2019) Bandwidth assessment of cooperative surveillance systems. In: 2019 3rd international conference on advanced information and communications technologies (AICT). IEEE, pp 1–6. https://doi.org/10.1109/aiact.2019.8847742
13. Svyd I, Obod I, Maltsev O, Tkachova T, Zavolodko G (2019) Optimal request signals detection in cooperative surveillance systems. In: 2019 IEEE 2nd Ukraine conference on electrical and computer engineering (UKRCON). IEEE, pp 1–5. https://doi.org/10.1109/ukrcon.2019.8879840
14. Siergiejczyk M, Krzykowska K, Rosiński A (2014) Reliability assessment of cooperation and replacement of surveillance systems in air traffic. In: Proceedings of the ninth international conference on dependability and complex systems DepCoS-RELCOMEX. June 30–July 4 2014, Brunów, Poland, pp 403–411. https://doi.org/10.1007/978-3-319-07013-1_39
15. Jackson D (2016) Ensuring honest behaviour in cooperative surveillance systems. The Centre for Doctoral Training in Cyber Security, Oxford
16. Ramasamy S, Sabatini R, Gardi A (2016) Cooperative and non-cooperative sense-and-avoid in the CNS + A context: a unified methodology. In: 2016 international conference on unmanned aircraft systems (ICUAS)
17. Bloisi D, Iocchi L, Nardi D, Fiorini M, Graziano G (2012) Ground traffic surveillance system for air traffic control. In: 2012 12th international conference on ITS telecommunications. IEEE, pp 135–139. https://doi.org/10.1109/itst.2012.6425151

18. Ahmadi Y, Mohamedpour K, Ahmadi M (2010) Deinterleaving of interfering radars signals in identification friend or foe systems. In: 18th Telecommunications forum Telfor, Telecommunications Society. Belgrade: ETF School of EE, University in Belgrade, IEEE Serbia & Montenegro COM CHAPTER, pp 729–733
19. Li W, Wei P, Xiao X (2009) A robust TDOA-based location method and its performance analysis. Sci China Ser F: Inf Sci 52(5):876–882
20. Trofimova Y (2007) Multilateration error investigation and classification error estimation. Transp Telecommun 8(2):28–37
21. Gaviria I (2013) New strategies to improve multilateration systems in the air traffic control. Editorial Universitat Politècnica de Valènci, Valencia
22. Naganawa J, Miyazaki H, Tajima H (2017) Detection probability estimation model for wide area multilateration. In: 2017 integrated communications, navigation and surveillance conference (ICNS). IEEE, pp 2B1-1–2B1-15. https://doi.org/10.1109/icnsurv.2017.8011897
23. Alia L, Italiano A, Pozzi F (2014) Advanced tools to analyze the expected performance of multilateration and wide area multilateration. In: 2014 Tyrrhenian international workshop on digital communications—enhanced surveillance of aircraft and vehicles (TIWDC/ESAV). IEEE, pp 82–86. doi:TIWDC-ESAV.2014.6945453
24. Naganawa J, Miyazaki H, Tajima H (2018) Localization accuracy model incorporating signal detection performance for wide area multilateration. In: IEEE transactions on aerospace and electronic systems, pp 1–1. https://doi.org/10.1109/tvt.2017.2699176
25. Garcia M, Dolan J, Hoag A (2017) Aireon's initial on-orbit performance analysis of space-based ADS-B. In: 2017 integrated communications, navigation and surveillance conference (ICNS), pp 1–28. https://doi.org/10.1109/icnsurv.2017.8011994
26. Revels M, Ciampa M (2018) Can software defined radio be used to compromise ADS-B aircraft transponder signals? J Transp Secur 11(1–2):41–52. https://doi.org/10.1007/s12198-018-0188-y
27. Naganawa J, Miyazaki H, Otsuyama T, Honda J (2018) Initial results on narrowband air-ground propagation channel modeling using opportunistic ADS-B measurement for coverage design. In: 2018 integrated communications, navigation, surveillance conference (ICNS). IEEE, pp 4F3-1–4F3-10 https://doi.org/10.1109/icnsurv.2018.8384895
28. Globus I (1972) Binary coding in asynchronous systems. Svyaz', Moscow
29. Kryvinska N (2010) Converged network service architecture: a platform for integrated services delivery and interworking. Electronic Business series, vol 2. International Academic Publishers, Peter Lang Publishing Group
30. Kryvinska N (2008) An analytical approach for the modeling of real-time services over IP network. Math Comput Simul 79(4):980–990. https://doi.org/10.1016/j.matcom.2008.02.016
31. Kryvinska N (2004) Intelligent network analysis by closed queuing models. Telecommun Syst 27:85–98. https://doi.org/10.1023/B:TELS.0000032945.92937.8f
32. Ageyev D, Al-Anssari A (2014) Optimization model for multi-time period LTE network planning. In: Proceedings of the 2014 first international scientific-practical conference problems of infocommunications science and technology (PIC S&T'2014). Kharkov, Ukraine: IEEE, pp 29–30. https://doi.org/10.1109/infocommst.2014.6992288
33. Al-Dulaimi A, Al-Dulaimi M, Asevev D (2016) Realization of resource blocks allocation in LTE downlink in the form of nonlinear optimization. In: 2016 13th international conference on modern problems of radio engineering, telecommunications and computer science (TCSET). IEEE, pp 646–648. https://doi.org/10.1109/tcset.2016.7452140
34. Pereverzev A, Ageyev D (2013) Design method access network radio over fiber. In: 2013 12th international conference on the experience of designing and application of CAD systems in microelectronics (CADSM), Polyana Svalyava: IEEE, pp 288–292
35. Ageyev D et al (2018) Classification of existing virtualization methods used in telecommunication networks. In: Proceedings of the 2018 IEEE 9th international conference on dependable systems, services and technologies (DESSERT), pp 83–86

36. Ageyev D, Al-Ansari A (2015) LTE RAN and services multi-period planning. In: 2015 second international scientific-practical conference problems of infocommunications science and technology (PIC S&T). Kharkov, Ukraine: IEEE, pp 272–274. https://doi.org/10.1109/infocommst.2015.7357334

37. Radivilova T, Kirichenko L, Ageiev D, Bulakh V (2020) The methods to improve quality of service by accounting secure parameters. In: Hu Z, Petoukhov S, Dychka I, He M (eds) Advances in computer science for engineering and education II. ICCSEEA 2019. Advances in intelligent systems and computing, vol 938. Springer, Cham

38. Bondarenko O, Ageyev D, Mohammed O (2019) Optimization model for 5G network planning. In: 2019 IEEE 15th international conference on the experience of designing and application of CAD systems (CADSM). IEEE, pp 1–4. https://doi.org/10.1109/cadsm.2019.8779298

39. Kirichenko L, Radivilova T, Bulakh V (2020) Binary classification of fractal time series by machine learning methods. In: Lytvynenko V, Babichev S, Wójcik W, Vynokurova O, Vyshemyrskaya S, Radetskaya S (eds) Lecture notes in computational intelligence and decision making. ISDMCI 2019. Advances in intelligent systems and computing, vol 1020. Springer, Cham

40. Kirichenko L, Radivilova T, Bulakh V (2018) Machine learning in classification time series with fractal properties. Data 4(1):5. https://doi.org/10.3390/data4010005

41. Kirichenko L, Radivilova T, Bulakh V (2018) Classification of fractal time series using recurrence plots. In: 2018 international scientific-practical conference problems of infocommunications. Science and technology (PIC S&T). IEEE, pp 719–724. https://doi.org/10.1109/infocommst.2018.8632010

42. Kryvinska N, Zinterhof P, van Thanh D (2007) An analytical approach to the efficient real-time events/services handling in converged network environment. In: Enokido T, Barolli L, Takizawa M (eds) Network-based information systems. NBiS 2007. Lecture notes in computer science, vol 4658. Springer, Berlin, Heidelberg

43. Kryvinska N, Zinterhof P, van Thanh D (2007) New-emerging service-support model for converged multi-service network and its practical validation. In: First international conference on complex, intelligent and software intensive systems (CISIS'07). IEEE, pp 100–110. https://doi.org/10.1109/cisis.2007.40

44. Obod I, Svyd I, Maltsev O, Vorgul O, Maistrenko G, Zavolodko G (2018) Optimization of data transfer in cooperative surveillance systems. In: 2018 international scientific-practical conference problems of infocommunications. Science and technology (PIC S&T). IEEE, pp 539–542. https://doi.org/10.1109/infocommst.2018.8632134

45. Svyd I, Obod I, Maltsev O, Vorgul O, Zavolodko G, Goriushkina A (2018) Noise immunity of data transfer channels in cooperative observation systems: comparative analysis. In: 2018 international scientific-practical conference problems of infocommunications. Science and technology (PIC S&T). IEEE, pp 509–512. https://doi.org/10.1109/infocommst.2018.8632019

46. Pavlova D, Zavolodko G, Obod I, Svyd I, Maltsev O, Saikivska L (2019) Comparative analysis of data consolidation in surveillance networks. In: 2019 10th international conference on dependable systems, services and technologies (DESSERT). IEEE, pp 140–143. https://doi.org/10.1109/dessert.2019.8770008

47. Pavlova D, Zavolodko G, Obod I, Svyd I, Maltsev O, Saikivska L (2019). Optimizing data processing in information networks of airspace surveillance systems. In: 2019 10th international conference on dependable systems, services and technologies (DESSERT). IEEE, pp 136–139. https://doi.org/10.1109/dessert.2019.8770022

Adaptive Semantic Analysis of Radar Data Using Fuzzy Transform

Svitlana Solonska and Volodymyr Zhyrnov

Abstract The development of a new adaptive system of radar data semantic analysis with their non-stationarity, which is based on both numerical and logical methods of multiscanning processing of signals and methods of artificial intelligence using fuzzy transformations of the universe of signals and signal images, is proposed. The possibility of its hardware and software implementation is considered. The results of computer modeling, theoretical and experimental researches with processing of real radar signals are presented. The elements of logical analysis and algebra of finite predicates (AFP) are selected as mathematical apparatus. As experimental studies show, AFP is an appropriate tool for logical-mathematical constructions, with which it's possible to describe the radar operator actions. The basic concepts of Boolean algebra and graph theory are also used. The practical value of the work is: a method for formalizing the processes of perception and transformation of signals and signal images, algorithms and software are intended for information radar systems with natural-language intellectual interface; also for support the design of information structures. Mathematical and software results can be used in the systems of automatic processing of radar information, particularly, in the intelligent radar and radio-electronic systems and complexes for monitoring of mobile air and ground objects.

Keywords Adaptive system · Semantic analysis · Radar data · Multiscanning signal processing · Artificial neural network · Algebra of finite predicates · Fuzzy-transform

S. Solonska (✉)
Kharkiv National Automobile and Highway University, Kharkiv, Ukraine
e-mail: svetsolo27@gmail.com

V. Zhyrnov
Kharkiv National University of Radioelectronics, Kharkiv, Ukraine
e-mail: nauka123@ukr.net

T. Radivilova et al. (eds.), *Data-Centric Business and Applications*, Lecture Notes on Data Engineering and Communications Technologies 48,
https://doi.org/10.1007/978-3-030-43070-2_9

157

1 Introduction

This work is devoted to the decision of the problem of radar detection and tracking of low-sized air objects in the conditions of non-stationary signal information for the effective control and monitoring of airspace. Modern active radar park doesn't provide the necessary level of adaptation to the complex signal interfering environment. Objects, such as unmanned aerial vehicles and Stealth technology, characterized by a small radar cross section (RCS) and detection and tracking of their radar marks are carried out in the presence of many streaming and non-stationary signal information. Therefore, the problem of increasing their reliable detection is relevant, both during the modernization of existing ones and the creation of new means of airspace control.

To solve this problem, we propose the development of a new adaptive system of semantic analysis of radar data with their non-stationarity, which is based on both numerical and logical methods and methods of artificial intelligence using fuzzy transform of the universe of signals and signal images. The possibilities of its software and hardware implementation are considered. The results of computer modeling, theoretical and experimental researches with processing of records of real radar signals are presented.

The elements of logical analysis and the algebra of finite predicates are chosen as a mathematical apparatus. These tools appropriate for logical-mathematical constructions, creating a complete description of the actions of a human-operator of radar systems. The basic concepts of Boolean algebra and graph theory are also used.

A model has been created for the operator of survey radar on the principle of neural networks and proposals for its software implementation. This model is similar to the well-known so-called model of an artificial neural network (ANN).

Such an ANN model is a system of simple processor actions interconnected and interacting with each other (elementary processors—artificial neurons). These processors are usually quite simple and deal only with signals that a neuron periodically receives and signals that it sends to other neurons (processors) and, nevertheless, being connected to a large network with controlled interaction, they are able to perform quite complex tasks. With regard to the surveillance radar airspace control it can be the following tasks:

1. From a machine learning point of view, the ANN can be a special case of the formation of virtual images of signal images for recognizing and detecting of aircraft marks and, in the case of positive results, taking on automatic tracking of this mark.
2. From a mathematical point of view, ANN learning is a multiparameter problem of nonlinear optimization in the "noise" of signals.
3. From the point of view of the nonstationarity of the signal environment—the ANN can be used in optimal control problems.
4. From the point of view of programming development, ANN is a way to solve the problem of efficient parallelism.

5. From the point of view of artificial intelligence, ANN is the main direction of modeling the methods of semantic analysis of radar information by a man—the operator of the radar.

Neural networks are not programmed, they are trained. Technically, training is finding the coefficients of connections (dependencies) between neurons or, in our case, the connection between periodically incoming signals, or between signals in the neighboring elements. In the process of learning, the neural network or the semantic analysis system is able to identify complicated dependencies between input and output data, as well as perform a generalization as a result of the semantic analysis. It can detect, recognize and track the signal mark of the air object.

These tools that in case of successful training, the network or system will be able to return the correct result based on virtual data that was previously absent in the training set, as well as incomplete and/or partially distorted data.

2 Learning ANN as a Special Case of the Formation of Virtual Images of Signal for the Unification of Procedures for Processing Radar Data Based on Semantic Analysis

The problems of processing signals and signal images in the intelligent radar systems, as well as approaches and technologies for their solution are analyzed [1–3]. It is determined that the prospects for the development of existing radar systems are to increase the level of automation of signal processing in the conditions of non-stationary data, including in the systems of radar recognition of air objects by the signal images. In addition, it is advisable to improve the technology of signal processing, when objects and their reflections are connected by complicated logical dependencies, which approximate the processes of perception and analysis of signals to human-operator logic.

This work is devoted to the problem solution of radar detecting and tracking of low-sized aerial objects in the conditions of non-stationary signal information for effective control and monitoring of airspace. It is planned to solve the following main tasks:

- development of methods of data structuring and formalization of processes of perception and transformation of signals and signal images based on algebra of finite predicates and the definition of the semantic components of radar signals;
- development of methods for training the processing system depending on the signal-interference situation on the basis of spatial-semantic and spectral-semantic signal models, which allow to reduce the number of processing procedures, taking into account non-stationary and non-linear dynamics of the process;
- using fuzzy transform to analyze radar data in the form of coordinates of marks and their virtual images, which are formed during training in the space of useful

and interfering signals, and are treated as fuzzy samples and sets. It is shown how this approach can be used to identify the functional (semantic) dependencies between radar data and image components;

- development of recommendations for the creation of a module for processing radar signals based on the developed information technology.

The proposed processing system is based on artificial intelligence methods and imitates the operator work. There are a lot of scientific papers in this area. In particular, in the paper [2] it is shown that primary stage of radar data processing is image analysis, and secondary stage—processing it at semantic level. This approach is based on modeling the human operator work when developing a technology for automated processing of signal information, which is expected to increase productivity of technological systems, in particular, intelligent radar systems. Analysis of the work on the mathematical modeling of adaptation in modern intelligent information systems [2, 4] shows their relevance to describe conditions and requirements of the tasks set by the project.

An essential role in building distributed systems is played by infocommunication networks [5]. In paper [6] studies the existing methods of virtualization of different resources. The positive and negative aspects of each of the methods are analyzed, the perspectivity of the approach is noted. It is also made an attempt to classify virtualization methods according to the application domain, which allows us to discover the method weaknesses which are needed to be optimized.

Modern studies of the processes occurring in technical systems have revealed their self-similar nature. So in works [7–10] a study of self-similar processes is carried out, both of signals of various nature and the properties of traffic in networks.

One of the directions in developing mathematical models for representing knowledge and processing of heterogeneous radio information is the adaptation models in the intelligent systems [2]. In [4] the methods of development of such models are shown due to the development of the mathematical apparatus for modeling intellectual functions for the formalization of the process in the construction of intelligent information systems. It is shown that to solve the problem, it is necessary to study models and methods for creating modern systems based on artificial intelligence methods [11, 12], in particular, methods for transforming into predicate algebra operations and developing a base of logical navigation rules for content adaptation and adaptive link hiding [13]. In the paper [14] it is emphasized that for the construction of information technology it is necessary to use algebra of finite predicates and predicate operations. The basic models of knowledge representation are analyzed; the basic methods of identification of knowledge are considered (classification, method of computer analysis). In [15], the mathematical tools on the base of algebra of finite predicate for the presentation of knowledge and strategy modeling in the intelligent systems with adaptation elements are defined in the construction of an integrated model of the intellectual adaptive system.

2.1 Research in the Field of the Formalization of the Processes of Forming the Pattern of Positional Behavior of Signals from Air Objects and Interferences in the Absence of Stationary Data

Based on the critical analysis of existing problems and tasks of radar signal processing [1–3], the purpose of the research is formulated and grounded. If the language of functions is required for the description of objects of the outside world, then a simulation of the semantic component of the signal information requires a more general language of relations and actions on them [4, 11]. The concept of a relationship is essentially equivalent to the notion of a predicate. The algebraic apparatus of predicates and predicate operations is most effective and appropriate for the study of intellectual activity of an operator and is considered to be the most promising in this field [11–13]. At present, logical mathematics is a collection of different algebras, the presence of each of which is determined by the relevant subject areas [14, 16, 17]. Primarily, we note the algebra of finite predicates, proposed in [4, 11]. It represents a generalization of the apparatus of boolean functions and the apparatus of multivalued logic. Using this algebra makes it possible to formalize the abstract concepts used by the survey radar operator.

Behavior analysis of the signal mark images of interfering reflections (IR) (clutters), discrete interfering reflections (angel echo) (AE) and aerial object reflections (AO) shows [14, 15, 18 20] that informative is view of picture around the analyzable (locatable) element of a space formed by accumulating information in time and space in the series of radar surveillances. Such a procedure will be called the artificial neural network training procedure. Analyzable (locatable) elementary cells of space are separate elements of neural network. Based on this, all possible types of image received during the training were divided into S_j, $j = 1, \ldots, N$ types, having different extent of "plausibility" for situations when only AO or IR or AE signals are received and for situations where they are received together in various combinations. At the same time, the basis for identifying AO on the background of AE or IR is a image analysis of the ordered with a certain speed movement of marks of aerodynamic objects, which differs significantly from the character of movement of DIR marks, including the angel-echo type.

Based on the analysis of differences in the positional behavior of AO and AE, is formed predicate attributes used to identify radar marks, i.e. their alternative relating to one of two classes: AO or AE marks (Fig. 1).

The degree of formalization of processes of perception and transformation of signals is determined by the semantic component.

The following designations are used: i, j are discrete numbers in distance R and azimuth β; Z_n is a sign of repetition in the signal processing element of the previous $k - 1$ and current k surveys; Z_{nc} is a sign of repetition in the element of signal processing of $k - 2$ and $k - 1$ surveys (for processing elements $i(\pm 1)j$, $i(j \pm 1)$ Z_{nc} is a sign of repetition in the neighboring signal processing element); Z_y is a sign of leaving the mark outwards the processing element during the survey period; AE, AO are angel-echo and aerial object marks respectively.

Fig. 1 Positional behavior
of signals

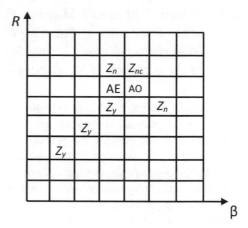

2.2 Algebraic Formalizing of the Semantics of Radar Signal Information in the Conditions of Non-stationary Data

Predicative representation of the semantic processing operations of radar signals as follows. Research shows that the mathematical apparatus of predicate algebra allows to formalize the main events (operations) for implementing intelligent algorithm for detecting and recognizing obtained by training images of interference marks and air objects based on the use of logical dependencies similar to the logic of a human operator [15, 20].

For formation of the picture of the positional behavior of signals from air objects and clutters, the following basic procedures of numerical and logical processing, based on semantic data analysis, are proposed:

- forming of maps of the signal intensities of the reflection marks for the radar field of view elements taking into account the multi-survey signal behavior in each element of processing or neuron in the neural network;
- forming of the elements of the radar coverage area, the predicate maps of the occurrence events, the repetition in this element, the repetition in the neighboring elements and the output of the signal from the processing element on the basis of interrogative mapping of the intensity map of the markings and predicates of events.;
- adaptive change (learning) of the semantic attributes Z_{ncij} and Z_{yij} and the value of the signal in the processing element (neuron) in the case of non-stationary data;
- selection and classification of air objects against the background of clutters or angel echo by solving predicate equations generated for the corresponding radar marks or neurons.

To perform these operations, an intelligent system (IS) algorithm, which is partially implemented in the unified module for multi-survey signal and information processing, has been developed [4, 11, 12]. The data array is a rectangular matrix

$\|A\|$ of $M \times N$, each element of which is a g-bit binary number. Moreover, the α bits of the binary number of each of the elements a_{ij} (RAM cells) are reserved for quantizing the signal intensities, the θ bits for storing information about predicates, and the $(g - \alpha - \theta)$ bits for storing information about the decision to detect a mark. Each element a_{ij} is connected with the corresponding section of the locatable zone (element of processing). The dimensions of the processing elements are chosen in such a way that air objects, moving with speeds from the range of the most probable speeds of movement, don't go out of its limits during the survey period.

A set $M = \{q_{11}, q_{12}, \ldots, q_{ij}, \ldots, q_{mn}\}$ is introduced processing elements of the radar field of view. Predicate $A(x)$ on the set M, corresponding to the set B of processing elements that exceed the threshold

$$A(x)^k = \bigvee_{i=1, j=1}^{mn} x_{ij}^{q_{ij}} \tag{1}$$

where expression $x_{ij}^{q_{ij}}$ is a form of event recognition, i, j are discrete numbers in distance R and azimuth β, k is survey number. At $x_{ij} = q_{ij}$ is determined $x_{ij}^{q_{ij}} = 1$. An example of positional behaviour of the signal is presented on the Fig. 1. For the formation of an intelligent image around element of the radar survey area currently being analyzed, a system of predicate attribute (functions) is introduced, which allows to formalize the changes $A(x)$ taking place in a series of radar surveys:

- predicate attribute Z_{nij}^k of signal repetition in the processing element a_{ij} of the previous $k - 1$ and current k surveys;
- predicate attribute Z_{ncij} of signal repetition in the adjacent processing element $k - 2$ and $k - 1$ surveys for processing elements $i(\pm 1)j$, $i(j \pm 1)$;
- predicate attribute Z_{yij} of mark's leaving outwards the processing element during the survey period.

Under these initial conditions, the formation of predicate features can be carried out according to the following algorithm:

$$
\begin{aligned}
&Z_{nij} = 1, \text{ at } A(x_{ij})^k > 0 \wedge A(x_{ij})^{k-1} > 0; \\
&Z_{ncij} = 1, \text{ at } Z_{n(i+l),(j+l)} = 1; \\
&Z_{yij} = 1, \text{ at } A(x_{ij})^{k-1} > 0 \wedge A(x_{ij})^k = 0 \wedge Z_{nij} = 0,
\end{aligned} \tag{2}
$$

The predicate attributes for each processing element determined to (2) are stored in RAM and, as the current radar information is received, the values of attributes and values of signals are updated in the surveys. The value of attributes either does not change or changes depending on the behavior of the signal in the processing element or and in the adjacent elements. According to the dynamics analysis of the predicate attributes behavior over time, the decision is made to identify the object's mark in the processing element.

Adaptation or extrapolation of semantic attributes in the case of non-stationary radar data is as follows:

$$Z_{nij} = 1, \text{ at } Z_{nij}^k = 1 \cup \left(Z_{nij}^{k-1} = 1 \cap A_{ij}^k = 0 \cap A_{ij}^{k-1} > 0 \right); \qquad (3)$$

$$Z_{yij} = 1, \text{ at } Z_{yij}^k = 1 \cup \left(Z_{yij}^{k-1} = 1 \cap A_{ij}^k = 0 \cap A_{ij}^{k-1} > 0 \right)$$

$$q_{ij} = q_{ij}^{k-1} - \Delta, \text{ at } q_{ij}^k = 0 \cap q_{ij}^{k-1} > 0 \qquad (4)$$

Adaptive change (learning) of semantic attributes Z_{nij} and Z_{yij}, and the signal in the processing element (neuron) is made according to (3) and (4) by overwriting them in the same memory cells in which they were recorded during previous surveys. The basis for saving by rewriting the value of the attribute is either its re-formation when the appropriate conditions (4) exist or the presence of a non-zero q_{ij} signal amplitude level in the considered cell, or in the case when the first two conditions are simultaneously.

The basis for the formation of an adaptive attenuating signal, decreasing from survey to survey by Δ value, is the fact that there is no signal coming in the current survey (4). In the case of a signal not being received in the current survey, the q_{ij} value recorded in the corresponding memory cell in the previous survey is reduced by a certain amount of Δ. Thus, the adaptation of attributes for a time is proportional to signal attenuation in the current processing element. The value Δ may be different depending on the type of semantic attribute for which adaptation is produced (Z_{nij} or Z_{yij}). The adaptation of the attribute Z_{yij} occurs automatically upon adaptation of the attribute Z_{nij} in the processing elements adjacent to the current element.

To increase the efficiency of small-sized air objects detection, the algorithm can be modified so that in case of signal repetition in adjacent surveys in one processing element (unit), their levels can be summed up, thereby increasing the accumulation energy of the useful signal and the time interval for attribute Z_{nij} adaptation.

$$q_{ij} = q_{ij}^k + q_{ij}^{k-1}, \text{ at } q_{ij}^k \cap q_{ij}^{k-1} > 0. \qquad (5)$$

For example, on the Fig. 2 a functional diagram of possible hardware and software device for implementing an algorithm for generating and extrapolating attribute Z_{nij} is shown.

The device works as follows. Data on the signal levels, that have passed the threshold processing in the input data converter (IDC), in binary parallel code arrive at the Switch and on the "OR-NOT" and first "OR" elements. These logic elements are intended respectively to determine the presence of zero (no signal) and non-zero (presence of signal) signal levels in each of the processing elements in the current survey. The second logical element "OR" is intended to determine the presence of a non-zero signal level in each of the processing elements in the previous survey. The data stored in the memory at the previous stage of processing (survey) are fed to it.

Logic elements "AND" perform functions of the comparison device. When signals are received at both inputs of the second logical element "AND", that is, under the condition $A_{ij}^k > 0 \cap A_{ij}^{k-1} > 0$ at its output an attribute Z_{nij} is generated, which

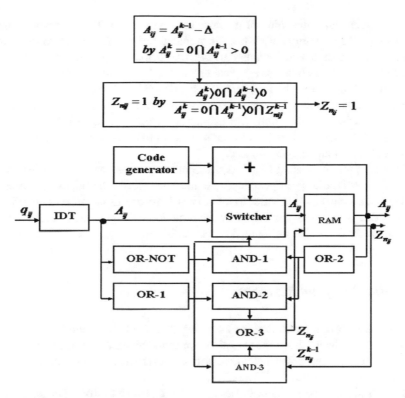

Fig. 2 Functional diagram of the software and hardware implementation of the algorithm for generating and extrapolating the Z_{nij} attribute

through the third logical element "OR" is recorded into the corresponding memory cell.

Extrapolation by rewriting an attribute Z_{nij} is performed when an attribute Z_{nij}^{k-1} is received on the third logical element "AND" recorded in the corresponding memory cell in the previous survey and there is a non-zero signal level stored in this cell, or when attribute Z_{nij} is updated.

The extrapolation time of attribute Z_{nij} and attribute Z_{yij} is determined by the value of the accumulated signal and the value Δ of the subtractor of the stored signal from survey to survey and is provided as follows. With a zero signal level in the ijth processing element in the current survey and at non-zero signal level stored in the corresponding RAM cell, at the output of the first logical element "AND" generates a control action transferred to the switch. Then there is a connection to the second input of the adder which generates a code (code of Δ value) and provides a reduction in the level of the signal recorded in RAM during the previous survey.

To generate attribute Z_{ncij}, data about the current survey signal levels in the binary parallel code is fed to the input of the OR element, which is intended to form a signal presence predicate in each processing element. In the coincidence circuit occurs element-wise matching of the current survey data with the previous survey data stored in the memory unit. When the predicates of events of nonzero signal level coincide in the current and previous surveys, the attribute Z_{nij} is generated that is fed to the second memory block as a logical unit level. As a result, an attribute matrix of signals repetition of the previous survey is generated in the second memory block.

The computer memory stores, with constant updating, data on the attributes Z_n of adjacent azimuthal directions (for example, three). If there attributes Z_n is in at least one of the eight processing elements adjacent to the ijth considered processing element, the attribute Z_{ncij} is formed.

Similarly, the processing elements adjacent to the considered one can be viewed while the attribute Z_{yij} is generated. At the same time, the adjacent processing elements are considered in order to identify at least one in the current overview of the situation which is characterized by the reducing the signal level recorded in the corresponding cell of the attribute Z_{nij} (condition 5).

2.3 Results of the Research

For such situation, the criterion for assessing the quality of identification of radar marks can be selected from the statistical approach given in [1]. It assumes that at one point in space (processing element) with a nonzero a priori probability there can be marks of various classes (AO, AE).

The solution to the problem of optimizing the procedure for radar marks identifying is based on the statistical theory methods of multi-alternative (in this case, two-alternative) recognition. In the presence of a two-alternative situation for the classical analysis procedure, the optimal system for radar marks identifying is determined by the value of likelihood ratio or its logarithm [1, 2]. At the same time $A = (A_1, A_2)$ is the alphabet of classes of identifiable marks, $P(A/b_i)$, $P(A/b_j)$ are conditional probabilities of classes, where b_i and b_j vectors of a set of attributes that characterize the belonging of radar marks to the different classes.

Then likelihood ratio is determined as

$$l(A) = \frac{P(A/b_i)}{P(A/b_j)} \geq \frac{P(b_j)}{P(b_i)} \tag{6}$$

where $P(b_i)$, $P(b_j)$ are priori probabilities of occurrence of sets of attributes b_i and b_j; $P(b_i)/ P(b_j)$ is threshold value of likelihood ratio.

The task of identifying marks is to identify two hypotheses: $H1$ is the mark refers to a class that is characterized by a combination of attributes b_i; $H2$ is mark refers to the class characterized by a set of attributes b_j. By comparing $i(A)$ with a threshold value, it is determined whether the event belongs to the hypothesis $H1$ or $H2$. Inequality (6) is a Bayesian criterion that minimizes the error in solving the problem of marks identifying. The probability of error determines measure of the separability of the different class's marks with vector A distribution data.

The total error, taking into account the errors of the first (non-detection) and second (false alarm) series, is defined as their weighted sum

$$P_{ou\Sigma} = \sum_{G_1} P(b_i)[1 - P(A/b_i)] + \sum_{G_2} P(b_j)[1 - P(A/b_j)] = P(b_i) \cdot P_1 + P(b_j) \cdot P_2,$$

(7)

where G_1, G_2 are possible A areas corresponding to the combination of attributes b_i and b_j; P_1, P_2 are errors of the first and second kinds, respectively.

P_1, P_2 can also be determined by integrating the likelihood ratio probability density $P(l/b_i)$ and $P(l/b_j)$.

$$P_1 = \int_0^T P(l/b_i)dl; \quad P_2 = \int_T^\infty P(l/b_j)dl$$

where $T = P(b_j)/P(b_i)$ is the identification threshold.

The calculation $P_{ou\Sigma}$ is essentially come down to integration in the n-dimensional space of the probability distribution of the analyzed classes.

In this case, the identification of radar marks is made by analyzing the attributes assigned to each of the marks based on a multiple-element element-by-element comparison of the positions of the signals on the generated map of the intensities of radar clutters. This raises the problem of identifying features that provide minimal error in the identification of radar marks from the entire set of generated attributes.

The considered algorithms allow to generate all possible types of images S_j, $j = 1, \ldots, N$ for each processing element, determined by the set of predicate features. In this case $N = 8$.

As a result of comparing the predicate equations with the reference ones, obtained on the basis of a priori and operational data about the classes of identifiable marks, information about the detection of air object signal is generated.

The algorithm for identifying types of images is described by the following equation

$$S_j = \left(\bigvee_{\gamma=0}^{1} Z_n^\gamma \right) \wedge \left(\bigvee_{\gamma=0}^{1} Z_{nc}^\gamma \right) \wedge \left(\bigvee_{\gamma=0}^{1} Z_y^\gamma \right).$$

(8)

Expression Z_n^γ is a form of recognition of the value of the predicate attribute; when $\gamma = 0$ $Z_n^\gamma = 0$, when $\gamma = 1$ $Z_n^\gamma = 1$.

As a result of the analysis of the set of b_i features, radar marks are identified, that is, they are assigned to one of two classes A_1 (AE marks), or A_2 (AO marks). For this case the decisive rule of maximum likelihood ratio is used, where the resolver identifies the mark with the class, the conditional probability of which is higher

$$b_i \Rightarrow A_1, \text{ if } P(A_1/b_i) > P(A_2/b_i);$$
$$b_i \Rightarrow A_2, \text{ if } P(A_1/b_i) < P(A_2/b_i). \tag{9}$$

The distribution density of features in each of the classes is estimated by computer simulating, provided that the model of the array of video data entering the processing device is adequate to the real AE characteristics.

To evaluate the effectiveness of algorithms for the selection of radar objects against the background of AE, an array of video data arriving at the primary processing device is simulated, when interfering clutters (angel echo) was applied to radar. In this case, the analysis of the probability characteristics of identifying radar marks was carried out. As a result of the simulation is:

- estimation of the probabilities of the distribution of Z_n, Z_{nc}, Z_y and their combinations in classes A_1 and A_2;
- estimation of the likelihood of errors in identifying the radar marks $P(L)$ using different combinations of attributes b_i;

When identifying radar marks in the model, the following decision rule is applied:

- if the predicate attributes Z_n, Z_{nc} correspond to the mark, or their combination, then the decision is made that the mark belongs to the class A_1;
- if the mark corresponds to the predicate attribute Z_y one or in combination with attributes Z_n, Z_{nc}, then the decision is made that the mark belongs to the class A_2.

Probability graphs of predicate attributes occurrence in various combinations in classes A_1 and A_2 are presented on the Fig. 3.

On the Fig. 4 presents the dependences of estimates of the probability of error identification AO $P(b_i/A_i)$ from b_i features. Estimates obtained for different speeds of radar objects and density of the DIR $\eta = 0.1$ (the ratio of the number of clutters to the total number of processing elements).

Fig. 3 Probability graphs of predicate attributes occurrence in classes A_1 and A_2

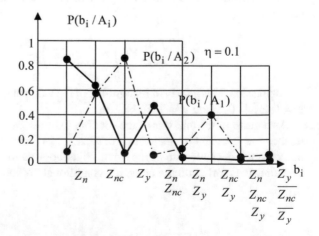

Fig. 4 Estimate of the error likelihood in the identification AO at different sets of predicate attributes b_i and velocity of radar objects movement

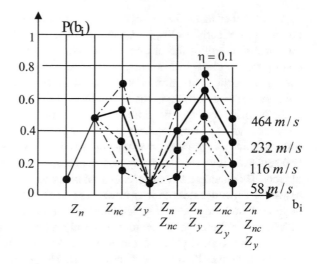

3 Optimization of Training in the Semantic Processing of Radar Data Under Conditions of Non-stationarity

It is proposed to use fuzzy transform to optimize the semantic analysis of radar data in the form of mark coordinates and their virtual images, which are formed during training on the space of useful and interfering signals, and are considered as fuzzy samples and sets. It is shown how this approach can be used to identify functional (semantic) dependencies between the components of radar data and images.

3.1 Using the Fuzzy Transform for Analyzing Radar Data in the Form of Mark Coordinates and Their Virtual Images

The traditional arsenal of mathematical methods for constructing models of uncertainty, based on the use of probability theory, fuzzy sets [21], inaccurate distribution densities of the analyzed random variable—random variables with their own sets [22] offers deterministic descriptions of these models. At the same time, when analyzing non-stationary radar data, including radar images, it becomes necessary to use more complex models in which the parameters of the model are uncertain too. In such systems, two or more level models may appear, for example, stochastic models in which the parameters are distribution densities; fuzzy models of fuzzy numbers, the parameters of which membership functions are fuzzy numbers. The use of such models gives rise to a number of problems associated with development of technologies for the practical use of these models as well as the study of systems which uncertain parameters are given by multilevel models.

On the other hand, the theory of inaccurate (coarse) sets, proposed by Zdzislaw Pawlak in the early 1980s, is capable of describing human capabilities in distinguishing a variety of the objects. As a rule, these possibilities are limited only in distinguishing classes of objects, and not the objects. Some elementary classes of such an indistinguishability relation may be incompatible, i.e. include objects that have the same description, but are assigned to different categories. Owing to the indistinguishability described above, it is impossible in general case to accurately describe a set of objects in terms of elementary sets of indistinguishable objects. In order to solve this problem, the concept of an inaccurate set is introduced as a pair of two sets—the lower and upper approximations, constructed from elementary sets of objects. This idea is a key to solving a lot of tasks, in particular, classification tasks, assessing dependencies between attributes and object classification, determining the degree of such dependence, calculating the importance of attributes, reducing the number of attributes and generating, deciding rules from the source data. The theory of inaccurate sets complements the theory of fuzzy sets and soft calculations, because deals with another kind of vagueness and inconsistency.

Fuzzy mathematics offers effective tools for building models of systems that operate under conditions of similar non-stationarity. Wide use of fuzzy mathematics is explained by the simplicity of determining the values that directly describe the non-stationarity of state of the environment or the system. At the same time, the necessary and sufficient way to determine these values is the assignment of the membership function.

Some useful definitions: The concept of a fuzzy set—this is attempt to mathematical formalize fuzzy information, for example, radar information, to build mathematical models. This concept is based on the idea that the components of a given set of elements, for example, radar marks with a common property, may have this property to various degrees and, therefore, belong to this set with various degrees. With this approach, statements such as "such a mark belongs to the given set" lose their meaning, since it is necessary to indicate "how strong" or with what degree a particular element (radar mark) satisfies the properties of this set.

Definition 1 A fuzzy set $\tilde{A}$ on a universal set U is a collection of pairs $(\mu_A(u), u)$, where $\mu_A(u)$ is the degree to which an $u \in U$ element belongs to the fuzzy set $\tilde{A}$. Membership is a number from the range [0, 1]. The higher the degree of belonging, the more the element of the universal set corresponds to the properties of a fuzzy set.

Definition 2 The membership function is a function that allows you to calculate the degree to which an element of a universal set belongs to a fuzzy set.

If a universal set consists of a finite number of $U = \{u_1, u_2, \ldots, u_k\}$ elements, then the fuzzy set $\tilde{A}$ is written like $\tilde{A} = \sum_{i=1}^{k} \mu_A(u)/u$. In the case of a continuous set U, this notation $\tilde{A} = \int_U \mu_A(u)/u$ is used. Note: the signs $\sum$ and $\int$ in these formulas mean a collection of pairs $\mu_A(u)$ and U.

In general, the F-transform (fuzzy transform) is a means of representing an infinite-dimensional function space using a finite-dimensional vector space by establishing the definite correspondence between the set of continuous real functions defined on a bounded interval and the set of vectors. Naturally, the inverse F-transform transforms obtained vectors into other continuous functions that approximate the original ones.

For example, in our case, there are radar data in the form of coordinates of marks, their angular size, as well as various types of virtual images formed on the basis of the primary processing of echo signal. The incoming data is a mixture of useful, various noise and interfering marks, that is, it is fuzzy samples and sets. Then it is possible to use fuzzy transform [4, 11] to detect functional dependencies among a mixture of marks and images. The dependencies are transformed into semantic attributes (Sect. 2.1): the Z_{nij}^k of signal repetition in the processing element a_{ij} (i, j—numbers of elements (discrete) in distance and azimuth; k—survey number, starting with the current survey); predicate attribute Z_{ncij} of signal repetition in the adjacent processing element; predicate attribute Z_{yij} of mark's leaving outwards the processing element during the survey period.

Fuzzy transform method. Let the transformed function of behavior of the radar mark or signal image in the form of $y = f(x)$ is a set on the limited interval [a, b], on which a set of fixed nodes $a \leq c_1 < c_2 < \ldots < c_j < \ldots < c_h \leq b$ is also set. Each of the nodes is associated with the membership function $\mu_j(x)$ so that their full set of $\mu_1(x) \ldots \mu_j(x) \ldots \mu_h(x)$ which can also be given in the vector form $\mu^h = (\mu_1(x) \ldots \mu_j(x) \ldots \mu_h(x))^T$, forms a fuzzy division of the interval [a, b].

In the theory of fuzzy transformation (F-transformation) of the membership function $\mu_j(x)$, which satisfies the requirements: $\mu_j(x)$: [a, b] $\rightarrow$ [0, 1], $\mu_j(c_j) = 1$; $\mu_j(x) = 0$ if x $\notin$ (c_{j-1}, c_{j+1}) i.e. membership functions have a compact base; $\mu_j(x)$ are continuous; $\mu_j(x)$ increases on the interval $[c_{j-1}, c_j]$ and decrease on $[c_j, c_{j+1}]$; and, finally, $\sum_{j=1}^{h} \mu_j(x) = 1$ are called basic.

Then the set of real numbers $F^h = \left(F^1, \ldots, F^j, \ldots, F^h\right)^T$ such that

$$F_j = \frac{\int_a^b f(x) \cdot \mu_j(x) \cdot dx}{\int_a^b \mu_j(x) \cdot dx}, j = 1, 2, \ldots, h \tag{10}$$

is called the integral fuzzy F- transformation of the function $f(x)$ relative to $\mu_1(x) \ldots \mu_j(x) \ldots \mu_h(x)$. This is function $f \in C[a, b]$ relative to $\mu_1 \ldots \mu_h$ designated $F_h[f]$. Then, according to the previously given definition, we can write $F_h[f] = [F_1, \ldots, F_h]$.

The elements $F_1, \ldots, F_h$ are called F-transform components. Figure 1 shows the component F_j of the fuzzy transform of the function $f(x)$, calculated from the basis function $\mu_j(x)$ of a triangular shape.

The useful properties of the F-transform components follow from [23–26]. One of the key properties concerns the problem: how accurate the original function f is represented by the F-transform? This can be shown by assumptions about the original function, by the F-transform components as a weighted average of the given function, where the weights are determined by the basic functions.

Consider the discrete case when the original function f is known (can be computed) only in the individual points $x(1), y(1), x(2), y(2), \ldots, x(k), y(k), \ldots, x(l), y(l), h \leq 1$.

In this case, it is appropriate to use a discrete transform in the form

$$F_j(l) = \frac{\sum_{k=1}^{l} y(k)\mu_j(x(k))}{\sum_{k=1}^{l} \mu_j(x(k))} = \frac{\sum_{k=1}^{l} f(x(k))\mu_j(x(k))}{\sum_{k=1}^{l} \mu_j(x(k))}, \ j = 1, 2, \ldots, h, \qquad (11)$$

Figure 5 shows the result of fuzzy transform of virtual image of radar data obtained from signal actual records.

Inverse F-transform is determined by the expression

$$y_F(k) = f_{F,h}(x(k)) = \sum_{j=1}^{h} F_j(l)\mu_j(x(k)) = (F^h(l))^T \mu^h(x(k)). \qquad (12)$$

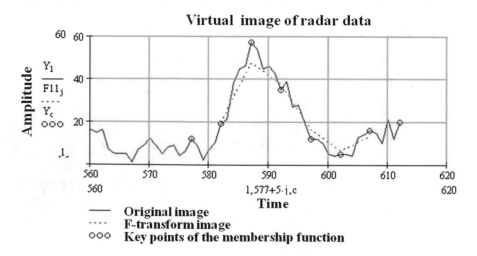

Fig. 5 The result of F-transform image based on real radar data

3.2 Adaptive Fuzzy Transform Based on the Artificial Neural Network Model

Note also that in the F-transform the nodes location c_j isn't connected with observations arrangement $x(k)$, and the transformation structure, can be referred to systems based on optimization, if it's considered from the position of learning neural networks.

An alternative to systems based on optimization training is so-called instantly trained systems (based on memory), at the heart of which is the concept of "neurons on data points". The most typical representative of such systems is the generalized regression neural network (GRNN) [27], which is widely used for the nonlinear objects identification and represents the problem of approximating functions from "noisy" data.

The basis of generalized regression neural networks is the idea of non-parametric estimation using Parzen windows [28], while for the training sample $\{x(k), y(k)\}$, $k = 1, 2, \ldots, l$ and the most popular cores Gauss network implements mapping

$$y_G(x) = \frac{\sum\limits_{k=1}^{l} y(k) \exp\left(-\frac{D^2(k)}{2\sigma^2}\right)}{\sum\limits_{k=1}^{l} \exp\left(-\frac{D^2(k)}{2\sigma^2}\right)}, \tag{13}$$

where $D(k) = |x - x(k)|$, σ is the width parameter chosen from empirical considerations.

Since, instead of Gaussians in the network, nuclear functions of the general form $\varphi(D(k), \sigma(k))$ can be used, the mapping (13) can be rewritten in a generalized form

$$y_G(x) = \frac{\sum\limits_{k=1}^{l} y(k) \varphi(D(k), \sigma(k))}{\sum\limits_{k=1}^{l} \varphi(D(k), \sigma(k))}, \tag{14}$$

where the functions $\varphi(\cdot)$ can have different width parameters. Learning GRNN is implemented very simply and consists in a single installation of the centers of kernel functions $\varphi(\cdot)$ at the observation points $x(k)$ (neurons at the data points). As noted in [3], the main advantages of GRNN are faster learning and convergence to an optimal non-linear regression curve.

In our case, when the number of cells in the radar or training sample l is large, the placement of neurons at data points makes the network too complex and can lead to dimensional curse. In this case, the centers of kernel functions $\varphi(\cdot)$ are proposed to be placed in prototype clusters, each of which is characterized by a certain radius of influence r. In this case, the number of clusters h can be much less than l.

The learning process of GRNN in this case includes the specification of cluster parameters in the form

$$
\begin{cases}
a_j(k) = a_j(k-1) + y(k), \ j = 1, 2, \ldots, h; \ k = 1, 2, \ldots, l, \\
b_j(k) = b_j(k-1) + 1, \\
c_j(k) = \frac{k-1}{k} c_j(k-1) + \frac{1}{k} x(k), \ |x(k) - c_j(k-1)| \leq r.
\end{cases}
\tag{15}
$$

$$
\begin{cases}
a_j(k) = a_j(k-1), \\
b_j(k) = b_j(k-1), \\
c_j(k) = c_j(k-1), \ |x(k) - c_j(k-1)| > r.
\end{cases}
\tag{16}
$$

If the characteristics of the approximated function change with time, then in the procedure (15) you can enter the forgetting procedure in the form

$$
\begin{cases}
a_j(k) = \frac{\tau-1}{\tau} a_j(k-1) + \frac{1}{\tau} y(k), \ j = 1, 2, \ldots, h; \ k = 1, 2, \ldots, l, \\
b_j(k) = \frac{\tau-1}{\tau} b_j(k-1) + \frac{1}{\tau}, \\
c_j(k) = \frac{\tau-1}{\tau} c_j(k-1) + \frac{1}{\tau} x(k), \ |x(k) - c_j(k-1)| \leq r,
\end{cases}
\tag{17}
$$

where τ is the time constant of the exponentially attenuating function. Procedure (16) remains unchanged.

In this case, the generalized regression neural network as a result of learning implements the mapping

$$
y_G(x) = \frac{\sum\limits_{j=1}^{h} a_j(l) \varphi_j(D_j, \sigma_j)}{\sum\limits_{j=1}^{h} b_j(l) \varphi_j(D_j, \sigma_j)} = \frac{\sum\limits_{j=1}^{h} a_j \varphi_j(D_j, \sigma_j)}{\sum\limits_{j=1}^{h} b_j \varphi_j(D_j, \sigma_j)},
\tag{18}
$$

where $D_j = |x_j - c_j|$, σ_j, the width parameter of the kernel function $\varphi(\cdot)$ of the jth cluster, and its architecture for scalar input and output signals is shown on Fig. 6.

When signal $x(p)$, not belonging to the training sample, is applied to the network input, a response appears at its output

$$
y_G(x(p)) = y_G(p) = \frac{\sum\limits_{j=1}^{h} a_j \varphi_j(D_j(p), \sigma_j)}{\sum\limits_{j=1}^{h} b_j \varphi_j(D_j(p), \sigma_j)},
\tag{19}
$$

where $D_j(p) = |x(p) - c_j|$ belongs to optimal regression curve approximating the original function.

Consider the architecture that implements the inverse F-transform (12), as shown on Fig. 7.

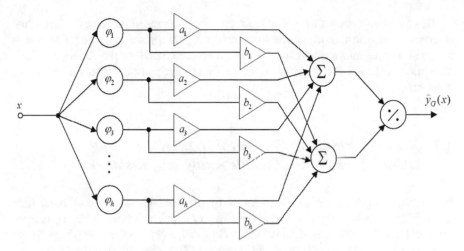

Fig. 6 The architecture of a generalized regression neural network (GRNN)

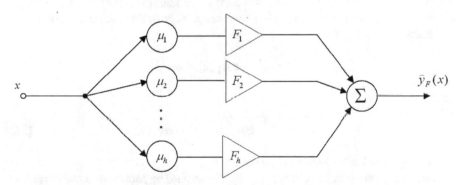

Fig. 7 Inverse F-transform architecture

Approximating the original function, but, naturally, different from (19). At the same time, it can be noted that if instead of expression (15), to refine the parameters of the clusters, use the

$$\begin{cases} a_j(k) = \frac{k-1}{k}a_j(k-1) + \frac{1}{k}y(k), \ j = 1, 2, \dots, h; \ k = 1, 2, \dots, l, \\ b_j(k) = \frac{k-1}{k}b_j(k-1) + \frac{1}{k}, \\ c_j(k) = \frac{k-1}{k}c_j(k-1) + \frac{1}{k}x(k), \ |x(k) - c_j(k-1)| \le r, \end{cases} \quad (20)$$

instead of nuclear functions $\varphi(\cdot)$, basic membership functions of $\mu_j(x)$, taking into account the fact that transformations (19) and (20) become structurally equivalent, although the first of them was obtained as a result of applying the procedure based on memory, and the second—as a result of solving the optimization problem.

This similarity allows us to construct an adaptive procedure for calculating the discrete F-transform, in which the components F_j are refined in real time as data is received, taking into account the possible non-stationarity of the functions $f(x)$. The initial data are the required accuracy of approximation ε and the radius of influence r.

3.3   Research Results of the Neural Network Learning Optimization in the Semantic Analysis of Radar Data

The neural network learning optimization in the semantic analysis of radar data is as follows. After the first radar data for two points of the training sample $x(1)$, $y(1)$, $x(2)$, $y(2)$, it is assumed that $x(1) = c_1$, $x(2) = c_2$ and two basic functions $\mu_1(x)$ and $\mu_2(x)$ are formed. Next point $x(3)$ is checked of belonging to the zone of influence c_1 and c_2. If this point is outside the influence zones c_1 and c_2 ($|x(3) - c_1| > r$, $|x(3) - c_2| > r$), the third basic function $\mu_3(x)$ is formed, but if $x(3)$ is in the influence zone of one of centers c_j, its position is refined according to expression

$$c_j(3) = (1 - \alpha)c_j(2) + \alpha x(3) \tag{21}$$

or for kth iteration

$$c_j(k) = (1 - \alpha)c_j(k - 1) + \alpha x(k), \tag{22}$$

where $0 \leq \alpha \leq 1$ is the smoothing parameter.

Generating of the basic functions $\mu_j(x)$, the smoothing parameter α and the radius of influence r takes into account current values of semantic attributes Z_{nij}, Z_{yij}, and value of the signal in processing element (neuron). The type of basic functions or membership functions $\mu_j(x)$ differs for the semantic attribute Z_{nij} (typical for clutter marks of "angel-echo" type) and for the semantic attribute Z_{yij} (typical for air object marks). The value of the smoothing parameter α and influence radius (cluster) r depends on measured and current value of the signal in processing element (neuron).

Adaptive change (learning) of semantic attributes Z_{nij} and Z_{yij}, the value of the signal in processing element (neuron) is made according to (3) and (4) by rewriting them in the same memory cells in which it was recorded in previous reviews. The basis for saving by rewriting the value of a sign is either its update when corresponding conditions (2) are or presence of non-zero signal amplitude level in the considered cell, or in the case when the first two conditions are simultaneously.

The basis for generating of an adaptive attenuating signal, decreasing from survey to survey by Δ value, is no signal coming in the current survey (4). In the case of no signal in the current survey, the value q_{ij} recorded in the previous survey is reduced by certain amount Δ. This ensures the adaptation of attributes for a time proportional to the attenuation of the signal in this processing element, value Δ depends on the type of semantic attribute.

This process continues the training set is finished, or continuously in real time. As a result of this training is obtained

$$\frac{b-a}{2r} \leq h \leq \frac{b-a}{2r} + 1 \tag{23}$$

When the training set is finished, the condition is checked.

$$\left| f(x(k)) - f_{F,h}(x(k)) \right| < \varepsilon, \tag{24}$$

where ε is a small positive number, i.e. the inverse F-transform approximates the original function arbitrarily accurately.

If expression (24) is true, then r and h are chosen correctly. If required accuracy is not provided, the radius of influence r is reduced and tuning is performed with a larger number of clusters h. Note also that with such an approach it's not necessary to set the interval $[a, b]$, but instead to use the minimum and maximum values from the training sample $x(k)$. This approach allows to associate the location of F-transform nodes c_j with the function behavior being approximated and to change their number h depending on the required accuracy ε. Here, a structural similarity is established between the discrete F-transform and the generalized regression neural network. An adaptive modification of F-transform is proposed, which allows sequentially processing radar data and changing a number of nodes in the learning process.

4 Conclusion

Thus, use of the proposed semantic technology in surveillance radars can increase their effectiveness to detect low-sized air objects. This technology is based on methods of artificial intelligence, which is based on semantic analysis of radar information by generating spatial-temporal images using algebra of finite predicates. This approach makes it possible to realize a technology of detecting low-visible objects in real time when radar information is compressed, while preserving its completeness and speed of semantic analysis.

The following results were obtained during the research:

- further development of data structuring methods and formalization processes of perception and transformation of signals and signal images based on algebra of finite predicates for the definition of semantic components of radar signals;
- development of methods for training the processing system depending on the signal-interference situation on the basis of spatial-semantic and spectral-semantic signal models, that allows to simplify the processing procedure taking into account non-stationary and non-linear dynamics of the process;

- using a fuzzy transform to analyze radar data in the form of marks coordinates and their virtual images, which are formed during training in the space of useful and interfering signals, and are considered as fuzzy samples and sets. It is shown how this approach can be used to identify the functional (semantic) dependencies between radar data and image components.

References

1. Skolnik MI (2008) Radar handbook, 3rd edn. McGraw-Hill, New York, USA
2. Jian L, Hummel R, Stoica P, Zelnio EG (2013) Radar signal processing and its applications. Springer
3. Guz' VI et al (2007) Peculiarities of airway processing when tracking air targets observed at small eleveation over underlying surface. Radioelectron Commun Syst 50(1):9–16. https://doi.org/10.3103/S0735272707010025
4. Bondarenko MF, Rusakova NE, Shabanov-Kushnarenko JuP (2010) O mozgopodobnyh strukturah [About brainlike structures]. Bionika intellekta 2:68–73 (In Russian)
5. Kryvinska N, Zinterhof P, van Thanh D (2007) New-emerging service-support model for converged multi-service network and its practical validation. In: First international conference on complex, intelligent and software intensive systems (CISIS'07). IEEE, pp 100–110. https://doi.org/10.1109/cisis.2007.40
6. Ageyev D, et al (2018) Classification of existing virtualization methods used in telecommunication networks. In: Proceedings of the 2018 IEEE 9th international conference on dependable systems, services and technologies (DESSERT), pp 83–86
7. Ageyev DV, Salah MT (2016) Parametric synthesis of overlay networks with self-similar traffic. Telecommun Radio Eng (English translation of Elektrosvyaz and Radiotekhnika) 75(14):1231–1241
8. Ageyev D, et al (2018) Method of self-similar load balancing in network intrusion detection system. In: 2018 28th international conference Radioelektronika (RADIOELEKTRONIKA). IEEE, pp 1–4. https://doi.org/10.1109/radioelek.2018.8376406
9. Kirichenko L, Radivilova T, Bulakh V (2020) Binary classification of fractal time series by machine learning methods. In: Lytvynenko V, Babichev S, Wójcik W, Vynokurova O, Vyshemyrskaya S, Radetskaya S (eds) Lecture notes in computational intelligence and decision making. ISDMCI 2019. Advances in intelligent systems and computing, vol 1020. Springer, Cham
10. Bulakh V, Kirichenko L, Radivilova T (2019) Classification of multifractal time series by decision tree methods. CEUR Work Proc 2105:457–460. http://ceur-ws.org/Vol-2105/10000457.pdf
11. Bondarenko MF, Shabanov-Kushnarenko JuP, Sharonova NV (2010) Situacionno-tekstovyj predikat [Situational-text predicate]. Bionika intellekta 3:20–25 (In Russian)
12. Aho AV, Hopcroft JE, Ullman JD (1974) The design and analysis of computer algorithms. Addison-Wesley, Reading, MA
13. Charniak E, Wilks Y (1976) Computational semantics: an introduction to artificial intelligence and natural language comprehension. North-Holland, Amsterdam
14. Chetverikov G, Leschinskaya I, Vechirskaya I (2009) Methods of synthesizing reversible spatial multivalued structures of language systems. Inf Sci Comput 15:32–39
15. Deschamps J, Bioul G, Sutter G (2006) Synthesis of arithmetic circuits. John Wiley, Hoboken (New Jersey)
16. Zhuravlev Y, Aslanyan L, Ryazanov V (2014) Analysis of a training sample and classification in one recognition model. Pattern Recognit Image Anal 24(3):347–352

17. Solonskaya S, Zhirnov V (2018) Intelligent analysis of radar data based on fuzzy transforms. Telecommun Radio Eng 77(15):1321–1329
18. Russell S, Norvig P (2010) Artificial intelligence: a modern approach. 3rd edn. Pearson
19. Luger G (2005) Artificial intelligence: structures and strategies for complex problem-solving. 4th edn. Williams
20. Chen K-M, Huang Y, Zhang J, Norman A (2000) Microwave life-detection systems for searching human subjects under earthquake rubble or behind barrier. IEEE Trans Biomed Eng 47(1):105–114
21. Pawlak Z (1985) Rough sets and fuzzy sets. Fuzzy Sets Syst 17(1):99–102
22. Solonskaya S, Zhirnov V (2018) Signal processing in the intelligence systems of detecting low-observable and low-doppler aerial targets. Telecommun Radio Eng 77(20):1827–1835. https://doi.org/10.1615/TelecomRadEng.v77.i20.50
23. Zadeh L (1965) Fuzzy sets. Inf Control 8(3):338–353
24. Perfilieva I (2006) Fuzzy transforms: Theory and applications. Fuzzy Sets Syst 157(8):993–1023. https://doi.org/10.1016/j.fss.2005.11.012
25. Specht D (1991) A general regression neural network. IEEE Trans Neural Netw 2(6):568–576
26. Parzen E (1962) On estimation of a probability density function and mode. Ann Math Stat 33(3):1065–1076. https://doi.org/10.1214/aoms/1177704472
27. Zhirnov V, Solonskaya S, Zima I (2014) Application of wavelet transform for generation of radar virtual images. Telecommun Radio Eng 73(17):1533–1539. https://doi.org/10.1615/TelecomRadEng.v73.i17.20
28. Zhirnov V, Solonskaya S, Zima I (2016) Magnetic and electric aspects of genesis of the radar angel clutters and their virtual imaging. Telecommun Radio Eng 75(15):1331–1341. https://doi.org/10.1615/TelecomRadEng.v75.i15.20

Research of Dynamic Processes in the Deterministic Chaos Oscillator Based on the Colpitts Scheme and Optimization of Its Self-oscillatory System Parameters

Andriy Semenov, Oleksandr Osadchuk, Olena Semenova, Serhii Baraban, Oleksandr Voznyak, Andrii Rudyk, and Kostyantyn Koval

Abstract The results of researching the modern methods and devices of deterministic chaos signal generation, which are constructed on the Colpitts scheme, for the infocommunication systems are given. The main variants for circuitry solutions of the Colpitts chaotic oscillators and their mathematical models are investigated. It is proposed to use voltage-controlled transistor capacitance equivalents on the basis of bipolar transistor structures with negative impedance as capacitive elements of oscillating systems of the Colpitts scheme-based deterministic chaos oscillators. The elements of the theory of the Colpitts scheme-based deterministic chaos oscillators with single-transistor and multi-transistor active elements are presented. The chaotic dynamics of generated electric oscillations and their statistical and informational properties arc investigated. Parameters of oscillating systems of the single-transistor and two-transistor Colpitts oscillators for the Kolmogorov-Sinai maximum were optimized. It is established that the maximum Kolmogorov-Sinai entropy is $H = 0.1292$

A. Semenov (✉) · O. Osadchuk · O. Semenova · S. Baraban · O. Voznyak · K. Koval
Vinnytsia National Technical University, Vinnytsia, Ukraine
e-mail: semenov.a.o@vntu.edu.ua

O. Osadchuk
e-mail: osadchuk.av69@gmail.com

O. Semenova
e-mail: Helene_S@ukr.net

S. Baraban
e-mail: serg@politex.org.ua

O. Voznyak
e-mail: alex.voz1966@gmail.com

K. Koval
e-mail: kkoval@vntu.edu.ua

A. Rudyk
National University of Water and Environmental Engineering, Rivne, Ukraine
e-mail: a.v.rudyk@nuwm.edu.ua

© The Editor(s) (if applicable) and The Author(s), under exclusive license to Springer Nature Switzerland AG 2021
T. Radivilova et al. (eds.), *Data-Centric Business and Applications*, Lecture Notes on Data Engineering and Communications Technologies 48,
https://doi.org/10.1007/978-3-030-43070-2_10

for the single-transistor Colpitts oscillator and H = 0.1642 for the two-transistor one with the fractal Hausdorff dimension of $d_F = 2.1123$ and $d_F = 2.6293$ respectively.

Keywords Deterministic chaos · Colpitts oscillator · Mathematical model

1 Introduction

The development of information and telecommunication technologies is carried out in two interrelated areas—hardware and software [1–4]. The current level of development of radio equipment and telecommunications (hardware) is due to the limit technology of the manufacture of semiconductor devices [5–7]. Therefore, considerable attention is paid to the study of the physical phenomena of semiconductor electronics and the creation of new and improved methods of constructing radioengineering devices and telecommunication facilities [7, 8].

A new direction in the development of infocommunication technologies, radioengineering and communications is the use of chaotic signals. Over the past 30 years, the number of scientific publications has expanded rapidly, which cover the theoretical aspects of nonlinear dynamic systems and the results of their practical application. Significant interest in the phenomenon of deterministic chaos is due to the following reasons: (1) the simplicity of the implementation of devices of deterministic chaos with a wide range of operating modes; (2) high sensitivity of devices of deterministic chaos; (3) small size, weight and cost [9–12].

Chaotic oscillations are characterized by the following basic properties [13–15]: (1) non-periodicity and wide band of spectrum frequencies; (2) weak correlation and orthogonality; (3) considerably greater noise immunity than in periodic oscillations; (4) high information capacity.

Chaotic signals are the type of signals that can be used as a media carrier in ultra-wideband communication media [16–18]. They are recommended by the IEEE 802.15.4a standard for use in UWB WPAN (Ultra Wide Band Wireless Personal Area Networks) and UWB WMAN (WMA Bandwidth) for use in ultra-wideband wireless personal networks [19–21]. Therefore, the development of devices for generating and generating signals of deterministic chaos is an actual scientific and technical task.

Devices for generation and formation of signals of deterministic chaos must have effective energy consumption. They must form signals of deterministic chaos in a given band of frequencies and have small radiation out of bandwidth [19–21]. Schematic solutions of generators of deterministic chaos should be such that in the long term they could be implemented as integral chips based on silicon technology [18].

2 Electric Circuits and Mathematical Models of Single-Transistor Generators of Deterministic Chaos Based on Colpitts Scheme

The most common and fully investigated transistor generator of deterministic chaos is the single-transistor Colpitts oscillator [22–33]. There are two main schemes of a single-transistor Colpitts generator—with a symmetrical (Fig. 1a) and asymmetric (Fig. 1b) power supply. The basic frequency of the generated signal is determined by the parameters of the reactive elements of the circuit of the single-transistor Colpitts oscillator [30].

$$\omega_0 = \frac{1}{\sqrt{L \frac{C_1 C_2}{C_1 + C_2}}}. \tag{1}$$

The main disadvantage of this generator is the low stability of the chaotic mode, which requires the use of precision capacitors. The paper proposes the use of voltage-controlled transistor capacitance equivalents based on a bipolar transistor structures with negative differential resistance, the schemes of which are shown in Fig. 2 [34].

The shorter system of Kirchhoff equations for the scheme of a single-transistor Colpitts oscillator with symmetrical power supply (Fig. 1a) has the following form [30]:

$$\begin{cases} C_{1eql} \frac{dV_{CE}}{dt} = I_L - I_C, \\ C_{2eql} \frac{dV_{BE}}{dt} = -\left(\frac{V_E + V_{BE}}{R_E} + I_L + I_B \right), \\ L \frac{dI_L}{dt} = V_C - V_{CE} + V_{BE} - I_L R_L. \end{cases} \tag{2}$$

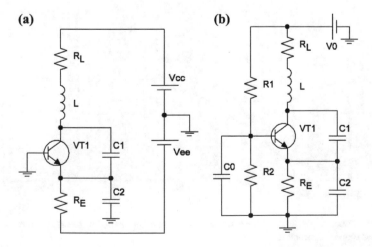

Fig. 1 Electrical circuits of a single-transistor Colpitts generator with a symmetric (**a**) [22, 30] and asymmetric (**b**) [31] power supply

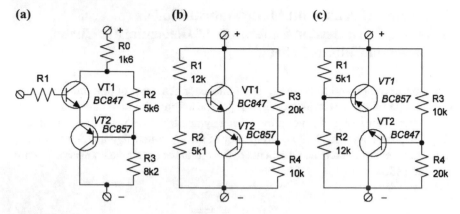

Fig. 2 Electrical circuits of voltage-controlled transistor capacitance equivalents on the basis of bipolar transistor structures with negative resistance: **a** similar to C_1 (Fig. 3.1a,b), **b** similar to C_2 (Fig. 3.1a), **c** similar to C_2 (rice 3.1b) [34]

where V_{CE}, V_{BE} is a collector-emitter and base-emitter voltages; I_L, I_C, I_B is are currents of inductance, collector and base of bipolar transistor, C_{1eql}, C_{2eql} is an equivalent capacitance of bipolar transistor structures in Fig. 2. Graphs of reactive properties of a bipolar transistor structure in Fig. 2a is shown in Fig. 3, and its equivalent capacity in Fig. 4 [34].

At low frequencies, the study of a bipolar transistor is carried out using a family of its volt-ampere characteristics. In high signal mode, the model of a bipolar transistor is reduced to the form of a two-segment voltage-controlled piece-to-line resistor and a linear current-controlled current source [30]:

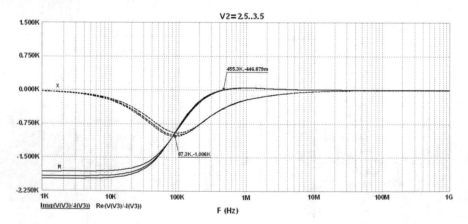

Fig. 3 Frequency dependence of the active and reactive components of the complete resistance of the bipolar transistor structure

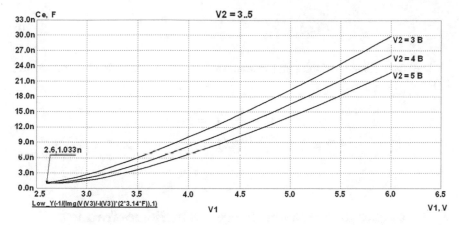

Fig. 4 Dependence of the equivalent capacitance of the bipolar transistor structure on the voltage V_1 at different voltage values V_2

$$I_C = \beta I_B, \quad I_B = \begin{cases} 0, & V_{BE} \leq V_0, \\ (V_{BE} - V_0)/R_1, & V_{BE} > V_0, \end{cases} \tag{3}$$

where V_0 is a threshold voltage of emitter p-n transition; R_1 is a resistance of emitter p-n transition in low signal mode; β is a current gain of bipolar transistor.

The mathematical model of a single-transistor Colpitts oscillator with symmetric power supply in physical variables (not normalized) has the form [30]

$$\begin{cases} \frac{dx_1}{dt} = \frac{1}{C_{1eql}} x_3 - \frac{\beta}{C_{1eql}} h(x_2), \\ \frac{dx_2}{dt} = -\frac{1}{C_{2eql} R_E} x_2 - \frac{V_E}{R_E C_{2eql}} - \frac{1}{C_{2eql}} x_3 - \frac{1}{C_{2eql}} h(x_2), \\ \frac{dx_3}{dt} = \frac{V_K}{L} - \frac{1}{L} x_1 + \frac{1}{L} x_2 - \frac{R_L}{L} x_3, \end{cases} \tag{4}$$

where $V_{CE} = x_1$, $V_{BE} = x_2$ and $I_L = x_3$.

The mathematical model of a single-transistor Colpitts oscillator with asymmetric power supply (Fig. 3.1b) in normalized (dimensionless) variables has the form [32, 33]

$$\begin{cases} \frac{dx_1}{dT} = x_2 - a \cdot F(x_3) - \sqrt{2Dn}(t), \\ \frac{dx_2}{dT} = c - x_1 - bx_2 - x_3, \\ \varepsilon \frac{dx_3}{dT} = x_2 - dx_3, \end{cases} \tag{5}$$

where $n(t)$ is a normalized source of white Gaussian noise, those parameters are $\langle n(t) \rangle = 0$, $\langle n(t)n(t - \tau) \rangle = \delta(\tau)$, D is an intensity level of noise in dimensionless quantities ($0 < D \leq 1$) [35], $F(x_3)$ is a function of approximation of collector current, that in the normalized variables has the form [32, 33]

$$F(x_3) = \begin{cases} e - 1 - x_3, & x_3 < e - 1, \\ 0, & x_3 \geq e - 1. \end{cases} \tag{6}$$

The standardized variables and system coefficients (5) are defined as follows [32, 33]

$$x_1 = \frac{V_{C1}}{V^*}, x_2 = \frac{\rho I_L}{V^*}, x_3 = \frac{V_{C2}}{V^*}, T = \frac{t}{\tau}, \dot{u} \equiv \frac{du}{dT}, \rho = \sqrt{\frac{L}{C_1}},$$
$$\tau = \sqrt{LC_1}, \varepsilon = \frac{C_{2eql}}{C_{1eql}}, a = \frac{\rho}{r}, b = \frac{R_L}{\rho}, c = \frac{V_0}{V^*}, d = \frac{\rho}{R_E}, e = \frac{R_2}{R_1 + R_2} c. \tag{7}$$

3 Investigation of Dynamic Processes in Single-Transistor Generators of Deterministic Chaos Based on the Colpitts Scheme

The results of simulation of the Colpitts generator with asymmetrical power supply on the bipolar transistor 2N3904 with the following nominal values of passive elements are given in Figs. 5, 6, and 7: L = 850 μH, $C_1 = C_2 = 470$ nF, $C_0 = 47$ μF, R = 36 Ω, $R_e = 510$ Ω, $R_1 = R_2 = 3$ kΩ, $V_0 = 15$ V. For such denominations, the normalized coefficients of the system of differential Eq. (5) have the following values: ε = 1, a = 30, b = 0.8, c = 20, d = 0.08, e = 10, D = 0 [32, 33].

When statistical processing of random signals observed over a period of time from 0 to T, an approximate mutual correlation function is used [36]

$$R_{X1X2} = \frac{1}{T - \tau} \int_0^{T-\tau} x_1(t) x_2(t + \tau) dt, \tag{8}$$

at $0 \leq \tau << T$. Approximately R_{X1X2} is calculated as the sum of the multiplies of discrete samples $X1_k$ and $X2_{k+n}$ random processes $X1(t)$ and $X2(t)$ [36]

$$R_{X1X2}(n\Delta t) = \frac{1}{N - n + 1} \sum_{k=0}^{N-n} X1_k X2_{k+n}, \tag{9}$$

where Δt is a sampling step, $n = 0, 1, 2, \ldots, M$ and $M << N$.

The dimensionless coefficient of mutual correlation is calculated by the formula [36]

$$\rho = \frac{R_{X1X2}(\tau)}{\sigma_{X1} \sigma_{X2}}, \tag{10}$$

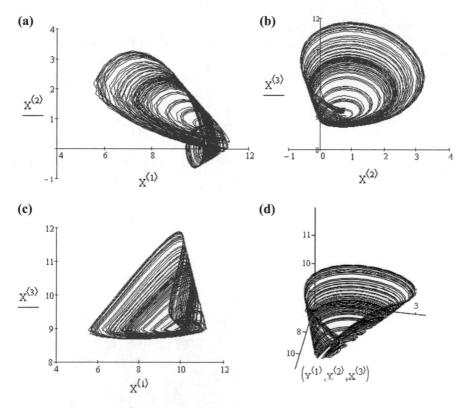

Fig. 5 Phase portraits of the Colpitts oscillator in the planes of dynamic variables x_1–x_2 (**a**), x_2–x_3 (**b**), x_1–x_3 (**c**) and within the limits of dynamic variables x_1–x_2–x_3 (**d**), where $X^{<1>}$, $X^{<2>}$, $X^{<3>}$ are column vectors of values corresponding to dynamic variables x_1, x_2, x_3 [33]

where σ_{X1} and σ_{X2} is a rejection of random variables $X1(t)$ and $X2(t)$ from their mathematical expectations, respectively.

Coefficients of mutual correlation of chaotic signals of a single-transistor generator of deterministic chaos based on Colpitts scheme are calculated using the standard function corr(X1, X2) of the program MathCad 15.0 and are [36]

$$\rho(X1, X2) = -0.69375, \ \rho(X1, X3) = 0.21144, \ \rho(X2, X3) = 0.06956. \quad (11)$$

Investigation of the dynamics of chaotic oscillations in the scheme of the Colpitts generator with symmetrical power supply was carried out by means of the virtual laboratory NI Multisim 10.1 (Fig. 8) [22, 37]. Results of simulation are presented in (Fig. 9).

Fig. 6 Diagrams of normalized dynamic variables of the oscillator system of the Colpitts generator with respect to the normalized time $T = t / \sqrt{LC_1}$: x_1 (**a**), x_2 (**b**), x_3 (**c**) [33]

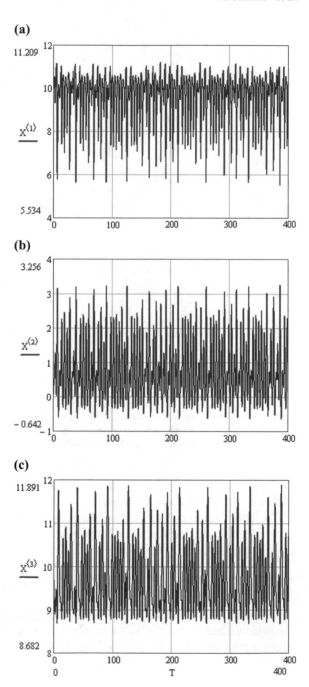

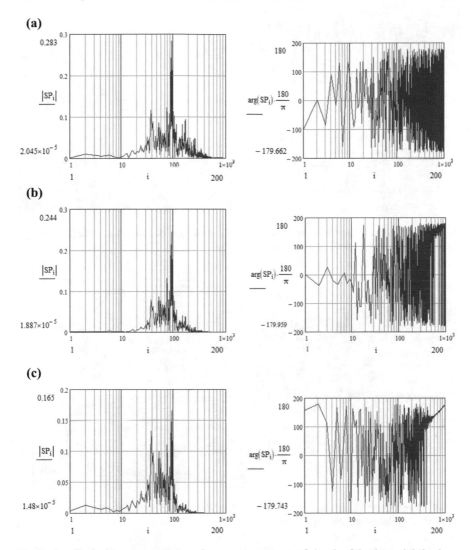

Fig. 7 Amplitude-frequency and phase-frequency spectrums of signals of the deterministic chaos of the single-transistor Colpitts oscillator [33]: **a** variable x_1, **b** variable x_2, **c** variable x_3 (on the x-axis i—is the harmonic number relative to the normalized frequency $\omega = 1/\sqrt{LC_{1eql}}$, along the axis of the ordinate $|SP_i|$—the module of the corresponding normalized dynamic variable and its argument in degrees for the AFS (to the left) and the PFS (to the right) respectively)

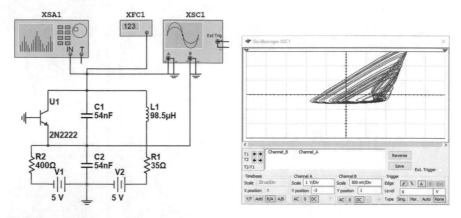

Fig. 8 Electrical scheme of the deterministic chaos Colpitts generator with symmetric power supply (left) [37] and its phase portrait (right)

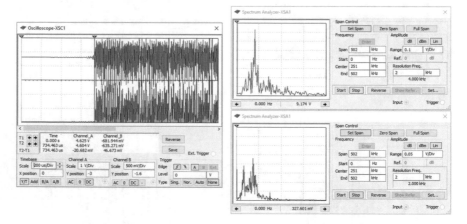

Fig. 9 Oscillograms of generated oscillations (left) of chaotic voltage v_{C1} (above) and v_{C2} (below) and their amplitude-frequency spectrum (right)

4 Optimization of the Parameters of a Oscillator System of Single-Transistor Generators of Deterministic Chaos Based on Colpitts Scheme for Kolmogorov-Sinai Maximum Entropy

In order to optimize the parameters of single-transistor generators of deterministic chaos based on Colpitts scheme, it is necessary to determine the basic statistical and informational properties of the attractor of its oscillator system. The analysis of the state of deterministic chaos oscillators is performed by methods of phase plane [18]. Phase portraits of generators of deterministic chaos in planes and limits of dynamic variables have the appearance of a strange attractor [18]. The exponential increase in the distance between two initially close trajectories is determined by the characteristic parameters of Lyapunov [37]. The rate of difference between a large number of infinitely close trajectories is determined by the entropy of the dynamic system. That is, entropy is the amount of information that is needed to determine the location of the system in some of its state, is the degree of ignorance of the system [38].

In the theory of information, the concept of entropy is introduced for systems that may be in different states x_i with different probabilities $p_i = p(x_i)$ [39]. According to Shannon, the state of uncertainty in the information system is characterized by entropy, as [39]

$$H = -\sum_i p_i \log p_i. \tag{12}$$

Entropy—is the average amount of information per message and is an informative measure of the source of the message. In addition to the classical Shannon's entropy, for the analysis of the information properties of chaotic systems, we use the generalized Reny entropy H_q in order q [39]

$$H_q = \frac{1}{1-q} \log\left(\sum_i p_i^q\right). \tag{13}$$

With a boundary condition provided $q \to 1$ Eq. (13) converted into a kind of Eq. (12). That is, the Shannon entropy is a partial case of the generalized Reny entropy [39]. From the general theory of systems of dynamic chaos, the following types of entropy are known [36, 39]: (1) metric entropy (Kolmogorov-Sinai entropy); (2) the entropy of the cascade; (3) generalized entropy (Reny entropy); (4) topological entropy.

The entropy of a dynamical system is directly related to the Lyapunov's characteristic indices. At the same time, the negative character characteristics of Lyapunov do not contribute to the total value of entropy. The entropy of chaotic systems is determined only by the positive parameters of Lyapunov [40, 41]

$$H = \int_A \left[\sum_{\lambda_i(x)>0} \lambda_i(x)\right] dA. \tag{14}$$

In Eq. (14) are added all positive characteristic values of Lyapunov, and the integral is taken over some invariant region of the phase plane of a dynamical system. Entropy is understood as some characteristic of one stochastic component of the system. In this case, the characteristic values of Lyapunov does not depend on the trajectory of the phase portrait, and therefore the integral in (14) is equal to the one [39]. Thus, the Kolmogorov-Sinai chaotic system entropy value is determined by the algebraic sum of positive characteristic parameters of Lyapunov [39]

$$H = \sum_{\lambda_i(x)>0} \lambda_i(x). \tag{15}$$

A fractal dimension is applied for processing wideband chaotic signals of a complex form [9–13]. The concept of fractal dimension attractor is closely related to Lyapunov dimension [39], which is determined by the formula

$$d_{L1} = j + \sum_{i=1}^{j} \frac{\lambda_i}{|\lambda_{j+1}|}, \tag{16}$$

where all characteristic indicators of Lyapunov are arranged by the grow

$$\lambda_1 \geq \lambda_2 \geq \ldots \geq \lambda_n,$$

n is a dimension of phase plane, and the number j is determined by such conditions

$$\lambda_1 + \lambda_2 + \cdots + \lambda_j \geq 0; \quad \lambda_1 + \lambda_2 + \cdots + \lambda_{j+1} < 0.$$

For the dynamical system (5), among the six parameters (7) we select the coefficients ε and b, the change of which corresponds to the control of the dynamics of the generator by changing the equivalent capacitances of the C_{1eql} and C_{2eql} and the resistance of the resistor R_L, respectively. Charts of Lyapunov's indices depending on the change of coefficients ε and b of the oscillator system of a single-transistor generator of deterministic chaos based on the Colpitts scheme are shown in Fig. 10. Charts in Fig. 10 are obtained through a specialized program DEREK 3.0 [42]. DEREK software calculates a part of Lyapunov exponent spectra (not more than 4 first ones) using the Benettin numerical iteration algorithm [43, 44].

As can be seen from Fig. 10, the maximum value of the higher Lyapunov exponent is provided at $\varepsilon = 1.07$ and $b = 0.93$. Under these conditions, older values of Lyapunov has values that are calculated using the program DEREK 3.0 [36]

$$\lambda_1 = 1.1324112 \times 10^{-1}, \ \lambda_2 = 1.5944613 \times 10^{-2}, \ \lambda_3 = -1.0087567 \times 10^{0}.$$

Kolmogorov-Sinai entropy of chaotic signals of a single-transistor generator of deterministic chaos based on Colpitts scheme is [36]

$$H = \sum_{\lambda_i(x)>0} \lambda_i(x) = \lambda_1 + \lambda_2 = 0.1292,$$

and it fractal dimension [36]

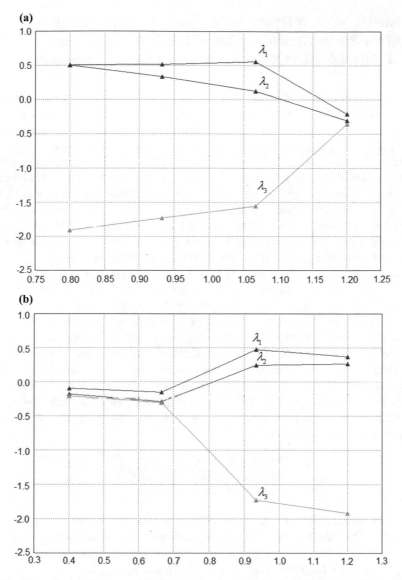

Fig. 10 Dependence of Lyapunov's indices on the change in the coefficients of a oscillator system of a single-transistor generator of deterministic chaos based on the Colpitts scheme: **a** ε (ε = 0.8 ... 1.2) and **b** b (b = 0.4 ... 1.2)

$$d_F = d_{L1} = 2 + \frac{\lambda_1}{|\lambda_3|} = 2, 1123.$$

5   Investigation of Dynamic Processes and Optimization of the Parameters of the Two-Transistor Generator of Deterministic Chaos Based on the Colpitts Scheme

The application of a chaotic Colpitts generator in the high frequency range imposes restrictions on the oscillation dynamics [33, 45]. The first studies of the chaotic mode of the Colpitts generator were conducted for the low-frequency range—tens of kilohertz [30]. Later, the study of dynamic processes in the Colpitts generator was carried out in the frequency range of HF—from 3 to 30 MHz [46, 47]. Chaotic oscillations with fundamental frequencies $f = 23$ MHz and $f = 26$ MHz were generated by the circuits using bipolar transistors 2N2222A and 2N3904 respectively [45]. The maximum frequency of these transistors is 300 MHz. When using microwave bipolar transistors AT41486 with a limiting frequency of 3 GHz and BFG520 with a limiting frequency of 9 GHz, chaotic oscillations were generated with the fundamental frequencies $f = 500$ MHz [45] and $f = 1.0$ GHz [47, 48] respectively.

Thus, experimentally it is confirmed that the main frequency of generated chaotic oscillations of the classical Colpitts generator is $f = 0.1 \cdot f_T$. Reduction of the influence of the parameters of the generator circuit and the load on the dynamics of microwave chaotic oscillations is ensured by the use of a gain cascade in accordance with the scheme of a common emitter [49]. Increasing the frequency of chaotic oscillations to the level $f = 0.3 \cdot f_T$ is provided by the use of a two-transistor active element [48, 50]. The electrical scheme of the two-transistor oscillator of the deterministic chaos of Colpitts is shown in Fig. 11 [50]. As the elements C_1–C_3 the author proposed to use transistor equivalent capacities on Fig. 2: C_1—Fig. 2a, C_2—Fig. 2c, and C_3—Fig. 2b. The main frequency of generation of a two-transistor chaos generator based on the Colpitts scheme (Fig. 11) is equal [50]

$$F = \frac{1}{2\pi} \sqrt{\frac{C_{1eql}C_2 + C_{1eql}C_{3eql} + C_{2eql}C_{3eql}}{L C_{1eql}C_{2eql}C_{3eql}} - \frac{R_L^2}{L^2}}. \tag{17}$$

The dynamics of the two-transistor Colpitts generator of the deterministic chaos is described by the system of the following ordinary differential equations [50]

$$\begin{cases} C_{1eql}\frac{dV_{C1}}{dt} = I_L - I_{EVT1}(r, V_{C2eql}, V_{C3eql}), \\ L\frac{dI_L}{dt} = V_0 - V_{C1eql} - V_{C2eql} - V_{C3eql} - R_L I_L, \\ C_{3eql}\frac{dV_{C3}}{dt} = I_L - I_{EVT2}(r, V_{C2eql}), \\ C_{2eql}\frac{dV_{C2}}{dt} = I_L - I_0. \end{cases} \tag{18}$$

When compiling the system (18) it was considered that [50]: current gain of the bipolar transistor $\alpha = 1$; differential resistance of the base-emitter transition R is a constant value (in practice, this can be ensured by selecting the bias current I_0).

In the normalized variables, the system of Eq. (18) will look [50]

Fig. 11 Electrical circuit of
a two-transistor chaos
generator based on Colpitts
scheme [50]

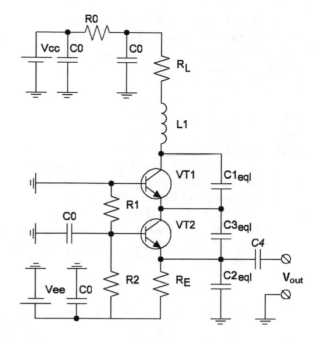

$$\begin{cases} \dot{x}_1 = x_2 - F_1(a, x_3, x_4), \\ \dot{x}_2 = -x_1 - x_3 - x_4 - bx_2, \\ \varepsilon_3 \dot{x}_3 = x_2 - F_2(a, x_4), \\ \varepsilon_2 \dot{x}_4 = x_2 - 1, \end{cases} \quad (19)$$

where

$$x_1 = \frac{V_{C1eql}}{\rho I_0}, \ x_2 = \frac{I_L}{I_0}, \ x_3 = \frac{V_{C3eql}}{\rho I_0}, \ x_4 = \frac{V_{C2eql}}{\rho I_0}, \ T = \frac{t}{\tau},$$
$$a = \frac{\rho}{r}, \ b = \frac{R_L}{\rho}, \ \rho = \sqrt{\frac{L}{C_{1eql}}}, \ \tau = \sqrt{LC_{1eql}}, \ \varepsilon_2 = \frac{C_{2eql}}{C_{1eql}}, \ \varepsilon_3 = \frac{C_{3eql}}{C_{1eql}}, \quad (20)$$

and the piecewise-linear functions of the VAC approximation of the base-emitter
transitions will look [50]

$$F_1(a, x_3, x_4) = \begin{cases} 1 - a(x_3 + x_4), & a(x_3 + x_4) < 1, \\ 0, & a(x_3 + x_4) \geq 1, \end{cases} \quad (21)$$

$$F_2(a, x_4) = \begin{cases} 1 - ax_4, & ax_4 < 1, \\ 0, & ax_4 \geq 1. \end{cases} \quad (22)$$

The experimental studies, the results of which are given in [50], has shown that the
main frequency of chaotic oscillations $f = 0.3 \cdot f_T$. In this case, the spectral density
graph of the signal has a greater uniformity in the frequency range from 250 МГц
to 1.1 ГГц [29, 51].

Further investigations of the modes of operation of the radiogenerator of deterministic chaos were carried out by the author on the basis of a mathematical model (19) and the approximations of current-voltage characteristic (21) and (22). All calculations and graphic dependencies were obtained by the author using the program package MathCad 15.0 [52].

As stated above, the proposed two-stage generator of deterministic chaos based on Colpitts scheme can operate in different modes depending on the values of system parameters (19). Research of dynamic processes is carried out in the mode of developed chaotic oscillations with the following values of parameters of an self-oscillator system a $= 11.5$, b $= 0.6$, $\varepsilon_2 = 4$, $\varepsilon_3 = 4$ for such initial conditions (1, 0, 0, 1) [52].

In Figs. 12 and 13 received phase portraits of a two-transistor generator of deterministic chaos based on the Colpitts scheme in different planes and limits of normalized variables are shown, respectively [52]. The time charts of the generated oscillations of the normalized variables of the dynamic system (19) of the two-transistor generator of deterministic chaos based on the Colpitts scheme and it amplitude-frequency spectrums are shown in Fig. 14 [52].

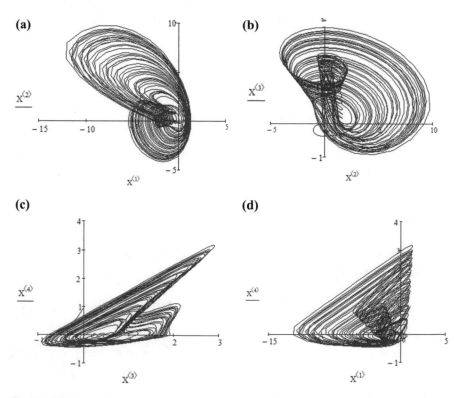

Fig. 12 Phase portraits of a two-transistor generator of deterministic chaos based on the Colpitts scheme in variable planes: **a** x_1-x_2, **b** x_2-x_3, **c** x_3-x_4, **d** x_1-x_4

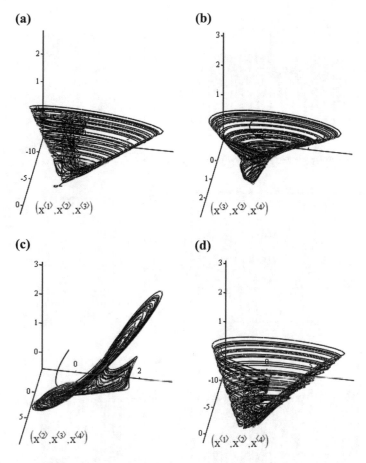

Fig. 13 Phase portraits of a two-transistor generator of deterministic chaos based on Colpitts scheme in different planes of normalized variables: **a** x_1–x_2–x_3, **b** x_3–x_2–x_4, **c** x_2–x_3–x_4, **d** x_1–x_2–x_4

Graphs of the highest Lyapunov's indexes of the dynamic system (19) obtained using DEREK 3.0 are shown in Figs. 15 and 16 [36]. Analysis of the graphs of Lyapunov's indicators confirms the steady work of the two-transistor Colpitts generator in the mode of developed chaos when changing the parameters of its oscillator system in wide range.

According to the results of the analysis of the Lyapunov's spectrums (Figs. 15 and 16), it can be concluded that the hyperchaotic mode in the oscillator system of the two-transistor Colpitts generator is impossible [53–56].

The optimization of the oscillator system of the two-transistor Colpitts generator is performed on the maximum of its information properties. Calculated using the DEREK 3.0 program, the higher figures of Lyapunov has corresponding values [36]

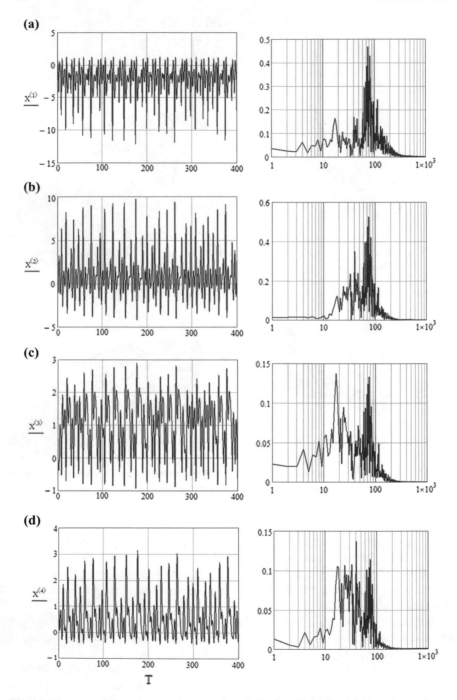

Fig. 14 Diagrams of generated chaotic oscillations (left) of normalized variables x_1 (**a**), x_2 (**b**), x_3 (**c**) and x_4 (**d**) in normalized time $T = t / \sqrt{LC_1}$ and it amplitude-frequency spectrums

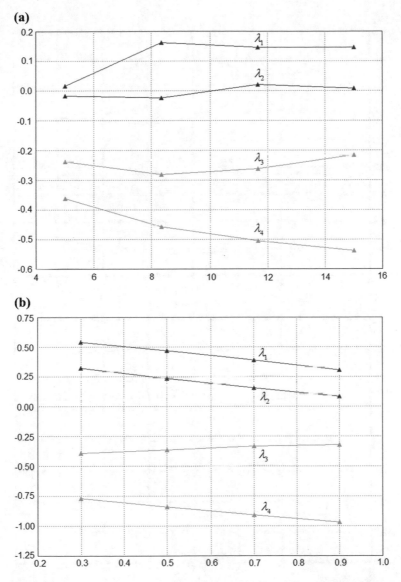

Fig. 15 Dependence of Lyapunov's indices on the change in the coefficients of a oscillator system of a two-transistor generator of deterministic chaos based on the Colpitts scheme (19): **a** a (a = 5.0 ... 15.0) and **b** b (b = 0.3 ... 0.9)

$$\lambda_1 = 1.4335192 \times 10^{-1}, \ \lambda_2 = 2.0865944 \times 10^{-2},$$
$$\lambda_3 = -2.6095239 \times 10^{-1}, \ \lambda_4 = -5.0326709 \times 10^{-1}.\text{'}$$

Kolmogorov-Sinai entropy of the two-transistor generator of deterministic chaos based on the Colpitts scheme is

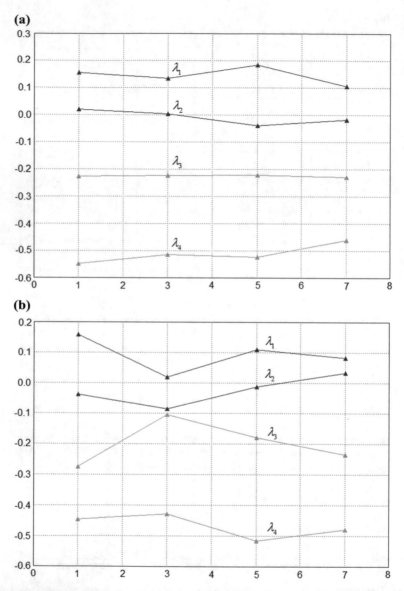

Fig. 16 Dependence of Lyapunov's indices on the change in the coefficients of a oscillator system of a two-transistor generator of deterministic chaos based on the Colpitts scheme (19): **a** ε_3 ($\varepsilon_3 = 1.0 \ldots 7.0$) and **b** ε_2 ($\varepsilon_2 = 1.0 \ldots 7.0$)

$$H = \sum_{\lambda_i(x)>0} \lambda_i(x) = \lambda_1 + \lambda_2 = 0.1642,$$

and the dimension of Lyapunov of its chaotic attractor is

$$d_{L1} = 3 + \frac{\lambda_1}{|\lambda_4|} + \frac{\lambda_2}{|\lambda_4|} = 3.3263.$$

Unlike chaotic systems of the third order, for which the numerical values of the Lyapunov's dimension and the fractal dimension coincide, for the systems of the fourth and higher orders, the Lyapunov's dimension gives the estimate of the maximum value of the fractal (Hausdorff) dimension [57–60].

$$d_F \le d_{L1},$$

the value of it equals

$$d_F = d_{L2} = 2 + \frac{\lambda_1}{|\lambda_3|} + \frac{\lambda_2}{|\lambda_3|} = 2.6293.$$

Calculated using the standard function corr(X1, X2) program MathCad 15.0 the coefficients of mutual correlation of chaotic signals of the two-transistor generator of deterministic chaos based on the Colpitts scheme are [36]

$$\rho(X1, X2) - -0.52975, \quad \rho(X1, X3) = 0.53036,$$
$$\rho(X1, X4) = 0.31557, \quad \rho(X2, X3) = -0.3676,$$
$$\rho(X2, X4) - 1.14945 \times 10^{-3}, \quad \rho(X3, X4) = 0.69758.$$

As follows from the obtained results, the smallest value of the correlation coefficient has the chaotic signals of the dynamic variables x_2 and x_4 (current of inductance and voltage on the capacity C_{2eql}, respectively) of the two-transistor generator of deterministic chaos based on the Colpitts scheme. Thus, the two-transistor generator of deterministic chaos based on the Colpitts scheme is capable of generating two information signals with a small correlation, and provided that it rapidly decay its autocorrelation functions, such signals are orthogonal [55, 56].

6 Conclusion

The paper presents the results of research of modern methods and devices for generating signals of deterministic chaos, which are constructed according to the Colpitts scheme. The main variants of circuitry solutions of chaotic generators based on the Colpitts scheme and their mathematical models are considered. The elements of the theory of generators of deterministic chaos based on the Colpitts scheme with one-transistor and two-transistor active elements are given.

The results of the calculation of statistical and informational parameters of chaotic signals of generators of deterministic chaos based on the Colpitts scheme shows that the entropy $H = 0.1642$ and the fractal dimension $d_F = 2.6293$ of the two-transistor generator of the deterministic chaos based on the Colpitts scheme are larger than the entropy $H = 0.1292$ and the fractal dimension $d_F = 2.1123$ one-transistor generator, with the chaotic electrical fluctuations of the current of inductance and voltage on the capacitor C_2 of the two-transistor generator has a low correlation coefficient, that allows to use the generator for the simultaneous generation of two orthogonal signals. The advantage of a single-transistor generator of deterministic chaos based on Colpitts scheme is that it is capable of generating signals at a higher base frequency than a two-transistor generator, while the last one is about 3 times the wider band of the amplitude-frequency spectrum of chaotic signals.

References

1. Skulysh M (2017) The method of resources involvement scheduling based on the long-term statistics ensuring quality and performance parameters. In: Proceedings of the 2017 international conference on information and telecommunication technologies and radio electronics, Odessa, Ukraine, 11–15 Sept 2017. https://doi.org/10.1109/ukrmico.2017.8095430
2. Globa L, Skulysh M, Reverchuk A (2014) Control strategy of the input stream on the online charging system in peak load moments. In: Proceedings of the 24th international crimean conference microwave and telecommunication technology, Sevastopol, Ukraine, 7–13 Sept 2014. https://doi.org/10.1109/crmico.2014.6959409
3. Skulysh M, Romanov O (2018) The structure of a mobile provider network with network functions virtualization. In: Proceedings of the 14th international conference on advanced trends in radioelecrtronics, telecommunications and computer engineering, Lviv-Slavske, Ukraine, 20–24 Feb 2018. https://doi.org/10.1109/tcset.2018.8336370
4. Globa L, Skulysh M, Zastavenko A (2015) The method of resources allocation for processing requests in online charging system. In: The experience of designing and application of CAD systems in microelectronics, Lviv, Ukraine, 24–27 Feb 2015. https://doi.org/10.1109/cadsm.2015.7230838
5. Pidchenko S, Taranchuk A (2016) Principles of quartz multifrequency oscillatory systems with digital compensation of temperature and vibrational instability frequency. In: Proceedings of the international conference radio electronics and infocommunications, Kiev, Ukraine, 11–16 Sept 2016. https://doi.org/10.1109/ukrmico.2016.7739620
6. Pidchenko SK (2013) Dual-frequency temperature-compensated quartz crystal oscillator. In: Proceedings of the 23rd international crimean conference microwave and telecommunication technology, Sevastopol, Ukraine, 8–14 Sept 2013, pp 669–670
7. Pidchenko S, Taranchuk A (2017) Synthesis of quartz measuring transducers with low Q— factor sensor element. In: Conference proceedings of 2017 IEEE 37th international conference on electronics and nanotechnology, Kyiv, Ukraine, 18–20 April 2017. https://doi.org/10.1109/elnano.2017.7939801
8. Kychak V et al (2010) Integrated microwave modulator based on induction of dynamic. In: Proceedings of the 2010 international conference on modern problems of radio engineering, telecommunications and computer science, Lviv-Slavske, Ukraine, 23–27 Feb 2010, p 72
9. Kennedy MP, Rovatti R, Setti G (2000) Chaotic electronics in telecommunications. CRC Press, Inc., Boca Raton, FL, USA

10. Stavroulakis P (2005) Chaos applications in telecommunications. CRC Press, Taylor and Francis, London
11. Jovic B (2011) Synchronization techniques for chaotic communication systems. Springer International Publishing, Switzerland
12. Lau FCM, Tse CK (2003) Chaos-based digital communication systems: operating principles, analysis methods, and performance evaluation. Springer Science and Business Media
13. Dmitriev AS, Kyarginsky BYe, Panas AI, Starkov SO (2003) Experiments on direct chaotic communications in microwave band. Int J Bifurc Chaos 13(6):1495–1507. https://doi.org/10.1142/s0218127403007345
14. Dmitriev AS et al (2003) Basic principles of direct Chaotic communications. Nonlinear Phenom Complex Syst 6(1):488–501. https://doi.org/10.1007/978-94-010-0217-2_3
15. Dmitriev AS, Panas AI, Zakharchenko KV (2003) Principles of direct chaotic communications. In: 2003 IEEE international workshop on workload characterization, vol. 2. Saint Petersburg, Russia, 20–22 Aug 2003, pp 475–483. https://doi.org/10.1109/phycon.2003.1236868
16. Vovchuk D, Haliuk S, Politanskii L (2014) Experimental research of the process of masking of digital information signals using chaotic oscillation. East Eur Sci J 3:245–253. https://doi.org/10.12851/EESJ201406C06ART09
17. Efremova EV (2018) Generation of dynamic chaos in a range of 10–30 GHz. J Commun Technol Electron 63(4):367–373. https://doi.org/10.1134/S1064226918040046
18. Anishchenko VS, Vadivasova TE, Strelkova GI (2014) Deterministic nonlinear systems. Springer International Publishing, Switzerland, A Short Course
19. IEEE Standard for Information Technology (2007) Telecommunications and in-formation exchange between systems. Local and metropolitan area networks. Specific requirements. Part 15.4: Wireless medium access control (MAC) and Physical layer (PHY) Specifications for low-rate wireless personal area networks (WPANs). Amendment 1: add alternate PHYs
20. Revision of part 15 of the commission's rules regarding ultra-wideband transmission systems, first report and order. Federal communications commission (FCC), ET Docket 98–153, FCC 02–48; Adopted: 14 Feb 2002. Released: 22 April 2002
21. Report from the commission to the european parliament and the council on the implementation of the radio spectrum policy programme, Brussels, 22 April 2014
22. Aissi C, Kazakos D (2006) A review of chaotic circuits, simulation and implementation. In: Proceedings of the 10th WSEAS international conference on CIRCUITS, Vouliagmeni, Athens, Greece, 10–12 July 2006, pp 125–131
23. Qiao S, Jiang T, Ran L et al (2007) Ultra-wide band noise-signal radar utilizing microwave chaotic signals and chaos synchronization. In: Progress in electromagnetics research symposium, Prague, Czech Republic, 27–30 Aug 2007, pp 503–506
24. Tsakiridis O, Syvridis D, Zervas E, Stonham J (2004) Chaotic operation of a Colpitts oscillator in the presence of parasitic capacitances. WSEAS Trans on Electron 1:416–421
25. Sarkar S, Sarkar S, Sarkar BC (2014) Dynamics of driven Colpitts oscillator in presence of co-channel tone interference: an experimental study. Int J Electron Appl Res 1:1–14
26. Buscarino A, Fortuna L, Frasca M et al (2010) Chaos control in inductor-based Chaotic oscillators. In: Proceedings of the 19th international symposium on mathematical theory of networks and systems, Budapest, Hungary, 5–9 July 2010, pp 2207–2210
27. Bonetti RC et al (2014) Super persistent transient in a master–slave configuration with Colpitts oscillators. J Phys A: Math Theor 47(40):405101, pp 1–12. https://doi.org/10.1088/1751-8113/47/40/405101
28. Kengne J, Chedjou JC, Kenne G et al (2012) Dynamical properties and chaos synchronization of improved Colpitts oscillators. Commun Nonlinear Sci Numer Simul 17(7):2914–2923
29. Prodyot KR, Arijit B (2003) A high frequency chaotic signal generator: a demonstration experiment. Am J Phys 71(1):34–37
30. Kennedy MP (1994) Chaos in the Colpitts oscillator. IEEE transactions on circuits and systems—I: fundamental theory and applications 11:771–774
31. Cenys A, Tamasevicius A, Baziliauskas A et al (2003) Hyperchaos in coupled Colpitts oscillators. J Chaos, Solitons Fractals 17:349–353

32. Li GH, Zhou SP, Yang K (2007) Controlling chaos in Colpitts oscillator. Chaos, Solitons Fractals 33:582–587
33. Semenov A (2016) Reviewing the mathematical models and electrical circuits of deterministic chaos transistor oscillators. In: Proceedings of the international siberian conference on control and communications, Moscow, Russia, 12–14 May 2016. https://doi.org/10.1109/sibcon.2016. 7491758
34. Osadchuk A et al (2008) Mathematical model of transistor equivalent of electrical controlled capacity. In: Proceedings of the international conference modern problems of radio engineering, telecommunication and computer science, Lviv-Slavske, Ukraine, 19–23 Feb 2008, pp 35–36
35. Semenov A (2016) The Van der Pol's mathematical model of the voltage-controlled oscillator based on a transistor structure with negative resistance. In: Proceedings of the XIII international conference modern problems of radio engineering, telecommunications, and computer science, Lviv-Slavsko, Ukraine, 23–26 Feb 2016, pp 100–104. https://doi.org/10.1109/tcset. 2016.7451982
36. Semenov A, Osadchuk O, Semenova O et al (2018) Signal statistic and informational parameters of deterministic chaos transistor oscillators for infocommunication systems. In: Proceedings of the 4th international scientific-practical conference problems of infocommunications science and technology, Kharkiv, Ukraine, 9–12 Oct 2018, pp 730–734. https://doi.org/10.1109/ infocommst.2018.8632046
37. Kuznetsov SP (2018) Simple electronic chaos generators and their circuit simulation. Izvestiya VUZ, Appl Nonlinear Dyn 26(3):35–61. https://doi.org/10.18500/0869-6632-2018- 26-3-35-61
38. Garrido PL (1997) Kolmogorov-Sinai entropy, Lyapunov exponents, and mean free time in billiard systems. J Stat Phys 88(3/4):807–824
39. Loskutov A (2010) Fascination of chaos. Phys Usp 53(12):1257–1280. https://doi.org/10.3367/ UFNr.0180.201012c.1305
40. Awrejcewicz J, Krysko AV, Erofeev NP et al (2018) Quantifying chaos by various computational methods. Part 1: simple systems. Entropy 20:175. https://doi.org/10.3390/e20030175
41. Radivilova T, Kirichenko L, Yeremenko O (2017) Calculation of routing value in MPLS network according to traffic fractal properties. In: Proceedings of the 2nd international conference on advanced information and communication technologies, Lviv, Ukraine, 4–7 July 2017, pp 250–253. https://doi.org/10.1109/aiact.2017.8020112
42. DEREK—research of dynamic systems. http://derek-ode.sytto.com. Accessed 18 Feb 2019
43. Bennetin G et al (1980) Lyapunov characteristic exponents for smooth dynamical systems and for Hamiltonian systems; a method for computing all of them. Part 1: theory. Meccanica 15(1):9–20. https://doi.org/10.1007/BF02128236
44. Bennetin G et al (1980) Lyapunov characteristic exponents for smooth dynamical systems and for Hamiltonian systems; a method for computing all of them. Part 2: numerical application. Meccanica 15(1):21–30. https://doi.org/10.1007/BF02128237
45. Wegener C, Kennedy MP (1995) RF Chaotic Colpitts oscillator. In: Proceedings of the 3rd international workshop on nonlinear dynamics of electronic systems, Dublin, Ireland, pp 255–258
46. Mykolaitis G, Tamaševičius A, Bumelienė S et al (2001) HF and VHF chaos oscillators. Electron Electr Eng 32:12–17
47. Tamaševičius A, Mykolaitis G, Bumelienė S et al (2001) Two-Stage Chaotic Colpitts Oscil Electron Lett 37(9):549–551. https://doi.org/10.1049/el:20010398
48. Bumelien S, Tamaševicius A, Mykolaitis G et al (2004) Hardware prototype of the two-stage Chaotic Colpitts oscillator for the UHF range. In: Proceedings of the 12th nonlinear dynamics of electronic systems, pp 99–102
49. Shi ZG, Ran LX (2006) Microwave chaotic Colpitts oscillator: design, implementation and applications. J Electromagn Waves Appl 20(10):1335–1349. https://doi.org/10.1163/ 156939306779276802
50. Mykolaitis G, Tamaševičius A, Bumelienė S et al (2004) Two-stage chaotic Colpitts oscillator for the UHF range. Elektronika Ir Elektrotechnika 53:13–15

51. Lindberg E, Murali K, Tamasevicius A (2008) The Colpitts oscillator family. In: Proceedings of the international symposium: topical problems of nonlinear wave physics, Nizhny Novgorod, NWP-1: Nonlinear Dynamics of Electronic Systems, NDES'2008, 20–26 July 2008, pp 47–48

52. Semenov A et al (2018) Mathematical modeling of the two-stage Chaotic Colpitis oscillator. In: Proceedings of the 14th international conference on advanced trends in radioelectronics, telecommunications and computer engineering, Lviv-Slavske, Ukraine, 20–24 Feb 2018, pp 835–839. https://doi.org/10.1109/tcset.2018.8336327

53. Zhang Z et al (2018) Design and nonlinear dynamical performance analysis of novel improved two-stage Colpitts wideband Chaotic oscillator. In: Proceedings of the 2018 IEEE 3rd international conference on image, vision and computing, Chongqing, China, 27–29 June 2018, pp 919–926. https://doi.org/10.1109/icivc.2018.8492812

54. Chen WL, Zheng LH, Song XX (2016) Design of two-stage chaotic Colpitts oscillator. In: Proceedings of the 2016 IEEE international conference on microwave and millimeter wave technology, Beijing, China, 5–8 June 2016, pp 1029–1031. https://doi.org/10.1109/icmmt.2016.7762523

55. Semenov A (2016) Mathematical simulation of the chaotic oscillator based on a field-effect transistor structure with negative resistance. In: 2016 IEEE 36th international conference on electronics and nanotechnology (ELNANO), Kyiv, Ukraine, 19–21 April 2016, pp 52–56. https://doi.org/10.1109/elnano.2016.7493008

56. Osadchuk VS et al (2010) Experimental research and modeling of the microwave oscillator based on the static inductance transistor structure with negative resistance. In: Proceedings of the 20th international crimean conference microwave and telecommunication technology (CriMiCo), Sevastopol, Ukraine, 13–17 Sept 2010, pp 187–188. https://doi.org/10.1109/crmico.2010.5632543

57. Kirichenko L, Radivilova T (2017) Analyzes of the distributed system load with multifractal input data flows. In: Proceedings of the 14th international conference the experience of designing and application of CAD systems in microelectronics, Lviv, Ukraine, 21–25 Feb 2017, pp 260–264. https://doi.org/10.1109/cadsm.2017.7916130

58. Radivilova T, Kirichenko L, Ivanisenko Igor (2016) Calculation of distributed system imbalance in condition of multifractal load. In: Proceedings of the third international scientific-practical conference problems of infocommunications science and technology, Kharkiv, Ukraine, 4–6 Oct 2016, pp 156–158. https://doi.org/10.1109/infocommst.2016.7905366

59. Ivanisenko I, Kirichenko L, Radivilova T (2016) Investigation of multifractal properties of additive data stream. In: Proceedings of the IEEE first international conference on data stream mining and processing, Lviv, Ukraine, 23–27 Aug 2016, pp 305–308. https://doi.org/10.1109/dsmp.2016.7583564

60. Ivanisenko I, Radivilova T (2015) The multifractal load balancing method. In: Proceedings of the second international scientific-practical conference problems of infocommunications science and technology, Kharkiv, Ukraine, 13–15 Oct 2015, pp 123–123. https://doi.org/10.1109/infocommst.2015.7357289

Approaches to Building a Chaotic Communication System

Mykola Kushnir, Dmytro Vovchuk, Serhii Haliuk, Petro Ivaniuk, and Ruslan Politanskyi

Abstract Nowadays are being held intensive researches involving the use of deterministic chaos in communication systems. Indeed chaotic oscillations in such systems serve as a carrier of information signals, and means of encryption, both in hardware and in software. This work is devoted to the complex analysis of the practical implementation of a chaotic secure communication system. The generalized synchronization as a synchronization response was selected. The circuit implementation of a modified Colpitts oscillator as a source of chaos was proposed. We also discussed questions of modeling chaotic oscillations and, in this context, analyzing the differences between chaotic, pseudo-chaotic and pseudo-random oscillations.

Keywords Deterministic chaos · Synchronization · Colpitts oscillator · Liu system · Lyapunov exponents · Pseudorandom sequences

1 Introduction

The discovery of deterministic chaos [1] (50 years ago) led to the fast growth of both fundamental and applied scientific investigations related to the description of the behavior of this nonlinear phenomenon [2–6]. When two important properties of chaotic oscillations had been discovered—namely synchronization and sensitive dependence on initial conditions [7], it became possible to investigate the usage of deterministic chaos in information systems [8–12]. At present, all works on using chaos in communication systems can be divided into the following four groups:

M. Kushnir · D. Vovchuk (✉) · S. Haliuk · R. Politanskyi
Yuriy Fedkovych Chernivtsi National University, Chernivtsi, Ukraine
e-mail: dimavovchuk@gmail.com

P. Ivaniuk
CheAI, Chernivtsi, Ukraine

T. Radivilova et al. (eds.), *Data-Centric Business and Applications*, Lecture Notes on Data
Engineering and Communications Technologies 48,
https://doi.org/10.1007/978-3-030-43070-2_11

207

- *generation and general properties of chaotic oscillations*—these are works devoted to both chaos generators of different dimensions (including hyperchaos) and basic research of new properties of deterministic (or dynamic) chaos;
- *synchronization and control of chaos*—these include studies that analyze the possibility of synchronous chaotic reviews without which it is impossible to realize a chaotic communication system. Nowadays the studies related to the release of unstable periodic orbits, which are also used in chaotic communication are developing intensively;
- *chaotic cryptography*—this is another important direction of the utilization of chaos in secure communication systems. First, chaotic encryption was performed on the logistic map based on the fact that a certain range of variation of chaotic amplitude corresponds to a particular character [13]. Modern methods of chaotic cryptography use two, three or more chaotic mappings and securely encrypt texts, images and videos in both on-line and off-line modes;
- *information systems using deterministic chaos*—here we should talk about several subgroups. First, all chaotic communication systems can be attributed to either analog or digital. In addition, the solutions can be represented both in hardware and software. Some groups include systems in integrated performance and chaotic communication based on symbolic dynamics. It is clear that all these subgroups can overlap.

The analysis of works on chaotic information systems of recent years makes possible to conclude several important outcomes. First, the predominant number of works devoted to information security via the chaos includes studies of chaotic cryptography implemented in software. Second, schematic implementations of both chaotic communication systems in general and their receiving and transmitting units in the integrated performance are intensively studied. And third, according to many researchers, direct chaotic communication systems are especially perspective, where the change of actually chaotic system parameter transmits the information.

And another important point, which concerns the nature of chaotic oscillations. The real chaotic oscillations can be realized only if the exact values of their chaotic time series are fully taken into account. Since all computing devices have limited accuracy, then when studying the chaotic systems we always deal with pseudo-chaotic fluctuations. It is clearly understood their fundamental difference from the pseudo-random oscillation, which is a sensitive dependence on initial conditions of chaotic mappings. Preliminary calculations show that in most cases if there is the accuracy of 16 decimal places in computer simulations than one can precisely talk about chaotic oscillations [14–16]. In this case, the repetition period is long enough and allows us to carry out safely all information transfer and receiving operations.

This work is a continuation of a series of works carried out at the Department of Radio Engineering and Information Security of Chernivtsi National University and devoted to the study of chaotic secure communication systems. The work has the following structure. The second part provides the results of the study of generalized synchronization for hidden information transfer. The third part deals with the modification of the Colpitts oscillator for a covert communication system. In the fourth part, the generation of stable chaotic oscillations was proposed that can be used in the transmitting-receiving parts of the communication systems. In the fifth part, we calculate a spectrum of Lyapunov exponents and the sixth part deals with chaotic cryptography.

2   Chaos Synchronization in Communication Systems

Synchronization is a universal phenomenon that occurs between two or more connected nonlinear oscillators. The ability of nonlinear oscillators to synchronize with each other is fundamental to understanding many processes in nature because synchronization plays a significant role in science [17].

An interesting type of chaotic synchronization for hidden transmission of information is generalized synchronization (GS) [1]. GS occurs if, after the end of transient processes, the functional relationship $Y = F(X)$ exists between the states of two chaotic systems, where X and Y are state vectors of driving and driven systems. The functional dependence of F can be complex or fractal, which complicates the detection of the information from the carrier signal. High noise resistance distinguishes GS among the other types of chaotic synchronization [18, 19].

The other advantage of using GS in chaotic communication systems is also the ability to synchronize nonidentical or parametrically different systems and systems with different dimensions of phase space.

The method for secure data transmission proposed in [18] is based on the properties of stability of boundaries of GS to noise and switching chaotic modes using small change parameters of the driving system. This method consists of the following. One or several control parameters of the driving generator of the transmitter are preliminarily modulated by a binary information signal. The obtained signal is transmitted via a communication channel. The receiver consists of two identical generators which can be in the mode of GS with a transmitter. Depending on the transmitted binary bit 0 or 1, the parameters of modulation of the control parameter of the transmitter should be chosen to lead to the presence or absence of generalized synchronization between the transmitter and receiver. Thus, it is possible to detect an occurrence of generalized synchronization or desynchronization between the connected systems, that is used to recover the information signal. To detect the GS one should use the method of an auxiliary system [19].

To use the generator of chaos in this system it is necessary that the minimum value of the coupling strength between systems (the boundary of synchronization) quickly changed at a small change of the control parameter of a generator. A sharp change in the GS border may be caused by two different ways of the occurrence of GS. In the first case, GS arises due to the suppression regime of natural oscillations of the driven system. Otherwise, GS is established, when the driven generator captures the fundamental frequency of the driving signal.

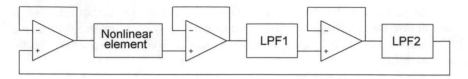

Fig. 1 Scheme of the chaos ring oscillator

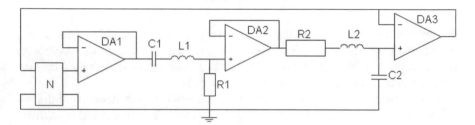

Fig. 2 Scheme of the modified generator

The investigation has shown that these properties are not inherent in most known chaotic systems. Therefore, the question appeared about the possibility of artificial systems constructing, which are characterized by the mentioned properties of GS.

The most suitable for solving this problem in terms of circuit implementation can be various modifications of the chaos ring oscillator, which consists of a filter and nonlinear element serially connected via buffer and closed by feedback. Simplest chaos ring oscillator consists of a first-order low-pass filter (LPF1), second-order low pass filter (LPF2) and nonlinear element (Fig. 1).

It should have a distinct frequency component in the signal spectrum for the occurrence of capture frequency between connected systems signals of chaotic systems. Therefore, in terms of the generator, there is necessary to include the narrowband filter and replace the first-order low-pass filter (Fig. 1) with the second-order narrowband one. The scheme of the modified generator is shown in Fig. 2.

To investigate synchronization let's consider two unidirectionally coupled generators, which are described by the following system of differential equations:

$$\dot{x}_1 = w_{11}^2(v_1 - y_1)$$
$$\dot{y}_1 = x_1 - by_1$$
$$\dot{z}_1 = -v_1$$
$$\dot{v}_1 = w_{12}^2(c(nl(y_1) - v_1) + z_1)$$
$$\dot{x}_{2,3} = w_{12,13}^2(v_{2,3} - y_{2,3}) \tag{1}$$
$$\dot{y}_{2,3} = x_{2,3} - by_{2,3}$$
$$\dot{z}_{2,3} = -v_{2,3} + e(v_1 - v_{2,3})$$
$$\dot{v}_{2,3} = w_{22,23}^2(c(nl(y_{2,3}) - v_{2,3}) + z_{2,3})$$

where x, y, z, w are the state variables; the indices $i = 1, 2, 3$ correspond to the signals of driving, driven and auxiliary systems respectively; e is coupling coefficient; $nl(*)$ is nonlinear function; $w_{1i} = 5$, $w_{2i} = 6.28$, $b = 1.38$, $c = 10$ are parameters of the system.

If $w_1 > 4.83$ the system (1) is in the chaotic regime. Changing the parameter w_{11} corresponds to the changing of the fundamental frequency oscillations of a driven system. Through numerical simulations, the dependence of synchronization error of GS on w_{11} for $w_{21} = 5$ was investigated. If the synchronization error is close to zero, the synchronization is established.

The dependence of synchronization error GS on coupling strength e and parameter w_{11} were calculated. As shown in Fig. 3, there exists a range of parameter values where synchronization error is small. This means that generators are synchronized if their parameters are picked up from this range.

The resistance of GS to noise in the channel was also studied. Here was found that GS is established if SNR > 1.5 dB (Fig. 4).

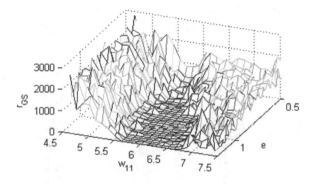

Fig. 3 The dependence of synchronization error r_{GS} on coupling strength e and parameter w_{11}

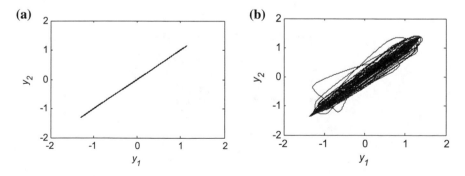

Fig. 4 GS of modified chaos ring generators with SNR $= 1.5$ dB: **a** synchronization for $w_{11} = 5.5$; **b** desynchronization for $w_{11} = 5.3$

Therefore, the method of modeling of chaos generator circuits for communication using chaos was proposed and the noise immunity of their generalized synchronization was investigated.

3   The Properties of Colpitts Oscillator Signals for Hidden Communication Systems

The development of HF hidden communication systems based on the usage of nonlinear dynamics requires the creation and study of chaotic generators the signals of which occupy the bandwidth of several hundred MHz. There are many scientific works were analog and digital communication systems use generators of chaotic oscillations [20–23]. However, the information transfer speed is very slow, because the simplest chaotic generators (Chua's circuit, Rossler, Lorentz systems etc. [24–26]) are used, which generate the signals whose spectrums are at the low frequencies range. Thus, Colpitts oscillator is more perspective in this case. It has a simple implementation and generates signals at a wide frequency range (Fig. 5a) [27, 28].

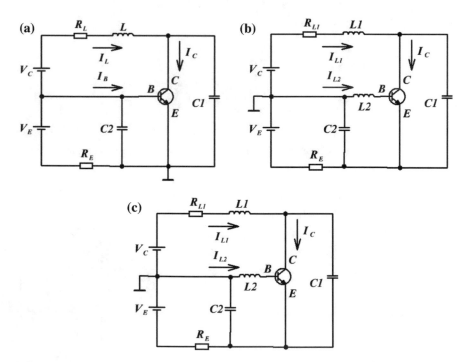

Fig. 5 Colpitts oscillators: **a** classic scheme with common emitter; **b** modified scheme with common base and inductor in base subcircuit; **c** modified scheme with common base and inductor in emitter subcircuit

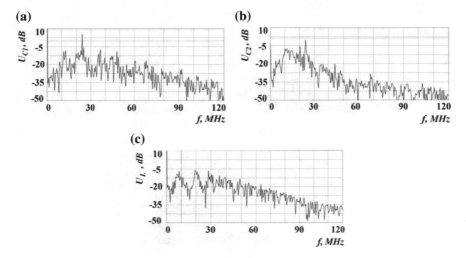

Fig. 6 Spectral characteristics of signals generated by Colpitts oscillator: **a** U_{C1}; **b** U_{C2}; **c** U_L

At the values of system parameters $R_L = 450\ \Omega$, $R_E = 2400\ \Omega$, $L = 1\ \mu H$, $V_C = V_E = 20\ V$, $C_1 = 150\ pF$ and $C_2 = 70\ pF$, spectrums of signals U_{C1}, U_{C2} and U_L occupy about 100 MHz ranges (Fig. 6).

For information secrecy, it is necessary that the chaotic signals were similar to white noise and it requires that the statistical characteristics were similar to normal distribution ones. However, the Colpitts oscillator signals are not characterized by properties similar to random variables, because the signals' distributions are not symmetry at least (Fig. 7), proving the values of the mathematical expectation μ and standard deviation σ and the values of kurtosis and skewness |E| and |A| are more than 0.25.

To solve this issue, the modification of the scheme was proposed. One more inductor was placed into one of the generator subcircuits. Several options of modification were considered, but only the signals U_{L2} of the scheme (Fig. 5b), and U_{L2} of the scheme (Fig. 5c), have statistical characteristics approximated to noise ones. The distributions of those signals and their spectral characteristics are depicted in Fig. 8.

The statistical characteristics of the signals are shown in Table 1. One can see that the Colpitts oscillator signals significantly differ from the normal distribution. The signal generated by the modified scheme (Fig. 5c) is the most similar to noise.

4 Method of Synthesis of Chaotic Signals Generators

An important task in the design of chaotic signals generators is a complex study of mathematical models that describe them. In particular, it is necessary to determine the exact boundaries of limits of system parameters and when they are changed various types of oscillatory modes can be formed, in particular—chaotic, hyperchaotic,

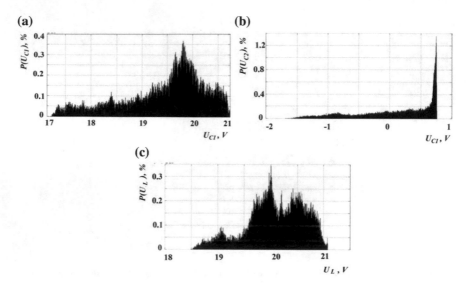

Fig. 7 Signals distribution generated by Colpitts oscillator: **a** U_{C1}; **b** U_{C2}; **c** U_L

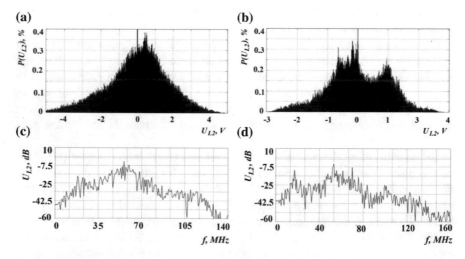

Fig. 8 The characteristics of signal U_{L2} of the scheme that is shown in Fig. 5b: **a** distribution; **c** spectrum; and the scheme that is shown in Fig. 5c: **b** distribution; **d** spectrum

periodic or quasi-periodic. This allows calculating the values of the Lyapunov exponent spectrum that characterizes the dependence of the dynamic system on initial conditions and determines the rate of divergence of trajectories in phase space.

Equally important tasks that help to determine the structure of the generated signals are mathematical modeling and numerical analysis of oscillatory modes of the chaotic systems under study, described by difference equations or systems of nonlinear differential equations of different orders.

Table 1 Statistical characteristics of studied chaotic generators

Scheme	Signal	μ	σ	$|E|$	$|A|$
Figure 5a	U_{C1}	19.43	0.86	3	0.75
	U_{C2}	0.07	0.64	0.31	0.63
	U_L	20.1	0.62	3	0.21
Figure 5b	U_{L2}	0	1.05	0.82	0.48
Figure 5c	U_{L2}	0	0.99	0.35	0.1

Modeling of the general-circuit solution of chaotic signals generators is made possible by using the designed electrical circuit and modern approaches to circuit simulation using specialized software, such as Micro-Cap or Multisim.

In the practical implementation of the electrical circuit of chaotic signals generator it is necessary to attribute the dynamic system parameters, its dynamic variables, and mathematical operations over them in the equations of the system to signals and their processing circuits and perform the following steps:

1. Develop a block diagram of chaotic signals generator based on the systems of nonlinear differential equations.
2. Develop an electrical circuit of chaotic signals generator based on the block diagram.
3. Calculate the value of the electric circuit components on the basis of systems of differential equations of an electrical circuit and the system of nonlinear differential equations.
4. Choose the base components according to the calculated values, considering the precision of circuit elements up to two decimal places.
5. Implement the electrical circuit of the chaotic signals generator.
6. Make adjustment of modes of chaotic signals generator according to the types of oscillatory modes that can be generated by changing their values of system parameters.

Liu chaotic system [29] was chosen as the basis for the realization of chaotic signals:

$$\dot{x} = a(y - x)$$
$$y = bx - hxz + \lambda w$$
$$\dot{z} = cx^2 - dz$$
$$\dot{w} = -ny. \tag{2}$$

Calculation of Lyapunov exponent spectrum, the results of mathematical modeling as time diagrams of dynamic variables and phase portraits, development of block diagram and electrical circuits, calculation of values of the electrical circuit components are given in [23, 30]. The results of mathematical modeling and modeling of general-circuit solutions of chaotic signals generator were confirmed by experimental results.

The final stage of designing is to adjust the prototype of chaotic signals generator and experimental studies of oscillatory modes which may arise therein. Figure 9 shows the prototypes of communication system based on chaotic signals generators for experimental studies of secure data transfer process.

Figure 10 shows the prototype of the communication system based on chaotic signals generators. Figure 11 reveals the phase portraits of the chaotic signals generator for chaotic and hyperchaotic modes of oscillations obtained by experimental studies.

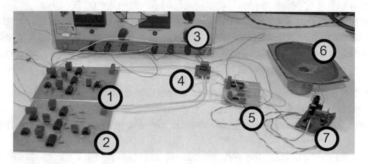

Fig. 9 Experimental studies of communication system based on chaotic signals generators: (1)—transmitter; (2)—receiver; (3)—power supply; (4)—scheme of synchronization; (5)—scheme of subtraction, addition of signals; (6)—audio speaker; (7)—audio amplifier

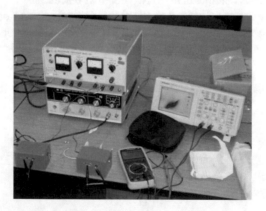

Fig. 10 The prototype of communication system based on chaotic signals generators

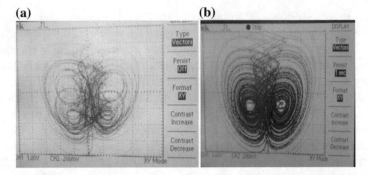

(a) **(b)**

Fig. 11 Experimental phase portraits of chaotic signals generator: **a** chaotic mode and **b** hyper-chaotic mode

5 Calculation of a Spectrum of Lyapunov Exponents

The nature of chaotic oscillations in continuous dynamic systems can be evaluated according to different criteria. The nature of chaotic oscillations in continuous dynamic systems can be evaluated according to different criteria. One of the main evaluation criteria is a spectrum of Lyapunov exponents, which is determined at each point in the phase space. The importance of calculating the spectrum of Lyapunov exponents for continuous dynamic systems is due to the fact that it is by spectrum that it is possible to establish ranges of values of the parameters of systems at which there are periodic, quasi-periodic, chaotic and hyper-chaotic oscillations.

Lyapunov exponents characterize the degree of dependence of the dynamic system on the initial conditions and determine the rate of differentiation of its trajectories in the phase space.

Table 2 shows the correspondence of the type of oscillations taking place in dynamic systems, signs, and values of Lyapunov exponents.

The number of Lyapunov exponent at each point for a given continuous system is determined by the number of differential equations that are analytically described. That is, for arbitrary chaotic system it is necessary to monitor the evolution of 3 vectors of perturbation along the phase trajectory.

The calculation of the Lyapunov spectrum is carried out by numerical methods, since analytical solutions to this problem can only be found for some systems.

Table 2 Correspondence of the type of oscillations of the chaotic system with signs and the value of Lyapunov exponent

Sign and values of Lyapunov exponent	Type of oscillations
$- - - 0$	Periodic
$- - 0\,0$	Quasi-periodic
$- - + 0$	Chaotic
$+ +$	Hyper-chaotic

There are several different algorithms for calculating Lyapunov exponent. According to the first algorithm, it is necessary to trace the nature of the evolution of the distance between certain two points of the phase trajectory in time of two copies of a dynamic system with similar initial conditions. According to the second one, it is necessary to numerically solve the system of differential equations and the system of equations in variations describing the evolution of a continuous dynamic system along the phase trajectory and infinitely small perturbation of this trajectory, respectively. The choice of the calculation scheme basically depends on the way the dynamic system is presented, namely, it is described by a system of differential equations or reflections. In calculating Lyapunov exponents for the continuous dynamic system under investigation (3.2), we used a second algorithm, since it is more adapted for continuous dynamic systems described by systems of differential equations and subjected to a numerical solution by applying a modified Benettin algorithm.

It should be noted that in order to calculate a certain number of Lyapunov exponents, it is necessary to monitor the evolution of the same number of vectors of perturbation along the phase trajectory. In our case, this number is 4.

The calculation of Lyapunov exponents begins with a numerical solution of a system of differential equations that describes the behavior of a continuous dynamic system in the interval of time during which the trajectory goes to a chaotic attractor. Starting from the arbitrarily chosen point of the phase trajectory, we trace the evolution of the trajectory and the four initial vectors of perturbation that are orthogonal to each other and normalized per unit. After a certain period of time, the trajectory will move to a point x_1. In this case, the vectors of perturbation will be changed accordingly by taking values $\tilde{x}_1, \tilde{y}_1, \tilde{z}_1, \tilde{w}_1$.

We renormalize them and orthogonalize them using the Gram-Schmidt method. We obtain a new set of orthogonal and normalized ones per unit of perturbation vectors $\tilde{x}_0^0, \tilde{y}_0^0, \tilde{z}_0^0, \tilde{w}_0^0$. The number of repetitions of the described procedure depends on the chosen interval of time during which the numerical solution of the system of differential equations is carried out.

At the same time, the calculation of the sum of the vectors of perturbation after orthogonalization is carried out:

$$S_1 = \sum_{i=1}^{M} \ln\|\tilde{x}\|; \tag{3.1}$$

$$S_2 = \sum_{i=1}^{M} \ln\|\tilde{y}\|; \tag{3.2}$$

$$S_3 = \sum_{i=1}^{M} \ln\|\tilde{z}\|; \tag{3.3}$$

$$S_4 = \sum_{i=1}^{M} \ln\|\tilde{w}\|. \tag{3.4}$$

The calculation of the first four Lyapunov exponents is carried out on the following expression

$$i = 1, 2, 3, 4. \tag{4}$$

Each system Eq. (4) corresponds to a system of four equations in variations:

$$
\begin{aligned}
\dot{x}_1 &= a(y_1 - x_1), \\
\dot{y}_1 &= bx_1 - hzx_1 - hxz_1 + \lambda w_1, \\
\dot{z}_1 &= 2cxx_1 - dz_1, \\
\dot{w}_1 &= -ny_1.
\end{aligned}
\tag{5}
$$

$$
\begin{aligned}
\dot{x}_2 &= a(y_2 - x_2), \\
\dot{y}_2 &= bx_2 - hzx_2 - hxz_2 + \lambda w_2, \\
\dot{z}_2 &= 2cxx_2 - dz_2, \\
\dot{w}_2 &= -ny_2.
\end{aligned}
\tag{6}
$$

The resulting complete system of twenty differential equations was solved by a numerical Runge-Kutta method using renormalization of perturbation vectors and their orthogonalization in the computation process for the following parameter values: $a = 15, b = 30, h = 1, \lambda = 1, c = 4, d = 2.5$. Lyapunov's parameters were calculated for whole parameter values $n \in [1; 100]$. The initial vectors of perturbation x_0, y_0, z_0, w_0 are orthogonal with each other and are normalized to one.

The orthogonalization of the perturbation vectors by the Gramm-Schmidt method and normalization to a fixed constant were carried out after the completion of each stage of the calculations.

From the obtained Lyapunov spectrum (Fig. 12), it follows that for any integer value n, the first and second indices change their sign, the third indicator is inaccurate, and the fourth is negative and accepts the minimum value $-18, 58$ at $n = 10$ and the maximum value of -14.78 at $n = 84$.

The obtained results of the study of the nature of oscillatory processes determined by the values of the Lyapunov exponents, depending on the integer value of the parameter n, are given in Table 3.

6 Determination of the Period of Pseudorandom Sequences for Cryptography Application

The development of new technologies related to the use of computerized devices connected to public networks, including the Internet, has led to the need to develop new methods of encryption, which is called light cryptography [31]. The field of

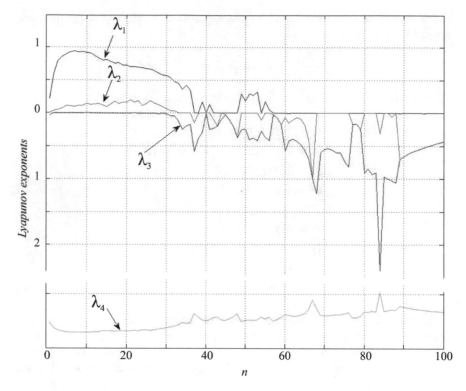

Fig. 12 The calculated spectrum of Lyapunov exponents for the studied system

Table 3 Types of oscillations of a chaotic system for different intervals of parameter values n

Parameter values n	Type of oscillations
$n \in (0, 33]$	Hyper-chaotic
$n \in (38; 42] \cup (48; 57]$	Chaotic
$n \in (42; 48] \cup (57; 68] \cup (76; 80] \cup (83; 100]$	Periodic
$n \in (68; 76] \cup (80; 83]$	Quasi-periodic

application of these devices is extremely wide: here we can refer to the Internet of things, automotive systems, sensor networks, monitoring systems for sick patients, distributed control systems, cyber-physics industrial systems, networks guided by artificial intelligence [20, 32].

The algorithms for encrypting information forming the technology of light cryptography are intended for use in devices with limited memory resources and processor time. Therefore, to provide acceptable speed, there is a need for encryption algorithms that use only fast operations: bitwise addition of XOR, cyclic shifts, etc. Proceeding from the above, we can conclude that the most attractive algorithm, with limited computing resources, is so-called streaming encryption, or a one-time notebook. Its action is based on performing the XOR operation over the bits of the source text and the bits of the encryption stream. To implement the bit sequence generator in the hardware encryption system, you can go two ways: implement some acceptable algorithm for the firmware of the encryption controller, or use removable external flash memory, the volume of which is sufficient to store a binary file with a written pseudo-random sequence.

The implementation of the first method, that is, when the code of the mobile device controller implementing this or that encryption algorithm is quite common. Examples of the use of the second algorithm are not that much. When comparing the advantages and disadvantages of both algorithms, the first method prevails from the point of view of ease of use, since it is not necessary to change the hardware configuration of the entire system. The application of the second algorithm assumes that the encryption key is the entire pseudo-random sequence and can be changed if, for example, the flash drive is changed. It also enables you to implement a pseudorandom sequence generation algorithm.

6.1 Features of Stream Ciphering

The feature of stream ciphers is the bitwise processing of information, and therefore encryption and decryption in stream ciphering may be interrupted at any time, and also be restored in the event of a continuation of the transmission process.

The most common algorithm of streaming encryption for a long time was the encryption sequence of a linear offset register. But it was shown that in the case of intercepting a 2 * N bit of the encryption sequence of a linear shift register; it is always possible to restore its internal structure. After that, such ciphers ceased to be reliable (the cipher is considered reliable if, during a long time, for example, a decade, no way was found to break it).

The basis of streaming encryption is the use of the XOR operation, and the cipher itself can be submitted as a machine that performs the following actions at each step of its work:

(1) generates one bit of encryption sequence according to some law;
(2) one bit of an open stream imposes one encryption bit applying a certain turnaround transformation.

In the field of operations on binary numbers, there are only two, which have the property of reversal: XOR and negation.

Attacks on gamma encodings in the case of an unknown encryption algorithm generate two main vulnerabilities:

(1) a significant deviation of their statistical characteristics from a truly random flow;
(2) re-use of some parts of the gamut in the process of encryption.

To prevent the vulnerability of the first type, a number of powerful statistical tests have been developed, which must match the generated sequence: NIST, DIE HARD, etc. The second type of vulnerability associated with possible repetition of parts of the gamma is much less explored due to the complexity of this problem.

Let's consider the problem of determining the period of a number of numbers represented in the form of decimal fractions generated by one-dimensional discrete reflections (7) that have the property of dynamic chaos:

$$x_{n+1} = f(x_n) \tag{7}$$

The numeric sequence generated by relation (7) is Markov since each subsequent term of the sequence x_{n+1} depends only on the current value of x_n. In order for the mapping (7) to be the property of dynamic chaos, it is necessary that its Lyapunov index be positive. From theory, it is well known that in this case, the generated sequence will never have a period (for example, for logistic mapping). But in any digital device or programming system, there is always a limitation of the number of digits representing the numbers obtained during the iterations (that is, the finite accuracy of the calculations). This is especially true for devices used in light cryptography.

Therefore, at some iteration stage, a number will be formed equal to the number generated earlier, and as a consequence of the fact that the sequence has the property of the Markov chain, a period will arise. For mappings possessing the property of dynamic chaos, it is impossible to determine the sequence period by analytical methods.

One of the most well-known standards for representing decimal numbers is the IEEE standard, which assumes that any mathematical operations are performed with an accuracy of 324 decimal places. This is an extremely high precision since the total number of 324-bit decimal digits that they can create is 10,324. In the description of the IEEE standard, data is given that this declared accuracy of the calculations is oriented or approximate. But we have been researching one of the systems development programs in the language C++, namely DEV C++. The theoretically, the limiting value of the maximum period of the resulting decimal pseudorandom sequences coincides with the maximum number of nonrepeatable numbers in a series of real numbers. Thus, after analyzing the problem of the number of the possible placement of decimal numbers with a given number of digits n, we conclude that the maximum period of the sequence of decimal numbers is determined by the formula (8).

$$N = 10^n \tag{8}$$

For example, consider the logistic map, setting the value of the parameter $\lambda = 4$.

It is well known [33] that for this value of this parameter, the sequence will never have a period for all possible initial values taken from the segment (0; 1), and the values of the elements of the series will be evenly distributed throughout the segment of possible values (0; 1) under one condition: if you do not limit the accuracy of finding the elements of the generated numerical series.

6.2   Methods of the Searching of the Period

There are relatively few studies that are designed to determine the periods of sequences shaped by decimal fractions with the finite number of discharges generated by chaotic reflections. And in existing researches are given results where the accuracy of calculations is limited to 10–15 decimal places.

But it is well known that any modern computing system (C++, Pascal) uses much more precision to perform mathematical operations over decimal fractions. Therefore, it is clear that for the study of numerical sequences whose elements are represented with such precision, it is necessary to find new algorithms for research. In the end, the direct method of pairwise comparison of elements of the numerical series is extremely ineffective in terms of computational complexity and the cost of machine time.

We describe the methods used by us. The first method is a method of searching for elements of a sequence in which a given number of decimals after a comma coincides, and the initial element of the numerical sequence is given.

In this case, the representation of numbers with the highest possible accuracy adopted by the standard IEEE-324 decimals, and which is implemented in most programming systems designed for the Windows operating system, was used for calculations.

Our research will allow us to determine the pseudo period of the generated sequence. The prefix "pseudo" is used to emphasize that in reality, only a certain number of decimal numbers that we choose for analysis will repeat. The period in the mathematical sense, that is, the absolutely exact repetition of the elements of the numerical sequence will take place in the case when taking into account all the significant figures of the decimal fraction generated by this computer system. In connection with this, that is, in connection with the pseudo-periodicity of the investigated sequence, we have developed another algorithm. The content of this algorithm is that after determining the coincidence, we continue the iteration of the logistic map, after which we will again have the coincidence of the selected number of decimal digits after the comma of the next iteration element and the decimal point digits, which is the first element of the numerical series. And so we continue several times, as in the future we say doing a few jumps. The statistically processed results for this improved algorithm are given in Table 4.

As can be seen from this table, the value of the sequence is repeated because of the number of iterations that exceeds the number 10n (the maximum number of digit numbers n), these lines are highlighted in Italics. The same is true for averaged values. Therefore, we conclude that the repetition of a given number of digits will occur earlier, but we do not know the value of this number (Fig. 13).

Based on Table 4 we computed the averaged values of the number of iterations for 2, 3, 4 and 5 significant digits for different initial values. Figure 1 shows the minimum, average and maximum values of the number of iterations, depending on the number of decimal values in the logarithmic scale.

Table 4 Definition of the pseudo period (5 jumps)

Initial value	The average number of iterations and standard deviation in percentage for different number of digits for which we determine the coincidence				
	k = 2	k = 3	k = 4	k = 5	k = 6
0.050000	109 ± 94%	282 ± 81%	8076 ± 115%	50,843 ± 69%	844,753 ± 88%
0.100000	70 ± 77%	456 ± 91%	8526 ± 91%	50,717 ± 17%	223,806 ± 66%
0.150000	119 ± 64%	917 ± 106%	6499 ± 126%	80,131 ± 118%	1,449,218 ± 100%
0.200000	60 ± 52%	1885 ± 169%	8248 ± 76%	332,538 ± 80%	997,276 ± 61%
0.260000	128 ± 62%	2930 ± 97%	6494 ± 59%	156,369 ± 87%	1,120,994 ± 90%
0.300000	126 ± 76%	887 ± 117%	19,451 ± 86%	122,667 ± 57%	679,801 ± 80%
0.350000	181 ± 58%	941 ± 94%	8709 ± 116%	166,147 ± 82%	1,038,490 ± 58%
0.400000	133 ± 65%	1302 ± 84%	8080 ± 86%	92,615 ± 80%	1,575,146 ± 68%
0.450000	82 ± 89%	1318 ± 43%	9309 ± 117%	224,633 ± 75%	3,536,094 ± 110%
0.510000	102 ± 84%	1031 ± 99%	31,166 ± 62%	162,095 ± 115%	614,639 ± 180%
0.550000	160 ± 65%	1007 ± 76%	19,968 ± 62%	56,730 ± 38%	329,029 ± 135%
0.600000	80 ± 53%	619 ± 73%	11,732 ± 91%	186,467 ± 77%	1,244,174 ± 133%
0.650000	224 ± 64%	1763 ± 39%	11,828 ± 98%	104,293 ± 69%	2,378,694 ± 66%
0.700000	129 ± 72%	1088 ± 65%	7860 ± 69%	64,460 ± 169%	1,673,257 ± 99%
0.760000	319 ± 116%	1477 ± 78%	13,446 ± 63%	84,940 ± 96%	1,281,827 ± 75%
0.800000	173 ± 93%	1052 ± 52%	4267 ± 83%	125,280 ± 184%	1,267,180 ± 86%
0.850000	95 ± 74%	484 ± 65%	7842 ± 97%	155,319 ± 126%	706,764 ± 147%
0.900000	51 ± 92%	311 ± 80%	13,987 ± 138%	79,053 ± 65%	615,664 ± 106%
0.950000	105 ± 64%	2085 ± 116%	5262 ± 92%	78,785 ± 67%	828,459 ± 76%
Average values	129	1149	11,092	124,952	1,179,224

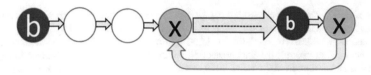

Fig. 13 "b"—initial value, "x"—the unknown value which can be repeated

Fig. 14 The number of iterations on a logarithmic scale versus number of digits

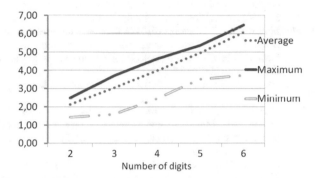

From the results obtained, it can be seen that the rate of growth of the number of iterations is practically unchanged at the logarithmic scale, but numerical analysis shows that the increase in the number of iterations is slowing down: about 10 to 3 digits, 100 to 4 digits, 950 for 5 digits and 14,500 for 6 digits.

Thus a general conclusion is that takes place almost linear dependence of the number of iterations on the number of coinciding numbers. The graph built on the calculated data is shown in Fig. 14.

7 Conclusions

This paper presents practical results on the implementation of a chaotic secure communication system. Particular attention is paid to the issue of generalized chaotic synchronization, the formation of the reliable generators of chaos and their hardware implementation. The current prototype of the chaotic communication system confirms the proposed algorithms of transfer and data protection using deterministic chaos. An important question about the frequency of pseudo-chaotic oscillations is also analyzed.

References

1. Lorenz EN (1963) Deterministic nonperiodic flow. J Atmos Sci 20:130–141
2. Ageyev DV, Salah MT (2016) Parametric synthesis of overlay networks with self-similar traffic. Telecommun Radio Eng (English translation of Elektrosvyaz and Radiotekhnika) 75(14):1231–1241

3. Kirichenko L, Radivilova T, Bulakh V (2018) Machine learning in classification time series with fractal properties. Data 4(1):5. https://doi.org/10.3390/data4010005

4. Pichugina O, Yakovlev S (2020) Quadratic optimization models and convex extensions on permutation matrix set. In: Shakhovska N, Medykovskyy M (eds) Advances in intelligent systems and computing IV. CCSIT 2019. Advances in intelligent systems and computing, vol 1080. Springer, Cham

5. Obod I, Svyd I, Maltsev O, Vorgul O, Maistrenko G, Zavolodko G (2018) Optimization of data transfer in cooperative surveillance systems. In: 2018 International scientific-practical conference problems of infocommunications. Science and Technology (PIC S&T). IEEE, pp 539–542. https://doi.org/10.1109/infocommst.2018.8632134

6. Matuszewski J, Dikta A (2017) Emitter location errors in electronic recognition system. In: 2017 XI conference on reconnaissance and electronic warfare systems, proceedings of SPIE— the international society for optical engineering, vol 10418, art. no. 104180C, pp C1–C8. https://doi.org/10.1117/12.2269295

7. Pecora LM, Carroll TL (1990) Synchronization in Chaotic systems. Phys Rev Lett 64(8):821–825

8. Shergin VL, Chala LE, Udovenko SG (2018) Fractal dimension of infinitely growing discrete sets. In: 2018 14th international conference on advanced trends in radioelecrtronics, telecommunications and computer engineering (TCSET). IEEE, pp 259–263. https://doi.org/10.1109/tcset.2018.8336198

9. Alsaleem NY, Moskalets M, Teplytska S (2016) The analysis of methods for determining direction of arrival of signals in problems of space-time access. East-Eur J Enterp Technol 4(9):82. https://doi.org/10.15587/1729-4061.2016.75716

10. Kirichenko L, Radivilova T (2017) Analyzes of the distributed system load with multifractal input data flows. In: 2017 14th international conference the experience of designing and application of CAD systems in microelectronics (CADSM). IEEE, pp 260–264. https://doi.org/10.1109/CADSM.2017.7916130

11. Kryvinska N, Zinterhof P, van Thanh D (2007) An analytical approach to the efficient real-time events/services handling in converged network environment. In: Enokido T, Barolli L, Takizawa M (eds) Network-based information systems. NBiS 2007. Lecture notes in computer science, vol 4658. Springer, Berlin, Heidelberg

12. Ageyev D, et al (2018) Method of self-similar load balancing in network intrusion detection system. In: 2018 28th international conference Radioelektronika (RADIOELEKTRONIKA). IEEE, pp 1–4. https://doi.org/10.1109/radioelek.2018.8376406

13. Baptista MS (1998) Cryptography with chaos. Phys Lett A 240:50–54

14. Kushnir M, Galiuk S, Rusyn V, Kosovan G, Vovchuk D (2014) Computer modeling of information properties of deterministic chaos. In: Proceedings of the 7th Chaotic modeling and simulation international conference. Lisbon, Portugal, 7–10 June 2014, pp 265–276

15. Semenov A (2016) Mathematical simulation of the chaotic oscillator based on a field-effect transistor structure with negative resistance. In: 2016 IEEE 36th international conference on electronics and nanotechnology (ELNANO). Kiev, Ukraine, 19–21 April 2016. https://doi.org/10.1109/elnano.2016.749300

16. Semenov A (2016) The Van der Pol's mathematical model of the voltage-controlled oscillator based on a transistor structure with negative resistance. In: 2016 13th international conference on modern problems of radio engineering, telecommunications and computer science (TCSET). Lviv, Ukraine, 23–26 Feb 2016. https://doi.org/10.1109/tcset.2016.7451982

17. Boccaletti S, Kurths J, Osipov G, Valladares DL, Zhou CS (2002) The synchronization of chaotic systems. Phys Rep 366(1–2):1–101

18. Moskalenko OI, Koronovskii AA, Hramov AE (2010) Generalized synchronization of chaos for secure communication: remarcable stability to noise. Phys Lett A 374:2925–2931

19. Abarbanel HDI, Rulkov NF, Sushchik MM (1996) Generalized synchronization of chaos: the auxiliary system approach. Phys Rev E 53(5):4528–4535

20. Vovchuk D, Haliuk S, Politanskii L (2013) Adaptation of the chaotic masking method for digital communication. East-Eur J Enterp Technol 2/4(62):50–55

21. Vovchuk D, Haliuk S, Politanskii L (2014) Research of the Process of masking of digital information signals using chaotic oscillation. East Eur Sci J 3:245–253
22. Yang T (2004) A survey of chaotic secure communication systems. Int J Comput Cignition 2(2):81–130
23. Ivanyuk P, Politansky L, Politansky R, Eliyashiv O (2012) Chaotic masking of information signals using generator based on the Liu system. Tekhnologiya i Konstruirovanie v Elektronnoi Apparature 3:11–17
24. Fortuna L, Fraska M, Xibilia M (2009) Chua's circuit implementations. Yesterday, today and tomorrow. World scientific series on nonlinear science, series A, vol 65, p 222
25. Chen H-C, Liau B-Y, Hou Y-Y (2013) Hardware implementation of Lorenz circuit systems for secure Chaotic communication applications. Sensors 13:2494–2505
26. Alsafasfeh QH, Al-Arni MS (2013) A new chaotic behavior from Lorenz and Rossler systems and its electronic implementation. Circuits Syst 2:101–105
27. Efremova EV (2008) Generatory haoticheskih kolebanij radio-I SVCh-diapazonov. Uspehi sovremennoj radiojelektroniki 1:17–31
28. Dmitriev AS, Klecov AV, Laktjushkin AM, Panas AI, Starkov SO (2008) Svcrhshirokopolosnye kommunikacionnye sistemy na osnove dinamicheskogo haosa. Uspehi sovremennoj radiojelektroniki 1:4–16 (In Russian)
29. Wang F-Q (2006) Hyperchaos evolved from the Liu chaotic system. Chin Phys 15(5):963–968
30. Ivanuyk PV (2011) Investigation of Chaotic processes generated Liu system. East-Eur J Enterp Technol 4/9(52):11–15
31. Naru ER, Saini H, Sharma M (2017) A recent review on lightweight cryptography in IoT. In: 2017 international conference on I-SMAC (IoT in social, mobile, analytics and cloud) (I-SMAC)
32. Kirichenko L, Radivilova T, Bulakh V (2020) Binary classification of fractal time series by machine learning methods. In: Lytvynenko V, Babichev S, Wójcik W, Vynokurova O, Vyshemyrskaya S, Radetskaya S (eds) Lecture notes in computational intelligence and decision making. ISDMCI 2019. Advances in intelligent systems and computing, vol 1020. Springer, Cham. https://doi.org/10.1007/978-3-030-26474-1_49
33. Patel ST, Mistry NH (2015) A survey: lightweight cryptography in WSN. In: 2015 international conference on communication networks (ICCN)

Radiomeasuring Optical-Frequency Converters Based on Reactive Properties of Transistor Structures with Negative Differential Resistance

Oleksandr Osadchuk⬤, Vladimir Osadchuk⬤, Andriy Semenov⬤,
Iaroslav Osadchuk, Olena Semenova⬤, Serhii Baraban⬤,
and Maksym Prytula

Abstract The paper presents the results of the investigation of the photoreactive effect in semiconductor diodes in the dynamic mode on the basis of the solution of the transport equation under the action of optical radiation, which gave the possibility to theoretically calculate the complete resistance of the base area and obtain an analytical dependence of its components on the power of optical radiation. The authors developed a model of an optical-frequency radiomeasuring converter of optical radiation, which describes the dependence of the complete resistance of the bipolar transistor structure that lies in the base of the converter of the optical radiation power. A photosensitive diode is used as a photosensitive element in a radiomeasuring optical-frequency converter. The proposed radiomeasuring optical-frequency converters have high sensitivity in the range of low optical radiation power values, which makes it possible to reliably measure even weak optical signals. The analytical expressions for transformation function and the sensitivity equation are obtained. The results of the simulation are confirmed by experimental studies, the error of the developed mathematical models is ±5%.

Keywords Photoreactive effect · Reactive properties · Negative differential resistance · Radiomeasuring optic-frequency converter · Optical radiation · Full resistance

1 Introduction

At the stage of development of scientific and technological progress, analytical instrumentation is one of the new areas of microelectronic use, which, with the help of

O. Osadchuk · V. Osadchuk · A. Semenov (✉) · I. Osadchuk · O. Semenova · S. Baraban ·
M. Prytula
Vinnytsia National Technical University, Vinnytsia, Ukraine
e-mail: semenov.a.o@vntu.edu.ua

O. Osadchuk
e-mail: osadchuk.av69@gmail.com

© The Editor(s) (if applicable) and The Author(s), under exclusive license to Springer Nature   229
Switzerland AG 2021
T. Radivilova et al. (eds.), *Data-Centric Business and Applications*, Lecture Notes on Data
Engineering and Communications Technologies 48,
https://doi.org/10.1007/978-3-030-43070-2_12

nanotechnology methods, allows the development and creation of primary converters for controlling environmental parameters [1, 2]. Creation of primary converters with improved metrological characteristics is one of the topical problems for today [3–6]. Improvement of systems of automatic control of various objects, processes, production, robotic processes is determined by achievements in this field [7–10]. The general trend in the development of measuring converters, including optical radiation sensors, is due to an increase in the accuracy requirements for it while complicating the operational requirements [11–14]. All this leads to the search and development of new principles for the construction of a variety of primary converters [15–19]. At the current level of development of functional microelectronics and nanoelectronics, its achievement can be used to create a new class of optical radiation converters with a frequency output signal [20–22].

One of the scientific directions in the development of primary measuring converters is the use of dependence of reactive properties and negative differential resistance of semiconductor devices on the influence of external physical quantities and the creation of a new class of microelectronic converters of optical radiation with a frequency output signal on this basis [23–25]. Devices of this type convert the power of optical radiation to a frequency that allows to make the sensors on integral technology, and provides an opportunity to increase the speed, accuracy and sensitivity, expand the range of measured values, improve reliability, noise immunity and stability of the parameters. The combination of a single-chip primary measuring converter with information processing schemes allows you to create a "smart sensor". Use as an informative, frequency parameter avoids the use of analog-to-digital converters in processing information, that reduces the cost of automatic control and control systems [26–28].

A separate direction of research in this area is development of wireless sensor networks based on the proposed sensors [29–31]. This is a relevant research problem. Wireless Sensor Network is a distributed and self-organizing network of multiple sensors and actuators [32, 33]. They are interconnected via radio channel [34, 35]. This paper do not consider the task of constructing a Wireless Sensor Network. Methods of constructing and organizing such networks are widely known [36]. Approaches to optimization of such sensor networks are considered in [37, 38]. Processing of information signal time series using fractal analysis approaches is perspective according to [39–41]. Methods of controlling such networks proposed in [42–44] are based on intellectual technologies. Approaches to studying and modeling devices and components are proposed in [45–47].

In addition to removing measurable information it is practically simultaneous processing, filtering, compression and adjustment. In this case, the advantage of using the frequency information signal of the primary converter over its analog form of voltage or current is due to the simplicity and precision of the frequency conversion into a digital code, its high impedance during transmission and the efficiency of switching in multichannel measuring systems [48, 49]. The transformation of the "power of optical radiation-frequency" can be accomplished using semiconductor structures containing a self oscillating device implemented in the form of circuits consisting of bipolar, field-effect or bipolar-field-effect transistor structures. As a photosensitive element in the circuits, the photodiode was used, due to its simplicity, high gain and a fairly linear dependence on the illumination. In photodiode mode, the sensitivity of the photodiode increases in comparison with the photovoltaic mode, as well as the input signal increases; the internal resistance of the photodiode to the alternating current also increases. In the scheme of the positive feedback of the self-oscillator converter, a photodiode on which optical radiation operates is included. Thus, the optical radiation is converted into a frequency output signal. To study the properties of such optical converters, it is necessary to consider the mechanism of interaction of optical radiation with a semiconductor diode in a dynamic mode, to develop a mathematical model that takes into account this effect and on its basis to obtain analytical dependences of the complete resistance of the base area of the diode. This work is devoted to consideration of these issues.

2 Mathematical Model of Photoreactive Effect in Semiconductor Diodes

Absorption of photons in the base area of a semiconductor diode is accompanied by the formation of electron-hole pairs and the appearance of excess charge carriers. The diffusion processes, as well as the bulk and superficial recombination of charge carriers, lead to a system of conduction band—the valence band in equilibrium state. The excess concentration of charge carriers diffuses to the transition point where it distribution by electric field occurs. The transport equation establishes the relationship between generation, recombination, diffusion and drift processes. Its solution allows us to obtain a mathematical expression for a complete photocurrent, that allows us to determine the voltage drop on the base area of the diode. We consider the solution of the transport equation for ordinary boundary conditions when the system of differential equations with boundary conditions is linear in relation to the concentration of charge carriers and it derivatives. As a result of this, it is possible to sum up the concentration of charge carriers under the action and the lack of lighting, as well as light and dark currents. If there is an electric field in the base area of the diode, it affects the process of transporting non-main charge carriers in such a way that the directional motion of charge carriers with constant velocity μE in the electric field is superimposed on the chaotic thermal motion.

Under the influence of monochromatic radiation on a semiconductor, the rate of generation of charge carriers per unit volume is determined [50].

$$G(x, \lambda) = \alpha(\lambda)\Gamma(x), \tag{1}$$

where $\alpha(\lambda)$ is an absorption coefficient of light, $\Gamma(x)$ is a photon flux density. Expression (1) is fair under the condition that the process of generation occurs with the participation of one photon and the quantum yield of this process is equal to one. The magnitude $G(x, \lambda)$, taking into account all the radiation contained in the solar spectrum, is 10^{21}–10^{22} cm^{-3} s^{-1}. The process of recombining excess charge carriers is described in accordance with the Shockley-Reed theory [51]. The velocity of bulk recombination is determined by the effective lifetime τ_n of non-main charge carriers, that does not depend on it concentration and coordinates in the volume of the semiconductor. When the conditions are fulfilled $n_p \ll p_p$ recombination rate

$$V(x) = \frac{n_p - n_{p0}}{\tau_n},\tag{2}$$

where n_{p0} is a concentration of non-main charge carriers under thermal equilibrium conditions. The rate of change n_p in time is determined by the equation [50, 51].

$$\frac{dn_p}{dt} = G(x) - V(x) = \alpha(\lambda)\Gamma(x) - \frac{n_p - n_{p0}}{\tau_n}.\tag{3}$$

Now it is possible to consider the equation of transport of charge carriers in a semiconductor. To do this, it must be taken into account that the current density consists of drift and diffusion part

$$\vec{J}_n = nq\mu\vec{E} + qD_n\nabla n,\tag{4}$$

$$\vec{J}_p = pq\mu\vec{E} - qD_p\nabla p.\tag{5}$$

The fulfillment of the charge storage condition in the infinitely small volume of the semiconductor can be represented as:

$$\frac{\partial P}{\partial t} + \nabla\vec{J} = 0,\tag{6}$$

$$\frac{\partial n}{\partial t} - \nabla\vec{J}_n/q = G_n - V_n,\tag{7}$$

$$\frac{\partial p}{\partial t} + \nabla\vec{J}_p/q = G_p - V_p,\tag{8}$$

where $\vec{J} = \vec{J}_n + \vec{J}_p$. If, as a result of inter-band transitions, electron-hole pairs are generated, then $G_n = G_p$. From the condition of maintaining electro-neutrality, when $\rho = 0$ and $\partial\rho/\partial t = 0$, and also using (6)–(8), can be written

$$V_n = V_p = \frac{n - n_0}{\tau_n} = \frac{p - p_0}{\tau_p},\tag{9}$$

where n_0 and p_0 are concentration of carriers at thermal equilibrium. When used (4), (5) and (9) we obtain one-dimensional transfer equations [50]

$$\frac{\partial n}{\partial t} = D_n \frac{\partial^2 n}{\partial x^2} + \mu_n E \frac{\partial n}{\partial x} + \mu_n n \frac{\partial E}{\partial x} - \frac{n - n_0}{\tau_n} + G_n(x), \tag{10}$$

$$\frac{\partial p}{\partial t} = D_p \frac{\partial^2 p}{\partial x^2} - \mu_p E \frac{\partial p}{\partial x} - \mu_p p \frac{\partial E}{\partial x} - \frac{p - p_0}{\tau_p} + G_p(x), \tag{11}$$

where D_n and D_p are coefficients of diffusion of electrons and holes, μ_n and μ_p are coefficients of mobility of electrons and holes, E is an electric field strength in the base area.

Equations (10) and (11) can be greatly simplified if considering a stationary case when $dn/dt = 0$, $dp/dt = 0$, except for Eqs. (10) and (11) can be solved independently of each other, when $G_n = G_p$ and $p \gg n$ (or $n \gg p$) with $dE/dx = 0$. Consequently, the transport equation takes the form [50, 52]

$$D_n \frac{d^2 n_p}{dx^2} + \mu_n E \frac{dn_p}{dx} - \frac{n - n_{p0}}{\tau_n} = -G(x). \tag{12}$$

Let us consider the solution of the equation of continuity (12) for a semiconductor diode, which operates optical radiation. The energy diagram is shown in Fig. 1.

We will assume that all optical radiation is absorbed in the quasineutral region of the p-type ($x_p \leq x \leq x_p'$). Under the action of an electric field, excess non-main charge carriers effectively leave an area that is located near the boundary of the depleted layer ($x = x_p$), which equal the condition, when $x \to x_p$ and with a zero displacement value n_p coming to n_{p0}. If a direct voltage is applied to the p–n transition U, then the concentration of charge carriers at $x = x_p$ according to Shockley's law, may be represented in the form [51]

Fig. 1 Power diagram of semiconductor diode in direct mode

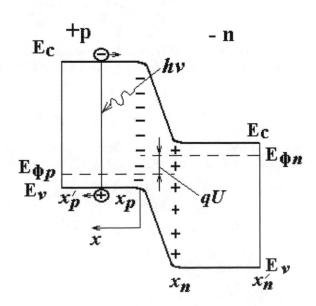

$$n_p = n_{p0} \exp(qU/kT).$$

We will assume that the thickness of the absorption layer is much larger than the diffusion length of charge carriers and the depth of light penetration for the considered light wavelength, that is, $x'_{p-}x_p \gg L_n$ and $x'_{p-}x_p \gg \alpha^{-1}$. We also consider that the electric field is absent in the quasineutral p-domain of absorption, i.e. $E = 0$. Taking into account these remarks, the transfer Eq. (12) takes the form

$$\frac{d^2 n_p}{dx^2} - \frac{n_p - n_{p0}}{D_n \tau_n} = -\frac{G(x)}{D_n}. \tag{13}$$

The magnitude $G(x)$ according to the theory of photoconductivity is determined [50, 52]

$$G(x) = G(0)e^{-ax}. \tag{14}$$

where $G(0) = \alpha\Gamma$, $\Gamma = \Gamma(\lambda)$ is a flux density of photons of monochromatic radiation at $x = x_p$. The boundary conditions after the solution of Eq. (13) has a form

$$n_p = n_{p0}e^{\frac{qU_0}{kT}} \text{ at } x = x_p, \tag{15}$$

$$n_p = n_{p0} \text{ at } x = \infty. \tag{16}$$

Taking into account (14) and $L_n^2 = D_n \tau_n$ Eq. (13) transformed into a form

$$\frac{d^2 n_p}{dx^2} - \frac{n_p - n_{p0}}{L_n^2} = -\frac{\alpha\Gamma}{D_n}e^{-\alpha(x-x_p)}. \tag{17}$$

The solution of Eq. (17) consists of two parts, respectively, for the right and left components. We first consider the solution for the left-hand side of Eq. (17), that has the form

$$n_p - n_{p0} = A_1 e^{\frac{(x-x_p)}{L_n}} + A_2 e^{\frac{-(x-x_p)}{L_n}}, \tag{18}$$

where A_1 and A_2 are determined from the boundary conditions (15) and (16). The solution for the right side of Eq. (17) in [20] has the form

$$n_p - n_{p0} = -\frac{\alpha\Gamma(\lambda)e^{-\alpha(x-x_p)}}{D_n\left[\alpha^2 - \frac{1}{D_n\tau_n}\right]}. \tag{19}$$

Thus, the general solution of Eq. (17) has the form

$$n_p - n_{p0} = n_{p0}\left(e^{\frac{qU_0}{kT}} - 1\right)e^{-\frac{x-x_p}{L_n}} - \frac{\alpha\Gamma(\lambda)e^{-\alpha(x-x_p)}}{D_n\left[\alpha^2 - \frac{1}{L_n^2}\right]}. \tag{20}$$

The first component in Eq. (20) characterizes the diffusion component of the concentration of non-main charge carriers in the dark when the diode is in direct mode, and the second component (20) determines the concentration of non-main charge carriers, that is due to the action of light. Equation (20) allows to determine the resistance of a diode under the action of optical radiation and at direct voltage. However, for a diode, except for the constant voltage and optical radiation, an alternating voltage operates and an electric field appears in the base region, that is connected either with a high level of injection or with strong optical radiation. For a case where a diode has a constant voltage and there is an electric field in the base region, the following transfer equation must be considered [50]

$$D_n\frac{d^2(n_p - n_{p0})}{dx^2} + \mu_E E\frac{d(n_p - n_{p0})}{dx} - \frac{n_p - n_{p0}}{\tau} = -G(x). \tag{21}$$

where μ_E is a bipolar drift mobility.
Denoting

$$\frac{\mu_E E}{D_n} = \frac{\mu_E E\tau_n}{D_n\tau_n} = \frac{l_E}{L^2}, \quad L^2 = D_n\tau_n.$$

Equation (21) is written in the form

$$\frac{d^2(n_p - n_{p0})}{dx^2} + \frac{l_E}{L^2}\frac{d(n_p - n_{p0})}{dx} - \frac{n_p - n_{p0}}{L^2} = -\frac{G(x)}{D_n}, \tag{22}$$

where $G(x)$ is described by Eq. (14). The boundary conditions for the solution of Eq. (22) has the form

$$n_p = n_{p0}e^{\frac{qU_0}{kT}} \text{ at } x = x_p, \tag{23}$$

$$n_p = n_{p0} \text{ at } x = W, \tag{24}$$

where W is the thickness of the diode base.
The solution of Eq. (22) consists of a general solution of a homogeneous equation and a partial solution of a nonhomogeneous equation. The solution of a homogeneous equation

$$n_p - n_{p0} = A_3 e^{K_1(x-x_p)} + A_4 e^{K_2(x-x_p)}, \tag{25}$$

where K_1 and K_1 is the roots of the square equation

$$K^2 + \frac{l_E}{2L^2} - \frac{1}{L^2} = 0, \tag{26}$$

that has values

$$K_{1,2} = -\frac{l_E}{2L^2} \pm \sqrt{\left(\frac{l_E}{2L^2}\right)^2 + \frac{1}{L^2}}. \tag{27}$$

Then the solution of the homogeneous Eq. (22) with substitution (27) takes the form

$$n_p - n_{p0} = A_3 e^{\left(-\frac{l_E}{2L^2} + \sqrt{\left(\frac{l_E}{2L^2}\right)^2 + \frac{1}{L^2}}\right)(x-x_p)} + A_4 e^{\left(-\frac{l_E}{2L^2} - \sqrt{\left(\frac{l_E}{2L^2}\right)^2 + \frac{1}{L^2}}\right)(x-x_p)}. \tag{28}$$

If we'll mark

$$\frac{1}{l_{1,2}} = \pm\sqrt{\left(\frac{l_E}{2L^2}\right)^2 + \frac{1}{L^2}} - \frac{l_E}{2L^2}, \tag{29}$$

then the solution of Eq. (22) can be written as

$$n_p - n_{p0} = A_3 e^{\frac{(x-x_p)}{l_1}} + A_4 e^{\frac{-(x-x_p)}{l_2}}. \tag{30}$$

Coefficients A_3 and A_4 define the boundary conditions (23) and (24). Having made the necessary transformations we will get the values A_3 and A_4, that is an expression (30)

$$n_p - n_{p0} = -\frac{n_{p0}\left(e^{\frac{qU_0}{kT}} - 1\right)e^{-\frac{(W-x_p)(l_1+l_2)}{l_1 l_2}}}{\left[1 - e^{-\frac{(W-x_p)(l_1+l_2)}{l_1 l_2}}\right]}e^{\frac{x-x_p}{l_1}} + \frac{n_{p0}\left(e^{\frac{qU_0}{kT}} - 1\right)}{\left[1 - e^{-\frac{(W-x_p)(l_1+l_2)}{l_1 l_2}}\right]}e^{-\frac{x-x_p}{l_2}}. \tag{31}$$

The partial solution of the nonhomogeneous Eq. (22) is sought in the form $\bar{y} = R_k(x)e^{\alpha x}$, where $R_k(x)$ is a polygon level k, if the right side of the equation has the form $f(x) = Q_k(x)e^{\alpha x}$ [53]. Thus, the partial solution of Eq. (22)

$$n_p - n_{p0} = -\frac{\Gamma(\lambda)\alpha e^{-\alpha(x-x_p)}}{D_n\left(\alpha^2 - \alpha\frac{l_E}{L^2} - \frac{1}{L^2}\right)}. \tag{32}$$

The general solution of Eq. (22) has the form

$$n_p - n_{p0} = A_3 e^{(x-x_p)/l_1} + A_4 e^{(x-x_p)/l_2} - \frac{\Gamma(\lambda)\alpha e^{-\alpha(x-x_p)}}{D_n\left(\alpha^2 - \alpha\frac{l_E}{L^2} - \frac{1}{L^2}\right)}. \tag{33}$$

In Eq. (33), the first two components describes the distribution of the concentration of charge carriers in the base region of the diode from the direct voltage, and the third component of the action of the optical radiation.

Since the diode works under alternating voltages and currents, it is necessary to determine the distribution of the concentration of charge carriers for this case. The one-dimensional transport equation for alternating current in a stationary mode under the action of light has the form

$$\frac{d^2 n_1}{dx^2} + \frac{l_E}{L^2}\frac{dn_1}{dx} - \frac{n_1(1 + j\omega\tau)}{L^2} = -\frac{G(x)}{D_n}, \tag{34}$$

where n_1 is a concentration of injected charge carriers due to the variable voltage at the emitter transition. It can be assumed that the electron concentration consists of a constant current component n_{E0} (function x) and alternating current component $n_1 e^{j\omega t}$ (function x and t). If the alternating voltage signal $U_{E1}(t)$ imposed on direct voltage U_{E0}, then the concentration of electrons in the p–n transition takes value

$$n_{E0} + n_{E1}(t) = n_{p0} \exp\left[\frac{q}{kT}(U_{E0} + U_{E1}(t))\right], \tag{35}$$

where $n_{E1}(t)$ is an electron concentration, that is determined by the alternating voltage applied to the average electron concentration caused by the direct voltage.

For a small signal case $q U_{E1}(t)/kT \ll 1$ and Eq. (35) can be greatly simplified in the expansion of a series of exponential functions while preserving the first two components of the decomposition. So, we can write

$$n_{E0} + n_{E1}(t) = n_{p0}e^{\frac{qU_{E0}}{kT}}\left[1 + \frac{qU_{E1}(t)}{kT}\right]. \tag{36}$$

On the basis of (36) we define the boundary conditions that must be used in solving Eq. (34)

$$n_1'(0, t) = n_{p0}e^{\frac{qU_{E0}}{kT}} + n_{p0}e^{\frac{qU_{E0}}{kT}}\frac{qU_{E1}(t)}{kT}, \tag{37}$$

$$n_1'(w, t) = n_{p0}. \tag{38}$$

The solution of the homogeneous Eq. (34) has the form

$$n_1(x,t) = A_5 e^{\frac{(x-x_p)C_6^*}{l_1}} + A_6 e^{-\frac{(x-x_p)C_6^*}{l_2}}, \tag{39}$$

where $C_6^* = \sqrt{1 + j\omega\tau}$, τ is a lifetime of electrons in the base, ω is an angular frequency, coefficients A_5 and A_6 are determined from the boundary conditions (37) and (38)

$$A_5 = \frac{n_1'(W,t) - n_1'(0,t)e^{-\frac{WC_6^*}{l_2}}}{e^{\frac{WC_6^*}{l_1}} - e^{\frac{WC_6^*}{l_2}}}, \quad A_6 = \frac{n_1'(0,t)e^{\frac{WC_6^*}{l_1}} - n_1'(W,t)}{e^{\frac{WC_6^*}{l_1}} - e^{\frac{WC_6^*}{l_2}}}. \tag{40}$$

Thus, the solution of the homogeneous Eq. (34) has the form

$$n_1(x,t) = \left[\frac{n_1'(W,t) - n_1'(0,t)e^{-\frac{WC_6^*}{l_2}}}{e^{\frac{WC_6^*}{l_1}} - e^{\frac{WC_6^*}{l_2}}}\right]e^{\frac{(x-x_p)C_6^*}{l_1}} + \left[\frac{n_1'(0,t)e^{\frac{WC_6^*}{l_1}} - n_1'(W,t)}{e^{\frac{WC_6^*}{l_1}} - e^{\frac{WC_6^*}{l_2}}}\right]e^{\frac{(x-x_p)C_6^*}{l_2}}.$$

$$\tag{41}$$

The partial solution of the nonhomogeneous Eq. (34) is determined by the expression (32). Thus, having obtained the general solution of the transport Eq. (34), we can determine the complete resistance of the base area of the diode. To do this, we'll use the general expression for current density

$$j_{gen} = q(\mu_n n + \mu_p p)E + q(D_n \nabla n - D_p \nabla p), \tag{42}$$

where

$$E = \frac{j_{gen} - q(D_n \nabla n - D_p \nabla p)}{q(\mu_n n + \mu_p p)}. \tag{43}$$

It is considered that in the base area there is a high level of injection, when $n_p \geq p_p$, as well as the condition of neutrality, from which follows that $\frac{\partial n}{\partial x} = \frac{\partial p}{\partial x}$ and $\frac{\partial n}{\partial t} = \frac{\partial p}{\partial t}$. Taking into account these remarks, Eq. (43) takes the form

$$E = \frac{j_{gen}}{q\mu_p n(b+1)} - \frac{(D_n/\mu_p)\nabla n}{n(b+1)} + \frac{kT}{q}\frac{\nabla n}{n(b+1)}, \tag{44}$$

where $b = \mu_n/\mu_p$. It should be noted that

$$j_{gen} = j_{c6} + j_{E0} + j_{E1}, \quad \nabla n = \nabla n_{c6} + \nabla n_{E0} + \nabla n_{E1}, \quad n = n_{c6} + n_{E0} + n_{E1},$$

then

$$E(x,\omega) = \frac{j_{c6}}{q\mu_p(b+1)(n_{c6} + n_{E0} + n_{E1})} + \frac{j_{E0}}{q\mu_p(b+1)(n_{c6} + n_{E0} + n_{E1})}$$

$$+ \frac{j_{E1}}{q\mu_p(b+1)(n_{c6} + n_{E0} + n_{E1})} - \frac{(D_n/\mu_p)\nabla n}{(b+1)(n_{c6} + n_{E0} + n_{E1})} + \frac{kT\nabla n}{q(b+1)(n_{c6} + n_{E0} + n_{E1})}.$$

$$\tag{45}$$

Contact voltage in the base area will be determined by the expression

$$U_{base} = -\int_0^W E(x,\omega)dx.$$ (46)

When substituting (45) in (46) we obtain

$$
\begin{aligned}
U_{base} = &-\int_0^W \frac{j_{c6}}{q\mu_p(b+1)(n_{c6}+n_{E0}+n_{E1})}dx - \int_0^W \frac{j_{E0}}{q\mu_p(b+1)(n_{c6}+n_{E0}+n_{E1})}dx \\
&-\int_0^W \frac{j_{E1}}{q\mu_p(b+1)(n_{c6}+n_{E0}+n_{E1})}dx + \int_0^W \frac{(D_n/\mu_p)\nabla n}{(b+1)(n_{c6}+n_{E0}+n_{E1})}dx \\
&-\frac{kT}{q}\int_0^W \frac{\nabla n}{q(b+1)(n_{c6}+n_{E0}+n_{E1})}dx.
\end{aligned}
$$ (47)

Complete resistance of the base area of the diode equals

$$Z_{base} = \frac{U_{E1}}{Sj_{E1}},$$ (48)

where U_{E1} is a voltage on the base area, S is a base area. When using the components that relate to the alternating voltage from the expression (47), we can write

$$
\begin{aligned}
Z_6 = &-\frac{2kT}{Sq^2 D_n(b+1)} \frac{\displaystyle\int_0^W \frac{A_5(C_6^*/l_1)-A_6(C_6^*/l_2)}{-\dfrac{G(0)e^{-\alpha(x-x_p)}}{D_n\left(\alpha^2-\dfrac{\alpha l_E}{L^2}-\dfrac{1}{L^2}\right)}+A_5 e^{\frac{(x-x_p)C_6^*}{l_1}}+A_6 e^{\frac{(x-x_p)C_6^*}{l_2}}}dx}{A_5\dfrac{C_6^*}{l_1}-A_6\dfrac{C_6^*}{l_2}} \\[2ex]
&+\frac{kT}{q(b+1)} \\
&*\int_0^W \frac{\alpha G(0)e^{-\alpha(x-x_p)}dx}{D_n\left(\alpha^2-\dfrac{\alpha l_E}{L^2}-\dfrac{1}{L^2}\right)qD_nS\left[\dfrac{A_5 C_6^*}{l_1}-\dfrac{A_6 C_6^*}{l_2}\right]\left[A_5 e^{\frac{(x-x_p)C_6^*}{l_1}}+A_6 e^{\frac{(x-x_p)C_6^*}{l_2}}-\dfrac{G(0)e^{-\alpha(x-x_p)}}{D_n\left(\alpha^2-\dfrac{\alpha l_E}{L^2}-\dfrac{1}{L^2}\right)}\right]} \\[2ex]
&+\frac{kT}{q(b+1)}\int_0^W \frac{A_5(C_6^*/l_1)e^{\frac{(x-x_p)C_6^*}{l_1}}-A_6(C_6^*/l_2)e^{\frac{(x-x_p)C_6^*}{l_2}}}{qD_nS\left[\dfrac{A_5 C_6^*}{l_1}-\dfrac{A_6 C_6^*}{l_2}\right]\left[A_5 e^{\frac{(x-x_p)C_6^*}{l_1}}+A_6 e^{\frac{(x-x_p)C_6^*}{l_2}}-\dfrac{G(0)e^{-\alpha(x-x_p)}}{D_n\left(\alpha^2-\dfrac{\alpha l_E}{L^2}-\dfrac{1}{L^2}\right)}\right]}dx.
\end{aligned}
$$ (49)

Fig. 2 Theoretical and
experimental dependences of
the active and reactive
components of the complete
resistance of the base of the
photosensitive diode

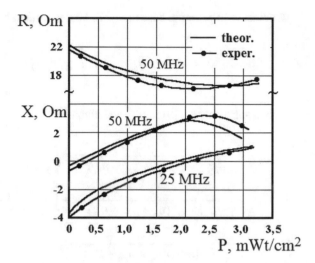

The solution of integrals in Eq. (49) is made on a personal computer in the computing environment "Mathlab 9.3" [54–56]. To verify the theoretical calculations, according to Eq. (49), experimental studies were carried out on the dependence of the complete resistance of the base of the diode, created on the basis of the integral transistor (base-collector transition: $I_{ss} = 0.4 \times 10^{-9}$ A; $C_D = 0.3 \times 10^{-12}$ F; $U_p = 12$ V). Theoretical and experimental dependences of the active and reactive components of the complete resistance against the power of optical radiation are given in Fig. 2.

As can be seen from the graph, the active component decreases with the increase in the power of optical radiation, that is explained by the decrease in the resistance of the base area. The reactive component with a capacitive character passes into an area with an inductive character, that initially sharply increases, and then saturation occurs, that is explained by the fulfillment of the condition of the approximate identity of the value of the photo-EMF voltage on the p–n transition at high optical radiation powers. Measurement of the complete resistance of the diode is made using the phase method [57].

3 Mathematical Model of Optical-Frequency Converter Based on Bipolar Transistors

In order to study the properties of a radiomeasuring optical-frequency converter, a mathematical model describing the dependencies of the active and reactive components of the complete resistance of the structure, of the sensitivity and frequency of generation on the power of optical radiation, as well as of carrying out experimental studies that would confirm the validity of the theoretical positions, should be described.

The scheme of a radiomeasuring optical-frequency converter based on a bipolar structure with a photosensitive diode is shown in Fig. 3.

In the circuit of the self oscillator converter, the oscillatory circuit is formed by the capacitive component of the complete resistance on the electrodes of the collector-collector of the bipolar transistors VT2 and VT4 and the inductance L. The cascaded activation of transistors VT1, VT2 and VT3, VT4 provides a high gain factor. Power supply is provided by voltage. Voltage is used to set the operating mode of the circuit (Fig. 4).

To determine the complete resistance of the optical-frequency converter based on the bipolar transistors, an equivalent circuit for alternating current is made (Fig. 5).

For the equivalent circuit of an optical-frequency converter (Fig. 5), the following symbols are used: $C_{be1}, C_{be2}, C_{be3}$ and C_{be4} are capacitances of transition base-emitter of the transistors VT1, VT2, VT3 and VT4 respectively; $C_{bc1}, C_{bc2}, C_{bc3}$ and C_{bc4} are capacitances of transition base-collector of the transistors VT1, VT2, VT3 and VT4

Fig. 3 Scheme of radiomeasuring optical-frequency converter on the basis of bipolar transistors

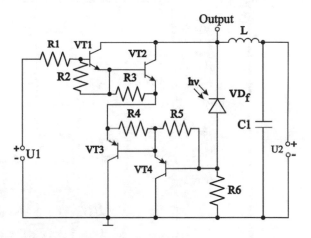

(a)   **(b)**

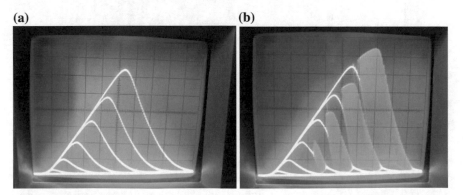

Fig. 4 Current-voltage characteristics of the optical-frequency converter on the basis of bipolar transistors: static (**a**) and dynamic (**b**)

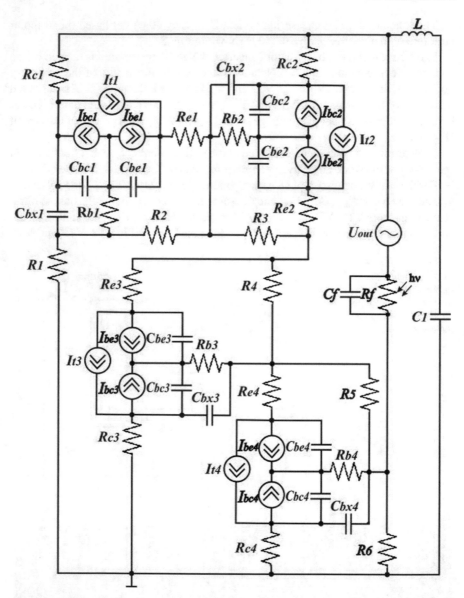

Fig. 5 Equivalent circuit of optical-frequency converter on the basis of bipolar transistors on alternating current

respectively; C_{bx1}, C_{bx2}, C_{bx3} and C_{bx4} are capacitances between the external output of the base and the collector of the transistors VT1, VT2, VT3 and VT4 respectively; R_f is a resistance of photosensitive diode R_f; C_f is a capacitance of a photosensitive diode; C_1 is a capacitance of the blocking capacitor; L is an inductance of oscillatory circuit; R_1, R_2, R_3, R_4, R_5, R_6 are resistances of the resistors R1, R2, R3, R4, R5,

R6 respectively; R_{c1}, R_{c2}, R_{c3} and R_{c4} are resistances of the transistor collectors VT1, VT2, VT3 and VT4 respectively; R_{e1}, R_{e2}, R_{e3} and R_{e4} are resistances of the transistor emitters VT1, VT2, VT3 and VT4 respectively; R_{b1}, R_{b2}, R_{b3} and R_{b4} are resistances of the transistor bases VT1, VT2, VT3 and VT4 respectively; I_{be1}, I_{be2}, I_{be3}, I_{be4}, I_{bc1}, I_{bc2}, I_{bc3}, I_{bc4}, I_{t1}, I_{t2}, I_{t3}, I_{t4} are currents of the transistors VT1, VT2, VT3 and VT4 respectively, which are determined by the equations [54]:

$$
I_{be} = I_s \exp\left(\frac{U_{be}}{NE \cdot V_t} - 1\right), \quad I_{bc} = I_s \exp\left(\frac{U_{bc}}{NC \cdot V_t} - 1\right), \quad I_t = \frac{I_{be} - I_{bc}}{Q},
$$

(50)

where $I_s = I_{ss} \exp\left(\frac{U_{js}}{NS \cdot V_t} - 1\right)$ is a saturation current; $V_t = \frac{kT}{q}$ is a reverse current of a p–n transition of the substrate, U_{be} is a base-emitter voltage, U_{bc} is a base-collector voltage, U_{js} is a contact potential difference of the collector-substrate area, NE is a coefficient of nonideality of emitter transition, NC is a coefficient of nonideality of collector transition; NS is a coefficient of nonideality of substrate area; Q is a charge in the base.

To perform calculations on a personal computer, we convert the scheme (Fig. 5) into a more convenient form (Fig. 6). Based on the equivalent scheme shown in Fig. 6 a system of Kirchhoff equations for an alternating current is made (51):

$$
\begin{cases}
I_{bc3} - I_{t3} = -\phi_1(Y_1 + Y_2 + Y_3) + \phi_2 Y_2 + \phi_{14} Y_3, \\
-(I_{be3} + I_{bc3}) = \phi_1 Y_2 - \phi_2(Y_2 + Y_4 + Y_5) + \phi_3 Y_5 + \phi_{14} Y_4, \\
I_{be3} - I_{t3} = \phi_2 Y_5 - \phi_3(Y_5 + Y_6) + \phi_{13} Y_6, \\
0 = -\phi_4(Y_7 + Y_8 + Y_9 + Y_{10}) + \phi_5 Y_8 + \phi_6 Y_9 + \phi_8 Y_{10}, \\
I_{t1} - I_{bc1} - U_{out} = \phi_4 Y_8 - \phi_5(Y_8 + Y_{11} + Y_{14}) + \phi_6 Y_{11}, \\
I_{bc1} + I_{be1} = \phi_4 Y_9 + \phi_5 Y_{11} - \phi_6(Y_9 + Y_{11} + Y_{12}), \\
-(I_{t1} + I_{be1}) = \phi_6 Y_{12} - \phi_7(Y_{12} + Y_{13}) + \phi_8 Y_{13}, \\
0 = \phi_4 Y_{10} + \phi_7 Y_{13} - \phi_8(Y_{10} + Y_{13} + Y_{15} + Y_{19} + Y_{23}) + \phi_9 Y_{19} + \phi_{10} Y_{15} + \phi_{13} Y_{23}, \\
I_{be2} + I_{bc2} = \phi_8 Y_{19} - \phi_9(Y_{16} + Y_{17} + Y_{19}) + \phi_{10} Y_{16} + \phi_{12} Y_{17}, \\
I_{t2} - I_{bc2} - U_{out} Y_{18} = \phi_8 Y_{15} + \phi_9 Y_{16} - \phi_{10}(Y_{15} + Y_{16} + Y_{18}), \\
U_{out}(Y_{14} + {}_{18} + Y_{20} + Y_{21}) = \phi_5 Y_{14} + \phi_{10} Y_{18} + \phi_{18} Y_{20}, \\
-(I_{t2} + I_{be2}) = -\phi_{12}(Y_{17} + Y_{22}) + \phi_9 Y_{17} + \phi_{13} Y_{22}, \\
0 = \phi_3 Y_6 + \phi_8 Y_{23} + \phi_{12} Y_{22} - \phi_{13}(Y_6 + Y_{22} + Y_{23} + Y_{24}), \\
0 = \phi_1 Y_3 + \phi_2 Y_4 + \phi_{13} Y_{24} - \phi_{14}(Y_4 + Y_{24} + Y_{24} + Y_{25} + Y_{26}), \\
I_{t4} + I_{be2} = \phi_{14} Y_{26} - \phi_{15}(Y_{26} + Y_{27}) + \phi_{16} Y_{27}, \\
-(I_{be4} + I_{bc4}) = \phi_{15} Y_{27} - \phi_{16}(Y_{26} + Y_{27} + Y_{28}) + \phi_{17} Y_{29} + \phi_{18} Y_{28}, \\
I_{bc4} - I_{t4} = \phi_{16} Y_{29} - \phi_{17}(Y_{290} + Y_{30} + Y_{31}) + \phi_{18} Y_{30}, \\
-U_{out} Y_{20} = \phi_{14} Y25 + \phi_{16} Y_{28} + \phi_{17} Y_{30} - \phi_{18}(Y_{20} + Y_{25} + Y_{28} + Y_{30}).
\end{cases}
$$

(51)

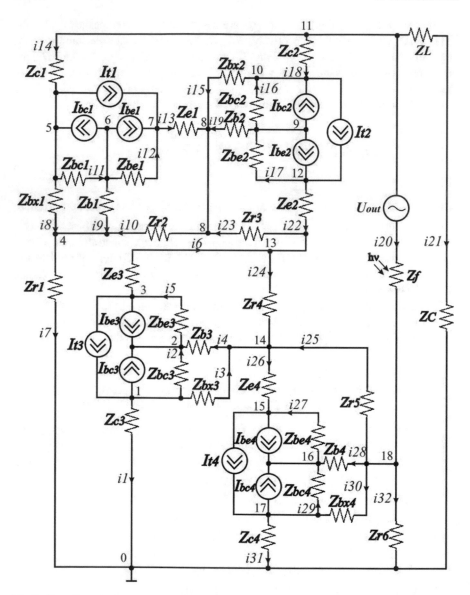

Fig. 6 Transform equivalent circuit of optical-frequency converter on the basis of bipolar transistors on alternating current

For the converted equivalent scheme (Fig. 6), the following notations are used:

$$Z_{r1} = R_1; \quad Z_{r2} = R_2; \quad Z_{r3} = R_3; \quad Z_{r4} = R_4; \quad Z_{r5} = R_5; \quad Z_{r6} = R_6;$$
$$Z_{b1} = R_{b1}; \quad Z_{b2} = R_{b2}; \quad Z_{b3} = R_{b3}; \quad Z_{b4} = R_{b4}; \quad Z_{c1} = R_{c1}; \quad Z_{c2} = R_{c2};$$
$$Z_{c3} = R_{c3}; \quad Z_{c4} = R_{c4}; \quad Z_{e1} = R_{e1}; \quad Z_{e2} = R_{e2}; \quad Z_{e3} = R_{e3}; \quad Z_{e4} = R_{e4};$$

$$Z_{bc1} = -j/(\omega \cdot C_{bc1}); \quad Z_{bc2} = -j/(\omega \cdot C_{bc2}); \quad Z_{bc3} = -j/(\omega \cdot C_{bc3});$$
$$Z_{bc4} = -j/(\omega \cdot C_{bc4}); \quad Z_{be1} = -j/(\omega \cdot C_{be1}); \quad Z_{be2} = -j/(\omega \cdot C_{be2});$$
$$Z_{be3} = -j/(\omega \cdot C_{be3}); \quad Z_{be4} = -j/(\omega \cdot C_{be4});$$
$$Z_{bx1} = -j/(\omega \cdot C_{bx1})l\,Z_{bx2} = -j/(\omega \cdot C_{bx2}); \quad Z_{bx3} = -j/(\omega \cdot C_{bx3});$$
$$Z_{bx4} = -j/(\omega \cdot C_{bx4}); \quad Z_C = -j/(\omega \cdot C_1); \quad Z_L = j\omega L;$$
$$Z_f = R_f/(1 + \omega^2 \cdot R_f^2 \cdot C_f^2) - j\omega R_f^2 C_f/(1 + \omega^2 R_f^2 C_f^2).$$

The conductivitys of the branches of the circuit (Fig. 6) are determined by the formulas:

$$Y_1 = 1/Z_{c3}; \quad Y_2 = 1/Z_{bc3}; \quad Y_3 = 1/Z_{bx3}; \quad Y_4 = 1/Z_{b3}; \quad Y_5 = 1/Z_{be3};$$
$$Y_6 = 1/Z_{e3}; \quad Y_7 = 1/Z_{r1}; \quad Y_8 = 1/Z_{bx1}; \quad Y_9 = 1/Z_{b1}; \quad Y_{10} = 1/Z_{r2};$$
$$Y_{11} = 1/Z_{bc1}; \quad Y_{12} = 1/Z_{be1}; \quad Y_{13} = 1/Z_{e1}; \quad Y_{14} = 1/Z_{c1}; \quad Y_{15} = 1/Z_{bx2};$$
$$Y_{16} = 1/Z_{bc2}; \quad Y_{17} = 1/Z_{be2}; \quad Y_{18} = 1/Z_{c2}; \quad Y_{19} = 1/Z_{b2}; \quad Y_{20} = 1/Z_f;$$
$$Y_{21} = 1/(Z_L + Z_C); \quad Y_{22} = 1/Z_{e2}; \quad Y_{23} = 1/Z_{r3}; \quad Y_{24} = 1/Zr_4;$$
$$Y_{25} = 1/Z_{r5}; \quad Y_{26} = 1/Z_{e4}; \quad Y_{27} = 1/Z_{be4}; \quad Y_{28} = 1/Z_{b4}; \quad Y_{29} = 1/Z_{bc4};$$
$$Y_{30} = 1/Z_{bx4}; \quad Y_{31} = 1/Z_{c4}; \quad Y_{32} = 1/Z_{r6}.$$

The system of Eq. (51) is solved by the Gaussian method using the software package Matlab 9.3. In Figs. 7 and 8 the theoretical and experimental dependences of the active and reactive components of the complete resistance against the power of optical radiation are given.

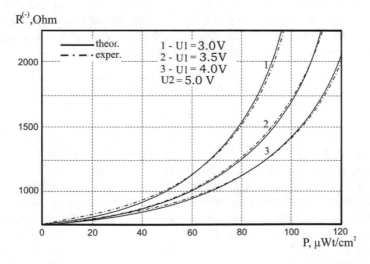

Fig. 7 Theoretical and experimental dependencies of the active component of the complete resistance against the power of optical radiation

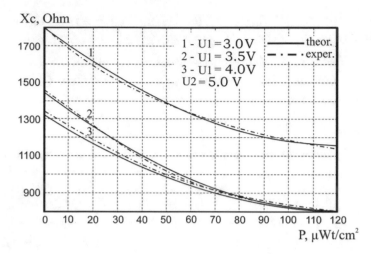

Fig. 8 Theoretical and experimental dependences of the reactive component of the complete resistance against the power of optical radiation

The reactive component of the complete resistance at a voltage of 3.0 V has a maximum value of 1800 Ω and decreases with an increase in the power of optical radiation and is 1165 Ω at 120 μW/cm^2.

The transformation function, that is, the dependence of the generation frequency on the change in optical radiation, is determined on the basis of the nonlinear equivalent scheme of the radiomeasuring converter. Initially, the reactive component of the complete resistivity of the transistor structure's drain–drain electrodes is determined, and then from the reactive component equivalent capacitance is determined that is depends against the optical radiation [58, 59]. Changing the equivalent capacitance determines the dependence of the generation frequency on optical radiation. The analytical expression of the transformation function has the form

$$F = \frac{1}{4\pi} \frac{\sqrt{2}\sqrt{R_f^2(P)C_f^2 + C_{bx1}R_f^2(P)C_f - LC_{bx1} - A_1}}{LC_{bx1}R_f^2(P)C_f^2}, \qquad (52)$$

where $A_1 = \sqrt{(R_f^2(P)C_f^2 + C_{bx1}R_f^2(P)C_f - LC_{bx1})^2 + 4C_{bx1}R_f^2(P)C_f}$.

Dependences of the generation frequency on the power of optical radiation, calculated by Eq. (52) and determined experimentally, are shown in Fig. 9.

From the graph, it is evident that with an increase in the radiation power a decrease in the generation frequency is observed, so at a control voltage of 3.0 V the frequency varies from 1745 kHz in the dark mode to 832 kHz with a power of optical radiation of 120 μW/cm^2. The difference between the theoretical and experimental components does not exceed $\pm 2.5\%$.

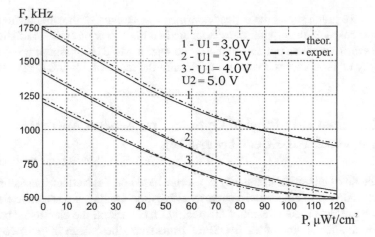

F, kHz

Fig. 9 Theoretical and experimental dependences of generation frequency against the power of optical radiation for different operating points of the optical-frequency converter

The sensitivity of the optical-frequency converter is determined by the formula:

$$S_P^F = \frac{1}{4} \frac{\sqrt{2}\left(\frac{\partial R_f(P)}{\partial P}\right)\left(R_f^2(P)C_f^2 - C_{bx1}R_f^2(P)C_f + \sqrt{A_2} + LC_{bx1}\right)}{\pi R_f^2(P)C_f\sqrt{A_2}\sqrt{-\frac{-R_f^2(P)C_f^2 - C_{bx1}R_f^2(P)C_f + \sqrt{A_2}\mid LC_{bx1}}{LC_{bx1}R_f^2(P)C_f^2}}}, \quad (53)$$

where

$$A_2 = R_f^4(P)C_f^4 + 2R_f^4(P)C_f^4 C_{bx1} + 2LC_{bx1}^2(P)C_f^2 + C_{bx1}^2 R_f^4(P)C_f^4$$
$$- 2C_{bx1}^2(P)R_f^2(P)R_f^2(P)C_f L + L^2 C_{bx1}^2.$$

Fig. 10 Dependence of the sensitivity of the optical-frequency converter from the change in the power of optical radiation

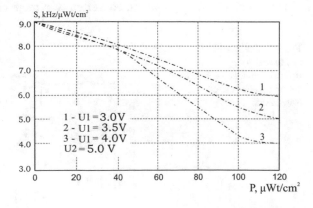

In Fig. 10 graph shows the dependence of the sensitivity of the optical-frequency converter based on the bipolar transistors against the power of optical radiation.

The sensitivity of an optical-frequency converter based on bipolar transistors at a frequency of 1750 kHz at a voltage of 5 V is in the range of 5.9–9.0 kHz/μW/cm^2.

4 Mathematical Model of Optical-Frequency Converter Based on Field-Effect Transistors

The scheme of an optical-frequency converter based on field-effect transistors with an active inductive element and a photosensitive diode, acting as a sensitive element, is shown in Fig. 11. As a result of the positive feedback on the electrodes between the drain and the source of the MOSFET transistors, there is an impedance which active component has a negative value (Fig. 12) and the reactive component has a capacitive character. The scheme (Fig. 11) contains a generator of electrical oscillations, formed by the capacitive component of the impedance on the fluxes of the MOSFET transistors VT1 and VT3, and the inductive component of the impedance on the electrodes of the emitter-collector of the bipolar transistor VT4 [60].

To creation a radiomeasuring optical-frequency converter as an integrated circuit, the design of passive inductance using thin-film technology is required. However, this inductance has low Q-factor and its size at frequencies 106 Hz not compatible with the size of integrated circuits [53, 57]. Thus, to solve this problem, it was proposed to use an active inductance, that has the inductive nature of the impedance on the outputs of the collector-emitter bipolar transistor VT4 as well as the phase-slip RC circuit [53, 61]. The phase-sliding RC-circle is realized on a resistor R3 and a capacitor C3.

To determine the impedance of a radiomeasuring optical-frequency converter based on field-effect transistors (Fig. 13), an equivalent circuit for alternating current is developed.

An equivalent circuit of an optical-frequency converter based on field-effect transistors on an alternating current, converted to a more convenient for calculations,

Fig. 11 Scheme of radiomeasuring optical-frequency converter based on field-effect transistors

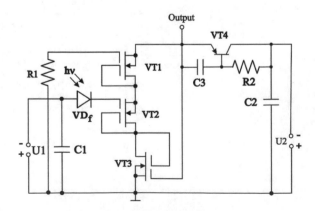

(a) **(b)**

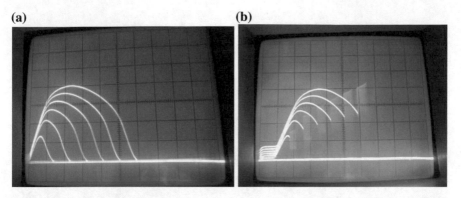

Fig. 12 Current-voltage characteristics of the optical-frequency converter based on field-effect transistors: static (**a**) and dynamic (**b**)

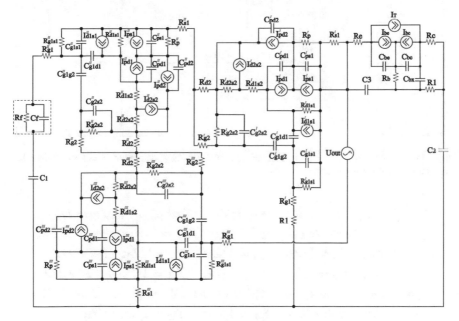

Fig. 13 Equivalent circuit of optical-frequency converter based on the field-effect transistors for alternating current

which is presented in Fig. 14. For the equivalent circuit of the optical-frequency converter (Fig. 13), the following symbols are used:

$$Y_1 = 1/Z_{46}; \quad Y_2 = 1/(Z_1 + Z_2 + Z_{29}); \quad Y_3 = 1/(Z_{32} + Z_{33});$$
$$Y_4 = 1/(Z_L + Z_{34}); \quad Y_5 = Z_{43}; \quad Y_6 = 1/Z_{44}; \quad Y_7 = Z_{45};$$
$$Y_8 = 1/Z_{38}; \quad Y_9 = 1/Z_{39}; \quad Y_{10} = 1/Z_{40}; \quad Y_{11} = 1/Z_{41}; \quad Y_{12} = 1/Z_{42};$$

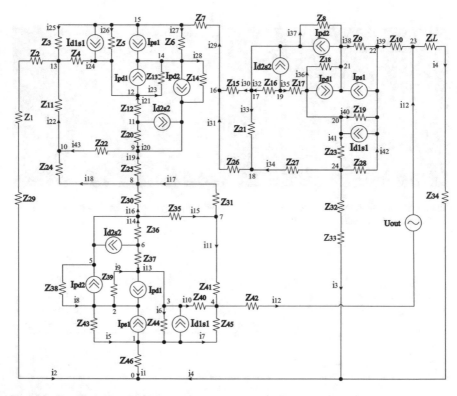

Fig. 14 Transduced equivalent circuit of optical-frequency converter based on field-effect transistors for alternating current

$$Y_{13} = 1/Z_{37}; \quad Y_{14} = Z_{36}; \quad Y_{15} = 1/Z_{35}; \quad Y_{16} = 1/Z_{30}; \quad Y_{17} = 1/Z_{31};$$
$$Y_{18} = Z_{24}; \quad Y_{19} = 1/Z_{25}; \quad Y_{20} = 1/Z_{20}; \quad Y_{21} = 1/Z_{12}; \quad Y_{22} = 1/Z_{11};$$
$$Y_{23} = 1/Z_{13}; \quad Y_{24} = 1/Z_{4}; \quad Y_{25} = 1/Z_{3}; \quad Y_{26} = 1/Z_{5}; \quad Y_{27} = 1/Z_{6};$$
$$Y_{28} = 1/Z_{14}; \quad Y_{29} = 1/Z_{7}; \quad Y_{30} = 1/Z_{15}; \quad Y_{31} = 1/Z_{26}; \quad Y_{32} = 1/Z_{16};$$
$$Y_{33} = 1/Z_{21}; \quad Y_{35} = 1/Z_{17}; \quad Y_{36} = 1/Z_{18}; \quad Y_{37} = 1/Z_{8}; \quad Y_{38} = 1/Z_{9};$$
$$Y_{39} = 1/Z_{8}; \quad Y_{40} = 1/Z_{19}; \quad Y_{41} = 1/Z_{23}; \quad Y_{42} = 1/Z_{28}; \quad Y_{43} = 1/Z_{22}.$$

On the basis of the equivalent scheme shown in Fig. 14 a system of Kirchhoff's equations for an alternating current is compiled (54):

$$
\left\{
\begin{aligned}
&I_{ps1} + I_{d1s1} = -\phi_1(Y_1 + Y_5 + Y_6 + Y_7) + \phi_2 Y_5 + \phi_3 Y_6 + \phi_4 Y_7, \\
&I_{pd2} - I_{pd1} - I_{ps1} = \phi_1 Y_5 - \phi_2(Y_5 + Y_8 + Y_9) + \phi_5 Y_8, \\
&I_{pd1} - I_{d1s1} = \phi_1 Y_6 + \phi_2 Y_9 - \phi_3(Y_6 + Y_9 + Y_{10} + Y_{13}) + \phi_4 Y_{10} + \phi_6 Y_{13}, \\
&0 = \phi_1 Y_7 + \phi_3 Y_{10} - \phi_4(Y_7 + Y_{10} + Y_{11} + Y_{12}) + \phi_7 Y_{11} + \phi_{23} Y_{12}, \\
&-I_{pd2} - I_{d2s2} = \phi_2 Y_8 - \phi_5(Y_8 + Y_{14} + Y_{15} + Y_{16}) + \phi_6 Y_{14} + \phi_7 Y_{15} + \phi_8 Y_{16}, \\
&I_{d2s2} = \phi_3 Y_{13} + \phi_5 Y_{14} - \phi_6(Y_{14} + Y_{13}), \\
&0 = \phi_4 Y_{11} + \phi_5 Y_{15} - \phi_7(Y_{11} + Y_{15} + Y_{17}) + \phi_8 Y_{17}, \\
&0 = \phi_5 Y_{16} + \phi_7 Y_{17} - \phi_8(Y_{16} + Y_{17} + Y_{18} + Y_{19}) + \phi_9 Y_{19} + \phi_{10} Y_{18}, \\
&-I_{pd2} - I_{d2s2} = \phi_8 Y_{19} - \phi_9(Y_{19} + Y_{20} + Y_{28} + Y_{43}) + \phi_{10} Y_{43} + \phi_{11} Y_{20} + \phi_{14} Y_{28}, \\
&0 = \phi_8 Y_{18} + \phi_9 Y_{43} - \phi_{10}(Y_{18} + Y_{22} + Y_{43}) + \phi_{13} Y_{22}, \\
&I_{d2s2} = \phi_9 Y_{20} - \phi_{11}(Y_{20} + Y_{21}) + \phi_{12} Y_{21}, \\
&I_{pd1} - I_{d1s1} = \phi_{10} Y_{22} - \phi_{12}(Y_{23} + Y_{24} + Y_{26}) + \phi_{13} Y_{24} + \phi_{14} Y_{23} + \phi_{15} Y_{26}, \\
&0 = \phi_{10} Y_{22} + \phi_{12} Y_{24} - \phi_{13}(Y_2 + Y_{22} + Y_{24} + Y_{25}) + \phi_{15} Y_{25}, \\
&-I_{ps1} - I_{pd1} + I_{pd2} = \phi_9 Y_{28} + \phi_{12} Y_{23} - \phi_{14}(Y_{23} + Y_{27} + Y_{28}) + \phi_{15} Y_{27}, \\
&I_{d1s1} + I_{ps1} = \phi_{12} Y_{26} + \phi_{13} Y_{25} + \phi_{14} Y_{27} - \phi_{15}(Y_{25} + Y_{26} + Y_{27} + Y_{29}) + \phi_{16} Y_{29}, \\
&0 = \phi_{15} Y_{29} - \phi_{16}(Y_{29} + Y_{30} + Y_{30}) + \phi_{17} Y_{30} + \phi_{18} Y_{31}, \\
&-I_{d2s1} - I_{pd2} = \phi_{16} Y_{30} - \phi_{17}(Y_{30} + Y_{32} + Y_{33} + Y_{37}) + \phi_{19} Y_{32} + \phi_{21} Y_{37}, \\
&0 = \phi_{16} Y_{31} + \phi_{17} Y_{33} - \phi_{18}(Y_{31} + Y_{33} + Y_{34}) + \phi_{24} Y_{34}, \\
&I_{d2s2} = \phi_{17} Y_{32} - \phi_{19}(Y_{31} + Y_{33} + Y_{34}) + \phi_{17} Y_{32} + \phi_{20} Y_{35}, \\
&I_{pd1} - I_{d1s1} = \phi_{19} Y_{35} - \phi_{20}(Y_{35} + Y_{36} + Y_{40} + Y_{41}) + \phi_{21} Y_{36} + \phi_{22} Y_{40} + \phi_{24} Y_{41}, \\
&I_{pd2} - I_{pd1} - I_{ps1} = \phi_{17} Y_{37} + \phi_{20} Y_{36} - \phi_{21}(Y_{36} + Y_{37} + Y_{38}) + \phi_{22} Y_{38}, \\
&I_{ps1} + I_{d1s1} - U_{out} Y_{39} = \phi_{20} Y_{40} + \phi_{21} Y_{38} - \phi_{22}(Y_{38} + Y_{39} + Y_{40} + Y_{42}) + \phi_{24} Y_{42}, \\
&0 = \phi_4 Y_{12} - U_{out}(Y_4 + Y_{12} + Y_{39}) + \phi_{22} Y_{39}, \\
&0 = \phi_{18} Y_{34} + \phi_{20} Y_{41} + \phi_{22} Y_{42} - \phi_{24}(Y_3 + Y_{34} + Y_{41} + Y_{42}).
\end{aligned}
\right.
\tag{54}
$$

For the equivalent circuit of an optical-frequency converter based on field-effect transistors (Fig. 13), the following symbols are used:

$$Z_1 = \frac{R_f}{1 + \omega^2 R_f^2 C_f^2} - j\frac{R_f^2 \omega C_f}{1 + \omega^2 R_f^2 C_f^2}; \quad Z_2 = R_{g1}^{II};$$

$$Z_4 = -j/\left(\omega C_{g1d1}^{II}\right); \quad Z_5 = R_{d1s1}^{II};$$

$$Z_3 = R_{g1s1}^{II}/(1 + \omega^2 (R_{g1s1}^{II})^2 (C_{g1s1}^{II})^2) - j\left((R_{g1s1}^{II})^2 \omega C_{g1s1}^{II}/(1 + \omega^2 (R_{g1s1}^{II})^2 (C_{g1s1}^{II})^2)\right);$$

$$Z_6 = R_p^{II}/(1 + \omega^2 (R_p^{II})^2 (C_{ps1}^{II})^2) - j\left((R_p^{II})^2 \omega C_{ps1}^{II}/(1 + \omega^2 (R_p^{II})^2 (C_{ps1}^{II})^2)\right);$$

$$Z_7 = R_{s1}^{II}; \quad Z_8 = -j/\left(\omega C_{pd2}^{II}\right);$$

$$Z_9 = R_p^{I}/(1 + \omega^2 (R_p^{I})^2 (C_{ps1}^{I})^2) - j\left((R_p^{I})^2 \omega C_{ps1}^{I}/(1 + \omega^2 (R_p^{I})^2 (C_{ps1}^{I})^2)\right);$$

$$Z_{10} = R_{s1}^{I}; \quad Z_{11} = -j/\omega(C_{g1g2}^{II}); \quad Z_{12} = R_{d1s2}^{II}; \quad Z_{13} = -j/\left(\omega C_{pd1}^{II}\right);$$

$$Z_{14} = -j/\left(\omega C_{pd2}^{II}\right); \quad Z_{15} = R_{d2}^{I}; \quad Z_{16} = R_{d2s2}^{I}; \quad Z_{17} = R_{d1s2}^{I};$$

$$Z_{18} = -j/\left(\omega C_{pd1}^I\right); \quad Z_{19} = R_{d1s1}^I; \quad Z_{20} = R_{d2s2}^{II};$$

$$Z_{21} = R_{g2s2}^I/(1 + \omega^2(R_{g2s2}^I)^2(C_{g2s2}^I)^2) - j\left((R_{g2s2}^I)^2\omega C_{g2s2}^I/(1 + \omega^2(R_{g2s2}^I)^2(C_{g2s2}^I)^2)\right);$$

$$Z_{22} = R_{g2s2}^{II}/(1 + \omega^2(R_{g2s2}^{II})^2(C_{g2s2}^{II})^2) - j\left((R_{g2s2}^{II})^2\omega C_{g2s2}^{II}/(1 + \omega^2(R_{g2s2}^{II})^2(C_{g2s2}^{II})^2)\right);$$

$$Z_{23} = -j/\left(\omega C_{g1d1}^I\right); \quad Z_{24} = R_{g2}^{II}; \quad Z_{25} = R_{d2}^{II}; \quad Z_{26} = R_{g2}^I;$$

$$Z_{28} = R_{g1s1}^I/(1 + \omega^2(R_{g1s1}^I)^2(C_{g1s1}^I)^2) - j\left((R_{g1s1}^I)^2\omega C_{g1s1}^I/(1 + \omega^2(R_{g1s1}^I)^2(C_{g1s1}^I)^2)\right);$$

$$Z_{27} = -j/\left(\omega C_{pd2}^I\right); \quad Z_{29} = -j/(\omega C_1); \quad Z_{30} = R_{d2}^{III}; \quad Z_{31} = R_{g2}^{III};$$

$$Z_{32} = R_{g1}^I; \quad Z_{33} = R_2; \quad Z_{34} = -j/(\omega C_2); \quad Z_{36} = R_{d2s2}^{III};$$

$$Z_{44} = R_{d1s1}^{III}; \quad Z_{46} = R_{s1}^{III};$$

$$Z_{35} = R_{g2s2}^{III}/(1 + \omega^2(R_{g2s2}^{III})^2(C_{g2s2}^{III})^2) - j\left((R_{g2s2}^{III})^2\omega C_{g2s2}^{III}/(1 + \omega^2(R_{g2s2}^{III})^2(C_{g2s2}^{III})^2)\right);$$

$$Z_{37} = R_{d1s2}^{III}; \quad Z_{38} = -j/\left(\omega C_{pd2}^{III}\right); \quad Z_{39} = -j/\left(\omega C_{pd1}^{III}\right); \quad Z_{40} = -j/\left(\omega C_{g1d1}^{III}\right);$$

$$Z_{41} = -j/\omega(C_{g1g2}^{III}); \quad Z_{42} = R_{g1}^{III};$$

$$Z_{43} = R_p^{III}/(1 + \omega^2(R_p^{III})^2(C_{ps1}^{III})^2) - j\left((R_p^{III})^2\omega C_{ps1}^{III}/(1 + \omega^2(R_p^{III})^2(C_{ps1}^{III})^2)\right);$$

$$Z_{45} = R_{g1s1}^{III}/(1 + \omega^2(R_{g1s1}^{III})^2(C_{g1s1}^{III})^2) - j\left((R_{g1s1}^{III})^2\omega C_{g1s1}^{III}/(1 + \omega^2(R_{g1s1}^{III})^2(C_{g1s1}^{III})^2)\right),$$

where R_f is a photodiode resistance, C_f is a photodiode capacities; C_{g1s1}^I, C_{g1s1}^{II}, C_{g1s1}^{III}, C_{g2s2}^I, C_{g2s2}^{II}, C_{g2s2}^{III} are capacities of a gate-source of the VT1, VT2 and VT3 transistors respectively; C_{pd1}^I, C_{pd1}^{II}, C_{pd1}^{III}, C_{pd2}^I, C_{pd2}^{II}, C_{pd2}^{III} are capacities of a body-drain of the VT1, VT2 and VT3 transistors respectively; C_{g1g2}^I, C_{g1g2}^{II}, C_{g1g2}^{III} are capacities between the first and the second gate of the VT1, VT2 and VT3; C_{bx} are capacity between a base and a collector terminal of the bipolar transistor VT4; C_1, C_2 and C_3 are capacities of the capacitors C_1, C_2 and C_3 respectively; C_{ps1}^{II}, C_{ps1}^{II}, C_{ps1}^{II} are capacities of a body-source of VT1, VT2 and VT3 transistors respectively; C_{g1d1}^I, C_{g1d1}^{II}, C_{g1d1}^{III} are capacities of a gate-drain of the VT1, VT2 and VT3 transistors respectively; C_c, C_e are capacities of a base–collector junction and a base–emitter junction of the bipolar transistor VT4 respectively; R_1 and R_2 are resistances of the resistors R_1 and R_2 respectively, R_{g1}^I, R_{g1}^{II}, R_{g1}^{III} are bulk resistances of gates of the VT1, VT2 and VT3 transistors respectively; R_{g1s1}^I, R_{g1s1}^{II}, R_{g1s1}^{III} and R_{g2s2}^I, R_{g2s2}^{II}, R_{g2s2}^{III} are bulk resistances of a gate-source of the VT1, VT2 and VT3 transistors respectively; R_{d1s1}^I, R_{d1s1}^{II}, R_{d1s1}^{III}, R_{d1s2}^I, R_{d1s2}^{II}, R_{d1s2}^{III}, R_{d2s2}^I, R_{d2s2}^{II}, R_{d2s2}^{III} are resistances of a drain-source of the VT1, VT2 and VT3 transistors respectively; R_{s1}^I, R_{s1}^{II}, R_{s1}^{III} are bulk resistances of a source of the VT1, VT2 and VT3 transistors respectively; R_{d2}^I, R_{d2}^{II}, R_{d2}^{III} are bulk resistances of a drain of VT1, VT2 and VT3 transistors respectively; R_{g1}^I, R_{g1}^{II}, R_{g1}^{III}, R_{g2}^I, R_{g2}^{II}, R_{g2}^{III} are bulk resistances of the first and second gates of the VT1, VT2 and VT3 transistors respectively; R_p^I, R_p^{II}, R_p^{III} are resistances of bodies of the VT1, VT2 and VT3 respectively; R_e, R_c, R_b are resistances of an emitter, collector and base of the transistor VT4 [62–64].

To determine the impedance of an active inductive element, a system of Kirchhoff's equations is developed for alternating current:

$$
\begin{cases}
U_1 = i_1 A_1 - i_2 Z_{c3} - i_4 Z_{r1}, \\
0 = -i_1 Z_{c3} + i_2 A_2 + i_3 Z_b + A_3, \\
0 = i_2 Z_b + i_3 A_4 + i_4 Z_{bx} + A_5, \\
0 = -i_1 Z_{r1} + i_3 Z_{bx} + i_4 A_6,
\end{cases}
\tag{55}
$$

where

$$
A_1 = Z_{c3} + Z_{r1}, \quad A_2 = Z_e + Z_{be} + Z_{c3} + Z_b, \quad A_3 = Z_{be}(I_{bc} - I_{be} - I_T),
$$
$$
A_4 = Z_{bc} + Z_{bx} + Z_b, \quad A_5 = Z_{bc}(I_{be} - I_{bc} + I_T), \quad A_6 = Z_{bx} + Z_c + Z_{r1}.
$$

The solution of the system of Eqs. (55) gives the impedance of an active inductive element:

$$
Z_L = U_1 / \left(\frac{U_1}{K_1} + \frac{A_5 A_6 Z_{c3} Z_b}{B_1 K_1 Z_{bx}(Z_{bx} - A_4 A_6/Z_{bx})} - \frac{Z_{c3} A_3}{K_1 B_1} + \frac{Z_{r1} A_5}{K_1(Z_{bx} - A_4 A_6/Z_{bx})} \right.
$$
$$
\left. - \frac{A_5 A_6 B_2 Z_b}{B_1 K_1 Z_{bx}(Z_{bx} - A_4 A_6/Z_{bx})} + \frac{A_3 B_2}{B_1 K_1} \right),
\tag{56}
$$

where

$$
B_1 = A_2 + Z_b^2 A_6/(Z_{bx}(Z_{bx} - A_4 A_6/Z_{bx})), \quad B_2 = Z_b Z_{r1}/(Z_{bx} - A_4 A_6/Z_{bx});
$$
$$
K_1 = A_1 - Z_{c3}^2/B_1 + Z_{c3} Z_b Z_{r1}/(B_1 Z_{bx}) + Z_{c3} Z_b Z_{r1} A_4 A_6 / \left(B_1 Z_{bx}^2(Z_{bx} - A_4 A_6/Z_{bx}) \right)
$$
$$
+ Z_{r1}^2 A_4/\left(Z_{bx}(Z_{bx} - A_4 A_6/Z_{bx})\right) + Z_{c3} B_2/B_1 - Z_b Z_{r1} B_2/(Z_{bx} B_1).
$$

The system of Eq. (54) is solved by the Gauss method using the software package Matlab 9.3. In Figs. 15 and 16 are given theoretical and experimental dependences of the active and reactive component of the complete resistance from the power of optical radiation.

The reactive component of the complete resistance at a voltage of 3.0 V has a maximum value of 2150 Ω and decreases with an increase in the power of optical radiation and at 120 $\mu W/cm^2$ is 1875 Ω.

The transformation function, that is the dependence of the generation frequency on the change in optical radiation, is determined on the basis of the nonlinear equivalent scheme of the radiomeasuring converter [25]. First, the reactive component of the complete resistance of the transistor structure on the drain–drain electrodes is determined, and then an equivalent capacitance is determined from the reactive component, that depends against the change of the optical radiation on the basis of the theory of Lyapunov's stability [65, 66]. Changing the equivalent capacitance determines the dependence of the generation frequency on the change in optical radiation

254

O. Osadchuk et al.

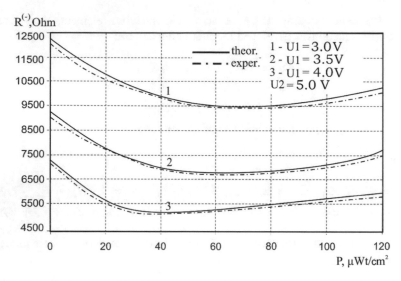

Fig. 15 Theoretical and experimental dependences of the active component of complete resistance against the power of optical radiation

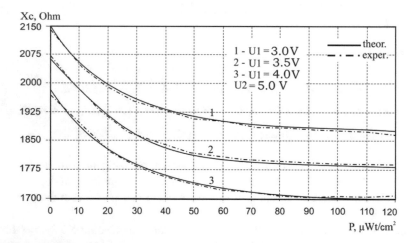

Fig. 16 Theoretical and experimental dependences of the reactive component of the complete resistance against the power of optical radiation

[67, 68]. The analytical expression of the transformation function has the form [69, 70]

$$
F = \frac{\sqrt{2}\sqrt{L_{ekv}C_4(-L_{ekv}C'_{pd2} + R_f^2(P)C_f^2 + R_f^2(P)C_f C'_{pd2} + A_1)}}{2L_{ekv}C_f C'_{pd2} R_f(P)}, \tag{57}
$$

where

$$A_1 = \sqrt{L_{ekv}^2 (C_{pd2}')^2 + 2L_{ekv} C_f^2 C_{pd2}' R_f^2(P) - 2L_{ekv}(C_{pd2}')^2 C_f R_f^2(P) + A_2},$$

$$A_2 = R_f^4(P)C_f^4 + 2R_f^4(P)C_f^3 C_{pd2}' + R_f^4(P)C_f^2(C_{pd2}')^2.$$

The sensitivity of the optical-frequency converter is determined by the formula:

$$
\begin{aligned}
S_P^F = \frac{1}{4}\sqrt{2}\Bigg(& 2R_f(P)C_f^2\left(\frac{\partial}{\partial P}R_f(P)\right) + 2R_f(P)C_f C_{pd2}'\left(\frac{\partial}{\partial P}R_f(P)\right) \\
& + \left(\frac{1}{2}\left(4L_{ekv}R_f(P)C_{pd2}'C_f^2\left(\frac{\partial}{\partial P}R_f(P)\right) - 4L_{ekv}R_f(P)\left(C_{pd2}'\right)^2 C_f\left(\frac{\partial}{\partial P}R_f(P)\right)\right.\right. \\
& + 4R_f^3(P)C_f^4\left(\frac{\partial}{\partial P}R_f(P)\right) + 8R_f^3(P)C_f^3 C_{pd2}'\left(\frac{\partial}{\partial P}R_f(P)\right) \\
& \left.\left. + 4R_f^3(P)C_f^2\left(C_{pd2}'\right)^2\left(\frac{\partial}{\partial P}R_f(P)\right)\right)\Big/ \sqrt{B_1}\right)\Big/ \left(\sqrt{-L_{ekv}C_{pd2}'(B_2 + \sqrt{B_1})}\right)\Bigg) \\
& - \frac{1}{2}\sqrt{2}\sqrt{L_{ekv}C_{pd2}'\left(B_2 + \sqrt{B_1}\right)}\left(\frac{\partial}{\partial P}R_f(P)\right)\Big/ \left(L_{ekv}C_{pd2}'C_f R_f^2(P)\right), \qquad (58)
\end{aligned}
$$

where

$$
\begin{aligned}
B_1 = & L_{ekv}^2\left(C_{pd2}'\right)^2 + 2L_{ekv}C_{pd2}'C_f^2 R_f^2(P) - 2L_{ekv}\left(C_{pd2}'\right)^2 C_f R_f^2(P) \\
& + R_f^4(P)C_f + 2R_f^4(P)C_f^3 C_{pd2}' + R_f^4(P)C_f^2\left(C_{pd2}'\right)^2.
\end{aligned}
$$

From the graph, it is evident that with an increase in the power of radiation, a decrease in the frequency generation is observed, so at a control voltage of 3.0 V the frequency varies from 2610 kHz in the dark mode to 2252 kHz with a power of optical radiation of 120 μW/cm^2 (Fig. 17). The difference between the theoretical and experimental components does not exceed $\pm 2.5\%$.

The sensitivity of the optical-frequency converter based on field-effect transistors at a frequency of 2600 kHz at a voltage of 5 V is in the range of 2.9–11.8 kHz/μW/cm^2 (Fig. 18).

5 Discussion and Conclusion

The authors proposed a method for constructing radiomeasuring optical-frequency converters on the basis of photoreactive effect in diode structures, which are included as sensitive elements in self oscillator devices based on bipolar and field-effect transistor structures with negative differential resistance, that created an opportunity for

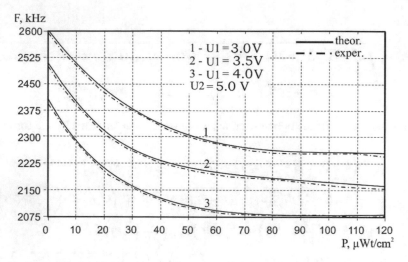

Fig. 17 Theoretical and experimental dependencies of the generation frequency against the power of optical radiation for different operating points of an optical-frequency converter based on field-effect transistors

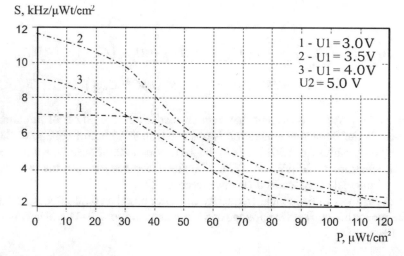

Fig. 18 Dependence of the sensitivity of the optical-frequency converter on the change in the power of optical radiation

the implementation of optical radiation converters with frequency output signal and microelectronic manufacturing technology.

In the paper, the investigation of the photoreactive effect in semiconductor diodes in a dynamic mode was carried out on the basis of the solution of the transport equation under the action of optical radiation, which allowed to obtain the distribution of charge carriers in the base area of the diode. It gave the opportunity to theoretically calculate the complete resistance of the base area and obtain the analytical dependence of its components against the power of optical radiation. The mathematical model of the optical-frequency converter with a photodiode as a photosensitive element is developed, based on the self oscillator generated by the capacitive component of the complete resistance of the bipolar structure and the inductance. Based on the model, analytical expressions for the transformation function and the sensitivity equation are derived. Theoretical and experimental dependencies has shown that the sensitivity of the developed optical-frequency converter at a voltage of 5 V lies in the range from 5.9 to 9.0 kHz/μW/cm^2. The mathematical model of the optical-frequency converter with a photodiode as a photosensitive element is developed, on the basis of the self oscillator created by the capacitive component of the complete resistance of the field-effect structure and active inductance. Based on the model, analytical expressions for the transformation function and the sensitivity equation are derived. Theoretical and experimental dependencies has shown that the sensitivity of the developed optical-frequency converter lies in the range from 2.9 to 11.8 kHz/μW/cm^2.

References

1. Bereziuk OV, Lemeshev MS, Bohachuk VV, Duk M (2018) Means for measuring relative humidity of municipal solid wastes based on the microcontroller Arduino UNO R3. Proc SPIE J 10808:108083. https://doi.org/10.1117/12.2501557
2. Bereziuk O, Lemeshev M, Bogachuk V et al (2019) Ultrasonic microcontroller device for distance measuring between dustcart and container of municipal solid wastes. Przegl Elektrotech J 4:146–150. https://doi.org/10.15199/48.2019.04.26
3. Yanenko A, Totsky A, Pidchenko S, Taranchuk A (2018) Experimental study of microwave radiation caused by the materials contacting with human body. J Telecommun Radio Eng (English translation of Elektrosvyaz and Radiotekhnika) 77(7):635–644. https://doi.org/10.1615/TelecomRadEng.v77.i7.60
4. Pidchenko S, Taranchuk A, Totsky A (2017) Multi-frequency quartz oscillating systems using digital compensation of frequency instability caused by variations of temperature and vibrations. J Telecommun Radio Eng (English translation of Elektrosvyaz and Radiotekhnika) 76(13):1193–1200. https://doi.org/10.1615/TelecomRadEng.v76.i13.70
5. Naumenko VV, Totsky AV, Pidchenko SK et al (2017) Multi frequency synthezier of bispectral triplet-signal designed for digital communication system. J Telecommun Radio Eng (English translation of Elektrosvyaz and Radiotekhnika) 76(2):147–155. https://doi.org/10.1615/TelecomRadEng.v76.i2.50
6. Taranchuk AA, Pidchenko SK, Khoptinskiy RP (2015) Dynamics of temperature-frequency processes in multifrequency crystal oscillators with digital compensations of resonator performance instability. J Radioelectron Commun Syst 58(6):250–257. https://doi.org/10.3103/S0735272715060023
7. Skulysh M (2017) The method of resources involvement scheduling based on the long-term statistics ensuring quality and performance parameters. In: Proceedings of the 2017 international conference on information and telecommunication technologies and radio electronics (UkrMiCo). https://doi.org/10.1109/ukrmico.2017.8095430
8. Globa L, Skulysh M, Reverchuk A (2014) Control strategy of the input stream on the online charging system in peak load moments. In: Proceedings of the 24th international Crimean conference microwave and telecommunication technology, 7–13 Sept 2014, pp 312–313. https://doi.org/10.1109/crmico.2014.6959409

9. Kryvinska N (2010) Converged network service architecture: a platform for integrated services delivery and interworking. In: Electronic business series, vol 2. International Academic Publishers, Peter Lang Publishing Group

10. Kryvinska N (2004) Intelligent network analysis by closed queuing models. Telecommun Syst 27:85–98. https://doi.org/10.1023/B:TELS.0000032945.92937.8f

11. Liang G, Shengchun L, Yi Zuowei et al (2011) Fiber-optic vibration sensor based on beat frequency and frequency-modulation demodulation techniques. J IEEE Photonics Technol Lett 23(1):18–20. https://doi.org/10.1109/LPT.2010.2089612

12. Huang CY, Wang WC, Wu WJ, Ledoux WR (2007) Composite optical bend loss sensor for pressure and shear measurement. IEEE Sens J 7(11):1554–1565. https://doi.org/10.1109/JSEN. 2007.908236

13. Samila A, Khandozhko V, Politansky L (2017) Energy efficiency increase of NQR spectrometer transmitter at pulse resonance excitation with noise signals. Solid State Nucl Magn Reson 87:10–17. https://doi.org/10.1016/j.ssnmr.2017.06.001

14. Osadchuk AV et al (2013) Noncontact infrared thermometer based on a self-oscillating lambda type system for measuring human body's temperature. In: 2013 23rd international Crimean conference "microwave and telecommunication technology". Sevastopol, Ukraine, 8–14 Sept 2013, pp 1069–1070

15. Vuitsik V, Hotra Z, Hotra O et al (2003) Mikroelektronni sensory fizychnykh velychyn (Microelectronic sensors of physical quantities). In: Hotra ZY (ed) vol 2. League-Press, Lvov

16. Krioukov E, Klunder DJW, Driessen A et al (2002) Sensor based on an integrated optical microcavity. J Opt Lett 27(7):512–514. https://doi.org/10.1364/OL.27.000512

17. Grattan KTV, Meggitt BT (eds) (1998) Optical fiber sensor technology, vol 3. Kluwer Academic Publishers, London

18. Michael KL, Taylor LC, Schultz SL et al (1998) Randomly ordered addressable high-density optical sensor arrays. J Anal Chem 70(7):1242–1248. https://doi.org/10.1021/ac971343r

19. Sharapova VM, Polishchuka ES (eds) (2012) Datchiki: Spravochnoye posobiye (Sensors: Reference book). Technosphere, Moscow

20. Osadchuk AV, Osadchuk IA (2015) Frequency transducer of the pressure on the basis of reactive properties of transistor structure with negative resistance. In: 2015 international Siberian conference on control and communications (SIBCON), 21–23 May 2015, Omsk, Russia, pp 1–3. https://doi.org/10.1109/sibcon.2015.7147168

21. Osadchuk VS et al (2010) Experimental research and modeling of the microwave oscillator based on the static inductance transistor structure with negative resistance. In: Proceedings of the 20th international Crimean conference microwave and telecommunication technology (CriMiCo). Sevastopol, Ukraine, 13–17 Sept 2010, pp 187–188. https://doi.org/10.1109/crmico.2010.5632543

22. Osadchuk VS, Osadchuk AV, Yushchenko YA (2008) Radiomeasuring thermal flowmeter of gas on the basis of transistor structure with negative resistance. Elektron Elektrotech J 84(4):47–52

23. Semenov A, Osadchuk O, Semenova O et al (2018) Signal statistic and informational parameters of deterministic chaos transistor oscillators for infocommunication systems. In: Proceedings of the international scientific-practical conference problems of infocommunications. Science and technology, 9–12 Oct 2018, pp 730–734. https://doi.org/10.1109/infocommst.2018.8632046

24. Schaumburg H (1992) Sensoren. BG Teubner, Stuttgart

25. Osadchuk OV, Osadchuk VS, Osadchuk IO et al (2017) Optical transducers with frequency output. Proc SPIE J 10445, Aug 2017. https://doi.org/10.1117/12.2280892

26. Novickiy PV, Knoring VG, Gutnikov VS (1970) Tsifrovyye pribory s chastotnymi datchikami (Digital devices with frequency sensors). Energiya, Leningrad

27. Skulysh MA (2019) Managing the process of servicing hybrid telecommunications services. Quality control and interaction procedure of service subsystems. Adv Intell Syst Comput 889:244–256. https://doi.org/10.1007/978-3-030-03314-9_22

28. Skulysh M, Romanov O (2018) The structure of a mobile provider network with network functions virtualization. In: Proceedings of the 14th international conference on advanced trends in radioelecrtronics, telecommunications and computer engineering, Lviv-Slavske, Ukraine, 20–24 Feb 2018. https://doi.org/10.1109/tcset.2018.8336370

29. Kryvinska N (2008) An analytical approach for the modeling of real-time services over IP network. Math Comput Simul 79(4):980–990. https://doi.org/10.1016/j.matcom.2008.02.016
30. Kryvinska N, Zinterhof P, van Thanh D (2007) An analytical approach to the efficient real-time events/services handling in converged network environment. In: Enokido T, Barolli L, Takizawa M (eds) Network-based information systems. NBiS 2007. Lecture notes in computer science, vol 4658. Springer, Berlin
31. Kryvinska N, Zinterhof P, van Thanh D (2007) New-emerging service-support model for converged multi-service network and its practical validation. In: First international conference on complex, intelligent and software intensive systems (CISIS'07), pp 100–110. https://doi.org/10.1109/cisis.2007.40
32. Ageyev D et al (2019) Infocommunication networks design with self-similar traffic. In: 2019 IEEE 15th international conference on the experience of designing and application of CAD systems (CADSM), pp 24–27. https://doi.org/10.1109/cadsm.2019.8779314
33. Ageyev DV, Salah MT (2016) Parametric synthesis of overlay networks with self-similar traffic. Telecommun Radio Eng (English translation of Elektrosvyaz and Radiotekhnika) 75(14):1231–1241
34. Pereverzev A, Ageyev D (2013) Design method access network radio over fiber. In: 2013 12th international conference on the experience of designing and application of CAD systems in microelectronics (CADSM). Polyana Svalyava, pp 288–292
35. Radivilova T et al (2018) Decrypting SSL/TLS traffic for hidden threats detection. In: Proceedings of the 2018 IEEE 9th international conference on dependable systems, services and technologies (DESSERT), pp 143–146. https://doi.org/10.1109/dessert.2018.8409116
36. Radivilova T, Kirichenko L, Ageiev D, Bulakh V (2020) The Methods to improve quality of service by accounting secure parameters. In: Hu Z, Petoukhov S, Dychka I, He M (eds) Advances in computer science for engineering and education II. ICCSEEA 2019. Advances in intelligent systems and computing, vol 938. Springer, Cham
37. Ageyev D, Yarkin D, Nameer Q (2014) Traffic aggregation and EPS network planning problem. In: Proceedings of the 2014 first international scientific-practical conference problems of infocommunications science and technology. Kharkov, Ukraine, pp 107–108. https://doi.org/10.1109/infocommst.2014.6992316
38. Ageyev D et al (2018) Method of self-similar load balancing in network intrusion detection system. In: 2018 28th international conference radioelektronika (RADIOELEKTRONIKA), pp 1–4. https://doi.org/10.1109/radioelek.2018.8376406
39. Bulakh V, Kirichenko L, Radivilova T (2018) Time series classification based on fractal properties. In: Proceedings of the 2018 IEEE second international conference on data stream mining and processing (DSMP), pp 198–201. https://doi.org/10.1109/dsmp.2018.8478532
40. Kirichenko L, Radivilova T, Tkachenko A (2019) Comparative analysis of noisy time series clusterin. In: CEUR workshop proceedings, vol 2362, pp 184–196. http://ceur-ws.org/Vol-2362/paper17.pdf. Accessed 15 May 2019
41. Kirichenko L, Radivilova T, Bulakh V (2018) Classification of fractal time series using recurrence plots. In: 2018 international scientific-practical conference problems of infocommunications. science and technology (PIC S&T). IEEE, pp 719–724. https://doi.org/10.1109/infocommst.2018.8632010
42. Semenova O, Semenov A, Voznyak O et al (2015) The fuzzy-controller for WiMAX networks. In: 2015 International Siberian conference on control and communications (SIBCON). Omsk, Russia, 21–23 May 2015. https://doi.org/10.1109/sibcon.2015.7147214
43. Semenov AA, Semenova OO, Voznyak OM et al (2016) Routing in telecommunication networks using fuzzy logic. In: 2016 17th international conference of young specialists on micro/nanotechnologies and electron devices (EDM). Erlagol, Russia, 30 June–4 July 2016. https://doi.org/10.1109/edm.2016.7538719
44. Semenova O, Semenov A, Koval K et al (2013) Access fuzzy controller for CDMA networks. In: 2013 international Siberian conference on control and communications (SIBCON). Krasnoyarsk, Russia, 12–13 Sept 2013. https://doi.org/10.1109/sibcon.2013.6693644

45. Brajlovskyj VV, Veryga AD, Gotra ZJ et al (2010) Simulation of circuit of autodyne sensor with field-controlled transistor. Radioelectron Commun Syst 53:550–554. https://doi.org/10.3103/S0735272710100055

46. Samila A (2017) Peculiarities of using s-simulation for parametric identification of multiplet [115]In NQR spectra in InSe. Meas J Int Meas Confeder 106:109–115. https://doi.org/10.1016/j.measurement.2017.04.035

47. Hotra O, Samila A, Politansky L (2018) Synthesis of the configuration structure of digital receiver of NQR radiospectrometer. Przegl Elektrotech 94(7/2018):58–61. https://doi.org/10.15199/48.2018.07.14

48. Rozensher E, Vinter B (2006) Optoelektronika (optoelectronics). Technosphera, Moscow

49. Vigdorvich EN, Winter B (2011) Fizicheskiye osnovy, proyektirovaniye i tekhnologiya optikoelektronnykh ustroystv (Physical fundamentals, design and technology of optoelectronic devices). MSTU Academic Publishers, Moscow

50. Pikhtin AN (2001) Opticheskaya i kvantovaya elektronika (Optical and Quantum Electronics). Vysshaya shkola, Moscow

51. Shalimova KV (1985) Fizika poluprovodnikov (Physics of semiconductors). Energiya, Moscow

52. Bonch-Bruevich VL, Kalashnikov SG (1990) Poluprovodnikovaya fizika (Semiconductor Physics). Nauka, Moscow

53. Osadchuk VS, Osadchuk AV (1999) Reaktyvni vlastyvosti tranzystoriv i tranzystornykh skhem (Reactive properties of transistors and transistor circuits). Universum-Vinnitsia, Vinnytsia

54. PSPICE User's Guide (1989) MicroSim corporation. La Cadensa Drive, Laguna Hills

55. Razevig VD (1992) Primeneniye programm P-CAD i PSpice dlya skhemotekhnicheskogo modelirovaniya na PEVM (Application of the P-CAD and PSPICE program for circuitry modeling on PEVM). Release 3, Radio i svyaz, Moscow

56. Potemkin VG (1999) Sistema inzhenernykh i nauchnykh raschetov Matlab 5.X (The system of engineering and scientific computations Matlab 5.X). Dialog-MIFI, Moscow

57. Osadchuk VS (1987) Induktivnyy effekt v poluprovodnikovykh priborakh (Inductive effect in semiconductor devices). Vyshcha shkola, Kiyev

58. Kayatskas AA (1988) Osnovy radioelektroniki (Fundamentals of radioelectronics). Vysshaya shkola, Moscow

59. Zi S (1984) Fizika poluprovodnikovykh priborov (Physics of semiconductor devices), vol 2. Mir, Moscow

60. Osadchuk OV, Osadchuk IO, Suleimenov B et al (2017) Frequency pressure transducer with a sensitivity of MEM capacitor on the basis of transistor structure with negative resistance. Proc SPIE 10445 August 2017. https://doi.org/10.1117/12.2280958

61. Osadchuk AV, Osadchuk IA (2015) Frequency transducer of the pressure on the basis of reactive properties of transistor structure with negative resistance. In: Proceedings of the 2015 international siberian conference on control and communications, 21–23 May 2015. Omsk, Russia. https://doi.org/10.1109/sibcon.2015.7147168

62. Zegrya GG, Perel VI (2009) Osnovy fiziki poluprovodnikov (Fundamentals of semiconductor physics). Fizmatlit, Moscow

63. Grundman M (2012) Osnovy fiziki poluprovodnikov. Nanofizika i tekhnicheskiye prilozheniya (Fundamentals of semiconductor physics. Nanophysics and technical applications). Fizmatlit, Moscow

64. Piter Y, Manuel Kardona (2002) Osnovy fiziki poluprovodnikov (Fundamentals of semiconductor physics). Fizmatlit, Moscow

65. Komar MV, Pryakhin AE (2010) Osnovy radioelektroniki: Kurs lektsiy (Fundamentals of radio electronics: a course of lectures). BGU Academic Publishers, Minsk

66. Deikova GM, Zhuravlyov VA, Maidanovsky AS et al (eds) (2006) Osnovy radioelektroniki: kompyuternyy laboratornyy praktikum (Bases of radio electronics: computer laboratory practice). Publishing House NTL, Tomsk

67. Semenov A (2016) The Van der Pol's mathematical model of the voltage-controlled oscillator based on a transistor structure with negative resistance. In: Proceedings of the XIII international conference modern problems of radio engineering, telecommunications, and computer science. Lviv-Slavsko, Ukraine, 23–26 Feb 2016, pp 100–104. https://doi.org/10.1109/tcset.2016.7451982

68. Semenov A (2016) Mathematical simulation of the chaotic oscillator based on a field-effect transistor structure with negative resistance. In: 2016 IEEE 36th international conference on electronics and nanotechnology (ELNANO), 19–21 Apr 2016. Kyiv, Ukraine, pp 52–56. https://doi.org/10.1109/elnano.2016.7493008

69. Tarnovskii NG, Osadchuk VS, Osadchuk AV (2000) Modeling of the gate junction in GaAs MESFETs. Russ Microlectron 29:279 283. https://doi.org/10.1007/BF02773276

70. Semenov A, Semenova O, Osadchuk O (2015) The UHF oscillators based on a HEMT structure with negative conductivity. In: Proceedings of the international Siberian conference on control and communications, 21–23 May 2015. Omsk, Russia. https://doi.org/10.1109/sibcon.2015.7147215

Implementation of Evolutionary Methods of Solving the Travelling Salesman Problem in a Robotic Warehouse

Andrii Oliinyk ⓘ **, Ievgen Fedorchenko** ⓘ **, Alexander Stepanenko** ⓘ **, Mykyta Rud, and Dmytro Goncharenko** ⓘ

Abstract An evolutionary method for solving the traveling salesman problem in the field of pharmacy business by optimizing the work of the drug delivery device is proposed in this paper. Modifications of three methods of initialization of the initial population of the genetic algorithm are developed. The software implementation is proposed to solve the problem of a sales-man in the pharmacy business by optimizing the process of drug delivery, using modified evolutionary methods. Unlike existing methods, the modified version of the evolution method allows to choose the original method of population initialization when solving the problem of traveling salesman, which, in turn, allows to generate more adapted chromosomes (chromosomes with better values of fitness functionality) at the stage of initialization and thereby improve the results algorithm. It is also possible to graphically monitor the process of solving the seller's problem, get the result in text and graphic forms. The principles of object-oriented programming, namely the use of classes, the principle of data encapsulation and inheritance, were used when writing the program. UML diagrams of classes, sequences, activities, states and cooperation were used to visualize the structure and functional relationships of the modules of the developed software.

Keywords Optimal way · Travelling salesman problem · Genetic algorithm · Evolutionary algorithm · Minimal distance · Graph

A. Oliinyk · I. Fedorchenko (✉) · A. Stepanenko · M. Rud · D. Goncharenko
Zaporizhzhia National Technical University, Zaporizhzhia, Ukraine
e-mail: evg.fedorchenko@gmail.com

A. Oliinyk
e-mail: olejnikaa@gmail.com

A. Stepanenko
e-mail: alex@zntu.edu.ua

M. Rud
e-mail: rudnike2@gmail.com

D. Goncharenko
e-mail: gdimk99@gmail.com

T. Radivilova et al. (eds.), *Data-Centric Business and Applications*, Lecture Notes on Data Engineering and Communications Technologies 48,
https://doi.org/10.1007/978-3-030-43070-2_13

263

1 Introduction

The problem of finding the shortest path is a classic optimization problem that is difficult to solve using traditional methods. Its purpose is to find the shortest way for a salesman who must visit N points. In a mathematical form it can be formulated as follows: let $K_n = (V, E)$ be a complete graph with n nodes and let c_e be the length $e \in E$ be the set of all Hamiltonian cycles (rounds) in K_n. Then we have:

$$min\{c(T)|T \in H\}, \tag{1}$$

where T is a Hamilton circuit.

This type of task exists in many forms, with several engineering applications that include the optimal location of the gas pipeline, the design of the feed system antenna, the configuration of transistors on boards, or sorting objects to obtain the most appropriate configuration. Much work on the salesman algorithm is motivated by its use as a platform for studying common methods, that can be applied to a wide range of discrete optimization problems [1].

Existing methods for solving this problem, for example, heuristic and greedy, have the disadvantages of finding a non-optimal solution, i.e. approximating the real value. That is why it is necessary to develop methods that will overcome these shortcomings.

2 Literature Review

Nowadays, scientists know the several methods of practical using of the salesman problem solution. Let's consider this task as one using way on automatic delivery of pills by a pharmacy robot to the clients, who need. Using these robots the issue of the solving salesman optimization arises when medicine is given by a automatic machine manipulator. There are many ways to automatize customer service, for example, the equipment of the Consis company (Fig. 1). These pharmacy robots, called robopharmacists, made by a German company, are ideally suited for pharmacies which want to robotize delivery of a part of their medicine in the range of 1000–1500 items. Available in 4 possible configurations Consis robots can contain from 8000 to 25,000 packages. One of its features is the high speed of delivery in just 7 s, which is achieved due to the thoughtful design of the robot and a high-speed manipulator. Such fast execution of orders allows serving 4 workplaces at peak times freely. Also, the compact storage capacity of 3500 packages per m^2 is very important—it is achieved by using angled shelves with an average package size of 100 × 30 × 60 mm [2–4].

Another tool that is worth mentioning is the robotic warehouse of the Italian Pharmathek company (Fig. 2)—it allows robotizing the most possible assortment with a various demand. Available in 2 standard sizes (and optionally—according

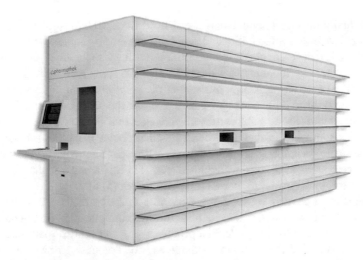

Fig. 1 Consis pharmacy robot

Fig. 2 Pharmatek robotic warehouse

to individually designed parameters) it can hold from 10,000 to 50,000 packages. Such a device provides flexibility and rapid adaptation to changes, since inside the device there is a different pharmacy assortment with various number of packages of each unit. Pharmacy robot chooses the right place to store each individual package by itself. The compact warehouse manipulator ensures dense storage of packages of various shapes up to 3500 packages per m^2. When the stock is changed, the robot automatically finds the best places to store packages in different demand. In addition, automatic loading of packages is provided—each device has a built-in function of semi-automatic package arrangement. It uses the space efficiently; the integrated intellectual system of robotic warehouses controls inside stocks, the shelf life of drugs and can make inventory in real time on request [5, 6].

An important element of the tool functioning is the algorithm of delivering the desired product to a customer, that is, the solution to the salesman optimization task, which in this case is to get the desired product using the shortest path possible. So the problem is to find the optimal path using as little computing power as possible.

Among the existing publications on selected topic, the following articles can be distinguished:

- Jean-Yves Potvin in his work "Genetic algorithms for the traveling salesman problem" (1996) analyzes the methods of solving the salesman optimization problem using heuristic algorithms and gives a detailed description how to use genetic algorithms to solve combinatorial problems, including the way of solving problems of permutation. In conclusion of the article the author states that genetic algorithms are able to compete with the best heuristic algorithms for medium-sized tasks (several hundreds points). However, they require a lot of time and cannot be successful in solving the problems of about a million points. On the other hand, they provide an environment for parallel implementation, because they work on a population of solutions [7].
- Woan Shin Tan, Siang Li Chua, Keng Who Yong, Tuck Seng Wu in their article "Impact of Pharmacy Automation on Patient Waiting Time: An Application of Computer Simulation" (2009) try to show the possibilities of using computer simulation to calculate the automated pharmacy distribution system influence on the time of customers' expectation and, in general, the potential of these systems as a routine process in pharmacy management. The conclusion of the article states that the use of automated distribution system increases the speed of service and allows reducing the number of workers by ~28% [8, 9].

In order to eliminate the identified drawbacks of known methods for solving the traveling salesman problem and their practical application, a method for solving the salesman problem with evolutionary search was developed.

The purpose of the research is to develop an evolutionary method for solving the traveling salesman problem as well as a software tool for finding an optimal path based on evolutionary optimization.

The function of cost for the simplest form of a task is the distance that the salesman passes in a given order (x_n, y_n), $n = 1, \ldots, N$ by the distance formula between two points, which is calculated using the Pythagorean theorem:

$$\cos t = \sum_{i=0}^{N} \sqrt{(x_i - x_{i+1})^2 + (y_i - y_{i+1})^2}, \tag{2}$$

where (x_n, x_{n+1}), (y_n, y_{n+1}) are the coordinates of the nth drug package [1].

2.1 Classification Analysis of the Functioning of the Investigated Methods

Simple Genetic Algorithm The methods of evolutionary search on the whole can be divided into binary and numerical. Both categories of methods are based on that model of genetic recombination and natural selection. First, present variables as coded binary to the term and work with them for minimization of function of cost, other work with numerical variables [1, 2].

On the whole a process to the robot ordinary EA can be described thus:

Cyclic bargaining to change a generation (a population is on a separate iteration) adjusted each of chromosomes by means of calculation of function of fitness, until then while the indicated limit of iterations will not be attained or a difference between two successive most optimal values of function of cost among generations will not be the less set number. On the rules of selection on every iteration there are only those individuals that is most adjusted—exactly they are used by the operator of crossing, for the receipt of descendants. Sometimes after crossing there are mutations—casual changes in a chromosome, that allow at the decision of tasks optimizations to go out from a local minimum [1–3, 9–12].

At the use GA for the decision of task of traveling salesman we must pay attention to that the operation of crossing can drive that the identical indexes of bales will appear to the way. Such situation is named the problem of transposition. She decides the use of the special operators of crossing and mutation [1, 2]. A numerical method was used in this work. For the selection of chromosomes for crossing the tournament operator of selection (Tournament of selection) was used. His algorithm consists of three steps.

1. By casual character to choose from a population individuals the amount of that equals the size of tournament.
2. Among select to choose the best and add to the resulting population.
3. To repeat previous steps so much times how many individuals in a population [3, 13–16].

The partly concerted operator (Partially of matched crossover, PMX) is used in quality to the operator of the two-point crossing. He offered Goldberg and Linglom for unhomologous numerical chromosomes. His algorithm is folded three steps.

1. Randomly select 2 crossing points.
2. Copy to the chromosome of the first descendant the segments of the first parent chromosome, located to the left and to the right of the crossing points and the middle of the second parent chromosome. Further, on the contrary, copy the lateral portions of the second parent chromosome and the middle of the first to the second offspring.
3. Replace the genes that are repeated in the lateral parts of each descendant on the genes from the opposite posterity at the duplicate positions [1, 17–22].

As a mutation operator, a classical mutation is used. His algorithm consists of three steps.

1. Create a chromosome offspring as a copy of the parent chromosome.
2. Select two non-identical index numbers from the set of chromosome genes.
3. Form a new chromosome by exchanging elements located at the positions in step 2 [3, 23–27].

The operation of a mutation is performed according to the formula:

$$position = ceil(rand(0.1) * N),\tag{3}$$

where rand () returns the value from the interval [0, 1], N—the number of individuals, ceil ()—rounds the result to an integer.

To generate random numbers, a uniform distribution is used that follows the formula:

$$P(i|a, b) = \frac{1}{b - a},\tag{4}$$

where a, b is the minimum and maximum value of the interval that the random variable takes.

2.2 Landscape Model

The island model is a parallel implementation of the GA and belongs to the category of distributed GA. Distribution is achieved using separate entities: for performing classical GA and distribution of individuals between islands. The main element (master, owner, manager) must carry out so-called migration with a given frequency and size, choosing the best individual from each island and replacing them with the worst individual on the other islands. The model can use different topologies like a star, exchange with a neighbor and others [28–30].

2.3 Activation Method

The activation method is intended to prevent the premature degeneration of the population of the genetic algorithm. That is, at the moment when the number of unique chromosomes is small, it must change part of the population, which may lead to the generation of a more optimal solution. Also an important element of this method is the use of the stop criterion—a condition or a set of them that allow you to choose the right moment to use the activation method [1, 2].

2.4 Comparative Analysis of Methods

The question of the difference between working with binary or numerical evolutionary methods is very relevant, and which method yields a better result. Practice shows that numerical methods are more effective. They do not need to convert the values of the variable to binary numbers and the need to monitor the number of bits needed to display the variable. Numerical genetic algorithms are also more compatible with other optimization algorithms, which makes them capable of combining or hybridizing [1, 2].

After a large number of binary and numerical GA comparisons, Mikhalevits (1992) said, "The experiments conducted show that numerical representation is faster, more consistent from start to launch, and provides greater accuracy (especially with a large plurality of where binary coding requires a much longer representation)." Inventors of GA in Europe in many cases used numerical variables in the algorithm [1].

To compare numerical and binary methods, I used a package of MATLAB applications. Each of the methods had to solve the problem of minimizing the two-dimensional function:

$$f(x_1, x_2, x_3, x_4) = 100 \cdot (x_2 - x_1^2)^2 + (1 - x_1)^2 + 90 \cdot (x_4 - x_3^2)$$
$$+ 10.1 \cdot (x_2 - 1)^2 + (x_4 - 1)^2 \tag{5}$$

To compare numerical and binary methods, I used a package of MATLAB applications. Each of the methods had to solve the problem of minimizing the two-dimensional function: Initial data of each of the limited methods were: population size—50, stop value—0.1, at $-20 <= x_1, x_2, x_3, x_4 <= 20$.

In the binary method, a proportional selection method, one-point cross-linking and random mutation were used. In the numeric Gauss mutation was used. In the process of performing 100 experiments, the resulting data were obtained. In Figs. 3 and 4 shows the resulting data for the binary algorithm.

In Figs. 5 and 6 shows the resulting data for a numerical algorithm.

In Table 1 the results of testing are given.

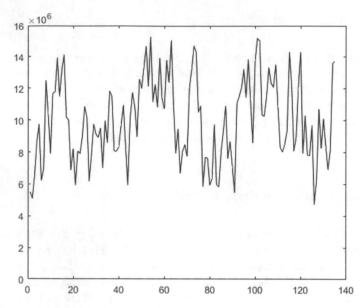

Fig. 3 Graph of values of the function of the binary method

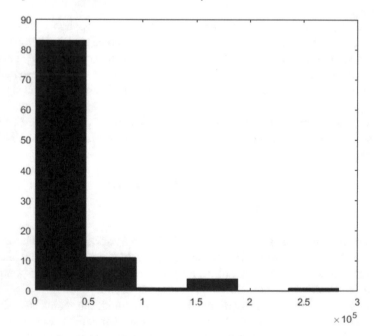

Fig. 4 Histogram distribution of the values of the binary method

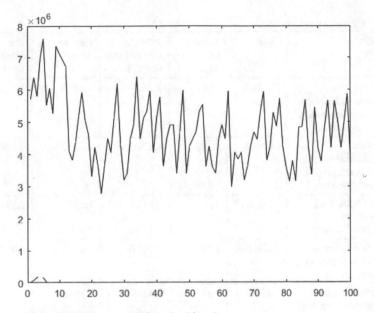

Fig. 5 Graph of values of the numerical method function

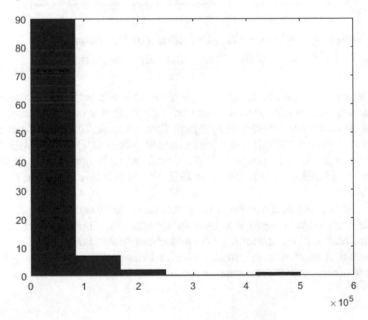

Fig. 6 Histogram distribution of the values of the numerical method

Table 1 Comparison of methods

Method	Type	Number of iterations	Average value of the function	Standard deviation	The coefficient of variation
Binary @proportionalselection, @crossoversinglepoint, @randommutation	Limited	135	2.6691e+04	4.3692e+04	1.9090e+09
Numerical @proportionalselection, @crossoversinglepoint, @mutationgaussian	Limited	100	4.3399e+04	9.5730e+04	9.1642e+09
@selectionroulette, @crossoverheuristic, @mutationgaussian	Unlimited	342	0.8175	1.5910	2.5314
@selectionuniform, @crossoversinglepoint, @mutationuniform	Unlimited	115	41.4078	54.9518	3.0197e+03
@selectionstochunif, @crossoverarithmetic, @mutationgaussian	Unlimited	400	0.4278	1.1695	1.3677

2.5 Choice of Software Development Tools. Overview of the Features of the Programming Language

In developing the program used language C++. This is a high-level programming language with support for several programming paradigms: object-oriented, generalized, and procedural. Developed by Bjarne Stroustrup at AT&T Bell Laboratories (Murray Hill, NJ) in 1979, it was initially named "Class C". Subsequently, Stroustrup renamed the C++ language in 1983. Based on the language of S. It was first described by ISO/IEC 14882: 1998, the ISO/IEC 14882: 2017 standard [31] is the most relevant.

In the 1990s, C++ became one of the most commonly used general-purpose programming languages. Language is used for system programming, software development, driver writing, powerful server and client applications, as well as for the development of entertainment programs such as video games. ++ has significantly influenced other, most popular today, programming languages: C# and Java [31].

2.6 Overview of Selected Libraries

Description of the Qt Quick Technology Qt Quick technology is a free framework that provides the ability to build a dynamic user interface with a declarative QML scripting language. The QML syntax allows you to use JavaScript to provide logic along with the C++ language. Qt Quick also uses the MOC (meta object compiler) to operate the Qt signaling/slot mechanism [32].

Description of the QWidget Class The QWidget class is the base for all user interface objects. The widget is an elementary object of the user interface: it receives mouse events, keyboards and other events from the window system and draws its image on the screen. Each widget has a rectangular shape, and they are all sorted by an overlay (Z-order). The widget is limited by its parent and other widgets located in front of him. A widget that is not embedded in a parent widget is called a window. Usually the window has a frame and a title bar, although using the appropriate window flags you can create a window without this external design. Qt QMainWindow and various subclasses QDialog are the most common types of windows [32].

Description of the QString Class The QString class is a string of Unicode characters. QString stores a string of 16-bit QChar, where each QChar retains the Unicode 4.0 character. Unicode is an international standard that supports most of the existing written systems. This is a superset with ASCII and Latin-1 (ISO 8859-1), and all ASCII/Latin-1 characters are available in the same code positions [32].

Implicitly, QString uses data sharing to prevent unnecessary memory allocation and unnecessary duplication of information. It also helps to reduce the inherent memory of 16-bit memory symbols compared to 8-bit characters [32].

QString converts const char * to Unicode using fromAscii (). By default, fromAscii () prefers characters with code greater than 128 such as the Latin-1 characters, but this can be changed by calling QTextCodec :: setCodecForCStrings () [32].

In addition to QString Qt also provides a QByteArray class for storing raw bytes and traditional null terminals. In most cases, QString is a class that is required for use. It is used throughout the Qt application programming interface, and Unicode support ensures that programs can be easily translated into another language, if at any time a programmer wants to increase their distribution market [32].

Description of the QLabel Class The QLabel widget allows you to display text or a picture. QLabel is used to display text or a picture. Does not support any user interaction features. The visual behavior of the label can be specified in different ways. The label can be used to task the mnemonic keys to get the focus by other widgets [32].

QLabel can contain the following types of data:

- plain text;
- formatted text;
- pixel card;
- animation;
- the number.

There are several ways to customize the look of QLabel. All QFrame settings are available for the taskbar widget task. The location of the content inside the QLabel widget area can be adjusted using setAlignment () and setIndent (). You can specify the transfer of text content by words using setWordWrap () [32].

Block Diagram of the Program The software (software) is developed using the principles of the PLO and consists of several interrelated modules—classes with their constituents—methods and data, so you can depict the structure of the program using the UML class diagram (Fig. 7).

The class diagram shows that one of the main classes is CGeneticAlgorithm. With him, the associative bond (composition) is associated with the CPopulation class. The

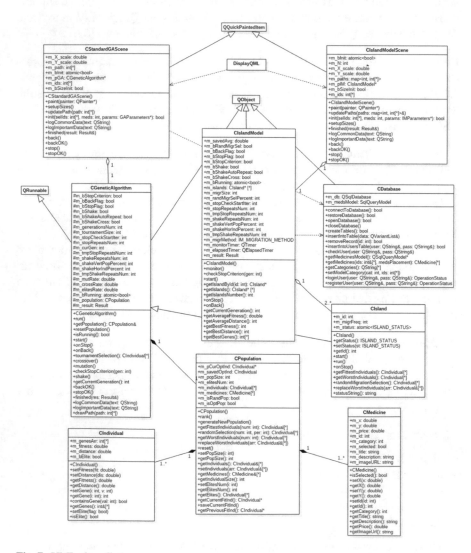

Fig. 7 UML class diagram

descendant of this class is the CIsland class that uses the simple GA functionality to apply it to the island model CIslandModel.

Also in the diagram is the QML-class Display—a graphical form of visualization of the algorithm, which is associated with C++ classes of graphical representation of the operation of a simple GA and island model—CStandardGAScene and CIslandModelScene, respectively.

The main classes also mimic the Qt library classes like QObject, QRunnable, QQuickPaintedItem.

For a more detailed review of the functional component of the system, it is appropriate to give UML diagrams of sequences, co-operation, states, activities.

In Figs. 8 and 9 depicts UML sequence diagrams.

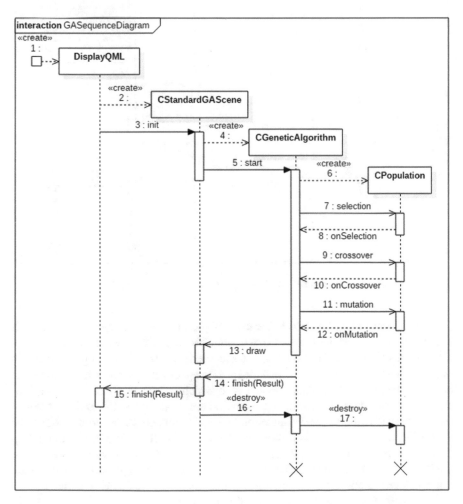

Fig. 8 UML sequence diagram of a simple GA

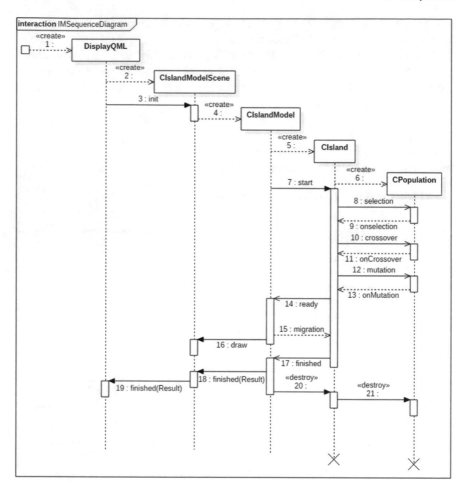

Fig. 9 The UML sequence diagram of the island model

The sequence diagram depicts the main objects of the developed program, their lifestyle and the existing links between them. This chart allows you to see in chronological order the sequence of creating the specified objects and exchanging messages between them. From Fig. 9 it is seen that the first parameter window is created—the object of the class Parameters, together with which objects of classes CIslandModelScene, CIslandModel, CIsland, CPopulation are created. The CIsland class object performs basic work on the optimization task (selection, crossing and mutation), the CIslandModel class object is responsible for executing the migration and updating the user interface. Upon completion of the work, the results are provided, an object of the Result class that contains data for constructing the resulting graphs and displaying important information (time of operation, best fitness functions and distance).

In Fig. 10 depicts the UML co-operation diagram.

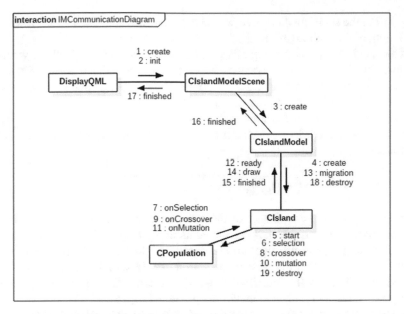

Fig. 10 The UML collaboration diagram of the island model

The cooperation diagram depicts the main objects of the developed program and the functional link between them. It provides an opportunity to follow the order and direction of communications between objects. This chart does not include the scale of the object's life, therefore it is more compact and optimal to look over the interaction of these objects. The diagram is an instant snapshot of the work of the island model—after executing the start () method. It shows events such as updating the user interface (draw), displaying the resulting information (finish (Result)), migration using the island model (migrate), selection, crossover, mutation.

The state diagram indicates a sequence of events that lead to a change in the abstractions of the values and connections of the system, that is, its states. The diagram shows that the change to some states, for example, "An algorithm started", requires that the system has been in the two previous states—"Chosen medications", "Chosen options". Finding a solution using an island model requires multiple repetition of migration operations and optimum searches.

3 Development of an Evolutionary Method for Solving the Traveling Salesman Problem

Let there be a set of medicines located in the device; all of them have their geometric coordinates in a two-dimensional coordinate system, which we denote $d_i(x_i, y_i)$.

Then, from the set of medicines A, we need to choose a subset B in such a way that the length of the path D{d_1, d_2, d_3, ...} should be minimal.

The standard way of solving this problem is the brute force method, when we count each of the possible ways and choose the best one. For n points exists (n − 1) possible ways. For example, for 10 points, the number of paths is more than 3 million, and for 20 points it is several quintillions [10, 11].

The cost function for the simplest form of a task is the distance traveled by the salesman in a given order (x_n, y_n); it is used to estimate the path, n = 1, ..., N by the formula:

$$\cos t = \sum_{i=0}^{N} \sqrt{(x_i - x_{i+1})^2 + (y_i - y_{i+1})^2},$$ (6)

where (x_i, y_i) are the coordinates of the n-th point, and the formula is the distance between the two points which is calculated using Pythagorean theorem [12, 13].

When using the genetic algorithm to solve the salesman problem, one must pay attention to the fact that the crossing operation may lead to the same indexes of the visited points in the path. This situation is called a permutation problem. It is solved using special cross-linking and mutation operators [14, 15].

To select the chromosomes for over-crossing in the developed method, the tournament selection operator (Tournament selection) was used. When using it is necessary at first to select random chromosomes from the population of N chromosomes; the number of them must be equal to the tournament size. Then, among the selected N chromosomes, we need to select the best one, namely the H_i-chromosome, with the minimum value of the fitness function f(H) and add it to the resulting population. Then it is necessary to repeat the previous steps as many times as the individuals in the population [16–18].

The ordering crossover operator (OX) is used as a crossover operator. It was proposed by D. Davis in 1985 for non-homologous numerical chromosomes. When using it is necessary at first to select randomly the point of crossing P1 and copy the chromosome segment P0–P1 of the first parent A1, located to the left of the crossing point, to the first child's D1 chromosome. Other genes in the descendant are copied from the second parent in an ordered form from left to right, except for the elements that have already entered the descendant. For example, the general case of arithmetic crossing is performed according to the formulas:

$$h_{1\pi i} = kh_{1i} + (1 - k)h_{2i},$$ (7)

$$h_{2\pi i} = kh_{2i} + (1 - k)h_{1i},$$ (8)

where k ϵ [0;1] is an actual coefficient which can depend on d (A1; A2) which is the distance between the chromosomes [19, 20].

A classical exchange mutation is used as a mutation operator. When using it is necessary at first to create a chromosome descendant D1 as a copy of the parent chromosome A1. Then it is necessary to select two different numbers-indices i1, i2 from the set of chromosome D1 genes and form a new chromosome by exchanging elements located on previously selected positions [21].

The mutation operation is performed according to the formula:

$$position = ceil(rand(0.1) * N), \tag{9}$$

where rand() returns a value from the interval [0, 1], N is the number of individuals, ceil () rounds the result to the whole number [22, 23].

To generate random numbers, uniform distribution is used according to the formula:

$$P(i \mid a, b) = \frac{1}{b - a}, \tag{10}$$

where a, b are the minimum and maximum value of the interval taken by a random variable, respectively [24].

It is well known that the effective use of evolutionary methods depends greatly on the formation of the initial set of solutions. Known evolutionary methods tend to use an approach that involves the random formation of an initial population and does not use known apriori information about a solvable practical problem. In many cases this leads to the fact that the decision points of the initial population are significantly remote from the area of global extremum and to the need for significant use of computer time and resources when using such methods [25–27].

In the developed evolutionary method of solving the salesman problem it is proposed to use modified initialization operators of the initial population to eliminate these shortcomings. Developed operators of initial population initialization provide the creation of an initial set of solutions, based on the features of solving the problem, which allows bringing the starting points closer to the area of global extremum and reduce the optimization time and use of computer resources.

- Standard method (Default). It is the method in which each of the population individuals is initialized by the same array of indices arranged in increasing order. It proposes to find the maximum point index (MAX_IDX) and change the size of the vector of points to the index found above. Then it is necessary to initialize the vector of point indexes—as the result we will get a vector of the form—1, 2, 3 ... MAX_IDX.
- Mixed method (Random). It is the method in which each population individual is initialized by a different array of indices located randomly. Here it is proposed to find the maximum point index—MAX_IDX and change the size of the vector of points to the index found above. Then it is necessary to initialize the vector of point indexes—as the result we will get a vector of 1, 2, 3 ... MAX_IDX type— and mix the content of the vector of point indexes—as the result we will get the vector of the form—4, 15, 5 ...;
- Mixed optimal (Random optimal). It is the method in which each population individual is initialized by the same array of indices, each of which is the optimal

path among the number of generated mixed paths specified by a user. It is proposed to find the maximum point index—MAX_IDX and change the size of the vector of points to the index found above. Then it is necessary to initialize the vector of point indexes—as the result we will get a vector of the form—1, 2, 3 ... MAX_IDX. Then it is necessary to mix the contents of the vector of point indexes (as the result we will get a vector of form—4, 15, 5 ...) and choose the most optimal path among the generated and initialize the entire population by it.

The work of the algorithm stops in case of reaching the maximum number of iterations, or in case of population degeneration the percentage of improvement of the value of the better chromosome F can be used. Let t_p denote the number of algorithm iterations, ρ_p denotes the threshold of the coefficient of the fitness function value improvement of a better chromosome. Starting from the cycle (t_{p+1}), the genetic operators calculate the value of the target function for each chromosome of the latest population, and the best value of the target function f_{best} t is selected from these values. From the previous T generations the best value of the target function $f_{best\,t-1}$ is selected. After that, the coefficient of improvement is calculated according to the formula:

$$\rho = \frac{f_{best\,t} - f_{best\,t-1}}{f_{best\,t-1}}. \tag{11}$$

To compare the results of the manipulator effectiveness for medicine delivery, we will use such indicators as the length of the path traversed and the time the manipulator passes this path.

$$t_i = \frac{L}{V}, \tag{12}$$

where t, s is time of passing the path from the beginning of the manipulator movement to the selection of medicine and the return path; L, m is the length of the path traversed; V, m/s is manipulator speed.

To determine a more efficient method, we will use the time difference:

$$\Delta t = t_{n+1} - t_n. \tag{13}$$

Unlike existing methods, a modified version of GA allows choosing the method of initializing the initial population in solving the traveling salesman problem, which in its turn allows generating more adapted chromosomes during initialization (chromosome with the best fitness function value) and thus improving the algorithm results.

The proposed modification can increase the speed significantly due to the fact that when using the standard method of initializing the initial population the initial value of fitness function and accordingly the distance is less if we use other initialization methods that will define the optimum problem more precisely.

Table 2 General description of the system model

The name of the entity	Description
Users	Contains registered users
Medicines	Contains drug information
Categories	Contains information on categories of drugs

Table 3 Description of the essence of "Users" ("Users")

Field name	Field name	Description
ID	Integer	User id
Username	Varchar(64)	User login
Password	Varchar(64)	User password

Table 4 Description of the essence of "Medication" ("Medicines")

Field name	Type of value	Description
ID	Integer	Medication ID
Title	Varchar(64)	The name of the medicine
Description	Varchar(128)	Description of medicines
Price	Real	The price of medicines
ImageURL	Varchar(128)	Link to image
Category	Integer	Category of medicines

Table 5 Description of the essence of the "Categories" ("Categories")

Field name	Type of value	Description
ID	Integer	Category ID
Name	Varchar(64)	Category name
Description	Varchar(128)	Category description
Rating	Integer	Category rating

3.1 Description Off the Developed Entities Database

In the process of development, a general description of the model of the system was created, shown in the Table 2. In Tables 3, 4 and 5 gives a description of the entities of the database.

3.2 Algorithm for the Functioning of the Developed Program

At the initial stage, the user must select the necessary parameters and input data.

The parameters include—the number of iterations, the size of the population, the size of the tournament, the elite coefficient, the crossing and the mutation. You can also select one of the available initialization methods.

Data in this case is a text file with the coordinates of the package of drugs for which the salesman's task with the help of GA must be solved.

After loading the data and running the algorithm, the search for the optimal solution will start. The search will be performed in a new thread, so the user can change the parameters to prepare for a new GA run.

After the GA is completed, the user can change the parameters or data and resume work again.

4 Experiments and Results of Researching Methods of Initialization of the Evolutionary Algorithm Initial Population

Three experiments describing the effectiveness of initialization methods for the genetic algorithm initial population were carried out. Each of methods started with the same parameters: the number of iterations (generations) is 1500, the size of the population is 100, the tournament size is 5, the elite coefficient is 0.02, the coefficient of crossing is 0.9, the mutation coefficient is 0.01, the number of optimal population is 200, the number of coordinate points is 95.

The following figures, which reflect the results of experiment 1, show the visualization of the found path (Fig. 11) and graphs describing dependence of iterations number on functions of cost and distance (Fig. 12). It can be seen that when using the standard method of the initial population initialization, the initial value of the function is 0.0019, the distance is 1.685. The experiment time is 8.104 s.

After conducting 3 experiments, it is worth noting that the best result is in the standard initialization method. This can be explained by the fact that in this problem the use of the initial population in the 1, 2, 3, 4, etc. form allows having lower initial value of the calculated fitness function and, accordingly, the distance. The first column of Table 6 gives us the opportunity to make sure that the standard population initialization method is appropriate. All 3 modifications showed roughly identical time scores, the results are shown in Table 6.

5 Discussing the Results of the Study of Initialization Methods of the Evolutionary Algorithm Initial Population

Figure 13 represents the visualization of the work of the developed algorithm.

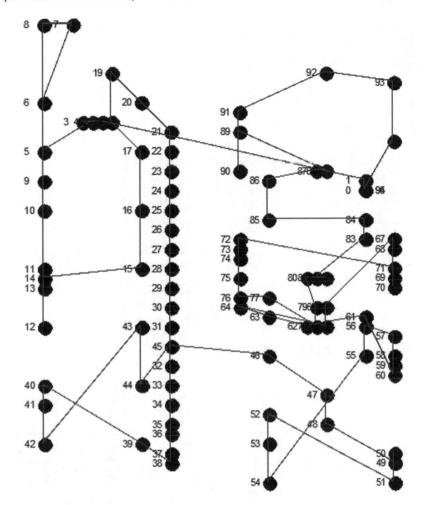

Fig. 11 The resultant path of the 1st experiment

From the visual analysis of Fig. 13, b, we can see that the manipulator path when using the modified version of the algorithm is optimal for use.

With the use of the original algorithm for the medicine delivery, the following results were obtained:

$$L1 = 5.2\,\text{m}, \quad t1 = 3.8\,\text{s}.$$

Using the modified method of finding the optimal path, the following results were obtained:

$$L2 = 3.7\,\text{m}, \quad t2 = 2.7\,\text{s}.$$

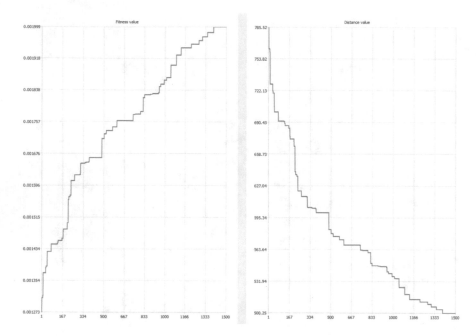

Fig. 12 The resultant graph of the 1st experiment

Table 6 Testing results

Method name	Initial distance value (m)	Resultant value of distance (m)	Runtime (s)
Standard	1.685	0.5	8.104
Mixed	1.784	0.521	8.258
Mixed random	1.671	0.529	8.28

Comparing the results of experiments, we get the difference:

$$\Delta t = 3.8 - 2.7 = 1.1\, \text{s}.$$

From Table 6 and comparison using formulas (12), (13) we can see that the proposed method allows increasing the speed and reducing the length of the path traversed by the manipulator.

(a)   **(b)**

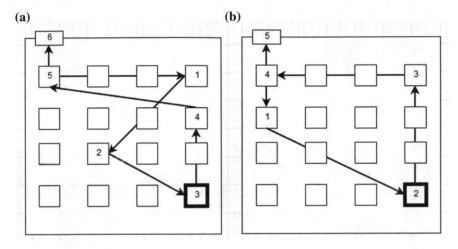

Fig. 13 The visualization of the manipulator path: **a** an original algorithm, **b** a modified algorithm

6 Features of Software Implementation of the System

We used C++ language to develop this program. This is a high-level programming language with support for several programming paradigms: object-oriented, generalized, and procedural [28]. Also the Qt library was used; it is designed to create cross-platform applications [29].

The software was developed using the principles of OOP; it consists of several interrelated modules—classes with their constituents—methods and data, so you can depict the structure of the program using the UML class diagram (Fig. 14).

The class diagram shows that the main class is the Widget one. The CGeneticAlghoritm, CMyGraphView classes are connected with it via the associative bond (composition), and the CresultForm class is connected with it via aggregation. The main classes also inherit the Qt library classes like QObject, QRunnable, QWidget, QgraphicsView.

For a more detailed review of the system functional component, it is appropriate to give UML diagrams of sequences, collaboration, states, activities depicted in Figs. 15, 16, 17 and 18.

The sequence diagram depicts the main objects of the developed program, their life lines and the links existing among them. This chart allows seeing in chronological order of the sequence of creating the specified objects and exchanging messages between them. It shows that the first object to be created is the main program window—the object of the Widget class, with which the objects of CGeneticAlghoritm, CMyGraphView classes are created. The object of the CGeneticAlghoritm class performs basic work on the optimization problem solution and updating the user interface. After completion, it gives the results of the object of the Widget class, which in its turn creates a form for their display—an object of CResultForm class.

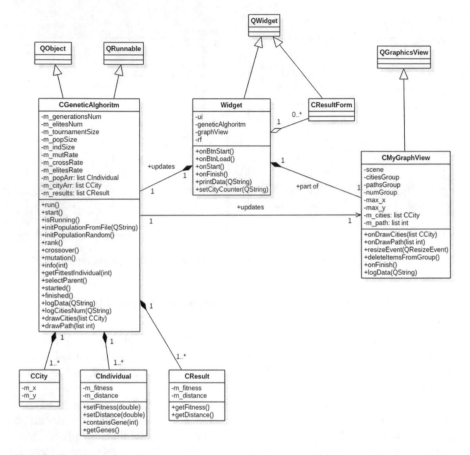

Fig. 14 Class diagram

The collaboration diagram depicts the main objects of the developed program and the functional link among them. It provides an opportunity to follow the order and direction of communications among objects. This diagram does not include the scale of the object's life, therefore, it is more compact and optimal to understand the interaction of these objects. The diagram is an instant snapshot of the GA operation— after executing the start() method. It shows events such as updating the user interface (printData (QString)), visualizing the algorithm operation (onDrawPath (list int)), displaying the resultant information (create Cresult-form).

The state diagram indicates system connections (its states) and a sequence of events that lead to a change in the value abstractions. The diagram shows that the change to some states, for example, "An algorithm started", requires the system to have been in two previous states—"Data are uploaded ", "Options are selected ".

The activity diagram depicts a sequence of steps that reflect the GA work; it focuses on operations and shows the control flow. The operations can be dis-played in parallel as "Searching Optimum," "Updating interface".

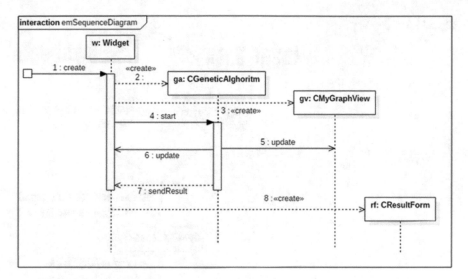

Fig. 15 Diagram of sequences

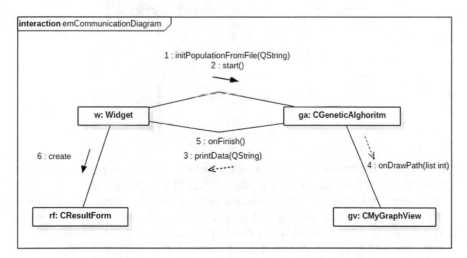

Fig. 16 Collaboration diagram

7 Conclusions

An evolutionary method for solving the traveling salesman problem has been developed. It can be used in the field of pharmacy business by optimizing the process of the medicine delivery equipment work. Unlike existing methods, the modified version of the evolutionary method allows choosing the initial population initialization method when solving the salesman problem, which in its turn allows generating

Fig. 17 State diagram

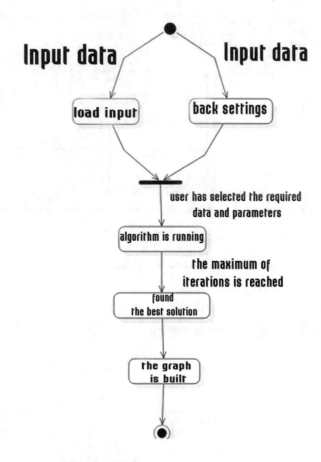

more adapted chromosomes (chromosomes with better fitness function values) at the initialization stage and thus improving the results of the algorithm work [30–59].

The actual scientific task of finding the optimal path is solved. The scientific novelty of the work is that the proposed modification of the genetic algorithm, which involves the use of three methods of initial population initialization. These methods allowed increasing the search speed and determining the problem optimum more precisely.

As suggestion for future research various ways of modification of existing initialization methods and genetic operators can be used, this includes modifications which are aimed on reducing algorithm work time and finding better optimal solution that for example can be done with application of distributed parallel genetic algorithms.

Fig. 18 Activity diagram

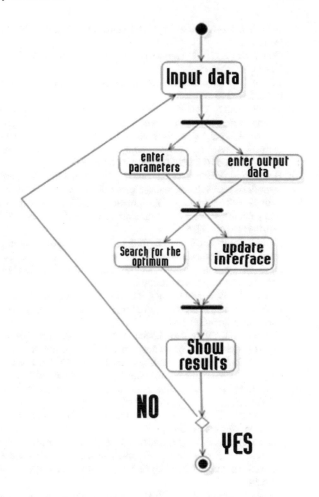

Acknowledgements The work was performed as part of the project "Methods and means of decision-making for data processing in intellectual recognition systems" (number of state registration 0117U003920) of National University "Zaporizhzhia Polytechnic".

References

1. Grötschel M (2009) The travelling salesman problem and its applications. Technisc he Universität, Berlin
2. Gen M, Cheng R (1997) Genetic algorithms and engineering design. Wiley, New Jersey, p 352
3. Haupt R, Haupt S, Haupt R (2004) Practical genetic algorithms, 2nd edn. Hoboken, New Jersey, p 261
4. Willach Pharmacy Solutions—UK (2018) https://www.willach-pharmacy-solutions.com/EN/index.php. Last accessed 06 July 2018

5. Pharmathek—Robot e magazzini automatizatti, http://www.pharmathek.com. Last accessed 06 June 2018
6. Gan Y, Dai X (2012) Human-like manipulation planning for articulated manipulator. J Bionic Eng 9(4):434–445
7. Potvin J (1996) Genetic algorithms for the traveling salesman problem. Ann Oper Res 63(3):337–370
8. Tan W, Chua S, Yong K, Wu T (2009) Impact of pharmacy automation on patient waiting time: an application of computer simulation. Ann Acad Med Singapore 5(38):501–507
9. Oliinyk A, Skrupsky S, Subbotin S, Korobiichuk I (2017) Parallel method of production rules extraction based on computational intelligence. Autom Control Comput Sci 51(4):215–223. https://doi.org/10.3103/S0146411617040058
10. Hoffman K, Padberg M, Rinaldi G (2013) Traveling salesman problem. In: Encyclopedia of operations research and management science, pp 1573–1578
11. Oliinyk AO, Skrupsky SY, Subbotin SA (2014) Using parallel random search to train fuzzy neural networks. Autom Control Comput Sci 48(6):313–323. https://doi.org/10.3103/S0146411614060078
12. Tsai C, Tseng S, Chiang M, Yang C, Hong T (2014) A high-performance genetic algorithm: using traveling salesman problem as a case. Sci World J 2014:1–14
13. Dopico J, Calle J, Sierra A (2009) Encyclopedia of artificial intelligence, 1st edn. Information Science Reference, New Jersey
14. Sanches D, Whitley D, Tinós R (2017) Improving an exact solver for the traveling salesman problem using partition crossover. In: Proceedings of the genetic and evolutionary computation conference on—GECCO
15. Kolpakova T, Oliinyk A, Lovkin V (2017) Improved method of group decision making in expert systems based on competitive agents selection. In: 2017 IEEE first Ukraine conference on electrical and computer engineering (UKRCON). Institute of Electrical and Electronics Engineers, pp 939–943. https://doi.org/10.1109/ukrcon.2017.8100388
16. Holland J (1992) Adaptation in natural and artificial systems, 2nd edn. MIT Press, Cambridge, p 223
17. Goldberg D (2012) Genetic algorithms in search, optimization, and machine learning, 1st edn. Addison-Wesley, Boston, p 432
18. Stepanenko O, Oliinyk A, Deineha L, Zaiko T (2018) Development of the method for decomposition of superpositions of unknown pulsed signals using the second-order adaptive spectral analysis. Eastern Eur J Enterp Technol 92(9):48–54. https://doi.org/10.15587/1729-4061.2018.126578
19. Nagata Y, Kobayashi S (2013) A powerful genetic algorithm using edge assembly crossover for the traveling salesman problem. INFORMS J Comput 25(2):346–363
20. Hussain A, Muhammad Y, Nauman Sajid M, Hussain I, Mohamd Shoukry A, Gani S (2017) Genetic algorithm for traveling salesman problem with modified cycle crossover operator. Comput Intell Neurosci 2017:1–7
21. Wang Y (2014) The hybrid genetic algorithm with two local optimization strategies for traveling salesman problem. Comput Ind Eng 70:124–133
22. Lin B, Sun X, Salous S (2019) Solving travelling salesman problem with an improved hybrid genetic algorithm. J Comput Commun 4(15):98–106
23. Cantú-Paz E (2001) Efficient and accurate parallel genetic algorithms, 2nd edn. Springer Science + Business Media, LLC, New York
24. Chambers L (2001) The practical handbook of genetic algorithms applications, 2nd edn. Chapman & Hall/CRC, Boca Raton, p 544
25. Oliinyk A, Skrupsky S, Subbotin SA (2017) Parallel computer system resource planning for synthesis of neuro-fuzzy networks. Adv Intell Syst Comput 543:88–96. https://doi.org/10.1007/978-3-319-48923-0_12
26. Oliinyk A, Subbotin S (2016) A stochastic approach for association rule extraction. Pattern Recognit Image Anal 26(2):419–426. https://doi.org/10.1134/S1054661816020139

27. Oliinyk A, Zayko T, Subbotin S (2014) Synthesis of neuro-fuzzy networks on the basis of association rules. Cybernet Syst Anal 50(3):348–357. https://doi.org/10.1007/s10559-014-9623-7

28. Stroustrup B (2015) The C++ programming language, 4th edn. Addison-Wesley, Upper Saddle River, p 1366

29. Shlee M (2015) Qt 5.5 professional'noe programmirovanie na C++, 3rd edn. BHV-Peterburg, Sankt-Peterburg

30. Oliinyk A, Fedorchenko I, Stepanenko A, Rud M, Goncharenko D (2018) Evolutionary method for solving the traveling salesman problem. Problems of infocommunications. In: 5th international scientific-practical conference PICST2018. Kharkiv National University of Radioelectronics, Kharkiv, pp 331–339. https://doi.org/10.1109/infocommst.2018.8632033

31. Lewis JP (1995) Fast template matching. In: Proceedings of the vision interface, 4th edn, pp 120–123

32. Sochman J, Matas J (2010) Center for machine perception. Czech Technical University, Prague, pp 1–17

33. Alsayaydeh JAJ, Shkarupylo V, Bin Hamid MS, Skrupsky S, Oliinyk A (2018) Stratified model of the internet of things infrastructure. J Eng Appl Sci 13(20):8634–8638

34. Oliinyk A, Fedorchenko I, Stepanenko A, Rud M, Goncharenko D (2019) Combinatorial optimization problems solving based on evolutionary approach. In: 2019 15th international conference on the experience of designing and application of CAD systems (CADSM), pp 41–45. https://doi.org/10.1109/cadsm.2019.8779290

35. Fedorchenko I, Oliinyk A, Stepanenko A, Zaiko T, Svyrydenko A, Goncharenko D (2019) Genetic method of image processing for motor vehicle recognition. In: CEUR workshop proceedings, vol 2353, pp 211–226 (2019). ISSN: 16130073

36. Fedorchenko I, Oliinyk A, Stepanenko A, Zaiko T, Korniienko S, Burtsev N (2019) Development of a genetic algorithm for placing power supply sources in a distributed electric network. Eastern Eur J Enterp Technol 5/3(101):6–16. https://doi.org/10.15587/1729-4061.2019.180897

37. Fedorchenko I, Oliinyk A, Stepanenko A, Zaiko T, Shylo S, Svyrydenko A (2019) Development of the modified methods to train a neural network to solve the task on recognition of road users. Eastern Eur J Enterp Technol 9(98):46–55. https://doi.org/10.15587/1729-4061.2019.164789

38. Stepanenko A, Oliinyk A, Fedorchenko I, Kuzmin V, Kuzmina M, Goncharenko D (2019) Analysis of echo-pulse images of layered structures the method of signal under space. In: CEUR workshop proceedings, vol 2353, pp 755–770. ISSN: 16130073

39. Oliinyk A, Fedorchenko I, Stepanenko A, Katschan A, Fedorchenko Y, Kharchenko A, Goncharenko D (2019) Development of genetic methods for predicting the incidence of volumes of emissions of pollutants in air. In: 2nd international workshop on informatics and data-driven medicine, IDDM 2019, pp 340–353. Lviv, Ukraine. ISSN: 16130073

40. Kryvinska N (2010) Converged network service architecture: a platform for integrated services delivery and interworking. Electronic business series, vol 2. International Academic Publishers, Peter Lang Publishing Group

41. Kryvinska N (2008) An analytical approach for the modeling of real-time services over IP network. Math Comput Simul 79(4):980–990. https://doi.org/10.1016/j.matcom.2008.02.016

42. Ageyev DV, Wehbe F (2013) Parametric synthesis of enterprise infocommunication systems using a multi-layer graph model. In: Proceedings of the 2013 23rd international crimean conference microwave and telecommunication technology (CriMiCo 2013), pp 507–508

43. Ageyev D, Al-Anssari A (2014) Optimization model for multi-time period LTE network planning. In: Proceedings of the 2014 first international scientific-practical conference problems of infocommunications science and technology (PIC S&T'2014). Kharkov, Ukraine, pp 29–30. https://doi.org/10.1109/infocommst.2014.6992288

44. Ageyev DV, Ignatenko AA, Wehbe F (2013) Design of information and telecommunication systems with the usage of the multi-layer graph model. In: Proceedings of the XIIth international conference the experience of designing and application of CAD systems in microelectronics (CADSM). Lviv Polytechnic National University, Lviv-Polyana, Ukraine, pp 1–4

45. Ageyev DV (2010) NGN network planning according to criterion of provider's maximum profit. In: 2010 international conference on modern problems of radio engineering. Telecommunications and Computer Science, Lviv-Slavske, p 256
46. Ageyev D et al (2018) Classification of existing virtualization methods used in telecommunication networks. In: Proceedings of the 2018 IEEE 9th international conference on dependable systems, services and technologies (DESSERT), pp 83–86
47. Karpukhin A et al (2017) Features of the use of software packages for modeling infocommunication systems. In: Proceedings of the 2017 4th international scientific-practical conference problems of infocommunications, pp 380–382. Science and technology (PIC S&T). https://doi.org/10.1109/infocommst.2017.8246421
48. Radivilova T, Kirichenko L, Ageiev D, Bulakh V (2020) The methods to improve quality of service by accounting secure parameters. In: Hu Z, Petoukhov S, Dychka I, He M (eds) Advances in computer science for engineering and education II. ICCSEEA 2019. Advances in intelligent systems and computing, vol 938. Springer, Cham
49. Andrushchak V. et al (2018) Development of the iBeacon's positioning algorithm for indoor scenarios. In: 2018 international scientific-practical conference problems of infocommunications, pp 741–744. Science and technology (PIC S&T), IEEE. https://doi.org/10.1109/infocommst.2018.8632075
50. Oliinyk A, Zaiko T, Subbotin S (2014) Training sample reduction based on association rules for neuro-fuzzy networks synthesis. Opt Memory Neural Netw Inf Opt 23(2):89–95. https://doi.org/10.3103/S1060992X14020039
51. Kirichenko L, Radivilova T, Bulakh V (2018) Machine learning in classification time series with fractal properties. Data 4(1):5. https://doi.org/10.3390/data4010005
52. Oliinyk AA, Subbotin SA (2015) The decision tree construction based on a stochastic search for the neuro-fuzzy network synthesis. Opt Memory Neural Netw Inf Opt 24(1):18–27. https://doi.org/10.3103/S1060992X15010038
53. Yarymbash D, Kotsur M, Subbotin S, Oliinyk A (2017) A new simulation approach of the electromagnetic fields in electrical machines. In: Proceedings of the international conference on information and digital technologies, pp 429–434. https://doi.org/10.1109/dt.2017.8024332
54. Kirichenko L, Radivilova T, Zinkevich I (2017) Forecasting weakly correlated time series in tasks of electronic commerce. In: 2017 12th international scientific and technical conference on computer sciences and information technologies (CSIT), pp 309–312. https://doi.org/10.1109/stc-csit.2017.8098793
55. Oliinyk AO, Oliinyk OO, Subbotin SA (2012) Software-hardware systems: agent technologies for feature selection. Cybern Syst Anal 48(2):257–267. https://doi.org/10.1007/s10559-012-9405-z
56. Kryvinska N (2004) Intelligent network analysis by closed queuing models. Telecommun Syst 27:85–98. https://doi.org/10.1023/B:TELS.0000032945.92937.8f
57. Kryvinska N, Zinterhof P, van Thanh D (2007) An analytical approach to the efficient real-time events/services handling in converged network environment. In: Enokido T, Barolli L, Takizawa M (eds) Network-based information systems. NBiS 2007. Lecture notes in computer science, vol 4658. Springer, Berlin
58. Oliinyk AO, Skrupsky SY, Subbotin SA (2015) Experimental investigation with analyzing the training method complexity of neuro-fuzzy networks based on parallel random search. Autom Control Comput Sci 49(1):11–20. https://doi.org/10.3103/S0146411615010071
59. Kryvinska N, Zinterhof P, van Thanh D (2007) New-emerging service-support model for converged multi-service network and its practical validation. In: First international conference on complex, intelligent and software intensive systems (CISIS'07), pp 100–110. https://doi.org/10.1109/cisis.2007.40

Unified Models of Gradation Image Correction

Kirill Smelyakov, Anastasiya Chupryna, Mykyta Hvozdiev, Denys Sandrkin, Igor Ruban, and Olena Voloshchuk

Abstract The efficiency of object recognition algorithms, computer vision systems and image analysis directly depend on the quality of image preprocessing. Gradation correction is one of the stages of this preprocessing. The paper discusses three the most common models of gradation image correction, which are able to work on any brightness scale. Also in this paper, criteria for the quality of gradation correction are formulated. Experiments have been carried out that confirm the operability and computational efficiency of the considered models.

Keywords Gradational correction · Digital image · Unified model · Offline machine vision system

1 Introduction

In the modern world, systems using machine vision are becoming increasingly popular, which requires the development of effective models and methods for processing images and video streams [1–4].

K. Smelyakov · A. Chupryna · M. Hvozdiev · D. Sandrkin (✉) · I. Ruban · O. Voloshchuk
Kharkiv National University of Radio Electronics, Kharkiv, Ukraine
e-mail: den.sandrkin@gmail.com

K. Smelyakov
e-mail: kirillsmelyakov@gmail.com

A. Chupryna
e-mail: nastya.chupryna@gmail.com

M. Hvozdiev
e-mail: nik4workgvozdev@gmail.com

I. Ruban
e-mail: ihor.ruban@nure.ua

O. Voloshchuk
e-mail: olena.voloshchuk@nure.ua

© The Editor(s) (if applicable) and The Author(s), under exclusive license to Springer Nature Switzerland AG 2021
T. Radivilova et al. (eds.), *Data-Centric Business and Applications*, Lecture Notes on Data Engineering and Communications Technologies 48,
https://doi.org/10.1007/978-3-030-43070-2_14

Many systems are autonomous and independently make decisions, without operator participation: drones, which analyze terrain, autonomous vehicle control systems, person identification systems [4–7]. The efficiency of those systems depends on the quality of the image [8–10].

Image preprocessing includes many algorithms and procedures: noise filtering, image smoothing, compressing and others. One of the most important stages, which soles several problems [11, 12] at once, is the gradation correction, which serves to tone correction of the image. This is especially important under conditions of insufficient illumination, or excessive brightness. An increase in the image contrast is also achieved by changing the informative range of the brightness. Moreover, the gradation correction is one of the factors that makes it possible to ensure the effectiveness of segmentation, vectoring and image recognition problems, as well as compression in order to ensure efficient storage and transmission of data in computer networks [13–18].

Analysis of the question allows us to conclude that the following tasks need to be solved:

- Selection of the function of gradation correction
- Definition of quality criteria
- Experimental confirmation of the adequacy of the selected functions and quality criteria.

The paper deals with halftone images, no reference to specific brightness scale.

2   Functions of Gradation Correction

At first, you need to define the general requirements for the function of gradation correction:

1. The input brightness range is determined by the interval $[a, \ldots, b]$, and the output by interval $[c, \ldots, d]$. Gradation function replaces input image pixel brightness x, $x \in [a, \ldots, b]$, with output brightness $f(x)$, $f(x) \in [c, \ldots, d]$. The range $[a, \ldots, b]$ is fully displayed on the range $[c, \ldots, d]$; for the ends of these brightness ranges the following conditions are met: $f(a) = c$, $f(b) = d$.
2. The gradation function $f(x)$ is continuous.
3. The function of the gradation correction cannot go beyond the rectangle formed by straight lines: $x = a$, $x = b$, $y = c$, $y = d$.
4. By default, we assume that the gradation correction function is increasing. That is, for any two brightness x_1 and x_2 next condition is met: $a \leq x_1 < x_2 \leq b \rightarrow f(x_1) \leq f(x_2)$.

Currently, many different types of functions are used, but most of them have several classes of functions in their basis. Most often it is the polynomial, logarithm and exponent, which are defined on the interval $[0, \ldots, 255]$.

2.1   *Linear Transformation*

This type of conversion is the most commonly used. Its unified model is written as follows:

$$p(x) = k \cdot (x - a) + c, \quad k = \frac{d - c}{b - a} \tag{1}$$

This transformation is a straight line equation that passes through points with coordinates (a, c) and (b, d).

Linear transformation (1) allows you to proportionally change the brightness of the image, which is not enough when solving some problems.

2.2   *Power Transformation*

Power transformation is used in cases where linear transformation, due to the property of proportionality, is not able to provide proper image quality.

$$y(x) = k \cdot x^\gamma, \quad \gamma > 0 \tag{2}$$

We transform this function to a unified form so that it meets the requirements formulated above. Also we make a shift along the coordinate axes by one unit to the right and one unit up and get

$$y(x) = k \cdot (x - a)^\gamma + c, \quad \gamma > 0 \tag{3}$$

Coefficient k is determined from condition $y(b) = d$

$$d = k \cdot (b - a)^\gamma + c \rightarrow k = \frac{d - c}{(b - a)^\gamma} \tag{4}$$

Substituting this coefficient into a function, we obtain a unified power transformation.

$$h_\gamma(x) = \frac{d - c}{(b - a)^\gamma} \cdot (x - a)^\gamma + c \tag{5}$$

For the convenience of solving the problems of gradation correction, instead of using private models, it is proposed to use one unified power model of gradation correction

$$F_\gamma(\lambda, x) = \begin{cases} \lfloor Z1 \rfloor & if\ 0 \le \lambda \le 1, \\ \lfloor Z2 \rfloor & if\ 1 \le \lambda \le 2, \\ \gamma > 1,\ 0 \le \lambda \le 2 \end{cases} \tag{6}$$

Which based on the use of inverse with respect to the straight lines (1) power functions of the gradation correction of the form (5)

$$\begin{cases} Z1 = (1 - \lambda) \cdot h_\gamma(x) + [1 - (1 - \lambda)] \cdot p(x), \\ Z2 = [1 - (\lambda - 1)] \cdot p(x) + (\lambda - 1) \cdot h_{1/\gamma}(x) \end{cases} \tag{7}$$

where $\lfloor \cdot \rfloor$—rounding operator, $F_\gamma(\lambda, x) \in [c, \ldots, d]$.

For a given level of significance γ using the unified model (6) allows you to vary only the weighting factor λ in order to determine the best combination of the parameters of the gradation correction (γ, λ).

A similar approach is also valid for the following exponential-logarithmic and sinusoidal models.

When $\lambda = 0$ we get $Z1 = h_\gamma(x)$. Under such conditions, model (6) degenerates into model (5), with the exponent $\gamma > 1$. This model is used for tone correction and enhancing the contrast of the lighted part of the image close to the end b of the interval $[a, \ldots, b]$.

When $\lambda = 1$ we get $Z1 = Z2 = p(x)$. That is, model (6) degenerates into a linear model (1), which serves to proportionally convert the brightness and contrast of an image pixel.

When $\lambda = 2$ we get $Z2 = h_{1/\gamma}(x)$. Under such conditions, model (6) degenerates into model (5), with the exponent $0 < \gamma < 1$. This model serves for tonal correction and enhancing the contrast of the darkened part of the image close to the beginning of the interval $[a, \ldots, b]$.

Power conversion allows nonlinear amplification of one and suppress other frequency ranges of the image.

2.3   Exponential-Logarithmic Transformation

Consider an exponential transformation based on a function of the form

$$y(x) = e^{kx} \tag{8}$$

We transform this function to a unified form so that it meets the requirements formulated above. Make a shift along the coordinate axes by a units to the right and $c - 1$ units up and get

$$y(x) = \left(e^{k(x-a)} - 1\right) + c \tag{9}$$

Coefficient k determine from condition $y(b) = d$

$$d = \left(e^{k(b-a)} - 1\right) + c \rightarrow k = \frac{\ln(d - c + 1)}{b - a} \tag{10}$$

Substituting the coefficient (10) into the function (9), we obtain the desired transformation

$$ef(x) = \left(e^{\frac{\ln(d-c+1)(x-a)}{b-a}} - 1\right) + c \tag{11}$$

Now consider the class of logarithmic functions, namely

$$y(x) = k \cdot \ln(x) \tag{12}$$

We transform this function to a unified form so that it meets the requirements formulated above. Make a shift along the coordinate axes by $a - 1$ units to the right and c units up and get

$$y(x) = k \cdot \ln(x - a + 1) + c \tag{13}$$

Coefficient k is determined from condition $y(b) = d$

$$d = k \cdot \ln(b - a + 1) + c \rightarrow k = \frac{d - c}{\ln(b - a + 1)} \tag{14}$$

Substituting the coefficient (14) into the function (13), we obtain the desired transformation

$$lf(x) = \frac{d - c}{\ln(b - a + 1)} \cdot \ln(x - a + 1) + c \tag{15}$$

For convenience of solving problems of gradation correction, instead of using private models, it is proposed to use one unified exponentially-logarithmic model of gradation correction of the form

$$F_{el}(\lambda, x) = \begin{cases} \lfloor Z1 \rfloor & if\ 0 \le \lambda \le 1, \\ \lfloor Z2 \rfloor & if\ 1 \le \lambda \le 2, \\ \gamma > 1,\ 0 \le \lambda \le 2 \end{cases} \tag{16}$$

Which based on the use of inverse with respect to the direct (1) exponential and logarithmic functions of gradation correction (11) and (15)

$$\begin{cases} Z1 = (1 - \lambda) \cdot ef(x) + [1 - (1 - \lambda)] \cdot p(x), \\ Z2 = [1 - (\lambda - 1)] \cdot p(x) + (\lambda - 1) \cdot lf(x) \end{cases} \tag{17}$$

where $\lfloor \cdot \rfloor$—rounding operator, $F_{el}(\lambda, x) \in [c, \ldots, d]$.

When $\lambda = 0$ we get $Z1 = ef(x)$. Under such conditions, model (16) degenerates into a pure exponential model (11). Such a model is used for tonal correction and enhancing the contrast of the illuminated part of the image close to the end b of the interval $[a, \ldots, b]$.

When $\lambda = 1$ we get $Z1 = Z2 = p(x)$. That is, model (16) degenerates into a linear model (1), which serves for proportional conversion of the brightness and contrast of an image pixel.

When $\lambda = 2$ we get $Z2 = lf(x)$. Under such conditions, model (16) degenerates into model (15). This model is used for tonal correction and enhancing the contrast of the darkened part of the image close to the beginning a of the interval $[a, \ldots, b]$.

The power and exponential-logarithmic models in spite of their similarity have a significant difference. The exponentially logarithmic model is steeper, especially near the boundaries of the interval $[a, \ldots, b]$. This allows you to use it for a significant (intermittent) correction of the brightness and contrast of the image near the borders of the interval $[a, \ldots, b]$.

The linear, power, and exponential-logarithmic transformations described above are applied in the vast majority of gradation correction tasks. At the same time, there is one more—a sinusoidal transformation, which is used in a special situation when it is generally required to softly and quasi-linearly convert the brightness of the image, but significant contrasting or brightness suppression is required near the boundaries of the brightness range $[a, \ldots, b]$.

2.4 Sinusoidal Transformation

The basic function of the sinusoidal gradation transform is written as follows

$$y(x) = \sin(x) \tag{18}$$

For the purpose of the gradation correction, we will consider the interval of the monotonous increase of the sine function from $-\pi/2$ to $\pi/2$.

Let's map the function $y(x) = \sin(x)$ from interval $[-\pi/2, \ldots, \pi/2]$ to interval $[a, \ldots, b]$ and get such a representation of it on the interval $[a, \ldots, b]$

$$y(x) = \sin\left(\frac{x-a}{b-a}\pi - \frac{\pi}{2}\right) \tag{19}$$

However, such a representation will not be adequately scaled along the ordinate axis. To eliminate this drawback, we rewrite the function (19) with two additional parameters $k1$ and $k2$, responsible for the shift and scaling of the function (19) along the ordinate axis so

$$s(x) = k1 \cdot \sin\left(\frac{x-a}{b-a}\pi - \frac{\pi}{2}\right) + k2 \tag{20}$$

Let's find the parameters $k1$ to satisfy the requirements above. To do this, we substitute into the function a pair of values $x = a$, $y = c$ and $x = b$, $y = d$ and obtain a system of two equations of the form

$$\begin{cases} c = -k1 + k2, \\ d = k1 + k2. \end{cases} \tag{21}$$

This system has the only solution

$$k1 = \frac{d-c}{2}, \quad k2 = \frac{d+c}{2} \tag{22}$$

In this situation, we obtain the following unified sinusoidal transformation

$$s(x) = \frac{d-c}{2} \cdot \sin\left(\frac{x-a}{b-a}\pi - \frac{\pi}{2}\right) + \frac{d+c}{2} \tag{23}$$

For the purpose of constructing the inverse sine function, we consider a combination of sinusoidal (23) and linear (1) transformations of the form

$$t(x) = p(x) + (p(x) - s(x)) \tag{24}$$

For the convenience of solving the problems of gradation correction, instead of using private models, it is proposed to use one unified sinusoidal model of gradation correction

$$F_s(\lambda, x) = \lfloor \lambda \cdot s(x) + (1-\lambda) \cdot t(x) \rfloor, \quad 0 \le \lambda \le 1 \tag{25}$$

where $\lfloor \cdot \rfloor$—rounding operator, $F_s(\lambda, x) \in [c, \ldots, d]$.

When $\lambda = 0$ we get $F_s(\lambda, x) = t(x)$. Under such conditions, model (25) degenerates into the inverse sinusoidal function (24). Such a model serves to alter the brightness more significantly and increase the contrast, the closer the value is to the end of the interval $[a, \ldots, b]$.

When $\lambda = 0.5$ we get $F_s(\lambda, x) = p(x)$. That is, model (25) degenerates into a linear model (1), which serves to proportionally convert the brightness and contrast of the image.

When $\lambda = 1$ we get $F_s(\lambda, x) = s(x)$. Under such conditions, model (25) degenerates into a sinusoidal model (23). This model serves to change the brightness more significantly and increase the contrast, the closer this value is to the middle of the brightness interval $[a, \ldots, b]$.

3 Quality Criteria of Gradational Image Correction

To improve the perception of objects in the picture, it is necessary to increase the dynamic range of brightness and increase the contrast of the object relative to the background.

When performing gradation correction, the brightness of all pixels of the image is converted [1, 12]. Thus, there is a general increase in the difference in brightness of the object relative to the background as a whole.

Two main indicators—global and local coefficients of change of the image contrast, will characterize the quality of the gradation correction.

Global image contrast enhancement ratio. Image sharpening can be expressed by an integral stretching factor c_s of an interval $[a, \ldots, b]$ for an interval $[c, \ldots, d]$

$$c_s = \frac{\nu}{\mu} = \frac{d - c + 1}{b - a + 1} \tag{26}$$

If the coefficient c_s is greater than one, in this case the narrower interval $[a, \ldots, b]$ is stretched to a wider interval $[c, \ldots, d]$. With an increase of coefficient c_s the contrast of the image will only increase. If a wider interval $[a, \ldots, b]$ is shrinked to a narrower interval $[c, \ldots, d]$, in this case the coefficient c_s will be less than one. With a decrease of coefficient c_s the contrast of the image will only decrease.

Local (object-oriented) object sharpening ratio. Gradual correction can be made in order to improve the overall quality of the image, and maybe - to improve the perception of objects of interest to us in the image. The coefficient c_s does not show a local change in the contrast of the object of interest relative to the background, depending on the applied gradation conversion model. To take this aspect into account, we will use a local coefficient c_v, that reflects the level of variation of the contrast of the object relative to the background

$$c_v = \frac{h'}{h} \tag{27}$$

where h' and h—new and previous estimates of the contrast of the object relative to the background [1, 2, 12].

Note that when using a linear transformation, the value of c_v is equal to coefficient c_s [it is also a coefficient in model (1)]. When using a nonlinear transformation, the coefficient c_v needs to be calculated, since it will depend on the frequency domain to which the brightness of the object pixels and the background belong. And for different objects in the picture the coefficient c_v can vary significantly.

4 Experiment Results

In order to compare the proposed functions of the gradation correction and the analysis of their quality, it is rational to carry out such an experiment:

- Select dark images
- Apply to them the proposed methods of gradation correction
- Evaluate the quality of the gradation correction using the coefficients c_s and c_v.

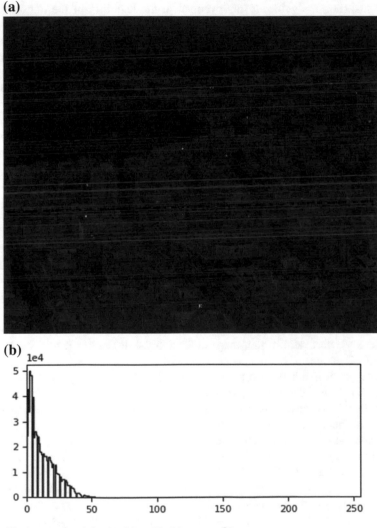

Fig. 1 The image of a night city (**a**) and its histogram (**b**)

Let's start with a snapshot of the night city (Fig. 1) [19]. As can be seen from the image histogram, the brightness values are shifted to the left and the range above 50 is not used, as a result of which it is very difficult to distinguish individual buildings or other objects. For this image we will use the following parameters constant $\gamma = 2$, $a = 0$, $b = 55$, $c = 0$, $d = 255$. Applying a linear transformation (Fig. 2), we obtain the coefficients $c_s = 4.64$ and $c_v = 4.69$. Buildings are clearly distinguishable in this picture, but the trees are still very poorly distinguishable. Applying a power transformation (Fig. 3) with parameter $\lambda = 1.75$ we obtain the coefficients $c_s = 4.64$ and $c_v = 5.62$. In this picture, according to the expert, an optimal quality of the correction was achieved. By applying an exponential-logarithmic transformation (Fig. 4) with parameter $\lambda = 2.0$ we obtain the coefficients $c_s = 4.64$ and $c_v = 7.08$. In this snapshot, a significant improvement in the legibility of the trees has been achieved, which is reflected in the maximum value of the coefficient c_v, but at the same time, the area with buildings according to the expert has a worse contrast than (Fig. 3). Applying a sinusoidal transform (Fig. 5) with parameter $\lambda = 0$ we obtain the coefficients $c_s = 4.64$ and $c_v = 5.62$. In this picture it is noteworthy that the coefficient c_v is the same as in (Fig. 3) but the overall image of the image is darker, which is normal because c_v can vary for different points in the image.

The next shot is a daytime shot of a city street (Fig. 6) [20]. As can be seen from the image histogram, a significant number of pixels is in the range of 50–200, which does not allow us to use too low b values. For this image, we will use the following parameters constant $\gamma = 2$, $a = 0$, $b = 200$, $c = 0$, $d = 255$. Applying a linear transformation (Fig. 7), we obtain the coefficients $c_s = 1.27$ and $c_v = 1.33$. This transformation practically did not improve the quality of image perception. Applying a power transformation (Fig. 8) with parameter $\lambda = 2.0$ we obtain the coefficients $c_s = 1.27$ and $c_v = 2.33$. In this picture, road signs and background objects are clearly distinguishable, but the noise level is elevated. By applying the exponential-logarithmic transformation (Fig. 9) with parameter $\lambda = 2.0$ we obtain the coefficients $c_s = 1.27$ and $c_v = 2.67$. In this picture, an unacceptable noise level is achieved, but at the same time, the best distinguishability of background objects. Applying a sinusoidal transformation (Fig. 10) with parameter $\lambda = 0$ we obtain the coefficients $c_s = 1.27$ and $c_v = 2.0$. According to the expert in this snapshot, it was possible to achieve the best ratio of the image noise and the distinctiveness of background objects.

The last shot is a shot of a poorly lit abandoned room (Fig. 11) [21], the histogram shows that there are a certain number of pixels with a brightness value greater than 100 that can be neglected. For this image we will use the following parameters unchanged $\gamma = 2$, $a = 0$, $b = 100$, $c = 0$, $d = 255$. Applying a linear transformation (Fig. 12), we obtain the coefficients $c_s = 2.55$ and $c_v = 2.55$. This transformation practically did not improve the quality of image perception. Applying a power transformation (Fig. 13) with parameter $\lambda = 2.0$ we obtain the coefficients $c_s = 2.55$ and $c_v = 3.0$. In this picture, pipes and irregularities on the walls in the shade are clearly distinguishable. By applying an exponential-logarithmic transformation (Fig. 14) with parameter $\lambda = 2.0$ we obtain the coefficients $c_s = 2.55$ and $c_v = 3.67$. In this picture, the best overall detail of the picture, but with a high level

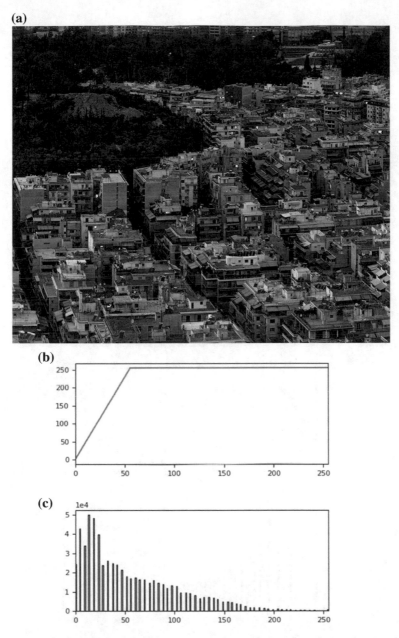

Fig. 2 Image of a night city with the linear transformation applied (**a**) the transform function (**b**) and brightness histogram (**c**)

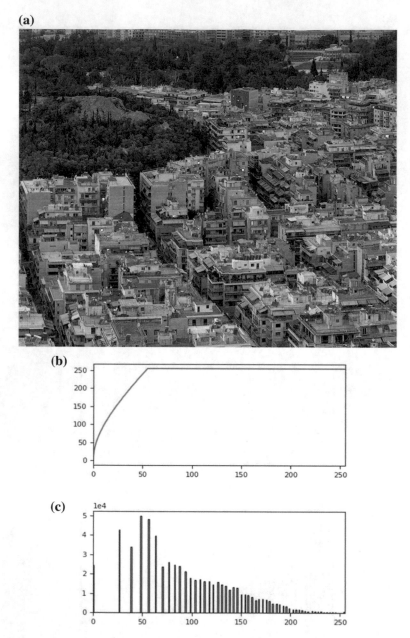

Fig. 3 Image of a night city with the power transformation applied (**a**) the transform function
(**b**) and brightness histogram (**c**)

(a)

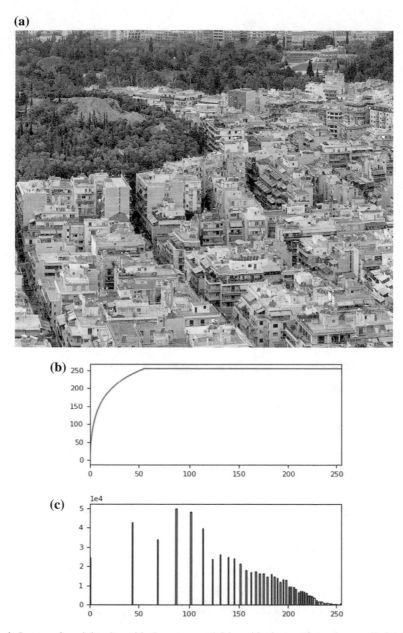

Fig. 4 Image of a night city with the exponential-logarithmic transformation applied (**a**) the transform function (**b**) and brightness histogram (**c**)

(a)

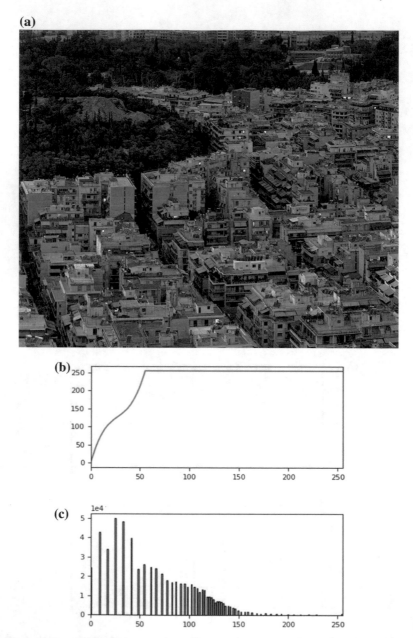

Fig. 5 Image of a night city with the sinusoidal transform applied (**a**) the transform function (**b**) and brightness histogram (**c**)

(a)

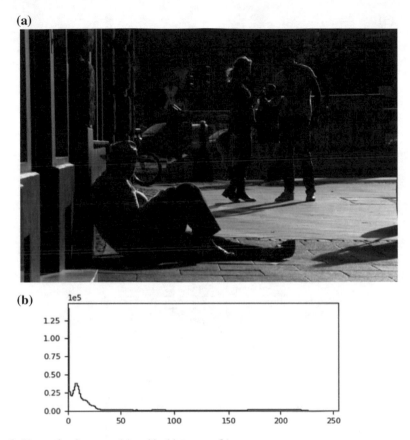

(b)

Fig. 6 Photo of a city street (**a**) and its histogram (**b**)

of noise. Applying a sinusoidal transform (Fig. 15) with parameter $\lambda = 0$ we obtain the coefficients $c_s = 2.55$ and $c_v = 3.33$. This transformation made it possible to achieve a high value of the coefficient in the region of interest without increasing noise.

5   Conclusions

Based on the analysis of the current state of the issue, the most common classes of functions used in the gradation correction are identified. Namely, linear, power, exponential, logarithmic, sinusoidal. For these classes of functions, the following models of gradation correction are proposed: power (6), exponential-logarithmic (16), and sinusoidal (25).

The quality indicators of the gradation image correction are described.

An experiment was conducted to confirm the performance of the proposed models and quality indicators. The experiment proved the practical importance of all the

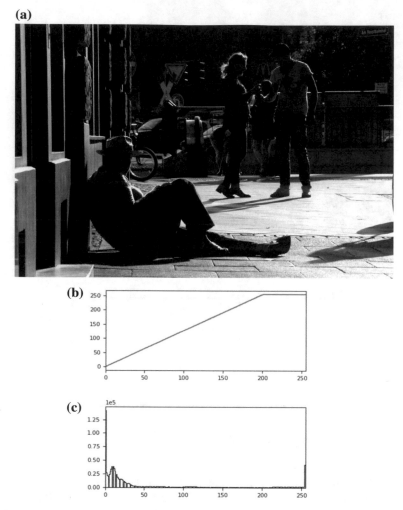

Fig. 7 Image of a city street with the linear transform applied (**a**) the transform function (**b**) and brightness histogram (**c**)

models presented, and also showed the optimal areas of application of the models, depending on the overall layout of the scene and the state of the object of interest. It is better to apply power transformation for a general improvement in the perception of a picture, when small details on the background are not so important. The exponential logarithmic transformation shows itself better when it is necessary to distinguish objects of the background, but at the same time some image details can be lost. Sinusoidal transformation shows good results when you need to achieve average indicators of improving the perception of the image, but with less noise. It is shown that quality indicators are not the only decisive factors when choosing a model, because in cases of exponential logarithmic transformation, the maximum values

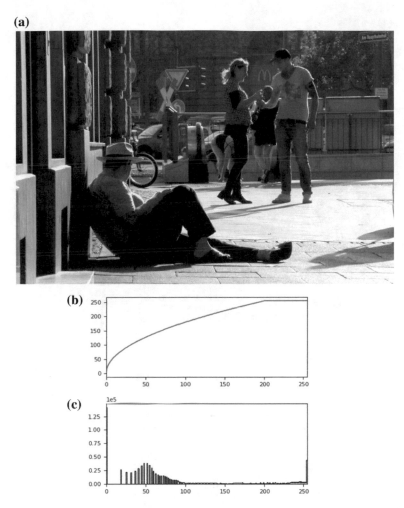

Fig. 8 Image of a city street with the power transform applied (**a**) the transform function (**b**) and brightness histogram (**c**)

of the quality criteria were obtained, but at the same time this conversion led to an excessive increase in noise. Thus, an expert opinion on the photorealism and noise level of the image is still required.

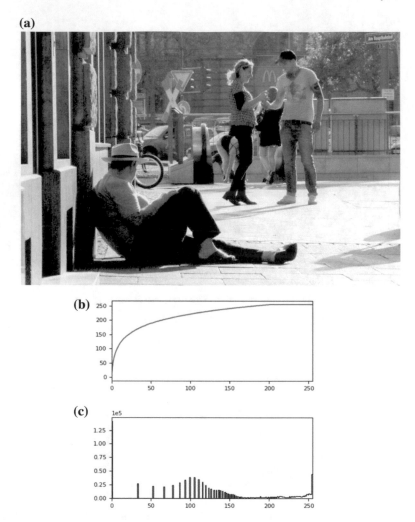

Fig. 9 Image of a city street with the exponential-logarithmic transform applied (**a**) the transform function (**b**) and brightness histogram (**c**)

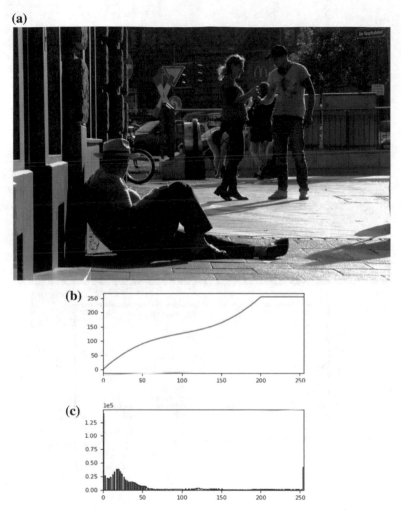

Fig. 10 Image of a city street with the sinusoidal transform applied (**a**) the transform function (**b**) and brightness histogram (**c**)

(a)

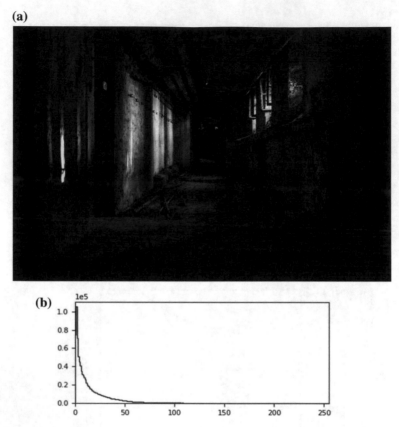

Fig. 11 An image of an abandoned room (**a**) and its histogram (**b**)

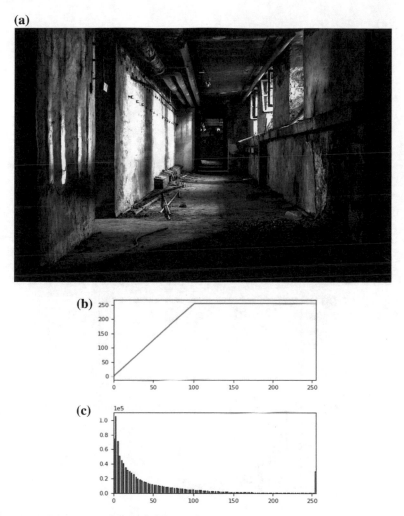

Fig. 12 An image of an abandoned room with the linear transform applied (**a**) the transform function (**b**) and brightness histogram (**c**)

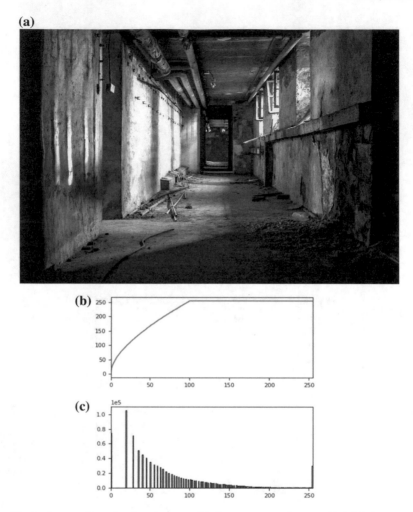

Fig. 13 An image of an abandoned room with the power transform applied (**a**) the transform function (**b**) and brightness histogram (**c**)

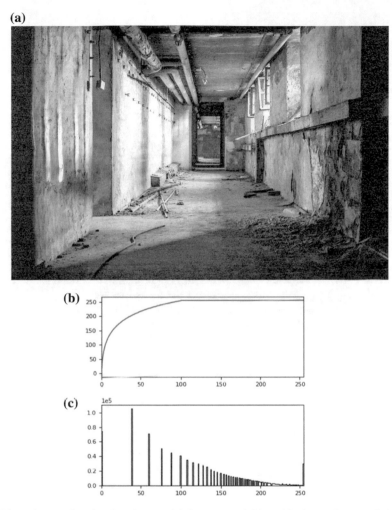

Fig. 14 An image of an abandoned room with the exponential-logarithmic transform applied (**a**) the transform function (**b**) and brightness histogram (**c**)

(a)

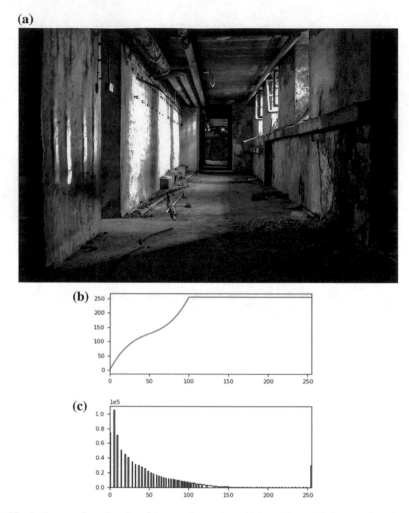

Fig. 15 An image of an abandoned room using a sinusoidal transform (**a**) the transform function (**b**) and brightness histogram (**c**)

References

1. Gonzalez R, Woods R (2018) Digital image processing, 4th edn. Pearson, New York
2. Forsyth D, Ponce J (2015) Computer vision: a modern approach, 2nd edn. Pearson, London
3. Corke P (2017) Robotics, vision and control. 2nd edn. Springer
4. Ruban I, Smelyakov K, Martovytskyi V, Pribylnov D, Bolohova N (2018) Method of neural network recognition of ground-based air objects. In: IEEE 9th international conference on dependable systems, services and technologies (DESSERT). IEEE, pp 589–592
5. Ageyev D et al (2018) Provision security in SDN/NFV. In: 2018 14th international conference on advanced trends in radioelecrtronics, telecommunications and computer engineering (TCSET). IEEE, pp 506–509. https://doi.org/10.1109/tcset.2018.8336252

6. Ruban I, Churyumov G, Tokarev V, Tkachov V (2017) Provision of survivability of reconfig-urable mobile system on exposure to high-power electromagnetic radiation. In: Selected papers of the XVII international scientific and practical conference on information technologies and security (ITS 2017). CEUR workshop processing, pp 105–111

7. Kirichenko L, Radivilova T, Bulakh V (2020) Binary classification of fractal time series by machine learning methods. In: Lytvynenko V, Babichev S, Wójcik W, Vynokurova O, Vyshe-myrskaya S, Radetskaya S (eds) Lecture notes in computational intelligence and decision making. ISDMCI 2019. Advances in intelligent systems and computing, vol 1020. Springer, Cham

8. Smelyakov K, Yeremenko D, Sakhon A, Polezhai V, Chupryna (2018) A Braille character recognition based on neural networks. In: IEEE second international conference on data stream mining & processing (DSMP). IEEE, pp 509–513

9. Smelyakov K, Ruban I, Sandrkin D, Martovytskyi V, Romanenkov Y (2018) Search by image. New search engine service model. In: 5th international scientific-practical conference problems of infocommunications. Science and technology (PIC S&T), pp 181–186

10. Ageyev DV, Salah MT (2016) Parametric synthesis of overlay networks with self-similar traffic. Telecommunications and Radio Engineering (English translation of Elektrosvyaz and Radiotekhnika) 75(14):1231–1241

11. Ionescu R, Popescu M (2016) Knowledge transfer between computer vision and text mining. Springer

12. Sonka M, Hlavac V, Boyle R (2014) Image processing, analysis, and machine vision. 4th edn. Cengage Learning

13. Ageyev D et al (2019) Infocommunication networks design with self-similar traffic. In: 2019 IEEE 15th international conference on the experience of designing and application of CAD systems (CADSM). IEEE, pp 24–27. https://doi.org/10.1109/cadsm.2019.8779314

14. Filatov V, Semenets V (2018) Methods for synthesis of relational data model in informa-tion systems reengineering problems. In: Proceedings of the international scientific-practical conference problems of infocommunications. Science and technology (PIC S&T), pp 247–251

15. Mukhin V, Romanenkov Y, Bilokin J, Rohovyi A, Kharazii A, Kosenko V, Kosenko N, Su Ju (2017) The method of variant synthesis of information and communication network structures on the basis of the graph and set-theoretical models. Int J Intell Syst Appl (IJISA) 9(11):42–51

16. Bielievtsov S, Ruban I, Smelyakov K, Sumtsov D (2018) Network technology for transmission of visual information. In: Selected papers of the XVIII international scientific and practical conference on information technologies and security (ITS 2018). CEUR workshop processing. Kyiv, pp 160–175

17. Lemeshko O, Arous K, Tariki N (2015) Effective solution for scalability and productivity improvement in fault-tolerant routing. In: Proceedings of second international IEEE conference problems of infocommunications. Science and technology (PICS&T-2015). Kharkiv, pp 76–78

18. Kryvinska N (2008) An analytical approach for the modeling of real-time services over IP network. Math Comput Simul 79(4):980–990. https://doi.org/10.1016/j.matcom.2008.02.016

19. Aristotle Roufanis photography, https://aristotle.photography/. Last accessed 10 Dec 2019

20. Pxhere free images, https://pxhere.com/en/photo/239908. Last accessed 21 Dec 2019

21. Pxhere free images, https://pxhere.com/en/photo/726866. Last accessed 21 Dec 2019

Study of Approaches to the Management of the Production of Entomophages

Vitaliy Lysenko⬥ and Irina Chernova⬥

Abstract The work is devoted to the study of approaches to the management of the production of entomophages, the processes of cultivation of entomophages from the positions of infocommunication, process and system approaches are considered; processes of production management using intelligent information processing algorithms—cognitive analysis, fuzzy logic and neural networks; the scientific novelty of the work is that, using intelligent algorithms for processing information to modeling complex systems, in order to increase the efficiency of production by analyzing and structuring large quantities and significant volumes of information flows in the production of entomophage *Habrobracon hebetor* proposed the method of cognitive analysis of information; mathematical model of control of caterpillar growing process *Ephestia kuehniella*, structural model of calculation of income, total electricity consumption and production profits *Ephestia kuehniella* and algorithm of control of the production of entomophages on the criterion of product quality using neural networks; the proposed approaches to the management of the production of entomophages allow the formalization of poorly structured insect cultivation processes, with sufficient accuracy to form management decisions in the conditions of incompleteness of information, automatically create knowledge bases, reduce energy costs for decision-making.

Keywords Production of entomophages · Management · Data mining

V. Lysenko
National University of Life and Environmental Sciences of Ukraine, Kyiv 03041, Ukraine

I. Chernova (✉)
Engineering and Technological Institute "Biotechnica" National Academy of Agrarian Sciences, Odesa Region 67667, Ukraine
e-mail: bioischernova@ukr.net

© The Editor(s) (if applicable) and The Author(s), under exclusive license to Springer Nature Switzerland AG 2021
T. Radivilova et al. (eds.), *Data-Centric Business and Applications*, Lecture Notes on Data Engineering and Communications Technologies 48,
https://doi.org/10.1007/978-3-030-43070-2_15

319

1 Introduction

The production of entomophages is a complex dynamic system with management, which contains a set of basic, auxiliary and service processes, which are linked by information interdependencies in order to obtain products of guaranteed quality [1]. The processes of the production of entomocultures by their nature are continuously periodic with a significant combination of discrete operations (preparation nutrient medium, premises, quality control of products, storage of products and other), stochastic (subject to the influence of a significant number of external perturbations—temperature and relative humidity of the environment, power cessation, partial loss of equipment efficiency etc.), have ambiguous behavior of the biological object, the number of input and output parameters are multidimensional, as well as those in which parameters change not only in time but also in space [1].

By optimizing the processes of cultivation of entomocultures, the use of innovative technologies that improve the quality of entomologic products, energy efficiency of production, increase in profits from sales through optimal production management is applied. The quality of entomological products, which is evaluated according to biological indicators, is formed progressively at all stages of the technological process of production and it depends on a large number of abiotic, technological and perturbing factors (temperature and relative humidity of air box for breeding entomocultures, photoperiod, lighting, quality and type of nutrient medium, height of the nutrient layer, type of insect-owners, etc.). In this case, the formalization of these dependencies in many cases is absent [1].

The quality of the insects received is of paramount importance for ensuring success with in strategies biological control, but technical and economic constraints hamper the development in cultivation of entomophages; with new directions being the development of production facilities, in particular, automation and product quality assurance [2].

The solution of the methodical issues of the management of production processes in the agroindustrial complex is carried out to a large extent in conditions of uncertainty, related to the inadequate level of knowledge about biological systems, the random nature of the processes occurring in them [3]. The main sources of manifestation of uncertainty in the tasks of management of biotechnological objects are the complexity of a formal description of a biotechnological object and management tasks; non-stationary parameters of the object [4].

The quality of production is directly related to the production technology. Violations of processes (physical wear of equipment, violation of climatic conditions of breeding entomocultures, prolonged breeding offspring of the initial population) cause deviation of the quality indicators of products from the normative values [5].

In order to increase the efficiency of production, the issue of quality control is relevant, which is related to the control of the parameters of significant impact on the quality of products. Solving this problem is possible using modern methods of intellectual analysis, in particular, cognitive models, fuzzy cognitive maps, neural networks, which allows to realize the visual representation of processes and simplify the formalization procedures of complex dependencies [1].

To date, is becoming widespread a cognitive approach to the modeling and management of weakly structured systems, aimed at the development of formal models and methods, supporting the intellectual process of problem solving due to consideration in these models and methods of cognitive capabilities (perception, representation, knowledge, understanding, explanation) of the subjects of management in the decision of managerial tasks [6].

The purpose of cognitive modeling of weakly structured systems is to find out the mechanism of the functioning of the system (the mechanism of phenomena and processes occurring in the system), forecasting the development of the system, managing by her, determining the possibilities of its adaptation to the external environment [7].

The methodology of cognitive modeling was proposed by R. Axelrod [8]. It is based on the modeling of the subjective representations of experts about the situation and contains: the methodology of structuring the situation; model of knowledge representation of the expert; methods of situations analysis [8].

Precondition for the rapid development of the cognitive approach was the inapplicability of precise models for analyzing and simulating problem situations, which arise in the process of development of complex systems, because of the need to consider a large number of factors, many of which proved to be difficult to formalize [9].

Today, the cognitive approach is actively developing in the research of complex technical systems [10, 11], on issues on energy efficiency management [12], intelligent decision-making systems [13], weakly structured processes [7].

At present, fuzzy logic is used to create different control systems, in particular temperature control, pH and saltiness of water when breeding shrimp [14], for controlling dissolved oxygen when growing shrimp indoors [15], fuzzy management for the cultivation of agricultural plants [16], management with the help of fuzzy logic for aquaculture of fresh water to control various stress factors on fish [17].

The use of neural networks provides some benefits to businesses that use them, in particular [18]:

– the possibility of use for problems with incomplete information when traditional mathematical models do not give the desired result;
– increase the accuracy of the decision and reduce its subjectivity;
– more detailed study of processes and situations;
– the ability to use a variety of analysis methods and a large number of algorithms;
– the ability to solve those tasks that have not been resolved before;
– accelerating the decision-making process, etc.

At the same time, in the context of increasing efficiency the production of entomophages has been investigated taking into account the infocommunication, process and system approaches as an infocommunication system consisting of a certain number of processes that are considered as a set of works, for which resources are used to convert inputs to outputs [19].

A characteristic feature of infocommunication systems is the continuous improvement of infocommunication technologies, development and implementation of progressive means of organization and management of databases, processing, analysis and presentation of information [20]. So, studies are being conducted today on the structural and parametric synthesis of modern information and communication systems [21], classification of virtualization methods for different resources [22], network intelligence [23, 24], ensuring the quality of services in information networks [25], optimizing resource utilization and reducing computation time [26].

The development of a management solution involves the necessary consideration of information communication for obtaining and using a certain amount of information about the state of the enterprise, personnel, its capabilities [27]. The completeness of the source information determines the division of managerial decisions into decisions that are taken in terms of certainty, risk and uncertainty [27, 28].

2   The Purpose of the Work

Increasing the efficiency of production by analyzing and structuring a large number and large volumes of information flows during the production of entomophage *Habrobracon hebetor*.

3   Materials and Methods of Research

The object of control is the processes of cultivating a caterpillar entomophage *Habrobracon hebetor* and its host insects *Ephestia kuehniella*. Use *Habrobracon hebetor* in the fight against harmful insects in agrocenoses is a promising area of biological protection of plants.

Methods of research—infocommunication, process and system approaches; intelligent information processing algorithms.

Applying infocommunication, process and system approaches to modeling complex systems [1, 5–7, 13, 19, 28–32], developed cognitive models of infocommunication aspects of entomophage production, the quality of production processes, production management processes, processes of entomophage *Habrobracon hebetor* growth in the form of events and a mathematical model of process control growing caterpillars *Ephestia kuehniella* on the criterion of maximizing profits on condition of minimization of energy consumption.

The main stages of creating cognitive models were the definition of target vertices, the factors of influence and the interrelationships between them.

On the basis of applying the cognitive approach and the theory of fuzzy logic [33, 34], using research results [35], a fuzzy cognitive map has been developed to control the effect of the height of the nutrient layer *Ephestia kuehniella* on her indicators quality. The main stages of developing a fuzzy cognitive map were the definition of the structure of the oriented graph (the input and output concepts and the connections between them); expert assessment of the degree of influence of input concepts on the weekend; calculation of the cognitive consonance for each of the input concepts; definition of the ranges of variation of input and output variables, term sets of each variable, type and parameters of the functions of the membership of the terms of each variable; construction of production rules of the knowledge base; calculation of the average error of approximation between the experimental data and the fuzzy conclusion; obtaining as a result of the fuzzy conclusion of the dependencies of output concepts from the input.

Using the ANFIS (Adaptive Network-Based Fuzzy Inference System)—the MAT-LAB editor, has developed a self-learning hybrid network in formation of controlling influences in production entomophage *Habrobracon hebetor* [36].

4 Research Results

Carried out the analysis of infocommunication aspects the production of ento-mophages, the result of which is a cognitive model (the model structure is shown in Fig. 1), consists of target, manager, information and communication vertices. The target node (infocommunication aspects of the production of entomophages) 1 consists of manager vertice (managerial decisions) 2, an information vertice (information resources) 7, a communication vertice (means of interaction) 18 and a controlling vertice (information flows of production control) 22. The controlling vertex 2 has the vertices that characterize managerial decisions in terms of certainty 3, under conditions of uncertainty 4, in terms of risk 5, in particular, the risk of production of low-quality products 6 [27, 28].

Information node 7 includes the dependences of the biological indices of the quality of entomocultures on the abiotic and technological parameters of technocenosis 8,

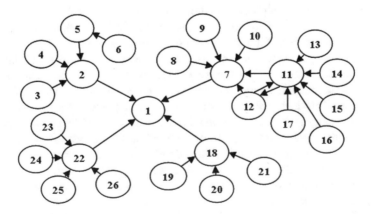

Fig. 1 Cognitive model of infocommunication aspects the production of entomophages

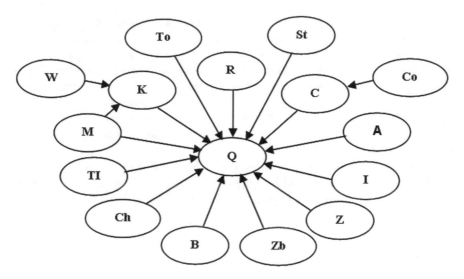

Fig. 2 Cognitive model of the quality of processes production entomophages

the knowledge base of the expert systems of the fuzzy and neuro-fuzzy conclusion 9, intelligent information processing algorithms 10, quality assurance system for entomological products 11, quality of production processes 12, technological support 13, methodology of quality management of entomological products 14, production base 15, personnel requirements 16, reference base 17 [1, 30].

The communication vertex 18 contains the user interface 19, the continuity of information 20 and the unimpeded information 21 [1, 27].

The controlling vertex 22 contains an inbound 23, selective 24, a production 25 and individual control 26 [29].

The structuring of the factors influencing the quality of the processes of production of entomophages is carried out, the result whose is the cognitive model on Fig. 2.

In model Q—quality of production processes; K—level of technological manageability (shows the flexibility of processes and the possibility of changing him parameters); W—management of the output of commodity entomological products; M—control of abiotic parameters of technocenosis (temperature and relative humidity of air box for insect breeding); TI—the level of technological intensity (determines the degree of use of material, energy and time parameters of processes); Ch—frequency of deviations of product quality indicators (the risk of producing poor quality products); B—the degree of continuity (the ratio of the duration of the technological part of the production cycle and the duration of the full production cycle); Z—coefficient of equipment loading (determined by the ratio of the estimated amount of equipment to the normative); I—coefficient of intensive use of equipment (correlation of actual production of products per unit time to technically grounded products output per unit of time); A—the level of automation (the ratio of the number of automated equipment to the total equipment); C—level of controllability; Co—the possibility of making corrections; R—the degree of rhythm; To—accuracy of maintaining given abiotic

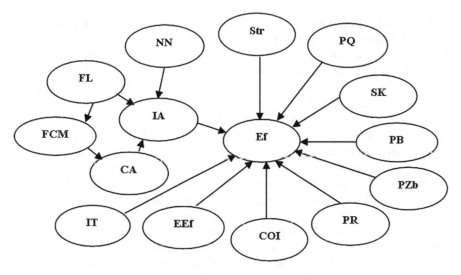

Fig. 3 Cognitive model of production management processes entomophages

parameters; St—stability of processes; Zb—account perturbations when managing abiotic parameters of technocenosis [28, 30, 37, 38].

Using the results of the analysis of the factors influencing the efficiency of the production of entomophages, a cognitive model of processes of production management, the structure of which is shown in Fig. 3.

The effectiveness Ef of production is influenced by: the application of information technology IT; energy efficient technologies EEf; speed of information processing COI; decision making in conditions of uncertainty of PR; decision making in conditions of perturbation PZb; profit from sales of products PB; situational management SK; measures to improve the quality of the entomologic products PQ; strategic management Str; the use of intelligent information processing algorithms IA; cognitive analysis of CA; Fuzzy Cognitive Cards FCM; fuzzy logic FL; neural networks NN.

The analysis of the factors influencing the efficiency of the production of entomophages makes it possible to allocate energy-efficient technologies, profit from sales of products and measures to improve its quality.

According to modern studies it should be noted that energy efficiency is:

- the use of less energy through the introduction of more advanced technologies or processes [39];
- characteristic of equipment, technology, production or system as a whole, indicating the degree of energy use per unit of final product [40];
- the state of the system, in which the achievement of goals and the fulfillment of its functions is ensured at a minimum energy cost [41];
- a set of indicators that makes it possible to compare different ware of the same function from the point of view of energy consumption [42].

Energy efficiency indicators of an enterprise depend on a large number of factors [43]: implementation of operating objectives; maintenance work; applicable technologies; achievement of balance of power, profit, energy consumption. At the same time, the indicators of the efficiency of maintaining the bioobject are: the cost of energy for technological processes; profit from the implementation of the management algorithm [44].

Formalized cognitive model of the production of entomophagus *Habrobracon hebetor* in the form of events, the structure which one is shown in Fig. 4.

In model A—preparatory work; B—checking the starting culture *Habrobracon hebetor*; C—preparation the host's insect nutrient medium (*Ephestia kuehniella*); D—inoculation of the nutrient medium with eggs *Ephestia kuehniella*; E—cultivation caterpillars *Ephestia kuehniella*; E1—assessment of the quality of entomological products according to biological indices (number of caterpillars, weight of caterpillars and weight of caterpillars of older age); F—detachment of caterpillars from nutrient medium; B1—assembling the imago *Habrobracon hebetor* for inoculation; I—preparation of the imago *Habrobracon hebetor* for infection; G—growing *Habrobracon hebetor* to the stage of the imago; H—infection of caterpillars *Habrobracon hebetor*; J—assessment of the quality of the infection *Habrobracon hebetor* caterpillars *Ephestia kuehniella*; K—growing *Habrobracon hebetor* to the

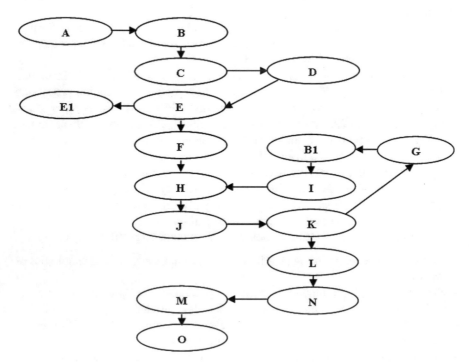

Fig. 4 Cognitive model for the production of entomophage *Habrobracon hebetor* in the form of events

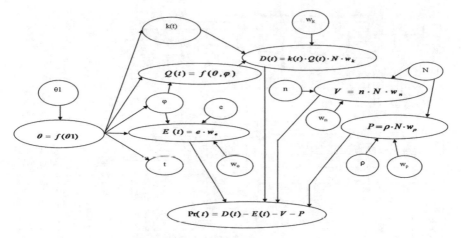

Fig. 5 Mathematical model of management the process of growing caterpillars *Ephestia kuehniella*

stage of pupae; L—quality assessment entomological products; N—accumulation of products; M—formation of commodity products; O—utilization of products.

Developed the mathematical model of management the process of growing caterpillars *Ephestia kuehniella* under the criterion of profit maximization under the condition of minimization of energy consumption [32], the structure of which is shown in Fig. 5.

In model θ—temperature of the box for insect breeding, °C; $\theta 1$—temperature of the environment, °C; $k(t)$—the number of caterpillars older age, pieces/cuvette; $Q(t)$—quality of entomocultures by the average weight of caterpillars of older age, g; $E(t)$—total electricity consumption for provision of set temperature and relative humidity of air box, UAH; e—electricity consumption, kWh; w_e—the cost of 1 kWh; UAH; φ—relative humidity of air box for breeding insects, %; t—the length of the cycle of caterpillars growing, days; $D(t)$—revenue from the implementation of entomoproduction, UAH; w_k—cost of 1 kg of caterpillars, UAH; N—number of cuvette, units; n—amount of nutrient medium, kg/cuvette; w_n—cost of 1 kg of grain, UAH; V—total expenditure of the nutrient medium (ground grain barley), UAH; ρ—is the number of eggs of *Ephestia kuehniella* introduced into grain, g/cuvette; w_ρ—cost of 1 g eggs, UAH; P—total expenses for grain inoculation eggs of *Ephestia kuehniella*, UAH; Pr—profit for the production of caterpillars in cultivation of the entomophage *Habrobracon hebetor*, UAH [32].

The practical implementation of the *Ephestia kuehniella* growing control process is accomplished by using fuzzy expert systems type Mamdani to estimate the profit of the production of entomocultures on the basis expansion pack of the Fuzzy Logic Toolbox for MATLAB [32]. As a result of the fuzzy conclusion, it is established that for $D(t) = (2254\text{-}2924)$ UAH, $E(t) = 131$ UAH, $\theta = 26.8$ °C, $\varphi = 70\%$, $V = 500$ UAH, $t = 30$ days, $P = 1020$ UAH profit $Pr(t)$ will be maximum—1440 UAH provided it minimizes energy costs [32]. The average error of the approximation for $Pr(t)$ was 6.62% (within the limits of permissible values [45]).

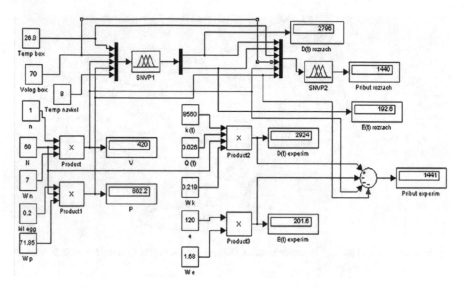

Fig. 6 Structural model of calculation of income, total electricity consumption and profits the production *Ephestia kuehniella* by results experimental studies taking into account perturbation

Using the software environment MATLAB/Simulink developed a structural model (Fig. 6), which allows you to calculate the income, total waste of electricity and profits the production of *Ephestia kuehniella* by the results of experimental studies taking into account perturbation (ambient temperature) [32].

In the model, the following notation is used: E(t) rozrach, D(t) rozrach, Pribut rozrach—respectively the value of total electricity consumption, revenue and profit by the fuzzy conclusion of the action on the biological object microclimate parameters (experimental temperature values Temp box and relative humidity of the air box Volog box) and ambient temperature (Temp navkol) and technological parameters of production (V, number of eggs of *Ephestia kuehniella*, brought in grain, kil egg, w_ρ). The designation Pribut experim, D(t) experim, E(t) experim is the experimental values of profit, income and total electricity consumption respectively.

For to control the influence of height of a layer of nutrient medium H (input concepts H1, H2, H3, H4 respectively 15 mm, 30 mm, 40 mm, 50 mm) of *Ephestia kuehniella* on its quality indicators—the number of caterpillars with 1 kg of nutrient medium (Y1); average mass of caterpillars, mg (Y2) and average mass of caterpillars of older age, mg (Y3) [35]—developed a fuzzy cognitive map in the form of a weighted oriented graph whose structure is shown in Fig. 7 [1].

An expert assessment extent of influence of height of the layer of the nutrient medium H of the *Ephestia kuehniella* on it quality indicators formalized as follows: 0.4 is weak; 0.7 is strong; 0.9 is very strong [1].

The cognitive consonance of KK [34] for each of the input concepts (Fig. 8) is calculated as a module of the sum of expert evaluations between H1 → Y1, H1 → Y2, H1 → Y3 (KK for H1); H2 → Y1, H2 → Y2, H2 → Y3 (KK for H2); H3 → Y1,

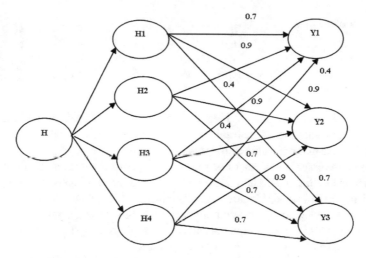

Fig. 7 Fuzzy cognitive map for to control the influence of height of a layer of nutrient medium of *Ephestia kuehniella* on its quality indicators [1]

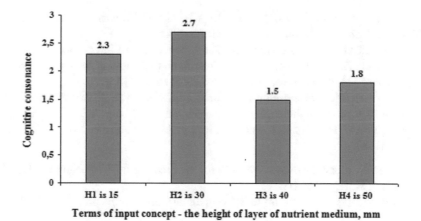

Fig. 8 The cognitive consonance fuzzy cognitive map [1]

H3 → Y2, H3 → Y3 (KK for H3); H4 → Y1, H4 → Y2, H4 → Y3 (KK for H4) and shows which one has the most impact on quality metrics *Ephestia kuehniella* [1].

Consonance is the maximum (2.7) for the concept of H2 (the average limit of the height of the layer of nutrient medium—30 mm), which agrees with the results of the studies given in [35].

The knowledge base of the fuzzy cognitive map is shown in Fig. 9 [1].

The production rules of the knowledge base have the form:

Fig. 9 View of the Rule
Editor [1]

1. If (H is H1) then (Y1 is Y12)(Y2 is Y23)(Y3 is Y31) (1)
2. If (H is H2) then (Y1 is Y13)(Y2 is Y24)(Y3 is Y32) (1)
3. If (H is H3) then (Y1 is Y11)(Y2 is Y21)(Y3 is Y31) (1)
4. If (H is H4) then (Y1 is Y11)(Y2 is Y22)(Y3 is Y31) (1)

R1: if H is «Lower boundary of height of the layer of the nutrient medium», then Y1 is «Standard number of caterpillars», Y2 is «High mass of caterpillars», Y3 is «Standard mass of caterpillars of older age»;

R2: if H is «Average boundary of height of the layer of the nutrient medium», then Y1 is «Higher than standard number», Y2 is «Higher than high mass of caterpillars», Y3 is «Higher than standard mass of caterpillars of older age»;

R3: if H is «Higher than average boundary of height of the layer of the nutrient medium», then Y1 is «Low number of caterpillars», Y2 is «Low mass of caterpillars», Y3 is «Standard mass of caterpillars of older age»;

R4: if H is «Upper bound of height of the layer of the nutrient medium», then Y1 is «Low number of caterpillars», Y2 is «Standard mass of caterpillars», Y3 is «Standard mass of caterpillars of older age».

The average error of approximation for Y1, Y2 and Y3 was respectively 1.78%, 2.32% and 1.68% [1] (within acceptable limits) [45].

As a result of the fuzzy conclusion, the dependences of the number [1] and the average mass of caterpillars *Ephestia kuehniella* on the height of the layer of nutrient (Figs. 10 and 11), which agree with the results of studies [35].

The mathematical description of a fuzzy Mamdani-type algorithm for a fuzzy cognitive map involves the definition of clear values of $Y1_0$, $Y2_0$, $Y3_0$, using the knowledge base (Fig. 9) and the clear values of H_0. He consisted of the following actions [1, 46]:

1. Fuzziness: found the degrees of truth for the preconditions of each rule: $H1(H_0)$, $H2(H_0)$, $H3(H_0)$, $H4(H_0)$.
2. Fuzzy conclusion: the cutoff levels were determined for the premise of each rule:

Fig. 10 The dependence of the number of caterpillars *Ephestia kuehniella* from 1 kg of nutrient medium from the height of its layer [1]

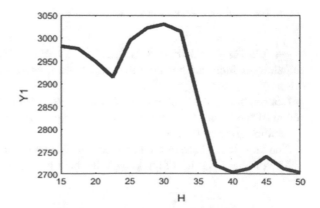

Fig. 11 The dependence of the medium weight of caterpillars *Ephestia kuehniella* from height of the layer nutrient medium

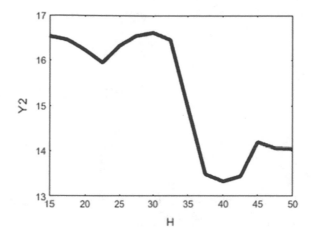

$$\alpha_1 = H_1(H_0), \tag{1}$$

$$\alpha_2 = H_2(H_0), \tag{2}$$

$$\alpha_3 = H_3(H_0), \tag{3}$$

$$\alpha_4 = H_4(H_0). \tag{4}$$

Then found truncated membership functions using the operation of the logical minimum (min):

$$Y12'(Y1) = (\alpha_1 \wedge Y12(Y1)), \tag{5}$$

$$Y23'(Y2) = (\alpha_1 \wedge Y23(Y2)), \tag{6}$$

$$Y31'(Y3) = (\alpha_1 \wedge Y31(Y3)), \tag{7}$$

$$Y13'(Y1) = (\alpha_2 \wedge Y13(Y1)), \tag{8}$$

$$Y24'(Y2) = (\alpha_2 \wedge Y24(Y2)), \tag{9}$$

$$Y32'(Y3) = (\alpha_2 \wedge Y32(Y3)), \tag{10}$$

$$Y11'(Y1) = (\alpha_3 \wedge Y11(Y1)), \tag{11}$$

$$Y21'(Y2) = (\alpha_3 \wedge Y21(Y2)), \tag{12}$$

$$Y31'(Y3) = (\alpha_3 \wedge Y31(Y3)), \tag{13}$$

$$Y11'(Y1) = (\alpha_4 \wedge Y11(Y1)), \tag{14}$$

$$Y22'(Y2) = (\alpha_4 \wedge Y22(Y2)), \tag{15}$$

$$Y31'(Y3) = (\alpha_4 \wedge Y31(Y3)). \tag{16}$$

3. Composition: using logical maximum, the combination of found truncated functions is performed:

$$\mu_\Sigma(Y1) = Y12' \vee Y13' \vee Y11' \vee Y11', \tag{17}$$

$$\mu_\Sigma(Y2) = Y23' \vee Y24' \vee Y21' \vee Y22', \tag{18}$$

$$\mu_\Sigma(Y3) = Y31' \vee Y32' \vee Y31' \vee Y31'. \tag{19}$$

4. Bringing to clarity is carried out is made, in particular, by the centroid method:

$$Y1_0 = \frac{\int_\Omega Y1 \mu_\Sigma(Y1) dY1}{\int_\Omega \mu_\Sigma(Y1) dY1}, \tag{20}$$

$$Y2_0 = \frac{\int_\Omega Y2 \mu_\Sigma(Y2) dY2}{\int_\Omega \mu_\Sigma(Y2) dY2}, \tag{21}$$

$$Y3_0 = \frac{\int_\Omega Y3 \mu_\Sigma(Y3) dY3}{\int_\Omega \mu_\Sigma(Y3) dY3}. \tag{22}$$

Approximation of the dependence of the quality of the production of the entomophage *Habrobracon hebetor* (the number of infested caterpillars *Ephestia kuehniella*) from the parameters of technocenosis (temperature and relative humidity of the air in the zone of entomocultur growth) was carried out using the hybrid network (Fig. 12) based on the neuro-fuzzy conclusion ANFIS (Adaptive Network-Based Fuzzy Inference System) [36].

The input variables (input1 and input2) of the network are respectively the temperature and relative humidity of the air, the output variable is the number of infected caterpillars *Ephestia kuehniella*. The generated network automatically created the knowledge base and the surface of the fuzzy conclusion (Fig. 13), which allows us to trace the dependence of quality biological index on the technocenose parameters [36]; the average error of approximation was 7.3% and was within the permissible limits (not exceeding 8–10%) [36, 45].

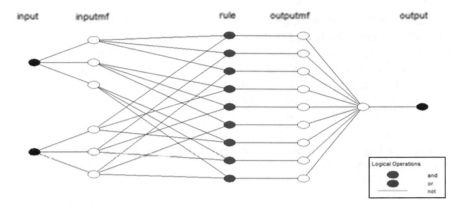

Fig. 12 Generated hybrid network [36]

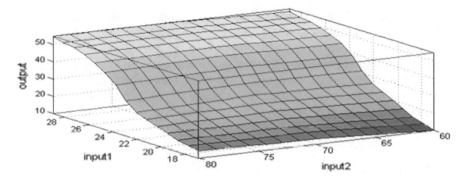

Fig. 13 The surface of the fuzzy conclusion of the hybrid network [36]

The analysis of the surface of the fuzzy conclusion suggests optimizing the micro-climate parameters in the entomoculture breeding zone in the range of temperature 26–29 °C and relative humidity of 60–80%, since the number of infected caterpillars meets the regulatory requirements [36].

A neuro-fuzzy network can be regarded as one of the varieties of fuzzy logic-type logic systems such as Sugeno [47]. At the same time, the functions of the membership of the synthesized systems are configured (trained) so as to minimize the deviation between the results of fuzzy simulation and experimental data [47].

The main stages of the fuzzy conclusion are [36, 48]:

– 'prod'—realization of the logical conjunction (operation and) by the method of algebraic product;
– 'probor'—implementation of logical disjunction (operations or) by the algebraic sum method;
– 'wtaver'—implementation of the de-phasification operation using the weighted average method.

According to the results of approximation, an algorithm for controlling the production of entomophages based on the product quality criterion using neural networks has been developed, which has the following order of action: the formation of a training sample based on the experimental dependencies of biological indicators of the quality of entomophages on the parameters technocenose, its loading into ANFIS—the editor of MATLAB, definition of the functions of the membership terms of input and output variables, definition of the structure of the network, its teaching by the chosen algorithm, determination of the network error from the number of training cycles, its testing, review of automatically generated rules, estimation of the surface of the neuro-fuzzy conclusion, determination of the average approximation error.

5 Discussion

The management of the production of entomophages is considered from the standpoint of infocommunication, system and process approaches, taking into account the peculiarities of production, with emphasis on increasing its efficiency, which is essentially related to energy efficiency, profitability, decision making under conditions of uncertainty and disturbances. The offered approaches are perspective in processes of creation of intellectual technologies of management of production of entomological means of protection of plants in Ukraine.

6 Conclusions

The proposed approaches to the management of the production of entomophages allow formalize the poorly structured processes of insect cultivation, with the with sufficient accuracy to form management decisions in the conditions of incomplete information, to automatically create knowledge bases, reduce energy costs for decision-making and production in general.

References

1. Lysenko V, Chernova I (2018) Intelligent algorithms of processing of information in the production entomophages. In: International scientific-practical conference problems of infocommunications. Science and technology (PIC S&T-2018). Conference proceedings. IEEE, Kharkiv, Oct 2018, pp 530–534. https://doi.org/10.1109/INFOCOMMST.2018.8632127
2. Simon Grenier (2012) Artificial rearing of entomophagous insects, with emphasis on nutrition and parasitoids—general outlines from personal experience. Karaelmas Fen ve Mühendislik Dergisi/Karaelmas Sci Eng J 2(2):1–12
3. Soloviev AI (2016) Theoretical essence of infocommunication providing management of agrarian production structures. Econ Enterp Manag 11:507–512

4. Lubentsova EV, Petrakov VA (2016) Intellectual technologies are a new base for building and developing automation systems for complex biotechnological processes. Sci Rev Tech Sci 5:64–80
5. Belchenko VM, Chernova IS (2015) The system of management of the quality of entomologic products using information technologies. Collection of scientific papers "plant protection". Minsk 39:262–267
6. Savchuk OV, Ladanyuk AP, Gritsenko NG (2009) Cognitive approach to modeling and management of weakly structured organizational and technological systems (situations). East Eur Mag Adv Technol 2/3(38):14–18
7. Makarova G (2013) Cognitive modeling in forecasting economic potential of the enterprise. J Natl Trade Econ Univ (Kyiv) 4:81–91
8. Mukhacheva NN, Popov DV (2009) The system-cognitive approach to the construction of ontological knowledge bases of information and intellectual resources. J Ryazan State Univ Radio Eng Ryazan 4(30):1–8
9. Feshchenko VV (2018) Cognitive management and cognitive modeling: principles, methods, functions. Sci J Econ Soc Right 2(10):54–60
10. Rudnichenko ND, Vychuzhanin VV (2013) Information cognitive model of technological interdependence of complex technical systems. Inf Math Methods Simul 3(3):240–247
11. León M, Rodriguez C, García MM et al (2010) Fuzzy cognitive maps for modeling complex systems. Mexican international conference on artificial intelligence MICAI: Advances in artificial intelligence. Springer, Berlin, pp 166–174
12. Margasov DV (2015) Development of information system structure on energy efficiency based on cognitive modeling. Management of the development of complex systems. Kyiv 24:97–105
13. Gorelova GV, Khlebnikova AI (2010) Cognitive modeling for intelligent transit trade management decision support system. Artif Intell 3:473–482
14. Wardhany VA, Yuliandoko H, Subono et al (2018) Fuzzy logic based control system temperature, pH and water salinity on vanammei shrimp ponds. In: International electronics symposium on engineering technology and applications (IES-ETA). Bali, Indonesia, Oct 2018. https://doi.org/10.1109/ELECSYM.2018.8615464
15. Yuswantoro D, Natan O, Angga AN et al (2018) Fuzzy logic-based control system for dissolved oxygen control on indoor shrimp cultivation. In: International electronics symposium on engineering technology and applications (IES-ETA). Bali, Indonesia, Oct 2018. https://doi.org/10.1109/ELECSYM.2018.8615569
16. Dias J, Coelho J, Gonçalves J (2015) Fuzzy control of a water pump for an agricultural plant growth system. In: Proceedings of the 7th international joint conference on computational intelligence (IJCCI 2015), vol 2. FCTA, pp 156–161
17. Rana DS, Rani S (2015) Fuzzy logic based control system for fresh water aquaculture: a MATLAB based simulation approach. Serb J Electr Eng 12(2):171–182
18. Cherednichenko AO, Shura NO (2015) Application of artificial neural networks as an effective mechanism for making effective management decisions at the enterprise. In: Global and national problems of the economy. Nikolayev National University named after V. O. Sukhomlynsky, vol 4, pp 628–630
19. DSTU ISO 9000: 2007 (2008) Quality management systems. Bases of the list of terms of terminology. Acting with 01 Jan 2008. Derzhspozhivstandart Ukraine, Kyiv, 29p
20. Soloviev AI (2017) Theoretical substantiation and development of the concept of infocommunication subsystem management of agrarian production structures. Econ Space 117:204–213
21. Ageyev DV, Salah MT (2016) Parametric synthesis of overlay networks with self-similar traffic. Telecommun Radio Eng (English translation of Elektrosvyaz and Radiotekhnika) 75(14):1231–1241
22. Ageyev D et al (2018) Classification of existing virtualization methods used in telecommunication networks. In: Proceedings of the 2018 IEEE 9th international conference on dependable systems, services and technologies (DESSERT), pp 83–86
23. Kryvinska N (2004) Intelligent network analysis by closed queuing models. Telecommun Syst 27:85–98. https://doi.org/10.1023/B:TELS.0000032945.92937.8f

24. Daradkeh YI, Kirichenko L, Radivilova T (2018) Development of QoS methods in the information networks with fractal traffic. Int J Electron Telecommun 64(1):27–32
25. Kirichenko L, Radivilova T, Zinkevich I (2017) Forecasting weakly correlated time series in tasks of electronic commerce. In: 2017 12th international scientific and technical conference on computer sciences and information technologies (CSIT). IEEE, pp 309–312. https://doi.org/10.1109/stc-csit.2017.8098793
26. Ivanisenko I, Radivilova T (2015) The multifractal load balancing method. In: 2015 second international scientific-practical conference problems of infocommunications science and technology (PIC S&T). IEEE, pp 122–123. https://doi.org/10.1109/infocommst.2015.7357289
27. Ozarko KS, Kudrya EN (2017) Consideration of directions of information communication in the management of communications enterprises. In: Fields of development of infocommunications: materials of sciences—practical. Seminar (Lviv, Dec 2017). Science. Information of Infocommunications. Lviv, pp 4–11
28. Kochegin AA (2011) Indicators of the quality of technological processes and systems: collection. In: Proceedings of the XVII international scientific and practical conference. Modern technology and technology, vol 3, Tomsk, pp 137–138
29. Chernova IS (2016) Basic approaches to the control of the production of entomological products. In: International scientific and practical conference natural science and education: the current state and prospects of development. Abstracts of the report, Kharkiv, Sept 2017, pp 54–55
30. Krutyakova VI, Chernova IS, Molchanova OD, Dolzhikova IV (2015) Basic approaches to ensuring the quality of entomological products. Mach Technol Agroindust Complex 11(74):30–31
31. Kryuchkova LP, Borisenko II (2017) Application of situational modeling in the management of technical systems. Communication 4:43–47
32. Lysenko VP, Chernova IS (2018) Intelligent control algorithm for energy-efficient entomophage growth. Autom Technol Bus Process 10(3):50–58. https://doi.org/10.15673/atbp.v10i3.1088
33. Mikhalev AI, Novikova EY (2006) A fuzzy-cognitive approach in the task of controlling the process of smelting FeSi. Adapt Autom Control Syst 9(29):133–139
34. Gozhii OP (2013) Construction of dynamic models based on fuzzy cognitive maps for solution of scenario planning tasks. J Lviv State Univ Life Saf (Lviv) 7:13–17
35. Molchanova OD, Kopko IA (2014) Breeding of *Ephestia kuehniella* for breeding ectoparasite bracon (*Habrobracon hebetor Say*). Agrar Bull South 1:131–134
36. Lysenko VP, Chernova IS (2017) On the issue of managing the production of entomophages. Power Eng Autom 3:15–24. http://journals.nubip.edu.ua/index.php/Energiya/article/view/9325/8398
37. Ivanova YV, Zhurukhin GI, Rutkauskas TK, Chuchkalova EI (2012) Applied economics: workshop. Russian State Vocational Pedagogical University, Ekaterinburg, 125p
38. Nedbay AA, Merzlikina NV (2008) Basics of qualimetry. In: Electronic resource. Institute for Advanced Studies Siberian Federal University, Krasnoyarsk, 126p. http://files.lib.sfu-kras.ru/ebibl/umkd/104/u_course.pdf
39. Khannik YM (2004) Energy-saving: a manual for university students. Thermodynamics. 4(1):205
40. Minyalenko IV, Poznyak YI (2014) Energy efficiency of production and its role in creating a competitive economy of Ukrainian regions. Eff Econ 11. http://www.economy.nayka.com.ua/?op=1&z=3579
41. Sukhodolya OV (2009) Types and objectives of managerial influence in the field of energy efficiency. Bull Natl Acad Publ Adm Under President Ukr 2:252–261
42. Ginzburg MD (2008) Terminology notes. What is energy efficiency? Market Install 5:54–56
43. Lakhov YA (2014) Determination of energy efficiency indicators of an oil refinery enterprise. Actual Probl Econ Manag 4(4):78–83
44. Lysenko VP (2014) The economic criterion for choosing a strategy for managing biotechnological objects. Bioresour. Nat. Manage. 6(3–4):174–179

45. Shalabanov AK, Roganov DA (2008) Econometrics. Teaching manual. Academy of Management "TISBI", Kazan 203p
46. Dyakonov V, Kruglov V (2001) Mathematical expansion packs MATLAB. Special handbook. Peter, St. Petersburg, 488p
47. Shtovba SD (2001) Introduction to the theory of fuzzy sets and fuzzy logic. In: Electronic resource. http://matlab.exponenta.ru/fuzzylogic/book1/index.php
48. Tarasyan V (2013) Package fuzzy logic toolbox for Matlab: tutorial. Ekaterinburg, 112p

Galois Field Augmentation Model for Training of Artificial Neural Network in Dentistry

Valeryi M. Bezruk, Stanislav A. Krivenko, and Liudmyla S. Kryvenko

Abstract In this paper, the authors consider how to label and save a large number of images that should be predicted in a single file. The technique of automatic labeling the data set with the finite element model for training of artificial neural network in tomography are proposed. A simple transparent example of thirty-two images to be predicted in a single HDF5 file training of artificial neural network in tomography show accuracy 100% for training set as well for the test set. Then this technique is able to build an information model of salivary immune and periodontal status and to evaluate the correlation between salivary immunoglobulin level, inflammation in periodontal tissues and orthodontic pathology. For this study, patients were divided into the following groups: 76 patients with chronic gingivitis and atopic diseases (group 1), among which the proposed treatment was used; 50 patients without clinical signs of gingivitis with atopic diseases (group 2) for which specific prevention was prescribed; 30 patients with chronic gingivitis and atopic diseases to which the standard treatment of gingivitis has been applied (group 3); 30 patients without clinical signs of gingivitis with atopic diseases using traditional preventive measures (group 4); 35 patients made control group with intact periodontal tissues without somatic pathology (group 5). This study was conducted to assess the state of lipid peroxidation in the oral liquid and the periodontal disease and to detect the correlation between the level of antioxidants in children and inflammation in periodontal tissues by means of regression analyses. The results showed changes in the antioxidant balance in children with atopy that were expressed in an increase in malondialdehyde level, a decrease in superoxide dismutase activity, and a level of reduced glutathione. These indicators can be considered as biological markers of the development of gingivitis at the preclinical stage in children against atopic diseases.

V. M. Bezruk · S. A. Krivenko (✉)
Kharkiv National University of Radio Electronics, Kharkiv, Ukraine
e-mail: stanislav.kryvenko@nure.ua

L. S. Kryvenko
Kharkiv National Medical University, Kharkiv, Ukraine
e-mail: lkryvenko@icloud.com

T. Radivilova et al. (eds.), *Data-Centric Business and Applications*, Lecture Notes on Data Engineering and Communications Technologies 48,
https://doi.org/10.1007/978-3-030-43070-2_16

339

Keywords Regression analyses · Atopic disease · Gingivitis · Convolutional
neural networks

1 Introduction

At nova days' dentists frequently make diagnosis "gingivitis" and all data can be
analyzed by methods of machine learning. A convolutional neural network for the
sequence-to-sequence learning task is a powerful tool. The dataset is split so that there
is no patient overlap between the training and validation sets. Two metrics to measure
model accuracy, using the dentist committee annotations as the ground truth are
used [1].

The prevention and treatment of allergic diseases in children is an extremely urgent
problem of medicine. According to the World Health Organization, the prevalence
of asthma in the world varies from an average of 4–8%, among the children's population is 2–15% [2–4]. According to an epidemiological study conducted in Kyiv in
the framework of the international forecasting of ISAAC, the prevalence of asthma
was 6.1–8.1%. In Ukraine, this percentage in the pediatric population is 5–10%.
According to the Ministry of Health of Ukraine, the disability of children due to res-
piratory diseases is 5.3% of the general level and is mainly due to bronchial asthma
[5]. The analysis of literature data proves the presence of functional and morpholog-
ical changes in the body of children with these diseases [6]. The described immune
metabolic changes, the trigger mechanism for which there is atopy, determine the
need to find a new concept of treatment of inflammatory diseases of the periodontal
disease in children with atopic diseases, based on early prediction biomarkers.

Deep learning usually involves a huge amount of data or a large number of pictures
(for example, several million pictures in ImageNet). In this case, it is not efficient to
download each picture separately from the hard drive, apply a preliminary processing
picture, and then transfer it to an artificial neural network for training, validating, or
testing. Preliminary processing takes a lot of time, but much more time is required to
read several images from the hard drive, compared to the case of storing them in one
file and reading them as a single data set. There are various data models and libraries
such as HDF5 and TFRecord. How to mark and save a large number of images that
you are trying to process in one HDF5 file, and then download them from a batch
file is a question. It does not matter how big the data is, or they are bigger than the
size of the memory. HDF5 provides tools organized for the collection, processing,
maintenance, use, sharing, dissemination, or disposition of data [7].

In our case, we collect all the images inside the train folder that has the
infrastructure of the well-known Dogs vs. Cats data set [8] for testing Python scripts.

Nowadays the wide spectrum of salivary elements provides valuable information
for clinical diagnostic applications. Whole saliva is most frequently used for the diagnosis of systemic diseases, because it can be easily collected and it contains most of
the serum constituents [9].

The rationale for the use of saliva as a diagnostic aid for diagnosis, treatment
planning, and an adjunctive aid for risk assessment has been the subject of numerous
studies and reviews in the literature [10].

On the other hand, logistic regression is a powerful tool, especially in epidemiologic studies, allowing multiple explanatory variables to be analyzed simultaneously, meanwhile reducing the effect of confounding factors [11].

The purpose of this research is to construct an informational model of salivary immune and periodontal status and to evaluate the correlation between salivary immunoglobulin level, inflammation in periodontal tissues and orthodontic pathology.

2 Materials and Methods

2.1 Labeled Data

Labeled data is a group of samples that have been tagged with one or more labels. Labeling typically takes a set of unlabeled data and augments each piece of that unlabeled data with meaningful tags that are informative. For example, labels might be indicating whether a photo contains a cat or a dog, whether the dot in an X-ray is inflammation in periodontal tissues and orthodontic pathology, etc.

Labels can be obtained by asking humans to make judgments about a given piece of unlabeled data, and are significantly more expensive to obtain than the raw unlabeled data.

After receiving a labeled dataset, machine learning models can be obtained to the data so that new unlabeled data can be presented to the model and a likely label can be guessed or predicted for that piece of unlabeled data.

2.2 Electrical Impedance Tomography

In this paper, images of electrical impedance tomography (EIT) are considered. The EIT uses an electric current generator and measures the voltage on the body surface to display the electrical properties of the inner tissues. Non-invasiveness and high temporal resolution are the benefits of the method, but it suffers from poor spatial resolution and sensitivity to electrodes and contact quality. Interpreting EIT-based activities is difficult, and describing ways of interpreting images are actual [12].

A mathematical data analysis technique of EIT that transforms a client request into an equivalent response of network with one more node based on the servers' applications writing with Python and MATLAB language is proposed, then, a database analysis technique to discover the edges of machine learning with additional constraints are implemented. Four element model of EIT a group with P positive instances and N negative instances of some condition (if the potentiometer is rotated more than halfway through its range) are considered [13]. A function returning a matrix of all possible binary condition vectors for a logical system with a finite number of inputs is known [14].

This function allows you to mark out the data set automatically without human intervention but does not take into account the repeatability of the labels when you rotate a planar finite element model. What is the most efficient way to shift a list in python? Numpy can do this using the roll command [15].

A technique of automatic labeling the data set with finite element model for training of artificial neural network in tomography is proposed that conjuncts the best features both functions.

2.3   Clinical Trial Methods

All clinical trial methods were performed by the dentist, according to the protocol of epidemiological examination recommended by WHO and according to the single algorithm for the treatment and prevention of gingivitis in periods of one, three, six, and twelve months [16]. To patients of all groups, the traditional method of examination of a dental patient with the additional registration of necessary indices and definition of certain clinical parameters was used. The analysis of the dental status began with an overview, paying attention to the anatomical features of the oral cavity, the presence of a pathological type of bite.

The level of gum inflammation was assessed using the SBI Bleeding Index (Mühlemann and Son). SBI was fixed on six dental surfaces. Results for SBI were evaluated on the following scale: 0: no bleeding, (1) bleeding in the sensation without changing the color and lack of edema, (2) bleeding when sensing with color change and no edema or macroscopic edema, (3) bleeding when sensing with a change in color and swelling of the edema, (4) bleeding during sensing, color changes due to inflammation, swelling with ulcers, (5) spontaneous bleeding, color changes, and significant edema with ulcers.

For this study patients were divided into the following groups: 76 patients with chronic gingivitis and atopic diseases (group 1), among which the proposed treatment was used; 50 patients without clinical signs of gingivitis with atopic diseases (group 2) for which specific prevention was prescribed; 30 patients with chronic gingivitis and atopic diseases to which the standard treatment of gingivitis has been applied (group 3); 30 patients without clinical signs of gingivitis with atopic diseases using traditional preventive measures (group 4); 35 patients made control group with intact periodontal tissues without somatic pathology (group 5).

The level of secretory immunoglobulin A, immunoglobulin A, G, M was determined on the immune enzyme analyzer "Labline-90" using commercial sets "Vector-Best" according to the method set by the set. For this purpose, unstimulated oral fluid was collected from patients from 8:00 to 12:00 with the volume of 2–5 ml on an empty stomach, patients were asked not to perform oral hygiene procedures (toothbrush, rinse aid) 60 min before the saliva collection procedure. Samples were transferred to the biochemical laboratory of the Kharkiv National Medical University, where they were stored at $-20\,°C$, after which an immunoanalysis was performed.

The state of prooxidant-antioxidant protection was determined by the level of malonic dialdehyde (MDA), catalase, glutathione and superoxide dismutase (SOD). Determination of the level of malonic dialdehyde was carried out by Uchiyma M., Michara M. in the modification of Volchegorsky I. A. and softened. on a test with thiobarbituric acid (TBC). The activity of catalase was determined by a method based on the ability of hydrogen peroxide to form a stable stained complex with molybdenum salts. The activity of superoxide dismutase was determined by the oxidation of quercetin in the modification of VA Kostyuk and co-authors.

Statistical analysis of data was performed using Microsoft Office 2016 software. Statistical analysis data were processed in Microsoft Excel 2016 using the t-test for distinguishing between groups and the least squares regression analysis, which is a statistical process for evaluating the relationships between variables. The values of P less than 0.05 were considered statistically significant [17]. Normality of data was tested before statistical analysis.

3 Analyses of the Inflammation Biomarkers Based on Regression Analyses

It was estimated the degree of inflammation of the gums with two indices—the PMA and the SBI. For regression analysis, we used both indexes for complex clinical picture and display the data. We conducted a regression analysis of data in the second group of patients who according to the clinical findings were almost without signs of gingivitis at the beginning, but allergists diagnosed atopic diseases. PMA index in this group of patients at the beginning was equal to $6.86 \pm 1.53\%$, immunoglobulin A levels were in the range 2.02 ± 1.45 g/l.

The results of the study of the relationship between the level of immunoglobulin A and the papillary-marginal-alveolar index and the regression analysis of data are presented in Tables 1, 2 and 3.

In order to evaluate the overall quality of the linear regression, a determination coefficient, designated as R-square, was used. This coefficient is an indicator of the quality of the regression equation and prediction accuracy using it.

Table 1 Regression statistics of levels of immunoglobulin A and the index of PMA

Regression statistics	Values
Multiple R	0.871753382
R-square	0.75995396
Normalized R-square	0.754953001
Standard error	0.755122086
Observation	50

Table 2 Dispersion analysis of regression statistics of immunoglobulin A levels and PMA-index

	df	SS	MS	F	Significance F
Regression	1	86.6499505	86.6499505	151.9616405	1.74997E−16
Residual	48	27.3700495	0.570209365		
Total	49	114.02			

Table 3 Determination of coefficients of regression statistics

	Coefficients	Standard error	t-statistics	P-meaning
Y-intercept	10.85519802	0.341234917	31.81151011	6.47787E−34
Variable X1	−1.637376238	0.132825511	−12.32727223	1.74997E−16

Table 4 Regression statistics of levels of immunoglobulin A and index of PMA in 6 months

Regression statistics	Values
Multiple R	0.963311458
R-square	0.927968965
Normalized R-square	0.926468319
Standard error	0.367869959
Observation	50

Since it is obvious from the regression analysis that the resulting coefficients are significantly larger than the standard error (coefficient 10.86 and standard error 0.34, coefficient −1.64 and standard error 0.13). Consequently, based on the data obtained, it is possible to construct a linear regression equation, which will have the following form: $y = -1.64 X1 + 10.86$.

According to the results of the regression analysis of Microsoft Excel 2016, the following conclusions can be drawn. The dependence of immunoglobulin A (Y) parameters on the level of inflammation according to the PMA index (variable X) is constructed in the form of linear regression. The linear regression coefficients are defined as minus 1.63 and 10.85. To assess the overall quality of the obtained linear regression equation, the determination coefficient is applied. The resulting value of 0.76 means that 76% of the variation of Y is associated with the variability of factor X. The other part of the Y variability (24%) is related to the influence of other factors. Thus, since the regression function is defined, interpreted and justified, and the quality of the regression equation satisfies the requirements, it can be assumed that the constructed model and predictive values have the necessary adequacy.

To assess the dynamics of changes in the indicators and to determine the possibility of using the level of immunoglobulin A as a non-invasive diagnostic tool in order to establish the periodicity of dispensary surveillance among children with atopic diseases, we conducted a regression analysis of the PMA index and immunoglobulin

Table 5 Dispersion analysis of regression statistics of immunoglobulin A levels and PMA index in 6 months

	df	SS	MS	F	Significance F
Regression	1	83.68424128	83.68424128	618.3794312	4.51951E−29
Residual	48	6.495758718	0.135328307		
Total	49	90.18			

Table 6 Determination of the coefficients of regression statistics after 6 months

	Coefficients	Standard error	t-statistics	P-meaning
Y-intercept	7.855325165	0.257659998	30.48717392	4.53811E−33
Variable X1	−0.992931197	0.039929295	−24.8672361	4.51951E−29

A in six months after the implementation of preventive measures. The results are given in Tables 4, 5 and 6.

It should be noted that the changes in the degree of gum inflammation in the second group of patients who were diagnosed with atopic diseases, however, there were almost no clinical signs of gingivitis, were negligible. At the beginning of the study, the PMA level was $6.86 \pm 1.53\%$, and in six months after the prophylactic measures were $1.58 \pm 1.36\%$. The level of immunoglobulin A was within 2.02 ± 1.45 g/l at the beginning of the study, after six months increased to 6.89 ± 0.66 g/l. Consequently, the dynamics of clinical and immunological indicators indicate the effectiveness of the proposed method of prophylaxis of periodontal disease, and regression analysis suggests that the level of immunoglobulin A can be a diagnostic and prognostic criterion for the prevention of chronic gingivitis in children with atopic diseases.

The constructed prognostic model showed an even greater degree of dependence of the indicators since the determination coefficient is equal to 0.92. The increase in the determination coefficient in this group of patients with atopic diseases is explained by the importance of the immunological correlation of the pathogenesis of chronic gingivitis in children with atopic diseases and the inclusion in the scheme of prophylaxis of a gingivitis drug with an immunomodulatory effect.

Regression analysis of data showed that even in the absence of clinical signs of gingivitis and minimal values of the PMA index there is a correlation between the level of inflammation and the level of immunoglobulin A. Thus, the level of immunoglobulin, which is determined in the oral fluid, can be used to predict periodontal disease, namely gingivitis, as well as for assessing the effectiveness of treatment and timely diagnosis in children with atopic diseases.

To confirm or refute the revealed pattern, we conducted a regression analysis of the data and in the group of children both with atopic diseases and the diagnosis of "chronic gingivitis" of moderate severity. As the data suggest, the dependence of the level of immunoglobulin A and the degree of inflammation in the gums is also observed, however, according to the determination coefficient, the dependence is somewhat weaker. The obtained data are presented in Tables 7, 8 and 9.

Table 7 Regression statistics of levels of immunoglobulin A and the index of PMA in a group of patients with gingivitis and atopic diseases

Regression statistics	Values
Multiple R	0.935855
R-square	0.875824
Normalized R-square	0.871389
Standard error	0.841048
Observation	30

Table 8 Dispersion analysis of regression statistics of immunoglobulin A levels and PMA index in a group of patients with gingivitis and atopic diseases

	df	SS	MS	F	Significance F
Regression	1	139.6939	139.6939	197.4858	3.29E−14
Residual	28	19.80612	0.707362		
Total	29	159.5			

Table 9 Determination of coefficients of regression statistics in a group of patients with gingivitis against the background of atopic diseases

	Coefficients	Standard error	t-statistics	P-meaning
Y-intercept	40.03061	0.357096	112.1004	1.08E−38
Variable X1	−3.77551	0.268663	−14.053	3.29E−14

To check the values of the regression coefficients according to the laws of statistical analysis, comparison of the absolute values of the coefficients (40.03061 and −3.77551) and their standard errors (0.357096 and 0.268663 respectively), and also analysis of the P-value was done. The obtained data testify to the reliability of the data obtained and the correctness of the constructed linear regression equation, which has the following form: $y = -3.78\,X1 + 40.03$.

According to the data obtained from the regression analysis, a regression model is constructed, the regression function is defined, interpreted and justified, and the quality assessment of the regression equation meets the requirements, thus it can be assumed that the constructed model and predictive values have the necessary adequacy. It is proved that 88% of the variability of immunoglobulin A level depends on the degree of gum inflammation according to the PMA index.

Since in both experimental groups (second and third), regardless of the severity of the clinical signs of gingivitis, the dependence of the degree of gingival inflammation according to the PMA index on the level of immunoglobulin A determined in the oral liquid, the determination coefficient and the coefficient of multiple correlation R are not significantly different between the groups, consequently, the use of the level of immunoglobulin A can be used as a non-invasive biomarker of inflammation of periodontal tissues in children with atopic diseases. It is important to note that according to the data of the regression analysis in the group of children with an average degree of severity of gingivitis, the dependence of immunological and periodontal indicators was expressed at 88% (12% falls on other etiological factors). In the second group of children who were ill with atopic diseases, however, clinical signs of gingivitis were hardly detected, the dependence of these parameters was determined by 76% (24% of the degree of gum inflammation varies with other factors).

In this regard, it is necessary to conclude that expressed parodontological symptoms, such as bleeding, edema, hyperemia, and others, are closely related to the deficiency of immunological defense, which is expressed in the decrease in the level of immunoglobulin A.

For further, more detailed analysis, it was also estimated the correlation between the assessment of the level of bleeding, and therefore inflammation in the tissues of the periodontal, and the level of secretory immunoglobulin A. In this case, our task was to investigate the level of inflammation in the group of children with established clinical diagnosis "Chronic gingivitis", an average degree of severity in a group of children with atopic diseases, and in a group of somatic healthy children. Comparison of these groups made it possible to determine the etiological significance of atopy and its role in the pathogenesis of inflammatory diseases of the periodontal disease, as well as to mathematically compare the severity of the correlation relationship.

According to the survey, the average level of immunoglobulin A in children with atopic diseases was 78.06 ± 9.32 μg/ml, index H. R. Muhlemann and S. Son were at 3.42 ± 0.42 points. To assess the results of the study the regression analysis was also chosen, which performs a linear regression analysis using the "least squares" method in the set of observations. This method has allowed determining how one dependent variable depends on the values of one or more independent variables. The results of the study of the relationship between the level of secretory immunoglobulin A and the index of gums bleeding and the regression analysis of data is presented in Tables 10, 11 and 12.

Table 10 Regression statistics of levels of secretory immunoglobulin A and the SBI index in the group of patients with gingivitis against and atopic diseases

Regression statistics	Values
Multiple R	0.933328946
R-square	0.871102921
Normalized R-square	0.865947038
Standard error	5.264184457
Observation	76

Table 11 Regression statistics of levels of secretory immunoglobulin A and the SBI index in the group of patients with gingivitis against and atopic diseases

	df	SS	MS	F	Significance F
Regression	1	4681.969735	4681.969735	168.9531934	1.2714E−12
Residual	74	692.79095	27.711638		
Total	75	5374.760685			

Table 12 Determination of coefficients of regression statistics

	Coefficients	Standard error	t-statistics	P-meaning
Y-intercept	−88.66873813	14.52809987	−6.103257752	2.2255E−06
Variable X1	50.62451495	3.894732842	12.99819962	1.2714E−12

In order to evaluate the overall quality of the linear regression equation, a determination coefficient, designated R-square, was used. This coefficient is an indicator of the quality of the regression equation and prediction accuracy using it.

According to the results of regression analysis of Microsoft Excel 2016, the following conclusions can be drawn. The dependence of the secretory immunoglobulin A (Y) on the level of inflammation according to the index H.R. Muhlemann and S. Son (variable X) in the form of linear regression. The linear regression coefficients are determined to be minus 88.7 and 50.6.

To assess the overall quality of the obtained linear regression equation, the determination coefficient is applied. The result obtained as a result of 0.87 means that 87% of the variation of Y is associated with the variability of factor X. The other part of the Y variability (13%) is associated with the influence of other factors. Thus, since the regression function is defined, interpreted and justified, and the quality of the regression equation satisfies the requirements, it can be assumed that the constructed model and predictive values have the necessary adequacy.

The level of inflammation in the fifth, control, group of patients where there were almost no clinical signs of chronic gingivitis and in the history of atopic diseases was not mentioned, was at the level of 0.48 ± 0.21 points, which significantly differed from those of the first group. At the same time, the level of secretory immunoglobulin A was at 123.57 ± 9.55 µg/ml, which also significantly differed from the indicators of the first group, where in the examined contingent of children the allergist doctors diagnosed atopic diseases.

A significant difference in the rates of secretory immunoglobulin A in the first and fifth groups alone shows in itself the lack of efficiency in the immune system in children with bronchial asthma, atopic dermatitis and allergic rhinitis. However, due to the regression analysis and the construction of the linear regression equation, we were able to determine that the variability in the level of bleeding and, therefore, inflammation, was associated with 87% with the level of secretory immunoglobulin A, and the determined coefficients allowed to construct a prognostic model of periodontal tissue inflammation.

The purpose of the regression analysis of the control group of patients was to determine the relationship between the variability of the level of secretory immunoglobulin A and its dependence on the index of SBI bleeding index. As we have seen with the example of regression analysis of the index of PMA and the level of immunoglobulin A described earlier, this kind of statistical analysis of data allows us to detect regularities even with the slightest changes in absolute values. The results of regression analysis are presented in Tables 13, 14 and 15.

Table 13 Regression statistics of levels of secretory immunoglobulin A and SBI index in the control group of patients

Regression statistics	Values
Multiple R	0.172584
R-square	0.029785
Normalized R-square	0.001249
Standard error	0.212147
Observation	36

Table 14 Dispersion analysis of regression statistics of secretory immunoglobulin A levels and SBI index in the control group of patients

	df	SS	MS	F	Significance F
Regression	1	0.046977	0.046977	1.043782	0.314158
Residual	34	1.530223	0.045007		
Total	35	1.5772			

Table 15 Determination of coefficients of regression statistics of the control group of patients

	Coefficients	Standard error	t-statistics	P-meaning
Y-intercept	0.005732	0.465559	0.012312	0.990249
Variable X1	0.003838	0.003757	1.021657	0.314158

To evaluate the overall quality of the linear regression equation we need to interpret the determination coefficient, which is designated as R-square. This coefficient is an indicator of the quality of the regression equation and prediction accuracy using it; since it is 0.03, so the dependence between the level of the SBI index and the level of secretory immunoglobulin A was not detected in the control group of children.

The reliability of the significance of the Fisher criterion (Significance F) is 0.31, which means that the model is significant. The degree of accuracy of the description of the model by the R-square process is 0.03, which suggests a low accuracy of approximation, which suggests that the constructed model can not be used to describe the process.

Thus, the conclusion of the regression analysis of data is the lack of dependence of the index of gut bleeding and the level of secretory immunoglobulin A in somatically healthy children. Comparison of data of regression analysis of secretory immunoglobulin A level in the first and fifth groups of patients has shown that against the background of atopic diseases, the immune link of pathogenesis of periodontal diseases is affected. In this case, the degree of inflammation depends on the values of the level of immunoglobulin. An important fact is that, according to the data obtained, both mucosal and general immunity, which is reflected in the condition of periodontal tissues, affects.

Another indicator of the state of the immune system is the level of immunoglobulin G. The purpose of the regression analysis of this indicator was to investigate the presence of inflammation of periodontal tissue infections according to the SBI index, as well as to compare the quality of the obtained linear regression equation with the immunoglobulin equations of other classes. For regression analysis, we selected the first group of patients, among which chronic gingivitis and atopic diseases were diagnosed. According to the obtained data, the average bleeding rate is clear by the index H.R. Muhlemann and S. Son were 3.42 ± 0.42 points, indicating a rather high rate of bleeding and, accordingly, the presence of inflammation in periodontal tissues. The mean IgG level was 1.72 ± 0.49 g/l. The results of the regression analysis data are presented in Tables 16, 17 and 18.

The determination coefficient R-square was 0.82, which makes it possible to use the results of regression analysis and forecast using the regression equation.

According to the results of the regression analysis of the gastrointestinal bleeding data on the level of the SBI index and immunological status by the level of immunoglobulin G, the following conclusions can be drawn. The dependence of immunoglobulin G (Y) parameters on the level of inflammation according to the index H.R. Muhlemann and S. Son (variable X) in the form of linear regression. The linear regression coefficients are determined to be minus 0.32 and 0.98, hence the

Table 16 Regression statistics of levels of immunoglobulin G and SBI index in the first group of patients

Regression statistics	Values
Multiple R	0.90689
R-square	0.82245
Normalized R-square	0.818119
Standard error	0.19452
Observation	76

Table 17 Dispersion analysis of regression statistics of levels of immunoglobulin G and SBI index in the first group of patients

	df	SS	MS	F	Significance F
Regression	1	7.186194	7.186194	18.99205	5.55E−17
Residual	74	1.551354	0.037838		
Total	75	8.737548			

Table 18 Definition of coefficients

	Coefficients	Standard error	t-statistics	P-meaning
Y-intercept	−0.32473	0.24433	−1.32906	0.191177
Variable X1	0.980734	0.071165	13.78116	5.55E−17

linear regression equation has the following form: $y = 0.98\,X1 + 0.32$. To determine the overall quality of the obtained linear regression equation, the determination coefficient was applied. The resulting value of 0.82 means that 82% of the variation of Y is associated with the variability of factor X. Another part of the variability Y (18%) is associated with the influence of other factors. Thus, since the regression function is defined, interpreted and justified, and the quality of the regression equation satisfies the requirements, it can be assumed that the constructed model and predictive values have the necessary adequacy.

In order to evaluate the immunoglobulin levels, we selected a regression analysis of the level of inflammation according to the PMA index and the level of immunoglobulin M. Some authors have reported a rather low informative significance of this level of immunoglobulin in the aspect of dental morbidity. We conducted our own assays, which showed a significant difference in the levels of immunoglobulin M in the control and experimental groups. An important fact was that even in groups with no pronounced clinical signs of chronic gingivitis, the level of immunoglobulin M was lowered, proving the importance of immune disorders in the pathogenesis of periodontal disease in children against atopic diseases. According to our results, the average level of immunoglobulin M in the third experimental group was 2.20 ± 0.46 g/l at the beginning of the study, the level of inflammation was $35.5 \pm 2.35\%$ according to the PMA index. The estimation of regression statistics is given in Tables 19, 20 and 21.

As can be seen from the results, the determination coefficient of the R-square was 0.69, which calls into question the possibility of using the results of regression analysis of the level of immunoglobulin M and the degree of gastritis and prognosis using the regression equation.

Table 19 Regression statistics of levels of immunoglobulin M and PMI index in the third group of patients

Regression statistics	Values
Multiple R	0.829052
R-square	0.687327
Normalized R-square	0.67616
Standard error	1.334587
Observation	30

Table 20 Dispersion analysis of regression statistics of immunoglobulin M levels and PMA index in the third group of patients

	df	SS	MS	F	Significance F
Regression	1	109.6286	109.6286	61.55033	1.52E−08
Residual	28	49.8714	1.781121		
Total	29	159.5			

Table 21 Determination of coefficients of regression statistics

	Coefficients	Standard error	t-statistics	P-meaning
Y-intercept	40.18569	0.645044	62.29916	1.4E−31
Variable X1	−2.94265	0.37508	−7.8454	1.52E−08

The rather high values of F Fisher's criterion confirm the rather high probability of improper prediction of the degree of gum inflammation based on the use of a prognostic value for immunoglobulin M. It was also conducted a regression analysis of data in a group of children with atopic diseases in history, but in which no chronic gingivitis was detected, to detect patterns or to prove their absence. To do this, the second group of patients parameters, identified at the beginning of the study, were analysed. As the results showed, the level of gingivitis according to the PMI index significantly differed from that of the third group, where the regression analysis was previously performed, and was $6.86 \pm 1.53\%$. It is important to note that the level of immunoglobulin M was not significantly different from the similar indicators of the third group, it was at 2.05 ± 0.52 g/l, which confirms the presence of immunological changes in children with atopic diseases and the existence of a stage of premedication of chronic gingivitis in children against the background of the burdened general-somatic status.

The obtained regression analysis data are presented in Tables 22, 23 and 24.

Regression analysis allowed determination of the determination coefficient, which was 0.43, and therefore the relationship between these parameters in the second group of patients is even weaker than in the third group.

Table 22 Regression statistics of levels of immunoglobulin M and PMI index in the second group of patients

Regression statistics	Values
Multiple R	0.65835
R-square	0.433425
Normalized R-square	0.421621
Standard error	1.160109
Observation	50

Table 23 Dispersion analysis of regression statistics of levels of immunoglobulin M and PMI index in the second group of patients

	df	SS	MS	F	Significance F
Regression	1	49.41907	49.41907	36.71953	2.03E−07
Residual	48	64.60093	1.345853		
Total	49	114.02			

Table 24 Determination of coefficients of regression statistics

	Coefficients	Standard error	t-statistics	P-meaning
Y-intercept	11.37319	0.762649	14.91276	1.21E−19
Variable X1	−1.68367	0.277848	−6.05966	2.03E−07

Since, according to the determination coefficient, only 43% of the variation of Y is, in this case, the level of immunoglobulin M is related to the variability of the variable X1, the PMA index, therefore, it is not possible to use this biomarker for diagnostic and prognostic purposes.

Another indicator that characterizes the state of the immune system of the body, and especially the state of mucosal immunity, and the ability of the macroorganism to withstand the processes of inflammation, is the level of lysozyme. As already established, the level of lysozyme was lowered at the beginning of the study in the major groups compared to the control. Thus, in patients of the first group at the beginning of the study, the level of lysozyme was 192.77 ± 23.12 cu/l, in the second group— 193.80 ± 20.30 cu/l, in the third group—184.48 ± 26.52 cu/l, in the fourth group— 189.06 ± 25.30 cu/l, in patients of the fifth control group—316.56 ± 15.87 cu/l.

The establishment of the dependence of lysozyme parameters on the degree of inflammation by the PMA index was made using regression analysis; the possibility of constructing a predictive model was determined by means of the determination coefficient; the accuracy of this forecast was determined using the F-criterion and the *P*-value. The obtained regression analysis data are presented in Tables 25, 26 and 27.

The conduct of regression analysis allowed determination of the determination coefficient, which was 0.48, which represents the weak severity of the relationship

Table 25 Regression statistics of lysozyme level and AMI index in the second group of patients

Regression statistics	Values
Multiple R	0.692919
R-square	0.480136
Normalized R-square	0.467457
Standard error	1.863449
Observation	50

Table 26 Dispersion analysis of regression statistics of lysozyme level and PMA index in the second group of patients

	df	SS	MS	F	Significance F
Regression	1	131.4904	131.4904	37.86684	2.62E−07
Residual	48	142.3701	3.472441		
Total	49	273.8605			

Table 27 Determination of coefficients of regression statistics

	Coefficients	Standard error	t-statistics	P-meaning
Y-intercept	42.41801	1.371294	30.93284	4.82E−30
Variable X1	−0.79191	0.12869	−6.1536	2.62E−07

between the level of lysozyme and the index of PMA, and therefore the degree of inflammation of periodontal tissues. The analysis of the given parameters, including F-value and P-value, proves the low informativeness of the values of lysozyme as a diagnostic and prognostic tool for periodontal tissue inflammation. Undoubtedly, according to the findings, lowering the level of lysozyme suggests a decrease in immunity, both local and mucosal, however, the use of its absolute values for the prognosis of inflammatory diseases of the periodontal disease is undesirable.

The analysis of the given parameters, including F-value and P-value, proves the low informativeness of the values of lysozyme as a diagnostic and prognostic tool for periodontal tissue inflammation. Undoubtedly, according to the findings, lowering the level of lysozyme suggests a decrease in immunity, both local and mucosal, however, the use of its absolute values for the prognosis of inflammatory diseases of the periodontal disease is undesirable.

Thus, due to the regression analysis method, we analyzed the data obtained during the clinical study of the data that characterized the immune competence of the body of the examined contingent of children with atopic diseases and somatic healthy children with marked clinical signs of gingivitis and without such. As a result of the regression analysis, the linear regression equations were constructed and their coefficients determined, determination coefficients were determined, which mathematically determined the degree of dependence of the change from one indicator to another.

According to the results of the research and data analysis, the most prognostically significant biomarkers were identified, which can be effectively used both for diagnosis and for the prognosis of periodontal diseases in children in the context of atopic diseases.

In order to detect inflammatory biomarkers that could be informative for the early diagnosis of periodontal inflammatory diseases, we conducted a regression analysis of clinical data and data characterizing the antioxidant status of children with atopic diseases and somatic healthy children. When analyzing the periodontal status at the level of the SBI index, it should be noted that the condition of periodontal tissues in patients of the first group in the index of bleeding was $3.42 \pm 0.42\%$, in patients with the second group with atopic diseases without manifestations of gingivitis $0.49 \pm 0.22\%$, among patients in the control group—0, $48 \pm 0.21\%$.

The average level of MDA of oral fluid in the control group was $3.83 \pm 0.83\,\mu m/l$. The median level in the first group was $6.87 \pm 0.91\,\mu m/l$, and in the second group it was $5.96 \pm 0.79\,\mu m/l$. The median MDA level of the control group was significantly lower compared to the groups of children with atopic diseases with and without clinical manifestations of gingivitis ($P < 0.05$). The average level of MDA of the first

Table 28 Evaluation and comparison of mean values of MDA and SOD in patient groups

Groups	MDA (μm/l)	SOD (cu/l)
First group	6.87 ± 0.91*	3.3 ± 0.41*
Second group	5.96 ± 0.79*	3.27 ± 0.4*
Control group	3.83 ± 0.83	5.26 ± 0.57

*P < 0.05, the difference is statistically significant in comparison with the control group and the second group of patients

and second groups did not have significant changes, the comparative characteristic is given in Table 28.

SOD activity was significantly lower in the first and second group of patients, compared with somatic healthy children. In the oral liquid, the concentration of superoxide dismutase was 3.3 ± 0.41 cu/l in the group of children in the background of atopy and gingivitis and 3.27 ± 0.4 cu/l in a group of children with atopy. There was no significant difference in the concentration of superoxide dismutase between these two groups, but a significant difference was observed in comparison with the control group.

The mean SBI index in the group of children with chronic gingivitis and atopic diseases was 3.42 ± 0.42. The average SBI level in a group of children without clinical signs of gingivitis against the background of atopic diseases was 0.49 ± 0.22. The level of the control group was 0.48 ± 0.21. The comparative characteristic of the inflammation rate according to the SBI index, which was carried out prior to the regression analysis, is given in Table 29.

A characteristic feature that distinguished the first and second group of patients was the presence of clinical signs of gingivitis. However, as shown by the analysis of indicators in dynamics, the level of inflammation of periodontal tissues gradually increased in the fourth group of patients without the use of pathogenetically substantiated correction of immune-metabolic status.

Comparison of mean values of SBI indicates a significantly higher level of inflammation in periodontal tissues in a group of patients with chronic gingivitis against the background of atopic diseases, compared to patients in the second group and control group, which completely coincides with the data of the clinical examination. A regression analysis was performed to assess the relationship between the SBI index and MDA level in the first group of patients (Tables 30, 31 and 32). Since R^2 is

Table 29 Evaluation and comparison of average values of the SBI index

Groups	SBI (points)
First group	3.42 ± 0.42*
Second group	0.49 ± 0.52
Control group	0.48 ± 0.21

*P < 0.05, the difference is statistically significant in comparison with the control group and the second group of patients

Table 30 Regression analysis of SBI and MDA ratio in the first group of patients

Regression statistics	Values
Multiple R	0.657539
R-square	0.432358
Normalized R-square	0.424687
Standard error	0.317763
Observation	76

Table 31 Regression analysis of SBI and MDA ratio in the first group of patients

	df	SS	MS	F	Significance F
Regression	1	5.691242	5.691242	56.36385	1.1E−10
Residual	74	7.472022	0.100973		
Total	75	13.16326			

Table 32 Determination of coefficients of regression statistics

	Coefficients	Standard error	t-statistics	P-meaning
Y-intercept	1.342471	0.279392	4.804968	7.9E−06
Variable X1	0.30261	0.040307	7.507586	1.1E−10

0.432358, there is a weak correlation between the level of malonic dialdehyde and the problems of periodontal disease.

In the second group of patients, the value of the determination coefficient R^2, which equals 0.000157, indicates that it is impossible to use regression analysis data to construct a prognostic model of periodontal tissue inflammation based on the level of malonic dialdehyde.

The regression analysis method shows that the regression function cannot be constructed in a group of patients with atopic diseases and gingivitis. The value of R^2, normalized by R^2, of multiple R, proves that inflammation in the periodontal tissues, which is measured by the SBI index, does not correlate with the level of malonic dialdehyde.

The regression analysis method shows that the regression function can not be constructed in a group of patients with atopic diseases and gingivitis. The value of R^2, normalized by R^2, of multiple R, proves that inflammation in the periodontal tissues, which is measured by the SBI index, does not correlate with the level of malonic dialdehyde. Regression statistics in the group of patients with atopic diseases and without the expressed signs of gingivitis indicates that there is no correlation between the SBI index and the level of malonic dialdehyde (Tables 33, 34 and 35).

In clinical practice, this means that the MDA level cannot be used to accurately predict SBI. An elevated level of malondialdehyde is observed at the preclinical stage

Table 33 Regression analysis of SBI and MDA in the second group of patients

Regression statistics	Values
Multiple R	0.012526
R-square	0.000157
Normalized R-square	−0.02067
Standard error	0.527224
Observation	50

Table 34 Dispersion analysis of regression statistics of SBI and MDA in the second group of patients

	df	SS	MS	F	Significance F
Regression	1	0.002093742	0.002093742	0.007532402	0.931199948
Residual	48	13.34230626	0.277964714		
Total	49	13.3444			

Table 35 Determination of coefficients of regression statistics

	Coefficients	Standard error	t-statistics	P-meaning
Y-intercept	0.50246508	0.575597987	0.87294447	0.387039511
Variable X1	0.008316302	0.095821614	0.086789413	0.931199948

of development of gingivitis, when such manifestations of inflammation have not yet been observed such as edema, hyperemia and bleeding.

This fact means that malonic dialdehyde can be used as an inflammation marker for early diagnosis and prophylaxis of periodontal inflammatory diseases, and normalization of its level may be a criterion for the effectiveness of programs for the prevention and treatment of gingivitis, but the construction of a prognostic model based on the values of MDA in a group of patients without clinical signs gingivitis is not possible, and in the group of patients with chronic gingivitis the correlation between the degree of inflammation and changes in the level of MDA is weak, therefore it can not be recommended for use in the clinical practice (Table 36).

Table 36 Regression analysis of SBI and SOD ratio in the first and second group of patients

Regression statistics		
Group of patients	First group	Second group
Multiple R	0.326201613	0.107837008
R-square	0.106407492	0.01162882
Normalized R-square	0.094331918	−0.008962246
Standard error	0.398690311	0.524190256

Table 37 Regression statistics of glutathione level and PMA index in the third group of patients

Regression statistics	Values
Multiple R	0.952755
R-square	0.907742
Normalized R-square	0.904447
Standard error	0.724943
Observation	30

As the results of the regression analysis showed, there was no correlation between the index of SBI bleeding index and superoxide dismutase activity (SOD), data are given in Table 37. In the group of patients with atopic diseases and chronic gingivitis (first group), R^2 is 0.106407492, multiple R: 0.326201613, normalized R^2: 0.094331918 [18]. A similar situation is observed in the group of patients without marked clinical signs of gingivitis against the background of atopic diseases: R^2 is 0.01162882, multiple R: 0.107837008, normalized by R^2: −0.008962246.

As a result of the evaluation of the determination coefficients in the first and second groups, it was concluded that there was no need to determine the coefficients of the linear regression function due to the low prognostic significance of the parameters obtained.

The evaluation of markers of antioxidant defense shows that MDA and SOD can serve as a potential diagnostic marker for children with atopy in the preclinical and clinical stages of gingivitis. The obtained data prove that in children with atopy there is an oxidative imbalance, since we have determined the increase of the level of MDA and decrease in the activity of SOD in the main groups compared with the control. At the same time, attention is drawn to the fact that the conducted regression analysis showed that there is no dependence between the SBI index and the level of antioxidants that could be used with a prognostic and diagnostic purpose. The MDA level is increased, and the level of SOD is lowered in the group of patients with atopy, and this fact does not depend on the presence of gingivitis in these groups.

Since there was no significant difference in the level of concentration of superoxide dismutase and malonic dialdehyde between a group of children with atopy and gingivitis and a group of children with atopy, we believe that antioxidant imbalance is primarily attributable to atopic disease and gingivitis develops at the background of disturbance of the prooxidant-antioxidant imbalance.

The next indicator of the prooxidant-antioxidant balance that was used to perform regression analysis, was the level of reduced glutathione and the degree of inflammation of the periodontal tissue from its level. According to biochemical studies conducted in all groups, glutathione levels were decreased in all major groups compared to control. Due to the fact that the level of glutathione was not significantly different in all major groups, for the implementation of regression analysis, we selected the third group of children with atopic diseases and chronic gingivitis.

According to the data, the average level of recovered glutathione among the examined contingent was 2.73 ± 0.58 mmol/l at the beginning of the study, and the level

Table 38 Dispersion analysis of regression statistics of glutathione and PMA index in the third group of patients

	df	SS	MS	F	Significance F
Regression	1	144.7848	144.7848	275.4963	5.06E−16
Residual	28	14.71517	0.525542		
Total	29	159.5			

Table 39 Determination of coefficients of regression statistics

	Coefficients	Standard error	t-statistics	P-meaning
Y-intercept	40.97408	0.355369	115.3	4.91E−39
Variable X1	−2.49199	0.150137	−16.5981	5.06E−16

of PMA was $35.50 \pm 2.35\%$, which corresponds to the average degree of severity of gingivitis. For the study of the dependence of the two indicators, we constructed linear regression equations, the results of regression analysis are given in Tables 37, 38 and 39.

It is noteworthy that the determination coefficient, denoted by the R-square, is 0.91, indicating a stable relationship between the value of the content of reduced glutathione in the oral liquid and the degree of periodontal tissue inflammation and proves that 91% of the variability of one parameter depends on the variability of another one. The multiple R is a correlation coefficient and confirms the dependence of the set parameters.

The definition of the standard error and the P-value indicates the reliability of the constructed linear regression equation. The given regression coefficients show that the dependence of the variable X (the degree of inflammation of the periodontal tissues) is inversely dependent on the values of Y (level of reduced glutathione).

Thus, regression analysis of data showed that the constructed prognostic model has the necessary adequacy, predictive values can be used as predictive for characterizing the degree of inflammation of periodontal tissues. As a result of the analysis, determination coefficient was determined, which showed that 91% of the variability of the level of glutathione depends on changes in the level of inflammation by the PMA index; 9% is influenced by other factors. Consequently, the level of reduced glutathione may be recommended as a predictive and diagnostic biomarker of inflammation of periodontal tissues in children with atopic diseases.

Another parameter that characterizes the state of the antioxidant system was the level of catalase that have been studied in the oral fluid. The level of catalase significantly did not differ among groups of children with atopic diseases, regardless of the presence of the expressed signs of chronic gingivitis in them. For regression analysis, we have processed the data of the second group of patients, in which the catalase level at the beginning of the study was 1.93 ± 0.54 cat/l, and the level of inflammation of the periodontal tissues was $6.86 \pm 1.53\%$.

The data of the regression analysis, namely the value of the determination coefficient, indicated that it was impossible to use the built-in prediction model and the absence of the need for determining the regression coefficients. The data obtained are given in Tables 40, 41 and 42.

As can be seen from the data presented, the determination coefficient, denoted by the R-square, is 0.42, indicating a weak correlation between the activity of catalase in the oral fluid and the degree of inflammation of the periodontal tissues. In connection with the definition of the determination coefficient, the use of other parameters of regression statistics, namely regression coefficients, is impossible, since the predictive values obtained as a result of the regression equation will not be significant.

Thus, the use of catalase as a prognostic and diagnostic marker of inflammation of periodontal tissues is inappropriate. Determination of the determination coefficient showed that this parameter of antioxidant defense has a weak prognostic value. The same weak diagnostic value, according to the results of regression analysis of data, has a level of urease. As previously described, urease activity was elevated in patients in this group and was 5.40 ± 1.47 μmol/min/l.

Determination of the determination coefficient for the level of PMA and urease is given in Table 43.

Table 40 Regression statistics of catalase level and PMA index in the second group of patients

Regression statistics	Values
Multiple R	0.65835
R-square	0.433425
Normalized R-square	0.421621
Standard error	1.160109
Observation	50

Table 41 Dispersion analysis of regression statistics of catalase and PMA index in the second group of patients

	df	SS	MS	F	Significance F
Regression	1	49.41907	49.41907	36.71953	2.03E−07
Residual	48	64.60093	1.345853		
Total	49	114.02			

Table 42 Determination of regression coefficients

	Coefficients	Standard error	t-statistics	P-meaning
Y-intercept	11.37319	0.762649	14.91276	1.21E−19
Variable X1	−1.68367	0.277848	−6.05966	2.03E−07

Table 43 Regression statistics of urease level and PMA index in the third group of patients

Regression statistics	Values
Multiple R	0.306916
R-square	0.094198
Normalized R-square	0.057965
Standard error	2.098678
Observation	30

Consequently, this indicator shows the presence of antioxidant imbalance in children with atopic diseases and chronic gingivitis, but can not be used to predict the level of inflammation of periodontal tissues.

4 Galois Field Augmentation Model

Linear-feedback shift register having many-to-many configuration is claim. It generates the Galois field different from finite field formed Galois linear-feedback shift register [19]. Finite field allows you to conjunction both functions shift a list and all possible binary condition vectors for a logical system with finite number of inputs.

4.1 Linear-Feedback Shift Register

In computing, a linear-feedback shift register (LFSR) consist of a shift register and combinational logic. The most commonly used combinational logic of single bits is XOR. In the Galois configuration, when the system is clocked, bits that are not taps are shifted one position to the right unchanged. The cycle is defined by power (m) of the feedback polynomial or reciprocal characteristic polynomial

$$x^m + x^n + 1 = pol \tag{1}$$

Case when number of terms more than three is not considered in this paper. The powers of the terms (the polynomial (1) index n, m) have to form the corresponding primitive feedback polynomial.

4.2 Logistic Regression Classifier

A logistic regression classifier for the recognition of inflammation was developed. The method of the neural network is considered in this chapter for the implementation of the classifier. The accuracy of training and testing to be close to 100% are expected. Logistic regression is a linear classifier. Loops aren't used in described code. The training algorithm includes the following elements: initialization of parameters; calculation of the function of value; an optimization of the gradient algorithm [20]. The main model function consists of the three functions listed above.

The cell of notebook Jupyter Anaconda Navigator is run to import all the five packages that we need during this research. Numpy is the fundamental package for scientific computing with Python. A dataset is processed by H5py package. Graphs in Python were plotted by Matplotlib. PIL and scipy tested proposed model with images of inflammation or non-inflammation. The advantages of these include the higher efficiency of using limited system resources [21]. This classifier has a very good performance in forecasting and order of magnitude smaller errors [22].

Problem Statement A training set of inTrain pictures labeled as inflammation (why = 1) or non-inflammation (why = 0). A test set of inTest pictures labeled as inflammation or non-inflammation.

Each image has a form (number_of_pixels, number_of_pixels, 3) where three is for the three RGB channels. Thus, each picture is rectangle (height_of_rectangle = number_of_pixels) and (width_of_rectangle = number_of_pixels). A picture-recognition algorithm that can correctly classify images as inflammation or non-inflammation was built for dataset ("data.h5") containing data above.

An array of picture was represented by each row of our train_set_eks_sa and test_set_eks_sa. A picture was visualized running the according code. The picture number was defined by index value.

Examples InTrain (number of training pictures). InTest (number of test pictures). Number_of_pixels.

Train_set_x_sa is a numpy array of shape (inTrain, number_of_pixels, number_of_pixels, 3). For convenience, images of shape (number_of_pixels, number_of_pixels, 3) were reshaped in a numpy-array of shape (three square of number_of_pixels, 1). After this, the training and test dataset is a numpy-array where each column represents a flattened picture. Column numbers is equal to inTrain or inTtest.

General Architecture of the Learning Algorithm A simple algorithm to separate inflammation images from non-inflammation images was designed on base a Logistic Regression, using a Neural Network mindset.

Sigmoid Function. Compute the sigmoid of z.

> Arguments: z—a scalar or numpy array of any size.
> Return: s—sigmoid(z).

Initialize with Zeros. This function creates a vector of zeros of shape (dim, 1) for w and initializes b to 0.

> Argument: dim—size of the w vector we want (or number of parameters in this case).

Returns: w—initialized vector of shape (dim, 1), b—initialized scalar (corresponds to the bias).

Propagation Function. Implement the cost function and its gradient for the propagation explained above.

Argument: w—weights, a numpy array of size (num_px * num_px * 3, 1), b—bias, a scalar, X—data of size (num_px * num_px * 3, number of examples), Y—true "label" vector (containing 0 if non-inflammation, 1 if inflammation) of size (1, number of examples).
Returns: cost—negative log-likelihood cost for logistic regression, dw—gradient of the loss with respect to w, thus same shape as w, db—gradient of the loss with respect to b, thus same shape as b.

Function of Optimize. This function optimizes w and b by running a gradient descent algorithm.

Arguments: w—weights, a numpy array of size (num_px * num_px * 3, 1), b—bias, a scalar, X—data of shape (num_px * num_px * 3, number of examples), Y—true "label" vector (containing 0 if non-inflammation, 1 if inflammation), of shape (1, number of examples), num_iterations—number of iterations of the optimization loop, learning_rate—learning rate of the gradient descent update rule, print_cost—True to print the loss every 100 steps.
Return: params—dictionary containing the weights w and bias b, grads—dictionary containing the gradients of the weights and bias with respect to the cost function, costs—list of all the costs computed during the optimization, this will be used to plot the learning curve.

Predict Function. Predict whether the label is 0 or 1 using learned logistic regression parameters (w, b).

Arguments: w—weights, a numpy array of size (num_px * num_px * 3, 1), b—bias, a scalar, X—data of size (num_px * num_px * 3, number of examples).
Return: Y_prediction—a numpy array (vector) containing all predictions (0/1) for the examples in X.

Model. Builds the logistic regression model by calling the function implemented previously.

Arguments: X_train—training set represented by a numpy array of shape (num_px * num_px * 3, m_train), Y_train—training labels represented by a numpy array (vector) of shape (1, m_train), X_test—test set represented by a numpy array of shape (num_px * num_px * 3, m_test), Y_test—test labels represented by a numpy array (vector) of shape (1, m_test), num_iterations—hyperparameter representing the number of iterations to optimize the parameters, learning_rate—hyperparameter representing the learning rate used in the update rule of optimize(), print_cost—Set to true to print the cost every 100 iterations.
Return: d—dictionary containing information about the model.

5   Results of Research

5.1   Primitive Polynomials

Technique of automatic labeling the data set is demonstrated by simple transparent example of five-element model of EIT. The feedback polynomial is

$$x^5 + x^2 + 1 = \text{pol5} \tag{2}$$

Thirty-two images for predicting or training of artificial neural network in tomography were saved in a single HDF5 file. Then HDF5 were loaded from the file in batch manner to a very simple Neural Network (Logistic Regression). Accuracy was scored for both train and test data set.

Overview of the Problem Set. Instance of inflammation image was plotted by running the designed script (see Fig. 1).

Matrix/vector dimensions were kept straight (446 * 446 pixels' image). The name of the axes is pixel number along the X and Y axes. Number of training instances: inTrain = 18. Number of testing instances: inTest = 8. Height of picture: number_of_pixels = 446. Width of picture: number_of_pixels = 446. Size of picture: (446, 446, 3). Train_set_eks shape: (18L, 446L, 446L, 3L). Train_set_why shape: (1L, 18L). Test_set_eks shape: (8L, 446L, 446L, 3L). Test_set_why shape: (1L, 4L). The data set was split into three parts: training (60%); validating (20%), and testing (20%).

Fig. 1 Example of a picture (index = 0, label = inflammation)

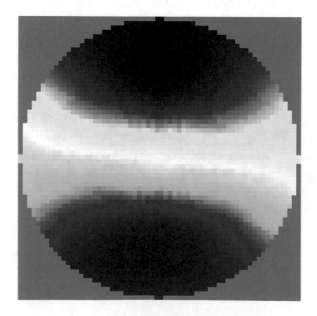

The results of automatic labeling the data set is in Table 44.

Table 45 shows output of model. Train accuracy is equal to 100%. Test accuracy is equal to 100%.

Figure 2 shows data set of first sixteen images.

Table 44 The results of automatic labeling the data

Dec	Label	Exponent	Condition vectors	x^0, x^1, x^2, x^3, x^4
0	Inflammation	Null	00000	00000
16	Non-inflammation	Beta (0)	00001	10000
8	Non-inflammation	Beta (1)	00010	01000
4	Non-inflammation	Beta (2)	00011	00100
2	Non-inflammation	Beta (3)	00100	00010
1	Inflammation	Bcta (4)	00101	00001
6	Inflammation	Beta (5)	00110	00110
3	Inflammation	Beta (6)	00111	00011
13	Non-inflammation	Beta (7)	01000	01101
10	Inflammation	Beta (8)	01001	01010
5	Inflammation	Beta (9)	01010	00101
14	Non-inflammation	Beta (10)	01011	01110
7	Non-inflammation	Beta (11)	01100	00111
15	Inflammation	Beta (12)	01101	01111
11	Non-inflammation	Beta (13)	01110	01011
9	Inflammation	Beta (14)	01111	01001
25	Inflammation	Beta (15)	10000	11001
24	Non-inflammation	Beta (16)	10001	11000
20	Non-inflammation	Beta (17)	10010	10100
18	Non-inflammation	Beta (18)	10011	10010
17	Non-inflammation	Beta (19)	10100	10001
28	Inflammation	Beta (20)	10101	11100
22	Inflammation	Beta (21)	10110	10110
18	Inflammation	Beta (22)	10111	10011
28	Non-inflammation	Beta (23)	11000	11101
26	Inflammation	Beta (24)	11001	11010
21	Inflammation	Beta (25)	11010	10101
30	Non-inflammation	Beta (26)	11011	11110
23	Non-inflammation	Beta (27)	11100	10111
31	Inflammation	Beta (28)	11101	11111
27	Non-inflammation	Beta (29)	11110	11011
25	Inflammation	Beta (30)	11111	11001

Table 45 Output of model

Number of iteration	Cost
0	0.693147
100	0.006366
200	0.003186
300	0.002124
400	0.001593
500	0.693147
600	0.006366
700	0.003186
800	0.002124
900	0.001593
1000:	0.000637
1100	0.000579
1200	0.000531
1300	0.000490
1400	0.000455
1500	0.000425
1600	0.000398
1700	0.000375
1800	0.000354
1900	0.000335

The training accuracy is equal to 100% as well as test accuracy is equal to 100%. Figure proofs it.

Regression analysis is a section of mathematical statistics, devoted to methods of analysis of the dependence of one value from another. In contrast to the correlation analysis it does not find out whether a significant relationship, but is engaged in the search for a model of this connection, expressed in the function of regression [10]. This type of analysis is much more accurate and precocious in compare to other one [14]. It makes it ideal tool for clinical investigations in medicine.

Immunoglobulins are usually measured in venous or capillary blood, however, alternative samples, including saliva, have also been used for children, given the non-invasive nature and ease of collection [15]. Therefore, salivary immunoglobulins detection that was made in our investigation is extremely important in medicine. Regression analysis showed that periodontal status depended on level of immunoglobulins in children with atopic diseases and orthodontic pathology.

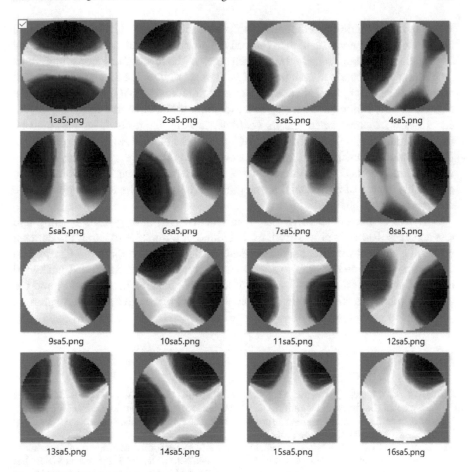

Fig. 2 The labeled data set of sixteen images

6 Conclusions

The present study has been done to build information model of salivary immunoglobulins and periodontal status and to evaluate the correlation between different types of immunoglobulin level in unstimulated saliva of children with atopy and inflammation in periodontal tissues. This study was conducted to assess the state of lipid peroxidation in the oral liquid and the periodontal disease and to detect the correlation between the level of antioxidants in children and inflammation in periodontal tissues by means of regression analyses. The results showed changes in the antioxidant balance in children with atopy that were expressed in an increase in malondialdehyde level, a decrease in superoxide dismutase activity, and a level of reduced glutathione. These indicators can be considered as biological markers of development of gingivitis at the preclinical stage in children against atopic diseases. However, the most valuable for prediction and diagnosis is the level of reduced glutathione in the oral liquid. Regression analysis of data showed the greatest dependence of inflammation of periodontal tissues, namely 91%, on changes in the level of reduced glutathione. According to the obtained results, the highest value at both the diagnostic stage and during the dispensary surveillance has immunoglobulins A and G, which showed the highest determination coefficient. For these immunoglobulins, the linear regression equations were constructed and its coefficients were determined, which could be used to predict inflammatory diseases of the periodontal disease in children against atopic diseases. On the basis of regression analysis data, the regression equation was constructed, its coefficients were determined, which allowed to mathematically predict the degree of periodontal tissue inflammation when the level of glutathione in the oral fluid is changed.

The aim of present research has been reached. We have got the training and test accuracy 100%. It proofs truth of technique of automatic labeling the data set with finite element model for training of artificial neural network in tomography. In this paper, we covered a brief overview of the forward propagation and backward propagation algorithm, an exploration of a node's behavior in a network, and the results of training a custom-built artificial neural network on EIT image recognition.

References

1. Bezruk V, Krivenko S, Kryvenko L (2017) Salivary lipid peroxidation and periodontal status detection in Ukrainian atopic children with convolutional neural networks. In: 2017 4th international scientific-practical conference problems of infocommunications. Science and technology (PIC S&T). Kharkov, Ukraine (2017)
2. Johansson SG (2003) Revised nomenclature for allergy for global use: report of the nomenclature review committee of the World Allergy Organization, Oct 2003. Johansson SG, Bieber T, Dahl RJ (2004) Allergy Clin Immunol 113(5):832–836
3. Hwang CY (2010) Prevalence of atopic dermatitis, allergic rhinitis and asthma in Taiwan: a national study 2000 to 2007. Hwang CY, Chen YJ, Lin MW et al (2010) Acta Derm Venereol 90:589–594
4. Steinbacher DM (2001) The dental patient with asthma. An update and oral health considerations. Steinbacher DM, Glick M (2001) J Am Dent Assoc 132:1229–1239
5. Akopian AZ (2000) Rasprostranennost allerhycheskykh zabolevanyi u detei. Akopian AZ (2000) Ukr Pulmonol Zhurnal 1:65–69
6. Ersin NK (2006) Oral and dental manifestations of young asthmatics related to medication, severity and duration of condition. Ersin NK, Gulen F, Eronat N et al (2006) Pediatr Int 48:549–554
7. Kazemi H (2019) Saving and loading a large number of images (data) into a single HDF5 file. In: Machine learning guru, 2 Jan 2019 [Online]. Available: http://machinelearninguru.com/deep_learning/data_preparation/hdf5/hdf5.html. Accessed 2 Jan 2019
8. Dogs vs. Cats (2014) Create an algorithm to distinguish dogs from cats. Kaggle, 2 Feb 2014 [Online]. Available: https://www.kaggle.com/c/dogs-vs-cats. Accessed 2 Jan 2019
9. Mittal S, Bansal V, Garg S, Atreja G, Bansal S (2011) The diagnostic role of Saliva—a review. J Clin Exp Dent 3(4):e314–e320
10. Malamud D (2006) Salivary diagnostics: the future is now. J Am Dent Assoc 137:284–286
11. Sperandei S (2014) Understanding logistic regression analysis. Biochem Med 24(1):12–18. https://doi.org/10.11613/BM.2014.003

12. Adler A, Boyle A (2017) Electrical impedance tomography: tissue properties to image measures. IEEE Trans Biomed Eng 64(11):2494–2504
13. Mandall NA (2005) Index of orthodontic treatment need as a predictor of orthodontic treatment uptake. Am J Orthod Dentofacial Orthop 128(6):703–707
14. Purrell (2017) Efficient way to shift a list in python. Stackoverflow, 4 July 2017 [Online]. Available: https://stackoverflow.com/questions/2150108/efficient-way-to-shift-a-list-in-python. Accessed 2 Jan 2019
15. Armstrong JS (2012) Illusions in regression analysis. Int J Forecast 28(3):689–694
16. World Health Organization (2013) Oral health surveys: basic methods, 5th edn, 125p
17. Aldrich J (2005) Fisher and regression. Stat Sci 20(4):417
18. Nazaryan R, Kryvenko L (2017) Salivary oxidative analysis and periodontal status in children with atopy. Interv Med Appl Sci 9(4):199–203. https://doi.org/10.1556/1646.9.2017.32
19. Krivenko SS, Krivenko SA (2014) Many-to-many linear-feedback shift register. In: 2014 IEEE 34th international conference on electronics and nanotechnology (ELNANO). Kyiv, Ukraine
20. Ng A (2017) Build your career in AI. Coursera, 21 Oct 2017 [Online]. Available: https://www.deeplearning.ai/. Accessed 23 Jan 2019
21. Ageyev D et al (2018) Classification of existing virtualization methods used in telecommunication networks. In: Proceedings of the 2018 IEEE 9th international conference on dependable systems, services and technologies (DESSERT), pp 83–86. https://doi.org/10.1109/dessert.2018.8409104
22. Kirichenko L, Radivilova T, Zinkevich I (2018) Comparative analysis of conversion series forecasting in E-commerce tasks. In: Shakhovska N, Stepashko V (eds) Advances in intelligent systems and computing II. CSIT 2017. Advances in intelligent systems and computing, vol 689. Springer, Cham, pp 230–242. https://doi.org/10.1007/978-3-319-70581-1_16

Deep Convolutional Neural Network for Detection of Solar Panels

Vladimir Golovko⬤, Alexander Kroshchanka⬤, Egor Mikhno⬤,
Myroslav Komar⬤, and Anatoliy Sachenko⬤

Abstract The article describes a method for detecting solar panels in satellite imagery. Due to the growing popularity of this technology, problems associated with the maintenance of solar panels are also becoming relevant. Many service companies are interested in obtaining information about potential customers. Thus, the analysis of photographs in order to identify solar panels and accumulate statistical information on energy production capacities, their territorial distribution is an urgent task. The development of the theory for learning the deep neural networks gave a powerful impact on the development of various versions of convolutional neural networks used to solve recognition and classification issues. The advent of GPU parallelization technology has made such training feasible in a reasonable amount of time. In this article authors solved the two independent problems. A first, the deep convolutional neural network recognizes the presence of solar panels in a photograph. This model was trained using a set of low-resolution photos from low-quality satellite images from Google and showed a high result of detecting the presence of an object in the image. A second, solar panels with obtaining the exact coordinates of their location were detected.

V. Golovko · A. Kroshchanka · E. Mikhno
Brest State Technical University, Brest, Belarus
e-mail: vladimir.golovko@gmail.com

V. Golovko
Państwowa Szkoła Wyższa im. Papieża Jana Pawła II, Biala Podlaska, Poland

M. Komar (✉) · A. Sachenko
Ternopil National Economic University, Ternopil, Ukraine
e-mail: mko@tneu.edu.ua

A. Sachenko
e-mail: sachenkoa@yahoo.com

A. Sachenko
University of Information Technology and Management, Rzeszow, Poland

© The Editor(s) (if applicable) and The Author(s), under exclusive license to Springer Nature   371
Switzerland AG 2021
T. Radivilova et al. (eds.), *Data-Centric Business and Applications*, Lecture Notes on Data
Engineering and Communications Technologies 48,
https://doi.org/10.1007/978-3-030-43070-2_17

Keywords Photovoltaic · Convolutional neural network · Deep learning ·
Determining the presence · Detection · Solar panel · Low-resolution photos ·
Images

1 Introduction

Between 1992 and 2018, the growth of photovoltaic energy production worldwide
was close to exponential [1]. During this period, photovoltaic technologies, also
known as solar photovoltaic systems, have evolved from a niche market for small
applications into a major source of electricity.

Improving the competitiveness of technology, built on the use of solar energy, in
comparison with other technologies enabled to increase essentially its share within
the total energy produced in the world.

By the end of 2018, the total aggregate installed capacity of photovoltaic plants
reached about 508 gigawatts (GW), of which about 180 GW (about 35%) were
utility-scale power plants [2]. This represents an increase of 27% from 2017 [3, 4].
That is enough to provide about 3% of the global electricity demand.

Bloomberg New Energy Finance predicts the growth of global solar installations in
2019, adding another 125–141 GW, resulting in a total capacity of 637–653 GW [5] by
the end of the year. According to the forecast published by the IEA, renewable power
capacity will expand by 50% between 2019 and 2024, led by solar PV (Fig. 1). By
2050, the IEA suggests that solar photovoltaic energy will reach 4.7 TW (4674 GW)
in its high-resiliency scenario, of which more than half will be deployed in China
and India, making solar energy the world's largest source of electricity [6, 7].

Fig. 1 Renewable capacity growth between 2019 and 2024 by technology [8]

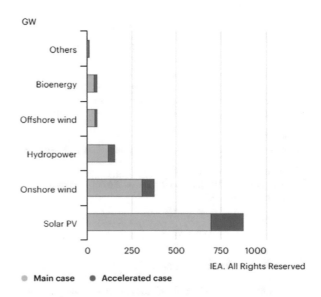

Due to the growing popularity of this technology, problems related to the maintenance of solar panels are also becoming relevant. Many service companies are interested in obtaining information about potential customers. Thus, the analysis of photographs in order to detect solar panels and accumulate statistical information is very important.

The development of machine learning methods [9] and the theory of deep neural networks [10–14] has led to the development of a great number of convolutional neural networks for images recognition and classification [15–17]. The advent of parallel computing technology using GPUs [18, 19] in turn made such learning workable in a reasonable amount of time.

We solved two independent problems of determining the presence of solar panels in the image and their detection with obtaining the exact coordinates of the location.

2 State-of-the-Arts

Convolutional neural networks are based on neocognitron models proposed by Fukushima [20] and networks with shared weights. The first classical model of a convolutional neural network was proposed by LeCun and received the name LeNet-5 [21, 22]. Convolutional neural networks have integrated three concepts, called the local receptor field, shared by weights and spatial subsample (Fig. 2).

At the moment, there are quite a few architectures of convolutional neural networks. Every year new articles, new architectures or existing ones are published. Consider the following architectures: DenseNet [23–25], InceptionV3 architecture [26], ResNet [27], AlexNet [28], Squeezenet [29, 30], MobileNet [31], etc. The last two were smaller and trained more quickly, since they were designed for use in devices with little memory and computing power.

Figure 3 depicts the architecture of DenseNet, as we see it consists of three blocks, an example of one such block is shown in Fig. 4.

Fig. 2 General architecture of the convolutional neural network

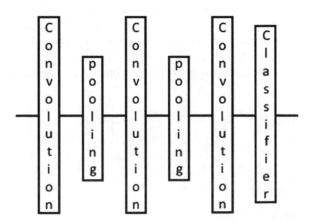

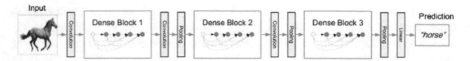

Fig. 3 DenseNet architecture [23]

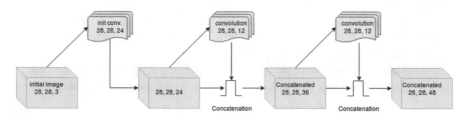

Fig. 4 The DenseNet architecture block [23]

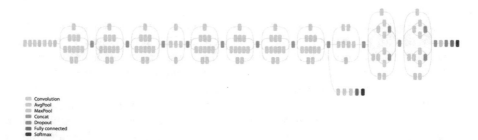

Fig. 5 InceptionV3 architecture [26]

At this time, this architecture shows one of the lowest errors in open CIFAR/SVHN data sets [32, 33].

Figure 5 depicts the InceptionV3 architecture [26], it was developed by Google, in front of DenseNet, and it also yields very good results on the same open CIFAR/SVHN data sets. As we see in the picture, the model consists of 11 modules, the output of each of which is given to the input of the next.

Figure 6 depicts the Squzeenet architecture [29]. The Squeezenet is characterized by a similar accuracy as AlexNet, but it is in 3 times faster and in 500 times smaller.

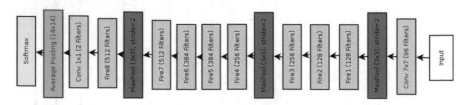

Fig. 6 Architecture Squzeenet [29]

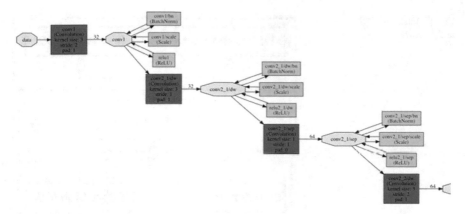

Fig. 7 MobileNet architecture [31]

Figure 7 depicts the MobileNet architecture [31], which shows similar results to the previous one. Usually These architectures are used commonly by mobile devices or other devices with a limited memory.

Classic implementations of CNN, which has become a breakthrough in the industry, are also LeNet5 and AlexNet [28] (Fig. 8).

All of these convolutional networks are used for constructing more complex architectures to solve object detection or image segmentations problems. Nowadays the Faster R-CNN [34] and SSD [35, 36] architectures are applied more often as others.

The Faster R-CNN (Faster Region-based Convolutional Neural Network) model consists of three parts (Fig. 9). The first part is the feature extractor-classifier (for example, ResNet-50/101, Inception), pre-trained on a specific dataset (e.g. COCO [37]). The second part is the RPN network that generates the candidate regions. The

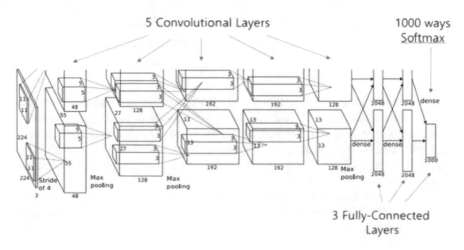

Fig. 8 Architecture AlexNet [28]

Fig. 9 Faster R-CNN architecture

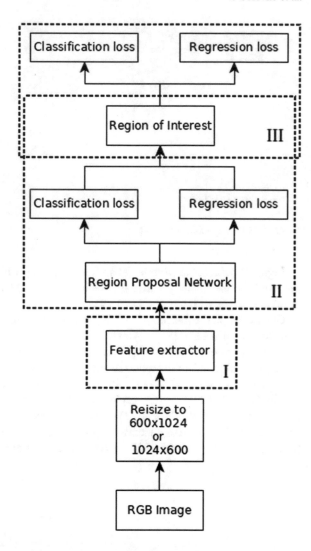

third part is a detector, which is represented by additional fully connected layers that generate the coordinates of bounding boxes containing the desired objects, and class labels per each such area. A key feature of this model is the RPN-network. Feature maps obtained by the preceding convolutional layer feed the RPN-network input. Due to this fact, the generation of applicants is faster than using the original full-size image.

This architecture provides the high performance indicators for solving problems of objects detection in systems where the total time, spent on processing and displaying results, is not critical.

Model SSD, like YOLO [38], belongs to the category of single-pass methods, which enable to solve the problem of object detection within a single network. The schematic design of the architecture is presented in Fig. 10.

The SSD model as well, as Faster R-CNN model consists of three parts. The first part is the feature extractor-classifier (Inception, MobileNet or other), which is pre-trained on a specific dataset and clipped to a completely convolutional neural network. The second part is the additional convolutional layers after feature extractor. They are used for reducing the dimension of feature maps for different detection and classification blocks. These blocks are the third part of this model that includes the default box generator, localization and confidence components. The Fast NMS algorithm is used to reduce the number of candidates.

SSD differs from other single shot detectors (in particular, YOLO) by the fact that each layer of the model takes a part in forming the information about objects and

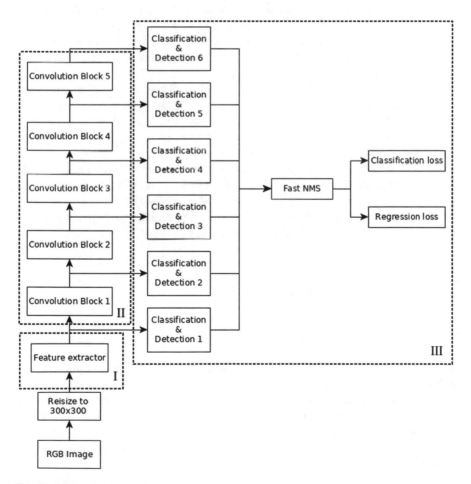

Fig. 10 SSD architecture

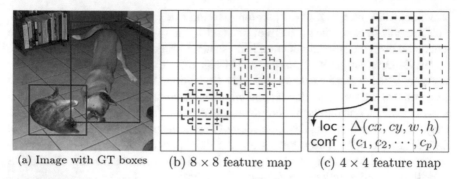

(a) Image with GT boxes   (b) 8 × 8 feature map   (c) 4 × 4 feature map

Fig. 11 Localization of objects on maps of signs of different sizes [35]

their location while taking into account the scale of these objects (each further layer processes objects larger than a size of the previous one) (Fig. 11).

Each element of feature maps forms a set of so-called default boxes (or Anchors), sides that are different in scale and ratio. The model is trained so that each anchor predicts correctly both its class and displacement.

Malof et al. [39] describes a method, based on SVM, to detect automatically the solar panels using high-resolution photos from satellite. Firstly, a pre-screening operation is used identifying the regions, which are processed to highlight the signs. After that, the obtained output information in a form of regions list and confidence values indicates how likely a solar panel is in a given area.

To detect direct the solar panels authors [40, 41] use the decision trees. The achieved indicator of panel's localization is 90% in the case of using the certain parameters of the algorithm, and the method itself consists of four stages. In addition, the Aerial Imagery Dataset is used here as a training sample, which includes images with a resolution of up to 5000 × 5000 pixels [42].

3 The Proposed Approach

3.1 General Scheme of Detection System

To solve the problem of determining the presence we proposed the approach, which differs from [36] in that the use of convolutional neural networks as a whole makes it possible to increase the detection accuracy. Another important feature is that we used the low-resolution photos to train the proposed model (for example, photos from Google Maps, often of poor quality, see Fig. 12).

To solve the detection problem and simplify the methods [40–42] we proposed to use the Faster R-CNN.

Let us divide the problem of solar panels detection into two tasks. The first task is to define the solar panels availability in images. The second one is to detect the solar

Fig. 12 Samples of poor quality images from the training set

panels from images, namely to generate the coordinates of the rectangular regions contained the solar panels and the class labels for each such region. The general scheme of proposed approach is shown in Fig. 13.

In order to define the solar panels availability we used the convolutional neural network with six layers and for solar panels detection the Faster R-CNN deep neural network, based on the ResNet-50 classifier is applied.

3.2 Building a Training Set

To form the training sample for solving the problems of determining the presence and detection of solar panels, we used Google Maps color images with a resolution of 200 × 200 pixels (Fig. 14).

Fig. 13 General scheme of
detection system

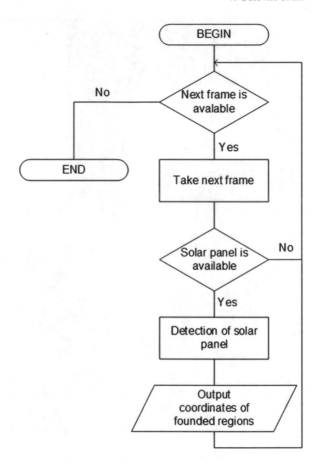

A sample of 3347 photographs was used (where 1643 images contained solar panels and 1704 did not contain), while 80% of the original sample formed a training subsample, and 20% a test one. For the detection task, we used a sample of 1000 images, 800 of which were assigned to the training, and 200 to the test subsample.

To determine the presence, the training sample preparation consisted of manually sorting the images and assigning them to two groups of images—containing solar panels and, not containing correspondingly. In a case of panel detection, sampling preparation was more laborious and consisted of manually sorting images with a definition per each characteristic of rectangular areas including panels (length, width, coordinates of the upper left corner).

When preparing a training sample for solving a detection problem, the selection of rectangular areas for some images does not seem appropriate, since the panels can be located at an angle to the horizontal axis of the image and have an elongated shape (Fig. 15).

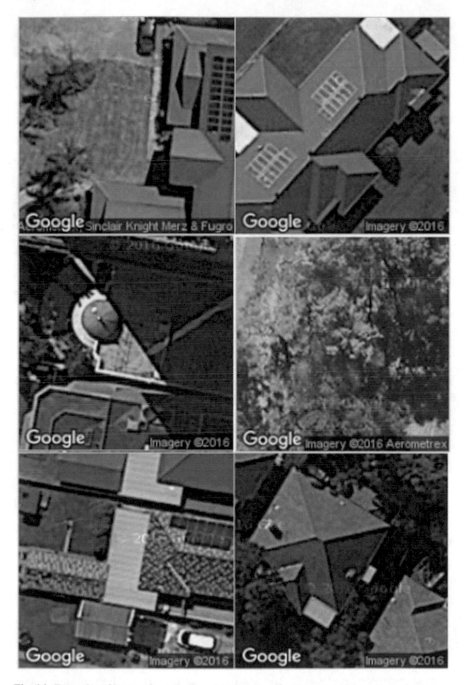

Fig. 14 Examples of images from the Google Maps sample

Fig. 15 Sample image with
a predominant background

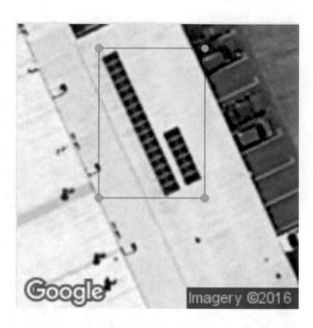

3.3 Definition of Solar Panels Availability

In order to define the solar panels availability in images, we proposed a convolutional neural network with the architecture presented in Fig. 16 [43].

The presented neural network consists of six layers. The first three layers are convolutional and perform low-level features. The last three layers are fully connected layers of neuronal elements that solve the classification problem.

For all layers, except for the latter, the ReLU activation function

$$y_j = F(S_j) = \begin{cases} S_j, & S_j > 0; \\ kS_j, & S_j \leq 0, \end{cases} \tag{1}$$

where S_j is a weighted sum of j-th neural unit, $k = $ const, $0 \leq k < 1$.

In the last fully connected layer, the sigmoid activation function

$$y_j = F(S_j) = \frac{1}{1 + e^{-S_j}}. \tag{2}$$

Additionally, we used a stride equal to 3 on the first convolutional layer to reduce the dimensionality of the image.

Images from the training sample were fed to the first layer of the convolutional network and passed down to the last layer. The last fully bound layer contains only one neuron, which returns the probability of having a solar panel in the image. The backpropagation algorithm was employed to train the convolutional neural network.

Fig. 16 Architecture of used convolutional neural network

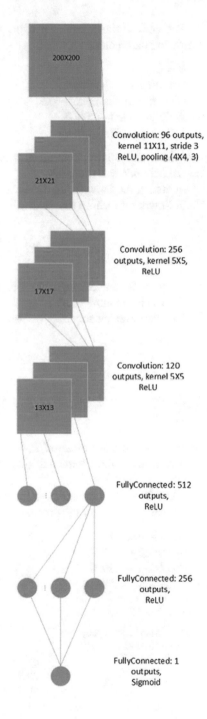

The convolutional neural network has been trained within 70 epochs using the following parameters:

- $k = 0$.
- Learning speed—0.001.
- Momentum parameter—0.9.
- Weight-decay—0.0005.
- Mini-batch size—20.

Additionally, a dropout with a probability of 0.5 for full connected layers of the neural network was used.

In testing mode, we used a simple conversion of output data in accordance with the following threshold function:

$$b_s = \begin{cases} 1, & y_j > 0.5; \\ 0, & \text{otherwise} \end{cases} \tag{3}$$

where y_s is the real output of the CNN network, b_s—binary output, $s = 1, 2, 3, \ldots,$ L where L is the number of images in the set.

To determine the accuracy we used the following formulas:

$$A = \frac{S}{L} * 100\% \tag{4}$$

$$S = \sum_{s=1}^{L} 1(b_s = e_s) \tag{5}$$

where $1()$—indicator function, e_s—reference value (label).

According to the obtained results, the accuracy of classification for this task was about 87.46% (Table 1).

The ROC curve, constructed for the trained classifier, is shown in Fig. 17, and other characteristics of the obtained classifier are indicated below:

- Recall—0.8339.
- Specificity—0.9104.
- Precision—0.8907.
- F-measure—0.8614.

Table 1 Matrix conjugacy for a trained model		Projected «No»	Projected «Yes»	Accuracy, %
	Waiting for «No»	325	32	91
	Waiting for «Yes»	52	261	83
	Total	377	293	87

Fig. 17 ROC curve for a trained binary classifier, AUC = 0.92

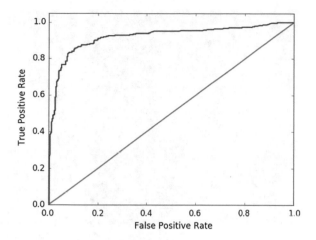

4 Case Study

The Faster R-CNN deep neural network, based on the ResNet-50 classifier, was used [44]. To assess the quality of the models, we employed the metric mAP (mean-average precision metric), which is the de facto standard of the metrics used for detection [45]. This metric is applied in conjunction with its modifications calculated for various IoU threshold values. Since there is one class of objects only for the problem under consideration, the mAP metric coincides with the AP.

The Faster R-CNN network has been trained for 5000 iterations (under iteration, here is the setting of parameters for one random image from the training sample) using the following parameters:

- Learning speed—0.0003.
- Momentum parameter—0.9.

After completing the training, we conducted a study of the generalization of the network using a 200-image test sample, the result is $AP = 0.9299$.

Figure 18 shows some results of detecting the solar panels in low-resolution images.

As it can be seen from above, the proposed method carries out objectively an accurate discovery. In Fig. 19 the results of detecting the arbitrary images (not from the test set) are shown.

5 Conclusion

A method for detecting the solar panels in satellite images based on convolutional neural networks is proposed.

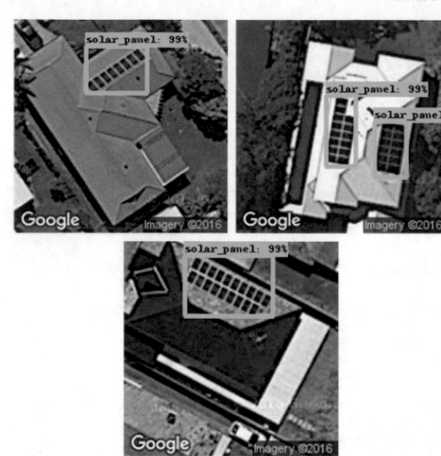

Fig. 18 Detection results for images from the test set

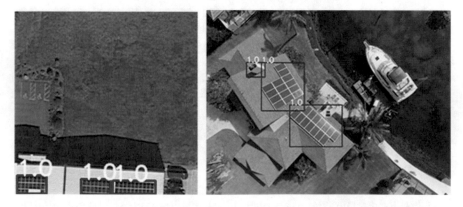

Fig. 19 Detection results for arbitrary images

A deep convolutional neural network recognizes the presence of solar panels in a photograph. The presented neural network consists of six layers. The first three layers are convolutional and perform high-level signs. The last three layers are full-knit layers of neuronal elements that solve the classification problem. For all layers, except for the latter, the ReLU activation function is used.

To solve the problem of detecting solar panels, the Faster R-CNN deep neural network, based on the ResNet-50 classifier, was used. This model consists of two parts. The first part is the ResNet-50 classifier, pre-trained on the COCO set. The second part is a detector, which is represented by a convolutional neural network generating the coordinates of the rectangular regions containing the objects being searched for and the class labels for each such region.

This model was trained using a set of low-resolution photos from low-quality satellite images from Google and showed a high result of detecting the presence of an object in the image. In addition, a trained convolutional neural network confirmed proper results in high-resolution images.

The obtained outcomes enable to detect solar panels and accumulate statistical information about the constructed energy production capacities and their territorial location based on Google satellite imagery analysis.

In the future we are going to explore ways to determine the boundaries of solar panels in photographs and develop a method for segmenting such objects.

Acknowledgements This work is performed under a grant by the Ministry of Education and Sciences, Ukraine, 2018–2019 as well as it's supported by the Belarusian State Research Program "Informatics and Space".

References

1. Growth of photovoltaics. https://en.wikipedia.org/wiki/Growth_of_photovoltaics. Accessed 12 Dec 2019
2. Utility-scale solar in 2018 Still growing thanks to Australia and other later entrants. https://wiki-solar.org/library/public/190314_Utility-scale_solar_in_2018.pdf. Accessed 12 Dec 2019
3. Clean Energy Investment Exceeded $300 Billion Once Again in 2018. https://about.bnef.com/blog/clean-energy-investment-exceeded-300-billion-2018. Accessed 12 Dec 2019
4. Trends in photovoltaic applications 2018. http://www.iea-pvps.org/fileadmin/dam/intranet/task1/IEA_PVPS_Trends_2018_in_Photovoltaic_Applications.pdf. Accessed 12 Dec 2019
5. Transition in energy, transport—10 predictions for 2019—2. Solar additions rise despite China. BNEF—Bloomberg New Energy Finance. https://about.bnef.com/blog/transition-energy-transport-10-predictions-2019. Accessed 12 Dec 2019
6. International Energy Agency. Technology roadmap: solar photo-voltaic energy. http://www.oregonrenewables.com/Publications/Reports/IEA_TechnologyRoadmapSolarPhotovoltaicEnergy_2014.pdf. Accessed 12 Dec 2019
7. Pozzebon S (2014) One chart shows how solar could dominate electricity in 30 years. https://www.businessinsider.in/one-chart-shows-how-solar-could-dominate-electricity-in-30-years/articleshow/43913904.cms?utm_source=contentofinterest&utm_medium=text&utm_campaign=cppst. Accessed 12 Dec 2019

8. Solar—fuels & technologies—IEA. https://www.iea.org/fuels-and-technologies/solar. Accessed 12 Dec 2019
9. Kirichenko L, Radivilova T, Bulakh V (2020) Binary classification of fractal time series by machine learning methods. In: Lytvynenko V, Babichev S, Wójcik W, Vynokurova O, Vyshemyrskaya S, Radetskaya S (eds) Lecture notes in computational intelligence and decision making. ISDMCI 2019. Advances in intelligent systems and computing, vol 1020. Springer, Cham
10. Hinton G, Osindero S, The Y (2006) A fast learning algorithm for deep belief nets. Neural Comput 18(7):1527–1554. https://doi.org/10.1162/neco.2006.18.7.1527
11. Bengio Y (2009) Learning deep architectures for AI. Found Trends Mach Learn 2(1):1–127. https://doi.org/10.1561/2200000006
12. Erhan D, Bengio Y, Courville A et al (2010) Why does unsupervised pre-training help deep learning? J Mach Learn Res 11:625–660. https://doi.org/10.1145/1756006.1756025
13. Golovko V, Kroschanka A (2016) The nature of unsupervised learning in deep neural networks: a new understanding and novel approach. Opt Mem Neural Netw 3:127–141. https://doi.org/10.3103/S1060992X16030073
14. Golovko V (2017) Deep learning: an overview and main paradigms. Opt Mem Neural Netw 26:1–17. https://doi.org/10.3103/S1060992X16040081
15. Komar M et al (2018) Deep neural network for image recognition based on the Caffe framework. In: Proceedings of the IEEE second international conference on data stream mining & processing (DSMP), Lviv, Ukraine, pp 102–106. https://doi.org/10.1109/dsmp.2018.8478621
16. Komar M et al (2018) Compression of network traffic parameters for detecting cyber attacks based on deep learning. In: Proceedings of the 9th IEEE international conference on dependable systems, services and technologies (DESSERT). Kyiv, Ukraine, pp 44–48. https://doi.org/10.1109/dessert.2018.8409096
17. Komar, M. et. al. (2018) Deep neural network for detection of cyber attacks. In: Proceedings of the IEEE first international conference on system analysis & intelligent computing (SAIC). Kyiv, Ukraine, pp 186–189. https://doi.org/10.1109/saic.2018.8516753
18. Korpała G, Kawalla R (2015) Optimization and application of GPU calculations in material science. https://library.wolfram.com/infocenter/Conferences/9346/1444771976.pdf. Accessed 12 Dec 2019
19. Dorosh V et al (2018) Parallel deep neural network for detecting computer attacks in information telecommunication systems. In: Proceedings of the 38th IEEE international conference on electronics and nanotechnology (ELNANO), Kyiv, Ukraine: TUU «Kyiv Polytechnic Institute», pp 675–679. https://doi.org/10.1109/elnano.2018.8477530
20. Fukushima K (1980) Neocognitron: a self-organizing neural network model for a mechanism of pattern recognition unaffected by shift in position. Biol Cybern 36(4):193–202
21. LeCun Y, Bottou L, Bengio Y et al (1998) Gradient-based learning applied to document recognition. Proc IEEE 86(11):2278–2324. https://doi.org/10.1109/5.726791
22. Golovko V, Krasnoproshin V (2017) Neural network data processing technologies. Minsk, Republic of Belarus (in Russian)
23. Huang G et al (2017) Densely connected convolutional networks. In: 2017 IEEE conference on computer vision and pattern recognition (CVPR). Honolulu, HI, USA, pp 4700–4708. https://doi.org/10.1109/cvpr.2017.243
24. Jégou S et al (2017) The one hundred layers tiramisu: fully convolutional DenseNets for semantic segmentation. In: 2017 IEEE conference on computer vision and pattern recognition (CVPR), Honolulu, HI, USA. https://doi.org/10.1109/cvprw.2017.156. https://arxiv.org/pdf/1611.09326.pdf. Accessed 12 Dec 2019
25. Zhu Y et al (2017) Densenet for dense flow. In: 2017 IEEE international conference on image processing (ICIP). https://doi.org/10.1109/icip.2017.8296389. https://arxiv.org/pdf/1707.06316v1.pdf. Accessed 12 Dec 2019
26. Evolution of neural networks for image recognition in Google: Inception-v3. https://habr.com/post/302242. Accessed 12 Dec 2019 (in Russian)

27. He K et al (2016) Deep residual learning for image recognition. In: 2016 IEEE conference on computer vision and pattern recognition (CVPR), Las Vegas, NV, USA, pp 770–778. https://doi.org/10.1109/cvpr.2016.90
28. Krizhevsky A, Sutskever I, Hinton G (2012) ImageNet classification with deep convolutional neural networks. Adv Neural Inf Process Syst 25(2):1097–1105. https://doi.org/10.1145/3065386
29. Iandola FN et al (2016) SqueezeNet: AlexNet-level accuracy with 50x fewer parameters and <0.5 MB model size. https://arxiv.org/pdf/1602.07360.pdf. Accessed 12 Dec 2019
30. Tsang S-H (2018) Review: SqueezeNet (image classification). https://towardsdatascience.com/review-squeezenet-image-classification-e7414825581a. Accessed 12 Dec 2019
31. Howard AG, Zhu M, Chen B et al (2017) MobileNets: efficient convolutional neural networks for mobile vision applications. https://arxiv.org/pdf/1704.04861.pdf. Accessed 12 Dec 2019
32. CIFAR-10 and CIFAR-100 dataset. http://www.cs.toronto.edu/~kriz/cifar.html. Accessed 12 Dec 2019
33. Examples of images from MNIST, CIFAR and SVHN datasets. https://www.researchgate.net/figure/Examples-of-images-from-MNIST-CIFAR-and-SVHN-Datasets_fig1_320564389. Accessed 12 Dec 2019
34. Ren S, He K, Girshick R et al (2017) Faster R-CNN: towards real-time object detection with region proposal networks. IEEE Trans Pattern Anal Mach Intell 39(6):1137–1149. https://doi.org/10.1109/TPAMI.2016.2577031
35. Liu W, Anguelov D, Erhan D et al (2016) SSD: single shot MultiBox detector. In: Leibe B, Matas J, Sebe N, Welling M (eds) Computer vision. In: ECCV 2016. Lecture notes in computer science, vol 9905. Springer, Berlin
36. Tsang S-H (2018) Review: SSD—single shot detector (object detection). https://towardsdatascience.com/review-ssd-single-shot-detector-object-detection-851a94607d11. Accessed 12 Dec 2019
37. Lin T, Maire M, Belongie S et al (2014) Microsoft COCO: common objects in context. In: Fleet D, Pajdla T, Schiele B, Tuytelaars T (eds) Computer vision. ECCV 2014. Lecture notes in computer science, vol 8693. Springer, Berlin
38. Redmon J et al (2016) You only look once: unified, real-time object detection. In: 2016 IEEE conference on computer vision and pattern recognition (CVPR), Las Vegas, NV. https://doi.org/10.1109/cvpr.2016.91
39. Malof J et al (2015) Automatic solar photovoltaic panel detection in satellite imagery. In: 2015 international conference on renewable energy research and applications (ICRERA), Palermo, Italy, pp 1428–1431. https://doi.org/10.1109/icrera.2015.7418643
40. Malof J, Bradbury K, Collins L et al (2016) Automatic detection of solar photovoltaic arrays in high resolution aerial imagery. Appl Energy 183:229–240. https://doi.org/10.1016/j.apenergy.2016.08.191
41. Malof J et al (2017) A deep convolutional neural network, with pre-training, for solar photovoltaic array detection in aerial imagery. In: 2017 IEEE international geoscience and remote sensing symposium (IGARSS), Fort Worth, TX. https://doi.org/10.1109/igarss.2017.8127092
42. Bradbury K, Saboo R, Johnson TL et al (2016) Distributed solar photovoltaic array location and extent dataset for remote sensing object identification. Sci Data 3:160106. https://doi.org/10.1038/sdata.2016.106
43. Golovko V et al (2017) Convolutional neural network based solar photovoltaic panel detection in satellite photos. In: Proceedings of the 9th IEEE international conference on intelligent data acquisition and advanced computing systems: technology and applications (IDAACS), Bucharest, Romania, pp 14–19. https://doi.org/10.1109/idaacs.2017.8094501
44. Golovko V et al (2018) Development of solar panels detector. In: Proceedings of the IEEE international scientific-practical conference problems of infocommunications. Science and technology (PIC S&T), Kharkiv, Ukraine, pp 761–764. https://doi.org/10.1109/infocommst.2018.8632132
45. Jonathan H (2018) mAP (mean average precision) for object detection. https://medium.com/@jonathan_hui/map-mean-average-precision-for-object-detection-45c121a31173. Accessed 12 Dec 2019

Designing Network Computing Systems for Intensive Processing of Information Flows of Data

Halina Mykhailyshyn, Nadia Pasyeka, Vasyl Sheketa, Mykola Pasyeka, Oksana Kondur, and Mariana Varvaruk

Abstract Systematic research of technologies and concepts used for designing and building distributed fault-tolerant web-systems is carried out. The general principles of design of distributed web applications and information technologies used in the design of web systems are considered. As a result of scientific research it became clear that data backup is a defining attribute of web systems serving a large number of customers. Therefore, the main role in building modern web applications comes down to their scaling. Scaling up in the distributed systems apply when performance of this or that operation demands a considerable quantity of computing resources. There are two variants of scaling, namely vertical and horizontal. Vertical scaling consists in increasing the performance of existing components in order to increase overall performance. However, horizontal scaling is used to build distributed systems. Horizontal scaling consists in the fact that the system is divided into small components and placed on various physical computers. This approach allows adding new nodes to increase the performance of the web system as a whole. However, this imposes certain limitations on the developers of software systems, namely, providing fault tolerance on each computer node as separately and as a whole in a distributed system.

H. Mykhailyshyn · N. Pasyeka · O. Kondur · M. Varvaruk
Vasyl Stefanyk Precarpathian National University, Ivano-Frankivsk, Ukraine
e-mail: galmuh60@gmail.com

N. Pasyeka
e-mail: pasyekanm@gmail.com

O. Kondur
e-mail: oxikon13@gmail.com

M. Varvaruk
e-mail: varvaruk.mariana@gmail.com

V. Sheketa · M. Pasyeka (⊠)
Ivano-Frankivsk National Technical University of Oil and Gas, Ivano-Frankivsk, Ukraine
e-mail: pms.mykola@gmail.com

V. Sheketa
e-mail: vasylsheketa@gmail.com

© The Editor(s) (if applicable) and The Author(s), under exclusive license to Springer Nature Switzerland AG 2021
T. Radivilova et al. (eds.), *Data-Centric Business and Applications*, Lecture Notes on Data Engineering and Communications Technologies 48,
https://doi.org/10.1007/978-3-030-43070-2_18

Keywords Up-diffused · Caching · Web-systems · Task distribution · Planning of
the systems

1 Introduction

Cloud-based software systems are becoming increasingly popular today. Many processes, which until recently were implemented on individual personal computers,
are now performed with the help of Web-browser and providing access to various
online services. The need for distributed software systems based on cloud technology over the past ten years has grown so much that the computational resources
of one server are becoming increasingly scarce. The Internet, and at the same time
distributed software systems have penetrated into all areas of human activity. Distributed Web-services or software applications can be found in various fields of
human activity, such as energy, medicine, education, engineering and others. Gradually, industrial society is moving away from the classical use of software systems
that can be performed only on limited architectural solutions and the available minimum computing performance. Mobility of software systems is the main need of most
users of the information society. Distributed software systems with the use of cloud
technology on independent computing platforms make it possible to work on any
modern information devices from any remote location, and therefore such systems
are devoid of the classic lack of software systems [1].

Such dynamic development of distributed software systems based on cloud technology was facilitated by the full use of processes in the development of multi-core
central processors, as well as the rapid growth in the availability of high-speed networks with access to the Internet. However, in today's world the need for computing is constantly growing, thus the performance of computing systems is becoming
increasingly scarce. In addition to the current trend of using information technologies, the number of potential users of distributed software systems based on cloud
technology is growing rapidly. Against the background of this problem in the computational resources and the urgent need for developed highly loaded distributed
software systems has arisen [2].

Distributed software system on the basis of cloud technology is a set of distributed
computing nodes, united in a single structure, which works as a single mechanism.
The main advantage of a highly loaded distributed software system based on cloud
technology is that such a system can easily increase computational resources by
simply adding new nodes. Development of highly fault-tolerant software systems is
a basic property of any system, which is to ensure that the developed software system
based on cloud technology to continue to function properly in case of abnormal errors
(disturbances) in one or more of its parts. When developing highly loaded, fault-
tolerant distributed software systems, as a rule, preference is given to such quality of
the system as fault tolerance as opposed to the performance of computing resources.
The main reason in choosing this criterion is that temporary partial decrease of
software system performance on the basis of cloud technology is more acceptable

criterion for a considerable number of users than absence of access at all and loss of information flows of data during even an insignificant period of time.

Let's consider architectural decisions on development of distributed fault-tolerant software systems on the basis of cloud technology (Fig. 1).

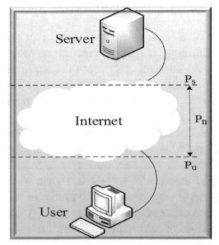

(a) Base of the software is subdivided system

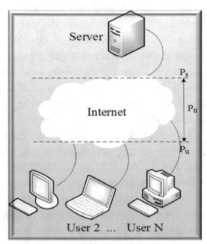

(b) Distributed user software system

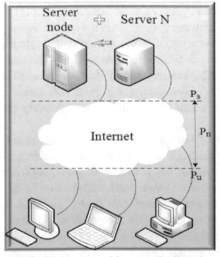

(c) Multiple user, multi-server distributed software system

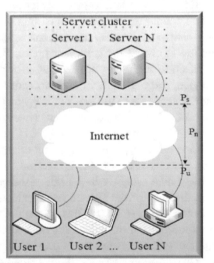

(d) A lot of custom cluster distributed software system

Fig. 1 Models of architectures of distributed software systems with regulation of computational load, where: P_n—moving the load on the server or to the user ("thin" or "thick client"); P_s—load on the server or cluster; P_u—load on the user's computer

At the primary level, it is only the messages of the user with a remote resource via the Internet (Fig. 1a). At the second level of development of distributed software systems, many user connections to one server are provided (Fig. 1b). At the third level of development of distributed software systems a lot of user connection to various services is provided (Fig. 1c), and at the fourth level, with many users connected servers are combined into one computing cluster, thus providing the union of computing capabilities of the node (Fig. 1d).

However, the scalability of development of a highly loaded fault-tolerant distributed software system based on cloud technology is that information resources or access to them are distributed among several computing servers (nodes). In the scientific work the basic principles of development of software systems using cloud technology on independent computing platforms, which make a significant impact on the ideology and design of large distributed software systems.

The architecture of a highly loaded distributed software system based on cloud technology should have a certain redundancy in its functionality for processing data flows. For example, if on one computing server there is only one copy of the file, the loss of capacity for work (failure) of this server will automatically mean the loss of this information file with data. The general way to avoid such problems is to create several information copies of this file on other computer nodes (pages). The same approach is used for functions of formation of information inquiries to cloud storehouses of the data [3].

If for a distributed software system based on cloud technology is the main criterion of functionality, the provision of simultaneous synchronization of several copies of the system or its versions can protect our software system from a short emergency failure of one of the computer nodes. For example, if in the technological process of business logic of processing of information data there are two copies of the same service, and one of them does not work or gives out obviously the worst result the system can pass to the working copy on the other computing resource. A failure of a distributed cloud-based software system can occur at any time, and its recovery can be either automatic or manual [4–11].

One more extremely important criterion of fault tolerance of highly loaded distributed software systems is creation of architecture with automatic redundancy of critical data arrays, where each of the computing nodes is able to perform certain functionality of the system independently of each other. Such architecture of distributed software system has no central computer node of state control and there is no coordination of actions for other computer nodes. This approach to the development of a software system provides operational scalability as new computational nodes can be increased without special conditions. Thus, the use of such architectural approaches to the development of distributed software systems on the basis of cloud technology provides sufficient stability of the system, because there is no single point of failure in it. Today we have to operate more and more often with large sets of information flows of data, and to speed up the processing we use algorithms of distribution of a certain part of this data on different computing servers. Also, there is often an urgent need when the computational algorithms require additional server capacity and to solve this problem they are combined into clusters. Therefore, to solve the needs in the lack of computing resources can be used vertical or horizontal scaling of distributed fault-tolerant software system based on cloud technology [12–14].

If an architect decides to use vertical scaling when developing a software system, it will mean adding additional computational resources when using a separate server. Proceeding from the aforesaid it becomes clear that for processing of the big set of the computing data it can be necessary to add more hard disks or replacement of existing ones with large ones, there providing the possibility of processing of information flows of data on one server. If the calculations require significant hardware resources, it is necessary to move the calculations to a more powerful server with a more powerful processor (increased number of computational cores, more frequency, more RAM, etc.). In each separate case, when developing software systems based on cloud technology, vertical scaling occurs by creating a new individual resource that is able to independently process the problems and tasks set before it. In contrast to the vertical, horizontal scaling is the action of adding other computing nodes to the software system based on cloud technology. For processing large data flows, the software architect can offer a physical addition of a server for storing and processing part of the data set, and at the level of computing resources it will mean the distribution of calculations on different servers or loading and processing between several nodes.

In order to fully use the advantages of horizontal scaling of a software system based on cloud technology, such an opportunity should be provided at the stage of development of internal design of the system architecture, otherwise this approach to the distribution of content between the computational nodes will be extremely difficult to implement on an already working software system. When architects come to the idea that it is time to scale horizontally, one of the most common approaches and methods is to divide the computational capabilities of the system into parts.

In order to fully use the advantages of horizontal scaling of a software system based on cloud technology, such a possibility should be provided at the stage of development of internal design of the system architecture, otherwise this approach to distribution of content between computing nodes will be extremely difficult to implement on the already working software system. When architects come to the idea that it is time to scale horizontally, one of the most common approaches and methods is to divide the computational capabilities of the system into parts. The computational parts of a software system based on cloud technology can be divided in such a way that any logical set of computational functionalities must be separate. Such an architectural solution in the development of software systems can be achieved by dividing separate computational nodes by geographical criterion or any other.

Of course, today there are still problems with distribution of computational data or their functionality on several servers. One of the main criteria for processing information flows of data is their location, when using distributed software systems based on cloud technology, the closer the data is to the operation or the point of calculation, the better the performance of the system as a whole. Thus, in the development of software systems a potential problem is the division of information flows of data into several servers, because at any time the information flows of data can flow through the network to the computing node. Along with the problems already mentioned, there can also be another problem in the form of inconsistency in the form of inconsistency, it is in the case when there are other service providers such as reads or writes from a shared information resource (storage or database). There is a critical possibility of conditions such as "races" where some part of the information flows to be processed must be updated, but the reading takes place before they are updated. Thus, the result of information data processing is not reliable and is not suitable for the next computational operations.

Innovative approaches to building distributed fail-safe cluster systems

In the conditions of total computerization of modern society, there are many possible variations in the implementation of highly loaded distributed fault-tolerant software systems based on cloud technology such as: clusters, mainframes, GRID-systems. However, despite such a variety of methods and approaches, as well as practical implementations in the design of distributed software systems, not all of them correspond to a high level of necessary and sufficient fault tolerance. Let's describe the basic requirements to development of highly loaded distributed fault-tolerant software systems based on cloud technology:

- increasing the performance of software systems and services through the effective use of computing resources on independent computing platforms and the overall increase in the computing power of the system as a whole;
- ensuring the availability of the software system for fault tolerance and availability of information flows of data in the 24/7 mode;
- implementation of automated centralized booking of information flows and their services;
- ensuring flexible management of information resources with significant changes in the load on computing nodes;
- ensuring the possibility of increasing the performance of the software system based on cloud technology without changing its overall architecture;
- minimization of labor costs and total cost of software system operation;
- anticipating the possibility of migration of the developed software system, services and data flows between different computing nodes or independent platforms.

Mixed clusters in software development

Mixed clusters are such an architecture of highly loaded distributed fault-tolerant software system, when storage and processing of significant data flows takes place both on **combined** and separate computing servers. Such architecture of software systems has a mixed structure in its design and is called cluster systems. The principle of their action is based on the distribution of information flows of requests through one or more input computing nodes, which in due time redirect significant data flows for processing by various computing nodes on independent platforms, based on the level of their maximum load. In such architectural decisions at designing of the highly loaded distributed fault-tolerant program systems high reliability of system as a whole without losses of its general productivity is reached.

From the user's point of view, the entire software system based on cloud technology is designed to form a single copy of the software system. The independent computing nodes included in the system work identically as separate servers, but they have the ability to ensure the balancing of the total load in automatic mode and the transfer of control when leaving the working condition of any of the modules of the software system. Software systems are built on the basis of clusters with such architecture, combining many different characteristics of high-performance clusters and clusters of extreme readiness, but, at the same time, such an architecture is not devoid of certain drawbacks. In contrast to high-performance software systems, which are focused on the computational nature of certain information flows of tasks, systems with mixed architecture provide high performance in the performance of input or output operations, at the edge of a critical indicator to assess the performance of the developed distributed fault-tolerant software system based on cloud technology and services that serve this system. Determining the performance of input or output computational operations in the developed software system makes it possible to determine the maximum number of simultaneously working users and the integrated response time of the software system to various user requests.

A significant part of the software modules of hot standby clusters is in the standby mode for processing data flows on independent computing platforms, in case any of the working clusters fails, it is replaced by another cluster, which are in the operational reserve. Consequently, with such an architecture of software systems design it is impossible to use standby clusters to solve other pressing problems, so the computational efficiency of the developed software system based on the cluster approaches of this architecture has little performance, and mixed-architecture clusters are deprived of this disadvantage [15–18].

As a rule, a typical architecture of mixed cluster systems consists of three computing nodes (servers), which interact with each other at the level of information and hardware resource sharing. The main elements to be divided in such computing systems are the drives that form disk information arrays. On one of the computational nodes of the developed software system on the basis of cloud technology a special program or service of the general load-balancer is installed, which provides automatic redistribution of load between these computational nodes of the cluster. In this case, the cluster management is performed with the help of independent cloud services of service messages transmitting information about the state of the computing node.

This cluster architecture has the following advantages:

- the possibility of uniform load balancing, which is provided by defining common rules and approaches when balancing computational queries between cluster nodes;
- ensuring redundant fault tolerance of the software system due to simultaneous operation of several computing nodes and backup channels for storage and processing of data flows;
- simple scaling of the developed software system by increasing the computing power of the cluster by simply adding a certain number of nodes

- the possibility of hot maintenance and replacement of computing nodes of the cluster system without stopping both its own and individual cloud services.

Mixed cluster architecture in software development

Development of a software system based on cloud technology using mixed type clusters can be divided into homogeneous (all computing nodes of the cluster have the same architecture and power) and heterogeneous, which use independent computing nodes with different architectures and capacities. As a rule, when it comes to the development of software systems based on cloud technology using high-performance computing clusters, the architects usually mean homogeneous clusters. Thus, when the number of clusters increases, it is necessary to use computing servers on independent computing platforms, which differ not only in power but also in their architecture. In this case, a homogeneous computing cluster can change its structure and move to the status of a heterogeneous cluster structure. In such an architecture of software systems development with the use of heterogeneous structure of cluster systems the following problems are created:

- diversity in computational power of nodes significantly complicates the task of distributing computational works between processors' work;
- diversity in the architecture of cluster systems requires a significant preparation of special executable files for the administration of individual nodes.

When developing distributed software systems based on cloud technology with the use of clusters and computing resources (RAM) can be divided into clusters with shared memory, distributed memory (UMA-systems), clusters with physically distributed and publicly available memory (hybrid systems, NUMA-systems).

The architecture of software systems using cloud technology on the basis of clusters with shared RAM usually has high bandwidth when transferring significant data flows between the processors of the computing platform, but provided that there is no simultaneous access to several processors of the same memory element. Clusters with distributed memory architecture are characterized by the presence of a significant number of fast data exchange channels, which connect individual parts of this RAM with individual processors on an independent computing platform. The developed software system based on the cluster with hybrid memory uses the available RAM, which is physically distributed in different parts of the system, but logically depleted and forms a single computational address space. Such memory on independent computing platforms is called logically shared memory.

Clusters with distributed resources consist of a common data storage system and cluster nodes that distribute access to common data streams. Architecture with shared disks is more efficient with high power of information flows storage system and work with tasks oriented to their processing. In this case, there is no need to keep several copies of information data and at the same time, in case of failure of the node, the tasks can be instantly available to other computing nodes on independent platforms. Based on previous statements, the growth of computing power of the software system can be achieved by using the horizontal scaling of the cluster, namely, increasing the total power of the distributed cluster system by adding new nodes. And vertical scaling, where there is a possibility of increasing the computational power of the system by upgrading the existing computational nodes of the cluster system or implementation of scaling, combining both options.

2   Caching and Load Balancing of Information Flows of the Software System

Avalanche similar growth of quantity of information streams of the data and inquiries of users causes two main problems of functioning of highly loaded fault tolerant program systems, such as, scaling of access to a server of a database or databanks. The method of caching information flows of data is based on the principle: "recently requested data", which is likely to be in demand again and again. Caching are used practically at all levels of processing of calculations, both hardware and operating systems, and also applied program means such, as Web-browsers, Web-appendices and so forth. Caching looks like a temporary storage of information data in the RAM, but it has a limited space, and usually the re-production of data is much faster process and contains fresh elements for the re-processing of data arrays. Caching can exist at all architectural levels of the developed program system, but it more often meets at the level nearest to processing where information processes of fast return of files of the data, without using other levels. Placement of information cache directly at the level of request processing allows to store these response data locally. Each time when the information inquiry is given to the certain service program the computing node will quickly find and return the local cached data if any (Fig. 2) [19–23].

If the information request is not in the cache, the computing node of the request requests data from the hard disk. The cache can be located both in the RAM (which is processed very quickly) and on the local disk of the computing node (which is much faster than the transition to the NAS). However, if this principle is extended for several computing nodes on independent computing platforms, an unexpected

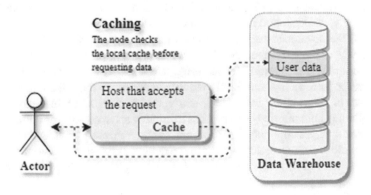

Fig. 2 Caching of data streams at the user request level

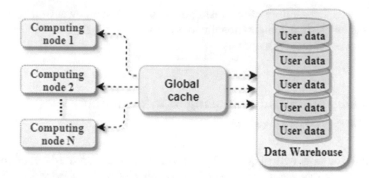

Fig. 3 Global cache, which is responsible for receiving data streams

situation may arise when each node will have its own cache of information data. Atypical cache processing errors may occur when the information load balancer accidentally distributes data requests from one user to different nodes. There are two ways to solve this problem: by organizing global or distributed caches.

Global cache

The global information cache works on a "one for all" basis, i.e. all computing nodes use the same cache space. This approach involves adding a computational server or file storage of a certain type of information data, which is processed much faster than the original data set. Each computational external node is capable to request a cache of the data as well as local. Such architectural approach to caching problems can be a little bit complicated. However, it is extremely effective in combination with specialized hardware and software. There are two general forms of global information cache organization, namely when the cached answer is not found in the cache, the cache itself becomes responsible for getting the missing data fragment from the database and database (Fig. 3).

However, when developing information flows of data on independent computing platforms there can be a problem situation when nodes independently query data flows if they were not found in the global cache (Fig. 4).

Most of the developed software systems that use global caches usually use the first type, where the information cache itself manages the receipt of data streams to prevent duplication of requests for the same user data. This approach may become a problem for software system developers in situations of unpredictable avalanche-like growth of information load which is a very negative factor and may cause an uncontrollable chain of refusals to process user requests in clusters or computational nodes. That's why at the stage of software systems development with the use of cloud technology it is necessary to carry out a detailed assessment of all the system's capabilities before introducing algorithms of using global caches of the first type [24].

Analyzing modern trends in the effective use of information caches, we can see that today the organization of distributed caches, where each of its nodes owns part of

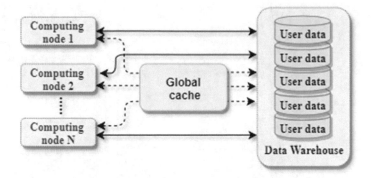

Fig. 4 Processing of data flows, where nodes independently carry out a user request

the cached data streams, is actively used. Usually at developed program systems, the cache is distributed between computing knots on independent platforms by means of consecutive function of caching. When a computational node searches for a certain fragment of data flows, it can quickly get information where to look for this fragment in the distributed cache to determine what data is available to it. In this case, each computational node has a small array of information cache, after which it will send a user request to another computational node before going to the initial state of data processing. Therefore, one of the main advantages of using distributed cache memory in the development of a software system is the significant space of cache memory, which the user can get by simply adding the computational nodes to the pool of requests (Fig. 5).

The generalized schemes of use of the developed program system with use of cloud technology on independent computing nodes with distribution of loadings (balancing). Load Balancing of the software system is used to optimize the performance of calculations using a distributed high-performance system, both on the user side and on the computer server or platform (Fig. 6) [25].

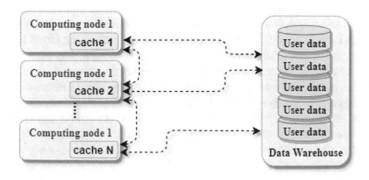

Fig. 5 Distributed cache usage scheme

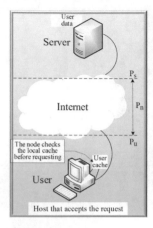 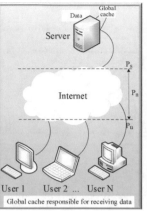 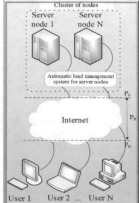

Fig. 6 Distributed computing schemes and use of caches for load balancing, where: P_n—moving the load on the server or to the user ("thin" or "thick client"); P_s—load on the server or cluster; P_u—load on the user's computer

Balancing the load on the computing nodes provides a uniform load of hardware and software systems on independent computing platforms. At occurrence of new technical tasks, the developed program system realizing load balancing should make a decision on where (on which the computing node) it is necessary to carry out the calculations connected with this new information task. Thus, the main task of balancing is to anticipate the process of migration (migration) of a part or all of the calculations from the busiest computational nodes to the less loaded ones.

3 Mathematical Models and Load Balancing Methods in Distributed Software Systems

On independent computing platforms, where the Web-server is located, which has a limited power resource, that is, it can perform only a certain number of operations per unit of time (second). Thus, architects of software systems using cloud technology should remember that in order to get an answer to a request you need to perform several computational operations. So, the Web-server can process for a unit of time only a finite number of user requests, the number of which is determined by the computing power of the Web-server. If for a certain unit of time the number of user requests received by the system for processing exceeds the computing power of the server, a certain number of these requests will remain unfulfilled or lost. In order to reduce the number of unfulfilled or lost user requests, it is necessary to increase the processing power of the Web-server. As a rule, to address the issue of lack of computing power Web-server is used linear increase in the power of the computing node by simply adding another Web-server. Such an architectural approach to solving the issue of lack of power of the computer node has a number of disadvantages: first, the power offered by the industry of computers is hardware limited, and second—the cost of computers is growing much faster than their performance, that is, the total cost of the computer, which has twice as much hardware power will increase significantly more than twice.

However, there is another approach to solving the above problems, which is to use for servicing (processing) user requests several Web-servers. With such an architectural approach to the development of software systems using cloud technology, the cost of management solutions will be directly proportional to the increase in its total power, while the upper limit on the computing power is not determined by the level of hardware technology used, and the total number of involved in the computing process Web-servers.

Traditionally, it has developed that a lot of computers, the association between themselves on a certain technological principle and jointly work out the user's requests, is called "Web-farm", and computers related to the "Web-farm", are called real servers processing information flows of data. According to the paradigm of the organization of the global Internet network, the information request of the user is sent to a certain address of the computer node, which can receive only one computer. Thus, in order for the user information requests sent to the Web server address to be able to process several real computing servers on the node that receives these requests, it is necessary to run a special service called "cloud load balancing service". The cloud load balancing service is designed to perform the following basic tasks:

- receiving and processing requests from the user;
- selection of a real Web-server, which will process this information request from the user;
- redirection of the information request from the user to the real Web-server;
- Reception of the answer of the real Web-server to the user's information request;
- redirection of the real Web-server response to the user's information request.

Thus, all information requests of users that fall into the "Web-farms", are first processed by the cloud Web-service of balancing the hardware and software computational load. Thus, to process a single information request from a user of the cloud-based Web-based load balancing service, a certain processor time is required. Therefore, the total capacity of the Web-server is limited not only by the computing power of the farm, but also by the capacity of a certain computing node, where the cloud load balancing service is launched. Proceeding from scientific researches it is possible to make a certain result how important it is to realize at development of the program system with use of cloud technology on independent computing platforms the most effective service for balancing of computing load.

By the type of hardware and software use we can distinguish two types of "Web-farms", such as symmetric and asymmetric. A symmetrical "Web-farm" can be called "Web-farm" if all its servers are able to process the same type of information requests of users. Asymmetric can be called "Web-farm" if each of its servers is specialized for processing of some one class of users' information requests. For example, if a user uses HTTP-service, one server with asymmetric architecture of "Web-farm" can specialize in processing of users' information requests for text html-pages, and another Web-server can specialize in developing user information requests for multimedia or graphic files.

For symmetric "Web-farms" cloud computing load balancing service should not determine which resource the user asks for, because all real Web-servers have the same set of software and hardware resources. Therefore, cloud computing load balancing service should accept the appropriate connection and, without waiting for the receipt of information request from the user, select any real Web-server for service. Despite the simplicity of the architectural solution of the organization of a symmetric "Web-farm", it is quite difficult to maintain. The administrator of the cloud platform must constantly ensure that all Web-servers have only the same information and software content, such an approach to the solution of cloud computing load balancing is a fairly complex process from a technical point of view. Thus, the proposed hardware-software solution requires additional long-term memory on independent computing platforms, as a copy of each information resource must be present on each Web-server.

Reasons for unbalanced load

The causes of problems in balancing the design load of a distributed system are:

- distributed system structure is different, different logical processes require different computational power [12];
- the structure of a computer computer's complex (e.g. cluster) is also different because different computing nodes have different computing powers;
- the structure of connections between computing nodes is different because all communication lines may have different processing power.

Static and dynamic balancing of the combing units

Static balancing of calculations is performed before starting a distributed computer system [26]. Often in the process of distributing calculations of logical processes among processors the experience of previous developments is used as well as genetic algorithms [27–32].

Previous experience of placing logical processes on all processors (computers)

It's becoming useless nowadays. It can be explained by a structural change in the computer environment where a program for a waveform gets into. Each use of the computing node may become invalid or be loaded with some other computer calculations, the number of which may be unjustified. In any case profit from distribution of logical processes between computational nodes for parallel processing becomes ineffective. Using dynamic balancing involves redistribution of computer computational load on the nodes during program system execution. At the same time, the system software that implements dynamic performance balancing determines:

- node loading on independent computing platforms;
- maximum throughput of computing lines;
- frequency of message exchange between logical computing processes, etc.

Based on the collected statistical data, either a distributed computing system or a computing environment makes an automatic decision to transfer logical processes from one independent node to another.

Setting tasks of dynamic balancing of computational load of a program system

According to the set of computational tasks, which includes processing of calculations, i.e. data transmission, as well as a network of fixed topology nodes, it is necessary to find such a distribution of computational tasks between independent nodes, which will provide almost uniform computational load of all nodes and minimal losses when transmitting significant amounts of data between them.

A distributed computer system on independent computing platforms can be represented as a graph:

$$G_p = \{V, E\}. \tag{1}$$

where

Gp graph of relationships between distributed system tasks on independent computing nodes;

V infinite number of computational nodes (tasks of a distributed system);

E set of arcs of a computational graph of a program system

The task of computer balancing on independent computing nodes is a mathematical description of the non-isomorphic coherent graph, $B: TM \rightarrow NG$, where TM—an infinite possible number of graph models presented, NG—an infinite possible number of graph configurations of the computer network. The graph $G \in NG$, $G = \{C, Ed\}$, is defined by the set of nodes on independent computational platforms C and the set of connective edges Ed, which initially define the connective lines of the computational process of the computer system. The NG computer distribution graph can be visualized as a super graph to which all possible subgraphs of G software system belong.

As a result, many graphs of computational tasks should be presented on many graphs of the distributed computer system correctly.

Methodology of practical solution of computational power balancing problem.

Usually practical and complete solutions of computational load balancing tasks on independent computing platforms include four basic steps [12]:

- Evaluation of independent computing nodes load level;
- Initiating the load balancing process on independent computing nodes;
- Conditions for making a managerial decision on balancing computational power;
- Redistribution of computational loops within the system.

The first stage is an approximate estimation of the loading power of each processor on an independent computing platform. The received information about the volume of loading is used as an information database for balancing process, firstly, for preliminary determination of the reasons of computational imbalance, and secondly, for the process of determination of a new object of power distribution of simulation model by mathematical calculation of the volume of potential work needed to move objects to other independent computing platforms. Thus, the quality of the performed work of balancing the computational load directly depends on the accuracy and completeness of the obtained information in the database [33].

Thus, such an information database may consist of two subtypes of data:

(1) Data on the CPU computing power usage (information about the CPU). These information data include CPU load, CPU idle time, CPU background load, data transfer rate through connection channels, etc.;
(2) Distributed application performance data. It includes: time to complete one task, downtime, data exchange rate, etc.

The communicative model of a computer system contains some important information for making a managerial decision about moving tasks while performing load balancing. If necessary, move the object from overloaded processor A on an independent computing platform to less loaded processor B.

Thus, it is reasonable to choose for moving such a computational object that interacts most intensively with tasks already placed into the container of processor B. In this case it is crucial to count two types of relations between computational objects: two-point and collective communication. During the distribution of computational tasks between processors their connection is estimated.

Two-point communication losses between two computational tasks can be mathematically calculated by the number of transmitted information data (b—number of messages in bytes) and the frequency of communication organization (n—number of messages for a certain period of time). Thus, it is possible to estimate total losses of information links between two computational tasks, using total processor losses for each message and each byte:

$$S = \alpha \times n + \beta \times b, \tag{2}$$

where

S   common ground between the two tasks;
α double-contact connections;
β single message size;
n   the total number of newsletters over time;
b   the total number of all n information messages over a given period of time

Summarizing the processor and object load on an independent computing platform can be done in several ways. One of them is the analytical one which is usually used in the static approach to load balancing and consists in approximate estimation of load of each object on the basis of statistical knowledge about the software system operation.

It may include a function from the size of information data, which integrally reflects the complexity of the algorithm used and the model of mutual relations between computational tasks on independent platforms [6, 11, 34].

The advantage of using the analytical method is a real possibility to estimate accurately the complexity of a computational task with variable productive load. Thus, instead of promptly informing about changes in the computational load only after its appearance, it can give forecasts about what possible future changes in the load are and react to them in advance. The disadvantage of changing this method is the necessity to apply huge efforts made by a software engineer who knows the algorithms used and that they may not be accurate enough if the developed speed estimation model is not.

Another way to collect information about loading is to measure the level of loading of processors and computational tasks directly. Most modern computers include runtime counters (accurate to microseconds), which can be used to measure the time of each computational task. This method can also potentially provide an automatic solution to a computational task to estimate the performance load of a software system.

The advantages of using it are that this method is precise and doesn't require considerable efforts of the developer of program systems on independent computational platforms. It should be noted that this method is not free of such disadvantages as balancing strategies based on the use of this method take into account the previous distribution of the program system's computational load. The results of using this method will not be exact enough if the loading of computational tasks is unpredictable [1].

Both methods of data collection described above can be combined and supplemented with a cognitive method based on measuring the performance of the computer system on an independent platform.

Engagement in the process of balancing the computational load can lead to a significant slowdown of the simulation model performance. The cost of the computational power balancing process itself can exceed the possible economic effect of its implementation. Thus, to balance the computational power performance on an independent platform it is necessary to define the moment of its initialization somehow. For this purpose, it is very important to determine the moment of imbalance of the program system's performance and the level of necessity of balancing, at that comparing possible financial profit and expenses on its implementation [2].

The unbalance of the computing load of a program system on an independent platform can be determined both synchronously and asynchronously.

During the process of synchronous unbalance determination all the nodes interrupt the work at certain moments of system synchronization and identify the unbalance of the computing load of the program system, comparing the load of a single node with the total average computational load.

During the asynchronous computational power balancing method, each node stores its boot history. With this approach, there are no computational load synchronization moments to determine the unbalance level. The mathematical calculation of the unbalance sum is performed in the background, which is executed in parallel.

Most dynamic load distribution strategies can be referred to the class of centralized or fully distributed [3].

During realization of the centralized strategy the special node collects all global information on a condition of all program system of calculation and makes decisions on moving of a computing problem for each separate node.

During the execution of a distributed strategy, computing load balancing algorithms are performed on each processor that exchanges status information with other nodes. In this case, the movement can be only between neighboring nodes.

The process of moving computing objects between nodes to achieve a new stage of load balancing is performed after the decision on balancing is made. In this case it is important to ensure the integrity of the state of the computational object when moving it. Usually additional program functions are used to move information data of the object, especially when there are complex data structures, such as lists of references or pointers.

To achieve this goal, it is necessary to have special software for performing the computational balancing process during simulation modeling. This software includes [24, 26, 35]:

- special software providing estimation of the state of the distributed simulation model and computational environment (analysis subsystem);
- special program which makes managerial decisions about the moment of implementing balancing of computational objects which should be moved from one processor to another;
- software that performs the movement of a computational object from one processor to another;
- visualization subsystem that displays the distribution of simulation components by computational nodes, communication environment, simulation model and changes in the state of the computational environment;
- an information database, which stores information about the computational components of the simulation model and the computational environment.

Distributed simulation model deduction load balancing

Balancing is also necessary during distributed modeling. The first purpose of parallel simulation modeling is quick execution of large and complex computational models [13]. For example, PDES (Parallel Discrete Event Simulation) tries to increase the speed of execution by distributing the computational model on some processors which work in parallel. But distribution of computational model objects has a great influence on the speed of simulation experiment execution because some processors (or computers in the network) are loaded too much and others are not loaded enough or even idle.

The way out of timely redistribution of computational load on less loaded nodes (computers) is the way out of this situation.

Most often, in parallelized imitation of modeling, the components are logical processes, that can function in parallel. Logical processes are distributed between physical processors; the synergy of processes are realized by sending messages from one process to another.

In most cases, the balancing algorithm is developed for an exact class of tasks. At SPEEDES, the balancing algorithm provides that software tools for modeling cover different branches of knowledge, and input changes in the code of user program are minor as well.

The dynamic balancing algorithm uses status characteristics of the system and chooses, what node should the work be moved from and what node it should be moved in during modeling. This approach gives an opportunity to react to the status changes of calculating machine or modeled system and perform balancing, if time, wasted on imitation run is increasing. But dynamic balancing guarantees some additional wastes on collecting statistical data about the status of calculating environment and the model, data analysis and decision-making.

Load moving (or moving an object or a process) is a mechanism that is used for moving the work from one node to another. If the balancing is connected with equal loading of processors, that load moving is connected with the moving a part of work on another node, and balancing is the purpose of moving the load.

Balancing and load moving are used for increasing the productivity of distributed modeling system. Because of heterogeneity of calculating environment, one algorithm may work well in distributed system, and badly in another one.

RCL—computational load distribution strategy

The RCL strategy uses a two-tier management decision making process that combines a centralized and decentralized approach. The two-tier management decision making process provides:

(1) Centralized process-coordinator makes a decision to move the computational load;
(2) Each computing process that sends an information message is responsible for selecting the load object to be moved to another computing resource.

At the first management level, the central coordinator is selected to make decisions among all data processing workstations. In the process of moving the computational load, all workstations stop working for some time, and the central coordinator collects information about the load of all available processors. The coordinator analyzes the obtained information, which includes data on the computational load on nodes and its distribution in the system. Then the coordinator checks the expediency of moving the computations from one node to another on the basis of the received information flow. If the state of the computing system shows that moving the load is necessary (the state corresponds to the criterion of moving the computational load), the central coordinator starts the procedure of redistribution of the load in accordance with the speed of processing nodes. If computing load transfer is not required, the node must send information messages about it to all other computers or nodes on independent platforms. In this case, the second level actions can be ignored, and each node continues to perform its local calculations. If, at layer 2, the computing stations sending information receive a load transfer signal from the central coordinator, they begin the process of selecting a portion of the load to move it to another computing node or platform. The RCL strategy is used in the process of selecting the computational object, and after selecting the parts to be loaded, the migration process begins. The sequence of steps that is used during the transfer of the computational load between two independent nodes includes data packing and sending, receiving and unpacking on the receiving node or platform. At the end of each migration step, the computing node that receives the load sends messages to all other nodes to complete the migration (transfer) procedure and to place new information data. After that each node independently continues processing the information data.

At the beginning of moving the information load all computing nodes stop working (data processing), and each of them receives information about the local load in a certain period of time. The local load information includes:

(1) Download quantity;
(2) Amount of time spent by the processor on processing;
(3) The amount of load that has been processed since its last move.

As soon as the coordinator receives information from all the nodes, he begins to analyze the general information about the load. The coordinator considers such characteristics as:

(1) The spread of the node load factor;
(2) the total amount of load that is pending processing;
(3) Distribution of the load that is expected to be processed.

The node load factor dispersion can show how loaded the nodes are. The higher the node load factor value, the less efficient the CPU resources are used. The total value of load, expected processing is the sum of all unfinished processes on each of the nodes. If this value is not high—there can be no profit from balancing, because the waste imposed when moving the load may exceed the profit from balancing. This is especially easy to see when the load is not large.

Multi-agent approach to balancing task

Currently, Web services have a much wider range of uses. They are used in a wide range of examples. The rapidly growing number of web services and their users need high-performance web servers to implement fast responses to requests that may appear at any second. It is also important to achieve some level of reliability for web servers.

Distributed web servers are used to implement high-performance and reliable web servers. Distributed Web servers are a set of web servers—duplicate resources for simultaneous access to services for many users. Input requests can be distributed among servers according to specific distribution strategies, and therefore these requests can be processed quickly. Distributed Web services can be organized in different ways:

(1) They can be integrated into a cluster of web servers connected by a local computer system to act as a single powerful web server;

(2) They can be used in different geographical locations through a global computing network.

The multi-agent approach is a software component that can automatically move from one node to another with its status and code (execution) and perform various operations on these nodes.

Advantages of using this method [14]:

(1) Agents can share functionality during the design of Web systems. In a traditional message-based load-balancing approach, the server service modules combine the core functionality of a web service with the support of functions such as load balancing. In some cases it is necessary to rewrite the server module when introducing a new load balancing policy. On the other hand, with the help of mobile agents the support functions can be separated from the service modules and performed by mobile agents. This approach is therefore based on mobile agents, which are more flexible when implementing a new load balancing policy for different web server systems.

(2) Mobile agents spend a small amount of traffic. In a data-balancing approach web servers have to periodically exchange messages with load information when making load-balancing decisions. Messaging reduces the bandwidth of the web system. Interaction instead of use eliminates direct messaging, which occurs between servers in both directions. In this way, bandwidth and communication losses can be significantly reduced.

(3) Mobile agents support asynchronous and autonomous operations. Servers can run mobile agents that move independently between servers while simultaneously performing different tasks. A mobile agent can run algorithms that contain different load balancing policies to make load balancing decisions.

(4) Using mobile agents increases the reliability of web systems because they can move client requests from invalid servers to active servers.

Client-Server approaches to load balancing

There are the following categories:

- client;
- based on DNS;
- dispatching;
- server.

The client approach selects a server from the client side. Clients can randomly select one of the Web-servers or choose the most suitable one using intelligent selection mechanisms.

Approach. Based on DNS, it is a solution from the system side. Domain Name Service is a routing mechanism for distributed web servers. It can select one of the web servers to process a request by displaying a unified resource pointer (URL) on the web server's IP address. But the DNS server can be a bottleneck in the routing process. There is some software for load balancing between multiple geographically distributed servers, in which the DNS server determines access to the servers, as well as the time delay in the network to select the optimal server that uses client-server technology [7].

To get full control over the routing of client requests, the dispatching approach completes the image of the address at the IP level. It identifies a group of web servers with a single virtual IP address, which is the dispatcher's IP address. The dispatcher works as a central scheduler and has full control over the routing of requests. The dispatcher identifies each server with its own private address and, if necessary, redirects the client-server packet by rewriting the IP address. It is based on a TCP router that acts as the dispatcher. The main drawback is the loss per address change.

The server approach uses a two-tier routing mechanism. DNS-server defines the server that receives the client's request. After that each server can override the request to another server. This is a decentralized load balancing strategy where each server is allowed to participate in the load balancing process. There is an implementation of this approach based on the Apache web server. Its implementation allows a smooth distribution of HTTP requests from overloaded to under loaded servers. The request is passed to a chain of decision-making functions called candidate functions. Each candidate function organizes a set of available servers and chooses the method of redirecting the request (redirecting HTTP or via HTTP proxy) based on resource information.

Web-System reliability and failover rating

The most important characteristic of a computer system is reliability—the ability to function properly over a long period of time. This property includes three components: the reliability itself, availability and ease of maintenance. The key to improving reliability is to prevent failures and errors by using components with a high level of integration, reducing the level of obstacles and simplifying modes of operation. Reliability is measured by the intensity of failures and their average response time. Reliability of distributed systems largely depends on the reliability of the switching system. Improved availability provides, to some extent, a reduction in the impact of failures and errors on system operation using error correction and monitoring tools, as well as tools for automatic restoration of information circulation after the definition of errors. Improved availability is a way to reduce system downtime.

The readiness criterion is the coefficient of readiness, which is equal to the fraction of the time when the system is able to work, and can be interpreted as the probability of finding the system able to work [19]. The availability factor is calculated as the ratio of the average time of fixing of an accident to the sum of this value itself and the average recovery time. Systems with a high level of readiness are also called fault tolerant. The main way to increase the level of resilience is to exceed, on the basis of which different versions of fault-tolerant architectures are executed [19]. Computing systems include a large number of elements of different types, and excess is necessary for each of the key elements to ensure its resistance to failure.

Switching from the main node to the backup node or vice versa can be done by an administrator in automatic mode or manually. Obviously, the automatic approach increases the availability of the system, as downtime will be significantly reduced in this case compared to the manual approach.

For the execution of automatic procedures of reconfiguration, the system has to have the intellectual communication nodes as well as the centralized management system, that helps the nodes to recognize the crashes and adequately react to them.

The high level of the readiness of a system can be in provided in the case, when the procedures of testing if the system elements are in working order and its transfer on reserve nodes are included in the communication protocols [24].

For example, if it fails, the communication is automatically reconfigured.

There is a wide variety of fault-tolerant systems. And here there are some common definitions [7]:

(1) High availability—describes systems that use redundant hardware and software;
(2) Failure-resistant—a feature of systems that have redundant hardware and software, and in addition, the recovery time in case of failure is not more than one second;
(3) Continuous availability—a feature of systems that also provide repair time within one second, but, unlike fault-tolerant systems, continuous availability has no degradation—the system must maintain a constant level of functionality and performance regardless of failure.

While systems serve a large number of customers at the same time, this fact should be taken into account when calculating the availability factor. The availability factor of the network must be equal to the proportion of time that the system uses to correctly perform its functions for all customers. Obviously, it is very difficult to provide a readiness factor close to one for large systems.

There is a fairly close link between performance and network reliability. This can be explained by the fact that failures and errors of communication channels and communication equipment cause loss or destruction of some parts of packets, and, as a consequence, communication protocols should organize retransmission of lost data [14]. While in LANs, transport or application layer protocols are responsible for recovering lost data that deals with timeouts, so performance loss due to low reliability can be hundreds of percent.

Evaluation of the reliability of the cluster calculation system

Reliability of corporate data processing systems, their ability to provide users with timely and reliable information in real time is one of the most important conditions for effective work. One of the things to achieve a high level of efficiency is the clustering of computing systems.

Creating fault-tolerant clusters makes it possible to use them as nodes for counting machines of highly reliable processors (servers) with double or multiple backup of all the main modules. Only the redundancy can provide the value of availability factor within 0.99999. Typical solutions are fault-tolerant systems (Fault Tolerance, FT) and high availability (HA). Currently, NA clusters are the most popular.

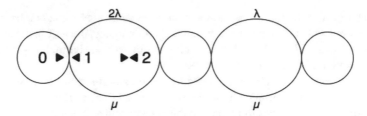

Fig. 7 A two-node cluster state diagram with an ideal control system

A cluster is a combination of two or more devices (modules) that are connected to each other and function as a single data processing node [26, 35]. The range of the proposed cluster solution is quite wide.

The global problem of reliability assessment of failure-resistant systems, including FT-systems and NA-clusters, is the problem of forecasting the reliability values of heavily loaded systems under development (design reliability assessment). In practice, a simplified approach to such assessment is used very often and the result is clearly overestimated reliability values.

According to two main fundamentally different areas of use of the counting systems—parts of information and accounting centers and real-time systems—independently formulate reliability requirements in different ways. In the first case, the reliability requirements are determined by the coefficient of performance reduction of the information-accounting complex due to equipment failures, and in the second case—by the probability of the necessary technological cycle performance for a certain period of time.

Reliability calculation of the duplicated group

Usually the duplicated group on nodes is taken into account in the process of cluster reliability calculation. In this case the failure is considered to be both nodes that cannot work. The model of the duplicated group with the same nodes is depicted on the (Fig. 7).

Duplicate group failures occur when one node fails while the other node is in the recovery process. Possible states:

- "0"—both nodes are fine;
- "1" is a failure of one node;
- "2"—failure of both nodes.

So, "0", "1"—system in working condition, "2"—system crash completed.

In the case when one of the group elements is not working, the node is repaired (restored) without system delay. Later it is included in the duplicated group after restoration in some period of time and distribution by exponential law with the parameter μ:

$$\mu = 1/T_b, \tag{3}$$

where

μ is the time of inclusion in the duplicate group;
T_b is the average repair time

Only one unit can be repaired at a time. For the mode of insufficient load redundancy with limited repair, the formula for calculating the availability factor:

$$K_g = \frac{\lambda\mu + \mu^2}{\lambda^2 + \lambda\mu + \mu^2},$$
(4)

where

K_g is the readiness factor;
μ the time of inclusion in the duplicate group;
λ load intensity

And with an unloaded reserve and unlimited repair:
For a redundant system with a loaded reserve and unlimited repair, the formula is as follows:

$$K_g = \frac{2\lambda\mu + \mu^2}{\lambda^2 + 2\lambda\mu + \mu^2},$$
(5)

where

K_g is the readiness factor;
μ the time of inclusion in the duplicate group;
λ load intensity

With a vanguard and an hour's worth of repairs:

$$K_g = \frac{2\lambda\mu + 2\mu^2}{\lambda^2 + 2\lambda\mu + 2\mu^2},$$
(6)

where

K_g is the readiness factor;
μ the time of inclusion in the duplicate group;
λ load intensity

Each element may contain hidden failures. The model described in Fig. 7 does not take into account the probability of failure detection, reliability of the reserve "switch", delay in switching the reserves, etc. E. In addition, a very significant characteristic of reliability is the apparent fault tolerance of software failures. Understandable resistance to server failures is the behavior when detecting hardware or software failures described below:

(1) the failure does not lead to loss or distortion of data in the server database;
(2) The server continues to run as normal, ignoring the failures;

(3) Clients of the server "do not notice" that the failures have occurred. The only acceptable actions from a normal client perspective are increased service time.

To overcome the consequences of failures, there is a technology based on the mechanism of control points (snapshots). According to this technology, the system must have a stable assembly and guarantee that its condition will not change when failures occur. The corresponding software periodically stores information about the state of processes in the stable node. In the event of a failure, the recorded information is used to repeat the calculation from the moment the information was recorded, in other words, to go back in time. The minimum amount of data that can be backed up in this way is called a checkpoint or snapshot. A stable node can contain one or more clusters. A combination of several types of backup systems is also used here. Unfortunately, the snapshot mechanism cannot be used directly to maintain clear stability due to undetermined behavior [7].

One of the methods to achieve clear failure resistance can be the creation of a hardware platform, which is presented as a multi-node cluster, where all nodes except one are basic and the only one is a backup. Each node has its own file system, but they are used to host shared collective data. Execution is performed on the nodes that are the primary nodes. If one of the master nodes is not in working condition, the backup will follow it.

Photo/recovery methods

This method is based on the mechanism of photo (snapshots). When using the snapshot/recovery method, the main node periodically captures the states, i.e. takes its own photo. A photo (snapshot) of the file system is created at the same time. After the snapshot is created, the nodes continue their work. The main node creates a history of all resources between two consecutive snapshots, which affects the determinism of behavior. In the event of a failure of the main node, state recovery (restoration) is performed on the backup node, so the last fixed node state is restored, the operating environment is moved to the state corresponding to the moment the snapshot was taken, and the file system is started from the moment the last snapshot was taken.

In addition, operations that the primary node performed when the last snapshot was created before the failure and operations that are fixed in the resource register are repeated on the backup node. Creating recovery and state snapshots based on a snapshot of a similar state on the standby node is a very complex task. In particular, it requires virtualization of the operating environment in which the node operates.

Distributed System Operation Problems

Despite all the advantages of distributed systems, compared to traditional centralized systems, they have several significant disadvantages. The main problems of distributed systems compared to the traditional ones are:

(1) Administration problems;
(2) Scalability problems of distributed system boundaries;
(3) Problems with unsupported software.

Problems of Web-system administration

Problems with Web-system administration include load balancing problems on the nodes of the system and data recovery problems in case of a failure. The fragmentation of resources in distributed systems requires flexible administration tools. As with global distributed systems, administration must be done automatically, so there are the following challenges in administering distributed systems:

(1) Balancing the load on the system nodes;
(2) Data recovery in case of failure;
(3) Collecting statistics from the nodes of the system;
(4) Updating the software on the nodes in automatic mode.

It should be noted that this list is generalized, so for each specific distributed system some atypical problems may occur. But practically data problems occur most often and you should pay special attention to them while designing. The last two problems are studied very well and there are a lot of software tools on the software market which provide statistics gathering and also software updating. Of scientific interest are methods of load balancing on system nodes and methods of data recovery **in case of failure**. **According to** the specifics of distributed systems and heterogeneity of architecture and equipment, there is no single design method that can solve all these problems.

Problems of Web computing load balancing

An essential problem in the design of a distributed high-performance system is to ensure effective distribution of computing load on the nodes of software systems. A correctly chosen strategy for balancing the computational load significantly affects the overall efficiency and speed of a distributed software system. Currently, there is a huge number of innovative approaches to solving this problem. In general, the classification of methods for balancing the computational load of nodes can be chosen as follows. By the nature of computational load distribution at nodes, there is dynamic alignment (redistribution) and static alignment. Static balancing of computational load in most cases is performed as a result of a priori analysis. The distributed computing system model is analyzed by resource allocation at the computing nodes to determine and apply the best balancing strategy. The structure of the distributed software system and the configuration of the computational nodes must be taken into account. The main drawback of this method of computational load balancing is the necessity to combine the nodes with another hardware configuration to calculate the complexity of a computer task, which is not always possible.

Dynamic balancing of a distributed system is the adaptation of the computational load on the nodes of the distributed system at work, which, in turn, allows more efficient use of available resources. The necessity of dynamic balancing of the computational load arises when at first it is impossible to determine the total load on a program system. These situations most often occur, for example, in tasks of mathematical modeling, when calculating on each iteration the complexity of forecasting increases and, consequently, the total time of calculation also increases. In addition, dynamic balancing makes it possible to use software that is invariant to the architecture of a distributed system.

The problem of data recovery in Web-systems

During the operation of distributed systems, the most common problem is failure monitoring and further data recovery. This situation can occur, for example, in case of a failure of one of the nodes. Automatic data recovery is a complex task that involves many problems. Data recovery requires finding out the nature of the error, classifying it and automatically recovering all the data. Not only the integrity of the data must be preserved, but also access to other data, because the recovery must be performed without blocking the main resources for read-write, i.e. the distributed system must function without any stops. Currently, there are many approaches to solving this problem. For example, one of the recovery methods in information distributed systems (distributed DBMS) is using a transaction register that stores all the information about all the changes that occurred in the database. In this exact case, the complexity is in a proper classification of mistakes and in fact, if methods of data restoring in automatic regime are used properly.

Scalable systems and the problems of their expanse

The scalability of distributed systems is one of the urgent tasks while designing. Distributed systems allowed to avoid the main drawback of the centralized systems— the limit of increasing the calculating power of the system. There are three main criterions of the system scalability [7]:

(1) *The scalability of a system according to its size.* The system is considered to be scalable according to its size, if it provides the simplicity of connection to its new nodes.
(2) *Geographic scalability.* The system is considered to be geographically scalable, if it's possible to connect new nodes to its network without exact geographic location (country, city, data center, etc.), i.e. globally distributed nodes.
(3) *The management scalability.* The system is considered to be scalable in terms of resource management, if the administrating the system isn't more complicated with increasing the general amount of nodes of the system.

It's not always possible to increase the amount of nodes of a system because of the limit of services, algorithms. Services often are set up on using some exact amount of equipment, for example, for using only one specific server, a specific architecture. So, it's the problem of centralization of resources as well as services. The problem of limited possibilities of a server, that accomplishes the data aggregation, collected from all system nodes in a general, global representation. The problem of limitation of data transfer networks. Whereas, according to the geographic scalability, the nodes of the distributed system can be located in geographically remote parts of the world, there are the problems of reliability of networks of data transfer while designing and operating the distributed systems. The decreasing of the general reliability and productivity is possible at low data transfer speed. The problem of limitation of data processing algorithms. It's necessary to use methods and algorithms of collecting data from system nodes, that overload the communication network marginally.

4 Conclusions

The conducted system analysis discusses the basic principles of designing computer-based high-performance distributed Web-systems using information technologies, which are most often used by architects of software systems in the design. It also became clear that the process of backup is an indispensable attribute in the development of most Web-systems, is highly productive with respect to user requests. The basic criterion at working out of program Web-systems is their scaling which apply when operations demand a considerable quantity of computing resources that considerably reduces productivity of system and demands from it the general increase in capacity. Features of development of program Web-systems with vertical and horizontal scaling are considered.

Also methods of an estimation of reliability and fault tolerance of the developed program Web-systems as any program system give into objective monitoring and can be predicted in the work are investigated. The basic aspects of operation of distributed fault-tolerant software Web-systems of their administration, and also restoration of working capacity at an exit from a working condition which can arise in the course of highly productive operation are considered.

Analyzing the technologies used in the development of Web-systems the methods of which the complex analysis of the computational load balancing has been carried out, which is the main task in the design of the software highly performing distributed fault-tolerant Web-systems with the use of cloud technology. To solve the problem of effective work of the developed software system algebraic methods of optimizing the work of balancing the computational load on nodes, as well as the network model of its distribution are used, thus creating a hybrid mechanism for the transfer of information flows of data.

Theoretical and practical bases providing effective functioning of the mechanism of failure hierarchy which will allow to provide effective calculation of computational loading of program modules and knots as a whole are considered. A thorough analysis of research has shown that the power of industrial computers is limited, and their cost grows much faster than the computational performance. When analyzing the high performance of distributed fault-tolerant Web-system software it was established that its basic criterion is the efficiency of processing data flows, which in general are calculated as a partial division of the sum of all output volumes of parameters by the sum of all input data flows. For each specific computational module or node its own efficiency value is determined. Comparison of the efficiency of computational nodes is carried out using the method of linear programming using different basic models and their variants. It is proposed to determine the number of involved computational modules or nodes by building a criterion of efficiency boundary, and for all other cases the criterion of their inefficiency degree.

Improved functionality of the network model differs significantly from the usual sequential and parallel models. The main difference of which is the possibility of aggregation and joint use of significant sets of heterogeneous computing resources for processing of information flows of data distributed among geographically separated territories. In many cases, this brings significant advantages, because developed software Web-system using cloud technology requires additional resources that are not available within a single computer node, and you can get the necessary resources from other nodes connected to the functional grid. In qualifying work the improved algorithm of transfer of computing load without application of reserve resources is offered the modified model of the organization of interaction of computing modules on processing of information streams of the data within the limits of one knot, and also the developed program Web-system as a whole. The section considers the method of determining the causes of failure of computational modules and nodes, and focuses on two types of major problems associated with the failure of the node from the working state (problems of software functioning and hardware and technical problems).

References

1. Zeeshan M, Mehtab Z, Waqas Khan M (2016) A fast convergence feed-forward automatic gain control algorithm based on RF characterization of software defined radio. In: International conference on advances in electrical, electronic and systems engineering, Putrajaya, Malaysia, pp 100–104. https://doi.org/10.1109/icaees.2016.7888017
2. Abbott ML, Fisher MT (2015) The art of scalability: scalable web architecture, processes, and organizations for the modern enterprise, 2nd edn, Kindle Edition. Addison-Wesley Professional, Boston, 618 p
3. Pasieka N, Sheketa V, Pasieka M, Domska U, Romanyshyn Y, Struk A (2019) Models, methods and algorithms of web system architecture optimization. In: Proceedings of the 2019 IEEE international scientific and practical conference problems of infocommunications science and technology—PIC S&T'2019. 08–11 Oct 2019, Kyiv, Ukraine, pp 147–153
4. Hassan A, Jamalludin Y (2016) Analysis of success factors of technology transfer process of the information and communication technology. In: International conference on advances in electrical, electronic and systems engineering (ICAEES) 2016 Putrajaya, Malaysia, https://doi.org/10.1109/icaees.2016.7888074
5. Leeuwen C, Gier J, Filho J, Papp Z (2014) Model-based architecture optimization for self-adaptive networked signal processing systems. In: Eighth international conference on self-adaptive and self-organizing systems, p 4
6. Pasyeka M, Sheketa V, Pasieka N, Chupakhina S, Dronyuk I (2019) System analysis of caching requests on network computing nodes. In: Proceedings of 3rd international conference on advanced information and communication technologies, AICT'2019, 2–6 July 2019, pp 262–269
7. Linling Q, Qingfeng W (2018) Research on automatic test of WEB system based on Loadrunner. In: 13th International conference on computer science & education, 8–11 Aug 2018, Sri Lanka, p 4. https://doi.org/10.1109/iccse.2018.8468852

8. Bandyra V, Malitchuk A, Pasieka M, Khrabatyn R (2019) Evaluation of quality of backup copy systems data in telecommunication systems. In: Proceedings of the 2019 IEEE international scientific and practical conference problems of infocommunications science and technology—PIC S&T'2019. 08–11 Oct 2019, Kyiv, Ukraine, p 7
9. Sheketa V, Chesanovskyy M, Poteriailo L, Pikh V, Romanyshyn Y, Pasyeka M (2019) Case-based notations for technological problems solving in the knowledge-based environment. In: Proceedings of the IEEE 2019 14th international scientific and technical conference on computer sciences and information technologies (CSIT), vol 1, 17–20 Sept 2019, Lviv, Ukraine, pp 10–15
10. Sheketa V, Vovk R, Romanyshyn Y, Pikh V, Pasyeka M (2019) Formal methods for solving technological problems in the infocommunications routines of intelligent decisions making for drilling control. In: Proceedings of the 2019 IEEE international scientific and practical conference problems of infocommunications science and technology, PIC S&T'2019, 08–11 Oct 2019, Kyiv, Ukraine, pp 29–34
11. Rashkevych Y, Peleshko D, Pasyeka M (2003) Optimization search process in database of learning system. In: IEEE international workshop on intelligent data acquisition and advanced computing systems: technology and application, pp 358–361
12. Brown M (2015) Learning Apache Cassandra—manage fault tolerant and scalable real-time data mat brown. Packt Publishing, Birmingham, 276 p
13. Pasyeka N, Mykhailyshyn H, Pasyeka M (2018) Development algorithmic model for optimization of distributed fault-tolerant web-systems. In: IEEE international scientific-practical conference «Problems of infocommunications. science and technology» (PIC S&T'2018), 9–12 Oct Kharkiv, pp 663–669
14. Riznyk O, Kynash Y, Povshuk O, Kovalyk V (2016) Recovery schemes for distributed computing based on bib-schemes. In: First international conference on data stream mining & processing (DSMP), pp 134–137
15. Kryvinska N (2004) Intelligent network analysis by closed queuing models. Telecommun Syst 27:85–98. https://doi.org/10.1023/B:TELS.0000032945.92937.8f
16. Ageyev DV, Salah MT (2016) Parametric synthesis of overlay networks with self-similar traffic. Telecommun Radio Eng (English translation of Elektrosvyaz and Radiotekhnika) 75(14):1231–1241
17. Ignatenko AA, Ageyev DV (2013) Structural and parametric synthesis of telecommunication systems with the usage of the multi-layer graph model. In: Proceedings of the 2013 23rd international Crimean conference microwave and telecommunication technology (CriMiCo 2013), pp 498–499
18. Ageyev D, Yarkin D, Nameer Q (2014) Traffic aggregation and EPS network planning problem. In: Proceedings of the 2014 first international scientific-practical conference problems of infocommunications science and technology, Kharkov, Ukraine. IEEE, pp 107–108. https://doi.org/10.1109/infocommst.2014.6992316
19. Tilley S (2013) Research directions in web systems evolution V: architecture. In: 15th IEEE international symposium on web systems evolution (WSE) 2013, p 1
20. Kryvinska N (2010) Converged network service architecture: a platform for integrated services delivery and interworking. Electronic business series, vol 2. International Academic Publishers, Peter Lang Publishing Group
21. Kryvinska N (2008) An analytical approach for the modeling of real-time services over IP network. Math Comput Simul 79(4):980–990. https://doi.org/10.1016/j.matcom.2008.02.016
22. Ageyev DV, Kopylev AN (2013) Modelling of multiservice streams at the decision of tasks of parametric synthesis. In: The 2013 23rd international crimean conference microwave and telecommunication technology (CriMiCo 2013). IEEE, pp 505–506
23. Radivilova T et al (2018) Decrypting SSL/TLS traffic for hidden threats detection. In: Proceedings of the 2018 IEEE 9th international conference on dependable systems, services and technologies (DESSERT). IEEE, pp 143–146. https://doi.org/10.1109/dessert.2018.8409116
24. Pasyeka N, Pasyeka M (2016) Construction of multidimensional data warehouse for processing students' knowledge evaluation in universities. In: 13th international scientific and technical conference, 23–26 Feb 2016, Lviv, pp 822–824

25. Romanyshyn Y, Sheketa V, Poteriailo L, Pikh V, Pasieka N, Kalambet Y (2019) Social-communication web technologies in the higher education as means of knowledge transfer. In: Proceedings of the IEEE 2019 14th international scientific and technical conference on computer sciences and information technologies (CSIT), vol 3, 17–20 Sept 2019, Lviv, Ukraine, pp 35–39

26. Gorbachuk M, Lazor A, Pasyeka M, Bandyra V, Yurchak I (2015) Method and parallelization algorithms of synthesis of empirical models taking into account the measurement errors. In: Proceedings of 13th international conference: the experience of designing and application of CAD systems in microelectronics, CADSM 2015, Lviv, pp 319–327

27. Radivilova T, Kirichenko L, Ageiev D, Bulakh V (2020) The methods to improve quality of service by accounting secure parameters. In: Hu Z, Petoukhov S, Dychka I, He M (eds) Advances in computer science for engineering and education II. ICCSEEA 2019. Advances in intelligent systems and computing, vol 938. Springer, Cham

28. Ageyev D et al (2019) Infocommunication networks design with self-similar traffic. In: 2019 IEEE 15th international conference on the experience of designing and application of CAD systems (CADSM). IEEE, pp 24–27. https://doi.org/10.1109/cadsm.2019.8779314

29. Kirichenko L, Radivilova T, Zinkevich I (2018) Comparative analysis of conversion series forecasting in e-commerce tasks. In: Shakhovska N, Stepashko V (eds) Advances in intelligent systems and computing II. CSIT 2017. Advances in intelligent systems and computing, vol 689. Springer, Cham, pp 230–242. https://doi.org/10.1007/978-3-319-70581-1_16

30. Kirichenko L, Radivilova T, Bulakh V (2020) Binary classification of fractal time series by machine learning methods. In: Lytvynenko V, Babichev S, Wójcik W, Vynokurova O, Vyshemyrskaya S, Radetskaya S (eds) Lecture notes in computational intelligence and decision making. ISDMCI 2019. Advances in intelligent systems and computing, vol 1020. Springer, Cham

31. Kirichenko L, Radivilova T, Bulakh V (2018) Machine learning in classification time series with fractal properties. Data 4(1):5. https://doi.org/10.3390/data4010005

32. Kryvinska N, Zinterhof P, van Thanh D (2007) New-emerging service-support model for converged multi-service network and its practical validation. In: First international conference on complex, intelligent and software intensive systems (CISIS'07). IEEE, pp 100–110. https://doi.org/10.1109/cisis.2007.40

33. Rashkevych Y, Peleshko D, Pasyeka M, Stetsyuk A (2002) Design of web-oriented distributed learning systems. Upravlyayushchie Sistemy i Mashiny, Issue 3–4, pp 72–80

34. Kryvinska N, Zinterhof P, van Thanh D (2007) An analytical approach to the efficient real-time events/services handling in converged network environment. In: Enokido T, Barolli L, Takizawa M (eds) Network-based information systems. NBiS 2007. Lecture notes in computer science, vol 4658. Springer, Berlin

35. Junwen L, Ziyan Z, Jiakai H (2017) The application of quantum communication technology used in electric power information & communication system confidential transmission. In: 19th international conference on advanced communication technology, p 5

Processing Signals in the Receiving Channel for the LoRa System

Dmytro Kucherov, Andrei Berezkin, Ludmila Onikienko, and Volodymyr Nakonechnyi

Abstract The chapter devoted studying some characteristics of signal used in receiving devices by LoRa technology. The main feature of this technology relates the ability to build wireless networks for transmitting short messages over long distances under the condition of the long-life battery. Our investigation focused on the impact of interference on the functioning of the receiving set. With this purpose, we studied the basic characteristics of the received radio signal, its frequency properties, the possibilities of spreading the spectrum, encoding and recovering the transmitted information, receiving a packet of individual pulse signals and the quality of packet processing as a whole, and the effect of multipath propagation. We also proposed an approach for studying a receiver in highly populated areas with an estimate of the calculating and experimental data. To check the quality of processing the received information the studies were carried out according to the "point-to-point" and "point-gateway" schemes under the remote server. The test results follow.

Keywords LoRa technology · Signal processing · Interference

D. Kucherov (✉)
Department of Computerized Control Systems, National Aviation University, Kiev, Ukraine
e-mail: d_kucherov@ukr.net

A. Berezkin
Pukhov Institute for Modelling in Energy Engineering, National Academy of Science, Kiev, Ukraine
e-mail: abis1999@ukr.net

L. Onikienko
Department of Communication Systems, Central Research Institute of Weapons and Military Equipment of Ukraine's Armed Forces, Kiev, Ukraine
e-mail: 9225foxtrot@gmail.com

V. Nakonechnyi
Department of Cyber Security and Information Protection, Taras Shevchenko National University, Kiev, Ukraine

© The Editor(s) (if applicable) and The Author(s), under exclusive license to Springer Nature Switzerland AG 2021
T. Radivilova et al. (eds.), *Data-Centric Business and Applications*, Lecture Notes on Data Engineering and Communications Technologies 48,
https://doi.org/10.1007/978-3-030-43070-2_19

1 Introduction

The section discusses the general principles of the construction of receiving devices built based on the technology of radio communication by short messages over long distances, called LoRa. The technology allows you to create a radio network with the transfer of information on both the router and the remote server. An attraction of technology, besides the greater range of transmission of messages (up to 20 km), is the low power consumption of transceivers, which makes it competitive with traditional Wi-Fi and LTE [1].

The intended purpose of the technology is the creation of a reasonable house, the digitization of the village, the city, inventory management (supply), agriculture is a promising application. It is believed that the control with the help of such sensors will allow rational maintenance, control costs in the operation of households, as well as improve the quality of life of residents of the city and village, monitor the operation of vehicles, and also save natural resources. Monitoring pets, collecting information from water, gas, and electricity consumption meters, assessing the pollution level of the air in the rooms and the state of garbage bins in everyday life are current potential applications of the technology.

Endpoint devices are combined into a star network architecture. Messages are double encrypted, which ensures network security. The use of a short information transfer time in a relatively long time interval, where the endpoint device is in the "sleep" mode, achieves its low power consumption. Information transfer can be performed in synchronous and asynchronous modes, then in the latter case, confirmation is necessary. Encrypted information generated at the end node of the network is transmitted through a gateway (hub) to the network server and then goes to the application server.

The LoRa technology is based on linear frequency modulated signals, integrated with error correction. This allows you to realize the main advantages of this technology, as well as to obtain the best sensitivity of the receiving device, high resistance to channel noise and insensitivity to frequency shifts of quartz resonators.

2 Literature Review

In the literature, the LoRa network protocol is presented in a large number of articles. The greatest interest among them is those in which the principal possibilities of receiving and transmitting data are described. This is partially considered in [2], and this paper is an extension of the results presented in it.

However, the presented work is not the only one in this area. So, for example, in [3], are studied the characteristics of LoRa systems depending on three important parameters of the technology: the transmission rate of the encoded message, the coefficient of expansion of the spectrum and the width of the spectrum. The paper shows that the first two factors have a contradictory effect on the accuracy of data transmission and the transmission distance. The presence of correction bits allows you to correct the received message, however, the transmission rate of the message decreases. Similarly, by expanding the spectrum of the signal, the data rate is reduced.

In [4], the estimated capacity limits of the LoRaWAN technology were simulated in the Matlab package for uplink and downlink. The obtained results can be useful in applications related to the transmission of measurement data, namely, they give an estimate of the average number of consumers per gateway with typical placement in urban environments. A probabilistic model of the transmission channel based on the Poisson model was obtained in [5]. The authors of [6] obtained estimates of the number of end devices and some other system indicators.

In papers [7–9], the possibilities of introducing technology into different fields of activity were studied. Thus, the practical application of LoRa technology for remote monitoring of end-users of electric power in an industrial area of 3.48 m^2 was investigated in [7]. Monitoring the temperature of the air on a grape farm in the interests of growing a quality product for wine production is presented in [8]. Control over the movement of public transport can be found in [9]. The study of technology for use in the city, suburb and for sparsely populated areas is given in [10].

The study of the influence of the main indicators of the system on the overall system performance was studied in [11–13]. For example, the authors of [11] studied the effect of the spreading factor on the packet time on air. In [12], a generalized probability density of a radio channel for portable devices was obtained. In [13], some limitations of the system are presented and a brief comparison with similar technologies, such as a system with channel separation by time, navigation devices, smart radio, and other meters.

Since LoRa is a free Standard, the natural desire of researchers is to overcome several limitations, primarily related to the quality of service. A variant of adaptive control of the duration of the access cycle in the environment is proposed in [14]. The adaptive mechanism for controlling the data rate was modeled by the authors [15] in the OMNeT++ environment. The authors of [16] achieve adaptive data rate control by dividing the network coverage area into regions and choosing the parameters of the optimal parameters of the nodes of a particular region, thus avoiding collisions and achieving fair in data transmission. In work [17], an adaptive MAC protocol is presented to ensure the transfer of information from sensors to a satellite information collection system based on an estimate of the communication channel budget.

An important direction of radio network modeling based on LoRa technology, as well as other computer networks, is their modeling in various software environments, for example, OMNeT++ [15], GNU Radio [18], Matlab [4, 11, 19, 20].

An alternative direction, representing the analysis of self-similar traffic in radio networks, is developing in [21–24].

3 Problem Statement

We will consider a system for collecting measurement data messages from end-users and transferring them to the server for further processing and, possibly, receiving responses to requests. Messaging between users and the server is carried out over the air in the free industrial-scientific-medical (ISM) range at frequencies below gigahertz. Each country uses its radio frequency range for this task. Popular ranges in various countries are shown in Table 1 [25].

A message is a packet consisting of a preamble, a header, and payload, the size of which depends on the expansion coefficient (SF) and varies from 51 to 256 bits. The data transfer rate can vary and is in the range of 22 bps–27 kbps also. The final value depends on range, bandwidth and spreading factor [25].

The emitted signal is a pulse signal with linear frequency modulation, which is currently considered a standard in the field of wireless short-range communication systems (IEEE 802.15.4a). The elementary signal obtained, for example, at the output of the dispersion of the ultrasonic delay line, it has the form

$$s(t) = \begin{cases} a(t)\cos(2\pi f t + ct^2) & \text{if } |t| \leq \tau_0/2; \\ 0, & \text{otherwise,} \end{cases} \tag{1}$$

In expression (1), $a(t)$ is the signal amplitude, $a = 0$, if $|t| > \tau_0/2$, where τ_0 is the pulse duration of elementary signal and t is the time; f is the initial frequency of the radiation, c is the parameter that sets the rate of change of frequency with time. A useful signal has a binary form, if it has a logical "1", which corresponds to an increase in frequency, and $c > 0$ in this case, and another case, if a logical "0", that it corresponds to a decrease in frequency, and $c < 0$ the other case. By the type of system organization and signals, it is assumed that it corresponds to the LoRa system. As usually this signal named chirp it means compressed high-intensity radar pulse. The waveform of the chirp signal is shown in Fig. 1.

A distinctive feature of this signal processing is the presence of various types of interference, for example, interference from external or adjacent emitters is possible. The signal at the input of the receiving device is as follows

$$y(t) = s(t) + n(t) + w(t), \tag{2}$$

where $n(t)$ is additive white noise with zero mean, $w(t)$ is interference due to reflection.

The main indicators of system performance are the signal-to-noise ratio (SNR) and the bit error rate (BER). The signal-to-noise ratio is determined by the expression [26]

Table 1 ISM varies for different continents	Countries	Europe	USA	Asia
	Range, MHz	863–868	902–928	779–787

Fig. 1 Waveform chirp
signal with $a = 1$ V, $f =$
7.5 Hz, $c = 15$, $\tau_0 = 2$ s

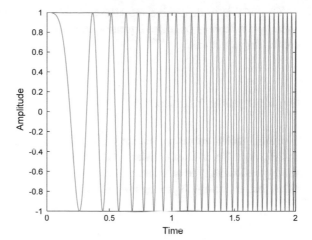

$$SNR = P_{signal}/P_{noise} = (a_{signal}/a_{noise})^2 \tag{3}$$

The type of signal for $SNR = 2$ is shown in Fig. 2.

Reception of a packet of length N bits is determined by the probability of receiving an error in at least one bit

$$p_p - 1 - (1 - p_e)^N, \tag{4}$$

where p_e is the probability of an error in the information bits or the BER, N is the number of bits in the packet. Thinking the value of p_e is small, we can write

$$p_p \approx p_e N. \tag{5}$$

Fig. 2 Chirp signal in noise,
signal ratio is 2

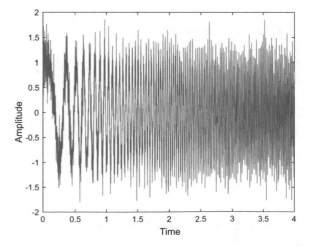

Our goal is to determine the effective tuning of the receiving part of the LoRa system in a densely populated urban area in the presence of natural interference and multipath reflections.

4 Reception Efficiency

The study of the receiving device will begin with an analysis of its susceptibility to interference. To do this, consider the spectrum of the chirp signal, the possibility of its expansion, as well as examine the losses in the transmission of data packets and the effect of multipath propagation of radio waves.

4.1 Spectrum

The chirp spectrum of the signal is completely determined by the phase modulation component and is a Fourier transform of the $s(t)$ signal, i.e.

$$N(f) = \int_{-\infty}^{\infty} s(t)e^{-j2\pi ft}dt. \tag{6}$$

For the signal $s(t)$, representing a rectangular pulse of unit amplitude, that is, if $a(t) = 1$, and the duration of the signal is τ_0, the expression (6) can be written as

$$N(f) = \int_{-\tau_0/2}^{\tau_0/2} e^{jct^2}e^{-j2\pi ft}dt. \tag{7}$$

Using the substitutions

$$x = \sqrt{\frac{2}{\pi}}\left(\sqrt{c}t - \frac{\pi f}{\sqrt{c}}\right), \quad dt = \sqrt{\frac{\pi}{2c}}dx \tag{8}$$

and applying it to Eq. (7), we get

$$N(f) = \sqrt{\frac{\pi}{2c}}e^{\frac{-j\pi^2 f^2}{c}}\int_{-x_1}^{x_2} e^{jx^2\pi/2}dx = \sqrt{\frac{\pi}{2c}}e^{\frac{-j\pi^2 f^2}{c}}\left(\int_{0}^{x_2} e^{\frac{jx^2\pi}{2}}dz - \int_{0}^{-x_1} e^{\frac{jx^2\pi}{2}}dx\right). \tag{9}$$

Now in Eq. (9), we introduce the notation

$$x_1 = -\sqrt{\frac{2c}{\pi}} \left(\frac{\tau_0}{2} + \frac{\pi f}{c} \right), \quad x_2 = \sqrt{\frac{2c}{\pi}} \left(\frac{\tau_0}{2} - \frac{\pi f}{c} \right). \tag{10}$$

Using in (9) for the Euler equations, the notation used in the Fresnel integrals

$$C(x) = \int\limits_0^x \cos\left(\frac{\pi}{2}t^2\right) dt, \quad S(x) = \int\limits_0^x \sin\left(\frac{\pi}{2}t^2\right) dt, \tag{11}$$

the expression (9) can be written in the form

$$N(f) = \sqrt{\frac{\pi}{2c}} e^{-\frac{j\pi^2 f^2}{c}} \{[C(z_2) + C(z_1)] + j[S(z_2) + S(z_1)]\}. \tag{12}$$

In (12), $C(-z) = -C(z)$, $S(-z) = -S(z)$ is taken into account. In the calculations for $z \gg 1$, you can use the following relations

$$C(z) \approx \frac{1}{2} + \frac{1}{\pi z} \sin\left(\frac{\pi}{2}z^2\right), \quad S(z) \approx \frac{1}{2} + \frac{1}{\pi z} \cos\left(\frac{\pi}{2}z^2\right). \tag{13}$$

Using (13), it is possible to obtain a graphical representation of the spectrum of the chirp signal of the form (1) with parameters $\tau_0 = 10$ μs and a bandwidth of 200 MHz, which is shown in Fig. 3.

In the receiving channel, the signal is compressed, resulting in a short signal, the duration of which is inverse to the width of the spectrum W_{ss}, i.e. $\tau = 1/W_{ss}$.

Fig. 3 A typical waveform of spectrum for chirp signal

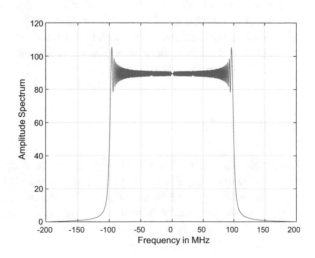

4.2 Chirp Spread Spectrum

As is known [13], the system efficiency in the conditions of interference with the use of spread spectrum signals is determined by the gain during processing, determined by the ratio

$$G_p = \frac{W_{ss}}{W_{min}}, \tag{14}$$

where W_{ss} is the bandwidth of the broadband signal spectrum, determined by the modulation rate of the chirp signal of duration τ_0, W_{min} is the bandwidth, depending on the bit rate. The higher the G_p coefficient, the harder it is to interfere.

Let the bandwidth of the signal spectrum W_{ss} for LoRa system is known and equal to a certain value, and the data transfer rate following [14] is determined by the spreading factor S_f, the coding rate R_c and the band of signal spectrum W_{ss}, under the expression

$$W_{min} = S_f \frac{R_c}{(2^{S_f}/W_{ss})}. \tag{15}$$

Then the processing gain for the LoRa system is

$$G_p = \frac{2^{S_f}}{S_f R_c}. \tag{16}$$

Taking into consideration the values of $S_f = 7 \ldots 12$, $R_c = 1 \ldots 4$ [12], the processing gain is in interval 3 ... 585. The larger values of this coefficient should be taken for large transmission distances. One should note that the width of the signal spectrum does not affect the quality of communication, but in this case, it is difficult to interfere with such a signal.

4.3 A Chirplet

The processing of the received chirp signal consists of optimal filtering and decoding the received message. The message is encoded by a set of some phase components. Thus, in addition to the optimal filtering, the receiver needs to find a match for the received phase component from the existing set. Since the transmitted signal is an element of some exponential signal, consider a Gaussian function, called a chirplet (by analogy with the wavelet), given by [27, 28]

$$g_c(t) = \frac{1}{\sqrt{\sqrt{\pi}\sigma}} \exp\left(-\frac{(t - t_c)^2}{2\sigma}\right) \exp(j2\pi f_c(t - t_c) + c(t - t_c)^2), \tag{17}$$

Fig. 4 A typical waveform of a chirplet

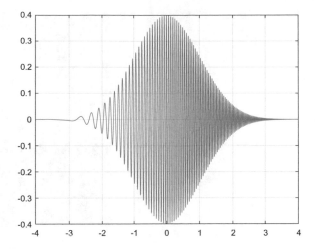

where t_c, f_c are the parameters of time and frequency of the function; σ is the variance depending on the duration of the chirp signal; c is the modulation rate. The shape of the chirplet is shown in Fig. 4.

For parameters $t_c = 0$, $\sigma = 1$, $c = 0$, $f_c = 0$, the function $g_c(t)$ takes the form

$$g_c(t) = \frac{1}{\sqrt[4]{\pi}} \exp\left(-\frac{t^2}{2}\right). \tag{18}$$

Expression (18) is called the base transformation function. Any modification (18) can be used to identify the parameters of the Gaussian function when it is represented by harmonic oscillation with frequency modulation of a given type. Since the Gaussian chirplet is a fundamental function, therefore, the received signal can be represented as a weighted sum of Gaussian chirplets.

The sequence of chirps represents a message packet consisting of characters, each of which includes a 2^{S_f} bit. Further processing consists of conducting a frequency-time analysis of the received bit of information. There are several approaches to performing this analysis. One of them is to determine the time convolution, where the signal is multiplied by shifted in time and frequency [7]. Unfortunately, this approach does not give a positive result, since the time-frequency transformation is interdependent, and the chirplet decomposition cannot be determined for arbitrary signal since the multiplied functions are not orthogonal [9]. A definite measurement is not possible. The type of signal in the time-frequency plane is shown in Fig. 5.

The situation deteriorates significantly when the signal is received in interference. Figure 6 shows a representation of the same signal measured in interference.

To calculate all four parameters of a received signal, it is necessary to perform a chirplet decomposition based on the calculation of the angular distribution of the signal energy in the time-frequency plane. Such a decomposition can be made based on the Wigner distribution or by the fractional Fourier transform.

Fig. 5 Time-frequency type for signal $a = 1$ V, $f = 15$ Hz, $c = 15$, $\tau_0 = 20$ s

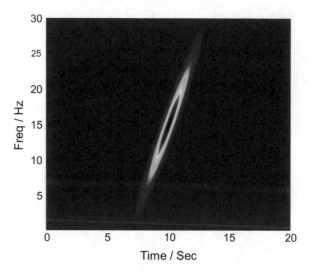

Fig. 6 Time-frequency representation for chirp signal in the noise, SNR $= 2$

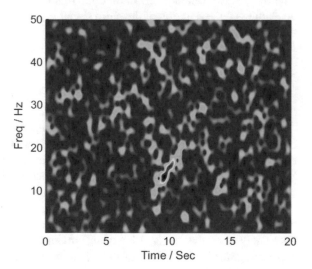

4.4 Operations with Chirplet

Let to the signal of the form (18), $s(t) \in R^+$, apply the Wigner distribution [29]

$$W(t, \omega) = \frac{1}{\sqrt{2\pi}} \int\limits_{-\infty}^{\infty} s\left(t - \frac{\tau}{2}\right) s^*\left(t + \frac{\tau}{2}\right) e^{-j\omega\tau} d\tau. \tag{19}$$

In Formula (19), the asterisk at the signal $s^*(t)$ means the complex conjugate signal $s(t)$, ω is the angular frequency equal to $2\pi f$.

Then scaling, shifting and rotating operations can be applied to (19). When scaling the signal (18) stretches in time (compressed in frequency). For the scale factor $l > 1$ (19) takes the form

$$W_l(t', \omega') = W\left(\frac{t}{l}, l\omega\right). \tag{20}$$

The delay of the signal at time τ leads to a time shift, then (19) is written in the form

$$W_\tau(t', \omega) = W(t - \tau, \omega). \tag{21}$$

and the shift to the f_c frequency is written like this

$$W_{f_c}(t, \omega') = e^{j2\pi f_c t} W(t, \omega). \tag{22}$$

Rotate the scaled function (18) by angle α has the form

$$W_g(t, w) = W_g\left(\frac{t\cos + w\sin}{s}, s(-t\cos + w\sin)\right). \tag{23}$$

It should be noted that the rotation for the Wigner distribution is equivalent to the fractional Fourier transform of the function $f(t)$ for the angle α

$$\Phi_\alpha[f(t)] = \int_{-\infty}^{\infty} f(t)K_\alpha(t, \omega)dt, \tag{24}$$

where

$$K_\alpha(t, \omega) = \begin{cases} \sqrt{\frac{1-j\cot\alpha}{2\pi}} \exp\left[j\left(\frac{t^2+\omega^2}{2}\cot\alpha - \omega t \csc\alpha\right)\right], & \text{if } 0 \leq \alpha \leq \pi, \\ \delta(t - \omega) \text{ if } \alpha = 2k\pi, k = 1, 2, \ldots, n, \\ \delta(t + \omega) \text{ if } \alpha + \pi = 2k\pi, k = 1, 2, \ldots, n. \end{cases} \tag{25}$$

In (25), $\delta(\cdot)$ is the Dirac function, $\csc(\cdot)$ is the reciprocal of sine, that is, $\csc(\cdot) = \sin^{-1}(\cdot)$.

However, there is a different approach to Chirplet decomposition for representing a complex signal in a compact form.

4.5 Chirplet Decomposition

The chirplet decomposition [30–32] determines the signal parameters $g(\gamma) = g[(l, \alpha, t, \omega)]$. After converting the signal into digital form, we obtain $g(\gamma_n) = g[(l_n, \alpha_n, t_n, \omega_n)]$, $n \leq N$. And γ_n is the set of possible values of the sampling parameters from the dictionary D, that is, $\gamma_n \in D$. Any function $s(t)$ can be represented by a set of atoms $g(\gamma)$. The algorithm that allows you to search for a suitable combination of data from the dictionary should provide the maximum search function

$$s(t) = \sum_{n=1}^{N} s(\gamma_n) g_{c_n}. \tag{26}$$

The parameters $s(t)$, which determine the maximum of the convolution (26), will determine the maximum approximation to the original signal. In this case, the l_n parameter defines the expansion of the time domain, and its inverse value is the signal compression in the frequency domain, the angle α_n of rotation of the ellipse corresponds to linear modulation of the center frequency of the signal, and the variables t_n, ω_n time and frequency of the central part of the signal. There is a need to develop a search algorithm $s(\gamma_n)$.

The proposed algorithm consists in the selection of unknown parameters of signals from a given dictionary D. The essence of the algorithm is to find the best coincidence of the projections of multidimensional data with a certain interval of a possible atom in the dictionary D. In this case, the desired signal $s(t)$ is approximated by a weighted sum of a finite set of functions (atoms) from the dictionary D, i.e.

$$s(t) \approx s_N(t) = \sum_{n=1}^{N} a_n g_{\gamma_n}, \tag{27}$$

where a_n is the weight coefficient for the atom $g_{\gamma_n} \in D$, and N is the number of atoms. Atoms from the dictionary are chosen to minimize the approximation error. Thus, the search is an iterative process that ends when the error e_N gets less then specified value ε, i.e.

$$e_N = |s(t) - s_N| \leq \varepsilon. \tag{28}$$

In (28) ε is a fairly small number.

The discrete-set search algorithm must be iterative to ensure the best structure of the desired signal, which can be achieved by calculating the scalar product of functions. The required parameters must satisfy the condition

$$|sg_0| = \sup_{\tilde{\gamma}} |sg_\gamma|. \tag{29}$$

Then the procedure for calculating the best signal is to calculate the remainder term, which at the beginning of the algorithm is equal to the signal itself

$$R_0 = s(t), \tag{30}$$

and after completing the following steps

$$R_1 = R_0 - |sg_0|g_0, \tag{31}$$

$$R_2 = R_1 - |sg_1|g_1 = R_0 - |sg_0|g_0 - |sg_1|g_1, \tag{32}$$

$$\ldots$$

$$R_i = R_{i-1} - |sg_{i-1}|g_{i-1} = R_0 - \sum_{k=1}^{i-1} |sg_k|g_k, \tag{33}$$

The expression (33) with the initial value (30) represents a recurrent procedure for finding the best representation of the signal $s(t)$. The stopping criterion is the ratio

$$\rho = \frac{\left\| \sum_{k=1}^{n} |sg_k|g_k \right\|}{\|R_n\|}. \tag{34}$$

The higher the value of this ratio, the better decomposition parameters are chosen. The proposed algorithm allows us to formulate the following statement.

Theorem *For the chirp signal $s(t)$ with the parameters $(l_n, \sigma_n, t_n, \omega_n)$ there is such a function*

$$g_c(t) = \frac{1}{\sqrt{\sqrt{\pi}\sigma}} \exp\left(-\frac{(t - t_c)^2}{2\sigma}\right) \exp(j2\pi f_c(t - t_c) + c(t - t_c)^2),$$

providing the best s_N approximation

$$e_N = |s(t) - s_N| \leq \varepsilon,$$

where ε is a sufficiently small number, by the algorithm

$$R_i = R_{i-1} - |sg_{i-1}|g_{i-1}$$

under condition

$$|sg_0| = \sup_{\tilde{\gamma}} |sg_\gamma|.$$

4.6 Packet Error Rate

As follows from expression (5) to reduce errors in the transmission of N information packets, the value of the bit error p_e should be reduced as well. To estimate the BER, we use the results of [14, 31], where the receiving signal is represented as an additive model of the useful signal of Gaussian white noise in the transmission channel. Calculate BER for wideband signals with binary phase shift keying (BPSK) modulation

$$p_e = 0.5 erfc\left(\sqrt{\frac{E_b}{N_0}}\right), \tag{35}$$

with 2FSK modulation

$$p_e = 0.5 erfc\left(\sqrt{\frac{E_b}{2N_0}}\right), \tag{36}$$

and BOK modulation

$$p_e = 0.5 erfc\left(\sqrt{\frac{E_b}{4N_0}}\right). \tag{37}$$

In Eqs. (35)–(37), E_b is the energy of a bit, N_0 is the spectral density of white noise, and

$$erfc(x) = \frac{2}{\sqrt{\pi}} \int_x^\infty e^{-t^2} dt. \tag{38}$$

Considering that the E_b/N_0 ratio is equivalent to the signal-to-noise ratio, the expressions confirm the physically obvious connection, when an increase in the E_b/N_0 ratio leads to a decrease in errors in the transmission of a data packet and, accordingly, increases with an increase in channel noise (Fig. 7). The calculation of the curves in Fig. 7 is made for BPSK modulation.

As can be seen from Fig. 7, an increase in the ratio E_b/N_0 in 5 times leads to a decrease in the probability of a bit error by a factor of 1000. A similar situation occurs in chirp modulation.

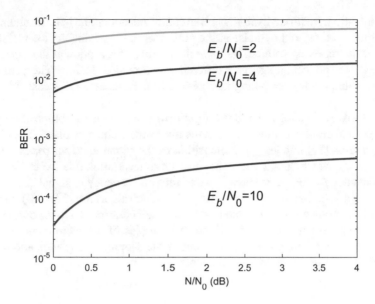

Fig. 7 BER for noise and different value E_b/N_0 ratio

4.7   Multipath Propagation

We shall consider the phenomenon is due to the reflection of antenna rays from surrounding objects, such as stationary structures, terrain elevations, trees, etc., which are perceived as well as interfering signals. Multipath propagation is also manifested in the effect of signal fading or it leads to a shift in the frequency of the main signal, which we shall see as an equivalent to the introduction of the Doppler additive into the signal (1).

These phenomena lead to a deterioration in the reception quality of the transmitted information due to the shift of the transmitted signal relative to the filter bandwidth also. A similar situation in the radar is solved by introducing the Doppler additive into the local oscillator frequency of the frequency converter or by a deviation of the frequency of the transmitted signal, which eliminates the frequency deviation of the received signal relative to the center passband of the receiving filter. Thus, the hardware implementation of the receiving part can be complicated to improve the quality of processing such signals.

4.8   Interference Processing

A natural way to increase the signal-to-noise ratio at the receiver output is to perform optimal filtering. However, multipath reflections are a serious problem in cellular communications, radio navigation, satellite communications and in LoRa technology. Exceeding the threshold by noise at the output of the receiving device violates the action of the receiving path. To improve the quality of reception at the input of the matched filter, protection against interference is provided. Unfortunately, the well-known technique for overcoming narrow-band interference, in this case, is of little use.

An effective technology for processing chirp signals is chirp decomposition, which can detect and eliminate interference in the joint time-frequency plane. The Chirplet decomposition [15] implies a decomposition of the signal into four parameters $\beta = (t_s, f_s, \sigma_T, c)$, which denote the signal energy concentration relative to the time $t_s \in R$, frequency $f_s \in R$, pulse spread σ_T and rate of frequency in signal c.

The basis for eliminating interference is the search algorithm (29)–(33). Since the reflections $w(t)$ are weakly correlated samples, their contribution to the output signal of the protection system decreases, and the samples of the main signal correlate strongly and, therefore, go on to the output of the suppression system, and then the main filtering of the receiving signal occurs.

5 Experimental Results

A study of the LoRa system was conducted by staging a series of experiments in a densely populated area of the city, where cellular communications and Wi-Fi devices interfered with the receiving channel.

The signal was transmitted according to the scheme "point-to-point", "point-gateway" with the subsequent processing on the cloud server. In the experiment, the LoRa system transceiver was used on the basis of the SX1276 chip to transmit a short message at a frequency of 868 MHz. The experiment used the message "hello" (Fig. 8); the transmitted packet was 60 bytes. The output power of the emitted signal did not exceed 20 mW [33].

Fig. 8 Board in an examination

5.1 Link Budget

The budget of the communication channel is considered, drawn up taking into account the components, including factors of increase and loss in the propagation channel of the target receiver. The budget of a network wireless link can be calculated as follows:

$$P_{RX}(\text{dBm}) = P_{TX}(\text{dBm}) + G_{system}(\text{dB}) - L_{system}(\text{dB}) - L_{channel}(\text{dB}) - M(\text{dB}), \tag{39}$$

where: P_{RX} is the expected power at the receiver input, P_{TX} is the transmitter power, G_{system} is the directional antenna gain, L_{system} is the loss in the system, for example, in the antenna power cable, the antenna itself, etc., $L_{channel}$ is the loss associated with the propagation channel, M is the attenuation level calculated or based on empirical data.

The estimated radio channel distance was determined for a LoRa-based transceiver utilizing the SX1276 chip, which operates at a frequency of 868 MHz, uses a bandwidth of 20.8 kHz, an expansion factor $S_f = 8$, a bit rate 1562 bps and sensitivity -128 dBm. For transmitter power of 1 mW, the ideal transmitting and receiving dipoles with a gain of 2.1 dB, the channel budget is 130.1 dB.

The maximum radio distance of direct visibility for the heights of the receiving and transmitting antennas in experimental conditions with normal refraction is approximately 25 km.

The calculation of the power of the receiving signal as a function of the distance r in conditions without interference was made using the Vvedensky formula and shown in Fig. 9.

The distance corresponding to the real sensitivity under the calculation conditions is 10,800 m.

5.2 Measurement

The receiver was installed on the 11th floor of a panel reinforced concrete building located in the city center of Kiev. Several Wi-Fi networks are deployed in the building, and mobile antennas of several operators are installed on the roof. The quality of the message received in the experiment was controlled by a mobile phone and an SDR receiver. During the experiment, the transmitter successively moved from the 11th floor to the first and further to the basement, which is lower than the first to two floors, in order to simulate the signal loss due to the multiple reflections. The scheme of the experiments is shown in Fig. 10.

In Fig. 10 we apply the abbreviation MS that means the designation of the antenna of the base stations of mobile operators, Tr is the transmitting device, Rc is the

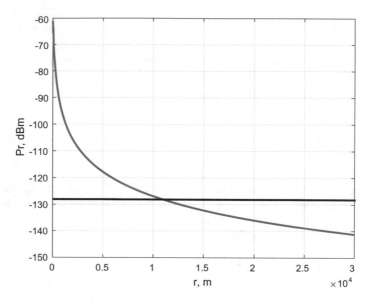

Fig. 9 Estimation of the power of the receiving signal depending on the distance

receiving device. An initial analysis of the interference situation is presented in Table 2.

Mobile transmitting antennas and Wi-Fi routers located next to the building interfere with reception. The signals of these devices create an interfering background, which is taken to be "white" noise $n(t)$. Cross noise $w(t)$ is created by multiple reflections from reinforced concrete structures inside the building. On each floor of the building, the signal and noise levels were recorded, and the quality of the message was monitored. For additional documentation of the measurement results, a panoramic receiver was used, which was considered to be the reference meter.

The results of measuring the number of error bits and the signal-to-noise ratio are shown in Figs. 11 and 12.

During the measurements, it was found that the message was received on all floors of the building and in the basement within 2 m from the stairwell, increasing the distance to 8–10 m from the stairs leading to a complete loss of communication.

The transmitter operates in the point-gateway mode is functionally similar to the point-to-point mode. The only difference was in the processing of data that were conducted on a remote server, so the results were obtained with a slight delay (several tens of seconds) in relation to the point-to-point mode.

Fig. 10 Scheme of the experiment

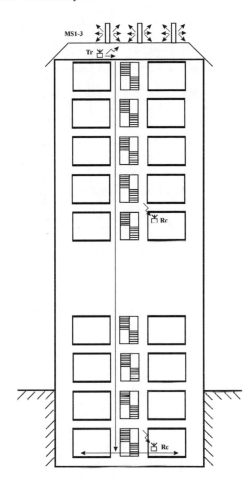

Table 2 Interference situation analysis

Sources of interference	F, MHz	P_{RX}, W
CDMA	800–900	2–60
LTE	900–2400	2–60
Wi-Fi	2400	0.1
Other	400–5000	<0.01

6 Conclusion

The paper is devoted to the consideration of one of the most poorly studied problems of the LoRa technology associated with the reception of messages in conditions of intense interference. Despite the advantages obtained by spreading spectrum techniques, message coding, and digital signal processing, the solution to this problem remains relevant in densely populated areas, the unconventional use of transceivers,

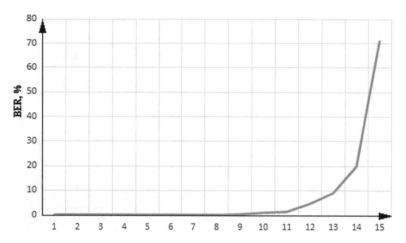

Fig. 11 BER in dependence on the number of the floor

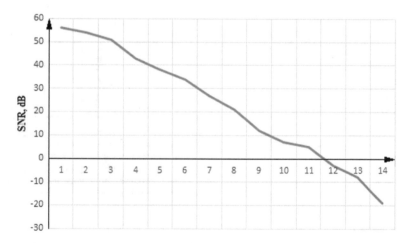

Fig. 12 SNR in dependence on the number of the floor

for example, in enclosed spaces such as a mine, a bunker, etc. Studies conducted in a multi-storey building in the center of Kyiv showed a reduction in transmission distance under conditions of signal attenuation due to multipath more than 10 times, which confirms the need for further research aimed at improving the quality of reception in such conditions. To protect against this kind of interference, in the processing of chirp signals, it is proposed decomposition useful signal into four components with a matching algorithm presented in the paper. A new method for estimating device parameters with limited conditions is proposed. Our further research plans to focus on studying the efficient processing of the chirp signal under interference conditions.

Acknowledgements Authors thank both the authorities of National Aviation University, National Academy of Science of Ukraine, Pukhov Institute for Modeling in Energy Engineering, and Central Research Institute of Weapons and Military Equipment of Ukraine's Armed Forces for their support during the preparation of this paper.

References

1. About LoRa alliance™. https://www.lora-alliance.org/about-lora-alliance. Accessed 19 Jan 2019
2. Kucherov D, Berezkin A, Onikienko L (2018) Detection of signals from a LoRa system under interference conditions. Paper presented at international scientific-practical conference problems of infocommunications, science and technology, Kharkov National University of Radio Electronics, Ukraine, 9–12 Oct 2018. https://doi.org/10.1109/infocommst.2018.8632135
3. Noreen U, Bounceur A, Clavier L (2017) A study of LoRa low power and wide area network technology. Paper presented at the international conference on advanced technologies for signal and image processing (ATSIP), Fez, Morocco, 22–24 May 2017, pp 1–7. https://doi.org/10.1109/atsip.2017.8075570
4. Varsier N, Schwoerer J (2017) Capacity limits of LoRaWAN technology for smart metering applications. Paper presented at the IEEE international conference on communications (ICC), Paris, France, 21–25 May 2017, pp 1–6. https://doi.org/10.1109/icc.2017.7996383
5. Bankov D, Khorov E, Lyakhov A (2017) Mathematical model of LoRaWAN channel access with capture effect. Paper presented at the IEEE 28th annual international symposium on personal, indoor, and mobile radio communications (PIMRC), Montreal, QC, Canada, 8–13 Oct 2017, pp 1–5. https://doi.org/10.1109/pimrc.2017.8292748
6. Mikhaylov K, Petäjäjärvi J, Hänninen T (2016) Analysis of the capacity and scalability of the LoRa wide area network technology. In: 22th European wireless conference, Oulu, Finland, 18–20 May 2016
7. Gören H, Alataş M, Görgün O (2018) Radio frequency planning and verification for remote energy monitoring: a LoRaWAN case study. Paper presented at the 26th signal processing and communications applications conference (SIU), Izmir, Turkey, 2–5 May, 2018, pp 1–4. https://doi.org/10.1109/siu.2018.8404166
8. Davcev D, Mitreski K, Trajkovic S et al (2018) IoT agriculture system based on LoRaWAN. Paper presented at the 14th IEEE International workshop on factory communication systems (WFCS), 13–15 June 2018, pp 1–4. https://doi.org/10.1109/wfcs.2018.8402368
9. James JG, Nair S (2017) Efficient, real-time tracking of public transport, using LoRaWAN and RF transceivers. Paper presented at the TENCON 2017–2017 IEEE Region 10 Conference, Penang, Malaysia, 5–8 Nov 2017, pp 2258–2261. https://doi.org/10.1109/tencon.2017.8228237
10. Sanchez-Iborra R, Sanchez-Gomez J, Ballesta-Viñas J et al (2018) Performance evaluation of LoRa considering scenario conditions. Sensors 18(772):1–19. https://doi.org/10.3390/s18030772
11. Lavric A, Popa V (2017) A LoRaWAN: long range wide area networks study. Paper presented at the international conference on electromechanical and power systems (SIELMEN), Iasi, Romania, 11–13 Oct 2017, pp 417–420. https://doi.org/10.1109/sielmen.2017.8123360
12. Catherwood PA, McComb S, Little M et al (2017) Channel characterisation for wearable LoRaWAN monitors. Paper Presented at the Loughborough antennas & propagation Conference (LAPC 2017), Loughborough, UK, 13–14 Nov 2017, pp 1–4. https://doi.org/10.1049/cp.2017.0273
13. Adelantado F, Vilajosana X, Tuset-Peiro P et al (2017) Understanding the limits of LoRaWAN. IEEE Commun Mag 55(9):34–40. https://doi.org/10.1109/MCOM.2017.1600613

14. Deng T, Zhu J, Nie Z (2017) An improved LoRaWAN protocol based on adaptive duty cycle. Paper presented at the IEEE 3rd information technology and mechatronics engineering conference (ITOEC), Chongqing, China, 3–5 Oct 2017, pp 1122–1125. https://doi.org/10.1109/itoec.2017.8122529
15. Slabicki M, Premsankar G, Francesco MD (2018) Adaptive configuration of LoRa networks for dense IoT deployments. Paper presented at the IEEE/IFIP network operations and management symposium (NOMS), Taipei, Taiwan, 23–27 April 2018. https://doi.org/10.1109/noms.2018.8406255
16. Abdelfadeel KQ, Cionca V, Pesch D (2018) Fair adaptive data rate allocation and power control in LoRaWAN. Paper presented at the IEEE 19th international symposium on "a world of wireless, mobile and multimedia networks" (WoWMoM), 12–15 June 2018. https://doi.org/10.1109/wowmom.2018.8449737
17. Deng T, Zhu J, Nie Z (2017) An adaptive MAC protocol for SDCS system based on LoRa technology. Adv Eng Res 118. http://scholar.google.com.ua/scholar_url?url=https%3A%2F%2Fdownload.atlantis-press.com. Accessed 19 Jan 2019
18. Knight M, Seeber B (2016) Decoding LoRa: realizing a modern LPWAN with SDR. Paper presented at the 6th GNU radio conference, Boulder, CO, 6 Sept 2016
19. Kucherov D (2017) Control of computer network overload. Paper presented at the CEUR workshop proceedings, 17th international scientific and practical conference on information technologies and security (ITS-2017), Kyiv, Ukraine, vol 2067, pp 69–75
20. Kucherov D, Berezkin A (2017) Identification approach to determining of radio signal frequency. Paper presented at the international conference on antenna theory and techniques, Kyiv, Ukraine, 24–27 May 2017, pp 1–4. https://doi.org/10.1109/icatt.2017.7972668
21. Kryvinska N (2010) Converged network service architecture: a platform for integrated services delivery and interworking. Electron business series, vol 2. International Academic Publishers, Peter Lang Publishing Group
22. Kryvinska N (2008) An analytical approach for the modeling of real-time services over IP network. Math Comput Simul 79(4):980–990. https://doi.org/10.1016/j.matcom.2008.02.016
23. Kryvinska N (2004) Intelligent network analysis by closed queuing models. Telecommun Syst 27:85–98. https://doi.org/10.1023/B:TELS.0000032945.92937.8f
24. Kryvinska N, Zinterhof P, van Thanh D (2007) An analytical approach to the efficient real-time events/services handling in converged network environment. In: Enokido T, Barolli L, Takizawa M (eds) Network-based information systems. NBiS 2007. Lecture notes in computer science, vol 4658. Springer, Berlin
25. AN1200.22. (2015) LoRa™ modulation basics. Semtech Corporation, Wireless Sensing and Timing Products Division. Rev. 2, May 2015, pp 1–26. https://www.semtech.com/uploads/documents/an1200.22.pdf. Accessed 19 Jan 2019
26. Sklar B (2002) Digital communications fundamentals and applications, 2nd edn. Prentice Hall PTR, USA
27. Mann S, Haykin S (1995) The chirplet transform: physical consideration. IEEE Trans Signal Proces 43(11):2745–2761. https://doi.org/10.1109/78.482123
28. Bultan A (1999) A four-parameter atomic decomposition of chirplets. IEEE Trans Signal Process 47(3):731–745. https://doi.org/10.1109/78.747779
29. Cui J, Wong W, Mann S (2005) Time-frequency analysis of visual evoked potentials using chirplet transform. Electron Lett 41(4):217–218. https://doi.org/10.1049/el:20056712
30. Xu C, Wang C, Gao J (2016) Instantaneous frequency identification using adaptive linear chirplet transform and matching pursuit. Shock Vib. https://www.hindawi.com/journals/sv/2016/1762010/
31. Aoi M, Lepage K, Lim Y et al (2015) An approach to time-frequency analysis with ridges of the continuous chirplet transform. IEEE Trans Signal Process 63(3):699–710. https://doi.org/10.1109/TSP.2014.2365756

32. Wang X, Fei M, Li X (2008) Performance of chirp spread spectrum in wireless communication systems. Paper presented at the 11th IEEE Singapore international conference on communication systems (SICCS), Guangzhou, China, 19–21 Nov 2008, pp 466–469. https://doi.org/10.1109/iccs.2008.4737227
33. SX1276/77/78—137–1050 MHz ultra low power long range transceiver. http://www.alldatasheet.com/datasheet-pdf/pdf/501037/SEMTECH/SX1276.html. Accessed 19 Jan 2019
34. Lavric A, Petrariu A (2018) LoRaWAN communication protocol: the new era of IoT. Paper presented at the 14th international conference on development and application systems, Suceava, Romania, 24–26 May 2018, pp 74–77. https://doi.org/10.1109/daas.2018.8396074

Information Technologies for Analysis and Modeling of Computer Network's Development

Nataliia Ivanushchak, Nataliia Kunanets, and Volodymyr Pasichnyk

Abstract One of the key problems to provide the secure management of complex computer networks is testing, which requires a functional depiction of such systems by a corresponding mathematical model. That's why it is needful to implement a statistical study of network ensembles, simulation their architecture and increase motions. The purpose of the work is to analyze the properties of computer networks of different Internet providers, develop new, improve and adapt existing methods and tools of mathematical simulation, which enable the study of their structure and parameters based on fragmentary observation data, modeling and forecasting processes for their development and structuring in Within the framework of the formalism of complex networks. Here a systematic analysis of methods and means of mathematical modeling of computer networks for prediction of their growth and clustering processes was carried out, the requirements for them based on the review of existing mathematical models were developed, on this basis a list of actual and unexamined tasks was developed for the purpose of further improvement and development of new methods scientifically grounded decisions. The developed method of modeling was used for analysis, evaluation and development of processes of stability of computer networks for directed hacker attacks and distribution of computer viruses in them.

Keywords Local area network · Mathematical modeling · Security mathematical model · Complex network theory

N. Ivanushchak (✉)
Department of Computer Systems and Networks, Yuriy Fedkovych Chernivtsi National University, Chernivtsi, Ukraine
e-mail: inm160286@gmail.com

N. Kunanets · V. Pasichnyk
Department of Information Systems and Networks, Lviv Polytechnic National University, Lviv, Ukraine
e-mail: nek.lviv@gmail.com

V. Pasichnyk
e-mail: icm.ikni@gmail.com

© The Editor(s) (if applicable) and The Author(s), under exclusive license to Springer Nature Switzerland AG 2021
T. Radivilova et al. (eds.), *Data-Centric Business and Applications*, Lecture Notes on Data Engineering and Communications Technologies 48,
https://doi.org/10.1007/978-3-030-43070-2_20

1 Introduction

One of the arrangements for assuring the security of complex computer networks is trial, which contains the amplification of a corresponding mathematical simulation. For a technical description of those models, it is required to implement a probabilistic observation of network ensembles, simulation their structure and evolution operations, anticipations the dynamics of their farther configuration, observing the stability to targeted attacks and dissemination malicious software. This process consists in repeated testing of the constructed probabilistic model and further statistical processing of the simulations results in order to determine the desired characteristics of the analyzed process in the form of estimates of its parameters. The accuracy of the estimates of these parameters determines the degree of approximation of the problem solution to the probabilistic characteristics. The investigation of formation and dynamics of complex networks' increasing permits us to compose circumstances for their efficient growing and defense.

2 Problem Setting

The evolution of methods and processing of probabilistic modeling of computer networks' development and configuration encourages the delight of explorers in research the complex networks theory, as established in [1–6], which suggests decisions to computationally complicated tasks that are standard for contemporary systems. The characteristic of many real biological, social, and computer networks notably vary from those with alike probable links between vertices, such as in the classical random graphs, that's why they are established on the joined constructions and power-law distributions, in accordance with [7–12]. The complex networks theory as an expanse of discrete mathematics investigates the properties of networks, inclusive of not only their structure, but also statistical phenomenon, allocation of vertices and connections' weights, stream effects, conductance etc. Unfortunately, the complex theory does not have a clear mathematical definition and can be characterized by the features of those systems and types of speakers that serve as the subject of its study [13–27]. The following basic properties of complex systems are distinguished:

- instability (complex systems tend to have many possible modes of behavior between which they wander as a result of small changes in parameters that control dynamics);
- indivisibility (complex systems act as a whole and can not be studied by dividing them into parts considered in isolation, that is, the behavior of the system is determined by the interaction of the components, but the reduction of the system to its components distorts most aspects inherent in systemic individuality);
- adaptability (complex systems often include a variety of decision makers and act based on partial information about the system as a whole and its environment, moreover, these agents can change their rules of behavior on the basis of such

partial information; in other words, complex systems have the ability draw hidden laws from incomplete information, learn from these laws and change their behavior on the basis of new incoming information);

- unpredictability (complex systems produce unexpected behavior that can not be predicted on the basis of knowledge of the properties of their constituents, if viewed in isolation).

In terms of the theory of complex systems, a more general heuristic value is inherent in the analysis of complex network high-tech and intellectually important systems. Such systems have a system-forming component, that is, their structure and dynamics actively influence those processes that are controlled by them. Their properties differ significantly from the properties of classical random graphs with evenly probable links among nodes, which until recently were considered as their basic mathematical model prototype, and therefore the construction of their models was proposed to implement [6, 7] using coherent structures and power law distributions.

The last decade of the development of the knowledge economy has led to a change in the paradigm of structural, functional and strategic positioning of modern enterprises. Vertically integrated corporations are replaced everywhere by distributed network structures (so-called business networks). Many of them instead of direct production now engaged in system integration.

Therefore, studying the topology and dynamics of network-based systems will optimize business processes and produce circumstance for their efficient evolution and defense.

In the complex networks theory three main directions are distinguished:

- study the statistical properties that characterize the behavior of networks;
- creation of network models;
- prediction of the behavior of networks when changing their structural properties.

The goal of the work is the task's decision that is substantial for performing new ways of mathematical simulation, i.e. the investigation and observation of approaches permitting the identification of the construction and characteristics of complex networks' models based on watching data; progressing of the means and methods of probabilistic simulation of the operations of the local area networks structures formalism.

3   Computer Network's Probabilistic Model

The local area network is represented as a graph G, which is determined as a set (V, E) of limited set of nodes V, $\dim(V) = N$, and a set of links E, forming of unordered pairs (u, v) where $u, v \in V$ and $u \neq v$. Each node is defined by its degree, i.e. the amount of incident links. A degree sequence is formed from an ordered list of degrees of nodes.

The integrant characteristic of a local area network is the degree distribution law p_k, which determines that degree k is implemented to the randomly selected vertex. The degree consistency for an undirected graph is easy to represent as $d = \left(k_1^{n_1}, k_2^{n_2}, \ldots, k_s^{n_s}\right)$, where numbers k_i are the degrees of vertices, and the index n_i defines the number of reiteration of k_i in the sequence. Discrete degree distribution of vertices p_k is associated with a degree sequence d in a form

$$p_k \stackrel{def}{=} P[x = k_i] = n_i/N \tag{1}$$

As it was shown in the random graphs model [3], the degree distribution of vertices p_k will be at large values of N binomial or Poisson's, because a link connected to two random nodes is absent or present with equal probability. But in real life the structure of most natural, biological and social networks is different from the topologies of random graphs that simulates the character of degree distribution of nodes, as recollected in [28–30]. Specifically, the experiential degree distribution of nodes is explained in terms of the power-law distribution $p_k = k^{-\gamma}$ in many real networks. It is defined by a determining parameter γ, which designates the rate of descent of the distribution "tail".

The development of technologies and methods of mathematical modeling of structuring and growth processes of real networks stimulates the interest of researchers of studying of complex networks theory, which approaches to solving computationally complicated problems that are standard for contemporary systems.

Complex network is a system, which consists of a large amount of components, admits "far-reaching" connections between components, has multiple scale (including spatial and temporal) variability. Such properties generate a number of specific issues of implementation of computer models of complex systems. A large number of components displayed on the cost-effectiveness of computing processes and increased requirements for RAM to store data structures. The multiple scale complexity of networks requires the use of parametric coupled models for their description, which describe the hierarchy of adjacent ranges of variability.

Complex networks are used for modeling objects and systems for which other methods of investigation (through observation and active experiment) are impractical or impossible.

Despite the fact that the extensive activities of the scientific community are deployed on the research of many real systems, which is inherent in the network structure (Internet, WWW, public transport networks, social networks), according to [7–12], the detailed analysis of local area networks from the standpoint of the theory of complex systems, the research of perspective directions of the use of methods and means of their study, predicting the formation and development at the moment is not yet implemented.

So, we investigated the degree distribution of vertices in local area networks in Chernivtsi in different time intervals.

The aim was to develop the basic mathematical model using the integral probabilistic characteristics of real computer networks of different Internet Support

Providers (ISP) and the apparatus of graph theory for modeling the structure of computer networks with the given volume and properties, which are in a steady state, and the justification of ways of identification of the parameters and structure of network simulations that found on supervisory data.

Understanding the structure of the local area network stems from the study of its evolution over time, topology and real location. Local networks are created for operation in a small geographic area. They allow multiple access to high-speed environments and management through local administration.

Analysis of a computer network development is based on the establishment of *influence factors* on the generation of nodes and conditions for the creation and joining of new vertices with their structure to the network.

Among the influence factors of the network growth in first, it is necessary to specify the size or length of the local network, which is determined by the number of computers connected to the network and the extent between the most distant stations, at which the collisions are clearly recognized from the normal functioning of the nodes. As the network grows, the number of collisions increases, and its useful bandwidth and signal transfer rate drops sharply, so you may need to use very expensive or rare equipment. Limiting the network length is a prerequisite for choosing a network structure, dividing it into separate parts, the appearance of the additional servers with a new network of connections. Multi-server connections allow increased stability, less delay, reduced access cost and improving the quality of communication for end users as a whole. There is a dynamics of the network, a kind of clustering, servers act as the centers of the formed clusters, spatial placement of the network component into a clear hierarchical structure.

The accession of new nodes is dictated by economic advantage, i.e. a resource costs, which depends on the geographic location of the users. Therefore, when designing the network equally important factor of its growth is consideration of efficiency relative price/quality.

Thus, various factors influence the formation of new relationships on the network, each of which can be assigned by its weight, and the nodes of the network—a certain amount of energy, which depends on the method of merging nodes in graphs.

The initial data of experiments were the properties of real networks "FoxNet", "Kabelne Telebachennia Mista" ("KTM") and "DSS-group" of Internet Support Providers (ISP) in Chernivtsi.

To determine the probabilistic characteristics of these networks, the real topological maps and the configuration schemes were investigated. They were derived from topological databases of routers using the Open Shortest Path First routing protocol (OSPF). Each router builds a graph of network connections, in which the vertices of the graph are routers and IP addresses and ribs are interfaces of routers. All routers share information with their neighbors about the topology of the network. These messages are called router links advertisement. Each router considers itself the center of the network and searches for the best route to each address known by him. Each route found in this way is remembered only one step—up to the next router, according to the principle of one-step routing. Data about this step fall into the routing table. The problem of finding the optimal path on the graph is quite complex. In the OSPF protocol, Deuter's iterative algorithm is used to solve it.

The data from topological databases were used to carry out the simulation operation of computer networks. These features were the amount of vertices, the amount of connections and the amount of degrees for every node that were established by means of a topology map. The degree of a vertex k sets the amount of links connected to a specific node, and n_k is the amount of nodes in a graph with a given k. The degree distribution of nodes has been calculated by these data.

Table 1 illustrates the characteristics of the computer networks "Fox Net" and "DSS-group". The table is incomplete, because the studies had been conducted at different time intervals; therefore we provide only partial information.

In Fig. 1 is shown the approximation of degree distribution of nodes in the explored networks and the characteristics of power-law distribution for them (e.g., $p_k = k^{-2.4}$ for the network "Fox Net", $p_k = k^{-2.6}$ for the network "KTM" and $p_k = k^{-2.1}$ for the network "DSS-group").

The analysis leads to the following conclusion: if the network is sufficiently structured, then in the course of its development the number of vertices n_k with a given degree k varies, such as the total amount of network users N, but the probability of joining these vertices p_k remains virtually unchanged, thus providing a power-law degree distribution of vertices $p_k = k^{-\gamma}$ with unchanged index of degree γ. The

Table 1 The characteristics of computer networks "Fox Net" and "DSS-group"

Network	"Fox Net"		"DSS-group"	
k	$N = 915$		$N = 2023$	
	n_k	p_k	n_k	p_k
1	614	0.671	1242	0.613
2	59	0.064	224	0.110
3	79	0.086	220	0.110
4	51	0.056	79	0.040
5	22	0.024	68	0.033
6	33	0.036	35	0.017
7	9	0.010	46	0.022
8	8	0.009	26	0.012
9	10	0.011	13	0.006
10	6	0.006	21	0.010
11	8	0.009	10	0.005
12	7	0.008	14	0.007
13	3	0.003	8	0.004
14	3	0.003	9	0.004
15	3	0.003	3	0.001

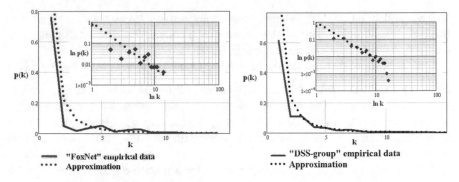

Fig. 1 The degree distribution of nodes approximation in explored networks

value γ of may be different. The larger—in less branched systems with a relatively small number of servers, switches and routers and with a large number of users with $k = 1$. The smaller—in more structured networks in the structure of which there are sufficient numbers of vertices with large degrees k, such as the network "DSS-group".

4 Computer Network's Evolution by Means of Probabilistic Model Algorithm

To realize a local area network simulation, which characteristics are analogous to topological maps researched in real, we submit the subsequent network structure. Supposing the network consists of N vertices (nodes); s—the number of classes of nodes; $i \in \{1, 2, 3, \ldots, s\}$ indicates a specific class of node; n_i—the number of nodes of the i-th class; k_i—the degree of the node i.

The procedure of the given scale-free network realization is based on the following algorithm, because the initial degree distribution of vertices p_k is assigned:

- let's permit a degree sequence d, by choosing s varieties of n_i in keeping with the given distribution p_k, where $i = \overline{1, s}$;
- for each node of the graph i, let's appoint k_i "billets" for upcoming links;
- the pairs of "billets" are randomly removed from the degree sequence. They are bounded by a link if the new link will not lead to the derivation of multi-edges or loops (edge cycles). From the degree sequence removed the appropriated indexes if the link is created;
- the previous step is repeated until the degree sequence becomes empty;
- a graph is randomly formed by allocation the nodes with the highest degrees in the centre of a graph, and setting the nodes with lower degrees radially from the center to the periphery in decreasing order of their degrees;
- the edges between vertices are filled gradually starting from the nodes with the highest amount of links.

The probabilistic model evolution algorithm ensures the association of all vertices in a single architecture of a stochastic graph considering the fact of obligatory connection of all users to a real computer network.

On the basis of distribution p_k, an arbitrary graph can be constructed by $\prod_i k_i!$ different ways, as "blanks" for future edges are indistinguishable. Thus, the process with equal probability generates an arbitrary possible configuration of the network with a given degree distribution of vertices p_k. The advantage of this algorithm is its versatility, it can be possible to build a network with random distribution of degrees of vertices.

Here is an example of constructing a graph for a formed degree sequence in the form

$$d = \left(3^2, 2^3, 1^4\right) \tag{2}$$

On the first step we numbered all vertices from a given sequence. In the center of the graph are placed vertices with the highest degree $k = 3$, around which are sequentially arranged vertices with lower degrees. The next step is edge generation E, i.e., the process of random connection of pairs of blanks of vertices. In this case, the formation of new ribs is monitored, so that the multi-edges do not appear in the graph. This is done because when generating the edge, the indices corresponding to it are removed from the degree sequence and the rules of prohibition are formed, i.e., for each vertex, the numbers of all vertices adjacent to it are stored in order not to allow the reconnection of the nodes between which the edge is already generated.

On the Fig. 2a each vertex of the graph i assigns k_i "blanks" for future edges, on the Fig. 2b is showing graph generated by one of the random implementations. Numbers in square brackets show the prohibition of connecting to the node of the corresponding number after the formation of the edge. A graph that corresponds to a step sequence (2) can be built in $1! \cdot 2! \cdot 3! = 12$ different ways.

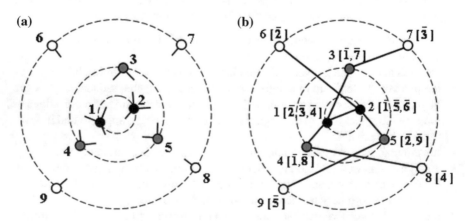

Fig. 2 Construction of the graph for a given degree sequence (2)

5 Implementation of the Algorithm

The outcome of the implementation of the sponsored algorithm is the virtual local area network, introduced as a stochastic graph with a definite amount of nodes and given probability distribution of their edges.

The sampling was carried out by the empirical distribution of degrees of vertices $p_k = k^{-\gamma}$. The process of network modeling with the subsequent possibility of comparing the results of simulation with the characteristics of the investigated networks (Table 1) was carried out. After the approximation of the degree distribution of vertices, which is illustrated in Fig. 1, we have been defining γ for various local computer networks, such as $\gamma = 2.4$ for the network "FoxNet", $\gamma = 2.6$ for the network "KTM" and $\gamma = 2.1$ for the network "DSS-group".

The performance of the simulation algorithm is demonstrated by means of creation a graph using the parameters of a real computer network "Fox Net" in Chernivtsi.

Figure 3 exemplifies a visualization sample of stochastic graph that detects the qualities of observed local area network.

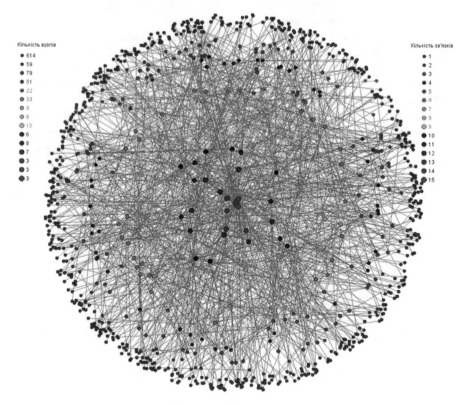

Fig. 3 The network structure "FoxNet" with distribution $p_k = k^{-\gamma}$: $\gamma = 2.4$, $N = 915$ presented by the stochastic graph

In accordance to the experiential degree distribution of nodes $p_k = k^{-\gamma}$, a probation was perpetrated, based on a network simulation with the following opportunity to juxtapose modelling results with the properties of the observed system and evaluated the accordance of the depiction of the real framework of simulation.

Vertices with different degrees of attachment k are depicted in different colors, their number in the generated network is placed on the left panel, and the number of links corresponding to each vertex is displayed in the panel on the right.

Studies have shown that for small values of the distribution parameter γ the network is merged into one larger connected cluster, rather than in the case of larger values of γ.

One of the most important tasks that need to be addressed when analyzing the development of local computer networks is receiving forecasts and assessments. This prediction is usually necessary to perform under uncertainty, in addition, you need to know and be able to model not only the structure of connections at the given time, but also the dynamics of the network with a specific distribution of vertices for a sufficiently large period of time. The algorithm proposed in the work allows to predict the development of the network, and the obtained values can be used not only to determine the number of potential users, but also for a certain approximate calculation of passive equipment.

As an example, by tracking the dynamics of the establishment of the "Fox Net" network over the past years, see Table 1, having carried out the simulation process according to the probability distribution defined by the integral law for it $p_k = k^{-2.4}$ and the annual increase of vertices of this network $\Delta N \approx 110$, we can calculate the predicted number of users, servers and switches in it (Table 2).

The results obtained by studying the model can be transferred to the real structure if the model describes it adequately. The question of the adequacy, accuracy and reliability of the simulated system and the real complex networks was studied by comparing and evaluating their numerical characteristics (number of nodes with different degrees k, probability of network integration). As the data distribution, the variance or mean square of the deviation was chosen (σ^2), which characterizes the deviation of random values from the mean value in this sample. The question of the accordance of the simulation is established by juxtaposing the estimates of the averaged approximated and real distributions of the nodes p_k in accordance with

$$\sigma^2 = \frac{1}{n} \sum_{k=1}^{n} (p_k - \bar{p}_k)^2 \tag{3}$$

Distribution of data for the studied networks were 7.5–3.7% variability of variation series.

Table 2 Estimated number of connections in network "Fox Net"

$N = 1245$															
k	1	2	3	4	5	6	7	8	9	10	11	12	13	14	15
n_k	751	271	89	45	26	17	12	8	6	5	4	3	3	3	2

6 Definition the Optimum Strategy of Local Area Networks' Defense

The main purpose of the above abstract model is to make predictions and assessments of different categories of offenders and cybercriminals who implement the most common network security threats, for example, random and target attacks on vertices. This forecast is not as resource intensive as the automated security analysis systems [18–20].

A random R-attack is characterizes by random choice of the attacking vertex (hardware failure, collapse, the effects of natural disasters). Successive demolition of the vertices with maximum connectivity is a definition of targeted I-attacks (DDOS-, DNS-attacks, swindling, diversion). The vertices and edges of local area networks of various structures and with different vulnerabilities in real life are subject to targeted and random attacks in different schemes, so, there is a more complex situation, as established in [31].

A general simulation of the network topology evolution has been developed under the circumstances of destabilization threats. *The threat model* and *the security model* are the main fragments of the expositive development of the LAN, in accordance with [32].

The threat model's input data is the designed stochastic graph, which considers the architecture of the real network. "Natural" liquidation of nodes and connections, i.e. random R-attacks, and liquidation nodes with the highest degrees, i.e. I-attacks are the main two scenarios of attacks over a stochastic graph. The part of the vertices including in the largest network cluster is the relative size of the maximum cluster S:

$$S = 1 - \sum_{k=1}^{k_{max}} \frac{k \cdot n_{del}(k)}{N} \qquad (4)$$

where $n_{del}(k)$ is an amount of deleted nodes of degree k, N is a total amount of nodes in the graph.

The evolution of this simulation allows model oneself on attacks in the form of complex threats (I-targeted and R-random) to defenseless vertices of the graph in straight (R-threats + I-threats) and converse (I-threats + R-threats) consistencies.

Figure 4 displays the outcomes of variation of the relative size of maximum cluster of scale-free local area networks "FoxNet", "KTM" and "DSS-group" in Chernivtsi, which requiring on the amount of attacked vertices. For any of the researched networks the attack scheme I-R is more efficient than R-I.

The security model believes that risk is the main security measure:

$$R = \rho_{th} \cdot \rho_v \cdot Pl \qquad (5)$$

where the risk is R, ρ_v and ρ_{th} are the appropriate probabilities of vulnerability and threat, Pl is the price of damages.

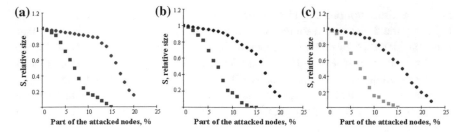

Fig. 4 The dependency of the maximum cluster relative size in the "FoxNet" (**a**), "KTM" (**b**) and "DSS-group" (**c**) networks on the amount of attacked vertices exposed to attacks in variety consistencies: I + R (black square) and R + I (black circle)

The simulation ensures an opportunity to carry out more complicated attacks on the network by describing sources of threats close to the real. The new approach allows for each node to enter a vulnerability (or security) parameter, in contradistinction to many known methods, for which the removing of a vertex or an edge is the ultimate stage of an attack. The system components obtain the value of their vulnerability, on the whole, with $\rho_v(i)$ different from 1, established by the amount of financial deposits F_i to the safety of the i-th element:

$$\rho_v(i) = f(F_i) \tag{6}$$

The cost of damages is the level of decrease of the graph implementation in case of the deleting of a vertex and all neighboring connections, established by choosing a parameter from a set of metrics M—those performance indexes, which determine the main goal properties of graph topologies. For instance, as plural elements M may use the value of the shortest paths, betweenness, clustering coefficient, maximum cluster size etc.

As a simulation model, the investigation of responses of the appropriated modeling stochastic graphs to the targeted threats of vertices was carried out. It is representing the topology of real computer networks "FoxNet", "KTM" and "DSS-group". It was found that the probability of a effective attack on a vertex varies only with the change of $\rho_v(i)$, which declines exponentially with an increment in the "thickening" of the barrier of defense d_i, which, alternatively, is established by the set of traditional safety steps: technical, physical, organizational, mathematical and legal. The variable d_i is solved by the number of spending, i.e. $d_i \sim F_i$, and in this case:

$$\rho_v(i) \sim \exp(-\mu F_i) \tag{7}$$

where μ is the multiplier, which determines the effectiveness of financial spending.

In fact, different network vertices and connections obtain financing for their safety in a diversity of modes. The highest levels of defense usually have the most momentous graph agents, while the defense of the leaves of simulated network (end-users) may be entirely not present. In accordance to the defense strategy, which is established by the division of resources between vertices, the network stability to threats was considered in the situation of lack of funding and three probable strategies for the localization of investments, which calculate on the degree of the vertex (for the same total volume): (a) no investments $F0_i = 0$; (b) low defense of all vertices in the network $F1_i = 1/\mu$; (c) average defense of vertices with high and medium degrees $F2_i = k_i/(C_1 \cdot \mu)$; (d) maximum level of defense for vertices with high degrees $F3_i = k_i^2/(C_2 \cdot \mu)$;

$$\sum_{i=1}^{1000} F1_i = \sum_{i=1}^{1000} F2_i = \sum_{i=1}^{1000} F3_i \tag{8}$$

Multipliers C_1 and C_2 play the role of normalizing constants and are determined in keeping with the formula $C_i = \frac{F_i}{N_k}$, where N_k is the amount of vertices with high and average degrees for C_1 or the amount of vertices only with high degrees for C_2.

To demonstrate the classic strategy of target attacks on the vertices of maximum connectivity a 1000-node test network was used. Figure 5 is shown these four defense strategies and the computation results of the transform of the relative size of the maximum cluster for them.

It is obvious that the strategy of over-defense of maximum connectivity vertices (i.e. $F_i \sim k_i^2$) is not optimal, because this scenario conducts to lack of funds for protection of the nodes of a lesser degree. In determining the optimal network security strategy, in addition to choosing the topology and distributing resources to defend its elements, the most efficient influence is the exact estimation of the strategies and actions of attacking side, in accordance with [33].

Figure 5 illustrates that the strategy of defending of the vertices with high and medium degrees is the optimal strategy for local area networks defending. Hence,

Fig. 5 Various defense strategies of investment distribution: $F0_i = 0$, $F1_i = 1/\mu$, $F2_i = 4.1k_i/\mu$, $F3_i = 28k_i^2/\mu$ and the dependency of the relative size of the maximum cluster in the test network on the amount of attacked vertices for them

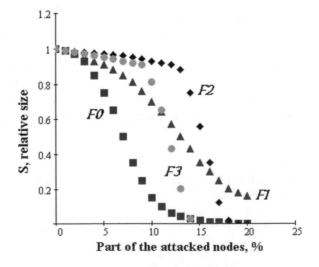

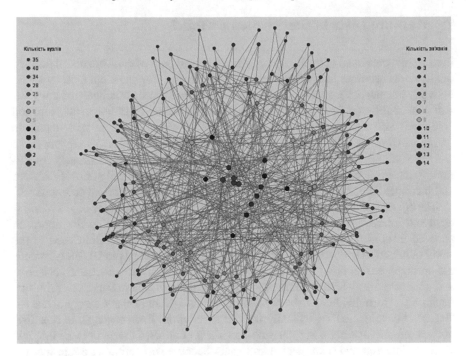

Fig. 6 Internet provider "FoxNet" backbone network

to explore the solidity of local area networks while defending from vulnerabilities, the localization algorithm of "backbone network" is implemented which forms in a multi-stage cutting of the original graph's leaves (Fig. 6).

Forming the backbone of an Internet support provider "FoxNet" consisted of the subsequent steps: during the first stage 614 leaves have been removed from the obtained graph, during the second stage—85 leaves, during the third stage—16, and during the fourth stage—3 leaves. The resulting backbone network consists of 197 vertices; each of them has a degree not less than 2. Backbone networks of variety Internet support providers were scale-free; the dependence of the degree distribution of their vertices is approximated by the power function with high precision.

The new samples of local area networks proceeding in the field of threats are researched and demonstrated, which makes suitable efficient arising to unleashing a wide range of stability problems of practically momentous network architectures.

The simulation algorithms proposed in this paper should become a means for developing approaches to the diagnosis of computer virus distribution processes in computer networks and studying the vulnerability and stability of the latter to targeted attacks.

7 Malware Distribution Dynamic Model

For Internet providers of local area networks, the research of malware distribution in this systems becomes particularly momentous (Trojan programs, computer viruses, network worms, etc.), according to [32, 34], because the insufficiency of antivirus software on the users' computers can lead to stream infection of users (the so-called epidemic), and consequent to network services' crashes. The simulated stochastic graphs can provide information about the structure and distribution of malware in the real network, after which we can propose an optimal immunization strategy (evolution of anti-virus projects), which is capable to restrain the distribution of malware and increment the reliability of the influence of network architectural components.

The epidemiological research of malware distribution has been the object of uninterrupted interest in scientific investigation [32], mainly following the approach adopted from biological epidemiology [10]. The standard simulation used in the study of infection with computer viruses is the epidemiological model SIS (susceptible–infected–susceptible). Each vertex in the network is an individual, and each edge is a connection in which the infection cans distribute to other users. The primitive depiction of the individuals in the population, which exist only in two discrete states— "healthy" or "infected" is typically for this simulation. Each susceptible (healthy) node is infected at a rate δ at each time step if it is connected to one or more infected vertices. Infected nodes are healed and again become susceptible at a rate ν at the same time determining the efficiency of the rate of infection $\lambda = \delta/\nu$. In the absence of uncertainty, we can install $\nu = 1$. In models with local connections (Euclidean lattice and middle-field model), the most important result is the generalized prediction of a non-zero epidemiological threshold λ_c. If the value λ is above the threshold $\lambda \geq \lambda_c$, the infection distributions and becomes permanent. If the value λ is lower than the threshold $\lambda \leq \lambda_c$, the infection dies out exponentially fast. The epidemic threshold is actually equivalent to the critical point in the nonequilibrium phase transition. From this point of view, the critical point distinguishes the active phase with the stationary density of infected vertices from the phase with only healthy nodes and zero activity. Studies [11, 32] have shown that the characteristic feature of scale-free networks is the lack of an epidemiological threshold $\lambda_c = 0$.

However, for the description of the distribution of computer viruses, the SIR-model (susceptible–infected–recovered) is a more appropriate model for the distribution of epidemics, which takes into account that some nodes return to the immune state and not infectious. For the local computer networks, immunization is achieved either by antivirus software or by simple recognition that avoids some viruses after some experience, such as the warning given in the message "I love you", etc.

Let's set an epidemiological algorithm in accordance with [35] in the plural of a pair $\{G, \Gamma\}$, where G is a graph which is determined as a cluster (V, E) of a set of nodes V, and a set of links E, and Γ is an evolution operator, which reflects the changing at discrete time intervals t of graph circumstances:

$$\langle V, E \rangle_{t+1} = \Gamma \langle V, E \rangle_t;$$

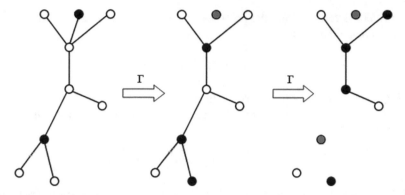

Fig. 7 Scheme of the virtual virus distribution in the graph. Gray circles indicate immunized vertices, black— infected vertices

$$\langle V, E \rangle_{t=0} \overset{def}{=} \langle V_0, E_0 \rangle. \tag{9}$$

The operation of the virtual virus distribution is neglected to research the stability of local area networks to malware attacks in simulative stochastic graphs. The three possible states for each vertex in the graph are susceptible, infected and recovered (immune) (Fig. 7). At each time step, each infected vertex transfers the virus to each of its neighbors with a definite probability (δ) after discrete intervals of time of implemented computations. At the same time step each infected vertex is reclaimed and gets immunity to the virus with a certain probability (v) after the transfer of the virus. This algorithm applies random variables to produce stochastic handling. A random variable U is chosen from a homogeneous distribution $\{0, 1\}$ every time before the possible infection of the vertex's neighbour. If $U > \delta$ infection does not take place, if $U < \delta$ the virus is transmitted. The same regulation is used for v.

We will suppose that the epidemic starts with one randomly chosen vertex to simplify the simulation.

Variable epidemiological characteristics:

- the efficiency of the virus, i.e. the probability of infection spreading to the connected neighbours of the vertex (δ);
- the probability of immunization/recovery (v);
- the launching vertex.

Simulating of the epidemic will continue until the amount of infected vertices becomes equal zero. Then an output file is generated which can be used for additional valuations. It includes the amount of infected nodes at each time step.

The variable n is used to determine the discrete time intervals $t = n\Delta t$, such as the model is discrete.

The chance of any susceptible vertex S_n at the $n + 1$ step not to have contact with any of I_n infected vertices is equal $(1 - \delta)^{I_n}$. Therefore, the probability to have at least one link and become infected is $1 - (1 - \delta)^{I_n}$.

All infected vertices from the previous time step recover at each next time step, so we can make the equation:

$$I_{n+1} = S_n(1 - (1 - \delta)^{I_n}) \tag{10}$$

The amount of susceptible vertices in the previous time step, except those that became infected at this time point is the amount of susceptible vertices during each time step:

$$S_{n+1} = S_n - S_n(1 - (1 - \delta)^{I_n}) = S_n(1 - \delta)^{I_n} \tag{11}$$

The amount of recovered nodes, taking into report the immunization parameter v, is:

$$R_{n+1} = R_n + v I_n \tag{12}$$

Thus, we get a system of equations:

$$S_{n+1} = S_n(1 - \delta)^{I_n};$$
$$I_{n+1} = S_n(1 - (1 - \delta)^{I_n});$$
$$R_{n+1} = R_n + v I_n. \tag{13}$$

To investigate the effectiveness of hindering the distribution of malware in a LAN, two immunization strategies of vertices are implemented: immunization of neighbors and random immunization. **Random immunization** provides the random input of immunized nodes into the graph. **Immunization of neighbors** is based on the hypothesis that a accidently chosen neighbor has more links than a randomly chosen vertex. The control operator is immunity Ω, which is defined by the relation of infected vertices ω to the total amount of vertices N and the amount of immunized vertices φ.

The outcome of the program simulation of the offered method permitted to diagnose the operation of malware distribution in local area networks and doing a lot of inferences (Fig. 8).

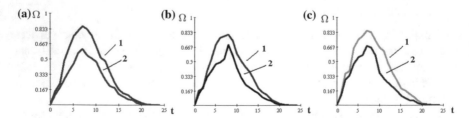

Fig. 8 Dependence Ω on time t in case of random immunization (1) and immunization of neighbors (2) in network "FoxNet" (**a**) "KTM" (**b**) and "DSS-group" (**c**) for values $\delta = 0.65$ and $v = 0.45$

For the three investigated networks of Internet providers "FoxNet", "KTM" and "DSS-group", immunization of neighbors was more efficient than random immunization, inasmuch as it leads to noticeably less infection of the vertices that is the cause for its recommendation in the expansion of antivirus projects and system security analysis of LAN.

The results of mathematical simulation make suitable to evaluate the amount of immunized vertices φ, which must be acceded into the graph to block the distribution of malware in it. It is defined by the created output file at the time t_{cr}, which represents the maximum of dependency $\Omega(t)$. For the moment of time $t > t_{cr}$, the distribution of the epidemic thanks to the operation of immunization in it slows down and declined (Fig. 8). Table 3 shows the minimum number of immunized nodes for the different networks, which is necessary to stop the distribution of viruses in their structures.

The distribution of the epidemic is affected by:

- the degree of the site that triggers the epidemic: large-scale causes major epidemics;
- the correlation between the probability of infection and the probability of recovery.

If $\delta \gg v$ there is an avalanche distribution of the epidemic, while immunization distributions at very slow pace, on the contrary, if $\delta \ll v$ the distribution of the epidemic is significantly slowed down, and it can be argued that such computer networks are resistant to viral infection (Fig. 9).

To define the values δ and v while simulating real networks, expert opinion and statistic data should be taken into examination to reach the highest similarity between the simulation and the network.

Table 3 Minimum required number of immunized nodes

	«FoxNet» $N = 915$		«KTM» $N = 1604$		«DSS-group» $N = 2023$	
	t_{cr}	φ	t_{cr}	φ	t_{cr}	φ
Random immunization	8	106	9	259	8	418
Immunization of neighbors	7	314	9	432	9	476

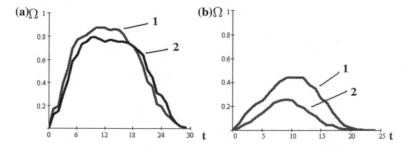

Fig. 9 1000-node test graph at values $\delta = 0.83$, $v = 0.16$ (**a**) and $\delta = 0.24$, $v = 0.79$ (**b**) and dependence Ω on time t in random immunization (1) and immunization of neighbors (2)

8   Conclusions

On the basis of the research, integral probability parameters of real local area networks are calculated in various intervals of time, their approximation is carried out for simulation.

The actual scientific task of developing methods and tools for mathematical modeling of the development processes of computer network topologies has been solved, which is important for increasing the efficiency of functioning of these networks and the defense of their components from targeted attacks.

A new mathematical model for generating the architecture of local area networks with a given density function of the distribution of vertices using the apparatus of the theory of complex networks has been developed, which enabled to reconstitute these networks as a stochastic graph with given probabilistic parameters. The reclaimed network obtained as a result of modeling in the form of a stochastic graph has characteristics that are close to the real network, which make suitable research the nature of the formation and growth of networks, the identification of new patterns in them, modeling of infection and immunization tasks, protection from network attacks.

On the basis of the modeling algorithm proposed in the work, the forecasting of the evolutionary dynamics of development and formation of real computer networks for the next years has been carried out.

The mathematical models of attack scenarios reproduced in the work were used to estimate the vulnerability of modeled stochastic graphs and to solve the problem of the stability of local area networks to random and targeted attacks. A general simulation of the network ensemble evolution under the circumstance of destabilization threats has been designed.

References

1. Golovach Y, Olmeskoy O, fon Ferber K et al (2006) Skladni merezhi (The complex networks). Zhurnal fizychnykh doslidzhen 10(4):247–289
2. Newman M (2003) The structure and function of complex networks. SIAM Rev 45(2):167–256
3. Erdős P, Renyi A (1960) On the evolution of random graphs. Publications of the Mathematical Institute of the Hungarian Academy of Sciences, vol 5, pp 17–61
4. Greenman (1977) Graphs and Markov Chains. Math Gazette 61(415):49
5. Watts D, Strogatz S (1998) Collective dynamics of 'small-world' networks. Nature 393(6684):440–442
6. Barabasi Albert R (1999) Emergence of scaling in random networks. Science 286(5439):509–512
7. Gao Z, Small M, Kurths J (2016) Complex network analysis of time series. EPL (Europhys Lett) 116(5):50001
8. Cencetti G, Bagnoli F, Battistelli G, Chisci L, Fanelli D (2017) Control of multidimensional systems on complex network. PLoS ONE 12(9):e0184431
9. Gershenson C, Niazi M (2013) Multidisciplinary applications of complex networks modeling, simulation, visualization, and analysis. Complex Adapt Syst Model 1(1):17

10. Monaco A, Monda N, Amoroso A et al (2018) A complex network approach reveals a pivotal substructure of genes linked to schizophrenia. PLoS ONE 13(1):e0190110
11. Leyffer S, Safro I (2013) Fast response to infection distribution and cyber attacks on large-scale networks. J Complex Netw 1(2):183–199
12. Schröter M, Paulsen O, Bullmore E (2017) Micro-connectomics: probing the organization of neuronal networks at the cellular scale. Nat Rev Neurosci 18(3):131–146
13. Kryvinska, N. (2010) Converged network service architecture: a platform for integrated services delivery and interworking. Electronic business series, vol 2. International Academic Publishers, Peter Lang Publishing Group
14. Kryvinska N (2008) An analytical approach for the modeling of real-time services over IP network. Math Comput Simul 79(4):980–990. https://doi.org/10.1016/j.matcom.2008.02.016
15. Kryvinska N (2004) Intelligent network analysis by closed queuing models. Telecommun Syst 27:85–98. https://doi.org/10.1023/B:TELS.0000032945.92937.8f
16. Kryvinska N., Zinterhof P, van Thanh D (2007) An analytical approach to the efficient real-time events/services handling in converged network environment. In: Enokido T, Barolli L, Takizawa M (eds) Network-based information systems. NBiS 2007. Lecture notes in computer science, vol 4658. Springer, Berlin
17. Ageyev DV, Salah MT (2016) Parametric synthesis of overlay networks with self-similar traffic. Telecommun Radio Eng (English translation of Elektrosvyaz and Radiotekhnika) 75(14):1231–1241
18. Ignatenko AA, Ageyev DV (2013) Structural and parametric synthesis of telecommunication systems with the usage of the multi-layer graph model. In: Proceedings of the 2013 23rd international Crimean conference microwave and telecommunication technology (CriMiCo 2013), pp 498–499
19. Ageyev D, Ali A-A, Nameer Q (2015) Multi-period LTE RAN and services planning for operator profit maximization. In: 2015 13th international conference on the experience of designing and application of CAD systems in microelectronics (CADSM). IEEE, pp 25–27. https://doi.org/10.1109/cadsm.2015.7230786
20. Radivilova T et al (2018) Decrypting SSL/TLS traffic for hidden threats detection. In: Proceedings of the 2018 IEEE 9th international conference on dependable systems, services and technologies (DESSERT). IEEE, pp 143–146. https://doi.org/10.1109/dessert.2018.8409116
21. Ageyev D et al (2018) Classification of existing virtualization methods used in telecommunication networks. In: Proceedings of the 2018 IEEE 9th international conference on dependable systems, services and technologies (DESSERT), pp 83–86
22. Karpukhin A et al (2017) Features of the use of software packages for modeling infocommunication systems. In: Proceedings of the 2017 4th international scientific-practical conference problems of infocommunications. Science and technology (PIC S&T). IEEE, pp 380–382. https://doi.org/10.1109/infocommst.2017.8246421
23. Radivilova T, Kirichenko L, Ageiev D, Bulakh V (2020) The methods to improve quality of service by accounting secure parameters. In: Hu Z, Petoukhov S, Dychka I, He M (eds) Advances in computer science for engineering and education II. ICCSEEA 2019. Advances in intelligent systems and computing, vol 938. Springer, Cham
24. Kirichenko L, Radivilova T, Tkachenko A (2019) Comparative analysis of noisy time series clustering. CEUR Workshop Proceedings, vol 2362, pp 184–196. http://ceur-ws.org/Vol-2362/paper17.pdf
25. Radivilova T, Kirichenko L, Yeremenko O (2017) Calculation of routing value in MPLS network according to traffic fractal properties. In: 2017 2nd international conference on advanced information and communication technologies (AICT). IEEE, pp 250–253. https://doi.org/10.1109/aiact.2017.8020112
26. Ivanisenko I, Radivilova T (2015) The multifractal load balancing method. In: 2015 second international scientific-practical conference problems of infocommunications science and technology (PIC S&T). IEEE, pp 122–123. https://doi.org/10.1109/infocommst.2015.7357289
27. Kryvinska N, Zinterhof P, van Thanh D (2007) New-emerging service-support model for converged multi-service network and its practical validation. In: First international conference on

complex, intelligent and software intensive systems (CISIS'07). IEEE, pp 100–110. https://doi.org/10.1109/cisis.2007.40

28. Albert R, Jeong H, Barabási A-L (2000) Error and attack tolerance of complex networks. Nature 406(6794):378–382
29. Alstott Bullmore E, Plenz D (2014) Powerlaw: A python package for analysis of heavy-tailed distributions. PLoS ONE 9(1):e85777
30. Da Silva D, Bianconi G, Da Costa R, Dorogovtsev S, Mendes J (2018) Complex network view of evolving manifolds. Phys Rev E 97(3)
31. Kitsak M, Ganin A, Eisenberg D et al (2018) Stability of a giant connected component in a complex network. Phys Rev E 97(1)
32. Yehezkel A, Cohen R (2012) Degree-based attacks and defense strategies in complex networks. Phys Rev E 86(6)
33. Duda O, Matsyuk O, Pasichnyk V, Kunanets N (2018) Kontsept «rozumne misto» ta informatsiyni tekhnolohiyi BigData (The concept of "smart city" and information technologies BigData). Paper presented at the 5th scientific and technical conference information models, systems and technologies, Teropil Ivan Pului National Technical University, Ternopil, 1–2 Feb 2018, p 30
34. Ivanushchak N, Kunanets N, Pasichnyk V (2018) Mathematical modeling and analysis of destabilization threats in computer networks. Paper presented at the international scientific and practical conference problems of infocommunications. Science and technology, Kharkiv National University of Radio Electronics, Kharkiv, 9–12 Oct 2018, pp 191–197
35. Sloot P, Ivanov S, Boukhanovsky A, van de Vijver D, Boucher C (2008) Stochastic simulation of HIV population dynamics through complex network modeling. Int J Comput Math 85(8):1175–1187

Adaptive Space-Time and Polarisation-Time Signal Processing in Mobile Communication Systems of Next Generations

Valeriy Loshakov⬮, Mykola Moskalets⬮, Dmytro Ageyev⬮, Abdnoure Drif⬮, and Konstantyn Sielivanov⬮

Abstract The possibilities of improving electromagnetic compatibility in mobile communication systems of next generations are analysed based on application of adaptive space-time signal processing in antennas of base stations. Features of solving the problem of interference suppression in adaptive antenna arrays are considered using recurrent adaptation algorithms realized on the basis of the minimum mean square error criterion. The results of modelling adaptation of signal processing and its analysis are presented. A promising method of interference suppression using polarisation-time processing of signals using optimal stochastic control is considered.

Keywords Electromagnetic compatibility · Adaptive antenna array · Weight vector · Recursive algorithm · Stability and convergence of adaptation algorithm

1 Introduction

Simultaneously with development of mobile networks, more and more attention is paid to their electromagnetic compatibility (EMC). Indicators of EMC are deteriorating due to increasing density of communication systems in cities, the growing number of subscribers of all types of networks and intensity of using radio frequency resource by all types of radio systems [1, 2]. With increase in the power of radio interference in modern conditions, the power of the transmitting devices should be increased. This leads to worsening the electromagnetic situation. Expanding the number of Base Stations (BS) and mobile terminals, as well as individual software applications, also leads to increase in time of using wireless networks and additionally complicates the interference situation [3–5]. One of the important directions

V. Loshakov · M. Moskalets (✉) · D. Ageyev · A. Drif · K. Sielivanov
Kharkiv National University of Radio Electronics, Kharkiv, Ukraine
e-mail: mykola.moskalets@nure.ua

© The Editor(s) (if applicable) and The Author(s), under exclusive license to Springer Nature   469
Switzerland AG 2021
T. Radivilova et al. (eds.), *Data-Centric Business and Applications*, Lecture Notes on Data
Engineering and Communications Technologies 48,
https://doi.org/10.1007/978-3-030-43070-2_21

Fig. 1 The scheme of space-time processing with an adaptive antenna arrays

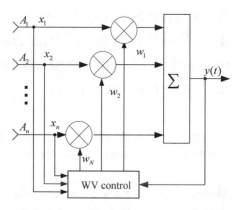

contributing to solving the EMC problem is the using of BS adaptive antenna arrays (AAA). The basis of most methods of space-time processing in systems with AAA (see Fig. 1) is the estimation of Weight Vector (WV) in reception channels and their modification by various algorithms [6, 7].

An important advantage of space-time processing in AAA systems is that it can be naturally combined with other time-frequency, code, and different organizational methods, successfully completing them and expanding signal reception/transmission capacity by expanding the decision-making space. In addition, the space-time processing has several advantages over other treatment methods. Thus, it does not take additional time and frequency resources. Moreover, it is possible to realize the reuse of operating frequencies when it is possible to receive different signals at the same frequency. The application of space-time processing makes it possible to leave unchanged the parameters of useful signals, the requirements for communication channels, the modes of communication without increasing the allocated frequency band.

However, the possibilities of using adaptive mechanisms for antennas in wireless communication systems of new generations have not been studied sufficiently. For practice, the values of reachable interference suppression levels, the value of the convergence coefficient, determining the adaptation time with sufficient stability of the algorithm, the possibility of forming the maximum of the main lobe of the directional pattern (DP) in needed direction are important.

2 Features of Solution for the Problem of Interference Suppression in the Adaptive Antenna Arrays of Base Stations

The problem of interference suppression in the BS AAA can be solved in different ways. For example, we can first evaluate the direction of interferences arrival and their polarisation, and then solve the task of controlling the weight coefficients of the

AAA to form a DP with dips towards the interference and a maximum toward the source of the useful signal. However, in practice, under the conditions of a limited time resource, it is more reasonable to directly estimate the weight coefficients wi. Currently, there are known solutions that differ both by the selected criteria for management effectiveness (adaptation) and the imposed restrictions. The most widely used are the following criteria: the maximum ratio of the useful signal to the sum of the interference at the output of the AAA; the minimum mean square deviation (MMSD) of the received signal from the reference (pilot) signal yr(t); the maximum interference suppression; the maximum likelihood ratio [8–13]. Each of these criteria has its limitations, advantages and disadvantages as well as areas of rational use. The MMSD criterion most closely corresponds to specificity of mobile communication tasks, since it is in good agreement with the traditional probabilistic criteria for radio communication systems that characterize the quality of information transmission. The MMSD criterion was proposed by Widrow [14]. The main limitation of using this criterion is the need have to know reference signal yr(t) on the receiving side (Fig. 2). However, for modern and perspective mobile systems it is not significant because almost all of them use pilot signals.

Using this criterion, as well as other criteria, the task of control in the AAA is reduced to providing the best quality of reception of a useful signal under the conditions of interference. Due to the weight coefficients, a complex weight vector is formed, which allows to change both the amplitude and the phase of the received signals

$$W^T(t) = (w_1(t), w_2(t), \ldots, w_N(t)) \tag{1}$$

The rate of changes of the weight coefficients should be consistent with the rate of change in the signal and interference situation, and the range is consistent with the dynamic range of signal and interference level changes as well as phase relationships in various AAA elements. In practice, based on the possibilities of technical feasibility and other reasons, it is necessary to limit these characteristics, which leads to corresponding decrease in the efficiency of the adaptation process [15, 16].

Fig. 2 The block diagram of the space-time signal processing with a reference signal

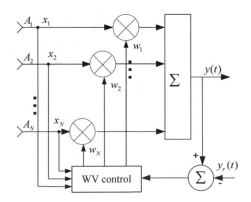

Unlike other problems of antenna technology, where the final result of the solved problems is the synthesis of DP with various restrictions on the design, dimensions, spectral composition of signals and interference and other parameters, the goal of using AAA is to provide the necessary characteristics quality of useful signals at the antenna output. This means that the output signal must correspond to the expression

$$y(t) = W^T(t)x^*(t) = w_1(t)x_1^*(t) + w_2(t)x_2^*(t) + \ldots + w_N(t)x_N^*(t), \qquad (2)$$

where $x_i(t) = s_i(t) + \sum_{j=1}^{J} n_{ij}(t) + v_i(t)$. In this case, the total DP of the AAA cannot be calculated at all, although being an intermediate characteristic, and also in the case of the solution of Space-Time Access (STA) problems, it is certainly interesting. If necessary, the DP of the AAA can be obtained by using the scalar product

$$F(\theta) = \left(W^T(t), f(\theta)\right), \qquad (3)$$

where $f^T(\theta) = \left(f_1(\theta), f_2(\theta)e^{i\varphi_1\theta_1}, \ldots, f_N(\theta)e^{i\varphi_N\theta_1}\right); \cdot f_i(\theta)$ is the not normalized mean square deviation of AAA elements; φ_i denotes phases of the envelope wave of the unit amplitude counted from the phase of the signal from the output of the *1st* element (at $\varphi_1 = 0$) fixed at the outputs of the receiving elements due to spatial differences.

2.1 Algorithm of Adaptive Antenna Arrays Work by Minimum Mean Square Deviation Criterion and Modelling Results

Let us consider features of AAA synthesis working algorithm based on the direct estimation weight vector $\hat{W}(t)$ over the MMSD criterion. In this case, it is important that when using MMSD, the residual component of the interference after the space-time signal processing (STSP) is an update process and can be approximated by the "white" noise [6, 8].

The basis of the STSP algorithm functioning according to the MMSD criterion, is a comparison after weighing the received signal realization X(t) with the reference signal y_r

$$v(t) = W^T X(t) - y_r(t) \qquad (4)$$

To calculate the mean square deviation (MSD) and determine the conditions for ensuring its minimum value, it is necessary to find the square of this *discrepancy* (4) and its mathematical expectation [14]:

$$v^2(t) = W^T X(t)X^T(t)W(t) - 2y_\partial W^T(t)X(t) + y_r^2; \qquad (5)$$

$$M[v^2(t)] = W^T(t)R_{xx}W(t) - 2W^T(t)r_{xy} + \overline{y_r^2(t)}, \qquad (6)$$

where

$$r_{xy} = (x_1(t)y_r(t); x_2(t)y_r(t); \ldots; x_N(t)y_r(t)) \qquad (7)$$

r_{xy} is covariance vector between the received and reference signals, $R_{xx} = X(t)X^T(t)$ is the correlation matrix of the signals received by the AAA.

By equating the derivative (5) to zero

$$dv^2/dW(t) = \nabla\omega\left(\overline{v^2(t)}\right) = 2R_{xx}W(t) - 2r_{xy} = 0 \qquad (8)$$

we come to the well-known vector-matrix Wiener-Hopf equation [8, 14], the solution of which gives the optimal value of the complex weight vector

$$\hat{W}_{opt} = R_{xx}^{-1}r_{xy} \qquad (9)$$

Knowing $\hat{W}_{opt}$ as an intermediate result, allows, if necessary, to calculate the optimal direction pattern of the AAA, for example, when it is needed to solve the space-time access problems,

$$F(\theta) = \left(\hat{W}_{opt}(t), f^*(\theta)\right). \qquad (10)$$

Let us proceed to the synthesis of the structure of the algorithm for performance of the BS AAA, which is optimal according the MMSD criterion. For real-time operation, the most appropriate is the class of algorithms based on the recurrent control and the direct calculation of the WV of the AAA based on the results of the determination of the discrepancy vk(t) at each kth sampling step yk(t). Recurrent adaptive algorithms, in comparison with asymptotic algorithms, do not require large expenditures for calculating matrices and are limited by the small volume of constant and operative memory.

Recurrent evaluation procedures $\hat{W}_{opt}$ begin with finding the extremum the gradient of a function containing the value $W(t)$ as its argument. This allows analysing and considering transient AAA mode as well as unsteadiness of $X(t)$. Recursive methods are based on a sequential iterative procedure of finding the solution for the WV in the direction opposite to the gradient of the quality index function

$$W(k) = W(k-1) - \mu\nabla(k), \qquad (11)$$

where k is the discrete time; μ is the step coefficient (step constant), taking into account the rate of the search for the extremum.

The algorithm (11) shows that the WV value at the kth step is equal to the WV value on the previous $(k-1)$th value with the adaptive additive $\mu\nabla(k)$, which depends on the *discrepancy* $y_r(k) - y(k)$. This procedure goes into a continuous one when the sampling period decreases $\Delta t = t_{k+1} - t_k \rightarrow 0$, which can be represented as a differential equation

$$dW(t)/dt = -\mu \nabla(t) \qquad (12)$$

To estimate the weight vector, we use the formalization of the Kalman filtering procedure for the case when the reference signal is used. For this case, the adaptation algorithm based on the criterion of minimum mean square deviation can be represented by the estimation equation [14]

$$\hat{W}(t) = A(t)W(t) + K(t)[W(t)x(t) - y_r(t)]x(t) \qquad (13)$$

The structural scheme of the AAA tracking algorithm with the reference is shown in Fig. 3.

The output Σ_r forms a *discrepancy*, which contains $\Delta s_\Sigma + \Delta n_\Sigma + v_\Sigma = v$ where $\Delta s_\Sigma \rightarrow 0$. Thus, the useful signal is compensated, and then the *residual* plays the role of a reference signal like a reference interfering signal in systems of auto-compensation interference [17].

In a discrete form, the algorithm for adapting to the MMSD criterion has the form

$$\hat{W}(k+1) = \hat{W}(k) + 2\mu \left[\hat{W}(k)x(k) - y_r(k) \right] x(k) \qquad (14)$$

where $y_r(k)$ is the reference signal at the time k; $x(k)$ is a vector of input influences; $1/\lambda_m > \mu > 0$; λ_m is the maximum eigenvalue of the covariance interference matrix [18, 19] (Fig. 4).

Fig. 3 The block diagram of the tracking adaptive antenna arrays algorithm

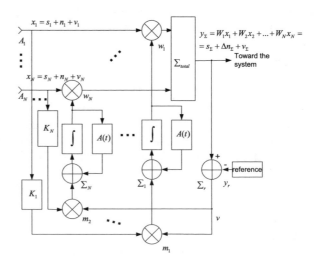

Fig. 4 The block diagram of the discrete adaptive antenna arrays algorithm

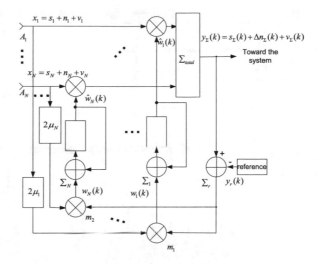

In practical implementation of this adaptive algorithm, the central question is the selection of the coefficient μ. This coefficient determines the rate of convergence (adaptation) of the algorithm and its stability and it can lie within the limits of [1]

$$\frac{1}{\sum_{i=1}^{N} D_i} > \mu > 0 \tag{15}$$

where D_i is the variance of the input process (taking into account all the interferences, signal and noise) in the ith element of the AA. If the value of this coefficient is too large, the algorithm will be unstable. If the coefficient is too small, then the algorithm will converge very slowly. In practice, the value of the coefficient is set approximately 10 times less than the maximum allowable value. Given that the level of interference from the zero direction is much higher than the level of the signal and noise, the requirement for the coefficient μ is simplified by the inequality $1/ND_1 > \mu$.

The results of studying the dependence between the convergence of the adaptation algorithm and the number of adaptation steps for the AAA with N = 5 under different values of the interference variance and the parameter μ are shown in Fig. 5, and the graphs of the directional patterns after adaptation are shown in Fig. 6. The graphs of the dependences of the adaptation errors on the difference in the angles of arrival of the signal and the interference as well as the dependence of the convergence of the adaptation algorithm on the variance of the interference are given in Figs. 7 and 8. From the analysis of simulation results (Figs. 5 and 7) it follows that the algorithm works well for large values of the variance of the interference, i.e. under strong interference. When the power of interference is reduced, the algorithm loses the ability to provide interference. Simulation also has shown that the adaptive antenna (AA) adapting to the pilot signal very deeply rejects the interference and receives signals with a maximum of DP in the direction of the arrival of useful signals. However,

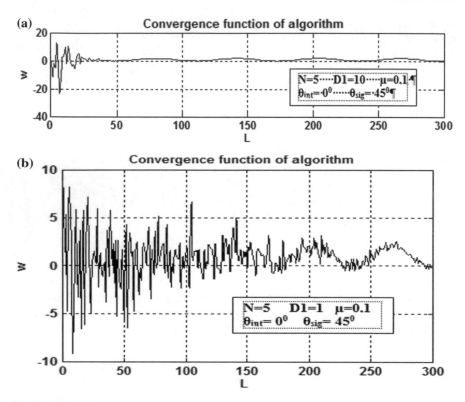

Fig. 5 Realizations convergence functions of algorithm from the number steps of adaptation

when the angles of the useful signal and interference approach, the adaptation gives an error in the angle of the useful signal (Fig. 6 and 8).

3   Polarisation-Time Signal Processing in Space-Time Access

3.1   Features of Polarisation in Communication Lines of Wireless Systems

Polarisation is a physical characteristic of an electromagnetic field, which determines the peculiarities of the spatial arrangement and changes in the time of its vector intensity. Linear, elliptic or circular polarisation in the plane perpendicular to the direction of distribution are distinguished depending on which geometric figure is formed by the end of the electromagnetic field intensity vector [20, 21].

Fig. 6 Patterns of the antenna array after adaptation at different angles of approaching of the signal and interference signals

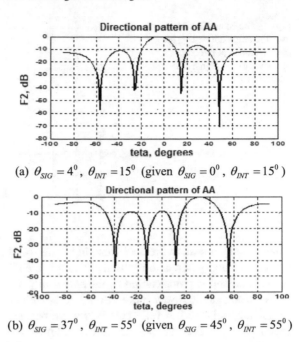

(a) $\theta_{SIG} = 4^0$, $\theta_{INT} = 15^0$ (given $\theta_{SIG} = 0^0$, $\theta_{INT} = 15^0$)

(b) $\theta_{SIG} = 37^0$, $\theta_{INT} = 55^0$ (given $\theta_{SIG} = 45^0$, $\theta_{INT} = 55^0$)

Fig. 7 Convergence of the adaptation algorithm depending on the variance of the interference

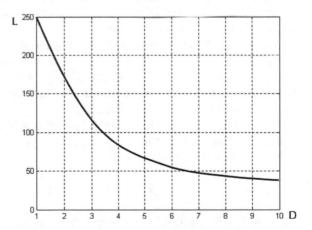

From the point of view of obtaining the necessary energy potential in communication lines, any polarisation does not have advantages, therefore in practice the linear polarisation is most often used, which is connected with the simplicity of the antenna design.

However, in some cases, due to the influence of the propagation medium or spatial displacement of carrier media, different variations in the orientation of the linear polarisation plane are possible, which requires a tracking system for these polarisation changes. The application of circular polarisation, invariant to the rotation of the

Fig. 8 The adaptation error over the angle depending on the difference in the angles of arrival of the signal and interference

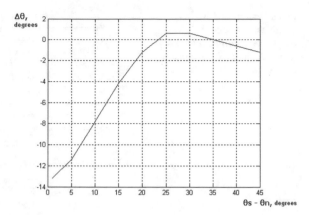

polarisation plane allows to avoid tracking systems. Therefore, in satellite communications, circular polarisation is used to exclude the influence of the Faraday rotation of the polarisation plane caused by the anisotropy of the ionosphere.

Electromagnetic waves emitted by antennas with fixed polarisation parameters are completely polarised. A fully polarised wave has a known position of the intensity vector $E(t)$, and for an unpolarised wave, all positions $E(t)$ are random and equally possible. The degree of polarisation in this case is defined as the ratio of the power of the fully polarised part of the wave to its entire power, which is equal to the sum of the polarised and unpolarised parts.

The change in the polarisation of the emitted waves can be performed by changing the spatial orientation of the antennas or amplitudes and phases (complex weight coefficients) of the currents supplying orthogonally polarised emitters: dipoles, spiral, round or square horns, or other antenna elements that form a two- or three-dimensional polarisation basis [20, 21].

With the appropriate selection of polarisation of the receiving antenna for any partially polarised wave, it is always possible to obtain the maximum signal power at the coordinated load. This power is

$$P_s = 0,5 \Pi A_{ef}(1 + m \cos \delta) \tag{16}$$

where m is the polarisation of a signal; δ is the angle between the points on the Poincare sphere, which determine the polarisation state of the signal and the antenna (in the case of their linear polarisation, δ is the angle between the vectors of the active antenna length and the field strength); A_{ef} is an effective area of the antenna; Π is the value of the Poynting's vector, which is characterized by the density of the power flow of the signal field at the receiving point.

Many problems of polarisation-time processing are directly reduced to obtaining extreme values of the power of received signals. For a fully polarised wave ($m = 1$), the signal strength is based on the expression (16) $P_s = A_{ef} \Pi \cos^2(\delta/2)$. When selecting polarisation of the antenna corresponding to the value $\delta/2 = n\pi$ (where $n =$

$0, 1, 2 \ldots$), we obtain an agreed polarisation method, which provides the maximum signal strength: $P_{smax} = A_{ef}\Pi$. The choice of $\delta/2 = (2n + 1)\pi/2$ allows to get $P_{smin} = 0$, that is, to reject the given signal. In this case, the efficiency of the polarisation-time processing depends on the accuracy of setting the values of the polarisation parameters for the antenna in relation to the polarisation of the signal field.

The main constraints that reduce the efficiency of polarisation-time processing can be attributed to:

- structural and technological errors of the antenna system;
- errors in determining polarisation parameters of received signals;
- errors in control devices for antenna polarisation parameters.

Let us analyse the effect of these constraints. The quality of the antenna system design is determined by the polarisation solution, which is characterized by the ratio of the levels of the main and the cross polarisation component of the received signals. When using serially manufactured nodes of antenna systems, the achievable level of polarisation is 25–35 dB. Various structural complications are used to improve it: polarising filters, two- and three-mirror antennas, corrugated horns and other devices that minimize the cross component and achieve a resolution of 40–50 dB or more in the frequency band of $\pm(10 \div 25) f_{mean}$.

Errors in determining the polarisation parameters of the received signals may have different nature. These errors may be due to the location uncertainty or changing parameters of the correspondent's radiation system (for example, under the connection with moving objects). At the same time, if the polarisation of the transmitted signal is determined, then at the receiving point it can be different. The main mechanisms that cause these differences are:

- the interaction of the main radiation in the direction of the field component correspondent with reflections from various objects or scattered on inhomogeneities of medium components in the intra-city connection, communication lines of centimetre or millimetre ranges in the event of the presence of hydrometeors or other finely dispersed formations on the radio track;
- the occurrence and interaction of ordinary and unusual components under distribution in anisotropic media, for example in the ionosphere;
- the displacement of the receiving point relative to the axis of the main radiation occurs when using the directional antennas, for example, in the tropospheric communication channel.

The polarisation deviations considered tend to randomly change over time. At the same time, the speed of these changes is determined by the correlation intervals for the ranges of the waves used, varying from a part of a second to several minutes, which makes it possible, with the help of various electronic methods, to accurately trace them and assume the existing degree of polarisation included in expression (16) close to one.

Errors in the case of statistical methods used to evaluate polarisation of signals depend both on the statistical structure of the random signal and the interference level of noise belonging to the type of own or another unpolarised "white noise". For optimal recurrent filtration of signals, the variance of estimate error is

$$\Delta^2 = 2P_c/(1 + \sqrt{1 + h^2}) \tag{17}$$

where h^2 is the ratio of the signal strength to the spectral density of the white noise power in the receiving channel.

For the Rice channel communication (for example, channels with the direct line of sight), the error variance value can be (30–40) dB in relation to the variance of the fluctuation component of the signal, that is, the accuracy of the estimation can be quite high.

Errors in the control devices of the polarisation parameters of the antennas depend on the errors in the estimation of signal parameters and structural errors of the antennas themselves. In this case, polarisation control, which is to establish the corresponding values of the complex weight coefficients of the antenna elements, can be made on the basis of evaluation of these values. An analysis of such algorithms shows that in real communication channels in such a way it is possible to choose polarisation of the receiving antenna in relation to the signal, and that the level of the orthogonal component will be lower than the fundamental one of 20–25 dB. These are the real possibilities of polarisation-time signal processing with suppressed interference and solving other problems. To successfully solve these problems, the antenna systems of communication means should provide the possibility of operative selection of the corresponding values for the polarisation parameters, that is, they must consist of two or three orthogonally polarised antenna elements that form a field-based basis and are equipped with amplitude control devices and phases (control devices for complex coefficients $w_i(t)$).

The polarisation and polarisation-time methods are most widely used in solving the following problems: interference suppression; provision of a consistent reception and minimization of the fading effect; frequency reuse; adjustment of antennas; ensuring electromagnetic compatibility; polarisation manipulations, etc. Let us dwell on the peculiarities of solving some of them.

3.2 Synthesis of Methods of Polarisation-Time Signal Processing Under Space-Time Access

Interference suppression through polarisation is a promising method of processing, especially in cases where interference occurs within the ML of the DP of the receiving antenna. Let us consider the solution of this problem from the standpoint of optimal stochastic control. Problems of polarisation-time processing as well as space-time can be solved in different ways. Thus estimating the polarisation of signals and noise, it is possible to construct a control system for the polarisation basis of antenna reception, which provides continuous orthogonalisation of the basis in relation to the noise. The solution of this problem can be complicated, especially in the nonlinear situation, when the conditions of the division theorem are not fulfilled.

In this case the above mentioned approach is considered more constructive when the evaluation is not subject to the parameters of the polarisation of the observed signals and interferences, and the values of the optimal weight coefficients $w_i(t)$ under $i = 1, 2$ for an antenna system consisting of two mutually orthogonal antenna elements included between the antennas elements and a common adder, and provide, for example, the MMSD of the received signal $y(t)$ from the reference one $y_r(t)$. Obviously, the choice of certain values of the weight coefficients $w_i(t)$ leads to the corresponding transformations of the polarisation basis of the system antenna, which, in turn, determines the level of partially polarised signals and interferences at the output of the common adder, but does not affect the level of unpolarised "white" noise, which falls into the band of reception frequencies. Equations of the weighting coefficient (WC) status, observation and estimation of $\hat{W}(k)$ in this case are similar to the corresponding equation for AAA (12), (13) and (14).

The block diagram of the device for estimation of weight coefficients, according to (15), which performs the polarisation-time processing, is presented in Fig. 9:

$$d\hat{w}_i(t)/dt = a_i(t)\hat{w}_i(t) + \sum_{j=1}^{N} 2K_{ij}(t)V_{obs}^1[y_e(t)-y(t)]x_i(t)$$

$$= a_i(t)\hat{w}_i(t) + 2V_{obs}^1 v(t) \sum_{j=1}^{N} K_{ij}(t)x_i(t) \qquad (18)$$

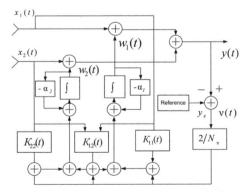

Fig. 9 The block diagram of the device for estimating weight coefficients that performs polarisation-time signal processing

It can be seen that the use of the chosen MMSD criterion leads to minimization of the interference strength of the system output, which for polarisation processing systems means orthogonalisation of the antenna polarisation in relation to noise.

Different generalisations and simplifications of the considered algorithm of polarisation-time processing are possible. In the presence of nonlinearities in the equations $x_i(t) = s_i(t) + \sum_{j=1}^{J} n_{ij}(t) + v_i(t)$ and (1) the values $K_{ij}(t)$ will be dependent on estimates of weight coefficients, which somewhat complicates the synthesised procedure. With a constant signal-to-noise situation, when the interferences are completely polarised, the equation of state (18) is simplified and brought to the form of $dw_i(t)/dt = 0$. Since here $A = 0$, the first term on the right-hand side (18) disappears, which allows the equation of estimation in the vector form to be written in the form

$$\frac{d\hat{w}(t)}{dt} = K(t)A^T(t)V_{\mathcal{H}}^{-1}v(t). \tag{19}$$

Equation (19) is a continuous version of the known Widrow-Hoff procedure. Thus, the synthesised algorithms can be considered as the generalisation of the procedure for a more complex statistical situation and general partially polarised fields.

The convergence rate of the considered evaluation procedures is quite high and is in the limit of the correlation interval. For discrete procedures, it is equal to 10–15 steps. For sub-optimal filter parameters, for example, when using the procedure (19), this time for an arbitrary signal-to-noise situation increases and can reach 300 steps or more.

These results are similar to those obtained in the problems for AAA and can be extended to space-polarisation-time problems. In addition, Eq. (19) corresponds to the algorithm for processing of space-time signals and interferences with a plane phase front and a point-source spatial spectrum. At fluctuation fronts, it is necessary to use procedures of [21]:

$$d\hat{w}_i(t)/dt = -a_i(t)\hat{w}_i(t) + \sum_{j=1}^{N} K_{ij}(t)F_j'(\hat{w}_t, t) \tag{20}$$

where $F_j'(\hat{w}_t, t) = dF_j(\hat{w}_t, t)/d\hat{w}_t$ is the N-dimensional column vector obtained through derivatives from the time of observation from the logarithm of the probability function.

Let us analyse the efficiency of the problems considered for different values of the angle δ, which is half the centre of the Poincare sphere and the point of the connection, which correspond to the values of signal polarisation and noise. For the analysis, we choose a coefficient η, which shows how the signal level related to the noise at the output of the system is greater than at the input:

$$\eta = \left(d_{s\_in}/(d_{int\_out} + d_{n\_out})\right)/\left(d_{s\_out}/(d_{int\_in} + d_{n\_in})\right) \tag{21}$$

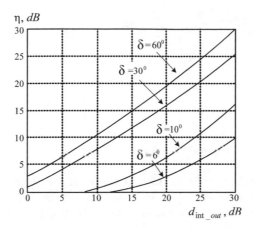

Fig. 10 Graphs of the dependence of the noise immunity coefficient η on the value of the difference in the angle of signal arrival and interference δ

where $d_{s\_out} = W^T R_c W$, $d_{int\_out} = W^T R_3 W$, $d_{n\_out} = W^T W$ are levels of signal, interference and noise itself normalised by the unit value of the spectral power density of the unpolarised "white" noise. We will determine the values of the optimal weight coefficients from the Wiener-Hopf matrix equation, $W = R_{xx}^{-1} r_{xy}$ where $R_{xx} = R_s + R_{int} + R_n$, and R_s, R_{int}, R_n are the correlation matrices of the signal, interference and noise respectively; r_{xy} is the matrix of mutual correlation between the vectors of received and reference signals.

Despite the fact that the analysis procedure is based on the Wiener filter equation, it can be used for this case, since the Kalman filter has the same efficiency. Dependence of η on d_{int} is shown in Fig. 10 where it is evident that with the decrease of the noise level the degree of interference suppression increases.

Efficiency also grows with increasing difference between signal polarisation and interference (with increasing of δ).

Specifically, even with a small difference in signal polarisations and interferences (at $\delta \leq 30°$) it is possible to achieve a suppression level of 20 dB or more, which respectively improves the quality of electromagnetic availability (EA) [21].

3.3 Methods of Space-Time Access with Polarisation-Time Signal Processing

The algorithms presented in the previous chapters refer to the cases of using either only space or only polarisation differences in the receiving signal and interference. In practice, however, it often happens that there are conditions for "blindness". The EA case for moving objects is especially characteristic. There is a natural desire to combine the possibility of space-time and polarisation processing in AAA controlled

Fig. 11 The AAA with two independent orthogonally polarised emitters

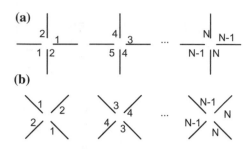

by polarisation. One of the first ideas was suggested by Compton. The essence of his work was the synthesis of AAA capable of responding to differences in signal polarisation and interference. At the same time, the interferences were assumed to be completely polarised. A more general solution suitable for suppressing partly polarised interferences is given in [6]. The idea of AAA controlled by polarisation is constructive also from the standpoint that in this case the complex solution to the problem of noise immunity is used. Moreover, in order to further improve the quality of radio-electronic means operation, it is necessary to involve the maximum number of methods for ensuring noise immunity, including not only space-polarised but also frequency-time, organizational ones, etc.

The antenna, which allows the synthesis of common space-time and polarisation-time processing algorithms, can be constructed in the form of a grid, in the nodes of which there are antenna elements consisting of two or three orthogonally polarised beams: electric (semiconductor dipoles, spirals, more complex constructions), magnetic (square, round horns, apertures, etc.) or combined (aperture and dipole) [20].

Let us consider AAA consisting of antenna elements with two independent orthogonally polarised emitters, an example of such a linear AAA consisting of turnstiles AEs is presented in Fig. 11.

The transition to the consideration of three-component emitters in this case does not cause fundamental difficulties. Choosing antenna elements of the array from a certain orthogonal polarisation, and the same direction, the number of equal degrees of freedom is doubled compared with the case when the array consists of elements of one polarisation. In this regard, the coordination of AAA with controlled polarisation is that placing in one electrical centre of two antenna elements reduces the overall dimensions of the aperture without losing the number of degrees of freedom of the array.

Methods of space-time processing, as well as polarisation-time, are in the corresponding weight ratio or phase-out of the output voltages of receiving antenna elements: $x_i(t)$ at $i = \overline{1, N}$.

Thus, the general structures of the algorithms of polarisation-time and space-time processing coincide [Figs. 4 and 9, Eqs. (12), (13), (14), (18), (19), (20)]. However, the input parameters, the values of the variables as well as the physical results are somewhat different. Thus, the state equation of type (12) corresponds to fully polarised signals and interferences that do not change the polarisation parameters over time. For partially polarised signals and interferences, which is typical for observing the signals of moving objects, with the connection in the decametre range, in lines of distant tropospheric propagation and in other cases, the state equation has the form (18), and in the case of nonlinearities it has the form of (2). The mathematical interpretation of the coefficients $a_i(t)$ and $\beta_{ij}(t)$ remains the same, that is, these are the coefficients of wear and diffusion.

The identity of the problems of the synthesis of space-time and polarisation-time algorithms is convenient not only from the methodological point of view, but also has a number of constructive advantages. For the considered common problem (or the problem of AAA with controlled polarisation synthesis), we will analyse the efficiency of the used indicators (21). Figures 12 and 13 show graphs of coefficients

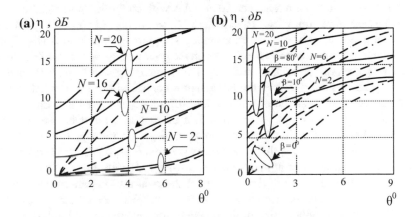

Fig. 12 **a** Dependence of η on $\theta°$ under the ratio $P_{int}/P_n = 10^2, P_s/P_n = 10$; **b** dependence of η on $\theta°$ under $\beta = \theta°$

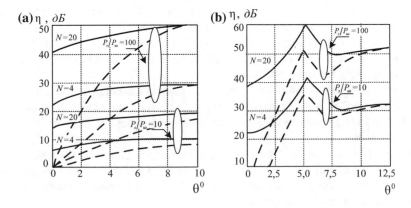

Fig. 13 **a** Dependence of η on $\theta°$ under $\beta = \theta°$; **b** dependence of η on $\theta°$ in case of two interferences effect

$\eta(t)$ where N characterizes an N-element grid in which one half of the elements have one polarisation, and the other half is orthogonal to it.

Figure 12 shows the graphs of the coefficient η, which is determined in the ratio $P_{int}/P_n = 10^2$; $P_s/P_n = 10$ for the different number of antenna elements N. Continuous lines correspond to $\beta = 30°$, intermittent lines correspond to $\beta = 0$. It is obvious that the coefficient η increases with the growth in the number of antenna elements. A noticeable decrease of η is noted when the interference $n(t)$ becomes closer to the signal by the space and polarisation parameters, and under $\theta = 0$ and $\beta = 0$ where the signal and interference coincide by these parameters, the situation is the worst: $\eta \rightarrow 0$ and the grid thus becomes "blind". In practice, to exclude such a situation, it is expedient to foresee the possibility of changing the polarisation of the transmitted signals, which allows to receive a significant positive effect even in the area of the coincidence of setting space spectra of signals and interferences (at $\theta \rightarrow 0$). This is an important advantage of AAA with controlled polarisation.

For the case of relations $P_{int}/P_n = 10^2$; $P_s/P_n = 10$, Fig. 13 shows graphs of the coefficient η indicating how much interference is suppressed in relation to the signal. Continuous lines correspond to $\beta = 80°$ when the polarised components of the signal and interference are significantly different. This situation ($\beta = 80°$) may correspond to the communication line with the polarisation control of the transmitted signals, which ensures maximization of polarisation differences in signal and interference at the receiving point. From the graphs it is seen that with the decrease of spatial differences between the signal and the noise (at $\theta \rightarrow 0$), the signal-to-noise ratio advantage is reduced by 3–8 dB. However, even under $\theta \rightarrow 0$ this gain is 40–80 dB, depending on the number of AAA elements. In this Figure, the dashed lines are graphs $\theta \rightarrow 0$ for the case of close values of the polarisation parameters between the signal and the interference ($\beta = 10°$). These graphs may reflect the situation where the control of polarisation of the transmission antenna is not implemented or is carried out with significant errors, for example, in the extended communication lines of the decametric range [20, 21].

The value of the gains here is 10–16 dB lower, however, when the spatial angles of the interference arrival change to 5–10° from the signal, this gain dramatically increases and little depends on the state of the polarisation of the interference. Intermitted lines show graphs in which the polarisation of the signal and the noise coincides ($\beta = 0°$). Curves for $N = 2$ show the effectiveness of polarisation-time processing using a single turnstile antenna.

Figure 12 shows graphs of η. Continuous lines correspond to $\beta = 80°$, dashed lines correspond to $\beta = 0°$. From the graphs, it is seen that the loss of efficiency of AAA with the coincidence of spatial parameters of the signal and interference are small (3–5 dB).

The graphs of η for the case of impact of two interferences, one of which comes with a fixed parameter $\theta_1 = 5°$, and the second parameter θ_2 varies, is shown in Fig. 13. Comparison of graphs on Figs. 12 and 13 shows a significant advantage of AAA with controlled polarisation.

4   Conclusions

Currently, active-passive antenna modules have emerged on the market, combining passive and active phased array antennas. The passive part replaces 2G and 3G antennas available to the mobile operators, and the active part solves the problem of adaptive space-time processing of 4G and 5G systems, thus significantly improving SINR. Such antenna systems can be implemented in base stations of next generation communications systems. Based on such antenna systems, algorithms for adaptive space-time and polarization signal processing using the criterion of minimum of the mean-square error compared to the reference (pilot) signal have been developed. This criterion, proposed by Widrow, is well-consistent with the traditional criteria for quality of information transmission of mobile radio systems. The main limitation of using this criterion is the need to have known reference signal on the receiving side. However, for modern and advanced mobile systems it is not significant as they use certain pilot signals.

The simulation confirmed the high efficiency of using antenna arrays with the algorithm of adaptation on pilot signal in mobile communication systems of new generations. The possibility of forming the maximum of directional pattern in the direction of the useful signal, determining the direction of the interference and its deep resection is confirmed. It has been found that this algorithm works best in difficult conditions with strong interference when its level approaches the level of the useful signal. The simulation also showed that the AAA, adapting by the pilot signal, sufficiently deeply suppress the interference and orientates the maximum of directional pattern in the direction of useful signal arrival. However, under convergence of directions between the useful signal and the interference adaptation gives an angle error relative to the real direction of the useful signal. The directional pattern is expanded compared to the case without adaptation, which causes some decrease in the gain of the antenna system and, accordingly, the signal level.

Adaptive polarization processing can largely eliminate this disadvantage. Thus, the use of AAA with controlled polarisation allows to reduce the linear dimensions of the array almost twice, which is especially important in the presence of significant constraints. At the same time, not only the quality of the processing is not reduced, but also the useful property is obtained, i.e. an opportunity to suppress interference, the spatial parameters of which are close or coincide with the useful signal, which significantly raises the reliability of electromagnetic availability.

References

1. Semenova O et al (2015) The fuzzy-controller for WiMAX networks. In: 2015 International Siberian conference on control and communications (SIBCON), Omsk, Russia, 21–23 May 2015, pp 1–4. https://doi.org/10.1109/sibcon.2015.7147214
2. Semenova O et al (2013) Access fuzzy controller for CDMA networks. In: 2013 International Siberian conference on control and communications (SIBCON), Krasnoyarsk, Russia, 12–13

Sept 2013, pp 1–2. https://doi.org/10.1109/sibcon.2013.6693644

3. Ageyev D, Al-Ansari A (2015) LTE RAN and services multi-period planning. In: 2015 Second international scientific-practical conference problems of infocommunications science and technology (PIC S&T). Kharkov, Ukraine: IEEE, pp 272–274. https://doi.org/10.1109/infocommst. 2015.7357334

4. Bondarenko O, Ageyev D, Mohammed O (2019) Optimization model for 5G network planning. In: 2019 IEEE 15th international conference on the experience of designing and application of CAD systems (CADSM). IEEE, pp 1–4. https://doi.org/10.1109/cadsm.2019.8779298

5. Ageyev DV, Salah MT (2016) Parametric synthesis of overlay networks with self-similar traffic. Telecommun Radio Eng (English translation of Elektrosvyaz and Radiotekhnika) 75(14):1231–1241

6. Naors Y, Anad A, Kashmoola MA, Moskalets M (2018) Analysis of the efficiency of spacetime access in the mobile communication systems based on an antenna array. East-Eur J Enterprise Technol 6(9(96)):38–47. https://doi.org/10.1587/1729-4061.2018.150921

7. Moskalets M, Samoylenkov O, Kulish K, Sielivanov K (2019) Analysis of restrictions influence in implementation of space-time access methods. Eskişehir Techn Univ J Sci Technol A Appl Sci Eng 20:149–156. https://doi.org/10.18038/estubtda.655714

8. Kohno R (1998) Spatial and temporal communication theory using adaptive antenna array. IEEE Pers Commun Mag 51:28–35

9. Shad F, Todd TD, Kezys V, Litva J (2001) Dynamic slot allocation (DSA) in indoor SDMA/TDMA using a smart antenna basestation. IEEE/ACM Trans Netw 9(1):69–81. https://doi.org/10.1109/90.909025

10. Stevanovic I, Skrivervik A, Mosig JR (2003) Smart antenna systems for mobile communications. Ecole Polytechnique Fédérale de Lausanne, Lausanne, Suisse, Tech. Rep. http://lemawww.epfl.ch. Accessed 23 Apr 2019

11. Godara LC (1997) Applications of antenna arrays to mobile communications, Part I: performance improvement, feasibility, and system considerations. Proc IEEE 85(7):1029–1030. https://doi.org/10.1109/JPROC.1997.611114

12. Okamoto G (2003) Developments and advances in smart antennas for wireless communications. Santa Clara University, Tech Rep. http://www.wmrc.com/businessbrieflng/pdf/wireless_2003/Publication/okamoto.pdf. Accessed 23 Apr 2019

13. Litva J, Lo T (1996) Digital beamforming in wireless communications. Artech House Publishers, Boston, MA

14. Widrow B (1989) Adaptive signal processing. Radio and Communication, Moscow

15. Kryvinska N (2010) Converged network service architecture: a platform for integrated services delivery and interworking. In: Electronic business series, vol 2. International Academic Publishers, Peter Lang Publishing Group

16. Kryvinska N, Zinterhof P, van Thanh D (2007) An analytical approach to the efficient real-time events/services handling in converged network environment. In: Enokido T, Barolli L, Takizawa M (eds) Network-based information systems. NBiS 2007. Lecture notes in computer science, vol 4658. Springer, Berlin

17. Loshakov V, Abdourahmane A (2017) Modeling adaptive communication system with MIMO and OFDM In: 2017 4th international scientific-practical conference problems of infocommunications science and technology (PIC S&T). Kharkov, Ukraine, 10–13 Oct 2017, IEEE, pp 581–592

18. Gooch R, Sublett B, Lonski R (1990) Adaptive beamformers in communications and direction finding Systems. In: 24th Asilomar conference on signals, systems and computers, vol 1, pp 11–15. REFERENCES 169

19. Compton RT (1988) Adaptive antennas: concepts and performance. Prentice Hall PTR, Upper Saddle River, NJ

20. Voskresensksy DI, Bakhraha LD (eds) (1989) Problems of antenna technology. Radio and Communication, Moscow

21. Popovskij V, Barkalov A, Titarenko L (2011) Control and adaptation in telecommunication system: mathematical foundations, vol 94. Springer Science & Business Media

Fast ReRoute Tensor Model with Quality of Service Protection Under Multiple Parameters

Oleksandr Lemeshko⬛, Maryna Yevdokymenko⬛, and Oleksandra Yeremenko⬛

Abstract The paper proposes a flow-based model of Fast ReRoute in a multiservice network with Quality of Service protection under multiple parameters, such as bandwidth, probability of packet loss, and average end-to-end delay. In the course of solving the task within the framework of this model, a result was obtained, the use of which contributes to the optimal use of the available network resource while ensuring a given level of Quality of Service and Quality of Resilience over both the primary and backup routes in the infocommunication network. The numerical example demonstrates the operability of the proposed Fast ReRoute flow-based model with detailing the procedures for geometrization of the network structure: the choice of space, coordinate systems and the formation of covariant transformation matrices of the introduced bases. The example covered the case of a possible failure of an arbitrary network router and (or) communication links incident to it. As a result of the problem solution, the primary and backup multipaths were obtained, along which a given level of Quality of Service was ensured in terms of bandwidth, the probability of packet loss and the average end-to-end delay.

Keywords Fast ReRoute · Tensor model · Quality of Service · Quality of Resilience · Protection · Bandwidth · Packet loss · Delay

1 Introduction

Construction of resilient networks is the response to the infocommunication industry regarding continuously increasing demands of users and network applications

O. Lemeshko · M. Yevdokymenko (✉) · O. Yeremenko
Kharkiv National University of Radio Electronics, Kharkiv, Ukraine
e-mail: maryna.yevdokymenko@ieee.org

O. Lemeshko
e-mail: oleksandr.lemeshko.ua@ieee.org

O. Yeremenko
e-mail: oleksandra.yeremenko.ua@ieee.org

© The Editor(s) (if applicable) and The Author(s), under exclusive license to Springer Nature Switzerland AG 2021

T. Radivilova et al. (eds.), *Data-Centric Business and Applications*, Lecture Notes on Data Engineering and Communications Technologies 48,
https://doi.org/10.1007/978-3-030-43070-2_22

489

for the Quality of Service (QoS) in the event of possible failures. In this case, the main reasons that may lead to a denial of service, as stated in [1–12], are hardware and software failures of network equipment. At the same time, science and practice increasingly use the Quality of Resilience (QoR) metrics to quantify the fault-tolerance of network solutions.

In order to ensure compliance with the requirements of the system approach to ensuring the resilience of infocommunication networks (ICN), the means of increasing fault-tolerance are technologically used at all Layers of the OSI (Open Systems Interconnection Basic Reference Model). At the same time, currently special attention [10, 13–18] is increasingly paid to hardware and software solutions, which are realized at the Network Layer through the protocols of fault-tolerant routing and technologies of fast rerouting. It should be noted that for the quick response of the network to probable failures of the switching equipment, the routing problem in its content and form is considerably complicated, since it is necessary to calculate not only the primary paths but also the set of backup routes with the protection of the network elements—the link/node/path/segment and their bandwidth. However, in multiservice infocommunication networks, for most multimedia flows, it is important to significantly expand the list of Quality of Service indicators, regarding the values of which the reservation of the network resource is performed. Network bandwidth provisioning, as a key indicator of the Quality of Service, should be supplemented by protection other key QoS indicators—the average end-to-end delay and the packet loss probability. Consideration of these requirements and aspects should take place at the level of mathematical models and methods, which are theoretical and algorithmic-software basis of perspective protocols of fast rerouting with protection of the Quality of Service level on a set of listed indicators.

2 Overview of the Existing Solutions and Requirements Formulation to Fast ReRoute in the Direction of Implementation of the Quality of Service Protection Schemes

Based on the analysis of existing results and prospective solutions for Fast-ReRoute (FRR), a classification of promising schemes for protecting the level of QoS in the infocommunication networks has been conducted:

1. For the first type of QoS^1–FRR, there is a solution for fast rerouting with the protection of one network performance (NP) indicator, for example, bandwidth [5–18].
2. The QoS^2–FRR provides the support of the QoS for two indicators of network performance, such that bandwidth and average delay or bandwidth and probability of packet loss [19, 20].

3. In addition, QoS^3–FRR, when the QoS is protected by an extended set of indicators, is based on three indicators of network performance, such as the bandwidth, the average delay and the probability of packet loss.

The traditional protection schemes for network elements during fast rerouting include the protection of the link, node, and path. In this regard, it should be noted that many works devoted to fast rerouting solutions are usually associated with the development of path protection schemes using the single path routing strategy [7, 10–12]. The main disadvantage of these solutions is the inefficient use of the available network resource. At the same time, only a few works are devoted to the protection of the bandwidth level [14–18], which can be attributed to the first type, QoS^1–FRR. The proposed approaches in [14–18] are aimed at the protection of specific network elements (for example, protection of border routers) and can be used for various types of networks [17, 18]. At the same time, the focus on implementing support for load balancing during fast rerouting also contributes to improving the Quality of Service and more efficient use of network bandwidth in general [16].

Among the solutions in the framework of QoS^2–FRR, when the protection of the Quality of Service level by two indicators is implemented, the following can be pointed out [19–23]. For example, in [19], an algorithm was proposed that forms the basis of the smart routing system, in which the level of Quality of Service is protected according to network delay and packet loss rates. Whereas in [20], the bandwidth and end-to-end delay protection is implemented. Both solutions are offered for using in SD-WAN.

In [23], a solution was proposed according to the QoS^2–FRR scheme, which is represented by a nonlinear fast rerouting flow-based model with the protection of two parameters of network bandwidth performance and packet loss probability. This model allows to consider the limitations of the queue buffer on the interfaces of the network routers and to control the probable overload of the link resource.

It is also important to note the result obtained in [22], where a tensor model of fault-tolerant routing with maintenance of Quality of Service in an infocommunication network was proposed, the novelty of which is to ensure the implementation of protection of Quality of Service level on the parameters of bandwidth and the average end-to-end packet delay. The presented solution is based on the tensor description of the fault-tolerant routing process, which allowed to present the required protection conditions in an analytical form, as well as to solve an optimization problem for the calculation of the primary and backup routes, along which the given level of Quality of Service was provided.

A promising direction in the development of solutions for fast rerouting is the support of the third by the complexity type of described schemes due to its suitability for protecting the level of QoS by the main NP indicators, which are critical for multimedia applications.

The analysis of known solutions in the field of fault-tolerant routing made it possible to formulate a list of key requirements, which should correspond to promising solutions in this area and, above all, mathematical models and methods on which they are based:

- considering the flow-based nature of traffic, which is a distinctive feature of most multimedia services and a compulsory moment when implementing bandwidth protection schemes and other indicators of network Quality of Service;
- problem statement as optimization task: the focus on optimal usage of available network resources;
- high scalability of solutions for fault-tolerant routing;
- support of basic protection schemes for network elements (link/node/path/bandwidth and QoS level for a set of indicators);
- coordinated solving of specific tasks for fault-tolerant routing, for example, default gateway protection, fast rerouting, etc.;
- enhancing the capabilities of existing solutions to support load balancing associated with the implementation of a multipath routing strategy with appropriate support for protection schemes for not a single path, but a multipath, that is, several paths in which packets of the same flow are transmitted;
- acceptable computational complexity of routing solutions.

A classification of advanced schemes for protecting the Quality of Service level that is important to implement during fault-tolerant routing, mainly multimedia flows, has shown that promising routing solutions should provide QoS protection for a variety of network performance indicators or Quality of Experience (QoE) simultaneously. All this requires the development of new or improved existing mathematical models and methods of fault-tolerant routing in accordance with the requirements. Thus, the problem with the design of the Fast ReRoute tensor model with Quality of Service protection under multiple parameters in accordance with the QoS^3–FRR scheme is relevant.

3 Flow-Based Model of Fast ReRoute with Considering Packet Loss and Delays in Infocommunication Network

Given the need to ensure and, in the future, protect the level of Quality of Service the user flows over a set of different indicators—bandwidth, average delay, and packet loss probability when they are fast rerouted, it is important to have an adequate mathematical model of calculations that takes into account these aspects of network operation. Based on the analysis of papers [22, 24–27], it can be concluded that in solving the problems of QoS routing and QoS fast rerouting, the tensor approach has proven itself as an effective means of an integral and multidimensional description of an infocommunication network. Tensor models are built on the basis of simultaneous and complementary use of information on the structural and functional construction of the network, which leads to an increase in the adequacy of its mathematical description and taking into account those patterns that would be hidden from the researcher when using other models [28].

In the framework of the proposed fast rerouting model, the structure of an infocommunication network is described using a one-dimensional simplicial complex (one-dimensional network) $S = (U, V)$, where $U = \{u_i, i = \overline{1, m}\}$ is a set of zero-dimensional simplexes—network nodes (routers), and $V = \{v_z = (i, j); z = \overline{1, n}; i, j = \overline{1, m}; i \neq j\}$ is a set of one-dimensional simplexes— network edges, where edge $v_z = (i, j)$ connects routers u_i and u_j. Thus, depending on the aspect of considering the processes occurring in the ICN, the characteristics and parameters related to the communication links of the network will be denoted by a single or double index. In the first case, when the link is considered as an independent object, continuous numbering of edges (links) will be applied, and in the second case, when it is important to take into account the location of the link in the network, numbering will be done via node numbers. For example, for each link modeled by the edge $v_z = (i, j) \in V$, the bandwidth measured in packets per second (1/s) will be denoted both as φ_z and $\varphi_{i,j}$ (1/s). Each network router has several interfaces through which it sends packets to its neighbor nodes. Moreover, the interface numbers for each individual node correspond to the numbers of adjacent nodes connected through them. Then $\varphi_{i,j}$ actually determines the bandwidth of the jth interface of the ith node.

To implement fast rerouting in the network [14–16], it is necessary to calculate two types of routing variables $x_{i,j}^k$ and $\bar{x}_{i,j}^k$, each characterizing the fraction of intensity of the kth flow sent from the node u_i to the node u_j via the link (i, j) that comprises the primary or backup path (multipath), respectively. In the future, the multipath will denote a set of paths between a given pair of source-destination nodes. When using a multipath routing scheme, the following restrictions are imposed on these routing variables:

$$0 < x_{i,j}^k \leq 1 \text{ and } 0 \leq \bar{x}_{i,j}^k \leq 1. \tag{1}$$

In the proposed fast rerouting model, the conditions of flow conservation on a network routers, which are part of the primary path, allow to take into account possible packet losses caused by queue buffer overload, unlike traditional conditions [14], and have the following form [24]:

$$\begin{cases} \sum\limits_{j:(i,j)\in V} x_{i,j}^k = 1, \ k \in K, \ i = s_k; \\ \sum\limits_{j:(i,j)\in V} x_{i,j}^k - \sum\limits_{j:(j,i)\in V} x_{j,i}^k (1 - p_{j,i}^k) = 0, \ k \in K, \ i \neq s_k, d_k; \\ \sum\limits_{j:(j,i)\in V} x_{j,i}^k (1 - p_{i,j}^k) = \varepsilon^k, \ k \in K, \ i = d_k, \end{cases} \tag{2}$$

where K is the set of flows in the network; s_k is the source node for the packets of the kth flow; d_k is the destination node for packets of the kth flow; ε^k is the fraction of the kth flow served by the network using the primary path (multipath), i.e. packets which are successfully delivered to the destination node; $p_{i,j}^k$ is the probability of packet loss of the kth flow on the jth interface of the ith node, when it is used by the primary path.

Constraints similar to conditions (2) are also imposed on routing variables of the backup path [14–16]:

$$\begin{cases} \sum_{j:(i,j)\in V} \bar{x}_{i,j}^k = 1, \ k \in K, \ i = s_k; \\ \sum_{j:(i,j)\in V} \bar{x}_{i,j}^k - \sum_{j:(j,i)\in V} \bar{x}_{j,i}^k (1 - \bar{p}_{j,i}^k) = 0, \ k \in K, \ i \neq s_k, d_k; \\ \sum_{j:(j,i)\in V} \bar{x}_{j,i}^k (1 - \bar{p}_{i,j}^k) = \bar{\varepsilon}^k, \ k \in K, \ i = d_k, \end{cases} \qquad (3)$$

where $\bar{\varepsilon}^k$ is the fraction of the kth flow served by the network using a backup path (multipath); $\bar{p}_{i,j}^k$ is the packet loss probability of the kth flow on the jth interface of the ith node, when it is used by the backup path.

As it is known [29, 30], each type of traffic and service discipline has its own interface operation model, represented by a particular queuing system, for example, M/M/1/N, M/D/1/N, SS/M/1/N, etc. In this paper, we consider the case when packet loss on the interface (in the communication link) is caused only by its overload. Then to determine the packet loss probability of $p_{i,j}^k$ and $\bar{p}_{i,j}^k$ of the kth flow on the jth interface of the ith node when using the primary and backup routes, respectively, for example, a queuing system with failures of M/M/1/N type will be used. Then the packet loss probability at the nodes interfaces of the primary $\left(p_{i,j}^k \right)$ and backup $\left(\bar{p}_{i,j}^k \right)$ routes can be calculated as:

$$p_{i,j}^k = \frac{\left(1 - \rho_{i,j}^k\right)\left(\rho_{i,j}^k\right)^N}{1 - \left(\rho_{i,j}^k\right)^{N+1}} \text{ and } \bar{p}_{i,j}^k = \frac{\left(1 - \bar{\rho}_{i,j}^k\right)\left(\bar{\rho}_{i,j}^k\right)^N}{1 - \left(\bar{\rho}_{i,j}^k\right)^{N+1}}, \qquad (4)$$

where $\rho_{i,j}^k = \frac{\lambda_{i,j}}{\varphi_{i,j}}$ and $\bar{\rho}_{i,j}^k = \frac{\lambda_{i,j}}{\varphi_{i,j}}$ is the link utilization (i, j) when it is used by the kth flow in the primary or backup route, respectively;

$\lambda_{i,j}$ and $\bar{\lambda}_{i,j}$ are the intensities of the aggregated flow sent to the link (i, j) when it is used by the kth flow in the primary or backup path, respectively, which, by analogy with [22], are calculated as:

$$\begin{cases} \lambda_{i,j} = \lambda_k^{\langle req \rangle} x_{i,j}^k + \sum_{p \in K, p \neq k} \lambda_p^{\langle req \rangle} \max[x_{i,j}^p, \bar{x}_{i,j}^p]; \\ \bar{\lambda}_{i,j} = \lambda_k^{\langle req \rangle} \bar{x}_{i,j}^k + \sum_{p \in K, p \neq k} \lambda_p^{\langle req \rangle} \max[x_{i,j}^p, \bar{x}_{i,j}^p], \end{cases} \qquad (5)$$

where $\lambda_k^{\langle req \rangle}$ is the average intensity of the kth packet flow arriving at the network for service, which specifies the QoS requirements for the packet transmission rate (bandwidth).

The physical meaning of expressions (5) is that the calculation of link utilization and the main indicators of Quality of Service, the packet loss probability (4), and further the average packet delay in communication links will be obtained for the worst case in terms of the aggregated flow intensity value. The second term on the right-hand side of expressions (5) is just introduced for this purpose: when determining the intensity of the aggregated flow, the maximum intensity value of each kth packet flow in an arbitrary link (i, j) is taken into account when it is used by the primary or backup path.

At the same time, it is important to understand that for each kth individual flow, its intensity of packets sent to the link (i, j) belonging to the primary or backup paths is determined by the expressions

$$\lambda_{i,j}^k = \lambda_k^{\langle req \rangle} x_{i,j}^k \text{ and } \bar{\lambda}_{i,j}^k = \lambda_k^{\langle req \rangle} \bar{x}_{i,j}^k. \tag{6}$$

Then the intensity of the dropped packets of the kth flow on the jth interface of the ith node belonging to the primary or backup path will be defined as

$$r_{i,j}^k = \lambda_k^{\langle req \rangle} x_{i,j}^k p_{i,j}^k \text{ and } \bar{r}_{i,j}^k = \lambda_k^{\langle req \rangle} \bar{x}_{i,j}^k \bar{p}_{i,j}^k. \tag{7}$$

4 Conditions for Protection of Network Elements and Level of Quality of Service According to Bandwidth and Packet Loss Probability During Fast ReRouting in Infocommunication Network

According to the results obtained in [14–16], we introduce a number of conditions that describe the implementation of protection schemes (reservation) of network elements and the level of Quality of Service in the network during fast rerouting of packets:

1. Conditions for protecting the communication link $(i, j) \in V$:

$$0 \le \bar{x}_{i,j}^k \le \delta_{i,j}^k, \tag{8}$$

where

$$\delta_{i,j}^k = \begin{cases} 0, & under \ (i, j) \in V \ link \ protection; \\ 1, & otherwise. \end{cases} \tag{9}$$

2. Protection conditions for the node $u_i \in U$ obtained by generalizing condition (8) in case of protection of multiple communication links incident to the protected node are:

$$0 \le \bar{x}_{i,j}^k \le \delta_{i,j}^k \text{ under } u_j \in u_i^*, \ j = \overline{1, m}, \tag{10}$$

where $u_i^* = \{u_j : (i, j) \neq 0; \ j = \overline{1, m}; \ i \neq j\}$ is the subset of routers that are adjacent to the router u_i, and the selection of values $\delta_{i,j}^k$ is performed similarly to condition (9).

3. Conditions for the implementation of the network bandwidth protection scheme as the main QoS indicator for fast rerouting are [25]:

$$\sum_{k \in K} \lambda_k^{\langle req \rangle} \max[x_{i,j}^k, \bar{x}_{i,j}^k] \leq \varphi_{i,j} \text{ under } (i, j) \in V. \tag{11}$$

4. The conditions for the protection of such a QoS indicator as the packet loss probability of the kth flow in the network based on (1)-(4), are as follows [24]:

$$1 - \varepsilon^k \leq p_k^{\langle req \rangle}; \tag{12}$$

$$1 - \bar{\varepsilon}^k \leq p_k^{\langle req \rangle}, \tag{13}$$

where $p_k^{\langle req \rangle}$ is QoS requirements regarding the allowed boundary values of the packet loss probability of the kth flow in the network.

To introduce protection conditions for such a QoS indicator as average end-to-end packet delay, which important for almost all types of traffic, it is advisable to apply tensor modeling functional of routing processes [24–27], which makes it possible to analytically describe the relationship between the previously introduced basic indicators of Quality of Service in the network.

5 Formalization of Conditions for Protecting Quality of Service Over Average Packet End-to-End Delay During Fast ReRouting in Infocommunication Network

In tensor modeling of a network, its structure determines an anisotropic space when calculating the primary multipath [22, 28]. The dimension of this space is determined by the total number of edges in the network and equals n. In addition, each independent path (edge, circuit or node pair) determines the coordinate axis in the space structure. As a rule, a network is modeled by a connected one-dimensional network, i.e. contains one connecting component, then the cyclomatic number μ and network rank ϕ respectively determine the number of basic circuits and node pairs, for which the following expressions are true:

$$\phi = m - 1, \quad \mu = n - m + 1, \quad n = \phi + \mu. \tag{14}$$

Base circuits and node pairs define a basis in n-dimensional space that corresponds to the network structure. Any other path of the network can be expressed through the basic paths and the algebraic sum of paths is the path that goes through all the sum components according to their orientation [28].

In the selected space, the infocommunication network in solving the problem of rerouting with the protection of the Quality of Service level can be represented by a mixed bivalent tensor

$$Q = T \otimes \Lambda, \tag{15}$$

where $\otimes$ is the tensor multiplication operator, and the components of the tensor Q are the univalent covariant tensor of average packet delays T and the univalent contravariant flow intensity tensor Λ in the coordinate paths of the network [24, 26].

In accordance with the results obtained in the articles [24–27], the tensor equation describing the infocommunication network in the introduced geometric space has the following form

$$\Lambda = GT, \tag{16}$$

where G is a doubly contravariant metric tensor in the introduced geometric space.

The infocommunication network within the framework of the tensor representation (15), (16) [24, 26] in order to obtain the desired protection conditions for the Quality of Service level will be described in two coordinate systems—network edges $\{v_l, l = \overline{1, n}\}$ as well as independent circuits $\{\pi_i, i = \overline{1, \mu}\}$ and node pairs $\{\eta_j, j = \overline{1, \phi}\}$. Then, in the coordinate system of network edges within the framework of the proposed model, when the network interface is modeled as a queuing system with failures of $M/M/1/N$ type, the average packet delay in the network communication link (i, j) is approximated by the expression [24]:

$$\tau_{i,j} = \frac{\rho_{i,j} - \rho_{i,j}^{N+2} - (N+1)\rho_{i,j}^{N+1}(1 - \rho_{i,j})}{\lambda_{i,j}(1 - \rho_{i,j}^{N+1})(1 - \rho_{i,j})}. \tag{17}$$

For the subsequent tensor generalization of the mathematical model of fast rerouting, taking into account the end-to-end link numbering, we write expression (17) in the form

$$\tau_z^v = \frac{\rho_z - \rho_z^{N+2} - (N+1)\rho_z^{N+1}(1 - \rho_z)}{\lambda_z(1 - \rho_z^{N+1})(1 - \rho_z)\lambda_z^v}\lambda_v^z, \quad v_z = (i, j) \in V, \tag{18}$$

where z is the number of the communication link in the network, and the index v indicates that all the parameters of expression (18) are assigned to the network edges; τ_z^v is average packet delay of the kth flow in the zth communication link; λ_v^z is the intensity of the kth flow in the zth communication link. For clarity, the index k near the QoS indicators τ_z^v and λ_v^z in expression (18) is omitted.

At the same time, in accordance with the postulate of the second generalization proposed by Kron [28], the system of expressions (18), referred to the kth flow, can be replaced by the following vector equation:

$$\Lambda_v = G_v T_v, \tag{19}$$

where $\Lambda_v = \begin{bmatrix} \lambda_v^1 & \dots & \lambda_v^z & \dots & \lambda_v^n \end{bmatrix}^t$ and $T_v = \begin{bmatrix} \tau_1^v & \dots & \tau_z^v & \dots & \tau_n^v \end{bmatrix}^t$ are the projections of the tensors Λ and T (15) in the coordinate system of the network edges, represented as vectors of flow intensities and average packet delays in the network edges of the size n, respectively; $[\cdot]^t$ is matrix transposition operation; $G_v = \left\| g_v^{zz} \right\|$ is a diagonal matrix of the size $n \times n$, the elements of the main diagonal of which are calculated according to expressions (18) related to the corresponding edges of the network, $\{ v_z, z = \overline{1, n} \}$ i.e.

$$g_v^{zz} = \frac{\lambda_z (1 - \rho_z^{N+1})(1 - \rho_z)\lambda_v^z}{\rho_z - \rho_z^{N+2} - (N + 1)\rho_z^{N+1}(1 - \rho_z)}. \tag{20}$$

The matrix G_v is a projection of the doubly contravariant metric tensor G of the introduced geometric space, and the vector-matrix Eq. (19) itself is a representation of the tensor Eq. (16) in the coordinate system of the network edges.

The considered coordinate systems of edges and independent circuits and node pairs of the network are orthogonal and are introduced for the same n-dimensional space. Therefore, there are unique rules for transforming the coordinates of any geometric objects when moving from one basis to another. For tensors, by definition [28], the coordinate transformation rules are linear and formalized using a nonsingular square matrix of the size $n \times n$:

$$\Lambda_v = C \, \Lambda_{\pi\eta}, \tag{21}$$

where $\Lambda_{\pi\eta}$ is the projection of the tensor Λ in the coordinate system of circuits and node pairs represented as a vector of the size $[n \times 1]$; C is the $n \times n$ matrix of contravariant transformation of tensors. At the same time, in the coordinate system of circuits and node pairs, the tensors projections of the flow intensity Λ and average delay T, represented by vectors $\Lambda_{\pi\eta}$ and $T_{\pi\eta}$ respectively, have the size n and the following components

$$\Lambda_{\pi\eta} = \left\| \begin{matrix} \Lambda_\pi \\ -- \\ \Lambda_\eta \end{matrix} \right\|; \; \Lambda_\pi = \left\| \begin{matrix} \lambda_\pi^1 \\ \vdots \\ \lambda_\pi^j \\ \vdots \\ \lambda_\pi^\mu \end{matrix} \right\|; \; \Lambda_\eta = \left\| \begin{matrix} \lambda_\eta^1 \\ \vdots \\ \lambda_\eta^p \\ \vdots \\ \lambda_\eta^\phi \end{matrix} \right\|, \; T_{\pi\eta} = \left\| \begin{matrix} T_\pi \\ -- \\ T_\eta \end{matrix} \right\|; \; T_\pi = \left\| \begin{matrix} \tau_1^\pi \\ \vdots \\ \tau_j^\pi \\ \vdots \\ \tau_\mu^\pi \end{matrix} \right\|; \; T_\eta = \left\| \begin{matrix} \tau_1^\eta \\ \vdots \\ \tau_p^\eta \\ \vdots \\ \tau_\phi^\eta \end{matrix} \right\| \tag{22}$$

where Λ_π, T_π are vectors of the intensities and average delays of the flow packets arriving to the network circuits; Λ_η, T_η, are vectors of intensities and average delays of flow packets arriving at its nodes or outgoing through them to the network users, respectively; λ_π^j is the flow intensity in the network circuit π_j; λ_η^p is the intensity of the external flow arriving the network or leaving the network through the node pair η_p; τ_j^π, τ_p^η are the average packet delays in the circuit π_j and between a pair of network nodes η_p, respectively.

The covariant nature of the delay tensor T determines the following law of coordinate transformation:

$$T_v = A\, T_{\pi\eta}, \tag{23}$$

where A is the covariant transformation matrix of the size $n \times n$ associated with the matrix C by the orthogonality condition $CA^t = I$; and I is the unit matrix of the size $n \times n$.

In the coordinate system of independent circuits and node pairs, the tensor Eq. (16) takes the form [24]:

$$\Lambda_{\pi\eta} = G_{\pi\eta} T_{\pi\eta}. \tag{24}$$

where $G_{\pi\eta}$ is the projection of the doubly contravariant metric tensor G in the coordinate system of independent circuits and node pairs represented by the matrix of the size $n \times n$.

Then the projection of the doubly contravariant metric tensor, when the coordinate system under consideration is changed, is transformed as follows [24, 28]:

$$G_{\pi\eta} = A^t G_v\, A. \tag{25}$$

Let us write the expression (24) with (22) in the following form

$$\left\| \begin{array}{c} \Lambda_\pi \\ \hline \Lambda_\eta \end{array} \right\| = \left\| \begin{array}{c|c} G_{\pi\eta}^{\langle 1\rangle} & G_{\pi\eta}^{\langle 2\rangle} \\ \hline G_{\pi\eta}^{\langle 3\rangle} & G_{\pi\eta}^{\langle 4\rangle} \end{array} \right\| \cdot \left\| \begin{array}{c} T_\pi \\ \hline T_\eta \end{array} \right\|, \tag{26}$$

where $\left\| \begin{array}{c|c} G_{\pi\eta}^{\langle 1\rangle} & G_{\pi\eta}^{\langle 2\rangle} \\ \hline G_{\pi\eta}^{\langle 3\rangle} & G_{\pi\eta}^{\langle 4\rangle} \end{array} \right\| = G_{\pi\eta}$, and $G_{\pi\eta}^{\langle 1\rangle}$, $G_{\pi\eta}^{\langle 4\rangle}$ are square submatrices of the size $\mu \times \mu$ and $\phi \times \phi$ respectively; $G_{\pi\eta}^{\langle 2\rangle}$ is the submatrix of the size $\mu \times \phi$; $G_{\pi\eta}^{\langle 3\rangle}$ is the submatrix of the size $\phi \times \mu$.

Based on the physical meaning of the components T_π, the following condition must be met:

$$T_\pi = 0, \tag{27}$$

which guarantees, firstly, the absence of loops in the routes, and secondly, the same average packet delay along each of them between the selected source-destination nodes.

Then from the expression (26) according to (27) there is a relation

$$\Lambda_\eta = G_{\pi\eta}^{\langle 4 \rangle} T_\eta. \tag{28}$$

Let us represent the vectors Λ_η and T_η from (28) in the following form

$$\Lambda_\eta = \left\|\begin{array}{c} \lambda_{(\eta)}^1 \\ --- \\ \Lambda_{\eta-1} \end{array}\right\|, \; \Lambda_{\eta-1} = \left\|\begin{array}{c} \lambda_\eta^2 \\ \vdots \\ \lambda_\eta^p \\ \vdots \\ \lambda_\eta^\phi \end{array}\right\|, \; T_\eta = \left\|\begin{array}{c} \tau_1^{(\eta)} \\ --- \\ T_{\eta-1} \end{array}\right\|, \; T_{\eta-1} = \left\|\begin{array}{c} \tau_2^\eta \\ \vdots \\ \tau_p^\eta \\ \vdots \\ \tau_\phi^\eta \end{array}\right\| \tag{29}$$

In addition, the coordinates of the vector Λ_η indicate the intensity of the flow between the reference node (usually the source node) of the network and other nodes of the network. Thus, the coordinate $\lambda_{(\eta)}^1$ determines the intensity of the flow, outgoing through the first node pair, in which the pair of nodes is the source-destination, i.e. $\lambda_{(\eta)}^1 = \lambda_k^{\langle req \rangle} \varepsilon^k$. In turn, the coordinate $\tau_1^{(\eta)}$ should not exceed the value of the average end-to-end packet delay in the network, i.e. $\tau_1^{(\eta)} \le \tau_k^{\langle req \rangle}$. For transit nodes that are not final recipients of packets, the coordinates of the subvector $\Lambda_{\eta-1}$ (29), i.e. values λ_η^j, should determine the total intensity of packet loss over all the interfaces for each individual node arriving the jth node pair. Hence, for example, if the number of the jth node pair coincides with the node number, then, taking into account (7), the intensity value of the dropped packets λ_η^j is calculated as

$$\lambda_\eta^j = \sum_{u_i \in u_j^*} r_{j,i}.$$

Then the expression (28) given (29) can be converted to the form

$$\left\|\begin{array}{c} \lambda_{(\eta)}^1 \\ --- \\ \Lambda_{\eta-1} \end{array}\right\| = \left\|\begin{array}{cc} G_{\pi\eta}^{\langle 4,1 \rangle} & G_{\pi\eta}^{\langle 4,2 \rangle} \\ --- + --- \\ G_{\pi\eta}^{\langle 4,3 \rangle} & G_{\pi\eta}^{\langle 4,4 \rangle} \end{array}\right\| \cdot \left\|\begin{array}{c} \tau_1^{(\eta)} \\ --- \\ T_{\eta-1} \end{array}\right\|, \tag{30}$$

where $G_{\pi\eta}^{\langle 4,1 \rangle}$ is the first element of the matrix $G_{\pi\eta}^{\langle 4 \rangle}$; $G_{\pi\eta}^{\langle 4,2 \rangle}$ is the submatrix of the matrix $G_{\pi\eta}^{\langle 4 \rangle}$ of the size $1 \times (\phi - 1)$; $G_{\pi\eta}^{\langle 4,3 \rangle}$ is the submatrix of the matrix $G_{\pi\eta}^{\langle 4 \rangle}$ of the size $(\phi - 1) \times 1$; and $G_{\pi\eta}^{\langle 4,4 \rangle}$ is the submatrix of the matrix $G_{\pi\eta}^{\langle 4 \rangle}$ of the size $(\phi - 1) \times (\phi - 1)$.

As a result of the mathematical transformation of expression (30), the condition for ensuring the Quality of Service in terms of bandwidth, average end-to-end delay and the packet loss probability for each kth flow transmitted along the primary multipath takes the following form similar to [24]:

$$\lambda_k^{\langle req \rangle} \varepsilon^k \leq G_{\pi\eta}^{\langle 4,2 \rangle} \left[G_{\pi\eta}^{\langle 4,4 \rangle} \right]^{-1} \Lambda_{\eta-1} + \left(G_{\pi\eta}^{\langle 4,1 \rangle} - G_{\pi\eta}^{\langle 4,2 \rangle} \left[G_{\pi\eta}^{\langle 4,4 \rangle} \right]^{-1} G_{\pi\eta}^{\langle 4,3 \rangle} \right) \tau_k^{\langle req \rangle}. \quad (31)$$

Thus, inequality (31), together with expressions (11)–(13), are the desired conditions for ensuring Quality of Service in terms of heterogeneous QoS indicators—rate ($\lambda_k^{\langle req \rangle}$), time ($\tau_k^{\langle req \rangle}$), and reliability indicators ($p_k^{\langle req \rangle}$) along the primary multipath. In addition, the requirement for equality to zero circuit delays (27) provides a minimum and equal for all calculated paths average packet delays [26], which helps to minimize jitter delay caused by the implementation of a multipath routing strategy.

The method of obtaining the desired conditions for ensuring the Quality of Service for a set of the mentioned indicators, but already along the backup multipath, is similar to the one described above, however, has several important features. To do this, it is also necessary to build a tensor model of an infocommunication network $\bar{S}$, in which the network structure already defines the discrete $\bar{n}$-dimensional space where $\bar{n}$ is the number of communication links in the network except for those links that are to be protected. In this case, the tensor model of the network (15), (16) is considered in two coordinate systems: network edges $\left\{ \bar{v}_l, l = \overline{1, \bar{n}} \right\}$ as well as linearly independent circuits $\left\{ \bar{\pi}_i, i = \overline{1, \bar{\mu}} \right\}$ and node pairs $\left\{ \bar{\eta}_j, j = \overline{1, \bar{\phi}} \right\}$ with

$$\bar{\mu}(\bar{S}) = \bar{n} - \bar{m} + 1 \text{ and } \bar{\varphi}(\bar{S}) = \bar{m} - 1 \quad (32)$$

which determine the number of basic circuits and node pairs in the network $\bar{S}$, respectively, when using the backup path.

The space metric will be represented by the tensor $\bar{G}$, the projections coordinates of which in the coordinate system of the edges are represented by the diagonal elements $\bar{n} \times \bar{n}$ of the matrix $\bar{G}_v$ and relate to each kth flow, which is transmitted via the backup multipath:

$$\bar{g}_v^{zz} = \frac{\bar{\lambda}_z (1 - \bar{\rho}_z^{N+1})(1 - \bar{\rho}_z) \bar{\lambda}_v^z}{\bar{\rho}_z - \bar{\rho}_z^{N+2} - (N+1)\bar{\rho}_z^{N+1}(1 - \bar{\rho}_z)}. \quad (33)$$

The relationship of the projections of the tensor $\bar{G}$ in the introduced coordinate systems is determined by a formula similar to (25):

$$\bar{G}_{\pi\eta} = \bar{A}^t \bar{G}_v \bar{A}, \quad (34)$$

where $\bar{A}$ is the $\bar{n} \times \bar{n}$ covariant coordinate transformation matrix, but for the space used in the calculation of the backup path; $\bar{G}_{\pi\eta}$ is the $\bar{n} \times \bar{n}$ matrix—projection of the tensor $\bar{G}$ in the coordinate system of circuits and node pairs, which can be represented in the form

$$
\bar{G}_{\pi\eta} = \left\| \begin{array}{c|c} \bar{G}_{\pi\eta}^{\langle 1 \rangle} & \bar{G}_{\pi\eta}^{\langle 2 \rangle} \\ --- + --- \\ \bar{G}_{\pi\eta}^{\langle 3 \rangle} & \bar{G}_{\pi\eta}^{\langle 4 \rangle} \end{array} \right\|, \ \bar{G}_{\pi\eta}^{\langle 4 \rangle} = \left\| \begin{array}{c|c} \bar{G}_{\pi\eta}^{\langle 4,1 \rangle} & \bar{G}_{\pi\eta}^{\langle 4,2 \rangle} \\ --- + --- \\ \bar{G}_{\pi\eta}^{\langle 4,3 \rangle} & \bar{G}_{\pi\eta}^{\langle 4,4 \rangle} \end{array} \right\|,
$$

where $\bar{G}_{\pi\eta}^{\langle 1 \rangle}$ and $\bar{G}_{\pi\eta}^{\langle 4 \rangle}$ are square submatrices of dimension $\bar{\mu} \times \bar{\mu}$ and $\bar{\phi} \times \bar{\phi}$, accordingly, $\bar{G}_{\pi\eta}^{\langle 2 \rangle}$ is the submatrix of size $\bar{\mu} \times \bar{\phi}$, $\bar{G}_{\pi\eta}^{\langle 3 \rangle}$ is submatrix of size $\bar{\phi} \times \bar{\mu}$; $\bar{G}_{\pi\eta}^{\langle 4,1 \rangle}$ is the first element of the matrix $\bar{G}_{\pi\eta}^{\langle 4 \rangle}$; $\bar{G}_{\pi\eta}^{\langle 4,2 \rangle}$ is the submatrix of the $\bar{G}_{\pi\eta}^{\langle 4 \rangle}$ matrix of the size $1 \times (\bar{\phi} - 1)$; $\bar{G}_{\pi\eta}^{\langle 4,3 \rangle}$ is the submatrix of the matrix $\bar{G}_{\pi\eta}^{\langle 4 \rangle}$ of the size $(\bar{\phi} - 1) \times 1$; $\bar{G}_{\pi\eta}^{\langle 4,4 \rangle}$ is submatrix of the matrix $\bar{G}_{\pi\eta}^{\langle 4 \rangle}$ of the size $(\bar{\phi} - 1) \times (\bar{\phi} - 1)$. Such modifications of the recording form and size caused by changes in the structure of the introduced geometric space (32) when calculating the backup path, also affected the vectors represented in expressions (22) and (29).

Then the conditions for ensuring QoS in a variety of indicators for each kth flow of packets, but already now the backup multipath, can be represented as follows:

$$
\lambda_k^{\langle req \rangle} \bar{\varepsilon}^k \leq \bar{G}_{\pi\eta}^{\langle 4,2 \rangle} \left[\bar{G}_{\pi\eta}^{\langle 4,4 \rangle} \right]^{-1} \bar{\Lambda}_{\eta-1} + \left(\bar{G}_{\pi\eta}^{\langle 4,1 \rangle} - \bar{G}_{\pi\eta}^{\langle 4,2 \rangle} \left[\bar{G}_{\pi\eta}^{\langle 4,4 \rangle} \right]^{-1} \bar{G}_{\pi\eta}^{\langle 4,3 \rangle} \right) \tau_k^{\langle req \rangle}. \quad (35)
$$

It is proposed to present the fast rerouting with QoS protection in an optimization form. Then, as a criterion for the optimality of the resulting routing solutions, it is advisable to choose a condition related to maximizing the overall performance of the infocommunication network:

$$
J = \sum_{k \in K} (c^k \varepsilon^k + \bar{c}^k \bar{\varepsilon}^k) \to \mathbf{max}, \quad (36)
$$

where c^k and $\bar{c}^k$ are weighting coefficients characterizing the importance (priority) of the kth flow. In this case, the condition $c^k > \bar{c}^k$ must be satisfied so that the QoS level for any flow along the primary path is not worse than the QoS level for the same flow along the backup path.

The constraints during the solution of the formulated optimization problem (36) were represented by the conditions for implementation of the multipath routing strategy (1), the conditions for the flow conservation (2), (3), the conditions for the protection of the link and the node (8)-(10), the conditions for preventing overload of the links—network bandwidth protection (11), conditions for ensuring and maintaining the Quality of Service level by the set of indicators (12), (13), (31), (35).

6 Calculation Example of Solving the Problem of Fast ReRouting with QoS Level Protection Based on Implementation of the Proposed Tensor Model

To assess the adequacy of the proposed model of fast rerouting (1)–(36) and the demonstrativeness of the obtained calculation results, we will solve this problem for a fragment of the infocommunication network, which is shown in Fig. 1. Let the network under investigation consist of sixteen routers and twenty-four communication links, indicating their capacity (1/s) in the gaps of the links. In the course of solving the problem of fast rerouting, it is needed to implement the protection scheme of the tenth router in case of its failure.

For clarity, the single flow case is considered, i.e. $k = 1$. Let the packet flow be transmitted between the first and sixteenth routers with the following QoS requirements:

$$\lambda_1^{\langle req \rangle} = 430 \, 1/s, \ p_1^{\langle req \rangle} = 0.015, \ \tau_1^{\langle req \rangle} = 150 \, \text{ms}.$$

For example, the operation of each of the interfaces of the network routers was simulated by the M/M/1/N queuing system, and the buffer capacity was ($N = 30$).

To calculate the primary multipath in the course of tensor formalization of the proposed model, the network S (Fig. 2) was determined by the n-dimensional space, where $n = 24$ and $m = 16$. Then the cyclomatic number was equal $\mu(S) = 9$, and

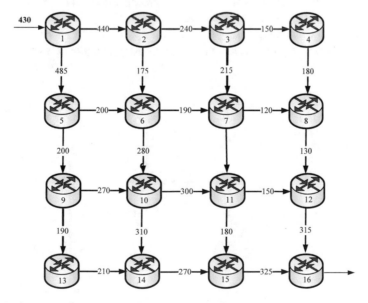

Fig. 1 The structure of the investigated infocommunication network

Fig. 2 An example of
selecting a set of basic
circuits and node pairs when
transmitting a flow of packets
along the primary multipath

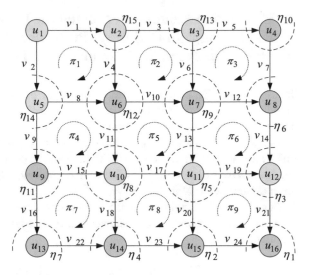

the network rank was $\phi(S) = 15$. The order of selecting a set of basic circuits and
node pairs is also presented in Fig. 2.

When transmitting a flow of packets through a backup multipath, the network
$\bar{S}$ defined a discrete $\bar{n}$-dimensional space, where $\bar{n} = 20$, and $\bar{m} = 15$ with the
exception of the node (in this example, the tenth router), which is protected, and
communication links incident to it. Then the cyclomatic number and rank of the
network $\bar{S}$ are equal $\mu(\bar{S}) = 6$ and $\phi(\bar{S}) = 14$, respectively. In this case, we used a
set of basic circuits and node pairs, shown in Fig. 3.

Then, when calculating the primary multipath for the selected skeleton
$\{v_1, v_2, v_3, v_4, v_5, v_6, v_7, v_9, v_{11}, v_{13}, v_{14}, v_{16}, v_{18}, v_{20}, v_{21}\}$ in the network, the chords

Fig. 3 An example of
selecting a set of basic
circuits and node pairs in the
transmission of a flow of
packets over a backup
multipath

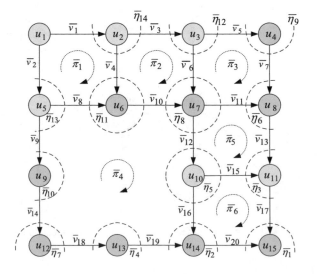

are v_8, v_{10}, v_{12}, v_{15}, v_{17}, v_{19}, v_{22}, v_{23}, v_{24}, which determine the basic circuits $\pi_1 \div \pi_9$. In turn, when calculating the backup multipath for a skeleton $\{\bar{v}_1, \bar{v}_2, \bar{v}_3, \bar{v}_4, \bar{v}_5, \bar{v}_6, \bar{v}_7, \bar{v}_9, \bar{v}_{12}, \bar{v}_{13}, \bar{v}_{14}, \bar{v}_{16}, \bar{v}_{17}, \bar{v}_{19}\}$ in the network $\bar{S}$, the chords are $\bar{v}_8$, $\bar{v}_{10}$, $\bar{v}_{11}$, $\bar{v}_{15}$, $\bar{v}_{18}$, $\bar{v}_{20}$, which define six basic circuits $\bar{\pi}_1 \div \bar{\pi}_6$. Then the rules for transforming the bases of edges and independent circuits and node pairs in the calculation of the primary and backup multipath are set respectively as follows:

$$
\begin{cases}
\pi_1 = -v_8; \\
\pi_2 = -v_{10}; \\
\pi_3 = -v_{12}; \\
\pi_4 = -v_{15}; \\
\pi_5 = -v_{17}; \\
\pi_6 = -v_{19}; \\
\pi_7 = -v_{22}; \\
\pi_8 = -v_{23}; \\
\pi_9 = -v_{24}; \\
\eta_1 = v_{21} + v_{24}; \\
\eta_2 = v_{20} + v_{23} - v_{24}; \\
\eta_3 = v_{14} + v_{19} - v_{21}; \\
\eta_4 = v_{18} + v_{22} - v_{23}; \\
\eta_5 = v_{13} + v_{17} - v_{19} - v_{20}; \\
\eta_6 = v_7 + v_{12} - v_{14}; \\
\eta_7 = v_{16} - v_{22}; \\
\eta_8 = v_{11} + v_{15} - v_{17} - v_{18}; \\
\eta_9 = v_6 + v_{10} - v_{12} - v_{13}; \\
\eta_{10} = v_5 - v_7; \\
\eta_{11} = v_9 - v_{15} - v_{16}; \\
\eta_{12} = v_4 + v_8 - v_{10} - v_{11}; \\
\eta_{13} = v_3 - v_5 - v_6; \\
\eta_{14} = v_2 - v_8 - v_9; \\
\eta_{15} = v_1 - v_3 - v_4.
\end{cases}
\qquad
\begin{cases}
\bar{\pi}_1 = -\bar{v}_8; \\
\bar{\pi}_2 = -\bar{v}_{10}; \\
\bar{\pi}_3 = -\bar{v}_{11}; \\
\bar{\pi}_4 = -\bar{v}_{18}; \\
\bar{\pi}_5 = -\bar{v}_{15}; \\
\bar{\pi}_6 = -\bar{v}_{20}; \\
\bar{\eta}_1 = \bar{v}_{17} + \bar{v}_{20}; \\
\bar{\eta}_2 = \bar{v}_{19} - \bar{v}_{20}; \\
\bar{\eta}_3 = \bar{v}_{13} + \bar{v}_{15} - \bar{v}_{17}; \\
\bar{\eta}_4 = \bar{v}_{18} - \bar{v}_{19}; \\
\bar{\eta}_5 = \bar{v}_{12} - \bar{v}_{15} - \bar{v}_{16}; \\
\bar{\eta}_6 = \bar{v}_7 + \bar{v}_{11} - \bar{v}_{13}; \\
\bar{\eta}_7 = \bar{v}_{14} - \bar{v}_{18}; \\
\bar{\eta}_8 = \bar{v}_6 + \bar{v}_{10} - \bar{v}_{11} - \bar{v}_{12}; \\
\bar{\eta}_9 = \bar{v}_5 - \bar{v}_7; \\
\bar{\eta}_{10} = \bar{v}_9 - \bar{v}_{14}; \\
\bar{\eta}_{11} = \bar{v}_4 + \bar{v}_8 - \bar{v}_{10}; \\
\bar{\eta}_{12} = \bar{v}_3 - \bar{v}_5 - \bar{v}_6; \\
\bar{\eta}_{13} = \bar{v}_2 - \bar{v}_8 - \bar{v}_9; \\
\bar{\eta}_{14} = \bar{v}_1 - \bar{v}_3 - \bar{v}_4.
\end{cases}
$$

Based on the obtained transformation rules, covariant coordinate transformation matrices are obtained for the primary (A) and backup ($\bar{A}$) multipaths:

$$A = \begin{Vmatrix}
0 & 0 & 0 & 0 & 0 & 0 & 0 & 0 & 0 & 0 & 0 & 0 & 0 & 0 & 0 & 0 & 0 & 0 & 0 & 0 & 0 & 0 & 1 \\
0 & 0 & 0 & 0 & 0 & 0 & 0 & 0 & 0 & 0 & 0 & 0 & 0 & 0 & 0 & 0 & 0 & 0 & 0 & 0 & 0 & 1 & 0 \\
0 & 0 & 0 & 0 & 0 & 0 & 0 & 0 & 0 & 0 & 0 & 0 & 0 & 0 & 0 & 0 & 0 & 0 & 0 & 0 & 1 & 0 & -1 \\
0 & 0 & 0 & 0 & 0 & 0 & 0 & 0 & 0 & 0 & 0 & 0 & 0 & 0 & 0 & 0 & 0 & 0 & 0 & 1 & 0 & 0 & -1 \\
0 & 0 & 0 & 0 & 0 & 0 & 0 & 0 & 0 & 0 & 0 & 0 & 0 & 0 & 0 & 0 & 0 & 1 & 0 & 0 & -1 & 0 & 0 \\
0 & 0 & 0 & 0 & 0 & 0 & 0 & 0 & 0 & 0 & 0 & 0 & 0 & 0 & 0 & 0 & 1 & 0 & 0 & 0 & -1 & 0 & 0 \\
0 & 0 & 0 & 0 & 0 & 0 & 0 & 0 & 0 & 0 & 0 & 0 & 0 & 0 & 1 & 0 & 0 & 0 & -1 & 0 & 0 & 0 & 0 \\
-1 & 0 & 0 & 0 & 0 & 0 & 0 & 0 & 0 & 0 & 0 & 0 & 0 & 0 & 0 & 0 & 0 & 0 & 0 & 0 & 1 & 0 & -1 \\
0 & 0 & 0 & 0 & 0 & 0 & 0 & 0 & 0 & 0 & 0 & 0 & 0 & 0 & 0 & 0 & 0 & 0 & 0 & 1 & 0 & 0 & -1 \\
0 & -1 & 0 & 0 & 0 & 0 & 0 & 0 & 0 & 0 & 0 & 0 & 0 & 0 & 0 & 0 & 1 & 0 & 0 & -1 & 0 & 0 & 0 \\
0 & 0 & 0 & 0 & 0 & 0 & 0 & 0 & 0 & 0 & 0 & 0 & 0 & 0 & 0 & 0 & 1 & 0 & 0 & -1 & 0 & 0 & 0 \\
0 & 0 & -1 & 0 & 0 & 0 & 0 & 0 & 0 & 0 & 0 & 0 & 0 & 0 & 1 & 0 & 0 & -1 & 0 & 0 & 0 & 0 & 0 \\
0 & 0 & 0 & 0 & 0 & 0 & 0 & 0 & 0 & 0 & 0 & 0 & 0 & 1 & 0 & 0 & 0 & -1 & 0 & 0 & 0 & 0 & 0 \\
0 & 0 & 0 & 0 & 0 & 0 & 0 & 0 & 0 & 0 & 0 & 1 & 0 & 0 & -1 & 0 & 0 & 0 & 0 & 0 & 0 & 0 & 0 \\
0 & 0 & 0 & -1 & 0 & 0 & 0 & 0 & 0 & 0 & 0 & 0 & 0 & 0 & 0 & 0 & 1 & 0 & 0 & -1 & 0 & 0 & 0 \\
0 & 0 & 0 & 0 & 0 & 0 & 0 & 0 & 0 & 0 & 0 & 0 & 0 & 0 & 0 & 1 & 0 & 0 & 0 & -1 & 0 & 0 & 0 \\
0 & 0 & 0 & 0 & -1 & 0 & 0 & 0 & 0 & 0 & 0 & 0 & 0 & 1 & 0 & 0 & -1 & 0 & 0 & 0 & 0 & 0 & 0 \\
0 & 0 & 0 & 0 & 0 & 0 & 0 & 0 & 0 & 0 & 0 & 0 & 1 & 0 & 0 & 0 & -1 & 0 & 0 & 0 & 0 & 0 & 0 \\
0 & 0 & 0 & 0 & 0 & -1 & 0 & 0 & 0 & 0 & 0 & 1 & 0 & -1 & 0 & 0 & 0 & 0 & 0 & 0 & 0 & 0 & 0 \\
0 & 0 & 0 & 0 & 0 & 0 & 0 & 0 & 0 & 0 & 1 & 0 & 0 & -1 & 0 & 0 & 0 & 0 & 0 & 0 & 0 & 0 & 0 \\
0 & 0 & 0 & 0 & 0 & 0 & 0 & 0 & 0 & 1 & 0 & -1 & 0 & 0 & 0 & 0 & 0 & 0 & 0 & 0 & 0 & 0 & 0 \\
0 & 0 & 0 & 0 & 0 & 0 & -1 & 0 & 0 & 0 & 0 & 0 & 1 & 0 & 0 & -1 & 0 & 0 & 0 & 0 & 0 & 0 & 0 \\
0 & 0 & 0 & 0 & 0 & 0 & 0 & -1 & 0 & 0 & 1 & 0 & -1 & 0 & 0 & 0 & 0 & 0 & 0 & 0 & 0 & 0 & 0 \\
0 & 0 & 0 & 0 & 0 & 0 & 0 & 0 & -1 & 1 & -1 & 0 & 0 & 0 & 0 & 0 & 0 & 0 & 0 & 0 & 0 & 0 & 0
\end{Vmatrix},$$

$$\bar{A} = \begin{Vmatrix}
0 & 0 & 0 & 0 & 0 & 0 & 0 & 0 & 0 & 0 & 0 & 0 & 0 & 0 & 0 & 0 & 0 & 0 & 0 & 0 & 1 \\
0 & 0 & 0 & 0 & 0 & 0 & 0 & 0 & 0 & 0 & 0 & 0 & 0 & 0 & 0 & 0 & 0 & 0 & 0 & 1 & 0 \\
0 & 0 & 0 & 0 & 0 & 0 & 0 & 0 & 0 & 0 & 0 & 0 & 0 & 0 & 0 & 0 & 0 & 0 & 1 & 0 & -1 \\
0 & 0 & 0 & 0 & 0 & 0 & 0 & 0 & 0 & 0 & 0 & 0 & 0 & 0 & 0 & 0 & 0 & 1 & 0 & 0 & -1 \\
0 & 0 & 0 & 0 & 0 & 0 & 0 & 0 & 0 & 0 & 0 & 0 & 0 & 0 & 0 & 1 & 0 & 0 & -1 & 0 & 0 \\
0 & 0 & 0 & 0 & 0 & 0 & 0 & 0 & 0 & 0 & 0 & 0 & 0 & 0 & 1 & 0 & 0 & 0 & -1 & 0 & 0 \\
0 & 0 & 0 & 0 & 0 & 0 & 0 & 0 & 0 & 0 & 0 & 1 & 0 & 0 & -1 & 0 & 0 & 0 & 0 & 0 & 0 \\
-1 & 0 & 0 & 0 & 0 & 0 & 0 & 0 & 0 & 0 & 0 & 0 & 0 & 0 & 0 & 0 & 0 & 1 & 0 & -1 & 0 \\
0 & 0 & 0 & 0 & 0 & 0 & 0 & 0 & 0 & 0 & 0 & 0 & 0 & 0 & 0 & 0 & 0 & 1 & 0 & 0 & -1 \\
0 & -1 & 0 & 0 & 0 & 0 & 0 & 0 & 0 & 0 & 0 & 0 & 0 & 1 & 0 & 0 & -1 & 0 & 0 & 0 & 0 \\
0 & 0 & -1 & 0 & 0 & 0 & 0 & 0 & 0 & 0 & 0 & 1 & 0 & -1 & 0 & 0 & 0 & 0 & 0 & 0 & 0 \\
0 & 0 & 0 & 0 & 0 & 0 & 0 & 0 & 0 & 0 & 1 & 0 & 0 & -1 & 0 & 0 & 0 & 0 & 0 & 0 & 0 \\
0 & 0 & 0 & 0 & 0 & 0 & 0 & 0 & 1 & 0 & 0 & -1 & 0 & 0 & 0 & 0 & 0 & 0 & 0 & 0 & 0 \\
0 & 0 & 0 & 0 & 0 & 0 & 0 & 0 & 0 & 0 & 0 & 1 & 0 & 0 & -1 & 0 & 0 & 0 & 0 & 0 & 0 \\
0 & 0 & 0 & 0 & -1 & 0 & 0 & 0 & 1 & 0 & -1 & 0 & 0 & 0 & 0 & 0 & 0 & 0 & 0 & 0 & 0 \\
0 & 0 & 0 & 0 & 0 & 0 & 0 & 1 & 0 & 0 & -1 & 0 & 0 & 0 & 0 & 0 & 0 & 0 & 0 & 0 & 0 \\
0 & 0 & 0 & 0 & 0 & 0 & 1 & 0 & -1 & 0 & 0 & 0 & 0 & 0 & 0 & 0 & 0 & 0 & 0 & 0 & 0 \\
0 & 0 & 0 & -1 & 0 & 0 & 0 & 0 & 0 & 1 & 0 & 0 & -1 & 0 & 0 & 0 & 0 & 0 & 0 & 0 & 0 \\
0 & 0 & 0 & 0 & 0 & 0 & 0 & 1 & 0 & -1 & 0 & 0 & 0 & 0 & 0 & 0 & 0 & 0 & 0 & 0 & 0 \\
0 & 0 & 0 & 0 & 0 & -1 & 1 & -1 & 0 & 0 & 0 & 0 & 0 & 0 & 0 & 0 & 0 & 0 & 0 & 0 & 0
\end{Vmatrix}.$$

In accordance with the presented initial data and QoS requirements, the solution of the fast rerouting problem was obtained using the proposed tensor model (1)–(36), the results of which are presented in Table 1.

Table 1 Results of solving the problem of fast rerouting

Link	$\varphi_{i,j}$, 1/s	QoS requirements: $\lambda_1^{\langle req \rangle} = 430$ 1/s, $p_1^{\langle req \rangle} = 0.015$, $\tau_1^{\langle req \rangle} = 150$ ms					
		Calculation results for the primary multipath			Calculation results for the backup multipath		
		$\lambda_{i,j}^1$, 1/s	$r_{i,j}^1$, 1/s	$\tau_{i,j}^1$, ms	$\bar{\lambda}_{i,j}^1$, 1/s	$\bar{r}_{i,j}^1$, 1/s	$\bar{\tau}_{i,j}^1$, ms
(1,2)	440	197.3368	0	4.1	192.8772	0	4
(1,5)	485	232.6632	0	4	237.1228	0	4
(2,3)	240	121.1699	0	8.4	153.3677	0.0001	11.5
(2,6)	175	76.1670	0	10.1	39.5094	0	7.4
(3,4)	150	49.5613	0	10	71.7225	0	12.8
(3,7)	215	71.6086	0	7	81.6451	0	7.5
(4,8)	180	49.5613	0	7.7	71.7225	0	9.2
(5,6)	200	102.6864	0	10.3	64.7351	0	7.4
(5,9)	200	129.9768	0.0001	14.3	172.3877	0.2788	34.4
(6,7)	190	0.2905	0	5.3	104.2445	0	11.7
(6,10)	280	178.5628	0.0001	9.9	–	–	–
(7,8)	120	26.0931	0	10.6	51.0906	0	14.5
(7,11)	140	45.8060	0	10.6	134.7991	2.3286	89.3
(8,12)	130	75.6544	0	18.4	122.8131	1.4880	86.9
(9,10)	270	99.1655	0	5.9	–	–	–
(9,13)	190	30.8112	0	6.3	172.1090	0.8748	47.1
(10,11)	300	134.1471	0	6	–	–	–
(10,14)	310	143.5811	0	6	–	–	–
(11,12)	150	95.7508	0	18.4	67.2401	0	12.1
(11,15)	180	84.2023	0	10.4	65.2303	0	8.7
(12,16)	315	171.4052	0	7	188.5652	0	7.9
(13,14)	210	30.8112	0	5.6	171.2342	0.0695	25.5
(14,15)	270	174.3923	0.0001	10.5	171.1647	0.0001	10.1
(15,16)	325	258.5945	0.0556	15	236.3950	0.0046	11.3

During fast rerouting, protection of a given QoS level was provided, since when using a primary or backup multipath, a given level of Quality of Service was ensured in terms of bandwidth, average end-to-end delay and packet loss probability. Thus, along each of the paths that are part of the primary multipath, the average end-to-end delay was 55.5 ms, and for the backup multipath it was 132.4 ms, which did not exceed the QoS requirements: $\tau_1^{\langle req \rangle} = 150$ ms. The total probability of packet loss when they were transmitted along the primary multipath was approximately equal 0.00013, and when using the backup multipath it was approximately equal 0.01173, which also met the QoS requirements: $p_1^{\langle req \rangle} = 0.015$.

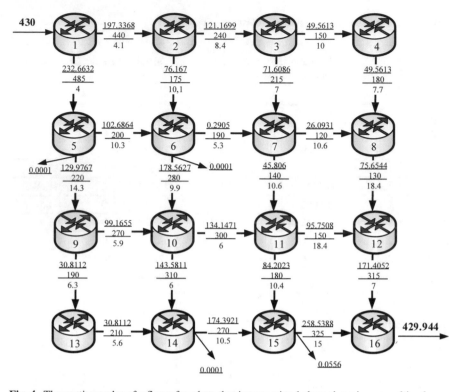

Fig. 4 The routing order of a flow of packets that is transmitted along the primary multipath

Figure 4 shows the routing order of the packet flow, which is transmitted through the primary multipath, and Fig. 5 shows the routing order over the backup multipath. In Figs. 4 and 5 in the gaps of the communication links, the following data is indicated (from top to bottom): the intensity of the packet flow in this communication link (1/s), the link capacity (1/s), the average delay of the packets in the link (ms). Near the overloaded interfaces, the arrow indicates the intensity of the dropped packets on $r_{i,j}^1$.

Thus, the presented calculation example confirmed the adequacy of the proposed fast rerouting model (1)–(36) in terms of the implementation of protection schemes for network elements (links, nodes) and the level of Quality of Service for a variety of indicators—bandwidth, average end-to-end delay, and packet loss probability.

7 Conclusions

In order to increase the level of resilience of infocommunication services provision in modern multiservice networks, a wide range of technological means is used, among which an important place is given to the protocols of fault-tolerant routing. At the level of the network transport (core), fault-tolerance is realized with the help of

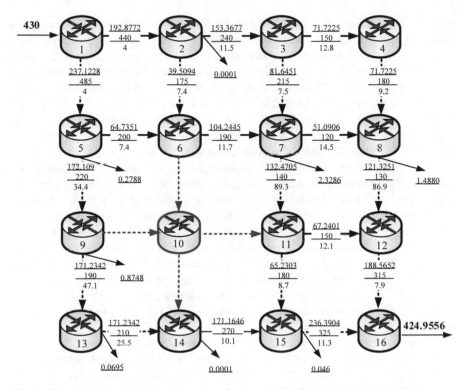

Fig. 5 The order of routing a flow of packets that is transmitted over the backup multipath

technologies of fast rerouting based on introduced resource redundancy when along with the calculation of the primary path, the backup route is also determined. The protection (reservation) of network elements and network bandwidth is provided, which allows to respond to the possible denial of service very quickly (~40 to 50 ms).

The efficiency of technological solutions regarding fast rerouting depends largely on the class of mathematical models used and the methods that will be the basis of the corresponding routing protocols. As shown by the analysis, the most effective is the use of the flow-based models of fast rerouting, which adequately take into account the flowing nature of modern predominantly multimedia traffic and adapted for the implementation of various protection schemes, including also such an important QoS indicator as bandwidth. However, given the multiservice nature of modern infocommunication networks, it is important during the fast rerouting that the level of Quality of Service is maintained over a set of indicators simultaneously, that is, to protect the bandwidth it is important to add and protect the permissible values of such QoS indicators as the average end-to-end packet delay and the probability of packet loss.

In this regard, the paper presents a flow-based model of fast rerouting in a multiservice network with the protection of the Quality of Service level for a variety of heterogeneous indicators at the same time: bandwidth, average end-to-end delay and probability of packet loss. Formulation of the conditions for the provision and protection of a given level of QoS according to these indicators (11), (12), (13), (31) and (35) in the analytical form was possible on the basis of the use of the tensor methodology for the research, when the infocommunication network was represented by a bivalent mixed tensor (15) in anisotropic space, which was determined by the structure of the network itself. The metric of this space depended entirely on the type of packet flows and the service disciplines on the interfaces of network routers (17). The use of coordinate systems of the network edges, independent circuits and node pairs allowed to provide an integral multidimensional network consideration.

In the framework of the proposed flow-based model (1)–(36), the technological problem of fast rerouting was formulated in the optimization form with the optimality criterion (36) and in which the constraints were the conditions for the implementation of the multipath routing strategy (1), the conditions for the flow conservation (2), (3), link and node protection conditions (8)–(10), conditions for preventing overload of communication links—network bandwidth protection (11), conditions for ensuring and providing the Quality of Service by a set of indicators (12), (13), (31), (35). Thus, the solution to the problem of fast rerouting, the use of which facilitates the optimal use of available network resources, while providing a predetermined level of the QoS and QoR along both the primary and backup routes in the infocommunication network, was obtained.

The numerical example demonstrates the performance of the proposed flow-based model of fast rerouting with detailed geometrization of the network structure: the choice of space, coordinate systems, and the formation of matrices of the covariant transformation for the introduced bases. An example was the case of a possible failure of an arbitrary network router and (or) incident communication links. As a result of the solution of the optimization problem, the primary and backup multipaths were obtained, along which a given level of Quality of Service was provided for the parameters of the bandwidth, the average end-to-end delay and the probability of packet loss.

The development of the proposed approach to the implementation of fast rerouting is seen in the extension of the scope of the results obtained for the case of QoE protection by indicators of Quality Rating (R) and Multimedia Quality (MMq) in the multiservice infocommunication network [31–37].

References

1. White R, Banks E (2017) Computer networking problems and solutions: an innovative approach to building resilient, modern networks. Addison-Wesley Professional
2. White R, Tantsura JE (2015) Navigating network complexity: next-generation routing with SDN, service virtualization, and service chaining. Addison-Wesley Professional
3. Monge AS, Szarkowicz KG (2015) MPLS in the SDN era: interoperable scenarios to make networks scale to new services. O'Reilly Media, Inc.
4. Stallings W (2015) Foundations of modern networking: SDN, NFV, QoE, IoT, and Cloud. Addison-Wesley Professional

5. Cholda P, Tapolcai J, Cinkler T, Wajda K, Jajszczyk A (2009) Quality of resilience as a network reliability characterization tool. IEEE Netw 23(2):11–19. https://doi.org/10.1109/MNET.2009. 4804331
6. Tipper D (2014) Resilient network design: challenges and future directions. Telecommun Syst 56(1):5–16. https://doi.org/10.1007/s11235-013-9815-x
7. Rak J (2015) Resilient routing in communication networks. Springer, Switzerland. https://doi. org/10.1007/978-3-319-22333-9
8. Mauthe A, Hutchison D, Cetinkaya EK, Ganchev I, Rak J, Sterbenz JP, Gunkelk M, Smith P, Gomes T (2016) Disaster-resilient communication networks: principles and best practices. In: Proceedings of 2016 8th international workshop on resilient networks design and modeling (RNDM). IEEE, pp 1–10. https://doi.org/10.1109/RNDM.2016.7608262
9. Rak J, Papadimitriou D, Niedermayer H, Romer P (2017) Information-driven network resilience: research challenges and perspectives. Opt Switching Netw 23(part 2):156–178. https://doi.org/10.1016/j.osn.2016.06.002
10. Pióro M, Tomaszewski A, Żukowski C, Hock D, Hartmann M, Menth M (2010) Optimized IP-based vs. explicit paths for one-to-one backup in MPLS fast reroute. In: Proceedings of 2010 14th international telecommunications network strategy and planning symposium (NETWORKS). IEEE, pp 1–6. https://doi.org/10.1109/NETWKS.2010.5624923
11. Gomes T, Tipper D, Alashaikh A (2014) A novel approach for ensuring high end-to-end availability: the spine concept. In: Proceedings of 2014 10th international conference on the design of reliable communication networks (DRCN). IEEE, pp 1–8. https://doi.org/10.1109/DRCN. 2014.6816142
12. Alashaikh A, Tipper D, Gomes T (2016) Supporting differentiated resilience classes in multilayer networks. In: Proceedings of 2016 12th international conference on the design of reliable communication networks (DRCN). IEEE, pp 31–38. https://doi.org/10.1109/DRCN.2016. 7470832
13. Hasan H, Cosmas J, Zaharis Z, Lazaridis P, Khwandah S (2016) Development of FRR mechanism by adopting SDN notion. In: Proceedings of 2016 24th international conference on software, telecommunications and computer networks (SoftCOM). IEEE, pp 1–72. https://doi. org/10.1109/SOFTCOM.2016.7772133
14. Lemeshko O, Arous K, Tariki N (2015) Effective solution for scalability and productivity improvement in fault-tolerant routing. In: Proceedings on 2015 second international scientific-practical conference problems of infocommunications science and technology (PIC S&T). IEEE, pp 76–78. https://doi.org/10.1109/INFOCOMMST.2015.7357274
15. Lemeshko AV, Yeremenko OS, Tariki N (2017) Improvement of flow-oriented fast reroute model based on scalable protection solutions for telecommunication network elements. Telecommun Radio Eng 76(6):477–490. https://doi.org/10.1615/TelecomRadEng.v76.i6.30
16. Lemeshko O, Yeremenko O (2017) Enhanced method of fast re-routing with load balancing in software-defined networks. Electr Eng 68(6):444–454. https://doi.org/10.1515/jee-2017-0079
17. Lemeshko O, Yeremenko O, Nevzorova O (2017) Hierarchical method of inter-area fast rerouting. Transp Telecommun J 18(2):155–167. https://doi.org/10.1515/ttj-2017-0015
18. Lemeshko O, Yeremenko O, Tariki N (2017) Solution for the default gateway protection within fault-tolerant routing in an IP network. Int J Electr Comput Eng Syst 8(1):19–26. https://doi. org/10.32985/ijeces.8.1.3
19. Golani K, Goswami K, Bhatt K, Park Y (2018) Fault tolerant traffic engineering in software-defined WAN. In: Proceedings on 2018 IEEE symposium on computers and communications (ISCC). IEEE, pp 01205–01210. https://doi.org/10.1109/ISCC.2018.8538606
20. Tomovic S, Radusinovic I, Prasad N (2015) Performance comparison of QoS routing algorithms applicable to large-scale SDN networks. In: Proceedings on 2015 IEEE international conference on computer as a tool (EUROCON). IEEE, pp 1–6. https://doi.org/10.1109/EUROCON.2015. 7313698
21. Tomovic S, Radusinovic I (2018) A new traffic engineering approach for QoS provisioning and failure recovery in SDN-based ISP networks. In: Proceedings on 2018 23rd international scientific-professional conference on information technology (IT). IEEE, pp 1–4. https://doi. org/10.1109/SPIT.2018.8350854

22. Lemeshko O, Yeremenko O, Yevdokymenko M (2018) Tensor model of fault-tolerant QoS routing with support of bandwidth and delay protection. In: Proceedings on 2018 XIIIth international scientific and technical conference computer sciences and information technologies (CSIT). IEEE, pp 135–138. https://doi.org/10.1109/stc-csit.2018.8526707

23. Lemeshko O, Yevdokymenko M, Yeremenko O, Hailan AM, Segeč P, Papán J (2019) Design of the Fast ReRoute QoS protection scheme for bandwidth and probability of packet loss in software-defined WAN. In: Proceedings on 2019 15th international conference the experience of designing and application of CAD systems in microelectronics (CADSM). IEEE, pp 3/72–3/76

24. Lemeshko AV, Evseeva OY, Garkusha SV (2014) Research on tensor model of multipath routing in telecommunication network with support of service quality by great number of indices. Telecommun Radio Eng 73(15):1339–1360. https://doi.org/10.1615/TelecomRadEng.v73.i15.30

25. Lemeshko O, Yeremenko O (2016) Dynamic presentation of tensor model for multipath QoS-routing. In: Proceedings on 13th international conference of modern problems of radio engineering, telecommunications and computer science (TCSET). IEEE, pp 601–604. https://doi.org/10.1109/TCSET.2016.7452128

26. Lemeshko OV, Yeremenko OS (2016) Dynamics analysis of multipath QoS-routing tensor model with support of different flows classes. In: Proceedings on 2016 international conference on smart systems and technologies (SST). IEEE, pp 225–230. https://doi.org/10.1109/SST.2016.7765664

27. Yeremenko OS, Lemeshko OV, Nevzorova OS, Hailan AM (2016) Method of hierarchical QoS routing based on the network resource reservation. In: Proceedings 2017 IEEE first Ukraine conference on electrical and computer engineering (UKRCON). IEEE, pp 971–976. https://doi.org/10.1109/UKRCON.2017.8100393

28. Kron G (1949) Tensor analysis of networks. Wiley

29. Kleinrock L, Gail R (1996) Queueing systems: problems and solutions. Wiley

30. Bose SK (2002) An introduction to queueing systems. Kluwer Academic/Plenum Publisher, New York

31. Lemeshko O, Yevdokymenko M, Naors Y, Anad A (2018) Development of the tensor model of multipath QoE-routing in an infocommunication network with providing the required Quality Rating. East-Eur J Enterp Technol 5/2(95), 40–46. https://doi.org/10.15587/1729-4061.2018.141989

32. ITU-T Y.1540 (2016) Internet protocol data communication service—IP packet transfer and availability performance parameters

33. ITU-T G.107 (2014) The E-model: a computational model for use in transmission n planning

34. ITU-T G.1011 (2015) Reference guide to quality of experience assessment methodologies

35. ITU-T P.806 (2014) A subjective quality test methodology using multiple rating scales

36. ITU-T G.1070 (2018) Multimedia Quality of Service and performance—generic and user-related aspects. Opinion model for video-telephony applications

37. ITU-T P.800.1 (2016) Methods for objective and subjective assessment of speech and video quality—Mean opinion score (MOS) terminology

The Development of Routing Flow Model in IEEE 802.11 Multi-radio Multi-channel Mesh Networks, Shown as a Konig Graph

Sergii Harkusha⊙ and Maryna Yevdokymenko⊙

Abstract An approach to the use of Konig graphs to model routing in IEEE 802.11 standard multi-radio multi-channel mesh networks, which enabled to describe more comprehensively and in more detail all possible configurations of the mesh network. Based on the representation planar Konig's of multi-radio multi-channel mesh network in the routing problem is solved, which is to identify those collision domains through which traffic is to be transmitted from the sender to the recipient to meet end-to-end performance requirements. As a mathematical model was proposed routing flow model of producing records of structural and functional features of multi-radio multi-channel mesh networks—the results of solving the problem of the distribution of frequency channels, bandwidth collision domains, traffic characteristics, order service packs on the mesh-station etc. The proposed model uses the optimality criterion, aiming to minimize the multi-radio multi-channel mesh network performance, i.e. intensity of the total network traffic catered with its priorities.

Keywords Multi-radio Multi-channel mesh network · Flow model · Routing · Hypergraph · Konig graphs

1 Introduction

The advent of Wireless Mesh Networks (WMNs), based on the IEEE 802.11 technology significantly changed the process of organizing wireless access networks and radio networks. So far there has been convincing evidence of the fact that the model of users' access with IEEE 802.11 standard infrastructure is indeed suitable for home and small corporate networks. These networks cover a limited area with

S. Harkusha
Poltava University of Economics and Trade, Poltava, Ukraine
e-mail: sergiy.garkusha@gmail.com

M. Yevdokymenko (✉)
Kharkiv National University of Radio Electronics, Kharkiv, Ukraine
e-mail: maryna.yevdokymenko@ieee.org

© The Editor(s) (if applicable) and The Author(s), under exclusive license to Springer Nature
Switzerland AG 2021
T. Radivilova et al. (eds.), *Data-Centric Business and Applications*, Lecture Notes on Data
Engineering and Communications Technologies 48,
https://doi.org/10.1007/978-3-030-43070-2_23

a limited number of users. In turn, the mesh network traffic model is suitable for the networks the high Access Point (AP) density, designed to connect users on a relatively large area. The mesh networking traffic architecture has various benefits, including reliability, scalability, profitability and easiness in building [1, 2].

Among the numerous requirements to the wireless mesh networks the key one is to provide high-performance with differentiated (or even guaranteed) quality of user request servicing. The studies on improving mesh networks productivity on the whole involve protocol means from the physical hardware to the network level of the Open Systems Interconnection Basic Reference Model (OSI). The analysis in [3, 4] showed that there is a number of ways to improve the performance of wireless networks: the use of intellectual antenna array technology, changing stations location, polarization-based signal spreading, the use of MIMO technology, multiplexing, the use of multi-radio multi-channel operation mode, ensuring efficient routing and traffic management. The highest result in improving a wireless network productivity level (up to 2–3 times) may be achieved by using Multi-Radio Multi-Channel Wireless Mesh Networks (MR MC WMN) [5, 6], implying the use of either one or more radio interfaces on each mesh station, set for different non-overlapping Radio Frequency Channels (RFCs). The specific feature of this type of networks is switching of one radio interface from the channel organizing user's access to the traffic channel, which enables to eliminate interferences, inherent in the use of single-channel solution. However, the use of the multi-radio multi-channel mode of WMN operation does not exclude its combined use along with other ways of performance enhancement as they belong to different OSI levels.

High declared potential capabilities of IEEE 802.11 MR MC WMN are ensured, on the one hand, by effective solutions of hardware and channel levels of the OSI Model, and, on the other hand, by rigid requirements to the efficiency to the network resources management and, in particular, to the traffic and routing management protocols. In wireless networks packet routing problem between an undefined pair of stations is much more complicated than in wired ones. The reason is that alongside general requirements, related to the support of the service quality and load balance, MR MC WMN routing protocols are to consider the randomized nature of created collision domains. Therefore, it appears topical to develop mathematical models and techniques of routing, which could be used as the basis for promising MR MC WMN routing protocols.

2 Analysis of Existing Routing Solutions in MR MC WMN

Despite the fact that the concepts of creating MR MC WMN were offered relatively recently, science and practice have already suggested quite a wide range of routing protocols [7–9]. Today, there are several approaches to the classification of routing protocols in mesh networks [10–21]. Classification is traditionally based on systematization and grouping into objects, which are analyzed in accordance with their common features, such as:

- by the principle of work: proactive, reactive, hybrid, flow-oriented and adaptive protocols [10–14]. In proactive protocols (AWDS, Babel, BATMAN, CGSR, DFR, DBF, DSDV, HSLS, Guesswork, HSR, IARP, LCA, MMRP, OLSR, TBRPF, WAR, WRP, STAR), the routing is based on tables, as a result of which also called table routing protocols.
- by the process of gathering information on the state of mesh-network [15–17]. It is possible to select protocols (AODV, DSDV, DVMRP, SEAD, AOMDV, SAODV, ARAN), using distance-vector algorithms by which the mesh-station sends its neighboring mesh-stations information about known routes to them.
- by the number of calculated routes. Single path routing protocols are selected, which, as a result of processing the mesh network topology information, look for the optimal path to the specified destination, and then enter it into the routing table (DVR, DSDV, WRP, ABR, HSR) [147, 153, 158]. Multipath routing protocols are also distinguished, which enter into the routing table more than one route to the required destination (AODV, AOMDV, SMR, MP-DSR, NDMR, DSR, TORA) [18–20].
- by the type of metric used to route the data. For example, protocols that use station power data (ISAIAH, PARO, EADSR, PAMAS, DSPRA, GAF) [21].

At the same time, as the analysis of routing protocols showed, they are all based on the use of the graph representation of WMN. The use of the MR MC WMN graph image is reasonable in case of applying beam antennas at mesh stations, since a radio channel is formed between them. It may be represented as a graph edge. However, provided non-beamed antennas are used at mesh stations, the MR MC WMN graph image does not ensure adequate consideration of their specifics due to forming a cluster mesh network structure. Good reputation has been gained by the solutions, which are also used in LTE and WiMAX technologies, based on the use of topological ideas [22–26] when mathematically describing the network structure. Papers [27, 28] and in recent studies [29, 30] show that when modelling MR MC WMN the approach based on representing a mesh network as a hypergraph proved its high performance for solving the tasks of distributing non-overlapping frequency channels. Therefore, the task, related to the development of the routing model in multi-radio multi-channel mesh networks, represented as hypergraphs is topical.

3 Multi-radio Multi-channel WMN Hypergraph Image

Using hypergraph imaging, as shown in works [25, 26], MR MC WMN may be represented as hypergraph $H(I, J; R)$, where I is a vertex set, J is an edge set, R is the predicate, determining adjacency of the stations with primary coverage areas. $I = \{n_i, i = \overline{1, N}\}$, where n_i is the element of set I, modelling MR MC WMN mesh stations, N is their total number in the mesh network. $J = \{z_j, j = \overline{1, Z}\}$, where z_j is the element of set J, modelling the primary coverage areas, Z is their total number in MR MC WMN. Predicate R, being incident or of hypergraph H,

Table 1 The number of non-overlapping frequency channels for different wireless network standards	Standard	The number of non-overlapping frequency channels, K
	IEEE 802.11a/n	12
	IEEE 802.11b/g/n	3 or 4
	IEEE 802.11ac	25

determines whether i station belongs to j primary coverage area. For example in case i mesh station is involved in forming the j primary coverage area, predicate $R(n_i, z_j)$ is true, i.e. equal to one, otherwise $R(n_i, z_j)$ is false, i.e. equal to zero. As a result the description of MR MC WMN may be made using final hypergraph $H(I, J; R)$, consisting of a pair of vertex sets $I = \{n_i, \ i = \overline{1, N}\}$ and edges $J = \{z_j, \ j = \overline{1, Z}\}$ along with binary predicate $R \Leftrightarrow R(n_i, z_j)$, defined for all as $n_i \in I$ and $z_j \in J$.

When MR MC WMN operates to eliminate interference the set of non-overlapping radio frequency channels $T = \{k_t, \ t = \overline{1, K}\}$ is used, where k_t is the element of the T set, modelling t non-overlapping frequency channels, K is their total number depending on the wireless network standard used (Table 1).

In the modelled MR MC WMN it is thought that non-overlapping frequency channels within given models [31–33] are pre-distributed between mesh stations radio interfaces. As a result the t frequency channels are connected to the i station, belonging to j primary coverage area. Thus, predicate $P(n_i, k_t, z_j)$ is found from:

$$P(n_i, k_t, z_j) = x_{n_i, k_t} R(n_i, z_j), \tag{1}$$

where x_{n_i, k_t} is a variable, characterizing the connection of frequency channel $k_t \in K$ to mesh station $n_i \in N$.

The connection of frequency channels to mesh stations radio interfaces of one primary coverage area leads to forming collision domains. Within a collision domain mesh stations use one and the same frequency channel. A MR MC WMN, built as a result of solving the task of distributing the frequency channels may also be shown as hypergraph $G(I, U; Q)$, where $U = \{d_u, \ i = \overline{1, D}\}$ is the set of formed collision domains, d_u is the element of U set, modelling collision domains, and D is the total number of collision domains in the MR MC WMN. Predicate $Q(n_i, d_u)$, in turn, is definitely determined by correspondence

$$Q(n_i, d_u) \Leftrightarrow P(n_i, k_t, z_j). \tag{2}$$

For example, if i mesh station, being a part of the j primary coverage area, is given t non-overlapped frequency channel $(P(n_i, k_t, z_j) = 1)$, the station participates in forming u collision domain and predicate $Q(n_i, d_u) = 1$. Otherwise, if i mesh station is not a part of the j primary coverage area, it is not given t non-overlapping frequency channel $(P(n_i, k_t, z_j) = 0)$, predicate $Q(n_i, d_u) = 0$. Thus, the use of i station in forming u collision domain is determined by predicate $Q(n_i, d_u)$.

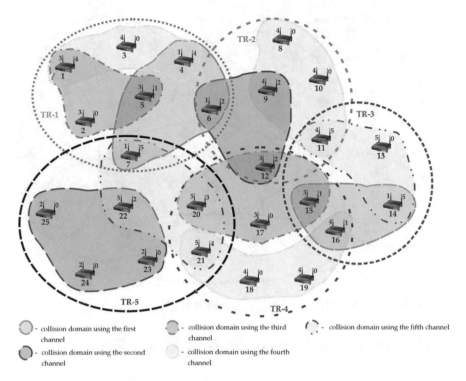

Fig. 1 A sample of MR MC WMN, using five non-overlapping frequency channels

As an example we can consider the MR MC WMN configuration, consisting of five primary coverage areas ($Z = 5$), formed by twenty-five mesh stations ($N = 25$) and operating with five non-overlapping frequency channels (K = 5). An example of this multi-radio mesh network is given in Fig. 1.

The above mesh network configuration is correspondent to hypergraph $G(I, U; Q)$, shown in Fig. 2, with vertex set $I = \{n_1, n_2, \ldots, n_{25}\}$, the set of used non-overlapping frequency channels $T = \{k_1, k_2, \ldots, k_5\}$ and predicate $Q(n_i, d_u)$, specifying the connection of a station to one of the collision domains $U = \{d_1, d_2, \ldots, d_{11}\}$.

Following the studies it was found out that the MR MC WMN hypergraph image enabled to describe potential configurations of the mesh network on the whole and its separate elements, shown as hypergraph vertices and edges more comprehensively and thoroughly. Accordingly, there emerges a need to formalize the routing task in the MR MC WMN, shown as a hypergraph. For further convenience the MR MC WMN hypergraph image will be transformed into Konig graph.

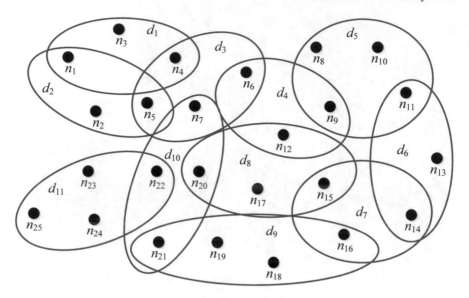

Fig. 2 Hypergraph image of the MR MC WMN, given in Fig. 1

4 Multi-radio Multi-channel WMN Image as a Konig Graph

To solve the traffic routing task in the configuration of a multi-radio multi-channel mesh networks let us refer Konig graph $K(G) = (I, U; Q)$ with vertex set $I \cup U$ [28–30] to hypergraph $G(I, U; Q)$, modelling the mesh network as collision domains.

Predicate $Q(n_i, d_u)$ specifies the adjacency of two vertices of two types so that vertices $n_i \in I$ and $d_u \in U$ in $K(G)$ are adjacent only when in G vertex i is incident to u edge, i.e. i station is involved in forming the u collision domain. At the same time graph $K(G)$ is called a Konig image of hypergraph G. Therefore, the possible configuration of WMN, shown in Fig. 1, corresponds to its Konig image (Fig. 3).

Transforming a Konig image of the MR MC WMN potential configuration, the mesh network may be shown as a flat Konig image and as a bipartite flow graph. An example of this transformation of the third collision domain, formed by mesh stations Nos. 4–7 (Fig. 1), is given in Fig. 4.

The flat Konig image of MR MC WMN configuration (Fig. 5) consists of vertex set $N = \{n_1, n_2, \ldots, n_{25}\}$, modelling mesh stations, as well as vertex set $D = \{d_1, d_2, \ldots, d_{11}\}$, modelling the collision domains formed. In turn, predicate $Q(n_i, d_u)$ specifies mesh station belonging to a collision domain.

Fig. 3 Konig image of MR
MC WMN potential
configuration

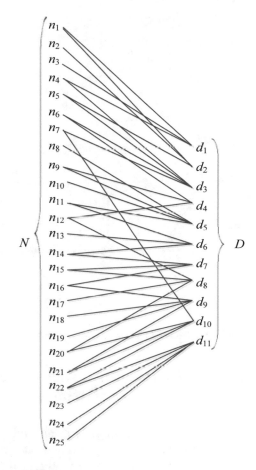

Fig. 4 The examples of
imaging a potential
configuration of a collision
domain (**a**) as a hypergraph
(**b**), as a flat Konig image (**c**),
and bipartite flow graph (**d**)

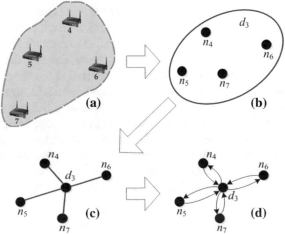

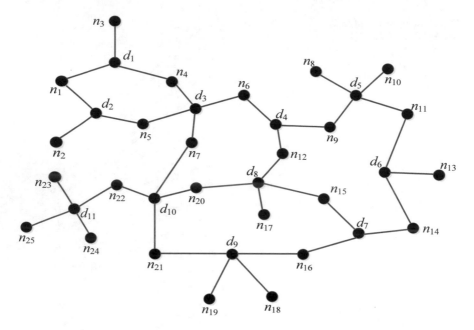

Fig. 5 The flat Konig image of WMN potential configuration

5 The Metrization of a Multi-radio Multi-channel WMN Shown as a Konig Graph

The key point in both setting ad solving routing tasks in telecommunication systems, especially in WMN, is to specify metrical features of the network elements. It is known that the key metrics when routing is often the function of the network element capacity. In this case these network elements are collision domains, whose properties and features fully identify the nature of mesh stations interaction. Thus, it appears important to solve the subtask, related to analytical formulation of an expression to calculate the collision domain capacity, namely, the current value of the available data transmission speed from i mesh station to other mesh stations of the u collision domain (c_{n_i,d_u}) and the current value of available data transmission speed to i mesh station from other mesh stations of the u-collision domain (c_{d_u,n_i}).

It is to be noted that according to the multiple access protocol with the Carrier Sense Multiple Access with Collision Avoidance (CSMA/CA) [32, 33] the access to the transmission medium of one WMN collision domain will be provided only to one mesh station at a time. Then, in case of high load on the WMN, i.e. when all the stations are seeking to transmit data, speeds c_{n_i,d_u} and c_{d_u,n_i} will be specified by expressions

$$c_{n_i,d_u} = \frac{Y_{d_u}}{\sum_{i=1}^{N} Q(n_i, d_u)}, \quad (u = \overline{1, D}), \tag{3}$$

Table 2 Output of IEEE 802.11 WMN family

Standard	Maximum output of u collision domain on physical level (Mbit/s)	Maximum output of u collision domain on network level, Y_{d_u} (Mbit/s)
802.11a	54	24
802.11b	11	4,8
802.11g	54	24
802.11n	300	60

$$c_{d_u,n_i} = Y_{d_u} \left(\frac{\sum_{i=1}^{N} Q(n_i, d_u) - 1}{\sum_{i=1}^{N} Q(n_i, d_u)} \right), \quad (u = \overline{1, D}), \tag{4}$$

where Y_{d_u} is the maximum output of the WMN u collision domain on the network level, supported in the applied wireless technology (Mbit/s) (Table 2); $\sum_{i=1}^{N} Q(n_i, d_u)$ is the number of mesh stations, involved in forming the u collision domain.

6 Model Routing Model in Multi-radio Multi-channel WMN

Basing on the flat Konig image, the routing solution is in specifying those collision domains, through which packages must be sent from the sender to the receiver. It is to be done so as to meet end-to-end capacity requirements. The suggested mathematical routing model will be a flow model, considering the specifics of MR MC WMN structural and functional building—the procedure of distributing frequency channels, the collision domain capacity, traffic features, package service procedure at mesh network stations (wireless routers) etc.

The development of the routing flow model requires the calculation of the set of route variables

$$\vec{X} = \begin{bmatrix} x_{n_i,d_u}^v \\ --- \\ x_{d_u,n_i}^v \end{bmatrix}, \tag{5}$$

where x_{n_i,d_u}^v is the share of intensity of the v flow of packages, transmitted from i mesh station to stations of the u collision domain, x_{d_u,n_i}^v is the share of intensity of the v flow of packages, transmitted from i mesh station from other stations of the u collision domain.

Under the model offered it is necessary to ensure flow maintenance [34, 35] in Konig graph vertices, modelling MR MC WMN, expressed as:

$$
\begin{cases}
\displaystyle\sum_{u:Q(n_i,d_u)=1} x^v_{n_i,d_u} = 1, v \in V, n_i = s_v; \\[2ex]
\displaystyle\sum_{u:Q(n_i,d_u)=1} x^v_{n_i,d_u} - \sum_{u:Q(n_i,d_u)=1} x^v_{d_u,n_i} = 0, v \in V, n_i \neq s_v, d_v; \\[2ex]
\displaystyle\sum_{i:Q(n_i,d_u)=1} x^v_{d_u,n_i} - \sum_{i:Q(n_i,d_u)=1} x^v_{n_i,d_u} = 0, v \in V; \\[2ex]
\displaystyle\sum_{u:Q(n_i,d_u)=1} x^v_{d_u,n_i} = 1, v \in V, n_i = d_v,
\end{cases}
\tag{6}
$$

where V is the set of flows in WMN; s_v is the sending station and d_v is the receiving station for the v flow packages.

Traditionally [34, 35], subject to on the additional restrictions on the controlling variables the task of routing may be classified in two ways. In case of restrictions

$$
0 \leq x^v_{n_i,d_u} \leq 1,
\tag{7}
$$

$$
0 \leq x^v_{d_u,n_i} \leq 1
\tag{8}
$$

the set task is classified as the task of multipath routing, and if there are restrictions

$$
x^v_{n_i,d_u} \in \{0, 1\},
\tag{9}
$$

$$
x^v_{d_u,n_i} \in \{0, 1\},
\tag{10}
$$

then single path way of serviced flow packages delivery is applied.

Restrictive conditions responsible for preventing the possibility to go beyond the current values of the available transmission speed are expressed as follows

$$
\sum_{v=1}^{V} r^v x^v_{n_i,d_u} \leq c_{n_i,d_u}, \quad (i = \overline{1, N}; u = \overline{1, D}),
\tag{11}
$$

$$
\sum_{v=1}^{V} r^v x^v_{d_u,n_i} \leq c_{d_u,n_i}, \quad (i = \overline{1, N}; u = \overline{1, D}),
\tag{12}
$$

where r^v is the average intensity of the v flow, coming into the mesh network to be serviced.

It should be noted that capacity value c_{n_i,d_u}, defined using expression (3), is true provided that all wireless network stations are seeking to transmit data. To record available flows, transmitted via the mesh station to the MR MC WMN in more detail, with the capacity of c_{n_i,d_u} may be defined based on expression

$$
c_{n_i,d_u} = \frac{Y_{d_u}}{\sum_{i=1}^{N-1} \sum_{k=1}^{K} Q(n_i, d_u)\lceil x^v_{n_i,d_u}\rceil + 1}, \quad (u = \overline{1, D}),
\tag{13}
$$

where $\lceil \cdot \rceil$ is the operation of rounding the number up.

In addition, to record not only the availability, but also the intensity of the flow, value c_{n_i,d_u} will be defined based on expression

$$c_{n_i,d_u} = \left[\frac{\sum_{v=1}^{V} x_{n_i,d_u}^{v}}{\sum_{j=1}^{N} \sum_{v=1}^{V} \left(x_{n_j,d_u}^{v} Q(n_j, d_u) \right)} \right] Y_{d_u}, \tag{14}$$

While solving the routing task we will opt for the condition below as the solutions optimality criterion:

$$\min_{x} \sum_{v=1}^{V} r^{v} \left(\sum_{u:Q(n_i,d_u)=1} x_{n_i,d_u}^{v} + \sum_{i:Q(n_i,d_u)=1} x_{d_u,n_i}^{v} \right), \tag{15}$$

Meeting this condition ensures load minimization of the elements (of stations and domains) and the MR MC WMN on the whole.

Depending on the selected option of using expressions (3)–(15) the optimization task may be classified as follows:

- (3)–(8), (11), (12), (15) are the class of Linear Programming tasks (LP);
- (3)–(6), (9)–(12), (15) are the class of Integer Linear Programming (ILP);
- (4)–(8), (11)–(13), (15) are the class of Nonlinear Programming (NLP);
- (4)–(8), (11), (12), (14), (15)—NLP;
- (4)–(6), (9)–(13), (15)—Mixed Integer Non-Linear Programming (MINLP);
- (4)–(6), (9)–(12), (14), (15)—MINLP.

7 A Sample of Solving the Routing Task in Multi-radio Multi-channel WMN

To visualize the obtained results it is reasonable to fulfil the task of multipath routing in MR MC WMN (Fig. 1), based on the suggested flow model (3)–(6), (9)–(12), (15). Here it is possible to take an example of a single product case, when the network services one flow, coming from the eleventh mesh station to the twentieth one. Suppose that within the considered MR MC WMN all mesh stations use IEEE 802.11g standard with 24 Mbit/s maximum capacity of each collision domain on the network level (Table 2). Figure 6 gives a flat Konig image of the possible MR MC WMN configuration as a bipartite flow graph, where the mesh stations transmission speeds (Mbit/s) are indicated in the arc breakings by collision domain, found from expressions (3) and (4).

Figures 7 and 8 provide sample solutions of routing under the flow intensity of $r = 13,47$ Mbit/s, coming into the WMN. The solution results are the values of the

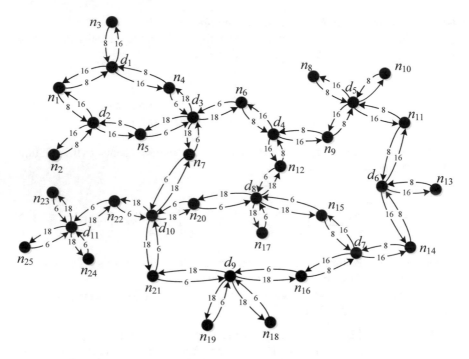

Fig. 6 The flat Konig image of MR MC WMN configuration, shown on Fig. 1 as a bipartite flow graph specifying mesh station capacity (Mbit/s)

flow intensity (Mbit/s) between mesh stations of different collision domains. The values of the flow intensity and their directions are specified in Fig. 7 along the arcs of the MR MC WMN flat Konig image.

As a result of solving the task of multipath routing by using the suggested model (3)–(6), (9)–(12), (15) four routes were built (Fig. 7):

- route $n_{11} \rightarrow d_5 \rightarrow n_9 \rightarrow d_4 \rightarrow n_6 \rightarrow d_3 \rightarrow n_7 \rightarrow d_{10} \rightarrow n_{20}$ services the flow with intensity 3,63 Mbit/s;
- route 2: $n_{11} \rightarrow d_5 \rightarrow n_9 \rightarrow d_4 \rightarrow n_{12} \rightarrow d_8 \rightarrow n_{20}$—services the flow with intensity 2,11 Mbit/s;
- route 3: $n_{11} \rightarrow d_6 \rightarrow n_{14} \rightarrow d_7 \rightarrow n_{15} \rightarrow d_8 \rightarrow n_{20}$—services the flow with intensity 2,32 Mbit/s;
- route 4: $n_{11} \rightarrow d_6 \rightarrow n_{14} \rightarrow d_7 \rightarrow n_{16} \rightarrow d_9 \rightarrow n_{21} \rightarrow d_{10} \rightarrow n_{20}$—services the flow with intensity 5,41 Mbit/s.

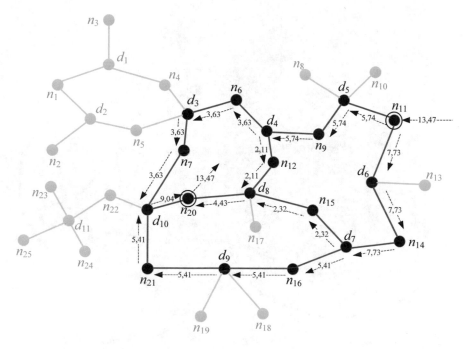

Fig. 7 A sample of calculating the routes when using suggested flow model, based on the flat Konig image of the mesh network

8 Conclusions

The work suggests an approach to using Konig graphs while modelling IEEE 802.11 standard multi-radio multi-channel mesh networks. This, in turn, enabled to describe in more comprehensively and more thoroughly possible configurations of the mesh network on the whole and its elements, shown as different vertices of the Konig graph. It was shown that the suggested mathematical tools may be efficiently used for routing in similar networks.

Based on the flat Konig image of the multi-radio multi-channel mesh network, the work solves the issue of routing, which is identifying the stations and collision domains, through which packages must be transmitted from the sender to the receiver so as to meet end-to-end capacity requirements. The flow model was offered as a mathematical routing model, considering the specifics of structural and functional building multi-radio multi-channel mesh networks—the results of solving the task of distributing radio frequency channels, collision domain capacities, graph features, the procedure of package servicing at network mesh stations (wireless routers) etc.

The offered model uses the optimality criterion, focused on minimizing the load of the multi-radio multi-channel mesh network. In addition, in the reduced setting the flow model may be formulated as several classes of optimization tasks depending

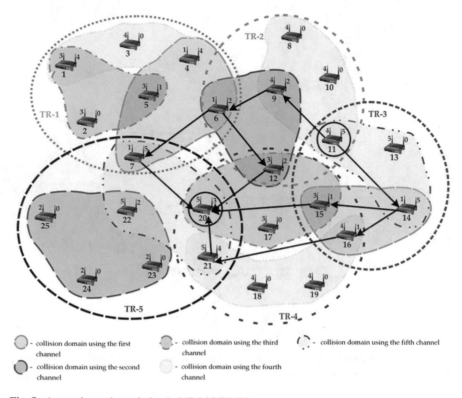

Fig. 8 A sample routing solution in MR MC WMN

on the number of formed paths and the extent of considering the characteristics of the mesh stations flows.

References

1. Smith C, Collins D (2013) Wireless networks. McGraw Hill Professional
2. Methley S (2010) Essentials of wireless mesh networking. Cambridge University Press, Cambridge
3. Akyildiz IF, Wang X, Wang W (2005) Wireless mesh networks: a survey. Comput Netw
4. Raniwala R, Gopalan K, Chiueh T (2004) Centralized channel assignment and routing algorithms for multi-channel wireless mesh networks. ACM SIGMOBILE Mob Comput Commun Rev 8(2):50–65
5. Chiang M (2005) Balancing transport and physical layers in wireless multihop networks: joint optimal congestion and power control. IEEE J Select Areas Commun 23(1):104–116
6. Skalli H, Ghosh S, Das SK, Lenzini L, Conti M (2007) Channel Assignment strategies for multiradio wireless mesh networks: issues and solutions. IEEE Comm Mag 45(11):86–95
7. Gogolieva M, Garkusha S, Abed AH (2011) Mathematical model of channel distribution in multichannel mesh networks 802.11. In: The experience of designing and application of cad systems in microelectronics (CADSM 2011). Ukraine, Lviv, pp 71–73

8. Garkusha SV (2012) Analysis the results of frequency planning in mesh networking standard IEEE 802.11. In: Modern problems of radio engineering, telecommunications and computer science: proceedings of the XI international conference (TCSET'2012), Slavske, Ukraine, Lviv, pp 285–286
9. Singh K, Behal S (2013) Review on routing protocols in wireless mesh networks. Int J Appl Innov Eng Manag (IJAIEM) 2(2):143–149
10. Harkusha S, Lemeshko O, Yevdokymenko M, Yeremenko O (2019) Model of structurally functional selforganization of multi-radio multi-channel mesh networks using hypergraphs. In: 8th international conference on "mathematics. information technologies. education", MoMLeT&DS-2019. Shatsk, Ukraine, pp 1–10
11. Núñez-Martínez J, Mangues-Bafalluy J (2013) A survey on routing protocols that really exploit wireless mesh network features. J Commun 5(3):211–231
12. Singh TP, Dua S, Das V (2012) Energy-efficient routing protocols in mobile Ad-Hoc networks. Int J Adv Res Comput Sci Softw Eng 5(1). http://www.ijarcsse.com/docs/papers/january2012/V2I1035A.pdf
13. Khetrapal A (2006) Routing techniques for Mobile Ad Hoc networks classification and qualitative/quantitative analysis. In: International conference on wireless networks, pp 251–257
14. Nageswara S, Rao S, Krishna YK, Nageswara Rao K (2013) A survey: routing protocols for wireless mesh networks. Int J Res Rev Wireless Sens Netw (IJRRWSN) 1(3):43–47
15. Qin L, Kunz T (2008) Adaptive MANET routing: a case study. In: 7th international conference on Ad-Hoc networks and wireless, Sophia-Antipolis, Springer Lecture Notes in Computer Science, France, pp 43–57
16. Marina MK, Das SR (2006) Ad hoc on-demand multipath distance vector routing. In: Wireless communications and mobile computing, pp 969–988
17. Adjih C, Baccelli E, Jacquet P (2003) Link state routing in wireless Ad-Hoc networks. Unité de recherche INRIA Rocquencourt. Institut National De Recherche En Informatique Et En Automatique
18. Tachtatzis C (2008) Performance evaluation of multi-path and single-path routing protocols for mobile ad-hoc networks. In: International symposium on performance evaluation of computer and telecommunication systems, pp 173–178
19. Rishiwal V, Yadav M, Verma S, Bajapai SK (2009) Power aware routing in Ad Hoc wireless networks. J Comput Sci Technol 9(2):101–109
20. Maghsoudlou A, St-Hilaire M, Kunz T (2011) A survey on geographic routing protocols for mobile ad hoc networks. Systems and Computer Engineering, Technical Report SCE-11-03, Carleton University
21. Garkusha S, Garkusha O, Ievdokymenko M (2015) Development of mathematical models of management bandwidth downlink LTE using Resource Allocation Type 1. In: Second international scientific-practical conference problems of infocommunications science and technology (PIC S&T). Ukraine, Kharkiv, pp 46–50
22. Garkusha S, Yevdokymenko M (2012) Classification and analysis of methods of the distribution channels in multichannel mesh networks IEEE 802.11. In: Modern problems of radio engineering, telecommunications and computer science: proceedings of the XI international conference (TCSET'2012), Slavske, Ukraine, Lviv, pp 273–274
23. Lemeshko O, Garkusha S, Abed AH (2012) Two-index mathematical model of channels distribution in multichannel mesh networks 802.11. In: Modern problems of radio engineering, telecommunications and computer science: proceedings of the XI international conference (TCSET'2012), Slavske, Ukraine, Lviv, pp 279–280
24. Al-Dulaimi AMK, Ievdokymenko M (2016) Mathematical model and method of the balanced management of time-frequency resources in LTE network. Problemi telekomunìkacìj 1(18):72–90. http://pt.journal.kh.ua/2016/1/1/161_aldulaimi_cqi.pdf
25. Lemeshko O, Al-Dulaimi AMK (2016) Priority based balancing model of resource allocation in LTE downlink. Scholars J Eng Technol 4(4):169–174
26. Lemeshko O, Al-Janabi HD, Al-Janabi HD (2016) Priority-based model of subchannel allocation in WiMAX downlink. Scholars J Eng Technol 4(4):200–206

27. Lemeshko OV, Yevseyeva OYu, Garkusha SV (2013) A tensor model of multipath routing based on multiple QoS metrics. In: International Siberian conference on control and communications (SIBCON), Krasnoyarsk, Russia, Siberian Federal University, pp 1–4
28. Garkusha S, Garkusha O, Abed AH (2014) Features of use of hypergraphs in the simulation of multi-channel mesh-networks IEEE 802.11. In: XII international conference modern problems of radio engineering, telecommunications, and computer science (TCSET'2014), Lviv-Slavske, Ukraine, pp 510–512
29. Berge C (1973) Graphs and hypergraphs. Elsevier, New York
30. Berge C (1989) Hypergraphs: the theory of finite sets, Amsterdam. North-Holland, Netherlands
31. R-I.T.U.R.M. 1450-4 (2010) Characteristics of broadband radio local area networks. ITU-T, Geneva
32. Seok Y, Lee Y, Choi Y (2001) Dynamic constrained multipath routing for MPLS networks. IEEE ICCCN, Scottsdale 2(1):348–353
33. Wang Y, Wang Z (1999) Explicit routing algorithms for internet traffic engineering. In: 8th international conference on computer communications and networks, Paris, pp 582–588
34. Lemeshko O, Yeremenko O, Hailan AM (2017) Two-level method of Fast ReRouting in software-defined networks. In: 2017 fourth international scientific-practical conference problems of infocommunications science and technology (PIC S&T). IEEE, pp 376–379
35. Mersni A, Ilyashenko AE (2017) Complex criterion of load balance optimality for multi-path routing in telecommunication networks of nonuniform topology. Telecommun Radio Eng 76(7):579–590

Processing of the Residuals of Numbers in Real and Complex Numerical Domains

Victor Krasnobayev⬤, Alexandr Kuznetsov⬤, Alina Yanko⬤,
Bakhytzhan Akhmetov⬤, and Tetiana Kuznetsova⬤

Abstract The chapter discusses the procedures for the formation and use of real residuals of real numbers on a real module, as well as complex and real residues of an integer complex number on a complex module. The chapter focuses on the processing of complex and real residuals of an integer complex number by a complex module. This procedure is based on using the results of the first fundamental Gauss theorem. The chapter of the proposed procedure provides examples of determining deductions in a complex numerical domain. On the basis of the considered procedure, an algorithm was developed for determining the real deduction of an integral complex number using a complex module in accordance with which the device was synthesized for its technical implementation. The device received a patent of Ukraine for the invention, which confirms the novelty and practical value of research results. The results obtained in the chapter are advisable to be used when implementing tasks and algorithms in real and complex numerical domains. In particular, the use of real numbers for cryptographic applications was considered.

V. Krasnobayev · A. Kuznetsov (✉) · T. Kuznetsova
V. N. Karazin Kharkiv National University, Svobody Sq., 4, Kharkiv 61022, Ukraine
e-mail: kuznetsov@karazin.ua

V. Krasnobayev
e-mail: v.a.krasnobaev@gmail.com

T. Kuznetsova
e-mail: kuznetsova.tatiana17@gmail.com

A. Yanko
Poltava National Technical Yuri Kondratyuk University, Poltava, Ukraine
e-mail: al9_yanko@ukr.net

B. Akhmetov
Abai Kazakh National Pedagogical University, Dostyk Ave., 13, Almaty 050010, Republic of Kazakhstan
e-mail: bakhytzhan.akhmetov.54@gmail.com

529

T. Radivilova et al. (eds.), *Data-Centric Business and Applications*, Lecture Notes on Data Engineering and Communications Technologies 48,
https://doi.org/10.1007/978-3-030-43070-2_24

Keywords Computer system · Modular arithmetic · Non-positional code structures · Numeral systems in residual classes · Positional numeral systems · Residual classes

1 Residual Class System Properties

1.1 Features of the Representation of Numbers in a Non-positional Numerical System in Residual Classes

The modern stage of the development of science and technology is characterized by increasingly complex tasks that require their solution. However, the complexity of the tasks being solved is ahead of the growth rate of the power of universal computers. In this aspect, the main areas of improvement of real-time processing of information processing systems (IPS) are the improvement of user productivity and faultless functioning, due to providing the required level of fault-tolerance.

Depending on the architectural decisions taken, the entire set of computing systems in positional numeral systems (PNS), usually binary, can be divided into four main groups: thus, the use of SISD architecture (single stream of instructions and single data stream) provides a dominant position in the classical von Neumann architecture. In such machines, information is processed sequentially, teams are executed one after another, with each instruction initiating, as a rule, one scalar operation. In this case, the use of parallel operation for the information input–output interface and the processor, the combination of operations performed by separate units and nodes of the arithmetic logic unit do not allow the effective realization of real-time parallel computing systems. Thus, the possibilities for increasing the speed of modern positional computers, which are based on the classical architecture of the successive implementation of operators, virtually reached their limit value; computer systems of the second group—MISD-architecture (multiple stream of commands and single data stream) did not get a lot of practical implementation; these tasks, in which several processors could effectively process one stream of data are still unknown to modern science and technology; the basis of the third group of computing systems consists of devices developed on the basis of SIMD-architecture (single stream of commands and multiple data streams); using SIMD-architecture allows to realize high-speed real-time IPS; with their help, the problems of vector and matrix calculations are effectively solved, the problem of determining the roots of systems of algebraic and differential equations, etc.; a special place is occupied by digital signal processing tasks, which are the most optimal for the SIMD structure. This architecture of the computer system is oriented on parallel-conveyer execution of the most laborious computational operations. To provide the limit for the given level of technology of the productivity of the computer system is possible only at the expense of the use of non-traditional arithmetic, in which the parallelization process is carried out at the level of arithmetic operations (microoperations); An alternative solution to the problem of solving real-time high-performance computing is the use of MIMD architecture (multiple stream of commands and multiple data streams). This class assumes that there are several command processing units in the computer system, united in a single complex and each with its own data and commands (multi-microprocessor, multi-machine, cluster and other similar computing systems). However, in spite of all the advantages mentioned above, such as the availability of memory for each processor element and the independence of the computing process, systems with mass parallelism generated a number of problems associated with the description and programming of process switching and their management. At the same time, the lack of a mathematical device, which allows solving the problem of increasing the productivity of computing systems, is a major deterrent to the widespread use of MIMD-systems with massive parallelism.

Thus, it is obvious that further progressive development of computer technology and information processing tools in PNS is directly related to the transition to parallel computing. This transition, of course, opens up new opportunities in the field of improving and developing computing devices [1–3].

Reserves for increasing reliability, fault tolerance and durability of functioning, as well as user-generated computing performance are the use of computing structures, specialized calculators and special processors, created on the principle of parallelization of the problem solved (algorithm) at the level of one micro-operation.

The concept of parallelism has long attracted the attention of specialists to its potential capabilities to increase the productivity and reliability of computing systems. The conducted theoretical, experimental and industrial developments in this direction have allowed to substantiate the basic principles of construction of parallel computing systems. It is precisely with such systems the current perspective of further increasing the computing power and reliability is connected.

In 2005, it turned out 50 years from the date of the publication of the article by the Czech engineer M. Valah, in which it was first proposed to apply for operations on computer numbers instead of operations of the rings of the residual modulo $M = 2^n$ the operation of the rings of the residual modulo $M = m_1 m_2 \ldots m_n$, where $m_1, m_2, \ldots, m_n$—pairwise mutually simple numbers. In computational practice, this was an outstanding idea, since all ring operations modulo $M = m_1 m_2 \ldots m_n$ were reduced to a homomorphic parallel implementation of the same operations on small modules $m_1, m_2, \ldots, m_n$. The well-known Chinese theorem on residual, which was previously treated as a structural theorem of abstract algebra, guaranteed this parallelism in calculations over integer numbers, provided that the result of ring operations belongs to the range of integers, which is determined by the product of the modules $M = m_1 m_2 \ldots m_n$. This idea attracted the attention of a large group of scientists. A new scientific direction arose—modular arithmetic [4–6].

Over the past 50 years, modular arithmetic (numeral systems in residual classes (NSRC), the class of residuals) has survived periods of rapid development and serious recessions. Currently, there is a progressive increase in interest in modular arithmetic among developers of complex systems related to the processing of signals and images, with cryptographic transformations, etc.

It is known that the non-positional numerical system in the system of residual classes (NSRC) is used in computer systems (CS) to increase the speed of the implementation of integer arithmetic operations. This is due to the presence and influence of the properties of the NSRC both in the structure of the data processing CS, represented in the integer-type, and on the principles of functioning (the numerical system influences the principles of the implementation of arithmetic operations more) of CS. In turn, the structure and principles of the functioning of the CS affect the characteristics of computing systems.

However, there are a number of factors, such as properties of the NSRC, influencing the structure of the CS, as well as the principles of the implementation of arithmetic operations. The use of the results of such research makes it possible to more accurately and precisely estimate the possibility of practical application of non-positional code structures (NPCS) in computational technology.

The creation of NPCS $A = (a_1 || a_2 || \ldots || a_n)$ in the NSRC is based on the use of the principles of parallelism and independence of the formation of residues a_i. These principles determine the three main properties of the NSRC. Let's analyze each of these properties [7–9].

1. Independence of each residual in the structure of the NPCS NSRC. This property enables to form the structures of the CS in the NSRC in aggregate form (by the number of n bases (modules) of the NSRC) of informatively independent, low-level computing paths (CP) that function independently of each other and in parallel in time.

    At the same time, note that:

    – in the general case, the time of execution of arithmetic operations by the CS in the NSRC is determined by the time of execution of operations for the largest bit grid CP;
    – the CS in the NSRC has a modular design consisting of separate independent CPs; it allows to carry out repairs and maintenance, as well as carry out operations of control, diagnostics and correction of data errors in the CP CS in the NSRC without interrupting the process of solving the problem, that is, without stopping the calculations;
    – errors occurring in the CP CS with base m_i, do not "multiply" to other CPs; It does not matter whether there was a single or multiple error in the CP with base m_i, or even a set of errors of length no greater than the binary digits $\left[\log_2(m_i - 1) + 1 \right]$ [10–12]. Thus, an error that arose in an arbitrary CP CS with base m_i, or persists in this CP before the end of the calculation, or in the process of subsequent calculations will be self-abolished (for example, if the erroneous value of residual a_i with base m_i of the first number is multiplied by zero second of the second on the same m_i NSRC basis);

2. The equality of residuals in NPCS NSRC. An arbitrary residual a_i of number $A = (a_1 || a_2 || \ldots || a_n)$ in the NSRC contains information about the entire output number. This makes it possible, subject to the equation $m_i < m_j$, to replace the

out of order CP of CS modulo m_i by the programmed methods, to workable CP modulo mi, without stopping the task solution [13–16].

This property is conditioned by the fact that the structure of the CS in the NSRC provides an opportunity to exchange, based on the application of program methods, such characteristics of the CS as the speed of execution of arithmetic operations, the accuracy of the execution of arithmetic operations and the reliability of the execution of arithmetic operations without stopping the calculations in the process of solving the problem. In this aspect, the CS in the NSRC can have different reliability depending on requirements, for example, the accuracy or speed of the calculations. In addition, the structure of the CS in the NSRC allows the organization of degradation of the computer system. This allows (subject to the refusal of the specified elements of the structure) to continue functioning without one or more functions of the CS, or to continue to function with deteriorated quality, for example, with a decrease in speed, or with a decrease in the accuracy of computing, etc.

The use of the first and second properties of the NSRC determines the availability of three types of reservation at the same time in the CS: structural, informational and functional. This, in turn, allows to synthesize mathematical models of fault-tolerance with a more accurate estimate of the required characteristics of the CS [7].

3. The low bitness of residuals, the totality of which determines the NPCS in the NSRC. This property allows to significantly improve the speed of the implementation of arithmetic operations both due to the low-level representation of the remains of the number in the NSRC, and due to the possibility (in contrast to the positional systems of the calculus) to use table arithmetic; application of the methods of table arithmetic allows to realize the basic arithmetic operations in the NSRC in fact for one cycle of the work of the CS.

1.2   The Influence of the Basic Properties of the System of Residual Classes on the Methods of Implementing Arithmetic Operations

The low bitness of the residuals of numbers in the NSRC provides a wide choice of variants of system-technical solutions to the tasks of implementing arithmetic modular operations, based on the following principles:

- adder principle (based on the use of low-level binary adder);
- tabular principle (based on the use of memory elements);
- the circular shift principle (based on the use of shift registers).
  Based on the use of the properties of the NSRC, the following (in comparison with positional numerical systems (PNS)) advantages become evident:

- possibility of realization of asynchronous arithmetic calculations at the level of decomposition of numbers, which increases the speed of CS computing;
- the possibility of organizing tabular (matrix) execution of arithmetic operations and sampling of the result of a modular operation in one cycle;
- the ability to create a CS and component with effective detection and correction of errors without interruption in calculations, as well as the ability to synthesize fault-tolerant digital devices;
- the possibility of creating a system for monitoring and correcting errors in the dynamics of the computational process CS;
- providing high active fault tolerance of computing structures based on operative reconfiguration of the structure of the CS;
- the possibility of increasing the reliability of the CS due to the effective use of passive and active fault tolerance.

The set of properties of NSRC determines the possible directions of its effective application. First of all, these are:

- modular and cryptographic integer transformations;
- signal processing;
- processing integer data of large (hundreds and thousands of bits) bitness in real-time;
- processing large data arrays represented in a matrix form;
- data processing in optoelectronic and neurocomputer scientific and technical areas;
- application of NSRC for implementation of algorithms of mobile communication processors, in which it is necessary to provide high speed of data processing with insignificant energy consumption; indeed, the data processing algorithms for mobile communication processors mainly consist of arithmetic operations of compilation and multiplication, thus, the fact of the absence of transfers in performing arithmetic operations in the NSRC allows to reduce the energy consumed by mobile communication processors.

The influence of the properties of the NSRC on the structure and principles of the implementation of arithmetic operations is as follows.

1. Increasing the speed of implementation of arithmetic operations is determined by the parallel structure of data processing in the NSRC and using the principle of tabular execution of basic arithmetic operations.
2. Improved reliability, fault-tolerance, and survivability of the CS is possible, firstly, by using the property of passive fault-tolerance, which a priori exists in the initial structure of the CS, which operates in the NSRC. In this case, the original natural structure of the CS in the NSRC has the form of a computing structure that is similar to the artificial permanent structural reservation in the PNS. Secondly, increasing the reliability of the CS can be due to the use of active fault-tolerance (dynamic structural redundancy) computing structures in the NSRC.
3. The use of the first and second properties of the NSRC causes more effective than the PNS, the simultaneous application of passive and active fault tolerance,

which makes the computer components of the CS a priori more suited to the control and diagnosis of data errors.

4. The influence of the first and second properties of the NSRC determines the possibility of the structure to adapt to the mode of operation of the CS. This circumstance allows directly, in the course of calculations, to perform exchange operations between the reliability of the functioning of the CS, the speed of the implementation of arithmetic operations and the accuracy of the solution of the problem.

The results of the study of the influence of the properties of the NSRC on the structure of the CS and the principles of implementation of arithmetic operations have shown that the use of NSRC can increase the speed of arithmetic operations and increase the failure-resistance of the functioning of the CS. In addition, the use of NSRC allows to create a unique system for monitoring, diagnosis and correction of data errors in CS without stopping calculations, which has no analogues in PNS.

1. Increasing the speed of implementation of arithmetic operations in the NSRC is achieved through the possibility of organizing parallel processing of data, as well as by using the tabular principle of the implementation of arithmetic operations.
2. The use of the properties of the NSRC determines the existence of three types of reservation at the same time in the CS: structural, informational and functional. This, in turn, allows us to increase the failure-resistance of non-positional computing structures in the NSRC through the application of methods based on passive (permanent structural reservation) and active (structural reserve replacement) fault-tolerance.
3. It was discovered that the CS and components in the NSRC belong to computational structures that are easy to control and diagnose. This feature facilitates the development of methods for effective control and diagnostics of data in the NSRC. Thus, the use of the properties of the NSRC makes it possible to create a unique system for monitoring and correcting data errors without stopping calculations, which is especially important for the CS, which function as part of complex technical real-time systems.

2 Processing of Data Presented in the Residual Classes System

2.1 Examples of Using a Residual Classes System in Data Processing Systems

In the present time there is a number of fields and directions of science and technology, where a need in fast, reliable and highly precise integer arithmetic calculations exists [17–29]. We can say, that in almost all fields of science the integer arithmetic calculations are used [22, 24, 26]. First of all, they are such fields of science as mathematics, physics, astronomy, technical science, cryptography and coding theory, geodesy and meteorology, seismology etc. [30–38]. Let's note the following directions in science and technology, where there exists the necessity in fast, reliable and highly precise integer arithmetic calculations: arithmetic operations with integer numbers and polynomials; integer linear programming; operations with numbers and sets, the solution of the multidimensional NP-complete problems; implementation of routing algorithms (algorithms for finding the shortest path); problems of ways and matrix multiplication; problems of fast Fourier transform and it's applications; the creation of artificial intelligence systems (neural network data processing system); tasks for military purposes; digital signal processing, digital image processing; cryptographic transformation; highly-precise integer arithmetic; the solution of problems related to the space research; highly-precise digital-to-analog and analog-to-digital conversions and so forth [39–43].

The results of the researches conducted during last few decades in the field of information technologies by different groups of scientists and engineers of methods of productivity improvement, reliability, survivability, and reliability of computer systems calculations and data processing means presented as integers (CSIDPM), showed that within the PNS, it is practically impossible to achieve it. First of all, it's caused by the main disadvantage of modern CSIDPM that operate in PNS: the presence of inter-bits links between the processed operands. These links significantly impact the architecture of the calculator and methods of implementation of arithmetic operations, implemented by CSIDPM; complicate the apparatus and limit the speed of the arithmetic operations of addition, subtraction and multiplication. In this regard, improving above mentioned characteristics of CSIDPM in PNS, is carried out, first of all, by increasing the clock frequency, development and application of methods and means of parallel data processing as well as by using different types of redundancy. This circumstance led to the need of finding the ways of increasing the effectiveness of CSIDPM functioning, for example, through the use of new architectural solutions by applying non-positional machine arithmetic, in particular, on the basis of non-positional numeral systems use in residual classes (NSRC). The well-known Chinese remainder theorem (the task of restoring the original number Ak by the aggregating of its remains (deductions) {ai} by dividing it into a series of natural numbers m1, m2, …, mn (modules) of NSRC), which was previously interpreted as a structural theorem of abstract algebra, guaranteed the specified parallelism in the calculations over integers, under the conditions that the result of ring operations belongs to the range of integers, defined by models product of NSRC. The results of conducted researches of the implementation of arithmetic operations methods in NSRC led to the creation of new machine arithmetic. Having its ideological roots of the classical works of Euler, Gauss and Chebyshev on the theory of comparisons, NSRC introduced new ideas in the development of creation methods of highly-productive and ultra reliable CSIDPM [4].

For the first time the results of theoretical studies devoted to the possibility of practical application of NSRC as a numeral system (NS) of CSIDPM, were published in 1955–1957 in the scientific works of Czech scientists M. Valaha and A. Svoboda.

Non-positional number system in NSRC is a NS where integers are presented as a set of non-negative deductions (residues) in the group of mutually pairwise prime numbers which are called bases or modules of NSRC. In this case there are no inter-bits relations between processed numbers residues, that gives opportunity to perform arithmetic operations excluding bit relations between numbers residues. The use of NSRC-based machine arithmetic allowed to create actually operating CSIDPM. In the 60s of the past century the team of scientists and engineers headed by the doctor of technical sciences, professor D. I. Yuditskii, created A-340A the world's first experimental computer and T-340A serial computers, functioning in NSRC. These computers were intended for regular polygon version of Dunay-3UP radar, which was the part of the USSR A-35 missile defense system. In the 70s of the past century for radar stations there were created such CSIDPM in NSRC as "Diamond" and 5E53 supercomputers.

However, in the 80s of the past century due to a number of objective and subjective reasons the interest to modular arithmetic (MA) is significantly reduced. It was primarily due to the death of the Director of the Microelectronics Center, developing the general theory and practical creation of a computer in NSRC located in Zelenograd, Moscow Region, the Director and the chief initiator of project Lukin Fedor Victorovich and therefore, the complete termination of practical works, connected with the use of MA. But then this direction was restrained by the imperfection of the existing at that time element base of computers, as well as the existing methodology of computer systems and components designing, principally focused at that time only on the binary system calculation.

Now the interest to the use of NSRC is increasing again. Ultimately it is caused by:

- the emergence of the numerous scientific and theoretical publications devoted to the theory and practice of the computer systems and components creating in NSRC;
- wide distribution of mobile processors that require high speed data processing at low energy consumption; the lack of inter-bits transfers during arithmetic operations of addition and multiplication of numbers in NSRC allows to reduce energy consumption;
- strong interest to NSRC is being shown by the banking structures, where it is necessary in real time to handle large amount of data safely and reliably, i.e. they are required highly-productive means for highly reliable computing with errors self-correction, that is typical to the NSRC codes;
- the elements density increasing on a single chip doesn't always allow to perform a complete and qualitative testing; in this case there is an increasing importance of providing failover operation of CSIDPM;
- the need for the use of the specialized CSIDPM to perform a large number of operations on vectors, which require high-speed performance of integer addition and multiplication operations (matrix multiplication problems, the problems of the scalar product of vectors, Fourier transformation, etc.);

- the widespread introduction of microelectronics into all spheres of human activity significantly increased relevance and importance of previously rare, and now so massive scientific and practical problems, as a digital signal and image processing, image recognition, cryptography, multi-bit data processing and storage, etc.; this circumstance requires enormous computing resources being in excess of the existing possibilities;
- the current level of microelectronics development is coming to its limits from the point of view of productive provision and reliability of existing and future computer systems and components of large data sets processing in real time;
- taking it over nanoelectronics, molecular electronics, micromechanics, bioelectronics, optical, optoelectronic and photonic computers and others are still rather far from the real industrial production and employment;
- the modern development of integrated circuit technology allows to have a fresh look at the principles of devices construction with modular arithmetic employment and provides wide opportunities to use new design techniques (such as the methodology of systems design on a chip-SOC) both in the development of individual computing units, and computer systems in general; integral technology enables more flexible design of computer systems and components and allows us to implement NSRC-based devices as effectively as on the basis of the binary system; furthermore at present in order to improve the effectiveness of computer devices development, automated design systems (ADS) are widely used; in this respect, the design of computer systems and components based on NSRC does not differ from the working with the help of ADS data of binary data-blocks in PNS;
- unfortunately, Ukraine today in contrast to the theoretical development, technologically is behind the foreign microelectronics of some leading countries; in this case, it is advisable to use the existing theoretical achievements and practical experience in the creation of effective computer systems and components in NSRC [4].

2.2 Residual Classes of Numbers in a Real Numerical Domain

In [1] it is given a definition of NSRC. In this case NSRC is considered a generalized version of NS, in which any natural number A, including zero, is represented as a set of the smallest positive residues (deductions) of the division of the original A number on preset $m_1, m_2, \ldots m_n$ natural numbers, called bases or NSRC modules. In literature it is often not entirely fair the term NSRC is identified with "residue class". In some cases, this circumstance can interfere the analysis of the results of solving the data processing problems presented in MA. In this regard it is important to consider the correlation between the notion of NSRC and RC. We'll give a definition to the notion "residue class". Let's consider the set {A} of all natural numbers, including zero. From the set of natural numbers we choose an arbitrary number (module) mi. While dividing any natural number on m_i module we can get the following set of residues: 0 (A number is divided into the m module integrally), 1, 2 ... m_{i-2} and m_{i-1}. All the set of natural numbers including zero, can be divided into m_i (0, 1, 2, ... m_{i-2} and m_{i-1}) of different groups of numbers (residue classes), including in each RC the numbers which, while dividing into the module mi, give the same remainder. It is considered, that these numbers are comparable with each other on module mi.

The residue class modulo mi of NSRC can be denoted by the symbol $RC_j^{(i)}$, where i—the number of the base of orderly ($m_i < m_{i+1}$) NSRC ($i = \overline{1, n}$); j—the RC number in the system of residues for a given module mi ($j = \overline{0, m_i - 1}$). In the general case, the residue class of $RC_j^{(i)}$ modulo mi we will call the set of all integers, including zero, which while dividing into the modules m_i give the same positive balance.

Taking into account the well-known correlation $(-A) \bmod m_i = (m_i \cdot k - A) \bmod m_i (k = 1, 2, 3, \ldots)$, all RC on arbitrary module m_i of NSRC can be represented in the form of residues:

Actually, there is an opinion [4], that it is possible for NSRC not to be called a number system. Indeed, NSRC bases are connected to each other so, that they are selected in a certain way and secured by the permanent modules for the given NS. Each residue modulo is informationally independent on other residues, however, during the implementation of arithmetic operations within each residue unitary or binary NS is generally used. Thus NSRC may be determined not as the number system, but as a special design code numeric data structure, that is specially encoded block of numerical data.

It should be noted that in the proposed approach the NSRC is not opposed to binary PNS, and serves as its extension that allows to solve effectively a certain class of problems. Therefore, the most effective in this case, is an approach that unites the use of a combined MA and binary PNS notation in constructing the control systems. Upon that, for example, control of the entire system can be carried out by the conventional binary commands and blocks; and data processing is performed on the basis of a modular representation of numbers. Thus, the use of the advantages and benefits of NSRC, along with the traditional binary method of control systems constructing can lead to the productivity increase of CSIDPM in general [7].

To answer the question of whether to use NSRC it's necessary to investigate the influence of the MA basic properties on the structure and operation principles of CSIDPM. Possible logical algorithm research diagram of NSRC effective application can be represented as follows:

- to indentify the areas and directions of science and technology where integer calculations are necessary; to show in which tasks and algorithms (specifically, to name and show the most important ones) integer calculations are used; first of all the tasks and algorithms, which include such operations as arithmetic operations of addition, subtraction and multiplication in a positive and negative number ranges, as well as arithmetic operation and algebraic comparisons of numbers;

- to justify the relevance requirements and the need to increase the speed of integer calculations, i.e. to justify the need to increase CSIDPM productivity in order to (to increase the speed of integer calculations it's necessary to create CSIDPM of increased (in comparison to the existing ones) productivity;
- to consider the existing and advanced methods for production increase of CDIDPM, operating in the PNS; possible conclusion: the existing and advanced methods of performance improving of CDIDPM in PNS do not always satisfy the increasing demands to the improved performance implementation of integer calculations (denote the main reason);
- to consider one of the possible (referred to in modern literature) options for creation of highly productive CDIDPM on the basis of NSRC; on the basis of the analysis of the NSRC properties and the results of the previous and up-to-date researches of theoretical and practical developments in the application field of non-positional number system, to justify the possibility of its effective application in order to improve the CSIDPM performance.

If the proposed algorithm research scheme is adopted, then the theoretical researches, devoted to the CSIDPM production increase on the basis of NSRC implementation can be carried out. Methods, models and data processing algorithms in NSRC are being developed. Comparative analysis of the achieved results are being conducted. Before defining a class of tasks and algorithms for which the mathematical apparatus of the numbers theory is effectively applied, it is necessary, on the basis of the results of the NSRC properties researches, to analyze the advantages and disadvantages of the MA use [7].

3 Methods for Implementing Arithmetic Operations in the System of Residual Classes

3.1 Tabular Method

The analysis of the methods for increasing of the efficiency of SEC in the HEC Jacobian allowed to theoretically substantiate and to practically demonstrate the dependence of the realization of the efficiency of SEC operations in the Jacobian of HEC upon the aggregate of the following basic characteristics—type of realization of cryptographic transformations (software, hardware and software-hardware); algorithm type of the SEC divisors; the prescribed base field, over which the given curve is set; the type of the curve; the values of the curve coefficients; the selected system of coordinates, in which the HEC Jacobian divisors (affinity, projective, weighted and mixed) are represented; the accepted method of arithmetical transformations etc. The known methods of realization of the SEC algorithm (the Quantor divisor summation method, the Kobliz method, the method of arithmetic transformations of divisors in the HEC

Jacobian of the second, the third and the fourth kinds, methods of summation of divisors with different weights, the Karatsuba method for multiplication and reduction of the polynomial functions by the module in the field, the method based upon several results of the Chinese remainder theorem etc.) do not always satisfy the requirements with respect to the efficiency of cryptographic transformations. At the same time, the reference sources [10] demonstrate high efficiency of the modular arithmetic (MA) codes, i.e., the system of computation in remainder classes (CRC) while solving separate problems of digital data processing (solving of filtering problems, problems of realization of FFT, DFT etc.) from the point of view of the high efficiency of their realizations. Thus, it is known that the Fourier transformation is related to calculation of the polynomial of the kind $P(x) = \sum_{i=1}^{n-1} \alpha_i x^i$. One of the applications of the Fourier transformation lies in calculation of the convolution $\sum_{i=1}^{n} \alpha_i \beta_i$ of two n-dimensional vectors $A = (\alpha_1, \alpha_2, \ldots, \alpha_n)$ and $B = (\beta_1, \beta_2, \ldots, \beta_n)$. In the given case the convolution operation is the complete analogue to the realization of arithmetic operations of multiplication of two numbers A and B in MA with consequent summation of the components of the kind $\alpha_i \beta_i (mod\ m_i) + \alpha_j \beta_j (mod\ m_j)$.

In the given aspect this phenomenon stipulates the importance and actuality of the search for the methods for increasing of the efficiency, reliability and validity of the public-key cryptographic transformations on the basis of the using the properties of the position-independent MA code structures.

The objective of the paper is to develop a highly efficient method for realization of public-key cryptographic transformations on the basis of the using the position-independent MA codes of position independent structures, i.e., CRC codes.

The influence of the CRC main parameters (independence, equality and short form of the operand-representing remainders) upon the structure and the principles of operation of the data processing system (DPS) in MA are considered in details in [11–13]. In particular, it is demonstrated that short form of the rests in representation of numbers in modular arithmetic provides for the possibility of wide selection between the options of system engineering solutions at realization of the modular arithmetic operations.

It is known that there exist four principles of realization of arithmetic operations in MA—the summation principles (SP) (on the base of short binary summators); the table principle (TP) (on the base of using ROM); the direct logical principle of realization of arithmetic operations based on description of module operations at the level of the systems of switching functions by means of which the values of binary digits of the resulting deductions are formed (it is reasonable to use systolic and programmable logical matrices as well as EPLD as the element base for technical realization of the given principle); the principle of ring shift (PRS) based on using of the ring shift register (RSR).

The absence of bit-to-bit associations (the absence of the transport process) between the binary digits in operands processed in DPS during the process of cryptographic transformations (at realization of module operations) on the basis of TP or PRS is one of the main and the most attractive particularities of modular arithmetic. Within the base notation system (BNS) the performance of an arithmetic operation assumes the subsequent processing of operands digits upon the rules determined by the contents of the given operation and cannot be finished up to the moment until the values of all of the intermediate results considering all the relationships between the bits, are sequentially determined. Thus, BNS in which the information is represented and processed in the present-day DPS, have a substantial drawback—the presence of bit-to-bit associations which impose their imprint upon the methods of realization of arithmetic operations; make the hardware more complicated, decrease the trustworthiness of calculations and restrict the computing speed of crypto-graphic transformations realization. Therefore, it is only natural to seek for the opportunities of creation of the kind of arithmetic, in which the bit-by-bit associations would be absent. In this connection it is worth to pay attention to the base notation system in the residual classes. The system of residual classes possesses a valuable parameter of independence of the remainders upon each other pursuant to the accepted system of bases. This independence opens up wide opportunities to the development of not only the new kind of machine arithmetic but also to the principally new structural realization of DPS, which, in its turn, is substantially extending the sphere of application of the machine arithmetic. In most of the reference sources it is noted that implementation of non-traditional methods for data representation and processing in the numerical systems with parallel structure and, in particular, within the so-called modular base notation systems possessing the maximal level of the internal parallelism in organization of the data processing procedures is one of the practical trends in increasing of the user efficiency of computing equipment. The position-independent computing system in the residual classes is also referred to the above systems.

Short form of the rests, which represent the operand, is one of the CRC properties. It is just this property that allows to substantially increase the computing speed at execution of the arithmetic operations due to the possibility of application (unlike in BNS) of the table arithmetic where the arithmetic operations of addition, deduction and multiplication are performed practically in one and the same cycle [10]. The search for the way of increasing of the data processing efficiency led to the necessity of development of the table method for realization of modular operations on the basis of PRS.

Thus, despite the difference in the digital structure of the tables of modular operations of summation deduction and multiplication there was created a new original table method for realization of arithmetic operations in MA. On the basis of the method it is possible to synthesize a structurally simple, highly reliable and super-efficient DPS in MA, the basis of which is formed by three separated switches each of them realizing only 0.25 part of the relevant complete table of modular operations of multiplication and deduction (the first switch is the II quadrant of the multiplication table; the second and the third switches are respectively I and II quadrants. In this sense the table multiplication code obtained a new quality and became the universal table code for performance of the three arithmetic operations in MA.

At the table option of realization of arithmetic operations the bit-to-bit associations between the processed operands are absent completely. However, for a quite large-digit grid of DPS (for larger in value modules of TMC) the number and complexity of equipment units in operation devices are sharply increased.

3.2 Ring Shift Method

It is important and actual to consider the intermediate option of realization of arithmetic operation in TMC based on application of the ring shift method (principle).

In [10] it is considered the principle of realization of arithmetic operations in TMC—the ring shift principle, the particularity of which lies in the fact that the result of the arithmetic operation $(\alpha_i \pm \beta_i) \, mod \, m_i$ upon the arbitrary module of TMC set by the aggregate of bases $\{m_j\}$ $(j = \overline{1, n})$ is determined only at the expense of cyclic shifts of the set digital structure. Actually, the well-known Kally theorem is setting the isomorphism between the elements of the finite Abelian group and the elements of the permutation group.

One of the consequences of the Kaly theorem is the conclusion that reflection of the elements of the Abelian group upon the group of all of the integer numbers is homomorphous. This circumstance allows to organize the process of determination of the result of arithmetic operations in TMC by means of using PRS.

Thus, the operand in MA is represented by the set of n remainders $\{\alpha_i\}$ formed by means of subsequent division of the initial number A by n mutually paired prime integers $\{m_i\}$ for $(i = \overline{1, n})$. In this case the aggregate of remainders $\{m_i\}$ is directly equaled to the amount n of prime Galois fields having the form of $\sum_{i=1}^{n} GF(m_i)$.

In order to consider the method of realization of arithmetic operations in TMC it would be sufficient to consider the option of for an arbitrary finite Galois field GF (m_i) at $i = const$, i.e., for the specific reduced system of deductions upon the module mi.

Let the Kally table is made for the set operation of modular summation $(\alpha_i + \beta_i) \, mod \, m_i$, in the field GF (m_i). From the existence of a neutral element in the field GF (m_i), which the elements of the given field are arranged in the ascending order. And from the fact that in the field of deductions GF (m_i) these elements are different (the order of the group is equal to mi) it follows that each row (column) of table contains all of the field elements exactly one time each. The use of the above particularities allows to realize the operations of modular summation and deduction in TMC by applying PRS with the help of n ring $M = m_i([\log_2(m_i - 1)] + 1)$—digit shifting registers.

Let the arbitrary algebraic system be represented in the form $S = \langle G, \otimes \rangle$, where G is the non-empty set; $\otimes$ is the type of operation determined for any of two elements $\alpha_i, \beta_i \in G$. The operation of summation $\oplus$ in the set of the classes of deductions R generated by the ideal J forms up a new ring called the class of deductions ring R/J. It can be represented in the form of Z/mi where Z is the set of integers 0, ± 1, ± 2, ... (If the base of TMC mi is the prime integer, then Z/mi is the field). As it is indicated above just this circumstance is stipulating the possibility of realization of the arithmetic operation of summation in TMC without any bit-to-bit transfers by means of the ring shift [10].

On the basis of the principle suggested in [10] there was developed the method of realization of arithmetic operations in TMC (the method of binary position and remainder encoding). The essence of the developed method lies in the fact that the

initial digital structure for each module (base) of TMS is represented in the form
of the contents of the first row (column) of the modular summation (deduction)
table $(\alpha_i \pm \beta_i)\, mod\; m_i$ of the form $P_{init}^{(m_i)} = \left[P_0(\alpha_0) \| P_1(\alpha_1) \| \dots \| P_{m_i-1}(\alpha_{m_i-1}) \right]$
where $\|$ is the operation of concatenation (gluing); $P_v(\alpha_v)$ is the k-bit binary code
correspondent to the value of the av-th remainder $(\alpha_v = \overline{0, m_i - 1})$ of the number
upon the module $m_i; k = \left[\log_2(m_i - 1) + 1\right]$. For the set specific module $m_i = 5$ the
initial digital structure of the RSR content has the following representation $P_U^{(5)} =$
$[000\|001\|010\|011\|100\;]$.

Thus, by means of the ring shift registers used in BNS it is easy to realize the
arithmetic operations in TMC, where the degrees of cyclic permutations as considered
on the basis of are determined by the following expressions:

$$\left[P_0(\alpha_0) \| P_1(\alpha_1) \| \dots \| P_{m_i-1}(\alpha_{m_i-1}) \right]$$
$$= \left[P_z(\alpha_z) \| P_{z+1}(\alpha_{z+1}) \| \dots \| P_0(\alpha_0) \| \dots \| P_{m_i-1}(\alpha_{m_i-1}) \right]^Z, \tag{1}$$

$$\left[P_0(\alpha_0) \| P_1(\alpha_1) \| \dots \| P_{m_i-1}(\alpha_{m_i-1}) \right]^{-z} = \left[P_{m_i-1-z}(\alpha_{m_i-1-z}) \| \dots \| P_{m_i-z}(\alpha_{m_i-z1}) \right.$$
$$\left. \| \dots \| P_0(\alpha_0) \| P_1(\alpha_1) \| \dots \| P_{m_i-z-2}(\alpha_{m_i-z-2}) \right]. \tag{2}$$

We note that $\left[P_0(\alpha_0) \| P_1(\alpha_1) \| \dots \| P_{m_i-1}(\alpha_{m_i-1}) \right]^{m_i} = \varepsilon$, i.e., at $z = m_i$ all the ele-
ments of the ordered set $\left\{P_j(\alpha_j)\right\}$ for $(j = \overline{0, m_i - 1})$ remain at the initial position.
During the technical realization of the given method the first operand a_i is determin-
ing the number α_i of the digit $P_{\alpha_i}(\alpha_{\alpha_i})$ with the content of the modular operation
result upon the module mi and the second operand β_i—the number of RSR digits
(of $\beta_i k$—bit binary digits) upon which it is necessary to perform the shifts of the
initial content of the RSR pursuant to the algorithms (1), (2). The main drawbacks
of the suggested method for realization of arithmetic operations in TMC include
comparatively larger time for execution of integer-number arithmetic modular operations that increases the efficiency of using of PRS. This drawback is stipulated
by the fact that the structure $P_{init}^{(m_i)}$ is represented by the set of initial remainders
of the first row of the matrix $(\alpha_i + \beta_i)\, mod\; m_i$, which are reflected by the binary
code. In this case the time for realization of the modular summation of two operands
$A = (\alpha_1, \alpha_2, \dots, \alpha_{n-1}, \alpha_n)$ and $B = (\beta_1, \beta_2, \dots, \beta_{n-1}, \beta_n)$ in TMC is determined
by the expression:

$$t = k\beta_{\max\, i}\tau, \tag{3}$$

where τ is the time of shift of one binary digit of RSR.

We consider the method for realization of arithmetic operations in TMC, which
is deprived of the above drawback. It is the method of unitary position and residual
encoding according to which the informational structure $P_{init}^{(m_i)}$ of the arbitrary module
mi, of TMC is represented in the form of a unitary $(m_i - 1)$–bit code.

$$P_{init}^{(m_i)} = \left[P(\alpha_{i-1}) \| P(\alpha_{i-2}) \| \ldots \| P(1) \| P(0)\right], \tag{4}$$

where $P(\alpha_j)$ is the binary digit of the digital structure (4), the unitary condition of which corresponds to the value of the operand α_j represented by a unitary code $(\alpha_j = \overline{0, m_i - 1})$. In this case the initial condition of RSR includes $m_i - 1$ binary digits and schematically can be represented in the form of.

As this takes place, the first operand $\{\alpha\}$ represented by a unitary code upon an arbitrary module m_i of TMC is entered into the jth digit of RSR, i.e., transfers the jth binary digit into the unitary condition. The second operand β_1 is pointing out to the number by the shift z of the RSR content determining the time of realization of arithmetic operations upon the module m_1 of TMC, i.e.,

$$t_{next} = \beta_i \tau. \tag{5}$$

We note that the time of realization of arithmetic operation A + B in TMC will be determined by the time of performing of the operation for the maximal value $\left(\beta_{max\ i}(i = \overline{1, n})\right)$ of the remainder from the set $\{\beta_i\}$ for the given operand $B = (\beta_1, \beta_2, \ldots, \beta_n)$, i.e.

$$t_{next} = \beta_i \tau. \tag{6}$$

Analysis of the expressions (5) and (6) demonstrates that the developed method of unitary representation reduces by $k = \left[\log_2(m_i - 1) + 1\right]$ times the time of performing of the arithmetic operations as compared to the method of binary encoding.

4 Processing of Data Presented in the Residual Classes System in the Complex Domain

4.1 Definition of Complex and Real Residues of Complex Numbers by a Complex Module

Let us consider one of the most interesting and important questions in the theory of whole complex numbers—the definition of the class of least residues and the related first fundamental Gauss theorem. This theorem establishes an isomorphism between the set of real and complex residuals of numbers.

The above material leads to the first fundamental Gauss theorem. This theorem establishes an isomorphism between complex and real residues.

Theorem *Given a complex module $\dot{m} = p + qi$, whose norm N is equal to $N = p^2 + q^2$ and for which p and q are mutually prime numbers, each integer complex* *number is $\dot{A} = a + bi$ by the complex module $\dot{m}$ is comparable to one and only one real residual from the $\overline{0, N-1}$ number series, i.e. we have $\dot{A} \equiv h(mod\,\dot{m})$ [11–13].*

Proof From the theory of numbers it is known that for two mutually simple numbers p and q you can find such two integers u and v, that the following condition is true:

$$u \cdot p + v \cdot q = 1. \tag{7}$$

First we show that the following identity exists:

$$i = u \cdot p - v \cdot q + \dot{m} \cdot (v + ui). \tag{8}$$

Then:

$$
\begin{aligned}
i &= u \cdot q - v \cdot p + (p + q \cdot i) \cdot (v + u \cdot i) \\
&= u \cdot q - v \cdot p + (p \cdot v + p \cdot u \cdot i + q \cdot v \cdot i + q \cdot u \cdot i^2) \\
&= u \cdot q - v \cdot p + (p \cdot v + p \cdot u \cdot i + q \cdot v \cdot i - q \cdot u) \\
&= u \cdot q - q \cdot u - v \cdot p + p \cdot v + p \cdot u \cdot i + q \cdot v \cdot i \\
&= (u \cdot p + v \cdot q) \cdot i.
\end{aligned}
$$

Let there be a complex number $\dot{A} = a + bi$. Then, taking (8) into account, we get:

$$
\begin{aligned}
a + bi &= a + b \cdot [u \cdot q - v \cdot p + \dot{m} \cdot (v + ui)] \\
&= a + (u \cdot q - v \cdot p) \cdot b + \dot{m} \cdot (v \cdot b + u \cdot bi)]. \tag{9}
\end{aligned}
$$

We denote by h the smallest positive real residue of number $a + (u \cdot q - v \cdot p) \cdot b$ modulo N, i.e.

$$h \equiv [a + (u \cdot q - v \cdot p) \cdot b]\,mod\ N. \tag{10}$$

In view of expression (7) we have that $i = i$. Thus, identity (8) is true. We write the expression (10) in the form:

$$
\begin{aligned}
a + (u \cdot q - v \cdot p) \cdot b &= h + s \cdot N = h + s(p + qi) \cdot (p - qi) \\
&= h + \dot{m} \cdot (p \cdot s - q \cdot si). \tag{11}
\end{aligned}
$$

Then, in view of (9), the following will be true:

$$
\begin{aligned}
a + bi &= h + \dot{m} \cdot (p \cdot s - q \cdot si) + \dot{m} \cdot (v \cdot b + u \cdot bi) \\
&= h + \dot{m} \cdot [p \cdot s + v \cdot b + (u \cdot b - q \cdot s)i], \tag{12}
\end{aligned}
$$

or in the following form:

$$(a + bi) \equiv h(mod\ \dot{m}). \tag{13}$$

Thus, it is proved that the smallest complex residue $x + yi$ of a complex number $a + bi$ is comparable modulo m_i to one and only one of the real numbers $0, 1, 2, \ldots, N - 1$.

Let us prove by contradiction that this number is unique. Assume that there are two comparisons:

$$(a + bi) = h_1(mod\ \dot{m}),$$
$$(a + bi) \equiv h_2(mod\ \dot{m}). \tag{14}$$

Based on the property of comparisons, we have $h_1 \equiv h_2(mod\ \dot{m})$ or $(h_1 - h_2) \equiv 0(mod\ \dot{m})$, i e.

$$(h_1 - h_2) = \dot{m} \cdot (e + f \cdot i). \tag{15}$$

From (15) we get the following:

$$(h_1 - h_2) = (p + qi) \cdot (e + fi),$$
$$(h_1 - h_2) \cdot (p - qi) = (p + qi) \cdot (p - qi) \cdot (e + fi),$$
$$(h_1 - h_2) \cdot (p - qi) = (p^2 + q^2) \cdot (e + fi),$$
$$(h_1 - h_2) \cdot (p - qi) = N \cdot (e + fi),$$
$$(h_1 - h_2) \cdot p - (h_1 - h_2) \cdot qi = N \cdot e + N \cdot fi.$$

which is equivalent to the following two real equalities:

$$\begin{cases} (h_1 - h_2) \cdot p = N \cdot e, \\ (h_1 - h_2) \cdot q = -N \cdot f, \end{cases} \tag{16}$$

since complex numbers are equal to each other, their real and imaginary parts are equal. Multiplying the first equality (16) by the value u and the second by the value v and adding them, we get:

$$(h_1 - h_2) \cdot (u \cdot p + v \cdot q) = N \cdot (e \cdot u - f \cdot v),$$

from where, taking into account the expression (7):

$$(h_1 - h_2) \equiv N \cdot (e \cdot u - f \cdot v),$$

or

$$(h_1 - h_2) \equiv 0(mod\ N). \tag{17}$$

So, as by assumption $h_1, h_2 < N$, the comparison (17) is possible only in the case when $h_1 = h_2$. Thus, the possibility of the existence of two different numbers h_1 and h_2, smaller than N, which would be comparable with the number $a + bi$ modulo $\dot{m}$. There is only one such number h, which is determined from the comparison:

$$[a + (u \cdot q - v \cdot p) \cdot b] \equiv h(mod\ N), \tag{18}$$

or

$$Z = (a + b \cdot \rho) \equiv h(mod\ N). \tag{19}$$

The expression $\rho = u \cdot q - v \cdot p$, by means of which a correspondence is established between the complex and real residue modulo $\dot{m} = p + qi$, is called the isomorphism coefficient (IC).

As an example, using Formulas (18) and (19), we define the values of real residues $Z_i \equiv h_i(mod\ N)$ $(i = \overline{0, N-1})$, corresponding to the smallest complex residues $x + yi$ modulo $\dot{m} = 1 + 2i$.

First, we define the value of the isomorphism coefficient $\rho = u \cdot q - v \cdot p = u \cdot 2 - v \cdot 1$. Values u and are determined from the relation known in number theory $u \cdot p + v \cdot q = 1$, i.e. $u \cdot 1 + v \cdot 2 = 1$. By selection (search) we determine that $u = -1$, and $q = 1$. Thus, $\rho = (-1) \cdot 2 - 1 \cdot 1 = -3$, or $(-3)\ mod\ 5 = 2$ $(N = p^2 + q^2 = 1^2 + 2^2 = 5)$.

Determine the values of the smallest real positive residues h_i, that are isomorphic to the smallest complex residues.

For $\dot{A} = 0 + 0i$. $Z_0 = a + b\rho = 0 + 0 \cdot \rho = 0$. $h_0 \equiv 0(mod\ 5)$.
For $\dot{A} = -1 + i$. $Z_1 = -1 + 1 \cdot (-3) = -4$. $h_1 \equiv 1(mod\ 5)$.
For $\dot{A} = i$. $Z_2 = 0 + 1 \cdot (-3) = -3$. $h_2 \equiv 2(mod\ 5)$.
For $\dot{A} = -1 + 2 \cdot i$. $Z_3 = -1 + 2 \cdot (-3) = -1 - 6 = -7$. $h_3 \equiv 3(mod\ 5)$.
For $\dot{A} = 2 \cdot i$. $Z_4 = 0 + 2 \cdot (-3) = -6$. $h_3 \equiv 4(mod\ 5)$.

Based on the results of the Gauss theorem, it is not difficult to show the following relation between the smallest complex and real residues. Assume that for two numbers $\dot{A}_1 = a_1 + b_1 i$ and $\dot{A}_2 = a_2 + b_2 i$ there are such values h_1 and h_2, $h_\pm$ and $h_\times$, that if $\dot{A}_1 \equiv h_1(mod\ \dot{m})$ and $\dot{A}_2 \equiv h_2(mod\ \dot{m})$, then the following are true $\dot{A}_1 \pm \dot{A}_2 \equiv h_\pm(mod\ \dot{m})$ and $\dot{A}_1 \cdot \dot{A}_2 \equiv h_\times(mod\ \dot{m})$. Then $h_\pm \equiv (h_1 \pm h_2)\ mod\ N$ and $h_\times \equiv (h_1 \cdot h_2)\ mod\ N$, where $N = p^2 + q^2$.

Let us give specific examples of the determination of the real deduction of an integer complex number by the complex module.

4.2 Examples of the Definition of Complex and Real Residues of Complex Numbers by a Complex Module

Example 1 Solve the comparison $(16+7i) \equiv h\,mod\,(5+2i)$. Since GCD $(5, 2) = 1$, the condition of the first fundamental Gauss theorem holds, therefore, there is a complete system of real residues modulo $N = p^2 + q^2 = 5^2 + 2^2 = 29$. The real deduction h is determined from the comparison (19), i.e.

$$16 + 7 \cdot \rho \equiv h(\,mod\,29).$$

The isomorphism coefficient ρ is $\rho = u \cdot q - v \cdot p = u \cdot 2 - v \cdot 5$, where the values u and v are determined from the condition of equality (7).

In this case we have that $u = 1$ and $v = -2$, i.e. $1 \cdot 5 + (-2) \cdot 2 = 5 - 4 = 1$.

In this case IC $\rho = 1 \cdot 2 - (-2) \cdot 5 = 2 + 10 = 12$. Therefore $16 + 7 \cdot \rho = = 16 + 7 \cdot 12 \equiv h(\,mod\,29)$ and $h \equiv 13(\,mod\,29)$. We can write that as $16 + 7i \equiv 13\,mod\,(5 + 2i)$.

Example 2 Solve the comparison $(1 + i) \equiv h\,mod\,(1 + 2i)$. Since GCD $(p, q) = (1, 2) = 1$. $N = p^2 + q^2 = 1 + 2^2 = 5$. $\dot{A} \equiv h(\,mod\,\dot{m})$. $h \equiv (a + b \cdot \rho)\,mod\,N$.

The value of IC is $\rho = u \cdot q - v \cdot p = u \cdot 2 - v \cdot 1$, and the values of u and v are determined from the relation (7). We get that $u = -1$, $v = 1$. In this case, the IC is $\rho = (-1) \cdot 2 - 1 \cdot 1 = -2 - 1 = -3 \equiv 2(\,mod\,5)$.

$$Z = a + b \cdot \rho = 1 + 1 \cdot (-3) = -2.$$
$$Z \equiv h(\,mod\,N).$$
$$(-2) \equiv h(\,mod\,5).$$
$$h = 3.$$
$$x + yi = 4 + 2i \sim h = 3,$$
$$(1 + i) \equiv 3\,mod\,(1 + 2i).$$

Consider the examples of determining the complex and real deductions of an integral complex number with respect to a complex module $\dot{m} = 1 + 2i$ with the control of the correctness of the solution of the problem.

Example 3 Determine the complex deduction $x + yi$ of complex number $\dot{A} = 1 + i$ by the complex modulus $\dot{m} = 1 + 2i$, i.e. find $\dot{A} \equiv (x + yi)mod\dot{m}$ $(a = 1, b = 1;$ $p = 1, q = 2; N = 5)$. By the Formula (13) we have that:

$$\begin{cases} (1 \cdot 1 + 1 \cdot 2) \equiv (x \cdot 1 + y \cdot 2) \, mod \, 5, \\ (1 \cdot 1 - 1 \cdot 2) \equiv (y \cdot 1 - x \cdot 2) \, mod \, 5. \end{cases}$$

$$\begin{cases} 3 = x + 2y, \\ -1 = -2x + y. \end{cases}$$

$$x = 3 - 2y,$$
$$-1 = -2 \cdot (3 - 2y) + y,$$
$$-1 = -6 + 4y + y,$$
$$5y = 5,$$
$$y = 1.$$
$$x = 3 - 2y = 3 - 2 = 1; \ x = 1.$$

Answer: the complex deduction $x + yi$ of complex number $\dot{A} = 1 + i$ in a complex modulus $\dot{m} = 1 + 2i$ is equal to a complex number $x + yi = 1 + i$.

Example 4 Determine the smallest deduction $x + yi$ of number $\dot{A} = 1 + i$ modulo $\dot{m} = 1 + 2i$, i.e. define the value of $\dot{A} \equiv (x + yi) \, mod \, (1 + 2i)$ $(a = 1, b = 1; p = 1, q = 2; N = 5)$. By the Formula (9) we have that: $R = (1 \cdot 1 + 1 \cdot 2) \, mod \, 5 = 3$; $R' = (1 \cdot 1 - 1 \cdot 2) \, mod \, 5 = (-1) \, mod \, 5 = 4$.

$$x + yi = \frac{3 \cdot 1 - 4 \cdot 2}{5} + \frac{4 \cdot 1 + 3 \cdot 2}{5}i = -\frac{5}{5} + \frac{10}{5}i = -1 + 2i.$$

Thus, $x + yi = -1 + 2i$ or $\dot{A} \equiv (-1 + 2i) \, mod \, (1 + 2i)$.

Example 5 Solve the comparison $\dot{A} \equiv h \, mod \, \dot{m}$, where $(1 + i) \equiv h \, mod \, (1 + 2i)$ $(a = 1, b = 1; p = 1, q = 2; N = 5)$; Formulas (7), (18) and (19):

$$u \cdot p + v \cdot q = 1, \ u = -1,$$
$$u \cdot 1 + v \cdot 2 = 1, \ v = 1.$$
$$\rho = u \cdot q - v \cdot p.$$
$$Z = a + b \cdot \rho \rightarrow Z \equiv h \, mod \, N.$$
$$\rho = (-1) \cdot 2 - 1 \cdot 1 = -2 - 1 = -3.$$
$$Z = [1 + 1 \cdot (-3)] \, mod \, 5 = (-2) \, mod \, 5 = 3.$$

Check. We will check the results. In Example 4, the smallest complex deduction $(-1 + 2i)$ was obtained, and in Example 5, a real deduction $h = 3$ was obtained. In accordance with the data of Table 2, we have that $(-1 + 2i) \sim 3$.

Example 6 Determine the complex deduction $x + yi$ of complex number $\dot{A} = 3 + 4i$ by complex module $\dot{m} = 1 + 2i$. $N = p^2 + q^2 = 1^2 + 2^2 = 5$.

In accordance with the well-known (13) rule, we compose a system of comparisons in the form:

$$\begin{cases} (3 \cdot 1 + 4 \cdot 2) \equiv (x \cdot 1 + y \cdot 2) \, mod \, 5, \\ (4 \cdot 1 - 3 \cdot 2) \equiv (y \cdot 1 - x \cdot 2) \, mod \, 5. \end{cases}$$

or

$$\begin{cases} 11 \equiv (x + 2y) \, mod \, 5, \\ (-2) \equiv (-2x + y) \, mod \, 5. \end{cases}$$

Based on the system of comparisons, we will compose a system of two linear equations:

$$\begin{cases} x + 2y = 11, \\ -2x + y = +3, \end{cases}$$

because $(-2) = 3 \, mod \, 5$.

$$x = 11 - 2y,$$
$$-2 \cdot (11 - 2 \cdot y) + y = 3,$$
$$-22 + 4y + y = 3,$$
$$5y = 25,$$
$$y = 5.$$
$$x = 11 - 2y = 11 - 10 = 1.$$

Thus, $x + yi = 1 + 5i$.

Example 7 Determine the smallest complex deduction $x + yi$ of complex number $\dot{A} = 3 + 4i$ by complex module $\dot{m} - 1 + 2i$; $N = 5$.

In accordance with the expression in (9), we have that the smallest complex deduction is equal to:

$$(x + yi) = \frac{R \cdot p - R' \cdot q}{N} + \frac{R' \cdot p + R \cdot q}{N}i.$$

Pre-define the values R and R':

$$R = (a \cdot p + b \cdot q) \, mod \, N = (3 \cdot 1 + 4 \cdot 2) \, mod \, 5 = 11(mod \, 5) = 1;$$
$$R' = (b \cdot p - a \cdot q) \, mod \, N = (4 \cdot 1 - 3 \cdot 2) \, mod \, 5 = (-2) \, mod \, 5 = 3.$$

In this case, we have that:

$$(x + yi) = \frac{1 \cdot 1 - 3 \cdot 2}{5} + \frac{3 \cdot 1 + 1 \cdot 2}{5} = -\frac{5}{5} + \frac{5}{5} = -1 + i.$$

So the smallest complex deduction is $-1 + i$.

Example 8 Determine the real deduction h of complex number $\dot{A} = 3 + 4i$ modulo $\dot{m} = 1 + 2i$; $N = 5$. Or you can formulate the task as follows. Solve the comparison $(3 + 4i) \equiv h \, mod \, (1 + 2i)$.

In accordance with expression (19), we have that $(a + b\rho) \equiv h(\bmod N)$, where the isomorphism coefficient ρ is $\rho = u \cdot q - v \cdot p$. Based on Formula (7), we define the values u and v, i.e. $u \cdot p + v \cdot q = 1$ or $u \cdot 1 + v \cdot 2 = 1$.

So, with the values $u = -1$ and $v = 1$, the condition (7) is true, i.e. $(-1)\cdot1+1\cdot2 = 1$.

Based on the calculations we get that $\rho = u \cdot q - v \cdot p = (-1) \cdot 2 - 1 \cdot 1 = -3$.

$$h = (a + b \cdot \rho) = (3 + 4 \cdot (-3)) \bmod 5,$$

or

$$(-9) \bmod 5 = 1.$$

Check. We will check the results. In Example 7, the smallest complex deduction $(-1 + i)$, was obtained, and in Example 8 we obtain a real deduction $h = 1$. In accordance with the obtained data, we have that $(-1 + i) \sim 1$.

5 Conclusion

In the presented chapter, the approach and the procedure for determining real residuals in a real number domain was considered. The main attention was paid to the description of the procedure for processing residuals of complex numbers by a complex module based on the use of the results of the first fundamental Gauss theorem. Many examples of determining the deduction of integer data in a complex number domain are given. Based on the procedure described, a device was developed for its technical implementation. The device for determining residuals of real and complex numbers in the system of residual classes (SRC) received a patent of Ukraine for invention. This confirms the global novelty and practical significance of the research results presented in this chapter. The examples of the implementation of real deductions for performing arithmetic modular operations in cryptography are shown. The results shown in the chapter are advisable to be used when solving problems and algorithms in the SRC in the complex numerical domain. The use of the considered procedure will help to increase the efficiency of the use of SRCs for the implementation of integer operations in a complex number domain.

The research results can be useful for various practical applications, in particular, for modeling real-time services [44], forecasting sequences [45], as well as in other applications of intelligent systems and computing [46, 47].

References

1. Akushskii IYa, Yuditskii DI (1968) Machine arithmetic in residual classes. Sov Radio, Moscow. [in Russian]
2. Wei S, Chen S, Shimizu K (2002) Fast modular multiplication using Booth recoding based on signed-digit number arithmetic. In: Asia-Pacific conference on circuits and systems, Denpasar, Bali, Indonesia, vol 2, pp 31–36. https://doi.org/10.1109/apccas.2002.1115104
3. Skavantzos A (1991) ROM table reduction techniques for computing the squaring operation using modular arithmetic. In: Conference record of the twenty-fifth Asilomar conference on signals, systems and computers, Pacific Grove, CA, USA, vol 1, pp 413–417. https://doi.org/10.1109/acssc.1991.186483
4. Yanko A, Koshman S, Krasnobayev V (2017) Algorithms of data processing in the residual classes system. In: 4th international scientific-practical conference problems of infocommunications. Science and technology (PIC S&T). Kharkov, pp 117–121
5. Suter BW (1974) The modular arithmetic of arbitrarily long sequences of digits. IEEE Trans Comput C-23(12):1301–1303. https://doi.org/10.1109/t-c.1974.223850
6. Bajard J, Imbert L, Plantard T (2005) Arithmetic operations in the polynomial modular number system. In: 17th IEEE symposium on computer arithmetic (ARITH'05), Cape Cod, MA, 2005, pp 206–213. https://doi.org/10.1109/arith.2005.11
7. Krasnobayev VA, Yanko AS, Koshman SA (2016) A method for arithmetic comparison of data represented in a residue number system. Cybern Syst Anal 52(1):145–150
8. Kasianchuk M, Yakymenko I, Pazdriy I, Zastavnyy O (2015) Algorithms of findings of perfect shape modules of remaining classes system. In: The experience of designing and application of CAD systems in microelectronics, Lviv, pp 316–318. https://doi.org/10.1109/cadsm.2015.7230866
9. Popov DI, Gapochkin AV (2018) Development of algorithm for control and correction of errors of digital signals, represented in system of residual classes. In: 2018 International Russian automation conference (RusAutoCon), Sochi, pp 1–3. https://doi.org/10.1109/rusautocon.2018.8501826
10. Krasnobayev VA, Koshman SA, Yanko AS (2017) Conception of realization of criptographic RSA transformations with using of the residue number system. In: Gorbenko ID, Kuznetsov AA (eds) ISCI'2017: information security in critical infrastructures. Collective monograph. LAP Lambert Academic Publishing, Omni Scriptum GmbH & Co. KG. ISBN: 978-3-330-06136-1. [Chapter № 3 in monograph, pp 81–92]. Germany (2017)
11. Yanko A, Koshman S, Krasnobayev V (2017) Algorithms of data processing in the residual classes system. In: 2017 4th international scientific-practical conference problems of infocommunications. science and technology (PIC S&T), Kharkov, pp 117–121. https://doi.org/10.1109/infocommst.2017.8246363
12. Krasnobayev V, Koshman S, Yanko A, Martynenko A (2018) Method of error control of the information presented in the modular number system. In: 2018 international scientific-practical conference problems of infocommunications. Science and technology (PIC S&T), Kharkiv, Ukraine, pp 39–42. https://doi.org/10.1109/INFOCOMMST.2018.8632049
13. Chen T, Yu B, Su J-H, Dai Z, Liu J-G (2007) A reconfigurable modular arithmetic unit for public-key Cryptography. In: 2007 7th international conference on ASIC, Guilin, pp 850–853. https://doi.org/10.1109/icasic.2007.4415764
14. Duan C, Liu Y, Chen Y (2009) A 3-stage pipelined large integer modular arithmetic unit for ECC. In: 2009 international symposium on information engineering and electronic commerce, Ternopil, pp 519–523. https://doi.org/10.1109/ieec.2009.115
15. Krasnobayev V, Kuznetsov A, Koshman S, Moroz S (2019) Improved method of determining the alternative set of numbers in residue number system. In: Chertov O, Mylovanov T, Kondratenko Y, Kacprzyk J, Kreinovich V, Stefanuk V (eds) Recent developments in data science and intelligent analysis of information. ICDSAI 2018. Advances in intelligent systems and computing, vol 836. Springer, Cham, pp 319–328, 05 Aug 2018. https://doi.org/10.1007/978-3-319-97885-7_31

16. Kasianchuk M, Yakymenko I, Pazdriy I, Melnyk A, Ivasiev S (2017) Rabin's modified method of encryption using various forms of system of residual classes. In: 2017 14th international conference the experience of designing and application of CAD systems in microelectronics (CADSM), Lviv, pp 222–224. https://doi.org/10.1109/cadsm.2017.7916120

17. Wahid KA, Dimitrov VS, Jullien GA (2003) Error-free arithmetic for discrete wavelet transforms using algebraic integers. In: Proceedings 2003 16th IEEE symposium on computer arithmetic, Santiago de Compostela, Spain, pp 238–244. https://doi.org/10.1109/arith.2003.1207684

18. Thomas JJ, Parker SR (1987) Implementing exact calculations in hardware. IEEE Trans Comput C-36(6):764–768. https://doi.org/10.1109/tc.1987.1676969

19. Luo Z, Martonosi M (2000) Accelerating pipelined integer and floating-point accumulations in configurable hardware with delayed addition techniques. IEEE Trans Comput 49(3):208–218. https://doi.org/10.1109/12.841125

20. Bautista JN, Alvarado-Nava O, Pérez FM (2012) A mathematical co-processor of modular arithmetic based on a FPGA. In: 2012 technologies applied to electronics teaching (TAEE), Vigo, pp 32–37. https://doi.org/10.1109/taee.2012.6235402

21. Shettar R, Banakar RM, Nataraj PSV (2007) Implementation of interval arithmetic algorithms on FPGAs. In: International conference on computational intelligence and multimedia applications (ICCIMA 2007), Sivakasi, Tamil Nadu, pp 196–200. https://doi.org/10.1109/iccima.2007.60

22. Stojčev MK, Milovanović EI, Milovanović IŽ (2012) A unified approach in manipulation with modular arithmetic. In: 2012 28th international conference on microelectronics proceedings, Nis, pp 423–426. https://doi.org/10.1109/miel.2012.6222892

23. Kuznetsov AA, Gorbenko YuI, Prokopovych-Tkachenko DI, Lutsenko MS, Pastukhov MV (2019) NIST PQC: code-based cryptosystems. Telecommun Radio Eng 78(5):429–441. https://doi.org/10.1615/telecomradeng.v78.i5.50

24. Shettar R, Banakar RM, Nataraj PSV (2006) Design and implementation of interval arithmetic algorithms. In: First international conference on industrial and information systems, Peradeniya, pp 328–331. https://doi.org/10.1109/iciis.2006.365747

25. Gorbenko I, Kuznetsov A, Gorbenko Y, Vdovenko S, Tymchenko V, Lutsenko M (2019) Studies on statistical analysis and performance evaluation for some stream ciphers. Int J Comput 18(1):82–88

26. Putra RVW, Adiono T (2015) Optimized hardware algorithm for integer cube root calculation and its efficient architecture. In: 2015 international symposium on intelligent signal processing and communication systems (ISPACS), Nusa Dua, pp 263–267. https://doi.org/10.1109/ispacs.2015.7432777

27. Kuznetsov A, Kiyan A, Uvarova A, Serhiienko R, Smirnov V (2018) New code based fuzzy extractor for biometric cryptography. In: 2018 international scientific-practical conference problems of infocommunications. science and technology (PIC S&T), Kharkiv, Ukraine, pp 119–124. https://doi.org/10.1109/infocommst.2018.8632040

28. Klotchkov IV, Pedersen S (1996) A codesign case study: implementing arithmetic functions in FPGAs. In: Proceedings IEEE symposium and workshop on engineering of computer-based systems, Friedrichshafen, Germany, pp 389–394. https://doi.org/10.1109/ecbs.1996.494565

29. Andres E, Molina MC, Botella G, del Barrio A, Mendias JM (2008) Area optimization of combined integer and floating point circuits in high-level synthesis. In: 2008 4th southern conference on programmable logic, San Carlos de Bariloche, pp 229–232. https://doi.org/10.1109/spl.2008.4547764

30. Gorbenko I, Kuznetsov A, Tymchenko V, Gorbenko Y, Kachko O (2018) Experimental studies of the modern symmetric stream ciphers. In: 2018 International scientific-practical conference problems of infocommunications. Science and Technology (PIC S&T), Kharkiv, Ukraine, pp 125–128. https://doi.org/10.1109/infocommst.2018.8632058

31. Martí-Campoy A (2016) Learning integer numbers representation by means of an Aronson's puzzle. In: 2016 technologies applied to electronics teaching (TAEE), Seville, pp 1–7. https://doi.org/10.1109/taee.2016.7528245

32. Gorbenko ID, Zamula AA, Semenko YeA (2016) Ensemble and correlation properties of cryptographic signals for telecommunication system and network applications. Telecommun Radio Eng 75(2):169–178. https://doi.org/10.1615/telecomradeng.v76.i17.40

33. Petryshyn M (2017) Modeling of the TIF processes in binary numeral systems based on the vector-branching diagrams. In: 2017 IEEE First Ukraine conference on electrical and computer engineering (UKRCON), Kyiv, Ukraine, pp 1078–1083. https://doi.org/10.1109/ukrcon.2017.8100416

34. Andrushkevych A, Gorbenko Y, Kuznetsov O, Oliynykov R, Rodinko MA (2019) A prospective lightweight block cipher for green IT engineering. In: Kharchenko V, Kondratenko Y, Kacprzyk J (eds) Green IT engineering: social, business and industrial applications. Studies in systems, decision and control, vol 171. Springer, Cham, pp 95–112. https://doi.org/10.1007/978-3-030-00253-4_5

35. Chamraz Š, Balogh R (2014) Control of the mechatronic systems using an integer arithmetics. In: 2014 23rd international conference on robotics in Alpe-Adria-Danube Region (RAAD), Smolenice, pp 1–6. https://doi.org/10.1109/raad.2014.7002269

36. Kuznetsov O, Potii O, Perepelitsyn A, Ivanenko D, Poluyanenko N (2019) Lightweight stream ciphers for green IT engineering. In: Kharchenko V, Kondratenko Y, Kacprzyk J (eds) Green IT engineering: social, business and industrial applications. Studies in systems, decision and control, vol 171. Springer, Cham, pp 113–137. https://doi.org/10.1007/978-3-030-00253-4_6

37. Bazelow AR, Raamot J (1983) On the Microprocessor solution of ordinary differential equations using integer arithmetic. IEEE Trans Comput C-32(2):204–207. https://doi.org/10.1109/tc.1983.1676207

38. Chornei R, Hans Daduna VM, Knopov P (2005) Controlled markov fields with finite state space on graphs. Stochast Models 21(4):847–874. https://doi.org/10.1080/15326340500294520

39. Shenoy AP, Kumaresan R (1988) Residue to binary conversion for RNS arithmetic using only modular look-up tables. IEEE Trans Circuits Syst 35(9):1158–1162. https://doi.org/10.1109/31.7577

40. Skavantzos A, Mitash N (1992) Computing large polynomial products using modular arithmetic. IEEE Trans Circuits Syst II: Analog Digital Sig Process 39(4):252–254. https://doi.org/10.1109/82.136577

41. Malina L, Hajny J (2011) Accelerated modular arithmetic for low-performance devices. In: 2011 34th international conference on telecommunications and signal processing (TSP), Budapest, pp 131–135. https://doi.org/10.1109/tsp.2011.6043757

42. Montgomery PL (1985) Modular multiplication without trial division. Math Comput 519–521 (1985)

43. Grochadl J, Avanzi RM, Savas E, Tillich S (2005) Energy-efficient software implementation of long integer modular arithmetic. In: Cryptographic hardware and embedded systems—CHES 2005, Springer, Berlin, pp 75–90

44. Kryvinska N (2008) An analytical approach for the modeling of real-time services over IP network. Math Comput Simul 79(4):980–990. https://doi.org/10.1016/j.matcom.2008.02.016

45. Kirichenko L, Radivilova T, Zinkevich I (2018) Comparative analysis of conversion series forecasting in E-commerce tasks. In: Shakhovska N, Stepashko V (eds) Advances in intelligent systems and computing II. CSIT 2017. Advances in intelligent systems and computing, vol 689. Springer, Cham, pp 230–242. https://doi.org/10.1007/978-3-319-70581-1_16

46. Kirichenko L, Radivilova T, Tkachenko A (2019) Comparative analysis of noisy time series clustering. In: CEUR workshop proceedings , vol 2362, pp 184–196. http://ceur-ws.org/Vol-2362/paper17.pdf

47. Radivilova T, Hassan HA (2017) Test for penetration in Wi-Fi network: attacks on WPA2-PSK and WPA2-enterprise. In: 2017 international conference on information and telecommunication technologies and radio electronics (UkrMiCo), IEEE, pp 1–4

Representation of Cascade Codes in the Frequency Domain

Alexandr Kuznetsov , **Andriy Pushkar'ov** , **Roman Serhiienko** , **Oleksii Smirnov** , **Vitalina Babenko** , **and Tetiana Kuznetsova**

Abstract The mathematical apparatus of the multidimensional discrete Fourier transform over finite fields is considered. Methods for the description of linear block codes in the frequency domain are investigated. It is shown that, in contrast to iterative codes (code-products), cascade codes in the general case cannot be described in the frequency domain in terms of multidimensional spectra. Analytic expressions are obtained that establish a one-to-one functional correspondence between the spectrum of a sequence over a finite field and the spectra of the corresponding words obtained by limiting this word to a subfield. A general solution of the problem of representation of cascade codes in the frequency domain is obtained, which allows constructing in the frequency domain using computationally efficient algorithms of encoding and decoding, and the derived analytic dependences of components of multidimensional spectra.

A. Kuznetsov (✉) · V. Babenko · T. Kuznetsova
V. N. Karazin Kharkiv National University, Svobody Sq., 4, Kharkiv 61022, Ukraine
e-mail: kuznetsov@karazin.ua

V. Babenko
e-mail: vitalinababenko@karazin.ua

T. Kuznetsova
e-mail: kuznetsova.tatiana17@gmail.com

A. Pushkar'ov
State Service of Special Communication and Information Pro-Tection, Solomianska 13 Str., Kyiv 03680, Ukraine
e-mail: push-karoff@ukr.net

R. Serhiienko
Hetman Petro Sahaidachnyi National Army Academy, 32 Heroes of Maidan Street, Lviv 79012, Ukraine
e-mail: romanserg69@gmail.com

O. Smirnov
Central Ukrainian National Technical University, Avenue University, 8, Kropivnitskiy 25006, Ukraine
e-mail: dr.smirnovoa@gmail.com

557

T. Radivilova et al. (eds.), *Data-Centric Business and Applications*, Lecture Notes on Data Engineering and Communications Technologies 48,
https://doi.org/10.1007/978-3-030-43070-2_25

Keywords Coding theory · Cascade codes · Code-products · Multidimensional discrete fourier transform over finite fields · Multidimensional spectra · Frequency domain

1 Challenge Problem in General and Previous Research

The mathematical apparatus of the discrete Fourier transform in Galois fields is used in the modern theory of noiseless coding both for describing the most important block codes in the frequency domain and for building new code constructions with improved properties [1–9]. Application of fast Fourier transform algorithms allows significantly reduce the computational complexity of coding and decoding algorithms, and also implement some computational processes in parallel [10–27].

The Fourier transforms over finite fields can be generalized to the multidimensional case. If code words of block codes could be represented as code polynomials of several variables and the corresponding code symbols could be written by some multidimensional matrix, then the mathematical apparatus of multidimensional spectra, as a rule, allows to specify codes in a multidimensional frequency domain. Codes that allow such a description include the simplest iterative codes (product codes) for which the transfer of calculations to a multidimensional frequency domain allows to increase the computational efficiency of the encoding-decoding algorithms and perform many operations in parallel [2, 3]. At the same time, the code interrelations of the iterative codes are far from optimal, which explains their small practical use [1–3].

The greatest spread in the technique of noise-immune encoding was obtained by the so-called cascade codes which have two codes in the design: the code of the inner (first) stage above the finite field $GF(q)$ and the code of the outer (second) stage over the extended field [1–3]. The resulting block code is defined over the field $GF(q)$. However, when generating codewords and decoding them, transformations are performed both over the field $GF(q)$ and over its extension $GF(q^m)$.

The code words of the cascade code are also represented as a matrix, however, the mathematical apparatus of multidimensional spectra is not applicable to cascade codes. It is also impossible to use fast multidimensional Fourier transforms, i.e. to obtain the effect provided by transformation in the frequency domain used in noise-immune coding technique. This article is devoted to the solution of this contradiction. Thus, the aim of the article is to develop a mathematical apparatus of multidimensional spectra for representing cascaded code structures in the frequency domain and implementing effective algorithms for encoding and decoding on their basis.

The work is structured as follows. In Sect. 2 the basic provisions and analytic relations for the discrete Fourier transform over finite fields and their connection with the polynomial description of block codes are given. In Sect. 3, the Fourier transforms are generalized to the multidimensional case. Using as example the iterative code the description of multidimensional codes in the frequency domain is given. It is shown that the corresponding representation cannot be obtained for cascade codes.

Section 4 is devoted to the solution of this problem. Through the introduction of a one-to-one correspondence of the code words of an arbitrary vector over a finite field and the spectra of its constraint words to an arbitrary subfield, it is possible to obtain analytical expressions that supplement the mathematical apparatus of multi-dimensional spectra. It allows describing cascade codes in the frequency domain. In item 5, the obtained results are generalized, their applied significance is discussed for the implementation of computationally efficient coding and decoding algorithms by cascade codes. All the statements provided on the text of the paper are supplemented with examples that clearly demonstrate the validity of the arguments above and simplify their comprehension.

2 Discrete Fourier Transform in Finite Fields

The Fourier transform plays an important role in the development of modern methods of the theory of information processing and transmission. In particular, the integral Fourier transform is used to process and study signals that are time-continuous and take real and complex values [6–17]. For digital processing of discretely-timed a discrete Fourier transform [3, 18–20] is used:

$$X_k = \sum_{j=0}^{n-1} e^{\frac{-2\pi i}{n} jk} x_j, \ k = 0, \ldots, n-1, \tag{1}$$

$$x_j = \frac{1}{n} \sum_{k=0}^{n-1} e^{\frac{2\pi i}{n} jk} X_k, \ j = 0, \ldots, n-1, \tag{2}$$

were: n—the number of signal values and the components of the expansion (spectrum); $x = \{x_j, j = 0, \ldots, n-1\}$—signal values in discrete time points with numbers $j = 0, \ldots, n-1$; $X = \{X_k, k = 0, \ldots, n-1\}$—$n$-complex amplitudes of the sinusoidal signals composing the original signal; k—frequency index.

For many sequence lengths the Fourier transform over Galois fields is also defined, which have the form of a developed analytic apparatus used to describe block codes in the frequency domain, to study their corrective properties, to construct computationally efficient algorithms for encoding and decoding [3, 18–20].

The core of the discrete Fourier transform in (1) and (2) is the complex nth root of unity $e^{\frac{-2\pi i}{n}}$.

Making an analogy with a finite field in which the element α of order n is nth root of unity, the following definition of the discrete Fourier transform over Galois fields was introduced in [3, 6–12].

Let $v = \{v_i, i = 0, \ldots, n-1\}$ is a vector over $GF(q)$ where n divides $q^m - 1$ with some m and let α is an element of nth order in field $GF(q^m)$. The Fourier transform in the Galois field of a vector v is defined as the vector over $GF(q^m)$ defined by equations [3, 6–12]:

$$c_j = \sum_{i=0}^{n-1} \alpha^{ij} v_i, \quad j = 0, \ldots, n-1. \tag{3}$$

The discrete index i in (1) is usually called as time, and v is a time function or signal. The corresponding j index is usually called as frequency, and c is the frequency function or spectrum [3, 6–12].

The order of element $\alpha \in GF(q^m)$ in (3) must be divisor of $q^m - 1$, therefore, unlike the field of complex numbers, the Fourier transform over a finite field is not defined for any length, but only for the corresponding divisors of $q^m - 1$.

The most important in applied respect is the choice $\alpha \in GF(q^m)$ as the primitive element with the maximum order of $n = q^m - 1$. In the general case, as an arbitrary length of n an arbitrary divisor $q^m - 1$ can be chosen for some positive integer m and an element $\alpha \in GF(q^m)$ of nth order as the transformation kernel. The spectrum in this case will be defined over the extension $GF(q^m)$, although it will contain only elements of nth order from this field.

Thus, over a field $GF(q)$ the vector v and its spectrum c are related by the equations [3, 6–12]:

$$c_j = \sum_{i=0}^{n-1} \alpha^{ij} v_i, \tag{4}$$

$$v_i = \frac{1}{n} \sum_{j=0}^{n-1} \alpha^{-ij} c_j, \tag{5}$$

where n it is interpreted as an element of a field $GF(q)$, $n | (q^m - 1)$, a $c_j \in GF(q^m)$.

Because the signal v is defined over the field $GF(q)$, and its spectrum c is defined over the extension $GF(q^m)$, not all the vectors over $GF(q^m)$ can be the spectra of any signals over $GF(q)$. The inverse Fourier transform of a signal c over $GF(q^m)$ is a vector v with components from $GF(q)$ if and only if the following equalities (conjugacy constraints) hold:

$$c_j^q = c_{qj \bmod n}, \quad j = 0, \ldots, n-1. \tag{6}$$

Indeed, according to the small theorem of Fermat,

$$c_j^q = \left(\sum_{i=0}^{n-1} \alpha^{ij} v_i \right)^q = \sum_{i=0}^{n-1} \alpha^{i(qj)} v_i = c_{qj \bmod n}, \quad j = 0, \ldots, n-1.$$

Let group the numbers $0, \ldots, n-1$ by $\bmod n$ on subsets [3, 6–12]:

$$A_j = \{j, jq, \ldots, jq^{m_j-1}\}, \tag{7}$$

where m_j is the smallest positive integer satisfying equality $jq^{m_j-1} = j \bmod n$, due the finiteness of the field such m_j always exists.

The set A_j allocates in the spectrum such a set of frequencies, called the chord, that if the signal takes values in a field $GF(q)$, then the value of the spectrum in one of the chord frequencies determines the values of the spectrum at all frequencies of this chord [3, 6–12]. In other words, in order to specify the signal through the inverse Fourier transform (5), it is sufficient to determine all the chords (7) with allowance for the constraints (6). The choice of the chord A_j corresponds to the definition of such a minimal polynomial $f_{\alpha^j}(x)$, whose roots are all elements α^{jq^s}, $s = 0, \ldots, m_j - 1$:

$$f_{\alpha^j}(x) = \prod_{s=0}^{m_j-1} (x - \alpha^{jq^s}).$$

The considered transformations are widely used in the theory of noise-immune coding for the description of codes in the frequency domain and for the study of their properties. For example, the most important applied polynomial codes are set in the frequency domain through certain zero components of the spectrum of their code words.

If the elements of the vector v are given in the form of the coefficients of the polynomial $v(x)$:

$$v(x) = v_0 + v_1 x + \ldots + v_{n-1} x^{n-1},$$

then, using the Fourier transform in the Galois field, it can be transformed into a polynomial

$$c(x) = c_0 + c_1 x + \ldots + c_{n-1} x^{n-1}$$

which is called a spectral polynomial or associated with a polynomial $v(x)$ [3, 6–12].

The properties of the spectrum are closely related to the roots of the polynomials [3, 6–12]:

- an element α^j is a root of a polynomial $v(x)$ if and only if the jth frequency component c_j is zero;
- an element α^{-i} is a root of a polynomial $c(x)$ if and only if the ith time component v_i is zero.

Indeed, computing the value of a polynomial $v(x)$ at a point α^j yields:

$$v(\alpha^j) = v_0 + v_1 \alpha^j + \ldots + v_{n-1} \alpha^{jn-1} = \sum_{i=0}^{n-1} \alpha^{ij} v_i = c_j,$$

in other words the vanishing of the frequency component c_j means that the corresponding element α^j is the root of the polynomial $v(x)$.

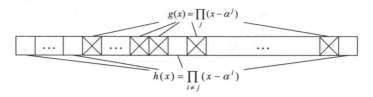

Fig. 1 Structure of code words of polynomial codes in the spectral domain

It follows that the polynomial codes given by the generating and / or checking polynomials are defined in the frequency domain by zero frequency components directly related to the roots of these polynomials. In particular, if the Bose–Chaud-huri–Hocquenghem (BCH) code over $GF(q)$ is given by its generating $g(x)$ and / or check $h(x)$ polynomials, i.e. if take into account of (6)–(8):

$$g(x) = LCM\left(\prod_j f_{\alpha^j}(x)\right) = \prod_j (x - \alpha^j),$$

$$h(x) = LCM\left(\prod_{i \neq j} f_{\alpha^i}(x)\right) = \prod_{i \neq j} (x - \alpha^i),$$

then the spectra of all the code words of such a code necessarily contain zeros in the components c_j and may not be zero in the components c_i (LCM—Least Common Multiple).

This is clearly demonstrated in Fig. 1, on which the zero components of the spectrum are schematically indicated—the roots of the generating polynomial $g(x)$ and the nonzero elements—the roots of the check polynomial $h(x)$.

The Fourier transform for functions defined over a real space is generalized in the form of a so-called multidimensional spectra [3, 6–12]. The corresponding multidimensional generalizations of the Fourier transform can also be introduced on discrete sequences from multidimensional spaces over finite fields.

3 Multidimensional Spectra and Their Application in the Theory of Coding

Let the lengths $n_1, n_2, \ldots, n_p$ be simultaneously divisors of the order of the multiplicative group of a finite field $GF(q^m)$ for some positive integer m, i.e.

$$n_1|(q^m - 1) \wedge n_2|(q^m - 1) \wedge \ldots \wedge n_p|(q^m - 1).$$

Then an arbitrary p-dimensional signal

$$v = \{v_{i_1,i_2,\ldots,i_p}, i_1 = 0, \ldots, n_1 - 1, i_2 = 0, \ldots, n_2 - 1, \ldots, i_p = 0, \ldots, n_p - 1\}$$

and its p-dimensional spectrum

$$c = \{c_{j_1,j_2,\ldots,j_p}, j_1 = 0, \ldots, n_1 - 1, j_2 = 0, \ldots, n_2 - 1, \ldots, j_p = 0, \ldots, n_p - 1\}$$

are related by appropriate transformations:

$$c_{j_1,j_2,\ldots,j_p} = \sum_{i_1=0}^{n_1-1} \sum_{i_2=0}^{n_2-1} \cdots \sum_{i_p=0}^{n_p-1} \alpha_1^{i_1 j_1} \alpha_2^{i_2 j_2} \ldots \alpha_p^{i_p j_p} v_{i_1,i_2,\ldots,i_p}, \tag{8}$$

$$v_{i_1,i_2,\ldots,i_p} = \frac{1}{n_1} \frac{1}{n_2} \cdots \frac{1}{n_p} \times \sum_{j_1=0}^{n_1-1} \sum_{j_2=0}^{n_2-1} \cdots \sum_{j_p=0}^{n_p-1} \alpha_1^{-i_1 j_1} \alpha_2^{-i_2 j_2} \ldots \alpha_p^{-i_p j_p} c_{j_1,j_2,\ldots,j_p}, \tag{9}$$

where $\alpha_1, \alpha_2, \ldots, \alpha_p$—elements of a finite field $GF(q^m)$ order of $n_1, n_2, \ldots, n_p$, respectively.

By analogy with the conjugacy constraints for one-dimensional spectra (6), we introduce the corresponding conditions for the multidimensional case:

$$c_{j_1,j_2,\ldots,j_p}^q = c_{q j_1 \bmod n, \, q j_2 \bmod n, \, \ldots, \, q j_p \bmod n},$$

$$i_1 = 0, \ldots, n_1 - 1, i_2 = 0, \ldots, n_2 - 1, \ldots, i_p = 0, \ldots, n_p - 1, \tag{10}$$

which can be is easily verified

$$c_{j_1,j_2,\ldots,j_p}^q = \left(\sum_{i_1=0}^{n_1-1} \sum_{i_2=0}^{n_2-1} \cdots \sum_{i_p=0}^{n_p-1} \alpha_1^{i_1 j_1} \alpha_2^{i_2 j_2} \ldots \alpha_p^{i_p j_p} v_{i_1,i_2,\ldots,i_p} \right)^q$$

$$= \sum_{i_1=0}^{n_1-1} \sum_{i_2=0}^{n_2-1} \cdots \sum_{i_p=0}^{n_p-1} \alpha_1^{i_1 (q j_1)} \alpha_2^{i_2 (q j_2)} \ldots \alpha_p^{i_p (q j_p)} v_{i_1,i_2,\ldots,i_p}$$

$$= c_{q j_1 \bmod n, \, q j_2 \bmod n, \, \ldots, \, q j_p \bmod n}, \quad i_1 = 0, \ldots, n_1 - 1, i_2 = 0, \ldots, n_2 - 1, \ldots,$$
$$i_p = 0, \ldots, n_p - 1.$$

Let break up the tuples of numbers

$$i_1 = 0, \ldots, n_1 - 1, i_2 = 0, \ldots, n_2 - 1, \ldots, i_p = 0, \ldots, n_p - 1$$

to subsets:

$$A_{j_1,j_2,\ldots,j_p} = \left\{ \begin{array}{c} \{j_1, j_1 q, \ldots, j_1 q^{m_{j_1}-1}\}, \\ \{j_2, j_2 q, \ldots, j_2 q^{m_{j_2}-1}\}, \\ \ldots, \{j_p, j_p q, \ldots, j_p q^{m_{j_p}-1}\} \end{array} \right\}, \tag{11}$$

where m_{j_s} is a least positive integer, satisfying the equality $j_s q^{m_{j_s}} = j_s \bmod n_s$, because of the finiteness of the field, this m_{j_s} always exists.

The set allocates $A_{j_1,j_2,\ldots,j_p}$ a multidimensional chord in the multidimensional spectrum, and if the time signal takes values in a field $GF(q)$, then the value of the spectrum in one of the chord frequencies determines the values of the spectrum at all frequencies of this chord [3, 6–12]. Thus, the signal can be specified through the inverse multidimensional Fourier transform (9) if we define chords (11) taking into account the constraints (10).

Multidimensional spectra are used in the theory of noise-immune coding for the description of so-called iterative codes (or code-products) in the frequency domain. Let us consider, without loss of generality, the simplest, two-dimensional case.

The information symbols $I = \{I_1, I_2, \ldots, I_k\}$ to be encoded with a two-dimensional iterative (n, k, d) code over $GF(q)$ are divided into k_2 sub-blocks containing by k_1 symbols in each, i.e. $k = k_1 k_2$. Let's write them in the form of a matrix of the size $k_1 \times k_2$, in which each column is a sub-block of k_1 symbols. Each row of the received matrix is encoded by a linear block (n_2, k_2, d_2) code over $GF(q)$, called the code of the second (outer) step. The encoding result yields a matrix containing n_2 columns of k_1 (n_2) characters in each.

Each of n_2 columns of the received matrix is encoded by a linear block (n_1, k_1, d_1) code over $GF(q)$, called the code of the first (internal) step. As a result of the last operation, we get a matrix of the size of $n_1 \times n_2$ symbols from $GF(q)$ in which each column is the code word of the first stage code, and each row is the code word of the second stage code (for the last $r_1 = n_1 - k_1$ rows this is a consequence of the linearity of the codes of the first and second stages). The resulting matrix is the code word of the iterative code with parameters: $n = n_1 n_2, k = k_1 k_2, d = d_1 d_2$.

To study the spectral properties of the iterative code, the mathematical apparatus of multidimensional spectra was used in [3, 6–12, 20–22]. Code word $v = \{v_{i_1,i_2}, i_1 = 0, \ldots, n_1 - 1, i_2 = 0, \ldots, n_2 - 1\}$ of two-dimensional code-product can be written as a polynomial in two variables:

$$v(x, y) = \sum_{i_1=0}^{n_1-1} \sum_{i_2=0}^{n_2-1} v_{i_1,i_2} x^{i_1} y^{i_2}, \tag{12}$$

where v_{i_1,i_2} are symbols from $GF(q)$ (signal components).

Using (8) and (9) for a given $p = 2$ signal which is a two-dimensional matrix v in the time domain, it is possible to calculate all the components of the spectrum $c = \{c_{j_1,j_2}, j_1 = 0, \ldots, n_1 - 1, j_2 = 0, \ldots, n_2 - 1\}$ and vice versa.

Fig. 2 The structure of code words of a two-dimensional code-product in the frequency domain (by analogy with Fig. 1)

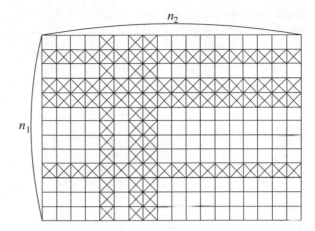

Thus, we can associate with its polynomial (12) its spectral polynomial

$$c(x, y) = \sum_{j_1=0}^{n_1-1} \sum_{j_2=0}^{n_2-1} c_{j_1, j_2} x^{j_1} y^{j_2}, \tag{13}$$

where c_{j_1, j_2} are symbols from $GF(q^m)$ which are components of the spectrum.

The representation of the code words of the iterative code in the form of a polynomial (12) and the associated spectral polynomial (13) is especially useful when polynomial codes are used in the first and second cascades, primarily BCH codes. The roots of the generating polynomials of such codes correspond to zero values in the spectral components, which for the two-dimensional case can be schematically represented in the form of Fig. 2.

The definition of the iterative code (multidimensional code-product) in the frequency domain allows us to use the developed apparatus of fast Fourier transform to reduce the computational complexity of the coding and decoding algorithms, and also perform some computational processes in parallel [3, 6–12]. For example, for the two-dimensional case considered above, the calculation of $n_2 - k_2$ the last words of the first-stage code in the time domain can be realized only after calculating all the words of the second-stage code. Using multidimensional Fourier transforms, the calculation of each code symbol v_{i_1, i_2} can be organized in parallel and independently from each other by computing the inverse transformation according to (9) with $p = 2$.

It should be noted, however, that the code interrelations of iterative codes are far from optimal, and for fixed parameters they tend to lose to other known constructions, for example, cascade codes.

Let consider a cascaded (Nn, Kk, Dd) code over $GF(q)$, formed from the (N, K, D) code over $GF(q^m)$ on the outer stage of the cascade and $(n, k = m, d)$ code over $GF(q)$ on the inner stage of the cascade.

Schematically the structure of the codeword of the cascade code is represented in Fig. 3 as a matrix of n rows and N columns, with the code symbols belonging to the

Fig. 3 Structure of the cascade code codeword

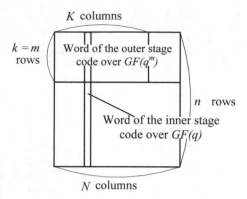

field $GF(q)$ written in the cells.

The top left area, consisting of $k = m$ rows and K columns, corresponds to information symbols from $GF(q)$. Each column in this area consisting of m symbols from $GF(q)$ is represented as one symbol from $GF(q^m)$:

$$V_i \Rightarrow \begin{pmatrix} v_{0,i} \\ v_{1,i} \\ \cdots \\ v_{m-1,i} \end{pmatrix}, i = 0, \ldots, K-1, \tag{14}$$

i.e. for the polynomial representation of the field elements, we have:

$$V_i(z) = v_{0,i} + v_{1,i}z + \ldots + v_{m-1,i}z^{m-1}, V_i(z) \in GF(q^m).$$

All K columns are thus represented as K symbols from $GF(q^m)$, which are processed as the information sequence

$$I = (V_0, V_1, \ldots, V_{K-1})$$

of the outer stage (N, K, D) code. The code word for such a code is represented as a sequence

$$V = (V_0, V_1, \ldots, V_{K-1}, V_K, V_{K+1}, \ldots, V_N), \tag{15}$$

where check symbols V_K, V_{K+1}, ..., V_N are formed in the process of coding by (N, K, D) code of the outer stage and written to the matrix in the form of the corresponding column vectors in the upper right region (see Fig. 3).

Thus, the first m rows in the matrix representation of the code word of the cascade code correspond to N symbols from $GF(q^m)$ of the code word of the outer step code. Each such symbol represented by a vector of m symbols

$$I_i = (v_{0,i}, v_{1,i}, \ldots, v_{m-1,i}), \, i = 0, \ldots, N-1$$

is processed as ith information sequence of the internal stage code.

The corresponding codeword is represented as a sequence

$$v_i = (v_{0,i}, v_{1,i}, \ldots, v_{m-1,i}, v_{m,i}, v_{m+1,i}, \ldots, v_{n-1,i}). \tag{16}$$

where check symbols $v_{m,i}, v_{m+1,i}, \ldots, v_{n-1,i}$ are formed in the process of coding by $(n, k = m, d)$ code of inner stage and are written in the matrix columns (Fig. 3).

Thus, the code word (signal) of the cascade code is represented as the following array of symbols from $GF(q)$:

$$v = \begin{pmatrix} v_{0,0} & v_{0,1} & \cdots & v_{0,K-1} & v_{0,K} & v_{0,K+1} & \cdots & v_{0,N-1} \\ v_{1,0} & v_{1,1} & \cdots & v_{1,K-1} & v_{1,K} & v_{1,K+1} & \cdots & v_{1,N-1} \\ \cdots & \cdots & \cdots & \cdots & \cdots & \cdots & \cdots & \cdots \\ v_{m-1,0} & v_{m-1,1} & \cdots & v_{m-1,K-1} & v_{m-1,K} & v_{m-1,K+1} & \cdots & v_{m-1,N-1} \\ v_{m,0} & v_{m,1} & \cdots & v_{m,K-1} & v_{m,K} & v_{m,K+1} & \cdots & v_{m,N-1} \\ v_{m+1,0} & v_{m+1,1} & \cdots & v_{m+1,K-1} & v_{m+1,K} & v_{m+1,K+1} & \cdots & v_{m+1,N-1} \\ \cdots & \cdots & \cdots & \cdots & \cdots & \cdots & \cdots & \cdots \\ v_{n-1,0} & v_{n-1,1} & \cdots & v_{n-1,K-1} & v_{n-1,K} & v_{n-1,K+1} & \cdots & v_{n-1,N-1} \end{pmatrix}. \tag{17}$$

Obviously, performing the two-dimensional Fourier transform of the matrix (17) by expression (9) with $p = 2$, we obtain corresponding two-dimensional spectrum

$$c = \begin{pmatrix} c_{0,0} & c_{0,1} & \cdots & c_{0,K-1} & c_{0,K} & c_{0,K+1} & \cdots & c_{0,N-1} \\ c_{1,0} & c_{1,1} & \cdots & c_{1,K-1} & c_{1,K} & c_{1,K+1} & \cdots & c_{1,N-1} \\ \cdots & \cdots & \cdots & \cdots & \cdots & \cdots & \cdots & \cdots \\ c_{m-1,0} & c_{m-1,1} & \cdots & c_{m-1,K-1} & c_{m-1,K} & c_{m-1,K+1} & \cdots & c_{m-1,N-1} \\ c_{m,0} & c_{m,1} & \cdots & c_{m,K-1} & c_{m,K} & c_{m,K+1} & \cdots & c_{m,N-1} \\ c_{m+1,0} & c_{m+1,1} & \cdots & c_{m+1,K-1} & c_{m+1,K} & c_{m+1,K+1} & \cdots & c_{m+1,N-1} \\ \cdots & \cdots & \cdots & \cdots & \cdots & \cdots & \cdots & \cdots \\ c_{n-1,0} & c_{n-1,1} & \cdots & c_{n-1,K-1} & c_{n-1,K} & c_{n-1,K+1} & \cdots & c_{n-1,N-1} \end{pmatrix}, \tag{18}$$

which, although having one-to-one correspondence with the time signal (17) above, not corresponds to the word (15) with the symbols from $GF(q^m)$.

In other words, the calculated spectrum (18) will correspond to such a time signal (17), which is the code word of some iterative code in which the second stage code is formed by constraining the word (15) with the symbols from $GF(q^m)$ in the subfield $GF(q)$. By limitation, hereinafter, is considered the formation of words

(a)   **(b)**   **(c)**

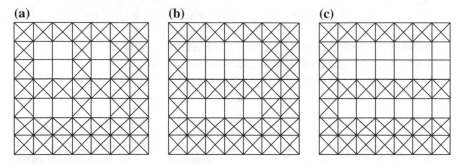

Fig. 4 Structure of code words of a two-dimensional code-product in the frequency domain

$$v_0 = (v_{0,0}, v_{0,1}, v_{0,2}, \ldots, v_{0,N-1}),$$
$$v_1 = (v_{1,0}, v_{1,1}, v_{1,2}, \ldots, v_{1,N-1}),$$
$$\cdots \tag{19}$$
$$v_{m-1} = (v_{m-1,0}, v_{m-1,1}, v_{m-1,2}, \ldots, v_{m-1,N-1}),$$

from the word (15) by applying the rule (14).

In practice this means that a direct two-dimensional Fourier transform of the matrix (17) will give such matrix signals (18) that *do not* correspond to the words of the cascade (Nn, Kk, Dd) code over $GF(q)$.

For clarity of the arguments above let's consider an example.

Example 1 First we consider an iterative binary (49, 9, 16) code formed by the product of two binary (7, 3, 4) BCH codes with a generating polynomial

$$g(x) = f_{\alpha^0}(x) f_{\alpha^3}(x) = (x - \alpha^0)(x - \alpha^3)(x - \alpha^5)(x - \alpha^6) = 1 + x + x^2 + x^4.$$

The two-dimensional spectrum of the code words of the iterative code defined in this way is schematically represented in Fig. 4a.

As can be seen in Fig. 4a in the two-dimensional spectrum there are nine non-necessarily zero spectral components. Using expression (11) for these components, we form three two-dimensional sets

$$A_{1,1} = \{\{1, 2, 4\}, \{1, 2, 4\}\}; A_{1,2} = \{\{1, 2, 4\}, \{2, 4, 1\}\};$$
$$A_{1,4} = \{\{1, 2, 4\}, \{4, 1, 2\}\},$$

which implement the conjugacy constraints (10) and point out in the two-dimensional spectrum three two-dimensional chords

$$c_{1,1}^2 = c_{2,2}, c_{2,2}^2 = c_{4,4}, c_{4,4}^2 = c_{1,1};$$
$$c_{1,2}^2 = c_{2,4}, c_{2,4}^2 = c_{4,1}, c_{4,1}^2 = c_{1,2};$$
$$c_{1,4}^2 = c_{2,1}, c_{2,1}^2 = c_{4,2}, c_{4,2}^2 = c_{1,4}.$$

By setting one (arbitrary) representative for each chord we compute the remaining spectral components of these chords according to the conjugacy rule. The corresponding two-dimensional signal is obtained through the inverse two-dimensional Fourier transform of the generated spectrum. And vice versa: having formed a signal according to the BCH coding rule by codes with a generating polynomial

$$g(x) = (x - \alpha^0)(x - \alpha^3)(x - \alpha^5)(x - \alpha^6)$$

and performing a direct two-dimensional Fourier transform over it we obtain a spectrum with zero components c_{j_1, j_2} for $j_1, j_2 \in \{0, 3, 5, 6\}$ and not necessarily zero spectral components c_{j_1, j_2} for $j_1, j_2 \in \{1, 2, 4\}$.

Now let's consider a binary cascade (49, 12, 16) code formed from the Reed-Solomon (RS) code over $GF(2^3)$ with parameters (7, 4, 4) on the outer stage and a binary BCH code with parameters (7, 3, 4) on the inner stage.

Let's define the RS code by the generating polynomial

$$G(X) = (X - \alpha^0)(X - \alpha^5)(X - \alpha^6) = \alpha^4 + \alpha^2 x + \alpha^3 x^2 + x^3.$$

On the inner stage of the cascade we will use the same binary BCH code given by the generating polynomial

$$g(x) = (x - \alpha^0)(x - \alpha^3)(x - \alpha^5)(x - \alpha^6) = 1 + x + x^2 + x^4.$$

First of all, we note that all words of the considered above iterative (49, 9, 16) code are contained in the words of the cascade (49, 12, 16) code, i.e. the iterative code is in this case the subcode of the cascade code. However, only by changing the encoding rule for a fixed code distance the relative speed of the code was increased by a third.

According to the value of the roots of the polynomial

$$G(X) = (X - \alpha^0)(X - \alpha^5)(X - \alpha^6),$$

the logical structure of the spectrum of code words of the cascade code would be the scheme represented in Fig. 4.b, i.e. matrix with zero columns corresponding to the roots of $G(X)$. However, the application the two-dimensional Fourier transform to the code words of the cascade code will lead to the representation of the words of the RS-code in the form of corresponding constraints on the binary subfield. Taking into account the conjugacy constraints (6), the binary words obtained by rule (18) form a binary (7, 6, 2) BCH code with a check polynomial

$$h(x) = (x - \alpha^1)(x - \alpha^2)(x - \alpha^4)(x - \alpha^3)(x - \alpha^5)(x - \alpha^6)$$
$$= 1 + x + x^2 + x^3 + x^4 + x^5 + x^6$$

and the generating polynomial

$$g(x) = (x - \alpha^0) = 1 + x,$$

respectively.

In other words, the non-zero spectral components $c_{i,3}$ of the two-dimensional spectrum can lead to nonzero values of all the spectral components inside the corresponding chords:

$$c_{1,3}^2 = c_{2,6}, c_{2,6}^2 = c_{4,5}, c_{4,5}^2 = c_{1,3};$$
$$c_{2,3}^2 = c_{4,6}, c_{4,6}^2 = c_{1,5}, c_{1,5}^2 = c_{2,3};$$
$$c_{4,3}^2 = c_{1,6}, c_{1,6}^2 = c_{2,5}, c_{2,5}^2 = c_{4,3}$$

and the spectrum of the codeword in general will have the form shown schematically in Fig. 4b.

Defining the codewords through the inverse two-dimensional Fourier transform of the spectrum corresponding to Fig. 4b, we get not a cascaded code, but some iterative code-product of binary and BCH codes.

Thus, the mathematical apparatus of multidimensional spectra *does not allow* describing the code words of cascade codes in the frequency domain. Accordingly, it is *impossible* to obtain for these codes the useful effect that fast multidimensional Fourier transforms give in the technique of noise-immune encoding. The next section of this paper is devoted to solving this contradiction.

4 Description of Cascaded Codes in the Frequency Domain

To solve the problem of representing cascade codes in the frequency domain it is necessary to analytically associate the values of the spectral components of the outer stage code with the corresponding spectral components of its restriction on the subfield. Then the mathematical apparatus of multidimensional spectra, taking into account this introduced analytic connection, will allow us to calculate the code word of the cascade code in the frequency domain.

Thus, it is necessary to solve the following *partial problems*:

(1) to express analytically the spectrum of the vector (15) (for given spectra of arbitrary sequences (19);
(2) to express analytically the spectrum of the sequences (19) according to a given spectrum of an arbitrary vector (15);
(3) to express analytically a multidimensional spectrum of the form (18) for given spectra of the sequences (19) and / or (15).

The solving of the 1st problem

Let's denote in general terms the spectrum of the sequence (15) in the form

$$C = (C_0, C_1, C_2, \ldots, C_{N-1}), \tag{20}$$

and also $V_i, C_j \in GF(q^m)$, i.e. the field of the signal symbols the and the signal's spectrum coincide.

Let's find the spectrum for each vector from (19), then we obtain

$$
\begin{aligned}
c_0 &= (c_{0,0}, c_{0,1}, c_{0,2}, \ldots, c_{0,N-1}), \\
c_1 &= (c_{1,0}, c_{1,1}, c_{1,2}, \ldots, c_{1,N-1}), \\
&\qquad\qquad \cdots \\
c_{m-1} &= (c_{m-1,0}, c_{m-1,1}, c_{m-1,2}, \ldots, c_{m-1,N-1}),
\end{aligned}
\tag{21}
$$

where spectral components $c_{i,j}$ for all m spectra $c_0, c_1, \ldots, c_{m-1}$ belong, as well as the components of the spectrum C, to extended field $GF(q^m)$.

Statement 1 The spectrum of an arbitrary time vector is a linear combination of the spectra of its constraint vectors on an arbitrary subfield.

Proof Let us find the spectrum of the signal V over $GF(q^m)$. The kernel of the Fourier transform in the Galois field is an element α of nth order equal to nth root of unity. We choose this element, for example, in the form $\alpha = z$. Then

$$V_i = v_{0,i} + v_{1,i}z + \ldots + v_{m-1,i}z^{m-1} = v_{0,i} + \alpha v_{1,i} + \ldots + \alpha^{m-1}v_{m-1,i}$$

and, respectively,

$$V = v_0 + \alpha v_1 + \ldots + \alpha^{m-1}v_{m-1}.$$

I.e. the vector V can be written as a linear combination of vectors $v_0, v_1, \ldots, v_{m-1}$ form (19) (the selection of another element $\alpha \in GF(q^m)$ of nth order will lead to an isomorphic representation of the element V, i.e. it will change only the form of this linear combination).

The Fourier transform is linear by definition and for finite fields can be written in the form of matrix multiplication of the signal V by the Vandermonde matrix, composed of all powers of the element (the transformation kernel), i.e.

$$C = VW, c_i = v_i W, i = 0, \ldots, m-1,$$

where

$$W^T = \begin{pmatrix} \alpha^0 & \alpha^0 & \alpha^0 & \ldots & \alpha^0 \\ \alpha^0 & \alpha^1 & \alpha^2 & \ldots & \alpha^{N-1} \\ \alpha^0 & \alpha^2 & \alpha^4 & \ldots & \alpha^{N-2} \\ \ldots & \ldots & \ldots & \ldots & \ldots \\ \alpha^0 & \alpha^{N-1} & \alpha^{N-2} & \ldots & \alpha^1 \end{pmatrix}, n = N = q^m - 1.$$

For the inverse transformation the inverse matrix should be used:

$$(W^{-1})^T = \begin{pmatrix} \alpha^0 & \alpha^0 & \alpha^0 & \ldots & \alpha^0 \\ \alpha^0 & \alpha^{N-1} & \alpha^{N-2} & \ldots & \alpha^1 \\ \alpha^0 & \alpha^{N-2} & \alpha^{N-3} & \ldots & \alpha^2 \\ \ldots & \ldots & \ldots & \ldots & \ldots \\ \alpha^0 & \alpha^1 & \alpha^2 & \ldots & \alpha^{N-1} \end{pmatrix}.$$

Using these expressions, i.e. property of linearity of the Fourier transform, we obtain:

$$C = VW = v_0 W + \alpha v_1 W + \ldots + \alpha^{m-1} v_{m-1} W = c_0 + \alpha c_1 + \ldots + \alpha^{m-1} c_{m-1}, \tag{22}$$

thus, *spectrum C of the arbitrary time vector V over $GF(q^m)$ is linear combination of spectra c_0, c_1, ..., c_{m-1} of vectors v_0, v_1, ..., v_{m-1}—constraints V on the subfield $GF(q) \subseteq GF(q^m)$*. The concrete form of this linear combination is obviously determined by the choice of the element α—the kernel of the Fourier transform.

The established assertion allows us to calculate the spectrum of the code word of the RS-code from the known spectra of its restriction on the subfield, that is, by the known spectra of the corresponding code words of some BCH code.

Example 2 Consider an arbitrary code word of the $(7, 4, 4)$ RS-code over $GF(2^3)$ from Example 1. Let it be, for example,

$$V = (\alpha^6, \alpha^4, \alpha^2, 0, 0, \alpha^4, \alpha^0).$$

The restriction (14) to the binary subfield gives three code words v_0, v_1 and v_2 of the binary $(7, 6, 2)$ BCH code (see Example 1). The concrete form of these vectors depends on the way the field $GF(2^3)$ elements are represented. Let's suppose, for example, that a field $GF(2^3)$ is constructed from a polynomial ring with operations modulo an irreducible polynomial $f(z) = 1 + z + z^3$, and a primitive element $\alpha = z$ of the field $GF(2^3)$ is a root of this polynomial.

Then the words v_0, v_1 and v_2 take the form:

$$v_0 = (1, 0, 0, 0, 0, 0, 1),$$

$$v_1 = (0, 1, 0, 0, 0, 1, 0),$$
$$v_2 = (1, 1, 1, 0, 0, 1, 0).$$

Let's find the spectrum of the vectors v_0, v_1, and v_2, usig the Fourier transform of the form (4) with the kernel $\alpha = z$, we obtain:

$$c_0 = (0, \alpha^2, \alpha^4, \alpha^5, \alpha^1, \alpha^6, \alpha^3),$$
$$c_1 = (0, \alpha^6, \alpha^5, \alpha^0, \alpha^3, \alpha^0, \alpha^0),$$
$$c_2 = (0, 0, 0, \alpha^6, 0, \alpha^3, \alpha^5).$$

Let's find the spectrum C of the vector V using (22):

$$C = c_0 + \alpha c_1 + \alpha^2 c_2 == \begin{pmatrix} 0 + 0 + 0 \\ \alpha^2 + \alpha^0 + 0 \\ \alpha^4 + \alpha^6 + 0 \\ \alpha^5 + \alpha^1 + \alpha^1 \\ \alpha^1 + \alpha^4 + 0 \\ \alpha^6 + \alpha^1 + \alpha^5 \\ \alpha^3 + \alpha^1 + \alpha^0 \end{pmatrix}^T = \begin{pmatrix} 0 \\ \alpha^6 \\ \alpha^3 \\ \alpha^5 \\ \alpha^2 \\ 0 \\ 0 \end{pmatrix}^T$$

The direct check shows that the spectrum of the vector

$$V = (\alpha^6, \alpha^4, \alpha^2, 0, 0, \alpha^4, \alpha^0),$$

calculated by the rule (4), is indeed equal to

$$C = (0, \alpha^6, \alpha^3, \alpha^5, \alpha^2, 0, 0).$$

The solving of the 2nd problem

Now we solve the inverse problem, i.e. the finding of the spectra (21) following on from the known spectrum (20). We use the notation introduced above. The following assertion is valid.

Statement 2 The components of the spectra of constraint vectors of an arbitrary time vector on an arbitrary subfield are a linear combination of the results of power mappings of the components of the spectrum of this vector.

Proof Using (22), note

$$
\begin{aligned}
C_0 &= c_{0,0} + \alpha c_{1,0} + \ldots + \alpha^{m-1} c_{m-1,0}, \\
C_1 &= c_{0,1} + \alpha c_{1,1} + \ldots + \alpha^{m-1} c_{m-1,1}, \\
C_2 &= c_{0,2} + \alpha c_{1,2} + \ldots + \alpha^{m-1} c_{m-1,2},
\end{aligned}
\tag{23}
$$

$$
\ldots
$$

$$
C_{n-1} = c_{0,n-1} + \alpha c_{1,n-1} + \ldots + \alpha^{m-1} c_{m-1,n-1},
$$

i.e. finding the components of the spectra (21) reduces to solving an underdefinite system of n linear equations from nm unknowns.

An underdefined system in the general form either has an infinite number of solutions, or does not have them at all, but in this case the system of equations can be somewhat simplified.

Indeed, the first equation while $\alpha = z$ takes the form

$$
C_0 = c_{0,0} + c_{1,0} z + \ldots + c_{m-1,0} z^{m-1},
$$

which, taking into account that $C_0 \in GF(q^m)$ and $c_{i,0} \in GF(q)$ for $i = 0, 1, \ldots, m - 1$, means, that elements $(c_{0,0}, c_{1,0}, \ldots, c_{m-1,0})$ are defined as base-q representation of C_0.

For the remaining equations we use the conjugacy constraints (10), which, for the two-dimensional case $(p = 2)$, can be rewritten as follows:

$$
(c_{i,j})^q = c_{i, jq \bmod (q^m - 1)}.
$$

Conjugacy constraints transform an underdefined system of linear equations (23) into a set of u certain subsystems with respect to u_s nonlinear equations (and u_s unknowns) in each subsystem, respectively:

$$
\begin{aligned}
C_s &= c_{0,s} + \alpha c_{1,s} + \ldots + \alpha^{m-1} c_{m-1,s}, \\
C_{sq \bmod N} &= (c_{0,s})^q + \alpha (c_{1,s})^q + \ldots + \alpha^{m-1} (c_{m-1,s})^q,
\end{aligned}
\tag{24}
$$

$$
\ldots
$$

$$
C_{sq^{u_s-1} \bmod N} = (c_{0,s})^{q^{u_s-1}} + \alpha (c_{1,s})^{q^{u_s-1}} + \ldots + \alpha^{m-1} (c_{m-1,s})^{q^{u_s-1}},
$$

where

u—the number of nontrivial conjugacy classes of the field elements $GF(q^m)$ (or, equivalently, the number of different nontrivial chords A_s in the spectrum of one-dimensional signals of length N over $GF(q)$);

u_s—the number of elements in sth class (or, equivalently, the number of elements in the chord A_s), s—a positive integer that runs through all powers of the primitive element from the expansion of the field $GF(q^m)$ on classes $\{\alpha^s, \alpha^{sq}, \ldots, \alpha^{sq^{u_s}}\}$ in a way that

$$\sum_s u_s = q^m - 2, \#s = u.$$

Let's find a solution for an arbitrary subsystem of nonlinear equations (24); i.e. for an arbitrary s. To do this, we use the following property of finite fields [3, 6–12, 20–22]:

$$(\alpha + \beta)^{q^b} = \alpha^{q^b} + \beta^{q^b},$$

valid for all $\alpha, \beta \in GF(q^m)$ and any positive integer b.

Raising the wth equation of subsystem (24)

$$C_{sq^w \bmod N} = (c_{0,s})^{q^w} + \alpha (c_{1,s})^{q^w} + \ldots + \alpha^{m-1} (c_{m-1,s})^{q^w}, \ w = 0, 1, .., u_s - 1$$

to the power q^{m-w} we obtain:

$$\left(C_{sq^w \bmod N}\right)^{q^{m-w}} = (c_{0,s})^{q^m} + \alpha^{q^{m-w}} (c_{1,s})^{q^m} + \ldots + \alpha^{(m-1)q^{m-w}} (c_{m-1,s})^{q^m},$$

that, according to a small theorem of Fermat, gives a linear equation:

$$\left(C_{sq^w \bmod N}\right)^{q^{m-w}} = c_{0,s} + \alpha^{q^{m-w}} c_{1,s} + \ldots + \alpha^{(m-1)q^{m-w}} c_{m-1,s}.$$

Let's denote the set of free terms on the left-hand side of system (24) in terms of

$$C^s = \{C_s, C_{sq \bmod N}, \ldots, C_{sq^{u_s-1} \bmod N}\}.$$

Then the functional correspondence

$$\left(C_{sq^w \bmod N}\right)^{q^{m-w}} = \phi\left(C_{sq^w \bmod N}\right), \ w = 0, \ldots, u_s - 1 \tag{25}$$

implements the power map $\phi : C^s \to \overline{C^s}$ of the set C^s to set

$$\overline{C^s} = \left\{(C_s)^{q^m}, \left(C_{sq \bmod N}\right)^{q^{m-1}}, \ldots, \left(C_{sq^{u_s-1} \bmod N}\right)^{q^{m-u_s+1}}\right\}.$$

By writing element-wise the result of the mapping ϕ, i.e. finding $\left(C_{sq^w \bmod N}\right)^{q^{m-w}}$ for all $w = 0, \ldots, u_s - 1$, we obtain a system of linear equations

$$(C_s)^{q^m} = c_{0,s} + \alpha^{q^m} c_{1,s} + \ldots + \alpha^{(m-1)q^m} c_{m-1,s},$$

$$\left(C_{sq \bmod N}\right)^{q^{m-1}} = c_{0,s} + \alpha^{q^{m-1}} c_{1,s} + \ldots + \alpha^{(m-1)q^{m-1}} c_{m-1,s}, \ldots$$

$$\left(C_{sq^{u_s-1} \bmod N}\right)^{q^{m-u_s+1}} = c_{0,s} + \alpha^{q^{m-u_s+1}} c_{1,s} + \ldots + \alpha^{(m-1)q^{m-u_s+1}} c_{m-1,s}, \tag{26}$$

whose solution gives the required components of the spectrum for the sth subsystem.

The solution found will be expressed by a linear combination of the free terms on the left side of the system, i.e. from the elements of the set $\overline{C^s}$—the results of the power map of the components of the spectrum C of the vector V.

Performing similar transformations for all u of the defined subsystems, we obtain, taking into account the conjugacy constraints (10), solutions for all unknown components of the spectra of the constraint vectors of the time-vector V on an arbitrary subfield. It is obvious that *the components of the spectra of constraint vectors of an arbitrary time vector on an arbitrary subfield, found in this way, will also be expressed by a linear combination of the results of power maps of the components of the spectrum of this vector.*

The statement formulated and proved allows analytical connecting the spectrum of constraint vectors of an arbitrary codeword with the spectrum of this codeword. For clarity, we give an example of computing the spectrum of code words of BCH-code from the known spectrum of the code word of RS-code from Example 2.

Example 3 Let's consider an arbitrary codeword $V = (V_0, V_1, V_2, V_3, V_4, V_5, V_6)$ of RS-code over $GF(2^3)$ from the previous example.

Let us write down its spectrum C with components from $GF(2^3)$ in a general form:

$$C = (C_0, C_1, C_2, C_3, C_4, C_5, C_6).$$

The spectrum of the corresponding code words of the BCH code (or, equivalently, the spectrum of constraint vectors v_0, v_1, v_2 of the word V on a binary subfield) is written as:

$$c_0 = (c_{0,0}, c_{0,1}, c_{0,2}, c_{0,3}, c_{0,4}, c_{0,5}, c_{0,6}),$$
$$c_1 = (c_{1,0}, c_{1,1}, c_{1,2}, c_{1,3}, c_{1,4}, c_{1,5}, c_{1,6}),$$
$$c_2 = (c_{2,0}, c_{2,1}, c_{2,2}, c_{2,3}, c_{2,4}, c_{2,5}, c_{2,6}).$$

Using expression (22), we obtain

$$C = c_0 + \alpha c_1 + \alpha^2 c_2,$$

that in the element-by-element notation (23) gives the following underdefined system of 6 equations and 18 unknowns

$$C_0 = c_{0,0} + \alpha c_{1,0} + \alpha^2 c_{2,0},$$
$$C_1 = c_{0,1} + \alpha c_{1,1} + \alpha^2 c_{2,1},$$
$$C_2 = c_{0,2} + \alpha c_{1,2} + \alpha^2 c_{2,2},$$
$$C_3 = c_{0,3} + \alpha c_{1,3} + \alpha^2 c_{2,3},$$
$$C_4 = c_{0,4} + \alpha c_{1,4} + \alpha^2 c_{2,4},$$

$$C_5 = c_{0,5} + \alpha c_{1,5} + \alpha^2 c_{2,5},$$
$$C_6 = c_{0,6} + \alpha c_{1,6} + \alpha^2 c_{2,6}.$$

Consider the first equation; note that $C_0 \in GF(2^3)$ and all $c_{i,0} \in GF(2)$. Then for $\alpha = z : C_0 = c_{0,0} + c_{1,0}z + c_{2,0}z^2$, i.e. elements

$$\left(c_{0,0}, c_{1,0}, c_{2,0}\right)$$

are defined as a binary representation of C_0.

Using constraints of conjugation

$$(c_{i,j})^2 = c_{i,2j \bmod (7)},$$

rewrite the remaining equations in the form:

$$C_1 = c_{0,1} + \alpha c_{1,1} + \alpha^2 c_{2,1},$$
$$C_2 = (c_{0,1})^2 + \alpha(c_{1,1})^2 + \alpha^2(c_{2,1})^2,$$
$$C_3 = c_{0,3} + \alpha c_{1,3} + \alpha^2 c_{2,3},$$
$$C_4 = (c_{0,1})^4 + \alpha(c_{1,1})^4 + \alpha^2(c_{2,1})^4,$$
$$C_5 = (c_{0,3})^4 + \alpha(c_{1,3})^4 + \alpha^2(c_{2,3})^4,$$
$$C_6 = (c_{0,3})^2 + \alpha(c_{1,3})^2 + \alpha^2(c_{2,3})^2.$$

Elements of the field $GF(2^3)$ form $u = 2$ of nontrivial conjugacy classes $\{\alpha^1, \alpha^2, \alpha^4\}$ and $\{\alpha^3, \alpha^6, \alpha^5\}$, i.e. according to (24) for $s = 1$ and for $s = 3$ we have two subsystems of $u_1 = u_3 = 3$ non-linear equations (and the same number of unknowns) in each subsystem:

$$C_1 = c_{0,1} + \alpha c_{1,1} + \alpha^2 c_{2,1},$$
$$C_2 = (c_{0,1})^2 + \alpha(c_{1,1})^2 + \alpha^2(c_{2,1})^2,$$
$$C_4 = (c_{0,1})^4 + \alpha(c_{1,1})^4 + \alpha^2(c_{2,1})^4;$$
$$C_3 = c_{0,3} + \alpha c_{1,3} + \alpha^2 c_{2,3},$$
$$C_5 = (c_{0,3})^4 + \alpha(c_{1,3})^4 + \alpha^2(c_{2,3})^4,$$
$$C_6 = (c_{0,3})^2 + \alpha(c_{1,3})^2 + \alpha^2(c_{2,3})^2.$$

For each $s = 1, 3$ using the functional correspondence (25)

$$(C_{s2^w \bmod 7})^{2^{3-w}} = \phi(C_{s2^w \bmod 7}), \quad w = 0, \ldots, 2,$$

we implement power maps $\phi : C^s \rightarrow \overline{C^s}$ of sets $C^s = \{C_s, C_{s2 \bmod 7}, C_{s4 \bmod 7}\}$ to sets

$$\overline{C^s} = \left\{ C_s, (C_{s2 \bmod 7})^4, (C_{s4 \bmod 7})^2 \right\},$$

the result will be written element-wise in the form (26).

We obtain two systems of linear equations:

$$C_1 = c_{0,1} + \alpha c_{1,1} + \alpha^2 c_{2,1},$$
$$(C_2)^4 = c_{0,1} + \alpha^4 c_{1,1} + \alpha c_{2,1},$$
$$(C_4)^2 = c_{0,1} + \alpha^2 c_{1,1} + \alpha^4 c_{2,1};$$
$$C_3 = c_{0,3} + \alpha c_{1,3} + \alpha^2 c_{2,3},$$
$$(C_5)^2 = c_{0,3} + \alpha^2 c_{1,3} + \alpha^4 c_{2,3},$$
$$(C_6)^4 = c_{0,3} + \alpha^4 c_{1,3} + \alpha c_{2,3},$$

whose solution has the form of linear combinations:

$$c_{0,1} = C_1 + (C_2)^4 + (C_4)^2,$$
$$c_{1,1} = \alpha^2 C_1 + \alpha(C_2)^4 + \alpha^4(C_4)^2,$$
$$c_{2,1} = \alpha C_1 + \alpha^4(C_2)^4 + \alpha^2(C_4)^2;$$
$$c_{0,3} = C_3 + (C_5)^4 + (C_6)^2,$$
$$c_{1,3} = \alpha^2 C_3 + \alpha(C_5)^4 + \alpha^4(C_6)^2,$$
$$c_{2,3} = \alpha C_3 + \alpha^4(C_5)^4 + \alpha^2(C_6)^2.$$

The remaining components of the spectrum are obtained from the conjugacy conditions (10):

$$c_{0,2} = (c_{0,1})^2 = (C_1)^2 + C_2 + (C_4)^4,$$
$$c_{0,4} = (c_{0,1})^4 = (C_1)^4 + (C_2)^2 + C_4,$$
$$c_{1,2} = (c_{1,1})^2 = \alpha^4(C_1)^2 + \alpha^2 C_2 + \alpha(C_4)^4,$$
$$c_{1,4} = (c_{1,1})^4 = \alpha(C_1)^4 + \alpha^4(C_2)^2 + \alpha^2 C_4,$$
$$c_{2,2} = (c_{2,1})^2 = \alpha^2(C_1)^2 + \alpha C_2 + \alpha^4(C_4)^4,$$
$$c_{2,4} = (c_{2,1})^4 = \alpha^4(C_1)^4 + \alpha^2(C_2)^2 + \alpha C_4,$$
$$c_{0,6} = (c_{0,3})^2 = (C_3)^2 + C_6 + (C_5)^4,$$
$$c_{0,5} = (c_{0,3})^4 = (C_3)^4 + (C_6)^2 + C_5,$$
$$c_{1,6} = (c_{1,3})^2 = \alpha^4(C_3)^2 + \alpha^2 C_6 + \alpha(C_5)^4,$$
$$c_{1,5} = (c_{1,3})^4 = \alpha(C_3)^4 + \alpha^4(C_6)^2 + \alpha^2 C_5,$$
$$c_{2,6} = (c_{2,3})^2 = \alpha^2(C_3)^2 + \alpha C_6 + \alpha^4(C_5)^4,$$
$$c_{2,5} = (c_{2,3})^4 = \alpha^4(C_3)^4 + \alpha^2(C_6)^2 + \alpha C_5.$$

Thus, in the general form, the solution of the system of equations for $0 < j \leq 6$ we write as a *linear combination of the results of power maps of the components of the spectrum C*:

$$c_{0,j} = C_j + (C_{2j \bmod 7})^4 + (C_{4j \bmod 7})^2,$$
$$c_{1,j} = \alpha^2 C_j + \alpha (C_{2j \bmod 7})^4 + \alpha^4 (C_{4j \bmod 7})^2,$$
$$c_{2,j} = \alpha C_j + \alpha^4 (C_{2j \bmod 7})^4 + \alpha^2 (C_{4j \bmod 7})^2.$$

Verification for the spectrum

$$C = (0, \alpha^6, \alpha^3, \alpha^5, \alpha^2, 0, 0)$$

gives

$$c_0 = (0, \alpha^2, \alpha^4, \alpha^5, \alpha^1, \alpha^6, \alpha^3),$$
$$c_1 = (0, \alpha^6, \alpha^5, \alpha^0, \alpha^3, \alpha^0, \alpha^0),$$
$$c_2 = (0, 0, 0, \alpha^6, 0, \alpha^3, \alpha^5),$$

which completely coincides with the data from Example 2.

The obtained analytical solutions of the first two problems connected with the search for a one-to-one correspondence between the spectrum of the vector (15) and the spectra of arbitrary sequences (19) make it possible in general to solve the problem of analytic representation of a multidimensional spectrum of the form (18) given spectra of the sequences (19) and/or (15).

The solving of the 3rd problem

Let us return to the code word of cascade code over $GF(q)$ in the form (17) and the corresponding spectrum (18) with components of $GF(q^m)$. If the first m rows of the matrix (17) are constraint vectors (19) of the code word (15) on the subfield $GF(q) \subseteq GF(q^m)$, then the spectrum of the words (19) is one-to-one functionally related to the spectrum of the word (15), which proves the previous assertions. We now consider the remaining $n - m$ rows of the matrix (17).

Statement 3 The code word of the cascade code is a linear combination of constraint vectors of the outer-stage code word.

Proof The structure of the cascade code (see Fig. 3) is such that after encoding by the outer stage code [forming the first m rows of the matrix (17)], each obtained column is considered as the information sequence of the $(n, k = m, d)$ code of the inner step. The corresponding code words are written along the columns of the matrix (17). For a linear code, this is equivalent to multiplication $v_i = (v_{0,i}, v_{1,i}, \ldots, v_{m-1,i})g$, where v_i is the code word (16) of the code of inner stage, written in the ith column of the matrix (17).

In other words, the process of forming the entire codeword (17) can be represented as multiplying $N \times k$ of the matrix $(k = m)$ formed by the elements of the vectors (19) by the generator $k \times n$ matrix g of the first stage code:

$$
v = \begin{pmatrix}
v_{0,0} & v_{1,0} & \cdots & v_{m-1,0} \\
v_{0,1} & v_{1,1} & \cdots & v_{m-1,1} \\
\cdots & \cdots & \cdots & \cdots \\
v_{0,N-1} & v_{1,N-1} & \cdots & v_{m-1,N-1}
\end{pmatrix} g.
$$

This is equivalent to forming rows of the matrix v, as a linear combination of vectors:

$$
v_0 = (v_{0,0}, v_{0,1}, v_{0,2}, \ldots, v_{0,N-1}),
$$
$$
v_1 = (v_{1,0}, v_{1,1}, v_{1,2}, \ldots, v_{1,N-1}),
$$
$$
\cdots
$$
$$
v_{m-1} = (v_{m-1,0}, v_{m-1,1}, v_{m-1,2}, \ldots, v_{m-1,N-1}),
$$

i.e. *linear combination of constraint vectors (19) of the code word of the outer stage to an arbitrary subfield.*

We now consider the spectrum of a two-dimensional word v. First, we note that the following statement holds for the general case of calculating the multidimensional spectra (8).

Statement 4 The multidimensional spectrum of a multidimensional word is the result of a multiple computation of a one-dimensional spectrum to all one-dimensional representations of this word.

Proof The multidimensional Fourier transform (8) is linear and can be written by repeated (p times, i.e. for each dimension of the matrix v with allowance for transposition) multiplications of the matrix v by the Vandermonde matrix W from Statement 1:

$$
c = \underbrace{\left(\left((vW)^T W \right)^T \ldots W \right)^T}_{p}.
$$

However, the result of the calculation vW in the row record gives a set of spectra corresponding to the rows of the matrix v, i.e. its one-dimensional representation. In other words, *the multidimensional spectrum of a multidimensional word v is the result of repeatedly computing a one-dimensional spectrum to all one-dimensional representations of a word v.*

Combining the previous two statements, we obtain the following.

Statement 5 The spectrum of the code word of the cascade code is, in a row-by-row notation, the set of results of a two-fold calculation of the one-dimensional spectrum to all linear combinations of constraint vectors of the outer step code word.

Proof The application of Statement 4 to the cascade code gives the following expression

$$c = (vW)^T W, \tag{27}$$

i.e. the spectrum of the codeword of the cascade code is, in a row notation, a set of spectra corresponding to the spectra of rows $(v_0, v_1, \ldots, v_{N-1})$ of matrix v.

However, as shown in Statement 3, the rows of the matrix v are linear combinations of constraint vectors $v_0, v_1, \ldots, v_{m-1}$ from (19) of the code word of the outer stage. The Fourier transform, by definition, is linear, hence the spectrum of the linear combination of the vectors $v_0, v_1, \ldots, v_{m-1})$ from (19) gives a linear combination of the corresponding spectra $(c_0, c_1, \ldots, c_{m-1})$ from (21). Then, the spectrum of the code word of the cascade code is, in a row notation, the set of spectra of the linear combinations of the spectra (21), or equivalently, *the set of results of the two-fold calculation of the one-dimensional spectrum to all linear combinations of constraint vectors of code words of the outer stage.*

The analytic connection of the spectrum of the codeword of the cascade code with the spectrum of words of the outer stage code gives the application of Statement 2 to the last result.

Statement 6 The components of the spectrum of an arbitrary code word of the cascade code are determined by a linear combination of the power map results of the components of the code-word spectrum of the outer-stage code.

Proof Indeed, if, according to Statement 2, the components of the spectra of constraint vectors of an arbitrary time vector on an arbitrary subfield are a linear combination of the results of power mappings of the components of the spectrum of this vector, then it follows from Statement 5 that the *components of the spectrum of the codeword of the cascade code are determined by a linear combination of the results of these mappings.*

Now les's give an example that clearly demonstrates the validity of the above considerations. As an initial data we will use Example 1. Recall that the use of two-dimensional spectra in Example 1 *did not allow* describing cascade codes in the frequency domain. We now show how, using the obtained analytical regularities, we can solve this problem and implement coding with cascade codes in the frequency domain.

Example 4 Consider a binary cascade (49, 12, 16) code formed from the RS-code over $GF(2^3)$ with parameters (7, 4, 4) on the outer stage and a binary BCH code with parameters (7, 3, 4) on the inner stage (see Example 1).

The spectrum of constraint vectors of the codeword of the outer step RS-code in general form gives the result of Statement 2. For the given case (see Example 2) the corresponding analytic expressions have the form:

$$c_{0,j} = C_j + (C_{2j \bmod 7})^4 + (C_{4j \bmod 7})^2,$$
$$c_{1,j} = \alpha^2 C_j + \alpha(C_{2j \bmod 7})^4 + \alpha^4(C_{4j \bmod 7})^2,$$
$$c_{2,j} = \alpha C_j + \alpha^4(C_{2j \bmod 7})^4 + \alpha^2(C_{4j \bmod 7})^2,$$

valid for all $j = 0, \ldots, N - 1$.

Suppose that on the inner stage of the cascade the encoding rule of the BCH code with $g(x) = 1 + x + x^2 + x^4$ is given in a systematic form through multiplication by the generator matrix g (see Statement 3):

$$g = \begin{pmatrix} 1\,0\,0\,1\,1\,1\,0 \\ 0\,1\,0\,0\,1\,1\,1 \\ 0\,0\,1\,1\,1\,0\,1 \end{pmatrix}.$$

Then the intermediate result vW in (27) in the row notation will be determined by linear combinations of vectors $c_{0,j}$, $c_{1,j}$, $c_{2,j}$ by the rule given by the matrix g, i.e. for all $j = 0, \ldots, 6$ we have:

$$c_{3,j} = c_{0,j} + c_{2,j} = \alpha^3 C_j + \alpha^5(C_{2j \bmod 7})^4 + \alpha^6(C_{4j \bmod 7})^2,$$
$$c_{4,j} = c_{0,j} + c_{1,j} + c_{2,j} = \alpha^5 C_j + \alpha^6(C_{2j \bmod 7})^4 + \alpha^3(C_{4j \bmod 7})^2,$$
$$c_{5,j} = c_{0,j} + c_{1,j} = \alpha^6 C_j + \alpha^3(C_{2j \bmod 7})^4 + \alpha^5(C_{4j \bmod 7})^2,$$
$$c_{6,j} = c_{1,j} + c_{2,j} = \alpha^4 C_j + \alpha^2(C_{2j \bmod 7})^4 + \alpha^1(C_{4j \bmod 7})^2.$$

Computation of the one-dimensional Fourier transform of the vector

$$c_j = \left(c_{0,j}, c_{1,j}, c_{2,j}, c_{3,j}, c_{4,j}, c_{5,j}, c_{6,j}\right)$$

for all $j = 0, \ldots, 6$ will give all the lines of the spectrum of the codeword of the cascade code. I.e. we write the general solution in the following form:

$$c_{0,j} = c_j(\alpha^i)^0 = 0; \ c_{j,0} = c_0(\alpha^i)^j = 0;$$
$$c_{1,1} = \alpha^6 C_1 + \alpha^0(C_2)^4 + \alpha^5(C_4)^2,$$
$$c_{2,2} = (c_{1,1})^2 = \alpha^5(C_1)^2 + \alpha^0(C_2)^1 + \alpha^3(C_4)^4,$$
$$c_{4,4} = (c_{2,2})^2 = \alpha^6(C_1)^4 + \alpha^0(C_2)^2 + \alpha^6(C_4)^1,$$
$$c_{1,2} = \alpha^6 C_2 + \alpha^0(C_4)^4 + \alpha^5(C_1)^2,$$
$$c_{2,4} = (c_{1,2})^2 = \alpha^5(C_2)^2 + \alpha^0(C_4)^1 + \alpha^3(C_1)^4,$$
$$c_{4,1} = (c_{2,4})^2 = \alpha^6(C_2)^4 + \alpha^0(C_4)^2 + \alpha^6(C_1),$$
$$c_{1,3} = \alpha^6 C_3 + \alpha^0(C_6)^4 + \alpha^5(C_5)^2,$$
$$c_{2,6} = (c_{1,3})^2 = \alpha^5(C_3)^2 + \alpha^0(C_6)^1 + \alpha^3(C_5)^4,$$
$$c_{4,5} = (c_{2,6})^2 = \alpha^6(C_3)^4 + \alpha^0(C_6)^2 + \alpha^6(C_5),$$

$$c_{1,4} = \alpha^6 C_4 + \alpha^0 (C_1)^4 + \alpha^5 (C_2)^2,$$
$$c_{2,1} = (c_{1,4})^2 = \alpha^5 (C_4)^2 + \alpha^0 (C_1)^1 + \alpha^3 (C_2)^4,$$
$$c_{4,2} = (c_{2,1})^2 = \alpha^6 (C_4)^4 + \alpha^0 (C_1)^2 + \alpha^6 (C_2),$$
$$c_{1,5} = \alpha^6 C_5 + \alpha^0 (C_3)^4 + \alpha^5 (C_6)^2,$$
$$c_{2,3} = (c_{1,5})^2 = \alpha^5 (C_5)^2 + \alpha^0 (C_3)^1 + \alpha^3 (C_6)^4,$$
$$c_{4,6} = (c_{2,3})^2 = \alpha^6 (C_5)^4 + \alpha^0 (C_3)^2 + \alpha^6 (C_6),$$
$$c_{1,6} = \alpha^6 C_6 + \alpha^0 (C_5)^4 + \alpha^5 (C_3)^2,$$
$$c_{2,5} = (c_{1,6})^2 = \alpha^5 (C_6)^2 + \alpha^0 (C_5)^1 + \alpha^3 (C_3)^4,$$
$$c_{4,3} = (c_{2,5})^2 = \alpha^6 (C_6)^4 + \alpha^0 (C_5)^2 + \alpha^6 (C_3),$$
$$c_{3,j} = 0, c_{5,j} = 0, c_{6,j} = 0.$$

For verification, we set the spectrum of the code word of RS-code as in Example 3: $C = (0, \alpha^6, \alpha^3, \alpha^5, \alpha^2, 0, 0)$, which gives

$$c_0 = (0, \alpha^2, \alpha^4, \alpha^5, \alpha^1, \alpha^6, \alpha^3),$$
$$c_1 = (0, \alpha^6, \alpha^5, \alpha^0, \alpha^3, \alpha^0, \alpha^0),$$
$$c_2 = (0, 0, 0, \alpha^6, 0, \alpha^3, \alpha^5)$$

and the corresponding spectrum of the code word of the cascade code, calculated from the derived analytic expressions, takes the form:

$$c = \begin{pmatrix} 0 & 0 & 0 & 0 & 0 & 0 & 0 \\ 0 & \alpha^2 & \alpha^6 & \alpha^4 & \alpha^5 & \alpha^6 & \alpha^1 \\ 0 & \alpha^3 & \alpha^4 & \alpha^5 & \alpha^5 & \alpha^2 & \alpha^1 \\ 0 & 0 & 0 & 0 & 0 & 0 & 0 \\ 0 & \alpha^3 & \alpha^6 & \alpha^4 & \alpha^1 & \alpha^2 & \alpha^3 \\ 0 & 0 & 0 & 0 & 0 & 0 & 0 \\ 0 & 0 & 0 & 0 & 0 & 0 & 0 \end{pmatrix}.$$

It is not difficult to see that the inverse two-dimensional Fourier transform of the matrix c gives the matrix

$$v = \begin{pmatrix} 1\,0\,0\,0\,0\,0\,1 \\ 0\,1\,0\,0\,0\,1\,0 \\ 1\,1\,1\,0\,0\,1\,0 \\ 0\,1\,1\,0\,0\,1\,1 \\ 0\,0\,1\,0\,0\,0\,1 \\ 1\,1\,0\,0\,0\,1\,1 \\ 1\,0\,1\,0\,0\,0\,0 \end{pmatrix}, \tag{28}$$

which is really the code word for the cascade (49, 12, 16) code.

Thus, Example 4 clearly demonstrates the implementation of encoding by cascade code through transformations in the frequency domain. This result, according to the authors' opinion, was obtained *for the first time*.

In conclusion, we give the expressions for obtaining components of the signal v over $GF(q)$ given in frequency-domain the spectrum components of the code word of the outer stage code over $GF(q^m)$. To this end, we state and prove the following statement.

Statement 7 The components of an arbitrary code word of the cascade code (in the time domain) are determined by a linear combination of the results of power mappings of the components of the code word spectrum of the outer stage code.

Proof It follows from (27) that to find the vector v it is necessary to perform a two-fold inverse one-dimensional Fourier transform over the rows of the matrix c with allowance for transposition. The intermediate result vW is already known from the statement (5). In the row-wise notation vW is the linear combinations of the spectra (21), which, by Statement 2, are determined by linear combinations of the results of power mappings (25) of the components of the spectrum (20).

Thus, to calculate the code word v it is sufficient to perform the inverse Fourier transform of all linear combinations of vectors (21), that is, taking into account the linearity of the transformation, *the codeword v is determined by a linear combination of results of the power mappings of the code-word spectrum components of the outer-stage code*.

Let's show the form of these linear combinations for the case considered in Example 4.

Example 5 Computing the inverse Fourier transform for all vectors c_j, $j = 0, \ldots, 6$ (see Example 4), we obtain:

$$v_{0,0} = \sum_{j=0}^{6} \left(C_j + (C_{2j \bmod 6})^4 + (C_{4j \bmod 6})^2\right),$$

$$v_{1,0} = \sum_{j=0}^{6} (\alpha^j)^{-1}\left(C_j + (C_{2j \bmod 6})^4 + (C_{4j \bmod 6})^2\right),$$

$$\cdots$$

$$v_{6,0} = \sum_{j=0}^{6} (\alpha^j)^{-6}\left(C_j + (C_{2j \bmod 6})^4 + (C_{4j \bmod 6})^2\right),$$

$$v_{0,1} = \sum_{j=0}^{6} \left(\alpha^2 C_j + \alpha(C_{2j \bmod 7})^4 + \alpha^4(C_{4j \bmod 7})^2\right),$$

$$v_{1,1} = \sum_{j=0}^{6} (\alpha^j)^{-1}\left(\alpha^2 C_j + \alpha(C_{2j \bmod 7})^4 + \alpha^4(C_{4j \bmod 7})^2\right),$$

$$\cdots$$

$$v_{6,1} = \sum_{j=0}^{6} (\alpha^j)^{-6}\left(\alpha^2 C_j + \alpha(C_{2j \bmod 7})^4 + \alpha^4(C_{4j \bmod 7})^2\right),$$

$$\cdots$$

$$v_{6,6} = \sum_{j=0}^{6} (\alpha^j)^{-6}\left(\alpha^4 C_j + \alpha^2(C_{2j \bmod 7})^4 + \alpha^1(C_{4j \bmod 7})^2\right).$$

A direct check with the spectrum $C = (0, \alpha^6, \alpha^3, \alpha^5, \alpha^2, 0, 0)$ from Example 4 yields the code word (28), which confirms the validity and adequacy of the above reasoning.

5 Conclusions

Thus, in the research the general solution of the problem of representation of cascade codes in the frequency domain has been obtained, which allows using deduced analytic dependences of components of multidimensional spectra to construct computationally efficient algorithms of encoding and decoding in the frequency domain. The most promising in this sense is the use of fast multidimensional Fourier transforms.

The solution of the problem of describing cascaded codes in the frequency domain required non-trivial abstract representations of the corresponding code words and their restrictions on the subfield. Nevertheless, the obtained result connects the spectrum of the code word of the outer stage with the code word of the cascade code (and / or its spectrum) in the form of simple analytic expressions (see Examples 4 and 5). The applied significance of this result is the possibility of the cascade code codeword constructing in the frequency domain through the corresponding components of the spectrum of the code word of the outer stage. Analyzing the results obtained, we should also note the specific structure of the final expressions. Indeed, the variables on the right-hand side of the equations (see Examples 4 and 5) are grouped according to the classes of conjugate elements; this indicates the direct influence of the group properties of the finite field. This observation obviously can also be the scope of further research aimed to reducing the computational complexity of the corresponding transformations.

The results can be used in various computer applications. In particular, it can be useful for modeling digital sequences of different nature, including telecommunication traffic of computer networks, etc. [28–31].

References

1. Sklar B (2016) Digital communications: fundamentals and applications. Prentice Hall communications engineering and emerging techno. Pearson Education
2. Clark GC, Cain JB (1981) Error-correction coding for digital communications. Springer, Berlin, 432 p
3. Blahut RE (1983) Theory and practice of error control codes. Addison Wesley Publishing Company, Inc., Reading, Massachusetts, 500 p
4. Wicker SB, Bhargava VK (1994) Reed-solomon codes and their application. IEEE Press, New York
5. Reed IS, Truong TK, Miller RL, Huang JP (1981) Fast transforms for decoding Reed-Solomon codes. IEE Proc 128(1):9–14
6. Winograd S (1978) On computing the discrete Fourier transform. Math Comput 32(141):175–199
7. Kuznetsov A, Serhiienko R, Prokopovych-Tkachenko D (2017) Construction of cascade codes in the frequency domain. In: 2017 4th international scientific-practical conference problems of infocommunications. Science and technology (PIC S&T), Kharkov, pp 131–136. https://doi.org/10.1109/infocommst.2017.8246366
8. Jeng JH, Truong TK (1999) On decoding of both errors and erasures of a Reed-Solomon code using an inverse-free Berlekamp-Massey algorithm. IEEE Trans Commun 47(10):1488–1494
9. Fedorenko SV, Trifonov PV (2002) Finding roots of polynomials over finite fields. IEEE Trans Commun 50(11):1709–1711
10. Fedorenko SV, Trifonov PV (2003) A method for Fast Computation of the Fast fourier transform over a finite field. Prob Inf Trans 39(3):231–238
11. Hatwar MA, Panse TG, Pothuri S (2015) Design of common sub expression elimination algorithm in fast Fourier transform. In: 2015 international conference on communications and signal processing (ICCSP), Melmaruvathur, pp 1703–1707. https://doi.org/10.1109/iccsp.2015.7322810
12. Bellini S, Ferrari M, Tomasoni A (2012) On the reduction of additive complexity of cyclotomic FFTs. IEEE Trans Commun 60(6)
13. Amerbaev VM, Solovyev RA, Stempkovskiy AL, Telpukhov DV (2014) Efficient calculation of cyclic convolution by means of fast Fourier transform in a finite field. In: Proceedings of IEEE East-West design and test symposium (EWDTS 2014), Kiev, pp 1–4. https://doi.org/10.1109/ewdts.2014.7027043
14. Smith SW (1997) The scientist & engineer's guide to digital signal processing. California Technical Pub
15. Stasev YuV, Kuznetsov AA (2005) Asymmetric code-theoretical schemes constructed with the use of algebraic geometric codes. Kibernetika i Sistemnyi Analiz (3):47–57
16. Bracewell R (1999) The Fourier transform & its applications. McGraw-Hill Science/Engineering/Math
17. Kuznetsov A, Lutsenko M, Kiian N, Makushenko T, Kuznetsova T (2018) Code-based key encapsulation mechanisms for post-quantum standardization. In: 2018 IEEE 9th international conference on dependable systems, services and technologies (DESSERT), Kyiv, Ukraine, pp 276–281. https://doi.org/10.1109/dessert.2018.8409144

18. Ting Y-R, Lu E-H, Lee C-Y (2001) A complex-valued cyclic code using fast Fourier transform. In: 2001 IEEE third workshop on signal processing advances in wireless communications (SPAWC'01). Workshop Proceedings (Cat. No.01EX471), Taiwan, China, pp 271–274. https://doi.org/10.1109/spawc.2001.923900

19. Kuznetsov A, Kiian A, Lutsenko M, Chepurko I, Kavun S (2018) Code-based cryptosystems from NIST PQC. In: 2018 IEEE 9th international conference on dependable systems, services and technologies (DESSERT), Kyiv, Ukraine, pp 282–287. https://doi.org/10.1109/dessert.2018.8409145

20. Blahut RE (1985) Fast algorithms for digital signal processing. Addison-Wesley, MA, p 441

21. Qi M, Zhong L, Guo Q (2010) Speeding up the arithmetic operations over optimal extension fields in the lagrange representation using DFT. In: 2010 international conference on innovative computing and communication and 2010 Asia-Pacific conference on information technology and ocean engineering, Macao, pp 39–42. https://doi.org/10.1109/cicc-itoe.2010.17

22. Stasinski R (1994) Fourier transform algorithm for a prime number blocklength Reed-Solomon code can be computationally efficient. In: Proceedings of 1994 IEEE international symposium on information theory, Trondheim, Norway, p 96. https://doi.org/10.1109/isit.1994.394852

23. Kuznetsov A, Pushkar'ov A, Kiyan N, Kuznetsova T (2018) Code-based electronic digital signature. In: 2018 IEEE 9th international conference on dependable systems, services and technologies (DESSERT), Kyiv, Ukraine, pp 331–336. https://doi.org/10.1109/dessert.2018.8409154

24. Deshmukh TP, Deshmukh PR, Dakhole PK (2015) Design of cyclotomic Fast Fourier Transform architecture over Galois field for 15 point DFT. In: 2015 international conference on industrial instrumentation and control (ICIC), Pune, pp 607–611. https://doi.org/10.1109/iic.2015.7150814

25. Kuznetsov AA, Kolovanova IP, Prokopovych-Tkachenko DI, Kuznetsova TY (2019) Analysis and studying of the properties of algebraic geometric codes. Telecommun Radio Eng 78(5):393–417. https://doi.org/10.1615/telecomradeng.v78.i5.30

26. Kuznetsov AA, Gorbenko YuI, Prokopovych-Tkachenko DI, Lutsenko MS, Pastukhov MV (2019) NIST PQC: code-based cryptosystems. Telecommun Radio Eng 78(5):429–441. https://doi.org/10.1615/telecomradeng.v78.i5.50

27. Deshmukh TP, Dewalkar VP (2014) The design approach for fast computation of fourier transform over a finite field. In: 2014 international conference on green computing communication and electrical engineering (ICGCCEE), Coimbatore, pp 1–4. https://doi.org/10.1109/icgccee.2014.6922465

28. Radivilova T et al (2018) Decrypting SSL/TLS traffic for hidden threats detection. In: Proceedings of the 2018 IEEE 9th international conference on dependable systems, services and technologies (DESSERT). IEEE, pp 143–146. https://doi.org/10.1109/dessert.2018.8409116

29. Radivilova T, Hassan HA (2017) Test for penetration in Wi-Fi network: attacks on WPA2-PSK and WPA2-enterprise. In: 2017 international conference on information and telecommunication technologies and radio electronics (UkrMiCo), IEEE, pp 1–4

30. Lyudmyla K, Vitalii B, Tamara R (2017) Fractal time series analysis of social network activities. In: 2017 4th international scientific-practical conference problems of infocommunications. Science and technology (PIC S&T). IEEE, pp 456–459. https://doi.org/10.1109/infocommst.2017.8246438

31. Kirichenko L, Ivanisenko I, Radivilova T (2016) Dynamic load balancing algorithm of distributed systems. In: 2016 13th international conference on modern problems of radio engineering, telecommunications and computer science (TCSET), IEEE, pp 515–518

The New Cryptographic Method for Software and Hardware Protection of Communication Channels in Open Environments

Yu. M. Penkin, G. I. Khara, and A. A. Fedoseeva

Abstract The chapter is devoted to presenting theoretical bases and features of practical realization the new cryptographic method for software and hardware protection of communication channels in open environments. In the theoretical part of the chapter is shown the possibility of modeling the modes of deterministic chaos in the oscillations of discrete structures formed in the form of special matrix forms of the Latin square. A mathematical model is proposed for such discrete systems, on the basis of which an analysis of the combination of conditions for discrete structures and the requirements for transformations of their evolution operators ensuring the achievement of such modes is carried out. The results of simulation modeling of oscillatory processes according to the cycles of states of a discrete system, forming similarities of attractor trajectories, are presented. The possibility of simulating oscillation beat modes in discrete cellular structures organized in the form of two-level matrix forms of Sudoku grids is substantiated. The results of simulation modeling of oscillation beats in cycles of varying states of a discrete system for two types of beats are presented: similar to the result of the superposition of harmonic oscillations at multiple frequencies in the theory of radio signals, as well as noise-like beats. Analyzed the principles of creating devices to implement a closed communication channel in an open data transmission environment. To solve the protection problem, it is proposed to use the AES standard cryptographic algorithm, supplemented by the author's system of dynamic generation of encryption keys. Is presented the description of the system for generating encryption keys, which based on the combinatorial properties of matrix structures in the Sudoku grids form. Are shown the results of external software testing of cipher text code sequences and computer tests of the proposed algorithm. Using the modified AES cryptographic algorithm, a working

Yu. M. Penkin (✉) · G. I. Khara
Pharmacoinformatics Department Kharkiv, National University of Pharmacy, Kharkiv, Ukraine
e-mail: penkin.yuriy@gmail.com

A. A. Fedoseeva
Kharkiv Radiotechnical College, Kharkiv, Ukraine
e-mail: fedoseeva@nuph.edu.ua

© The Editor(s) (if applicable) and The Author(s), under exclusive license to Springer Nature
Switzerland AG 2021
T. Radivilova et al. (eds.), *Data-Centric Business and Applications*, Lecture Notes on Data
Engineering and Communications Technologies 48,
https://doi.org/10.1007/978-3-030-43070-2_26

model of the system for protecting a remote control of a mobile object over the radio channel has been built.

Keywords Data protection · Cryptoalgorithm · Discrete structure · Sudoku grid · Deterministic chaos · Nonlinear operations of matrix transformations · Dynamic key generation · Distance control

1 Introduction

The mobile devices with distance control—is a one of the most progressive modern technologies. This technology allows solving problems in various fields of human activities. The wired and wireless networks used in the distance control systems. Recently, a significant preference is given to wireless networks. Also used high-frequency radiochannels, they built according to their communication protocols.

Distance control for many of the devices used is very critical with regard to unauthorized interference with their work. For example, for most modern technologies, which used the distance control in critical systems, unauthorized interference with their work can be dangerous. Statistics show that along with the growing number of mobile applications, the number of unauthorized access and hacking applications is also increasing. This fact creates growing demand for the development of more secure technologies, sophisticated authentication algorithms and data encryption methods. That is why ensuring the security and protection of distance object management systems is a priority and urgent task for mobile application developers.

It is considered that the implementation of any information channel, first of all, runs into two types of risks: unauthorized interception code in an open environment (external threat) and code capture by cracking special databases in computer networks (internal threat). The one of the key, in fact, issues for implementing any encryption system is the method of storing the private key. When using software and hardware protection [1], according to most experts, the most reliable storage for this key is an autonomous (from network) flash memory microcontroller. For hack this memory requires physical capture of the controller itself with the software and also complex and expensive equipment. In addition, even after the intended reading of the content of this memory, serious work is needed on cryptanalysis of the captured information. Thus, the use of autonomous units based on microcontrollers removes the risk of threats of an internal type. For combat external attacks in software and hardware encryption can be used the principle of storing not the key itself, but only information about the procedure for its generation. Such information (about keys generation) convenient to present in the form of two components, the first of them stored in security flash memory. The second is changed programmatically. This should provide an easy way to replace the permanent information component in case of loss of the encryption device.

The goal of this chapter is to present the results of initiative project to create software and hardware data protection, which transmitted over open channels of remote control of mobile devices, coding systems which are based on the new theory of deterministic chaos in oscillations of discrete structures of the matrix type, which proposed by professor Penkin Yu. M.

Here it is reasonable to recall that systems are called dynamic whose states change with time in accordance with the established rules of the evolution operator. Traditionally separated systems are systems with continuous and discrete time evolution (streams and cascades). In the case of discrete time, the concept of time slicing is sometimes used. The first type of systems in the literature is defined as continuous systems, the second (considering the not quite correct analogy in relation to the method of describing time)—discrete. In addition to the discrete time of evolution, a truly discrete system must also consist of spatially and functionally concentrated elements (that is, it is a discrete structure).

Dynamic system means that can specify such a set of values, called dynamic variables and characterizing the state of the system, their values at any subsequent time can be determined from the original using the evolutionary operator rules. If the state of the system is given by a set of N values, then the dynamics of the change in its state over time can be represented as the movement of a point along a trajectory in an N-dimensional phase space. This space called the phase trajectory. Initially, a purely Newtonian content was embedded in the concept of a dynamic system, bearing in mind a set of bodies connected by force interactions and obeying a system of differential equations. The modern concept of a dynamic system is broader and implies the possibility of defining an evolution operator in any way. In particular, for systems with discrete time, it can be represented as recurrent mappings or difference equations (using various discrete transforms), and for discrete systems—through the rules of mutual permutations of localized elements of the structure.

It should be noted that according to the Poincaré—Bendixson theorem, a continuous dynamical system on a plane cannot have chaotic dynamics. Among continuous systems, only non-planar spatial systems can be characterized by chaotic behavior in time, for which at least three dynamic variables and (or) non-Euclidean geometry are required. However, it is known that discrete dynamical systems at some stage of their evolution can exhibit chaotic behavior even in one-dimensional or two-dimensional spaces. Unfortunately, the conditions under which such behavior of discrete systems can be observed are not generalized in the literature, which caused the authors to desire additional study.

Discrete structures are one of the types of dynamic systems, to describe which it is necessary to use representations of their specific states. In this case, the phase space of the system is understood as the totality of all admissible states of the system in the dynamics of their changes over time. Thus, a discrete structure is characterized by its initial state and a law (given by the evolution operator), according to which the system passes from the initial state to the next. If the evolution operator on the time scale is set so that the structure, after a cycle of changes, periodically returns to the same state, then we can speak about the observation in the system of the oscillatory process.

The study of various types of oscillations (self, forced, self-oscillations, parametric) in systems of different nature has always been given much attention. It is believed that in the series of these studies, work [2] marked a turning point in the theory of oscillations and marked the beginning of the development of their general non-linear theory, on the basis of which modern interdisciplinary science was formed—non-linear dynamics. In nonlinear dynamics, nonlinear models are used to study the properties of dynamic systems, usually described by differential equations for systems with continuous time and recurrent (or discrete) mappings for systems with discrete time. However, dynamic models of discrete structures, in which evolution operators are specified in the form of algorithms (rather than formula relations), are practically not represented in modern literature.

In the first half of the theoretical part of this chapter, the authors first proposed a model of a simulator of oscillations in discrete structures built in the form of Sudoku grids, the dynamics of changes in which in discrete time is described by specially introduced operators of the evolution of group shifts of the linear and vortex types. Based on the simulation, the conditions for the manifestation of modes of deterministic chaos in the oscillations of such discrete structures were justified. The final part of the theoretical foundations presents a generalization of the author's approach to discrete cellular structures organized in the form of two-level matrix forms, in which it is possible to observe the effect of beating vibrations. The results of simulation modeling of oscillation beats in cycles of varying states of a discrete system are presented for two types of beats: similar to the result of a superposition of harmonic oscillations at multiple frequencies in radio signal theory, and noise-like beats. Later in the chapter, based on the theoretical results obtained, a description is given of the principles of creating device-encoders for implementing a closed communication channel in an open data transmission environment. To solve the protection problem, it is proposed to use the AES cryptographic algorithm, supplemented by the author's system for dynamic generation of encryption keys. A description of the system for generating encryption keys based on the combinatorial properties of matrix structures such as Latin squares is presented. The competitive advantages of the proposed cryptographic method are formulated. The results of external software testing of cipher text code sequences and computer tests of the proposed coding algorithm are shown. With using the modified AES cryptographic algorithm, a working mockup of security devices of the remote control system of a mobile object over the radio-channel.

2 Deterministic Chaos in the Oscillations of Discrete Structures of Matrix Type

2.1 Theoretical Bases of Observing Determined Chaos in Matrix Structures

2.1.1 The Paradigm of Determined Chaos

The pioneers of a separate branch of the theory of dynamical systems are: the French physicist and philosopher Henri Poincaré, who proved the return theorem, Soviet mathematicians A. N. Kolmogorov and V. I. Arnold, and the German mathematician Yu. K. Moser, who constructed a theory of small perturbations of conditionally periodic dynamics in Hamiltonian systems, called KAM (Kolmogorov-Arnold-Moser theory). Note that this theory introduces the concept of attractors (KAM-tors, like attracting cantor structures)—stable orbits of the phase trajectories of the system and allows us to show that even an arbitrarily small perturbation can lead to a strong qualitative reorganization of the phase portrait of a nonlinear system.

In 1963 in the magazine "Journal of the Atmospheric Sciences" was published in article by E. Lorenz "Deterministic Nonperiodic Flow" (the Russian translate [3]), launched a new direction in the qualitative theory of dynamical systems. In the work, the notion of "strange attractor" (from the English. Attract—to attract, attract), as a mathematical image of deterministic chaotic oscillations. This discovery was the beginning of the research boom, during which the number of publications devoted to the study of deterministic chaos grew exponentially, reaching several thousand a year. Strange attractors began to be found in the systems of the most diverse nature. The phenomenon of deterministic chaos, which had only recently appeared to be mathematical exotics, turned out to be typical, and inevitable under certain conditions, for most nonlinear systems.

It turned out that the phenomenon of deterministic chaos is universal and is observed at all levels of the organization of matter. Thus, a new scientific paradigm was created, covering all scientific fields. Of course, it is impossible to characterize all meaningful results obtained within the paradigm framework, we only note that in the phase space of dissipative systems, strange attractors are complex sets that demonstrate an increasingly fine structure at different levels of solving the problem of dynamics, and are fractals. This approach allowed us to establish the main scenarios of the transition to chaos (bifurcation sequences): through the period doubling cascade (M. Feigenbaum scenario), intermittency and quasi-periodic modes. Also by a group of researchers from the University of Maryland (USA) in 1984 it was established [4] that the scenario of transition to chaos through quasi-periodic regimes may include, as an intermediate stage, the formation of strange nonchaotic attractors. In the Maryland group interpretation, the term "strange" refers to the geometric structure of the attractor, which is a fractal object, and the term "non-chaotic" indicates the absence of a sensitive dependence of the dynamics on the initial conditions: among the Lyapunov indicators of the attractor there are no positive.

One of the key components of the general paradigm of deterministic chaos is the requirement of nonlinearity of its evolution operator. Usually for dynamic systems with continuous time, this operator is represented by a system of nonlinear differential equations, and for systems with discrete time, by difference equations or logistic-type mappings (discrete analogs of the continuous logistic Verhulst equation). Along with this, examples of definitions of nonlinear evolution operators for discrete structures, as well as requirements for the interpretation of their initial conditions, are unknown to the authors.

2.1.2   Latin Square and Sudoku Rule

The Latin square of the Nth order is called an N × N table filled with N elements of the set M in such a way that in any row and in any column of the table each element of M is exactly found. The classic example of a Latin square of the 3rd order for M set $M = \{A, B, C\}$ given at Fig. 1.

Usually like M set many natural numbers are taken $\{1, 2, ..., N\}$ or set $\{0, 1, ..., N - 1\}$. However, L. Euler, when examining such tables, used the letters of the Latin alphabet, whence the Latin squares got their name. Latin squares exist for any N, whereby the exact formula for the number $L(N)$ of Latin squares N-s order unknown. Exact numbers $L(N)$ defined for N values from 1 to 11. So, for example, from [5] this value is $L(9) = 5,524,751,496,156,892,842,531,225,600 \approx 5.525 \times 10^{27}$.

There are a number of games that use Latin squares, the most common of which— "Sudoku" (from Japan—数独—separate figure). It requires a partially filled table 9 × 9 (Fig. 2) add up to the Latin square of the 9th order, which has an additional property: all nine of its adjacent 3 × 3 sub-squares contain all natural numbers from 1 to 9 once. We will discuss the role of this additional rule below.

Exploring Sudoku as a mathematical object, use the accepted terms for matrix forms, calling the game table a grid. Namely: a grid line, a grid column, a bar—a block line of three adjacent small squares horizontally and a stack—a block column of three adjacent small squares vertically. Of course, the number of exact values Ls(9) of different Sudoku kinds will be less than the number for Latin squares L(9), if only because of L(9), as contradicting the additional rule Sudoku, it is necessary to exclude the number of matrices containing small squares in the form of normalized (or reduced) Latin squares, in which the first row and first column are filled

Fig. 1 The classical
example of Latin square

(a)

A	B	C
C	A	B
B	C	A

(b)

A	B	C
B	C	A
C	A	B

Fig. 2 Example of the
original Sudoku game field

5	3			7				
6			1	9	5			
	9	8					6	
8				6				3
4			8		3			1
7				2				6
	6					2	8	
			4	1	9			5
				8			7	9

in accordance with the order of the M set (Fig. 1b). Given the third (additional) rule, the number of different Sudoku grids, according to [6], decreases to Ls(9) = 6,670,903,752,021,072,936,960 ≈ 6.671×10^{21}.

But quite often we can get one Sudoku grid from another using simple transforms. Two Sudoku grids are called equivalent if we can convert one of them to another by applying one (or more than one) of the symmetries of the set G. If none of the symmetry sequences transforms one of the grids into another, the grids are called essentially different. It is assumed that the symmetry group G is generated by the transformations of the following types.

(1) reassignment of nine digits;
(2) shuffling three stacks;
(3) rearrangement of three bands;
(4) rearrangement of three columns in a stack;
(5) permutation of three lines in a strip;
(6) all secular reflections and turns from the set of symmetries of a square (rotate 0° (neutral element); 90° clockwise rotation; 180° clockwise rotation; 270° clockwise rotation; reflection relative to the horizontal axis passing through the center of the square; отражениеrelative to the vertical axis, also passing through the center of the square; reflection relative to the diagonal of the square from its lower left corner to the upper right; the reflection relative to the diagonal of the square from its upper left corner to the lower right).

Using the methods of group theory taking into account the specified set of symmetries G, the number Lb(9) of essentially different Sudoku grids was found in [6], which is significantly less than the number Ls(9) and is Lb(9) = 5,472,730,538 ≈ 5.473×10^9. Оценки Ls(9) and Lb(9) of fundamental interest in the analysis of possible transformations of matrix structures obeying the Sudoku rules.

Two more types of tasks were also actively studied: the minimum number of initial prompts (numbers pre-arranged on the grid), which leads to a single variant of its filling, and the construction of an optimal algorithm for filling the mesh based on the given prompts. Without considering these types of problems in detail, we point out that mathematicians from Ireland, on the basis of long-term computer simulations, found out [7] that the Sudoku puzzle with a single solution should have at least 17 tips. As for the methods of filling the grid, there are many proposed on the basis of different approaches, most of which are based on computer modeling. One of the most simple methods is presented in the paper [8]. It is interesting to note that in some papers, for example [9], the chaotic behavior of the decision-finding algorithms was observed, which was associated with the rigidity of defining the Sudoku grid structure.

2.1.3 Description of the Mathematical Model of the Discrete System

We will assume that the structure of a discrete dynamical system is given on the Sudoku grid. In this case, the numbers from 1 to 9 can be considered as indicators of an indicator of a quality factor (color, sound, density, heat, area, number of defects or micro-objects, probability, etc.). Figure 3 presents examples of three-dimensional discrete structure models (Fig. 3b) and a puzzle breakdown of a uniform array of objects (Fig. 3c) on a fragment of a small Sudoku square (Fig. 3a).

In our case, as the initial structure, we will consider a square matrix, the type of filling of which represent in Fig. 4. Here the lines are filled with ranked data, taking into account the closure of the sequence of numbers in the cycle, in which, after 9, 1, 2, 3 again goes … This arrangement of numbers clearly demonstrates the strict ordering of a given structure.

Of course, in the case of "digital" modeling, the Sudoku condition expressed in terminological form: "all rows, all columns and all nine adjacent small squares of 3×3 structures contain exactly one positive integer from 1 to 9" in the form of mathematical ratios:

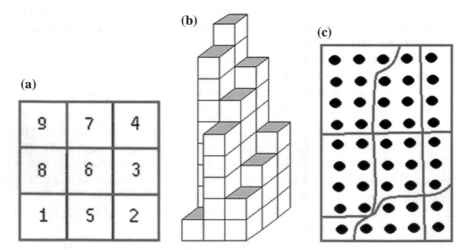

Fig. 3 Examples of discrete structure models

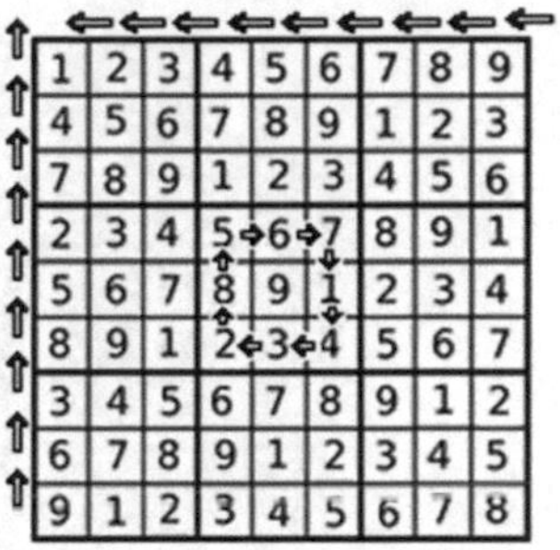

Fig. 4 Examples of discrete
structure models

1	2	3	4	5	6	7	8	9
4	5	6	7	8	9	1	2	3
7	8	9	1	2	3	4	5	6
2	3	4	5→6→7			8	9	1
5	6	7	8	9	1	2	3	4
8	9	1	2←3←4			5	6	7
3	4	5	6	7	8	9	1	2
6	7	8	9	1	2	3	4	5
9	1	2	3	4	5	6	7	8

$$\sum_{i=1}^{9} a_{ij} = 45, \quad \text{at } j = 1, 2 \ldots 9$$
$$\sum_{j=1}^{9} a_{ij} = 45, \quad \text{at } i = 1, 2 \ldots 9; \tag{1}$$
$$\sum_{i,j} a_{ij} = 45$$

where sum of nine elements of each small square, where a_{ij}—are matrix elements.

The three requirements for the structure of a discrete system (1) can be interpreted as a similarity to a system of three differential equations for dynamic systems with continuous time, formulated with respect to three dynamic parameters. Recall that it is in such continuous dynamic systems, in contrast to systems with two dynamic parameters, observations of deterministic chaos modes are possible. Therefore, in the case of the considered discrete structure, in contrast to the classical Latin squares, we can expect the manifestation of a chaotic nature.

2.1.4 Types of Transformation of the Operator of Evolution

To create a full-fledged model of a dynamic system, a given initial matrix structure (Fig. 4), being one of the possible forms of the set $L_s(9)$, it must be supplemented by the introduction of the evolution operator. That is, it is necessary to formulate the rule of dynamics, according to which a discrete system in time steps will change the form of its structural filling to a new form from the set $L_s(9)$, essentially meaning its different states. In other words, the trajectory of changes in the original Sudoku grid in discrete time must be specified. There are also a sufficiently large number of possible implementations for the introduction of the evolution operator. For example, it can be an established sequence of transformations from the symmetry group G, or, on the contrary, the trajectory along the indicated substantially different grids from the set $Lb(9)$, either hybrid versions of these scenarios. In our model, the evolution operator will be specified in the form of group (simultaneous for all digits filling the matrix) displacements of the matrix elements, which will ensure the transformation of the matrix structure at the next step of its evolution. In this case, two types of such

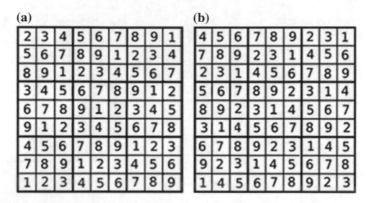

Fig. 5 Type of structure as a result of linear shifts

operators can be realized: linear-type shifters and vortex-type shifters. Consider them with specific examples.

Without loss of generality, we assume that the elements of the first row of the original matrix (Fig. 4) are shifted to the left in the cycle by one position, as shown in Fig. 4 top row of arrows. In this case, digit 2 moves to the first cell of the line, digit 3 to the second, and each subsequent cell to the left of it. In place of 9, there will be 1. Such a cyclic permutation of elements will be called a linear-type shift. It is convenient to describe it in the form of a vector, whose components will be pairs of numbers. The first digit will be the one to be changed, and the second is the digit that is inserted in place of the variable. In our case, such a transformation vector will have the following form: $T\{(1, 2); (2, 3); (3, 4); (4, 5); (5, 6); (6, 7); (7, 8); (8, 9); (9, 1)\}$. We emphasize that the group permutation described by the vector T, applies to all elements of the matrix structure, not only to the elements of the first row. As a result, we obtain a new form of the matrix presented in Fig. 5a.

As you can see, the new matrix structure differs from the original one only by rearranging the first (left) column of the original matrix to its right edge. This is one of the symmetry transformations from the group G, which leads us to the equivalent grid for the original. In fact, such a result of shifting elements in a row was easily predictable. Not quite obvious is the result of cyclic shift in the column. For example, we will select the first column in the initial structure and make a cyclic shift of the elements in it upwards, as shown in Fig. 4 side arrows. In this case, the transformation vector will be: $T\{(1, 4); (4, 7); (7, 2); (2, 5); (5, 8); (8, 3); (3, 6); (6, 9); (9, 1)\}$, and the form of the transformed matrix as in Fig. 5b. Analyzing the structure obtained, it can be noted that it differs from the original permutation of the first (left) stack of the original matrix to its right edge and the (left) stack of the original matrix to its right edge and the following two remapping numbers: 2 and 1, 3 and 1. That is, for the linear shift operator in the column, we obtain a certain sequence of symmetry transformations from the group G, which also leads the original matrix to the equivalent grid.

(a)

7	3	4	1	8	5	6	2	9
1	8	5	6	2	9	7	3	4
6	2	9	7	3	4	1	8	5
3	4	1	8	5	6	2	9	7
8	5	6	2	9	7	3	4	1
2	9	7	3	4	1	8	5	6
4	1	8	5	6	2	9	7	3
5	6	2	9	7	3	4	1	8
9	7	3	4	1	8	5	6	2

(b)

4	8	2	3	6	7	1	5	9
3	6	7	1	5	9	4	8	2
1	5	9	4	8	2	3	6	7
8	2	3	6	7	1	5	9	4
6	7	1	5	9	4	8	2	3
5	9	4	8	2	3	6	7	1
2	3	6	7	1	5	9	4	8
7	1	5	9	4	8	2	3	6
9	4	8	2	3	6	7	1	5

Fig. 6 View of structures as a result of vortex shifts

The second type of evolution operator is defined on the basis of vortex shifts. Without loss of generality, we assume that the elements of the central small square of the initial structure move clockwise around the perimeter of the square around the Fig. 9, as shown in Fig. 4 internal arrows. Here, the transformation vector has the form: $T\{(1, 7); (2, 3); (3, 4); (4, 1); (5, 8); (6, 5); (7, 6); (8, 2); (9, 9)\}$, and the form of the transformed matrix is shown in Fig. 6a. Analysis of the obtained matrix structure shows that it should be attributed to the set $L_b(9)$ of significantly different grids with respect to the original. Note that when choosing any of the 9 small squares of the original matrix to activate the vortex evolution operator, a similar result of the matrix transformation will be provided.

Based on a number of analyzed simulations (when changing rows and columns of activation of the linear shift operator, with a different choice of small squares and directions of activation of the vortex shift operator, as well as changing the matrix representation of the initial discrete structure), it can be stated that:

(1) the introduced operators of the evolution of linear group shifts lead to matrix transformations of Sudoku grids from $[L_s(9) - L_b(9)]$ set of equivalent grids with respect to the original and can be defined as linear operators;

(2) the introduced operators of evolution of group shifts of the vortex type lead to matrix transformations of Sudoku nets from the set $L_b(9)$ substantially different grids with respect to the original and can be defined as nonlinear operators.

Of course, by requiring the execution of a given sequence of entered operators, you can set a specific scenario for the dynamic change of structure states on a discrete time scale. One of the variants of such scenarios of fundamental interest is the implementation of a sequence of cyclic group shifts in the structure. In this case, it becomes possible to build a model of oscillatory processes in discrete structures. Let us verify this by the example of analyzing the sequence of linear shifts of the first line of the initial structure in Fig. 4. We believe that the above transformation with the vector $T\{(1, 2); (2, 3); (3, 4); (4, 5); (5, 6); (6, 7); (7, 8); (8, 9); (9, 1)\}$ will represent the first step in a dynamic scenario, which leads to a simple permutation of the first (left) column of the original matrix on its right edge. Next, we perform successively 8 transformations defined by the same vector, ensuring at each step the implementation of rearranging the left column of the matrix on its right edge. As a result of this cycle of 9 permutations, we return to the original matrix. Further we will execute a cycle from 9 consecutive shifts in the opposite direction with a transformation vector $T\{(2, 1); (3, 2); (4, 3); (5, 4); (6, 5); (7, 6); (8, 7); (9, 8); (1, 9)\}$, which will provide permutations of the right column of the original matrix on its left edge. As a result of a complete cycle of 18 steps, we again return to the original structure. If we further repeat the sequence of application of evolution operators on the time grid several times, we obtain a model of periodic oscillations in a discrete structure.

2.1.5   Observation of Determined Chaos in a Discrete System

To observe the chaotic manifestations in the time change of states of discrete systems, it is impossible to introduce an analog of the phase space, traditionally used in the analysis of systems with continuous time. Moreover, taking into account the uniform scattering of elements (figures) in a discrete structure established by the Sudoku rules, here the probabilistic analysis approaches become ineffective. Therefore, in this work, the principle of preserving the ranking of digital sequences 1.2 … 9.1.2 … in the rows of matrix structures, which was provided when filling out the initial matrix, will be used to indicate the ordering of the structure (Fig. 6). Moreover, one or another change in the order of numbers in the ranked sequences will be interpreted as a violation of the order in the discrete system as a whole, and their unordered distribution over the elements of the matrix structure as a manifestation of deterministic chaos.

According Fig. 5 the linear evolution operators introduced do not change the general order of the original structure (Fig. 5a), either these changes are local (at Fig. 5b the position of the elements with the 1). It can also be argued that when modeling oscillatory processes in discrete structures using linear-group group shift operators, no manifestations of a chaotic nature are observed. These conclusions were confirmed by a series of sampling simulations of oscillatory processes, the schemes of which are shown in Fig. 7 (Fig. 7a—oscillations in the rows of the matrix structure, Fig. 7b—oscillations in the columns of the matrix structure).

The situation is completely different when non-linear group shift operators are used. For example, analyzing the structure in Fig. 6 we no longer observe in the rows of the matrix the fulfillment of the principle of ranking digital values. However, it is clear that if in the first step we apply the transformation vector $T\{(1, 7); (2, 3); (3, 4); (4, 1); (5, 8); (6, 5); (7, 6); (8, 2); (9, 9)\}$, and then (clockwise) continue the sequence of 8 transformations defined by the same vector, we will return to the original ordered matrix form. Moreover, at any step (except the original form), the principle of ranking digital values will not be implemented in the matrices. As an example in Fig. 6b shows the matrix structure at the 8th (last but one in a cycle) step. Therefore, it can be argued that when modeling oscillations of the vortex type (Fig. 7c)

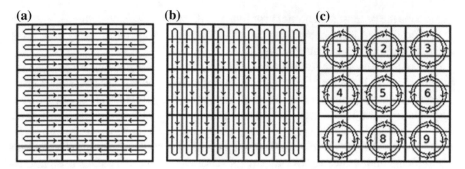

Fig. 7 Simulation of oscillatory processes in a discrete structure

in a discrete structure, the relative order is lost and a manifestation of deterministic chaos is observed. Moreover, 9 cyclic sequences of essentially different Sudoku grids in this case will be similar to the trajectories of strange attractors on a discrete-time scale. Since these attractors cannot be sensitively dependent on the initial conditions, most likely, they must be attributed (using the terminology of researchers from the Maryland group) to the type of strange non-chaotic attractors.

Of course, the theoretical results presented in the section are fundamental, but they can be directly used to develop effective methods for generating keys for block encryption in practical cryptography. It should be borne in mind that the effectiveness of the matrix approach is based primarily on the possibility of simultaneously obtaining permutations (generation round) of a set of equivalent keys for one round. We also emphasize that although only Sudoku grids of 9 × 9 sizes were considered, the results can be easily generalized to similar matrix structures of arbitrary sizes.

2.2 Monitoring of the Effect of Vibrations Beating in Matrix Structures

2.2.1 Description of the Mathematical Model of the Discrete System

In communication schemes, chaos can be used as a carrier of information, as a dynamic process that ensures the transformation of information to a new type, and, finally, as a combination of both. Device, which converts with using chaos information flow in the transmitter from one type to another, called chaotic coder. With it, you can change the information in such a way that it will be inaccessible to an outside observer, but at the same time it will be easily returned to its original form by a special dynamic system—chaotic decoder, located on the receiving side of the communication information channel.

However, the functionality of the use of controlled chaotic algorithms can be significantly extended if any (even for special cases) principles of their synchronization and resynchronization are known [10]. Therefore, the purpose of this section is to generalize the proposed theoretical approach to discrete cellular structures organized in the form of two-level matrix forms, in which the possibility of observing the effects of beating vibrations is manifested.

We assume that the initial structure of a discrete dynamic system is specified on the Sudoku grid in the ranked form shown in Fig. 4. The operators of the evolutionary structure must be transformed into a matrix in which the permutations of its elements (digit) correspond to scenarios of group shifts of a linear or vortex type. Recall that, using the entered evolution operators, simulation of the oscillatory process of a given type for a discrete structure in the form of a chain of cyclic changes of its states can be implemented. However, in order to observe the effect of beating oscillations, according to the general theory of signals, two oscillations must be simultaneously excited in a dynamic system at different frequencies. In this regard, it is necessary to introduce into consideration the cellular structure of a more complex organization.

It is proposed to use as a model of such a discrete structure a layer of cubic cells enclosed between two complementary Sudoku nets of the following type:

Here we will call a pair of grids Sudoku (Fig. 8), which superimposed on each other provide in each cell, located in a layer between them, sum of numbers (or number of inclusions) equal to 10, like in Fig. 9. In initial state the cell structure will be characterized by a uniform distribution of inclusions throughout the cells of the entire layer.

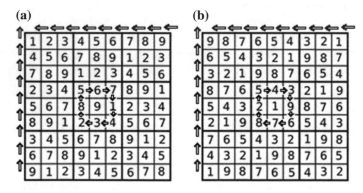

Fig. 8 The initial additional matrix structure

Fig. 9 Fragment of a two-layer matrix form

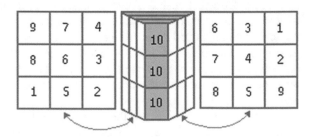

In considered option a two-tier structure, the upper grid is selected in the form given in Fig. 8a, down grid (complementary to the upper grid)—like in Fig. 8b. Then the model will provide the ability to analyze two independent oscillations in each grid structure, as a result of their imposition according to estimates of changes in the homogeneity of filling the cells of the layer between them.

2.2.2   Observing the Effect of Beating Vibrations in a Discrete Structure

Before proceeding to the analysis of the effect of beats for different types of oscillations, it should be noted that the characteristic property of the model used for a discrete structure is com. Without resorting to computer simulation, it is easy to verify that if the same type of oscillations are excited synchronously in time in both boundary grids (that is, described by the same group shift operators), the uniformity of cell filling of the model layer is preserved. Indeed, in this case, the transformed lower grid remains complementary to the transformed upper boundary grid at each discrete time step of their cyclic changes.

To observe the beats, we first consider two oscillations defined on the basis of the linear group shift operator of the first row of the matrix, as shown at Fig. 8a top row of arrows. That is, the evolution operator is applicable to the initial upper grid of the model (Fig. 1a) Tv{(1, 2); (2, 3); (3, 4); (4, 5); (5, 6); (6, 7); (7, 8); (8, 9); (9, 1)}, and to the bottom (Fig. 8b)—operator Tn{(9, 8); (8, 7); (7, 6); (6, 5); (5, 4); (4, 3); (3, 2); (2, 1); (1, 9)}. With that, in order to provide different frequency modes of grid oscillations, here we assume that transformations in the upper grid occur on each discrete interval P of the time scale, and conversions in the lower grid on each pair of such intervals (that is, two times slower). Of course, at the 18th evolutionary step (at $P = 18$), the discrete structure will return to the state of uniform filling of the model layer.

Deviations from the homogeneous state (when exactly 10 objects are in each cell of the structure) will be observed cellular, realizing that the same number of inclusions will be found in the cells of the two-level structure layer with the same upper digit. Therefore, it turns out to be sufficient to observe changes in the number of objects in a 9-layer layer located under the numerical values $N = 1, 2 \ldots 9$ on the upper grid. It is convenient to present the results of observations for analysis in the form of a two-entry table. So in Table 1 shows the number of inclusions in the cells of the two-level structure in time steps $P = 1.2 \ldots 18$ for the considered case of the addition of oscillations.

For clarity some results from the Table 1 are given in a Fig. 10 like temporary dots diagrams. As you see at Table 1 and Fig. 10 it turns out that in the cell structure under $N = 1$ in the first time step, the number of inclusions decreases from 10 to 2, and then gradually increases to 10. Moreover, the increase in the number of inclusions occurs in a multiple-time discrete 2P, and the chart steps are formed by pairs of identical values of the observed number of inclusions. A similar type of oscillations in the number of inclusions is observed in the cell under $N = 9$, where this number gradually increases from 10 to 18, and again at the last two time intervals 17P and

Table 1 Number of inclusions in the cells of the two-level structure

N/P	1	2	3	4	5	6	7	8	9	10	11	12	13	14	15	16	17	18
1	2	2	3	3	4	4	5	5	6	6	7	7	8	8	9	9	10	10
2	11	11	3	3	4	4	5	5	6	6	7	7	8	8	9	9	10	10
3	11	11	12	12	4	4	5	5	6	6	7	7	8	8	9	9	10	10
4	11	11	12	12	13	13	5	5	6	6	7	7	8	8	9	9	10	10
5	11	11	12	12	13	13	14	14	6	6	7	7	8	8	9	9	10	10
6	11	11	12	12	13	13	14	14	15	15	7	7	8	8	9	9	10	10
7	11	11	12	12	13	13	14	14	15	15	16	16	8	8	9	9	10	10
8	11	11	12	12	13	13	14	14	15	15	16	16	17	17	9	18	10	10
9	11	11	12	12	13	13	14	14	15	15	16	16	17	17	18	18	10	10

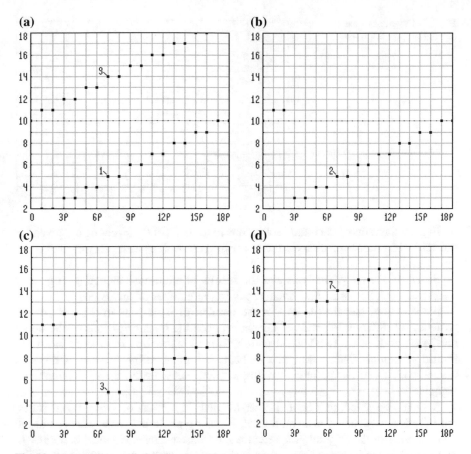

Fig. 10 Multiplicity oscillation diagrams 2P

18P it becomes equal to 10. These two types of oscillations, by analogy with the general theory of signals, can be defined as the main (envelope time dependencies for the amplitude of the resulting signal). In the remaining cells of the structure, oscillations of the hybrid type are observed. So, for the cells under $N = 2$ (Fig. 10b) on the first three temporary oscillation intervals correspond to the mode in the cell under $N = 9$, and on the rest—the mode of oscillation in the cell under $N = 1$.

Next for the cells under $N = 3$ (Fig. 10c) in the first five time samples, the oscillations correspond to the mode in the cell under $N = 9$, and on the rest— synchronized with the oscillation mode in the cell under $N = 1$. Similarly, oscillation modes and other hybrid types, for example, at cell under $N = 7$ (Fig. 10d).

As the simulation results showed, the principle of oscillation formation under the considered evolution operators is also preserved under other variants of the lower grid oscillation delays relative to the evolution rate of the oscillations in the upper grid of the two-level structure. In this case, the diagrams of the main oscillations are complicated with the manifestation of periodic fragments in their form. For example

Fig. 11 Diagrams of basic
oscillations of multiplicity
3P

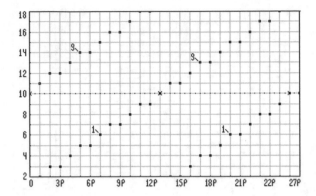

on Fig. 11 diagrams of the main oscillations in the cells of the structure are given for
the case when transformations in the lower grid occur three times slower than in the
upper grid.

Based on the above analysis and the results of additional modeling, a general
conclusion can be made, that fluctuations in the number of inclusions in the cells
of the structure, that fluctuations in the number of inclusions in the cells of the
structure, at different intervals of delays between the transformations of the upper
and lower grids of the two-level structure, formed in the form of ordered basic and
hybrid vibrations. Such a structure of beats is similar to the process of beats as
a superposition of oscillations at multiple frequencies in the theory of harmonic
signals.

It should be noted, that Sect. 1.2 established: as a result of group linear shifts,
the introduced operators of evolution of a linear type do not make changes to the
general order of the original grid structures, or these changes are local in nature. In
this regard, it can be argued that since in the case considered, the original matrices
were filled with ranked sequences of digits (Fig. 8), then ordered oscillations were
observed in the cells of the two-level structure.

When considering oscillations specified for boundary grids by vortex-type evolution operators, we should expect the implementation of more complex scenarios
for fluctuations in the number of inclusions in the structure cells, since, in this case,
elements of deterministic chaos may appear. It means that with the initial filling of
boundary grids with ranked sequences of numbers, one should expect unordered
(chaotic) forms of oscillation beats. These assumptions are fully confirmed by the
results of multiple simulations. For example in Table 2 are given the simulation
results for the same source grids (Fig. 8) in the mode of oscillations given by two
vortex-type group shift operators. This was supposed to, for top grid transformation
vector has the form: Tv{(5, 8); (6, 5); (7, 6); (8, 2); (9, 9); (1, 7); (2, 3); (3, 4); (4, 1)},
for bottom grid: Tn{(5, 2); (1, 5); (3, 4); (2, 8); (1, 1); (9, 3); (8, 7); (7, 6); (6, 9)}. Get
sight of that under the action of both operators, elements of the central small squares
of the original grids are shifted clockwise around the central numbers 9 and 1 along
the perimeter of the squares, respectively. It was also assumed that transformations in
the upper grid occur on each discrete interval P of timeline, and the transformations

Table 2 The simulation results for the same source grids

N/P	1	2	3	4	5	6	7	8	9	10	11	12	13	14	15	16
1	7	7	8	8	9	9	3	3	6	6	5	5	4	4	10	10
2	4	4	7	7	6	6	5	5	11	11	8	8	9	9	10	10
3	11	11	5	5	8	8	7	7	6	6	12	12	9	9	10	10
4	11	11	12	12	6	6	9	9	8	8	7	7	13	13	10	10
5	9	9	8	8	14	14	11	11	12	12	13	13	7	7	10	10
6	9	9	15	15	12	12	13	13	14	14	8	8	11	11	10	10
7	16	16	13	13	14	14	15	15	9	9	12	12	11	11	10	10
8	13	13	12	12	11	11	17	17	14	14	15	15	16	16	10	10
9	10	10	10	10	10	10	10	10	10	10	10	10	10	10	10	10

in the lower grid on each pair of such intervals (that is, two times slower). Since here the central figures of the small squares of the boundary grids are fixed, the discrete structure returns to the state of uniform filling of the model layer at the 16th step of evolution (at $P = 16$).

As seen from Table 2 fluctuations in the number of inclusions in all cells of the structure and at all steps of evolution, really, are disordered and their beats are chaotic. Like the general theory of signals, such oscillation beats can be defined as noise-like.

Despite the general theoretical nature of the concepts of beats in matrix structures, These results can be directly applied in cryptographic applications. For example, in solving problems: authentication of subscribers, organization of electronic databases, storage for secret keys, keys generation with a "double secret" or hidden key transfer via an open communication channel.

3   The New Method of Cryptographic Protection of Information Channels

The task of cryptographic protection of information channels was considered as part of an initiative project with the working title HiSNeC (High Speed Network Coder), the main results of which will be presented in this section. The project can be attributed to the development of technologies of the so-called Lightweight Cryptography (LWC). Moreover, the statement of the problem was based on the following a priori provisions:

(1) in the system of protection of the information channel should be involved autonomous program-controlled transceiver devices that perform encryption, decryption and other (necessary to protect the exchange of information) functions;

(2) the information security system should be built on the basis of the generally accepted, reliable, which has withstood the critical study by experts and has passed the time-tested AES standard. Advanced Encryption Standard (original name Rijndael)—an advanced encryption standard based on a symmetric block-encryption algorithm, established by the National Institute of Standards and Technology (NIST) in 2001;

(3) the private keys used in the process of information exchange should be stored in the program memory of the devices, and be inaccessible to persons using them. Compliance with this requirement is provided by most modern processors, which allow, after programming, to close access to the program code (reading the code is impossible even if the device is hacked);

(4) when transferring conditionally static (slowly changing over time) data blocks, it should be possible to synchronously modify keys in the data exchange system. Moreover, the modifier transmitted for key synchronization must not exceed one percent of exchange traffic;

(5) if it's necessary, to protect against theft of devices or their loss, additional means
     of user authentication (fingerprint identification, voice identification, etc.) may
     be included in the device composition.

Based on the analysis of these provisions, the main goal of further research was developed: the creation and practical implementation of a key generation method for symmetric block fast coding algorithms used in software and hardware security tools and fully compatible with the AES standard formats.

3.1   Description of the Method of Key Generation for Symmetric Block Algorithms for Encryption

It is advisable to recall that most of the existing block encryption algorithms are actively used for generating data encryption keys: substitutions (replacing some data elements with others when establishing a one-to-one correspondence between the sets containing these data) and permutations (changing the order of the data elements).For example, the method proposed in the patent of the Russian Federation № 2309549 [11] is based on the hardware implementation of controlled permutation operations—cryptographic primitives, which makes it possible to significantly speed up the encryption processes. However, it should be borne in mind that it is characteristic of such high-speed ciphers to use precomputations, including performing and expanding the secret key. In this case, the requirements for the frequency change of keys conflict with the high-speed use of ciphers based on precomputations, since the need to repeatedly execute the latter introduces significant limitations in speed. In this connection, it becomes very important to reduce the complexity of predictions (or reject them) while maintaining the high cryptographic strength of the transformations. Thus, the development of high-speed ciphers of the new generation, based on new ways of generating dynamic keys, allowing both cost-effective hardware implementation and retaining high encryption speed with frequent key changes, is practically significant. It should also be borne in mind that, due to the specifics of information representation in digital devices, block ciphers are of the greatest interest, allowing encoding information without changing its structure, that is, allowing to encrypt information that is pre-structured according to any protocol protection method.

Of course, any block cipher must comply with the three basic principles:

- diffusion, in which a change in any character of the plaintext or key affects a
  large number of ciphertext characters, which hides the statistical properties of the
  plaintext;
- confusion, that is, the use of transformations that make it difficult to obtain
  statistical dependencies between ciphertext and plaintext;
- enabling work with blocks and keys of different lengths.

In this regard, the disadvantages of the known prototypes include:

- the fixed structure of the network of control units of operations, which corresponds to a specific key size (data block). When changing the key size, the development of a new structure is required;
- this is a fixed-line size (data block). The structure of a new structure is required;
- the need to store information about the current round keys and the schedule for their use;
- complex organization of the network of operations management units and a special selection of information that controls the operation of this network to achieve the required strength of the encryption procedure, determined by the uniform distribution of codes and non-linearity of multi-round permutations.

Considering the listed requirements and eliminating the indicated disadvantages allows the new method proposed by the authors to build cryptographic primitives on the basis of operations of nonlinear transformation of matrix structures, described in theoretical part. Note that this method was protected by the patent of Ukraine [12]. Next, we emphasize the key points of its software implementation.

Let there be two finite sets Ω_1 and Ω_2, each of which contains n different elements. The permutation of σ is a one-to-one mapping Ω_1 at Ω_2. For example, if $\Omega_1 = (1, 2, 3)$, $\Omega_2 = (7, 11, 9)$, then one of the six possible substitutions will be:

$$\sigma = \left(\frac{1, 2, 3}{7, 11, 9} \right) \tag{2}$$

If the number of elements in each of the sets—n, then the number of possible substitutions is equal to factorial n (that is, each element of the Ω_1 set can match any element from Ω_2 set). In the particular case of the set Ω_1 and Ω_2 may match. A one-to-one mapping of a set onto itself will be called *rearrangement* (π). For example, if $\Omega = (1, 2, 3)$, then exist 6 (or 3!) transpositions:

$$\pi_1 = \left(\frac{1, 2, 3}{1, 2, 3} \right), \quad \pi_2 = \left(\frac{1, 2, 3}{2, 3, 1} \right), \quad \pi_3 = \left(\frac{1, 2, 3}{3, 1, 2} \right) \tag{3}$$

$$\pi_4 = \left(\frac{1, 2, 3}{1, 3, 2} \right), \quad \pi_5 = \left(\frac{1, 2, 3}{2, 1, 3} \right), \quad \pi_6 = \left(\frac{1, 2, 3}{3, 2, 1} \right) \tag{4}$$

Define a Ω set like set of natural numbers 1, 2, ..., n. Latin square L will be represented as a square table $n \times n$, in each row and column can in each row and in each column of which every element of the Ω set occurs exactly once. Choose the n value of so that $n = k^2$. Then the Latin square L with size $n \times n$ can be dividing into n adjacent, non-intersecting squares of size $k \times k$. We additionally require that in each small square $k \times k$ any element of the set Ω also occurs exactly once. Such a Latin square is called S-square. If $k = 3$, then we get the Sudoku grid squares 9×9 (with 9th small squares 3×3), characteristics which were given in the previous section. Small squares S-square have size $k \times k$ and contain $n = k^2$ elements.

Let built S-square order $n = k^2$. Select one of the small squares and assume that it is filled as shown at Fig. 12a. This arrangement is described by a permutation.

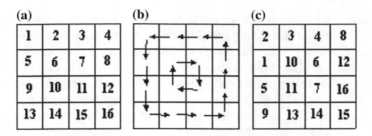

Fig. 12 DAn example of the permutation of small square elements for $k = 4, n = 16$

$$\pi_1 = \left(\frac{1, 2, 3, 4, 5, 6, 7, 8, 9, 10, 11.12, 13, 14, 15, 16}{1, 2, 3, 4, 5, 6, 7, 8, 9, 10, 11.12, 13, 14, 15, 16} \right), \tag{5}$$

If you rearrange the elements of the square by the vortex-shift algorithms as shown at Fig. 12b, we get the permutation shown at Fig. 12c:

$$\pi_2 = \left(\frac{1, 2, 3, 4, 5, 6, 7, 8, 9, 10, 11.12, 13, 14, 15, 16}{2, 3, 4, 8, 1, 10, 6, 12, 5, 11, 7, 16, 9, 13, 14, 15} \right).$$

Here permutations of cyclic type are used, which allow to provide various essentially matrix transformations.

At Fig. 12b can be distinguished of two cyclic permutations: the first is counter clock wise advancement of border cells and clockwise advancement of the central cells of the table. Matching the bottom lines of permutations π_1 and π_2 sets the rules for reassigning the numbers of a small square when performing a permutation π_2. Indeed, 1 changes by 2, 2—by 3, …, 11—by 7, …, 16—by 15. Having simultaneously carried out the same replacement of numbers in all small squares, we obtain a new matrix form of a large S-square.

Further, we assume the initial cell numbering of small squares of the matrix structure in accordance with a Fig. 12a. Then any permutation of the elements of a small square can be described by a vector of dimension 16. We agree that the record $\pi = (2, 3, 4, 8.1, 10, 6, 12, 5, 11, 7, 16, 9, 13, 14, 15)$ limits the following actions: the element from cell 1 is rearranged to cell 2, from cell 2 to cell 3, …, from 6 to 10, from 7 to 6, …, from 16 to 15. Thus, the sequence number of the vector coordinate indicates from which cell the element should be taken for permutation. The vector coordinate value determines which cell the selected element should be placed in.

In order to flesh out the presentation, we will assume that the encryption key is generated for the 128-bit version of the AES cryptoalgorithm. In this regard, we choose as the initial S-square some square $S_0(n = 16)$. We define the set of permutation operations by defining permutation vectors of dimension 16. We call these permutations basic. The number of such vectors is advisable to choose equal to degree 2. In the experiments conducted by the authors, this number m is 32. The operations table (T_{op}) above S-square swill have $n \times m$ size (in the author's version of the simulation used Table 16 × 32). In the general case, the structural components (rows, columns, or small squares) of S-squares matrices obtained as a result of these operations can be viewed as generating new sets of coding keys or their components. Here it is important to emphasize that for one round of collective permutations, the possibility is realized of the simultaneous generation of not one, but many ("bundles") of encryption keys. Moreover, these generations are carried out without using any additional mathematical procedures. Moreover, knowing only the sequence (schedule) of operations used T_{op} eliminates the need to store tables S-squares after each operation performed, unlike the currently used multi-key generation technologies. The listed features of the key generation method, first of all, allow declaring its significant advantages over prototypes in the speed of implementation of the coding and decoding processes of information flows.

In order to "disguise" the transmission of coded sequences (albeit random) from the open channel of communication with constantly used natural numbers (from 1 to 16), square tables are additionally formed R_0 and R_i with n^2 size. Moreover, the matrix R_0 must be pre-formed using a random number generator (with a given uniformity of distribution) from the range 0 to 255. For further receipt R_i from R_0 matrix the same table of operations will be used T_{op}, which will now determine the specific permutations of the locations of the cells of this matrix along with the random numbers contained in them.

Thus, in the present embodiment, the proposed key generation algorithm combines the following sequence of steps:

(1)   calculate the composition M_r of matrix R_0 and R_i (256 modulo 256 operations are used);
(2)   calculate the sums s_r of M_r table elements modulo m;
(3)   using s_r like a pointer, a permutation vector is selected from the table T_{op};
(4)   the selected permutation operation is performed on the S_0-square, as a result of which a new S_i-square is formed;
(5)   according to a given rule, from S_i, either a row (one of 16 rows), a column (one of 16 columns), or a small square (one of 16) is selected. The resulting 16 numbers form a permutation vector. Note that in the model implementation, the authors used only these 48 options, although there is no reason to use other options as well;
(6)   to form a sample of the matrix M_r an auxiliary function is built $G(i, R)$, the result of which is the assigned sequence of 16 matrix elements M_r, which form the current dynamic key for the AES encryption algorithm.

Under the conditions specified above, in accordance with the above algorithm, 48 $\times$ 256 = 12,288 encryption keys can be generated. Further, the following sr pointer to the T_{op} table can be used, and repeating steps 3–6. This will ensure the generation of the next set of 12,288 keys. Thus, the use of 32 permutation vectors contained in the T_{op} table generates a set of 48 $\times$ 256 $\times$ 32 = 393,216 essentially different encryption keys, which allows encrypting 393,216 blocks of 16 bytes each (over 6 MB of information). To use the considered algorithm, it is necessary to "close" the initial S_0-square and operation tables T_{op}. The random value stable R_0 is stored in the RAM of processors and can be updated again if necessary. Of course, the matrix

S_0, T_{op} and R_0 must be the same (specified) for the entire group of devices that will exchange coded information.

In the practical implementation of the protection of the transmission channel information between two devices T_t (transmitter) and T_r (receiver) the following sequence of procedures was carried out:

(1) transmitter T_r turns to the receiver T_r, sending a transfer request;
(2) upon receipt from T_t permission to transfer T_r generates a table of random numbers R_0 and transmits it to the receiver;
(3) the receiver and transmitter simultaneously calculate the matrix M_r;
(4) the receiver and transmitter, in accordance with the above algorithm, prepare the synchronous generation of identical keys: T_t for coding, and T_r for decoding information flow. We emphasize that in this case the devices of different groups (with different S_0, T_{op}, R_0) will not be able to share information. The same can be said about the attempt of unauthorized access to the communication session, since obtaining valid keys in real time is possible only when using the original S_0, T_{op} and R_0. These parameters, in accordance with the foregoing, are inaccessible not only to the hypothetical "attacker", but also to the user of the communication channel security device.

3.2   Competitive Advantages of the Proposed Cryptographic Method

The choice of a specific encryption method is a key element in the implementation of cryptographic information protection technology. The key, in the sense that it is he who is the determining factor of all the functional capabilities of the coding technology being implemented: program-procedural, resource, technical, economic and other. In this regard, the presentation of the specific advantages of the chosen encryption method (and the technology itself) becomes systemic. For this reason, the authors decided to put this presentation for *HiSNeC* technology into a separate paragraph, which will allow the interested reader to form a more comprehensive assessment of the proposed innovation.

Comprehensive description of *HiSNeC* technology implementation features and its differences from the existing prototypes we will present in the following three areas.

1. The procedural nature questions:

   (1.1) the encryption algorithm is based on the principles of a new, previously not used in prototypes, theory of discrete mathematics, developed for the study of the manifestation of deterministic chaos in the structures of the matrix type;

(1.2) the method of generating encryption keys realizes the possibility of simultaneous generation of a set of keys, is carried out without using any additional mathematical operations and eliminates the need to store the entire amount of current data, in contrast to the known multi-round technologies. These features allow you to declare the significant advantages of the new method of generating keys over prototypes in the speed of implementation of the processes of encoding and decoding information;

(1.3) key generation method allows you to get keys with different sizes and requires any additional modifications;

(1.4) the cryptographic strength of the encoding algorithm is based on the use of a huge scale of significantly different matrix forms, a priori ensuring uniformity of diffusion and mixing of data in code sequences, as well as the possibilities of frequent dynamic change of significantly different keys;

(1.5) performing encoding (and decoding) of information in autonomous block devices eliminates any adaptation requirements for the software environment used in communication networks;

(1.6) the technology can be used for any formats of information flows in conjunction with the original protocols used to protect their transmission without any modification.

2. The resource and technical nature questions:

(2.1) the autonomy of the functioning of microcontrollers in the encoder modules eliminates the need to allocate additional amounts of processor resources belonging to communication subscribers;

(2.2) the use of cryptographic protection technology in existing communication channels does not require additional configuration and procedures for the technological regulation on the part of network providers;

(2.3) application of the technology ensures the profitability of simply replacing the encoder modules with new ones in the event of their emergency failure.

3. The economic efficiency questions:

(3.1) key generation method can be directly used to generate cipher keys Rijndael (AES standard). This eliminates the introduction of cryptographic technology in existing information systems (or management systems) the need to spend money on additional licensing procedures or standardization of the original products;

(3.2) software closeness of the encoder modules when implementing the technology does not require additional retraining of network support personnel;

(3.3) the technology can be successfully implemented on the existing element base of mass-produced microprocessors, which, when implemented, does not require significant production reorganizations;

(3.4)   technological simplicity of manufacturing autonomous encoder modules
and the price availability of the element base is the basis of the low cost
of their production.

4   The Testing Results

This section of chapter will present the results of test modeling and mock research, confirming the possibility of successful implementation in practice of *HiSNeC* cryptographic technology.

4.1   The Results of the Preliminary Statistical Testing of Code Sequences

Here, the most common primary statistical test method for block ciphers was used, which consists in generating pseudorandom sequences using an encoder, which are then monitored using a set of statistical tests.

For the considered encoder, in each test control of the generated sequence (as a result of encoding electronic materials of a training manual with different format fragments), the problem of testing the hypothesis of the randomness of this sequence was solved. The fact that this sequence is random was declared as a null hypothesis, and the opposite was considered as an alternative hypothesis—that the sequence would not be random. At the same time, a significance level was set equal to 0.05 (that is, the probability of a false-negative result). In each of the tests conducted on the basis of the data, the probability P of the event was calculated, which consisted in the fact that the ideal random sequence generator would generate a "less random" sequence than the control. This comparison was made using the statistics of the test series, and the probability P was determined using the additional error function. Note that the testing used code sequences with a length of 1,000,000 bits (as recommended by the NIST STS).

300 code sequences were subjected to test controls, of which 296 successfully passed the comparison test. It turned out that the obtained probability value (in each case) of the probability P is uniformly distributed over the entire testing scale. Both of these facts together allow us to conclude that the proposed encoder has successfully coped with the primary statistical test control.

A test simulation of the encryption and decryption process was carried out using a computer with a 4-core Core i7 processor. Encryption and data transfer via ETHERNET (server and client on the same computer) was used. Multi-threaded programs were used for client and server. In one of the streams, data was transmitted (received), and in the other, dynamic key generation and block encryption. An experiment was also conducted with data encryption and transmission over wireless

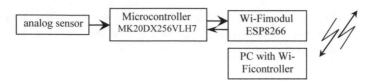

Fig. 13 Experiment on the implementation of the closed channel communication in an open computer network

communication. The scheme of the experiment is shown in Fig. 13. Data processing was performed using the controller MK20DX256VLH7 and communicational Wi-Fi controller ESP8266. Held repeatedly daily experiments showed high reliability of the closed communication channel and the failure-free operation of the data protection algorithm.

Also, an experiment was conducted to determine the time of data encryption using the proposed method for generating keys in conjunction with the AES algorithm. The experiment showed an insignificant (0.3%) increase in CPU time when using the proposed method of generating keys and dynamically changing the key when encrypting each 16-byte block compared to using the AES algorithm with a permanent private key.

4.2   The Model Experiments Results

With the aim of comprehensive experimental testing of the technology *HiSNeC* real prototyping of the radio-channel crypto-protection system of the vehicle model control for the game was carried out (as a variant of the mobile object of small robotics) at Fig. 14.

The command coding module was located in the control panel, and the decoding module was built into the frame of the car model. Both encoder modules were made on the basis of microprocessors of the AT91SAM7S128 type by firm ATMEL.

Selected microprocessors are characterized by a maximum clock frequency of 50 MHz, the amount of RAM 16 Kbytes and flash-memory 128 Kbytes. Micro modules NRF24L01 by Nordic Semiconductor firm used for functional radio channel support of strip type antennas prototyping at 2.4 GHz. At this frequency, which is standard for communication with low- or medium-power devices, a data transfer rate of 2 Mbit/ec was performed for the radius of action of the radio channel (in closed areas) of the order of 40–50 m. The micromodules and microprocessors connection implemented via SPI interfaces.

The operator control panel was equipped with a push-button joystick with the usual set of 6 commands for moving the model: forward-straight, forward-left, forward-right, back-straight, back-left and back-right. Note, that when coding commands, the R_0 matrix was required to be updated each time the radio channel was turned on, and the dynamic update of the command code sequences for each newly received

Fig. 14 Model of radio-controlled car for games

command (even for multiple commands of the repeated motion of the model). Also in the experimental sample was implemented the possibility of online visualization of existing code sequences on the PC screen.

A series of experiments fully confirmed the correctness of the functioning of the technology *HiSNeC* implemented in practice, the system of software and hardware cryptographic protection of the radio channel management of the mobile object.

5 Conclusion

This chapter presents the basics of a new theoretical approach for describing deterministic chaos in the oscillations of discrete structures (expanding the capabilities of modern discrete mathematics), describes the characteristics of the implementation of a cryptographic method for software and hardware to protect communication channels in open environments (based on this theoretical platform), as well as practical results of modeling and prototyping, carried out in the framework of the author's initiative project (named by *HiSNeC*).

In the theoretical part, the possibility of manifestation of modes of deterministic chaos in the oscillations of discrete structures formed in the form of special matrix forms is investigated (like variants of Latin square)—Sudoku grids. Are introduced the mathematical model for describing the evolution of the states of such discrete systems is proposed, the concepts of the shift operators of the transformation of matrix structures. Based on this model are analyzed the requirements for discrete structures and transformations of their evolution operators, which ensure the achievement of deterministic chaos regimes. In particular, it was justified to use for this purpose vortex type shift operators. The results of simulation modeling of oscillatory processes according to the cycles of states of a discrete system, forming similarities of attractor trajectories, are presented. The possibility of simulating oscillation beat modes in discrete cellular structures organized in the form of two-level matrix forms is shown. The results of such a simulation of oscillation beats in cycles of varying states of a discrete system for two types of beats are presented: similar to the result of the superposition of harmonic oscillations at multiple frequencies in the theory of radio signals, as well as noise-like beats. The latter can be used in solving critical problems of cryptographic systems: hidden authentication of subscribers, organization of electronic databases-storage for secret keys, generation of keys with a "double secret" or hidden transfer of a key via an open communication channel.

The chapter presents the principles of creating coding devices (of LWC) for implementing a closed communication channel in an open data transmission medium. To solve the protection problem, the cryptoalgorithm of the AES standard was used, supplemented by the author's system of dynamic generation of encryption keys. The description of the encryption key generation system based on the combinatorial properties of matrix structures in the form of Sudoku grids is given. Are given the characteristics implementation of the new cryptographic method and its main competitive advantages over the existing prototypes. At the same time, a comprehensive description of the advantages of the author's method is presented in three areas: program-procedural issues, issues of resource-technical and economic efficiency. Note that here the comparisons are qualitative, since their quantitative assessments can be obtained only from the analysis of the specific task of implementing the cryptographic protection system.

In order to verify the cryptocorrectness and performance of the proposed technology, both computer tests of the encryption algorithm and the actual prototyping of the radio channel control system of the car model for the game were carried out (as a variant of a small robot mobile object). The results obtained in test studies and in a series of experiments completely confirmed the correctness of the functioning of the *HiSNeC* technology with the practical implementation of software and hardware cryptographic protection.

References

1. Mołdovyan AA et al (2002) Cryptography: high-speed ciphers. BHV-Peterburg, 496 p. (in Russian)
2. Andronov AA, Vitt AA, Haykin SE (1981) Oscillation theory. Moscow, Nauka, 568 p. (in Russian)

3. Lorenz E (1981) Deterministic non-periodic flow. In: Strange attractors, Moscow, Mir, pp 88–117. (in Russian)
4. Grebogi C, Ott E, Pelikan S, Yorke JA (1981) Strange attractors that are not chaotic. Phys D, Nonlinear Phenom 261–268
5. McKay BD, Wanless IM (2005) On the number of Latin squares. Ann Combin 335–344
6. Russel, Jarvis AF (2019) Mathematics of Sudoku II. www.afjarvis.staff.shef.ac.uk/sudoku/russell_jarvis_spec2.pdf, last accessed 2019/11/21
7. McGuire G, Tugemann B, Civario G (2013) There is no 16-Clue Sudoku: solving the Sudoku minimum number of clues problem, 43
8. Crook JF (2009) A pencil-and-paper algorithm for solving Sudoku puzzles. Not Am Math Soc 460–468
9. Ercsey-Ravasz M, Toroczkai Z (2012) The chaos within sudoku. Sci Rep 2:725. https://doi.org/10.1038/srep00725
10. Pikovsky A, Rosenblum M, Kurths J (2002) Synchronization: a universal concept in nonlinear sciences. Cambridge University Press, 411 p
11. RF patent No. 2309549 dated 10.27.2007. The method of cryptographic conversion of digital data
12. Patent of Ukraine for utilitiy model No. 129836 dated November 12, 2018 "A method for generating keys for symmetric block encryption alogrithms"

Output Feedback Encryption Mode: Periodic Features of Output Blocks Sequence

Alexandr Kuznetsov⦿, Yuriy Gorbenko⦿, Ievgeniia Kolovanova⦿, Serhii Smirnov⦿, Iryna Perevozova⦿, and Tetiana Kuznetsova⦿

Abstract We investigate periodic characteristics of sequence of output blocks in the output feedback encryption mode. The model of random homogeneous substitution is used for an abstract description of this formation. This property is directly related to the periodic properties of output feedback encryption mode, since it characterizes the probabilistic distribution of output blocks with certain period appearance, provided that the assumption is made that the properties of the block symmetric cipher are consistent with certain properties of the random substitution. Also in the work specific practical tasks are solved, namely recommendations are being developed for the application of the outbound feedback on the encryption threshold, certain requirements and limitations are justified.

Keywords Encryption mode · Random substitution · Periodic characteristics of output blocks · Output feedback

A. Kuznetsov (✉) · Y. Gorbenko · I. Kolovanova · T. Kuznetsova
V. N. Karazin Kharkiv National University, Svobody Sq., 4, 61022 Kharkiv, Ukraine
e-mail: kuznetsov@karazin.ua

Y. Gorbenko
e-mail: gorbenkoU@iit.kharkov.ua

I. Kolovanova
e-mail: e.kolovanova@gmail.com

T. Kuznetsova
e-mail: kuznetsova.tatiana17@gmail.com

S. Smirnov
Central Ukrainian National Technical University, Kropyvnytskyi, Ukraine
e-mail: smirnov.ser.81@gmail.com

I. Perevozova
Ivano-Frankivsk National Technical University of Oil and Gas, Ivano-Frankivsk, Ukraine
e-mail: perevozova@ukr.net

© The Editor(s) (if applicable) and The Author(s), under exclusive license to Springer Nature Switzerland AG 2021
T. Radivilova et al. (eds.), *Data-Centric Business and Applications*, Lecture Notes on Data Engineering and Communications Technologies 48,
https://doi.org/10.1007/978-3-030-43070-2_27

1 Introduction

One of the most common block symmetric encryption modes used to provide confidentiality services is Outbound Feedback (OFB) [1–7]. The OFB has several advantages [8–15]: firstly, the output block can be formed in advance, even before the message, which can greatly speed up the process of protecting information; and secondly, in this mode, as in the mode Electronic Codebook—ECB, errors that can arise when transmitting ciphertext over the communication channels, are localized in the block, not extending to the neighboring, and in the OFB mode only changed bits will be false (in the ECB mode the entire block will change). Thirdly, the cryptographic properties of the output block do not depend on the open text, they are determined only by the properties of the base cryptographic transformation, and, possibly, by the value of the initialization block, which determines the specific form and frequency of the output block [16–23]. This work is devoted to study of the periodic properties of output block in OFB mode, because the occurrence of output block repetition is the most dangerous case, it gives an attacker the possibility to violate the established mode of message confidentiality.

We derive a formula for estimating the probability of a cycle occurrence for an arbitrary fixed value of a set of transformations. This property is directly related to the periodic properties of the output blocks, since it characterizes the probabilistic distribution of the output block with certain period appearance, provided that the assumption is made that the properties of the block symmetric cipher are consistent with certain properties of the random substitution [24–35]. Also in the work specific practical tasks are solved, namely recommendations are being developed for the application of the Outbound Feedback on the encryption threshold, certain requirements and limitations are justified. The conclusions summarize and concretize our results, discuss possible ways of further research.

2 Output Feedback Encryption Mode

The output feedback encryption mode is intended to provide a confidentiality service. This mode is based on encryption of the initialization vector to generate a sequence of output blocks that are added to the normal text to form encrypted text and, conversely, to the ciphertext to decrypt it. This mode requires unique initialization vector for each application with the provided (fixed) key. Let's consider the specification of output feedback encryption mode.

The parameters of the mode are encryption key K, $|K| = k$, and initialization vector S, $|S| = l$. Additional requirements for initialization vector are not imposed. When encrypting the message ($|M| \geq 1$) is presented as a sequence of blocks:

$$M = m_1||m_2|| \ldots ||m_n, \ |m_i| = l \text{ for } i = 1, 2, .., n - 1, \ 1 \leq |m_n| \leq l.$$

The initial value of the output block γ_0 ($|\gamma_0| = l$) is calculated as

$$\gamma_0 = T_{l,k}^{(K)}(S). \tag{1}$$

Each ciphertext block is calculated according to the ratio

$$c_i = m_i \oplus L_{l,|m_i|}(\gamma_{i-1}) \text{ for } i = 1, 2, .., n, \tag{2}$$

and

$$\gamma_i = T_{l,k}^{(K)}(\gamma_{i-1}) \text{ for } i = 1, 2, .., n-1. \tag{3}$$

The result of message encrypting is ciphertext $C = c_1||c_2|| \ldots ||c_n$.

When decrypting, the ciphertext C ($|C| \geq 1$) is presented in the form of a sequence of blocks:

$$C = c_1||c_2|| \ldots ||c_n, \ |c_i| = l$$

for $i = 1, 2, .., n-1, 1 \leq |c_n| \leq l$.

The initial value of the output block γ_0 ($|\gamma_0| = l$) is calculated as

$$\gamma_0 = T_{l,k}^{(K)}(S).$$

Each message block is calculated according to the ratio

$$m_i = c_i \oplus L_{l,|m_i|}(\gamma_{i-1}) \text{ for } i = 1, 2, .., n, \tag{4}$$

and

$$\gamma_i = T_{l,k}^{(K)}(\gamma_{i-1}) \text{ for } i = 1, 2, .., n-1.$$

The result of the decryption of ciphertext is the message $M = m_1||m_2|| \ldots ||m_n$.

The encryption and decryption scheme in the output feedback encryption mode is shown in Fig. 1. This scheme formally depicts the sequence of execution of transformations (2), (4) for all values of the cyclic variable $i = 1, 2, .., n$.

Let's consider the periodic characteristics of output feedback encryption mode. First, we note that, by definition, the output block consists of the initial value of the block γ_0 and the remaining blocks γ_i, which are calculated for (1), (3) for each value of the cyclic variable $i = 1, 2, .., n-1$. That is, the task of the research is precisely in determining the period of the sequence of blocks γ_i, $i = 0, 1, .., n-1$.

Each output block γ_i is the result of encrypting the previous block γ_{i-1}, where the initial value γ_0 is equal to the result of the initialization vector encryption. If you use the terminology of the substitutions theory [36–38] and present the basic encryption transformation $T_{l,k}^{(K)}$ initiated by the secret key K as some substitution s acting on the set of open texts, then the period of the sequence of output blocks $\gamma_0, \gamma_1, \ldots, \gamma_{n-1}$ will correspond to one of the cycles $s_i = (y, s_i(y), s_i^2(y), \ldots, s_i^{l_i-1}(y))$ of the substitution s where the initial value of the cycle y is equal to the initialization vector value S.

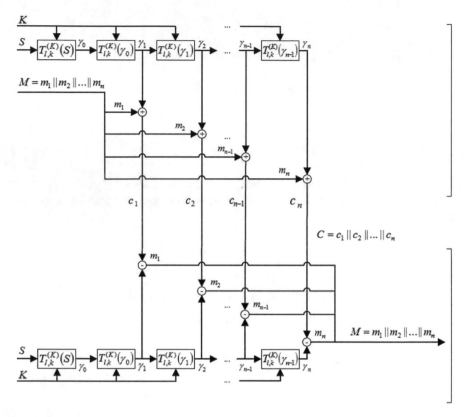

Fig. 1 Scheme of encryption and decryption in the output feedback encryption mode

Each next element $s_i^j(y)$ of the cycle s_i with the length l_i is the result of multiple encoding of the initialization vector:

$$y = S;$$

$$s_i(y) = T_{l,k}^{(K)}(S) = \gamma_0;$$
$$s_i^2(y) = T_{l,k}^{(K)}\left(T_{l,k}^{(K)}(S)\right) = \gamma_1;$$

$$\cdots$$

$$s_i^{l_i-1}(y) = \underbrace{T_{l,k}^{(K)}\left(T_{l,k}^{(K)} \ldots T_{l,k}^{(K)}(S)\right)}_{l_i-1 \text{ times}} = \gamma_{l_i-2};$$

$$s_i^{l_i}(y) = \underbrace{T_{l,k}^{(K)}\left(T_{l,k}^{(K)} \ldots T_{l,k}^{(K)}(S)\right)}_{l_i \text{ times}} = \gamma_{l_i-1} = \gamma_0;$$

$$\cdots$$

$$s_i^n(y) = \underbrace{T_{l,k}^{(K)}\left(T_{l,k}^{(K)} \ldots T_{l,k}^{(K)}(S)\right)}_{n \text{ times}} = \gamma_{n-1}.$$

Thus, the investigation of the periodic properties of the sequence of output blocks γ_i, $i = 0, 1, .., n$ is to study the cyclic structure of the substitution s, namely, in estimating the distribution of the length l_i of the cycles s_i for different initial values $y = S$ of the basic ciphering transformation $T_{l,k}^{(K)}$. Such investigations will make it possible to estimate the length l_i of output blocks period for different $y = S$ and K or to determine the probability of forming the output blocks of a certain period for an arbitrary fixed value of the initialization vector $y = S$ and a randomly selected secret key K.

Let us consider some of the provisions of the theory of substitutions and their relation to the properties of BSC, in particular, we introduce the basic concepts and definitions associated with certain properties of symmetric block crypto-transformations (the distribution of the number of cycles, magnifications and inversions, etc.).

3 Special Provisions of the Theory of Substitutions

By the definition BSC is a key-parameterized function of a bijective mapping of a set of plaintexts into a set of ciphertexts $V_l \rightarrow V_l$, $K \in V_k$ [9, 10]. In general, for any l-bit block cipher there are $2^l!$ possible permutations of plaintext. These transformations, called permutations of degree 2^l, form a group under the operation of performing sequential transformations. Such a group is called a symmetric group of permutations of degree 2^n and is denoted by S_{2^l} [36–38]. In practice, it means that the number of bits of the key, which is necessary to obtain all possible permutations, is about $\ln 2^l! \approx l \cdot 2^l$ bits (by the Stirling formula $\ln(x!) = x \ln(x) - x - O(\ln(x))$). For example, for $l = 128$, we have $2^{128}! \approx 2^{128 \cdot 2^{128}}$ possible permutations of 128-bits blocks, of which, depending on the length of the key, only 2^{128} or 2^{256} transformations are used. Thus, the basic transformation of the cipher is essentially a subset of the complete set of all possible substitutions acting on a set of processed data blocks. The basic assumption that is adopted in substantiating the stability of symmetric cryptographic transformation is in preserving the probabilistic properties of random substitution. It is assumed that while encrypting and applying a limited set of substitutions from S_{2^l}, however, certain distribution probabilities of these subset elements correspond to the properties of randomly selected substitutions from the whole set S_{2^l} [39–44].

Let's consider the basic concepts and definitions of the theory of substitutions [36–38] and associate them with cyclic properties of BSC. For this we consider the set of all bijective transformations of the set $Y = \{y_1, y_2, \ldots, y_n\}$ to itself, forming a symmetric group S_n with the power $n!$ of all possible substitutions of degree n. By definition of the symmetric group [36–38], each substitution $s \in S_n$ corresponds to a unique substitution $s^{-1} \in S_n$, such that

$$s^{-1} \cdot s(y) = s \cdot s^{-1}(y) = e(y), \ y \in Y,$$

where $e(y) \in S_n$ is the unit substitution, i.e. $e(y) = y$ for all $y \in Y$.

Let's use the following symbols:

$$s \cdot s \cdot \ldots \cdot s = s^k, \; s^{-1} \cdot s^{-1} \cdot \ldots \cdot s^{-1} = s^{-k},$$

where products contain k multipliers.

Accordingly, we have

$$s^k \cdot s^{-k} = s^{-k} \cdot s^k = s^0 = e.$$

The set of substitutions of degree n, which is locked in relation to the multiplication and inverse computation operation for $s \in S_n$ an element $s^{-1} \in S_n$, is called substitution group. Each such group is a subgroup of the symmetric group S_n [36–38].

Consider some substitutions $s \in S_n$ that act on a set Y. We will define a binary relation on a set Y, while we will assume that $y \sim y'$ for $y, y' \in Y$ if there exists such j that $y' = s^j(y)$. This binary relation is reflexive, symmetric and transitive, i.e. the relation is equivalence. Indeed, according to [36–38] we have:

- $y \sim y$, because $y = s^0(y) = e(y)$;
- it follows from condition $y \sim y'$ that $y' \sim y$, because it follows from equality $y' = s^j(y)$ that $y = s^{-j}(y')$;
- it follows from $y \sim y'$ and $y' \sim y''$ that $y \sim y''$, because from the equalities $y' = s^j(y)$ and $y'' = s^i(y')$ it follows that $y'' = s^i(s^j(y)) = s^{i+j}(y)$.

The cycle s_i of substitution $s \in S_n$ with the length l_i is defined as follows:

$$s_i = (y, s_i(y), s_i^2(y), \ldots, s_i^{l_i-1}(y)),$$

where $s_i^{l_i}(y) = y$.

An arbitrary substitution $s \in S_n$ can be expanded into the corresponding cycles [36–38]:

$$s = (y_1, s_1(y_1), s_1^2(y_1), \ldots, s_1^{l_1-1}(y_1)) \ldots (y_k, s_k(y_k), s_k^2(y_k), \ldots, s_k^{l_k-1}(y_k)). \quad (5)$$

Elements y_i and y_{i+1} in substitution $s \in S_n$ form increment, if $s(y_i) > s(y_{i+1})$ it is assumed that an element y_1 always preceded by increment. A pair of elements y_i and y_j in substitution $s \in S_n$ forms an increment if $s(y_i) > s(y_j)$, $i < j$.

For example, the substitution s of degree 4

$$s = \begin{pmatrix} y_1 & y_2 & y_3 & y_4 \\ s(y_1) & s(y_2) & s(y_3) & s(y_4) \end{pmatrix} = \begin{pmatrix} 1 & 2 & 3 & 4 \\ 1 & 4 & 3 & 2 \end{pmatrix}$$

can be given in the form of a schedule for 3 cycles:

$$s_1 = (y_1) = (1), \, l_1 = 1;$$
$$s_2 = (y_2, s_2(y_2)) = (2, 4), \, l_2 = 2;$$
$$s_3 = (y_3) = (3), \, l_3 = 1.$$

We have the following schedule:

$$s = (y_1)(y_2, s_2(y_y))(y_3) = (1)(2, 4)(3).$$

In this substitution there are two increments (the element $y_1 = 1$ always preceded by one increment, and one more increment forms the elements $y_1 = 1$ and $y_2 = 2$, because $s(y_1) = 1 > s(y_2) = 4$) and three inversions (they form pairs of elements y_2 and y_3, y_2 and y_4, y_3 and y_4, because there are inequalities $s(y_2) > s(y_3)$, $s(y_2) > s(y_4)$, $s(y_3) > s(y_4)$, respectively).

On the set of all permutations of the symmetric group S_n, we give a uniform probabilistic distribution, i.e. for each selected substitution $s \in S_n$ we put in correspondence the probability of its selection equal to $1/n!$. According to modern views of symmetric cryptography, such a set of equivalence mappings corresponds to the idea of an "ideal" cipher. After all, if the random selection of a separate substitution $s \in S_n$ is associated with the value of the entered encryption key, then the resulting conversion will match the random and evenly selected ciphertext for each open text with any key, i.e. in all possible variants of open text mapping in the ciphertext.

We will investigate the probabilistic properties of random substitution, in particular, the probabilities of a cycle of a certain length in a randomly selected substitution, since this particular event will correspond to the case when the output block y_i of a certain period is formed for an arbitrary fixed value of the initialization vector S.

4 Probability Evaluation of a Certain Length Cycle in Randomly Selected Substitution

Consider a random value ξ_n equals to the number of cycles in a randomly chosen substitution $s \in S_n$. The substitution $s \in S_n$ refers to a cyclic class $\{1^{\alpha_1} 2^{\alpha_2} \ldots n^{\alpha_n}\}$ if it contains α_1 cycles of length 1, α_2 cycles of length 2, and so on:

$$s = (y_1)(y_2) \ldots (y_{\alpha_1})(y_1', y_1'')(y_2', y_2'') \ldots (y_{\alpha_2}', y_{\alpha_2}'') \ldots,$$
$$1\alpha_1 + 2\alpha_2 + \ldots + n\alpha_n = n.$$

Denote by $C(\alpha_1, \alpha_2, \ldots, \alpha_n)$ the number of substitutions in the cyclic class $\{1^{\alpha_1} 2^{\alpha_2} \ldots n^{\alpha_n}\}$, and by $C(n, k)$ the number of substitutions of degree n that have k cycles. Then we have [36–38]:

$$C(\alpha_1, \alpha_2, \ldots, \alpha_n) = \frac{n!}{1^{\alpha_1} 2^{\alpha_2} \ldots n^{\alpha_n} \alpha_1! \alpha_2! \ldots \alpha_n!},$$

$$C(n, k) = \sum_{\substack{1\alpha_1 + 2\alpha_2 + \ldots + n\alpha_n = n \\ \alpha_1 + \alpha_2 + \ldots + \alpha_n = k, \ \alpha_i \geq 0}} \frac{n!}{1^{\alpha_1} 2^{\alpha_2} \ldots n^{\alpha_n} \alpha_1! \alpha_2! \ldots \alpha_n!} = |s(n, k)|, \tag{6}$$

where $s(n, k)$ are Stirling numbers of first kind, which is determined by the ratio

$$(x)_n = x(x - 1) \ldots (x - n + 1) = \sum_{k=0}^{n} s(n, k) x^k,$$

where $(x)_n = x(x - 1) \ldots (x - n + 1)$ is the common designation of the declining factorial (the symbol of Pohgammer).

Formula (6) implies the expression for the exact probability distribution of a random event $\xi_n = k$, in the case where randomly selected substitutions will observe exactly k cycles (see expression (5)).

Using the formula for computing Stirling numbers of the first kind, we have [36–38]:

$$P(\xi_n = k) = \frac{C(n, k)}{n!} = \frac{|s(n, k)|}{n!}, \ k = 0, 1, \ldots, n.$$

In [36, 37] we obtain the expected value $M\xi_n$ and variance $D\xi_n$ of the random variable ξ_n:

$$M\xi_n = \sum_{j=1}^{n} \frac{1}{j} = \ln n + C + o(1), \ D\xi_n = \sum_{j=1}^{n} \frac{1}{j} - \sum_{j=1}^{n} \frac{1}{j^2}$$

$$= \ln n + C + o(1), \ C = 0,5772 \ldots,$$

in addition, it is shown that when $n \to \infty$ a random variable $\xi_n' = (\xi_n - \ln n)/(\ln n)$ is distributed asymptotically with parameters $(0, 1)$

$$\lim_{n \to \infty} P(\xi_n' < u) = \frac{1}{\sqrt{2\pi}} \int_{-\infty}^{u} e^{-y^2/2} dy$$

For random variables ζ_n and η_n, which are equal to the number of increments and inversions in randomly selected substitution $s \in S_n$, the corresponding expected value and variance have the form [36–38]:

$$M\zeta_n = \frac{n(n - 1)}{4}, \ D\zeta_n = \frac{2n^3 + 3n^2 - 5n}{72}, \ M\eta_n = \frac{n}{2}, \ D\eta_n = \frac{n}{12},$$

in this case random variables

$$\zeta_n' = (\zeta_n - M\zeta_n)/(D\zeta_n)$$

and

$$\eta'_n = (\eta_n - M\eta_n)/(D\eta_n)$$

when $n \to \infty$ are also asymptotically distributed with parameters $(0, 1)$.

Empirical distributions of the probability of occurrence of a certain number of cycles, increments and inversions in randomly selected substitutions from a certain subset $V \subset S_n$, whose elements are substitutions implemented by the use of the encryption function on reduced cipher models, are investigated in [45–47]. It is established that the obtained empirical distributions are very close to the theoretical distributions under consideration, i.e. it can be argued that the reduced models of BSC on these criteria are similar to the properties of random substitution from S_n.

At the same time, to evaluate the probability of forming output blocks γ_i of a certain period for an arbitrary fixed value of the initialization vector S another characteristic of random substitution is required, namely, the distribution of the number of cycles of a given length. In accordance with [36–38], this characteristic in random substitution is determined as follows.

We denote $\chi_{n,L}$ as the number of cycles of length L in an arbitrary equivalence arbitrary substitution of degree n. Obviously that

$$\xi_n = \chi_{n,L=1} + \chi_{n,L=2} + \cdots + \chi_{n,L=n}.$$

The probability distribution of a random event $\chi_{n,L} = k$ is defined as [36–38]:

$$P(\chi_{n,L} = k) = \frac{1}{L^k k!} \sum_{j=0}^{[n/L]-k} \frac{(-1)^j}{L^j j!}, \quad k = 0, 1, \ldots, [n/L]. \qquad (7)$$

When $n \to \infty$ a random variable $\chi_{n,L}$ has a Poisson distribution with parameters $\lambda = 1/L$, i.e.

$$\lim_{n \to \infty} P(\chi_{n,L} = k) = \frac{1}{L^k k!} e^{-1/L}, \quad k = 0, 1, \ldots. \qquad (8)$$

We use the formula for the exact distribution of probabilities of a random event $\chi_{n,L} = k$ in the form (7) [36–38]. The value $n! P(\chi_{n,L} = k)$ corresponds to the number of substitutions containing k cycles with the length L. We are interested in the number of such substitutions $s \in S_n$, which for an arbitrary fixed $y_i \in Y$ will necessarily have cycles $s_i = (y_i, s_i(y_i), s_i^2(y_i), \ldots, s_i^{l_i-1}(y_i))$ of lengths $L = l_i$. Consider the case when $L = 1$, i.e. we will count the number of such substitutions from S_n which, for an arbitrary fixed $y_i \in Y$ will necessarily have a cycle (y_i) of length $L = l_i = 1$. In cryptography, when considering block symmetric cryptographic transformations, such cases are called fixed points of substitution [45–47].

Taking into account that $L = 1$, formula (7) takes the form

$$P(\chi_{n,L=1} = k) = \frac{1}{k!} \sum_{j=0}^{n-k} \frac{(-1)^j}{j!}, \quad k = 0, 1, \ldots, n.$$

and for each $k = 1, \ldots, n$ each of $n!P(\chi_{n,L=1} = k)$ cases for an arbitrary fixed $y_i \in Y$ will be observed precisely

$$\frac{C_{n-1}^{k-1}}{C_n^k} = \frac{(n-1)!}{(k-1)!(n-k)!} \frac{k!(n-k)!}{n!} = \frac{k}{n} \tag{9}$$

times, i.e. the number of substitutions containing one fixed point of a specific form (cycle (y_i)) will be determined by the formula:

$$\sum_{k=1}^{n} n!P(\chi_{n,L=1} = k) \frac{C_{n-1}^{k-1}}{C_n^k} = \sum_{k=1}^{n} \left(\frac{(n-1)!}{(k-1)!} \sum_{j=0}^{n-k} \frac{(-1)^j}{j!} \right) = (n-1)!, \tag{10}$$

and the corresponding probability of the appearance of such fixed point (for given initial value $y_i \in Y$) in the randomly chosen substitution of the degree n will look like:

$$\sum_{k=1}^{n} P(\chi_{n,L=1} = k) \frac{C_{n-1}^{k-1}}{C_n^k} = \frac{(n-1)!}{n!} = \frac{1}{n}. \tag{11}$$

Let's explain the formula (9). In total, there are exactly C_n^k methods for simultaneously choosing values $y_i, y_{i_1}, y_{i_2}, \ldots, y_{i_{k-1}} \in Y$, $y_i \neq y_{i_1} \neq y_{i_2} \neq \cdots \neq y_{i_{k-1}}$ which uniquely determine cycles (y_i) (y_{i_1}) $(y_{i_2}) \ldots (y_{i_{k-1}})$ of length $L = 1$. But for each fixed $y_i \in Y$ there are exactly C_{n-1}^{k-1} ways for choosing the remaining values $y_{i_1}, y_{i_2}, \ldots, y_{i_{k-1}} \in Y$. i.e. from the total number $n!P(\chi_{n,L=1} = k)$ of substitutions containing k cycles of length $L = 1$, only

$$n!P(\chi_{n,L=1} = k) \frac{C_{n-1}^{k-1}}{C_n^k} = n!P(\chi_{n,L=1} = k) \frac{k}{n}$$

substitutions will necessarily contain a cycle (y_i).

The last formula (11) can be obtained much simpler from trivial combinatorial considerations. Indeed, if on the set $Y = \{y_1, y_2, \ldots, y_n\}$ we will fix m elements, then there are $(n - m)!$ ways for permutations of the remaining elements. i.e. on the all set of substitutions from S_n with the random probability distribution, the probability of choosing a substitution with m fixed points is equal to

$$\frac{(n-m)!}{n!} = \frac{1}{(n-m+1)(n-m+2)\ldots n} = \frac{1}{(n)_m}, \tag{12}$$

Which for $m = 1$ coincides with (11).

Formulas (9–12) were obtained in [48] when studying Galois/Counter Mode and GMAC—GCM & GMAC, which allowed us to estimate the probability of a zero hash subkey, i.e. the probability of such event, when encryption of zero open text will get zero value of ciphertext. We extend the result obtained earlier to an arbitrary value of the length of cycle $L = l_i \in \{1, 2, \ldots, n\}$ in the study of periodic properties of the output blocks in OFB mode.

Consider the case of an arbitrary length of a cycle, we will calculate the number of such substitutions s from S_n which, for an arbitrary fixed $y_i \in Y$ will necessarily have a cycle

$$(y_i, s_i(y_i), s_i^2(y_i), \ldots, s_i^{L-1}(y_i))$$

of length $L = l_i \in \{1, 2, \ldots, n\}$.

For fixed lengths and quantities of cycles (L and k) there are exactly C_n^{kL} ways to simultaneously select values

$$\begin{aligned}
&y_i, y_{j_i}, y_{u_i}, \ldots, y_{v_i}, \\
&y_{i_1}, y_{j_1}, y_{u_1}, \ldots, y_{v_1}, \\
&y_{i_2}, y_{j_2}, y_{u_2}, \ldots, y_{v_2}, \\
&\qquad\qquad \cdots \\
&y_{i_{k-1}}, y_{j_{k-1}}, y_{u_{k-1}}, \ldots, y_{v_{k-1}},
\end{aligned} \tag{13}$$

which collectively determine k cycles of length L each:

$$\begin{aligned}
&(y_i, y_j = s_i(y_i), y_u = s_i^2(y_i), \ldots, y_v = s_i^{L-1}(y_i)), \\
&(y_{i_1}, y_{j_1} = s_{i_1}(y_{i_1}), y_{u_1} = s_{i_1}^2(y_{i_1}), \ldots, y_{v_1} = s_{i_1}^{L-1}(y_{i_1})), \\
&(y_{i_2}, y_{j_2} = s_{i_2}(y_{i_2}), y_{u_2} = s_{i_2}^2(y_{i_2}), \ldots, y_{v_2} = s_{i_2}^{L-1}(y_{i_2})), \\
&\qquad\qquad \cdots, \\
&(y_{i_{k-1}}, y_{j_{k-1}} = s_{i_{k-1}}(y_{i_{k-1}}), y_{u_{k-1}} = s_{i_{k-1}}^2(y_{i_{k-1}}), \ldots, y_{v_{k-1}} = s_{i_{k-1}}^{L-1}(y_{i_{k-1}})),
\end{aligned} \tag{14}$$

and all elements from (13) are unique, since the set of cycles (14) is included in the decomposition of the same substitution.

Of the C_n^{kL} ways of simultaneously choosing the values (13) for each fixed set $y_i, y_{j_i}, y_{u_i}, \ldots, y_{v_i} \in Y$ there are exactly C_{n-1}^{kL-1} ways for choosing the remaining values

$$\begin{aligned}
&y_{i_1}, y_{j_1}, y_{u_1}, \ldots, y_{v_1}, \\
&y_{i_2}, y_{j_2}, y_{u_2}, \ldots, y_{v_2}, \\
&\qquad\qquad \cdots, \\
&y_{i_{k-1}}, y_{j_{k-1}}, y_{u_{k-1}}, \ldots, y_{v_{k-1}},
\end{aligned}$$

because the selection of set $y_i, y_{j_i}, y_{u_i}, \ldots, y_{v_i} \in Y$ is determined by selecting only one element $y_i \in Y$, and from the remaining $n - 1$ elements possible different combinations of $kL - 1$ elements.

Thus, for each $k = 0, 1, \ldots, [n/L]$ of the total number $n! P(\chi_{n,l} = k)$ of substitutions containing k cycles of length L, only

$$n! P(\chi_{n,L} = k) \frac{C_{n-1}^{kL-1}}{C_n^{kL}} = n! P(\chi_{n,L} = k) \frac{(n-1)! kL! (n-kL)!}{n! (kL-1)! (n-kL)!}$$

$$= n! P(\chi_{n,L} = k) \frac{kL}{n} = (n-1)! kL P(\chi_{n,L} = k)$$

substitutions will necessarily contain a cycle

$$(y_i, s_i(y_i), s_i^2(y_i), \ldots, s_i^{L-1}(y_i)).$$

Summing up the last expression for all $k = 0, 1, \ldots, [n/L]$, taking into account (7), we obtain an exact formula for determining the number of substitutions s from S_n which, for an arbitrary fixed $y_i \in Y$ will necessarily have a cycle

$$(y_i, s_i(y_i), s_i^2(y_i), \ldots, s_i^{L-1}(y_i))$$

of length $L = l_i \in \{1, 2, \ldots, n\}$:

$$\sum_{k=1}^{[n/L]} n! P(\chi_{n,L} = k) \frac{C_{n-1}^{kL-1}}{C_n^{kL}} = \sum_{k=1}^{[n/L]} \frac{(n-1)!}{(k-1)! L^{k-1}} \sum_{j=0}^{[n/L]-k} \frac{(-1)^j}{L^j j!} = (n-1)!, \quad (15)$$

and the corresponding formula for calculating the probability of randomly choosing a substitution s from S_n with the following cycle:

$$\sum_{k=1}^{[n/L]} P(\chi_{n,L} = k) \frac{C_{n-1}^{kL-1}}{C_n^{kL}} = \frac{(n-1)!}{n!} = \frac{1}{n}. \quad (16)$$

It is obvious that the last analytic expression for $L = 1$ completely coincides with the formula (8) in [48] with the corresponding statement.

The resulting analytic expression (16) can also be considered as a combinatorial identity (simplified formula) for the sum of the members of formula (7) with the corresponding proportional coefficients

$$\frac{C_{n-1}^{kL-1}}{C_n^{kL}} = \frac{kL}{n},$$

or even for the Poisson distribution (8).

The probability estimate (16) of a cycle of a certain length can be obtained by another, in a much simpler way, using simple combinatorial considerations.

We fix some arbitrary value y_i from the set $Y = \{y_1, y_2, \ldots, y_n\}$. There are totally $n!$ substitutions on the set Y, of which only

$$\frac{n!}{n} = (n-1)!$$

substitutions will contain a cycle (y_i) of length $L = 1$ in their cyclic schedule.

In addition, from $n!$ substitutions of the symmetric group

$$\frac{n!}{n(n-1)}(n-1) = (n-1)!$$

substitutions will contain a cycle $(y_i, y_{j \neq i})$ of length $L = 2$,

$$\frac{n!}{n(n-1)(n-2)}(n-1)(n-2) = (n-1)!$$

substitutions will contain a cycle $(y_i, y_{j \neq i}, y_{u \neq i, j})$ of length $L = 3$ and so on.

That is, for an arbitrary fixed value $y_i \in Y$ the number of substitutions from S_n containing the cycle $(y_i, s_i(y_i), s_i^2(y_i), \ldots, s_i^{L-1}(y_i))$ is defined as $(n-1)!$, and the corresponding probability of randomly choosing a substitution containing such a cycle is defined as

$$\frac{(n-1)!}{n!} = \frac{1}{n},$$

regardless of the length of the cycle $L = l_i \in \{1, 2, \ldots, n\}$, nor its own value y_i from $Y = \{y_1, y_2, \ldots, y_n\}$.

Thus, the probability of a cycle of a certain length is determined only by the degree n of substitution. For example, for $n = 4$ from $n! = 24$ substitutions of the symmetric group for any fixed y_i from $Y = \{y_1, y_2, y_3, y_4\}$ we have $(n-1)! = 6$ substitutions each that necessarily contain cycles of different lengths (or (y_i), or $(y_i, y_{j \neq i})$, or $(y_i, y_{j \neq i}, y_{u \neq i, j})$, or $(y_i, y_{j \neq i}, y_{u \neq i, j}, y_{v \neq i, j, u})$), respectively). Consequently, the probability that a randomly selected substitution from S_4 contained cycle of length $L = l_i \in \{1, 2, \ldots, 4\}$ will be equal to $1/n = 1/4$ independent from either y_i no $L = l_i$.

Since the obtained analytical expressions (15) and (16) are rather complicated and cumbersome, especially the order of their output, we illustrate the example of calculating the probabilities of a cycle of given length in a randomly chosen substitution from a symmetric group S_4. An example will be supplemented by explanations showing the validity of the formulas obtained and the combinatorial arguments presented.

5   Example for Symmetric Group S_4

Consider an example of all biective transformations of the set $Y = \{y_1, y_2, y_3, y_4\}$ to itself, i.e. the set with $n! = 24$ substitutions of degree $n = 4$.

Table 1 shows all substitutions that make up the symmetric group S_4 (the results of each substitution are given, the order of each substitution for cycles, total number of cycles and the distribution of the number of cycles of a certain length, each substitution is numbered for convenience).

Table 2 shows the distribution of the number of values $\xi_n = k$ and $\chi_{n,L} = k$ for different k (the symbol # (x) indicates the number of cases (x) for all substitutions from S_4).

5.1   Estimation of the Probability of the Cycle (y_i)

Consider the case when for an arbitrary fixed $y_i \in Y$ randomly chosen substitution s from S_4 will necessarily contain a cycle (y_i) of length $L = 1$.

First we consider substitutions with $k = 1$ cycles of length $L = 1$. We have 8 such substitutions (Table 3). But each individual $y_i \in Y$ generates a cycle of length $L = 1$ only twice, that is, for each arbitrary fixed $y_i \in Y$ there are precisely 2 substitutions that contain a cycle of length $L = 1$ of the form (y_i). For example, for $y_1 \in Y$ these are 4th and 5th substitutions, for $y_2 \in Y$ these are the 16th and 21st substitutions, etc. The number of substitutions, which for an arbitrary fixed $y_i \in Y$ containing k $= 1$ cycle of length $L = 1$ of form (y_i) are calculated as follows. In total, there are exactly $C_{n=4}^{kL=1} = 4$ ways to select a value $y_i \in Y$.

This choice is no longer limited, because for each $y_i \in Y$ cycle (y_i) is defined unambiguously (according to the formula there are $C_{n-1=3}^{kL-1=0} = 1$ options for choosing a value $y_{j \neq i} \in Y$). That is, the total number of substitutions containing only k = 1 cycle of length $L = 1$ (such 8 substitutions) must be multiplied by value $\frac{C_{n-1=3}^{kL-1=0}}{C_{n=4}^{kL=1}} = \frac{1}{4}$. Thus, for an arbitrary fixed $y_i \in Y$ the number of substitutions containing only one cycle (y_i) is equal to two.

Consider now the substitutions containing $k = 2$ cycles of length $L = 1$. We have 6 substitutions (Table 4), of which three substitutions in a cyclic decomposition contain cycles (y_1) $(y_{i \neq 1})$, three substitutions contain cycles (y_2) $(y_{i \neq 2})$, three substitutions contain cycles (y_3) $(y_{i \neq 3})$ and three substitutions contain cycles (y_4) $(y_{i \neq 4})$. It is clear that one and the same substitution can be assumed to different ways, i.e. it can, in its cyclic decomposition, be treated to substitutions containing cycles (y_i) $(y_{j \neq i})$ and substitutions containing cycles (y_j) $(y_{i \neq j})$. For example, the sixth substitution has a cyclic decomposition (y_1) (y_2, y_4) (y_3), it should be considered as a substitution with cycles (y_1) $(y_{j \neq 1})$ and with substitutions with cycles $(y_{i \neq 3})$ $(y_{i \neq 3})$.

The number of substitutions containing k = 2 cycles (y_i) $(y_{j \neq i})$ of length $L = 1$ for fixed $y_i \in Y$ are calculated as follows. In total, there are exactly $C_{n=4}^{kL=2} = 6$ ways to simultaneously select values $y_i, y_{j \neq i} \in Y$. But for fixed $y_i \in Y$ there are

Table 1 The set of substitutions from S_4 and their cyclic properties

№	Result of substitution				Decomposition of substitution to cycles	Number of cycles, ξ_n, $s(y_1)$	Number of cycles of length L, $\chi_{n,L}$			
	$s(y_1)$	$s(y_2)$	$s(y_3)$	$s(y_4)$			$s(y_1)$	$s(y_2)$	$s(y_3)$	$s(y_4)$
1	y_1	y_2	y_3	y_4	$(y_1)(y_2)(y_3)(y_4)$	4	4	0	0	0
2	y_1	y_2	y_4	y_3	$(y_1)(y_2)(y_3,y_4)$	3	2	1	0	0
3	y_1	y_3	y_2	y_4	$(y_1)(y_2,y_3)(y_4)$	3	2	1	0	0
4	y_1	y_3	y_4	y_2	$(y_1)(y_2,y_3,y_4)$	2	1	0	1	0
5	y_1	y_4	y_2	y_3	$(y_1)(y_2,y_4,y_3)$	2	1	0	1	0
6	y_1	y_4	y_3	y_2	$(y_1)(y_2,y_4)(y_3)$	3	2	1	0	0
7	y_2	y_1	y_3	y_4	$(y_1,y_2)(y_3)(y_4)$	3	2	1	0	0
8	y_2	y_1	y_4	y_3	$(y_1,y_2)(y_3,y_4)$	2	0	2	0	0
9	y_2	y_3	y_1	y_4	$(y_1,y_2,y_3)(y_4)$	2	1	0	1	0
10	y_2	y_3	y_4	y_1	(y_1,y_2,y_3,y_4)	1	0	0	0	1
11	y_2	y_4	y_1	y_3	(y_1,y_2,y_4,y_3)	1	0	0	0	1
12	y_2	y_4	y_3	y_1	$(y_1,y_2,y_4)(y_3)$	2	1	0	1	0
13	y_3	y_1	y_2	y_4	$(y_1,y_3,y_2)(y_4)$	2	1	0	1	0
14	y_3	y_1	y_4	y_2	(y_1,y_3,y_4,y_2)	1	0	0	0	1
15	y_3	y_2	y_1	y_4	$(y_1,y_3)(y_2)(y_4)$	3	2	1	0	0
16	y_3	y_2	y_4	y_1	$(y_1,y_3,y_4)(y_2)$	2	1	0	1	0
17	y_3	y_4	y_1	y_2	$(y_1,y_3)(y_2,y_4)$	2	0	2	0	0
18	y_3	y_4	y_2	y_1	(y_1,y_3,y_2,y_4)	1	0	0	0	1
19	y_4	y_1	y_2	y_3	(y_1,y_4,y_3,y_2)	1	0	0	0	1

(continued)

Table 1 (continued)

№	Result of substitution				Decomposition of substitution to cycles	Number of cycles, ξ_n, $s(y_1)$	Number of cycles of length L, $\chi_{n,L}$			
	$s(y_1)$	$s(y_2)$	$s(y_3)$	$s(y_4)$			$s(y_1)$	$s(y_2)$	$s(y_3)$	$s(y_4)$
20	y_4	y_1	y_3	y_2	$(y_1, y_4, y_2)\,(y_3)$	2	1	0	1	0
21	y_4	y_2	y_1	y_3	$(y_1, y_4, y_3)\,(y_2)$	2	1	0	1	0
22	y_4	y_2	y_3	y_1	$(y_1, y_4)\,(y_2)\,(y_3)$	3	2	1	0	0
23	y_4	y_3	y_1	y_2	(y_1, y_4, y_2, y_3)	1	0	0	0	1
24	y_4	y_3	y_2	y_1	$(y_1, y_4)\,(y_2, y_3)$	2	0	2	0	0

Table 2 Distributions of quantities of values $\xi_n = k$ and $\chi_{n,L} = k$ for all substitutions from S_4

k	0	1	2	3	4
# $(\xi_n = k)$	0	6	11	6	1
# $(\chi_{n,L=1} = k)$	9	8	6	0	1
# $(\chi_{n,L=2} = k)$	15	6	3	0	0
# $(\chi_{n,L=3} = k)$	16	8	0	0	0
# $(\chi_{n,L=4} = k)$	18	6	0	0	0

Table 3 Substitutions containing 1 cycle of length 1 and one cycle of length 3

№	Result of substitution				Decomposition of substitution to cycles
	$s(y_1)$	$s(y_2)$	$s(y_3)$	$s(y_4)$	
4	y_1	y_3	y_4	y_2	$(y_1)\,(y_2, y_3, y_4)$
5	y_1	y_4	y_2	y_3	$(y_1)\,(y_2, y_4, y_3)$
9	y_2	y_3	y_1	y_4	$(y_1, y_2, y_3)\,(y_4)$
12	y_2	y_4	y_3	y_1	$(y_1, y_2, y_4)\,(y_3)$
13	y_3	y_1	y_2	y_4	$(y_1, y_3, y_2)\,(y_4)$
16	y_3	y_2	y_4	y_1	$(y_1, y_3, y_4)\,(y_2)$
20	y_4	y_1	y_3	y_2	$(y_1, y_4, y_2)\,(y_3)$
21	y_4	y_2	y_1	y_3	$(y_1, y_4, y_3)\,(y_2)$

Table 4 Substitutions containing 2 cycle of length 1 and one cycle of length 2

№	Result of substitution				Decomposition of substitution to cycles
	$s(y_1)$	$s(y_2)$	$s(y_3)$	$s(y_4)$	
2	y_1	y_2	y_4	y_3	$(y_1)\,(y_2)\,(y_3, y_4)$
3	y_1	y_3	y_2	y_4	$(y_1)\,(y_2, y_3)\,(y_4)$
6	y_1	y_4	y_3	y_2	$(y_1)\,(y_2, y_4)\,(y_3)$
7	y_2	y_1	y_3	y_4	$(y_1, y_2)\,(y_3)\,(y_4)$
15	y_3	y_2	y_1	y_4	$(y_1, y_3)\,(y_2)\,(y_4)$
22	y_4	y_2	y_3	y_1	$(y_1, y_4)\,(y_2)\,(y_3)$
2	y_1	y_2	y_4	y_3	$(y_1)\,(y_2)\,(y_3, y_4)$
3	y_1	y_3	y_2	y_4	$(y_1)\,(y_2, y_3)\,(y_4)$

exactly $C_{n-1=3}^{kL-1=1} = 3$ ways for choosing a value $y_{j \neq i} \in Y$. That is, the total number of substitutions containing $k = 2$ cycles of length $L = 1$ must be multiplied by the value $\frac{C_{n-1=3}^{kL-1=1}}{C_{n=4}^{kL=2}} = \frac{3}{6}$, we obtain the desired value, for an arbitrary fixed $y_i \in Y$ the number of substitutions containing $k = 2$ cycles $(y_i)\,(y_{j \neq i})$ of length $L = 1$, is equal to three. For example, for fixed value $y_1 \in Y$ the second, third, and sixth substitutions contain k = 2 cycles of length $L = 1$ with cyclic decompositions: $(y_1) (y_2) (y_3, y_4)$, $(y_1) (y_2, y_3) (y_4)$ та $(y_1) (y_2, y_4) (y_3)$.

Consider substitutions containing $k = 4$ cycles of length $L = 1$ (no substitution can have k = 3 cycles of length $L = 1$). We have 1 substitution (first in Table 1), its cyclic decomposition has the form: $(y_1) (y_2) (y_3) (y_4)$. Applying the same formula, we have $\frac{C_{n-1=3}^{kL-1=3}}{C_{n=4}^{kL=4}} = \frac{1}{1} = 1$, that is, the total number of substitutions containing four cycles of length 1 coincides with the number of substitutions with cyclic decomposition (y_i) $(y_{j\neq i})$ $(y_{u\neq i,j})$ $(y_{v\neq i,j,u})$, as it should be.

We will calculate the number of substitutions from S_4 (Table 1), which for fixed $y_i \in Y$ necessarily contain a cycle (y_i) of length (y_i). To do this, we must summarize the number of substitutions that for fixed $y_i \in Y$ in their cyclic decomposition contain a different number of cycles of length $L = 1$, namely, 2 substitutions with k = 1 cycle (y_i), three substitutions with $k = 2$ cycles (y_i) $(y_{j\neq i})$ and one substitution with $k = 4$ cycles (y_i) $(y_{j\neq i})$ $(y_{u\neq i,j})$ $(y_{v\neq i,j,u})$. In general, we have 6 substitutions from the total of 24 substitutions of the symmetric group. That is, at randomly equal probable selection of substitutions from S_4 the probability that in it for arbitrary fixed $y_i \in Y$ will be observed cycle (y_i) of length $L = 1$ is equal to $6/24 = 1/4$.

5.2 Estimation of Probability of Occurrence of a Cycle $(y_i, y_{j\neq i})$

Consider the case when for arbitrary fixed $y_i \in Y$ randomly chosen substitution s from S_4 will necessarily contain a cycle $(y_i, y_{j\neq i})$ of length $L = 2$.

First we consider substitutions containing $k = 1$ cycle of length $L = 2$. We have 6 such substitutions, which are given in Table 4 (if the substitution from S_4 contains two cycles of length 1, then it necessarily contains one cycle of length 2). We will calculate the number of substitutions, which for an arbitrary fixed $y_i \in Y$ contain k = 1 cycle of length $L = 2$ of form $(y_i y_{j\neq i})$. In total, there are exactly $C_{n=4}^{kL=2} = 6$ ways to simultaneously select values $y_i, y_{j\neq i} \in Y$. But for each $y_i \in Y$ there is exactly $C_{n-1=3}^{kL-1=1} = 3$ ways to choose a value $y_{j\neq i} \in Y$. That is, the number of substitutions, which for an arbitrary fixed $y_i \in Y$ in a cyclic decomposition contain k = 2 cycles (y_i) $(y_{j\neq i})$ of length l = 1, is equal to $6 \cdot \frac{C_{n-1=3}^{kL-1=1}}{C_{n=4}^{kL=2}} = 3$. For example, for $y_1 \in Y$ the seventh, fifteenth and twenty second substitutions contain k = 1 cycles of length $l = 2$ each.

Let's also consider substitutions containing $k = 2$ cycles of length $L = 2$. In total there are 3 such substitutions in S_n, this (see Table 1):

- the eighth substitution with a cyclic decomposition $(y_1, y_2) (y_3, y_4)$;
- seventeenth substitution with cyclic decomposition $(y_1, y_3) (y_2, y_4)$;
- last, twenty fourth substitution with cyclic decomposition $(y_1, y_4) (y_2, y_3)$.

The number of substitutions, which for any fixed $y_i \in Y$ contains $k = 2$ cycles $(y_i, y_{j \neq i})$ and $(y_{u \neq i,j}, y_{v \neq i,j,u})$ of length $L = 2$, is calculated in the same way. In total, there are exactly $C_{n=4}^{kL=4} = 1$ ways to simultaneously select values $y_i, y_{j \neq i}, y_{u \neq i,j}, y_{v \neq i,j,u} \in Y$. This choice is no longer limited, because for selected $y_i, y_{j \neq i}, y_{u \neq i,j}, y_{v \neq i,j,u} \in Y$ corresponding $k = 2$ cycles $(y_i, y_{j \neq i})$ та $(y_{u \neq i,j}, y_{v \neq i,j,u})$ are defined unambiguously (according to the formula there are $C_{n-1=3}^{kL-1=3} = 1$ variants). That is, the number of substitutions, which for an arbitrary fixed value $y_i \in Y$ have cyclic decomposition $(y_i, y_{j \neq i})$ $(y_{u \neq i,j}, y_{v \neq i,j,u})$ coincides with the total number of substitutions with $k = 2$ cycles of length $L = 2$, i.e. equal to 3.

No substitution from S_4 can have $k = 3$ and $k = 4$ cycles of length $L = 2$. We immediately proceed to calculate the probability of occurrence of the cycle $(y_i, y_{j \neq i})$ in randomly selected substitutions. We sum up the number of substitutions, which for an arbitrary fixed $y_i \in Y$ in their cyclic decomposition contain a different number of cycles of length $L = 2$, namely, we have 3 substitutions with $k = 1$ cycle $(y_i, y_{j \neq i})$, and three substitutions with $k = 2$ cycles $(y_i, y_{j \neq i})$ $(y_{u \neq i,j}, y_{v \neq i,j,u})$. In general, we have 6 substitutions from the total of 24 substitutions of the symmetric group, that is, at randomly equal probable selection of substitutions from S_4 the probability that in it for arbitrary fixed $y_i \in Y$ will be observed cycle $(y_i, y_{j \neq i})$ of length $L = 2$ is equal to $6/24 = 1/4$.

5.3   Estimation of Probability of Occurrence of a Cycle $(y_i, y_{j \neq i}, y_{u \neq i,j})$

Similar to the above, we will calculate the number of substitutions, which for arbitrary fixed $y_i \in Y$ have cycle $(y_i, y_{j \neq i}, y_{u \neq i,j})$. In an arbitrary substitution $s \in S_4$, there can be no more than one such cycle, i.e. cases with $k > 1$ are impossible. Each cycle of length $L = 3$ in the substitution decomposition is combined with a cycle of length 1, i.e. all eight of these substitutions are given in Table 3.

We will calculate the number of substitutions which for arbitrary fixed $y_i \in Y$ necessarily contain a cycle $(y_i, y_{j \neq i}, y_{u \neq i,j})$. In total, there are exactly $C_{n=4}^{kL=3} = 4$ ways to simultaneously select values $y_i, y_{j \neq i}, y_{u \neq i,j} \in Y$. But for each $y_i \in Y$ there are exactly $C_{n-1=3}^{kL-1=2} = 3$ ways to choose of values $y_{j \neq i}, y_{u \neq i,j} \in Y$. That is, the number of substitutions, which for an arbitrary fixed value $y_i \in Y$ necessarily have a cycle $(y_i, y_{j \neq i}, y_{u \neq i,j})$, is defined as $8 \cdot \frac{C_{n-1=3}^{kL-1=2}}{C_{n=4}^{kL=3}} = 6$. For example, for $y_1 \in Y$ these are 6 substitutions (№ 9, 12, 13, 16, 20, 21) with cyclic decompositions (see Table 3): (y_1, y_2, y_3) (y_4), (y_1, y_2, y_4) (y_3), (y_1, y_3, y_2) (y_4), (y_1, y_3, y_4) (y_2), (y_1, y_4, y_2) (y_3) and (y_1, y_4, y_3) (y_2).

Consequently, at randomly equal probable selection of substitutions from S_4 the probability that in it for arbitrary fixed $y_i \in Y$ will be observed cycle $(y_i, y_{j \neq i}, y_{u \neq i,j})$ of length $L = 3$ is equal to $6/24 = 1/4$.

Table 5 Substitutions containing 1 cycle of length 4

№	Result of substitution				Decomposition of substitution to cycles
	$s(y_1)$	$s(y_2)$	$s(y_3)$	$s(y_4)$	
10	y_2	y_3	y_4	y_1	(y_1, y_2, y_3, y_4)
11	y_2	y_4	y_1	y_3	(y_1, y_2, y_4, y_3)
14	y_3	y_1	y_4	y_2	(y_1, y_3, y_4, y_2)
18	y_3	y_4	y_2	y_1	(y_1, y_3, y_2, y_4)
19	y_4	y_1	y_2	y_3	(y_1, y_4, y_3, y_2)
23	y_4	y_3	y_1	y_2	(y_1, y_4, y_2, y_3)

5.4 Estimation of Probability of Occurrence of a Cycle $(y_i, y_{j \neq i}, y_{u \neq i,j}, y_{v \neq i,j,u})$

In any substitution $s \in S_4$ there can be only one cycle of length $L = 4$. Such substitutions are given in Table 5.

Obviously, all such substitutions necessarily have in their single cycle all elements from set $Y = \{y_1, y_2, y_3, y_4\}$, i.e. for each $y_i \in Y$ in each substitution of Table 5 there will be a cycle $(y_i, y_{j \neq i}, y_{u \neq i,j}, y_{v \neq i,j,u})$. Check the above formula: $\frac{C_{n-1=3}^{kL-1=3}}{C_{n=4}^{kL=4}} = 1$, indeed, all substitutions from Table 5 will necessarily contain a cycle $(y_i, y_{j \neq i}, y_{u \neq i,j}, y_{v \neq i,j,u})$ for each arbitrary fixed $y_i \in Y$. Consequently, the probability of a cycle $(y_i, y_{j \neq i}, y_{u \neq i,j}, y_{v \neq i,j,u})$ in a randomly selected substitution $s \in S_4$ is equal to $6/24 = 1/4$.

6 Interpretation of the Received Results to the Properties of BSC

The obtained analytic expressions (15) and (16) allow us to estimate the number of substitutions from the symmetric group, which for a certain element of the set of transformations necessarily contain a cycle of a certain length with this element, and the corresponding probability to randomly select substitution with such cycle.

We use these formulas to study periodic properties of OFB mode. In this case, we assume that the probabilistic properties of substitutions generated by the encryption function correspond to certain properties of random substitution, i.e. they correspond to our representations of such "ideal" BSC, which at any given encryption key randomly and equally compares any encrypt text to any open text. In this way, the l-bit BSC will implement a subset of the symmetric group of substitutions of degree 2^l, selecting a particular substitution s from S_{2^l} associated with the entered encryption key. On the all set of cipher keys we can choose substitution, which for an arbitrary fixed $y_i = S \in Y = \{y_1, y_2, \ldots, y_{2^l}\}$ necessarily has a cycle

$(y_i, s_i(y_i), s_i^2(y_i), \ldots, s_i^{L-1}(y_i))$ of length $L = l_i \in \{1, 2, \ldots, n\}$. The probability of this event is determined by (16).

In practice this means that the probability of a cycle of any length in randomly chosen substitution from a symmetric group S_{2^l} for an arbitrary fixed element of a set does not depend on either this element or the length of the cycle. It depends only on the order $n = 2^l$ of the substitutions of the symmetric group S_n and is defined as the inverse value, that is equal to $1/n$. For any fixed value of the introduced initialization vector $y = S$, the probability that the corresponding output block y_i, $i = 0, 1, .., n - 1$ formed in the OFB mode will have a period of length $L = l_i \in \{1, 2, \ldots, n\}$ does not depend on either the value of this initialization vector or the length of the period. It is determined only by the degree of substitution, that is, in this case, the digit of the cipher and it is equal to 2^{-l}.

Determine the probability that the period of formed output blocks will be not less than 2^m blocks, i.e. the probability of such an event, when for fixed value of the initialization vector $y_i = S$ the corresponding output blocks will not be repeated during 2^m iterations by the formulas (1) and (3) when forming the output blocks y_i:

$$P(\forall i, j \in \{1, 2, \ldots, 2^m\} : y_i \neq y_j |_{i \neq j})$$

$$= 1 - \sum_{L=1}^{2^m} \sum_{k=1}^{[2^l/L]} P(\chi_{2^l,L} = k) \frac{C_{2^l-1}^{kL-1}}{C_{2^l}^{kL}}$$

$$= \sum_{L=2^m+1}^{2^l} \sum_{k=1}^{[2^l/L]} P(\chi_{2^l,L} = k) \frac{C_{2^l-1}^{kL-1}}{C_{2^l}^{kL}}. \tag{17}$$

Taking into account (16) we obtain:

$$P(\forall i, j \in \{1, 2, \ldots, 2^m\} : y_i \neq y_j |_{i \neq j})$$

$$= 1 - \sum_{L=1}^{2^m} 2^{-l} = \sum_{L=2^m+1}^{2^l} 2^{-l} = 1 - 2^{m-l}. \tag{18}$$

Thus, when the assumption about the correspondence of certain probabilistic properties of the cipher to the properties of a random substitution is correct, the probability of non-repetition of output block at a certain length is a function of this length. This fact determines the main limitation of the use of the OFB mode, it is directly derived from the results of the research. The main limitation on the use of the OFB mode for BSC "Kalyna" [49] is specified in Appendix G.2 "Limit on the total length of messages protected by the use of one key" namely:

- when the block size is equal to 128 bits, it is recommended to limit the number of blocks protected by the single key to value 2^{60} (16 million TB);
- when the block size is equal to 256 bits, it is recommended to limit the number of blocks protected by the single key to value 2^{124};

- when the block size is equal to 512 bits, it is recommended to limit the number of blocks protected by the single key to value 2^{251}.

Since, as shown in Sect. 2, the essence of protecting an informational message according to the OFB mode specification is in addition of output blocks to it, then the restrictions specified in appendix G.2 relate to the restrictions on the length of output blocks, which are formed by multiple encryption of the same non-secret initialization vector by the formulas (1), (3). That is, the implementation of the restrictions recommended by the specification of the national standard provides certain probabilistic indicators of non-periodic output blocks, namely:

- when the block size is $l = 128$ bits, and when performing the recommended limitation of the number of blocks protected by a single key with the size $2^m = 2^{60}$, the probability of non-repetition of the output blocks $1 - 2^{m-l} = 1 - 2^{-68} > 1 - 2^{-64}$ will be ensured;
- when the block size is $l = 256$ bits, and when performing the recommended limitation of the number of blocks protected by a single key with the size $2^m = 2^{124}$, the probability of non-repetition of the output blocks $1 - 2^{m-l} = 1 - 2^{-132} > 1 - 2^{-128}$ will be ensured;
- when the block size is $l = 512$ bits, and when performing the recommended limitation of the number of blocks protected by a single key with the size $2^m = 2^{251}$, the probability of non-repetition of the output blocks $1 - 2^{m-l} = 1 - 2^{-261} > 1 - 2^{-256}$ will be ensured.

More often, in the theory of information security, the inverse value is used, it is the probability that a formed output blocks with a length that does not exceed a certain limit will have at least one repetition. Taking into account (17) and (18), this probability will be determined as:

$$P_{l,m} = 1 - P(\forall i, j \in \{1, 2, \ldots, 2^m\} : \gamma_i \neq \gamma_j |_{i \neq j})$$

$$= \sum_{L=1}^{2^m} \sum_{k=1}^{[2^l/L]} P(\chi_{2^l, L} = k) \frac{C_{2^l-1}^{kL-1}}{C_{2^l}^{kL}} = 2^{m-l}. \tag{19}$$

Figure 2 shows the dependence $P_{l,m}$ and m for different l.

From the dependencies shown in Fig. 2, it is evident that increasing the length of the output blocks leads to an increase in the probability of any number of repetitions of the output blocks. These graphs can be used to justify certain restrictions, for example, if you want to reduce the probability of any number of repetitions of the output blocks, you must reduce its length.

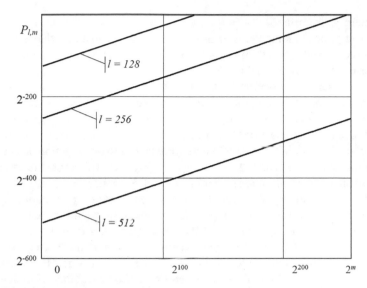

Fig. 2 The dependence of the probability that there will be at least one repetition in output blocks γ_i, $i = 0, 1, .., n - 1$ with the length not more than $2^m = 2^{60}$ blocks

7 Conclusions

Based on the obtained results, we can do conclusions that are important in practical terms.

1. Properties of modern symmetric cryptotransformations depend not only on the characteristics of BSC, but also on the mode of application. Therefore, the National Standard "Information technology. Cryptographic protection of information. The algorithm of symmetric block transformation" provides 10 modes of cryptotransformations: simple substitution (Electronic Codebook—basic transformation), Counter, Cipher Feedback, Symmetric Key Block Cipher-Based Message Authentication Code, Cipher Block Chaining, Output Feedback, Galois/Counter Mode and Galois Message Authentication Code, Counter with Cipher Block Chaining-Message Authentication Code, XOR Encrypt XOR (XEX) Tweakable Block Cipher, Key Wrapping.

2. In OFB mode, which is used to provide a privacy service, the output message is protected by addition of output blocks, which are formed by multiple encryption of the same non-secret initialization vector. If we assume that the properties of the cipher correspond to certain properties of random substitution, then the periodicity of the output blocks will be determined by the presence of cycles in randomly selected substitutions from the symmetric group, and the selection of the substitution is set to the value of the secret key.

3. Investigation of the properties of randomly selected substitution s from the symmetric group S_n has shown that the probability of cycle $s_i = (y_i, s_i(y_i), s_i^2(y_i), \ldots, s_i^{l_i-1}(y_i))$ of any length $L = l_i$ for an arbitrary fixed

element y_i from the set $Y = \{y_1, y_2, \ldots, y_n\}$ does not depend on either this element y_i or the length $L = l_i$ of the cycle. This probability depends only on the order n of the substitutions of the symmetric group S_n and is defined as an inverse value, that is equal to $1/n$.

4. Thus, the periodic properties of the output blocks in OFB mode are determined by the distribution of the probabilities of the number of cycles of random substitution. The selection of a secret key that parameterizes the encryption function corresponds to the selection of a particular substitution from the symmetric group; selecting the initialization vector value corresponds to the selection of an element y_i from the set of elements $Y = \{y_1, y_2, \ldots, y_n\}$ over which substitution occurs. But neither the actual value of the initialization vector nor the length of the output blocks period has any effect on the probability of obtaining an output blocks of a certain period. This probability is determined only by the degree $n = 2^l$ of substitution, that is, by the size l of the basic cipher transformation.

5. From the point of view of the practical application of symmetric cryptographic transformations to output blocks, the requirements of homogeneity are proposed at a length not exceeding the established limit. The probability of such event is determined by $1 - 2^{m-l}$ where 2^m is the length restriction of the output blocks. For example, for a 128-bit cipher "Kalyna", when we have limiting the length of a output blocks to 2^m in the OFB mode, the probability that the gamma blocks never coincide equals to $1 - 2^{m-l} = 1 - 2^{-68}$, i.e. it is much more than $1 - 2^{-64}$. The probability that at length no more 2^m blocks of output blocks in the OFB mode will coincide at least once, will equal $P_{l,m} = 2^{m-l}$. This dependence can be used to substantiate the restrictions on the length of the output blocks when the upper limit of probability $P_{l,m}$ is set.

6. The specification of BSC "Kalyna" [49, 50] recommended certain restrictions on the total length of messages protected by the use of one key. As for OFB mode such restrictions should be considered as requirements for the maximum length of output blocks, which with a certain probability will not be repeated. The above recommendations are fundamental because the occurrence of output blocks repetition is the most dangerous case when using the OFB mode, since in this case, the attacker will almost certainly violate the established privacy mode of the messages. For example, if the probability of a repetition of the output blocks is equal to 2^{-64}, then the length of output blocks for 128-bits BSC "Kalyna" should not exceed 2^{64} blocks (in the standard this restriction is more stringent and equal to 2^{60}).

7. The estimations of the probability of forming the output blocks with given period can be considered as a criterion for selection of cryptographic primitives, or criterion of the statistical test. Indeed, if the studied cryptographic primitive in OFB mode with equal probability forms output blocks with any period and this probability is determined by inverse to the degree of substitution, then by the cyclical properties this cryptographic primitive responsible to probabilistic properties of random substitution and by this criterion can be adopted for use. Actually such studies, particularly on nonlinear replacement nodes or reduced

BSC models are promising direction for further work. Disseminate the results to other modes is perspective.

It should be noted that research results can be useful not only in cryptography. In particular, the obtained analytical relationships can be used for other important practical applications in computer science, for forecasting and calculating telecommunication characteristics, etc. [51–54].

References

1. National Institute of Standards and Technology, Specification for the Data Encryption Standard (DES), Technical report NIST FIPS PUB 46-3, Department of Commerce, Oct 1999
2. National Institute of Standards and Technology, Specification for the Advanced Encryption Standard (AES), Technical report NIST FIPS PUB 197, Department of Commerce, Nov 2001
3. Anon, Information technology. Security techniques. Modes of operation for an n-bit cipher. Available at: http://dx.doi.org/10.3403/30062954
4. Elkamchouchi HM et al (2018). A new image encryption algorithm combining the meaning of location with output feedback mode. In: 2018 10th international conference on communication software and networks (ICCSN). Available at: http://dx.doi.org/10.1109/iccsn.2018.8488233
5. Kuznetsov A, Kolovanova I, Kuznetsova T (2017) Periodic characteristics of output feedback encryption mode. In: 2017 4th international scientific-practical conference problems of infocommunications science and technology (PIC S&T). Available at: http://dx.doi.org/10.1109/infocommst.2017.8246378
6. Kuznetsov O, Gorbenko Y, Kolovanova I (2016) Combinatorial properties of block symmetric ciphers key schedule. In: 2016 3rd international scientific-practical conference problems of infocommunications science and technology (PIC S&T). Available at: http://dx.doi.org/10.1109/infocommst.2016.7905334
7. Gorbenko I et al (2017) The research of modern stream ciphers. In: 2017 4th international scientific-practical conference problems of infocommunications science and technology (PIC S&T). Available at: http://dx.doi.org/10.1109/infocommst.2017.8246381
8. Heys HM (2003) Analysis of the statistical cipher feedback mode of block ciphers. IEEE Trans Comput 52(1):77–92. Available at: http://dx.doi.org/10.1109/tc.2003.1159755
9. Menezes A, van Oorschot P, Vanstone S (1996) Handbook of applied cryptography. Discrete mathematics and its applications. Available at: http://dx.doi.org/10.1201/9781439821916
10. Ferguson N, Schneier B, Kohno T (2015) Introduction to cryptography. In: Cryptography engineering, pp 23–39. Available at: http://dx.doi.org/10.1002/9781118722367.ch2
11. Moskovchenko I et al (2018) Heuristic methods of hill climbing of cryptographic boolean functions. In: 2018 international scientific-practical conference problems of infocommunications science and technology (PIC S&T). Available at: http://dx.doi.org/10.1109/infocommst.2018.8632017
12. Gorbenko I et al (2018) Experimental studies of the modern symmetric stream ciphers. In: 2018 international scientific-practical conference problems of infocommunications science and technology (PIC S&T). Available at: http://dx.doi.org/10.1109/infocommst.2018.8632058
13. Kuznetsov A et al (2017) Analysis of block symmetric algorithms from international standard of lightweight cryptography ISO/IEC 29192-2. In: 2017 4th international scientific-practical conference problems of infocommunications science and technology (PIC S&T). Available at: http://dx.doi.org/10.1109/infocommst.2017.8246380
14. Andrushkevych A et al (2018) A Prospective Lightweight Block Cipher for Green IT Engineering. In: Studies in systems, decision and control, pp 95–112. Available at: http://dx.doi.org/10.1007/978-3-030-00253-4_5

15. Jueneman RR (1983) Analysis of certain aspects of output feedback mode. Advances in cryptology, pp 99–127. Available at: http://dx.doi.org/10.1007/978-1-4757-0602-4_10

16. Altman J (2000) Telnet encryption: CAST-128 64 bit output feedback. Available at: http://dx.doi.org/10.17487/rfc2949

17. Ts'o T (2000) Telnet encryption: DES 64 bit output feedback. Available at: http://dx.doi.org/10.17487/rfc2953

18. Kuznetsov A et al (2018) Evaluation of algebraic immunity of modern block ciphers. In: 2018 IEEE 9th international conference on dependable systems, Services and Technologies (DESSERT). Available at: http://dx.doi.org/10.1109/dessert.2018.8409146

19. Kuznetsov OO et al (2018) Algebraic immunity of non-linear blocks of symmetric ciphers. Telecommun Radio Eng 77(4):309–325. Available at: http://dx.doi.org/10.1615/telecomradeng.v77.i4.30

20. Dong X (2008) Output feedback sliding mode control for a class of mismatched uncertain systems. In: 2008 27th Chinese control conference. Available at: http://dx.doi.org/10.1109/chicc.2008.4605032

21. Alsultanny YA (2008) Testing image encryption by output feedback (OFB). J Comput Sci 4(2):125–128. Available at: http://dx.doi.org/10.3844/jcssp.2008.125.128

22. Kuznetsov A et al (2018) Periodic properties of cryptographically strong pseudorandom sequences. In: 2018 international scientific-practical conference problems of infocommunications science and technology (PIC S&T). Available at: http://dx.doi.org/10.1109/infocommst.2018.8632021

23. Asaad R et al (2017) Advanced encryption standard enhancement with output feedback block mode operation. Acad J Nawroz Univ 6(3):1–10. Available at: http://dx.doi.org/10.25007/ajnu.v6n3a70

24. Gorbenko I et al (2018) Strumok keystream generator. In: 2018 IEEE 9th international conference on dependable systems, Services and Technologies (DESSERT). Available at: http://dx.doi.org/10.1109/dessert.2018.8409147

25. Biryukov A, Chosen plaintext and chosen ciphertext attack. In: Encyclopedia of cryptography and security, pp 77–77. Available at: http://dx.doi.org/10.1007/0-387-23483-7_61

26. Meyer CH (1978) Ciphertext/plaintext and ciphertext/key dependence vs. number of rounds for the data encryption standard. In: Proceedings of the 1978 national computer conference, AFIPS Press, Montvale

27. Kuznetsov A et al (2018) Research of cross-platform stream symmetric ciphers implementation. In: 2018 IEEE 9th international conference on dependable systems, Services and Technologies (DESSERT). Available at: http://dx.doi.org/10.1109/dessert.2018.8409148

28. Blakley GR (1979) Safeguarding cryptographic keys. In: Proceedings of the national computer conference, 1979. AFIPS Press, vol 47, pp 313–317

29. Hellman ME, Reyneri JM, The distribution of drainage and the DES. In: Advances in cryptography; proceedings of CRYPTO 82. Plenum Publishing Corp., 233 Spring Street, New York, NY 10013

30. Gait J (1977) A new non-linear pseudo-random number generator. IEEE Trans Softw Eng SE-3(5):359–363

31. Kuznetsov O et al (2018) Lightweight stream ciphers for green IT engineering. Studies in systems, decision and control, pp 113–137. Available at: http://dx.doi.org/10.1007/978-3-030-00253-4_6

32. Davies DW, Parkin GIP (1983) The average cycle size of the key stream in output feedback encipherment. In: Advances in cryptology, pp 97–98. Available at: http://dx.doi.org/10.1007/978-1-4757-0602-4_9

33. Kuznetsov O, Lutsenko M, Ivanenko D (2016) Strumok stream cipher: Specification and basic properties. In: 2016 3rd international scientific-practical conference problems of infocommunications science and technology (PIC S&T). Available at: http://dx.doi.org/10.1109/infocommst.2016.7905335

34. Campbell C (1978) Design and specification of cryptographic capabilities. IEEE Commun Soc Mag 16(6):15–19. Available at: http://dx.doi.org/10.1109/mcom.1978.1089775

35. Orceyre M, Heller R (1978) An approach to secure voice communication based on the data encryption standard. IEEE Commun Soc Mag 16(6):41–50. Available at: http://dx.doi.org/10.1109/mcom.1978.1089785

36. Sachkov VN, Kolchin V (1996) Combinatorial methods in discrete mathematics. Available at: http://dx.doi.org/10.1017/cbo9780511666186

37. Sachkov VN, Vatutin VA (1997) Probabilistic methods in combinatorial analysis. Available at: http://dx.doi.org/10.1017/cbo9780511666193

38. Newman SC (2012) A classical introduction to galois theory. Available at: http://dx.doi.org/10.1002/9781118336816

39. Lisitskaya I, Grinenko T, Bezsonov S (2015) Differential and linear properties analysis of the ciphers rijndael, serpent, threefish with 16-bit inputs and outputs. East-Eur J Enterp Technol 54(77):50. Available at: http://dx.doi.org/10.15587/1729-4061.2015.51701

40. Li R, Sun B, Li C (2011) Impossible differential cryptanalysis of SPN ciphers. IET Inf Secur 5(2):111. Available at: http://dx.doi.org/10.1049/iet-ifs.2010.0174

41. Krasnobayev V et al (2018) Improved method of determining the alternative set of numbers in residue number system. In: Recent developments in data science and intelligent analysis of information, pp 319–328. Available at: http://dx.doi.org/10.1007/978-3-319-97885-7_31

42. Lisickiy K, Dolgov V, Lisickaya I (2017) Block cipher with improved dynamic indicators of the condition of a random substitution. In: 2017 4th international scientific-practical conference problems of infocommunications science and technology (PIC S&T). Available at: http://dx.doi.org/10.1109/infocommst.2017.8246424

43. Zhang K, Guan J, Hu B (2016) Some properties of impossible differential and zero correlation linear cryptanalysis on TEA family-type ciphers. Secur Commun Netw 9(18):5746–5755. Available at: http://dx.doi.org/10.1002/sec.1733

44. Biryukov A, Cannière C, Linear cryptanalysis for block ciphers. In: Encyclopedia of cryptography and security, pp 351–354. Available at: http://dx.doi.org/10.1007/0-387-23483-7_233

45. Dolgov VI, Lisitska IV, Lisitskyi KY (2017) The new concept of block symmetric ciphers design. Telecommun Radio Eng 76(2):157–184. Available at: http://dx.doi.org/10.1615/telecomradeng.v76.i2.60

46. Lisickiy K, Dolgov V, Lisickaya I (2017) Cipher with improved dynamic indicators of the condition of a random substitution. In: 2017 4th international scientific-practical conference problems of infocommunications science and technology (PIC S&T). Available at: http://dx.doi.org/10.1109/infocommst.2017.8246425

47. Rodinko M, Oliynykov R (2017) Open problems of proving security of ARX-based ciphers to differential cryptanalysis. In: 2017 4th international scientific-practical conference problems of infocommunications science and technology (PIC S&T). Available at: http://dx.doi.org/10.1109/infocommst.2017.8246385

48. Kuznetsov OO, Ivanenko DV, Kolovanova IP (2014) Analysis of collision properties of galois message authentication code with selective counter. Bull V. Karazin Kharkiv Natl Univ 1097(23):55–71 (Mathematical Modelling, Information Technology, Automated Control Systems) (In Russian)

49. DSTU 7624 (2014) Information technologies. Cryptographic data security. Symmetric block transformation algorithm. Available at: http://shop.uas.org.ua/ua/informacijni-tehnologii-kriptografichnij-zahist-informacii-algoritm-simetrichnogo-blokovogo-peretvorennja.html

50. A New Encryption Standard of Ukraine: The Kalyna Block Cipher. Cryptology ePrint Archive: report 2015/650. Available at: https://eprint.iacr.org/2015/650.pdf

51. Ageyev D et al (2018) Method of self-similar load balancing in network intrusion detection system. In: 2018 28th international conference radioelektronika (RADIOELEKTRONIKA). IEEE, pp 1–4. https://doi.org/10.1109/radioelek.2018.8376406

52. Radivilova T, Hassan HA (2017) Test for penetration in Wi-Fi network: attacks on WPA2-PSK and WPA2-enterprise. In: 2017 international conference on information and telecommunication technologies and radio electronics (UkrMiCo), IEEE, pp 1–4

53. Lyudmyla K, Vitalii B, Tamara R (2017) Fractal time series analysis of social network activities. In: 2017 4th international scientific-practical conference problems of infocommunications science and technology (PIC S&T). IEEE, pp. 456–459. https://doi.org/10.1109/infocommst.2017.8246438
54. Kirichenko L, Ivanisenko I, Radivilova T (2016) Dynamic load balancing algorithm of distributed systems. In: 2016 13th international conference on modern problems of radio engineering, telecommunications and computer science (TCSET), IEEE, 2016, pp 515–518

Information-Measuring System of Polygon Based on Wireless Sensor Infocommunication Network

Igor Shostko⬤, Andriy Tevyashev⬤, Mykhaylo Neofitnyi⬤, and Yuliia Kulia⬤

Abstract The work is devoted to the development of methods for constructing an optical information-measuring system of a landfill for the accompany of air objects moving along a ballistic or linear trajectory at high speed. The basic idea is to use small-scale optical-electronic stations that are located on the range according to the task of trajectory measurements. All optical-electronic stations are combined into a single information and measuring system. Each opto-electronic station in its area of responsibility is programmed to support the target in the predicted trajectory. The programming process is automated and carried out at the same time for everyone stations. Determining the location of each optoelectronic station on the ground carried out by the GPS monitoring. An alternative method for determining the location of each optoelectronic station on a landfill using the technology of wireless sensor networks.

Keywords Information-measuring system · Detection technology and high precision object tracking · Wireless sensor infocommunication network · Optoelectronic station of trajectory measurements

I. Shostko (✉) · A. Tevyashev · M. Neofitnyi · Y. Kulia
Kharkiv National University of Radio Electronics, 14 Nauky Ave, Kharkiv, Ukraine
e-mail: ihor.shostko@nure.ua

A. Tevyashev
e-mail: tad45ua@gmail.com

M. Neofitnyi
e-mail: mykhailo.neofitny@nure.ua

Y. Kulia
e-mail: yuliia.kulia@nure.ua

T. Radivilova et al. (eds.), *Data-Centric Business and Applications*, Lecture Notes on Data Engineering and Communications Technologies 48,
https://doi.org/10.1007/978-3-030-43070-2_28

1 Introduction

At the final stage of development of new samples of ammunition, it is necessary to test them at the landfill. Therefore, there is a need for a special opto-electronic station for trajectory measurements (OESTM). OESTM can be used for certification of various aviation and rocket artillery systems. OESTM provides trajectory measurements in real-time. The result of the test is to record the flight characteristics of the ammunition at the control sections of the trajectory. The resulting video information is used for further analysis of the technical characteristics of the object tested. If it is important to observe the ammunition at many control sections of the trajectory, it is necessary to combine several OESTM into a single information-measuring system (IMS) of the landfill.

The analog of the IMS is a multi-point mobile measuring complex, manufactured by VAT NPIC «Armint» (Russia). The complex includes: digital theodolites designed to determine angular coordinates; high-speed recording module; information processing system; central control point. The digital theodolites at the landfill are located along the ballistic track so as to provide observation of the object in its various parts. There are also other developments in the IMS and in separate structures of optical-electronic stations (OES) made in the USA, Germany, Britain, Israel and Canada. But at this time, the capabilities of the known analogs of optical IMS do not allow to meet the growing requirements, such as:

- accuracy of support of air objects moving along a ballistic or linear trajectory at high speed;
- registration of the results of trajectory measurements, processing and transmission of video stream in real-time.

Thus, there is a problem concerning the insolvency of the testing ground to provide testing of modern models of missile and artillery equipment. To solve this problem, we propose a new solution for the methodology of construction of IMS, the software implementation of which is new and world-class.

2 Relevance of Research

The main objective of the testing ground is measuring the external-trajectory parameters of the flight of objects (coordinates, velocity vector, angular position in space, etc.). The values of the above parameters evaluate the quality of the functioning of objects. Doppler radars and OESTM are used to measure these parameters. Doppler radars are used to map the flight path of a target. OESTM is used when it is necessary to visually evaluate the quality of functioning of the test object from the time of departure of artillery ammunition and for rockets or jet ammunition, respectively, from the moment of descent from the launcher to the point of a hit. The initial speeds of various rocket artillery products can vary from 50 to 2000 m/s, and their flight distances—depending on the size of the landfill.

There are currently many different optical tracking systems for moving objects, but they are unsuitable for accompanying aerial objects moving along a ballistic or linear trajectory at high speed. This is due to the following reasons:

1. The considerable magnitude of the inertia of the support-rotary device in the accompaniment of ammunition in flight complicates, and in some cases, does not ensure the registration of the real motion parameters of modern ammunition.
2. The large mass and size of the complex do not allow speaking about its mobility in the conditions of the landfill. The inconvenience of transportation and complexity of maintenance (in many respects as a result of large masses and dimensions) in the field conditions at landfill tests makes it impractical, and sometimes—it is fundamentally impossible to use it.
3. The high cost of the equipment makes it virtually impossible to purchase it in the right amount. In addition, the high cost of the equipment does not allow it to be exposed to the risk of possible irreversible damage when testing ammunition.

Therefore, given these shortcomings, the development of specialized OESTM and their integration into a single IMS landfill is required.

Thus, the tasks that this research is aimed at are relevant.

3   Aims and Objectives of the Research

Object of study. The process of trajectory measurements by optical means.

Subject of study. Methods for automating and adapting the IMS based OESTV to accompany the air objects moving on a ballistic trajectory at high speed.

The purpose of the work is to develop recommendations for the construction of optical IMS, which provides support for air objects moving along a ballistic or linear trajectory at high speed).

In order to achieve this goal a solution of the following scientific and technical tasks is necessary:

1. Development of detection methods and high-precision tracking of air objects moving along a ballistic or linear trajectory at high speed.
2. Development of management, synchronization, and correction methods for targeting for each OESTM in the course of aerial object support.
3. Merging of OESTM in IMS.

4   Optical-Electronic Station of Trajectory Measurements

The problem of detecting and tracking aerial objects moving along a ballistic or linear path at high speed in a video stream is extremely complex. One of the ways to solve this problem for the motion of essentially non-stationary objects in a substantially non-stationary environment is proposed by the authors on the basis of information-analytical technology (IAT) detection and high-precision tracking of air objects [1]. IAT is an orderly sequence of procedures for obtaining, processing, analyzing video information, making and implementing decisions to achieve a given goal in the context of high a priori uncertainty about the behavior of the object of observation and the environment. IAT is a hierarchically ordered software package of libraries of methods for optimal processing of video stream, obtained from one or multiple video cameras.

OESTM should include:

- supporting and rotating device with a horizontal system;
- opto-electronic module (wide-angle television or thermal imaging camera, narrow-angle ultra-fast television camera);
- Digital Imaging Module (DIM), taking into account two modifications:

    - PC-based (Intel Core i7-8700K3.7 GHz/12 Mb processor, memory: 8G DDR3, PCI-E 6 Gb GeForce GTX 1060);
    - based on the board—Xilinx Zynq UltraScale + MPSoC ZCU104.

The Xilinx Zynq UltraScale + MPSoC ZCU104 board (Fig. 1) provides capture camera image and hardware processing to minimize delays and increase the speed of the OESTM digital image processing module (DIPM).

To compare the two modifications of the DIPM, their testing was performed. Testing of the algorithm of tracking the aerial objects and assessing its performance was carried out in the laboratory on dynamic scenes using a simulated moving air object, placed on the background, which allows simulating different conditions of visibility (contrast) of the accompanying object. In the experiment, the movement of the object was recorded using a video camera, real-time video stream was processed using Field-Programmable Gate Array (FPGA), video image after processing was displayed on the monitor screen. As a result of the video stream processing, the coordinates of the object are calculated and the object is maintained and held in the frame.

The algorithm performance was evaluated by measuring the time spent processing each frame in sequence. As a result of testing, it was found that when executing the FPGA algorithm, the time spent processing the frame does not depend on the object configuration, frame filling, and background characteristics.

The results of comparing the performance of the calculations to implement the Harris algorithm on a personal computer (PC) and FPGA are shown in Table 1.

The table shows that the FPGA-based approach shows much better performance than the PC. For example, the speed of an algorithm when executed on an FPGA at a frame size of 1920 × 1080 exceeds the speed of execution on a PC more than 20 times.

Fig. 1 Xilinx Zynq UltraScale + MPSoC ZCU104 board

Table 1 The results of the comparison

Resolution video frame	Processing time using a PC, ms	The processing using FPGA, ms
640 × 480	17.5	0.98
1280 × 480	68.2	4.3
1920 × 1080	272.3	13.6
1920 × 1920	613.5	46.7

5 Structure of the Information-Measuring System

It is proposed to use wireless sensor network (WSN) for data collection and transmission in IMS [2–17], which will unite all small-sized OESTM (Fig. 2), which are located in the territory of the landfill Fig. 3.

All OESTM are merged into a single IMS. Each OESTM in its area of responsibility is programmed to support the target in the predicted trajectory. The programming process is automated and carried out simultaneously for all OESTM. The work of all OESTM will be synced. The objectives of OESTM, which will be the first to accompany the target:

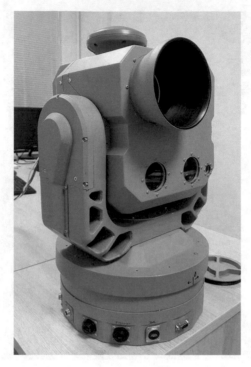

Fig. 2 Optical-electronic station of trajectory measurements

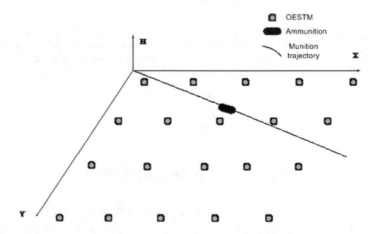

Fig. 3 Location of optical-electronic stations of trajectory measurements

- identify inconsistencies in the parameters of its predicted and real trajectory;
- provide guidance on pointing correction for all other stations.

Based on these data, the coordinates of the expected ammunition capture point of each OESTM are adjusted. The number of OESTM in IMS is determined depending on the test objectives.

It is problematic to measure the range of a target moving at a speed of 50–2000 m/s with a laser rangefinder. Therefore, it is recommended that trajectory measurements be made by the pelenhatsiynym method. Meanwhile, two OESTM simultaneously monitor the movement of one object and determine its coordinates as a function of time (Fig. 4). OESTM work is synchronized by a single time service.

The network structure is constructed in such a way that two switches synchronously serve the OESTM pair and then alternately the next pair and so on (Fig. 5) [22]. The survey sequence is set according to the order of work of each OESTM couple on the target.

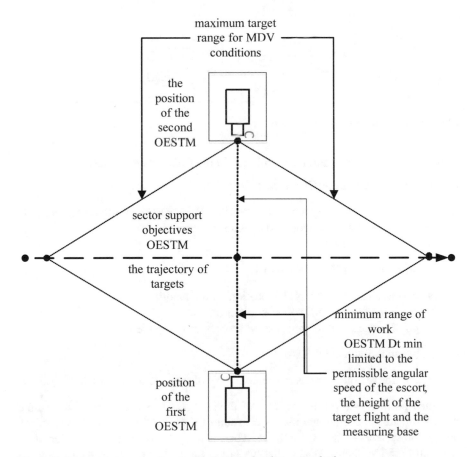

Fig. 4 Trajectory measurements using by the pelenhatsiynym method

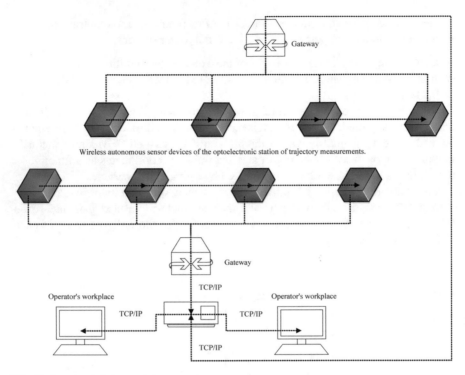

Fig. 5 Structural diagram of the network

The following information is circulating in the IMS sensor network:

(1) From command post (CP) to each OESTM:

- Targets for targeting the accompanying flight path at a predetermined point;
- team prohibition of work for the chosen purpose;
- time synchronization.

(2) From each OESTM to CP landfill:

- coordinates of the OESTM location on the landfill;
- confirmation of readiness to support the target;
- confirmation that the goal support has been completed.

(3) Between individual OESTM:

- corrective amendments to the trajectory of target, accompanied.

When placing IMS on the territory of the polygon, it is necessary to determine the coordinates and orientation on the ground for each OESTM according to the navigation systems—GPS, GLONASS. The accuracy of positioning and orientation for each OESTM should ensure that the target on the CP targeting data is uncovered, with correction from the first OESTM. For this orientation mean square error determination must not exceed 1/10 of the minimum angle camera search. With a minimum angle of view of the search camera 2.3°, the square root error of orientation determination should not exceed 0.23°. Accordingly, with a maximum detection distance of 5 km, the mean square error of the definition of the OESTM should not exceed 20 m.

For the stationary location of the IMS, it is proposed to determine the location of each OESTM using a portable GPS monitoring module (Figs. 6, 7 and 8).

Fig. 6 GPS monitoring system kit

Fig. 7 Fixing GPS sensor on OESTM

Fig. 8 Measurement of coordinates of the point location of OESTM

For the mobile location of the IMS, the GPS monitoring module is located at the first OESTM. To determine the position of other OESTM it is proposed to use distance measurement methods between OESTM using wireless sensor network technologies. Based on the known coordinates of the location of the base OESTM and measurement results of distances between each OESTM calculated coordinates of their location.

6 Method for Determining the Distance Between OESTM Using the Wireless Sensor Network Technology

In wireless networks, there are a number of methods for measuring communication range, but we would like to highlight two main ones, this RSSI method (Received Strength Signal Indication), based on the determination of the received signal strength, and ToF (Time of Flight), based on measuring the time between the passage of the signal between the nodes [18–22]. Under the nodes, we will further name the receiving and transmitting modules based on ZigBee technology, which we install on each OESTM. There are many different technologies and protocols that build on the WSN. The standards governing the operation of the network are described in the family of specifications IEEE 802.15.4. These specifications are the base for many wireless technologies, but when building ZigBee WSN is most prevalent. The main feature of ZigBee technology is that it at low power consumption supports not only simple network topologies, but also a self-organizing and self-healing mesh topology with relaying and message routing. Let's consider the methods of measuring the range of communication more detail.

6.1 RSSI Method

The simplest method for determining the range from object to node is to indicate the level of the received signal. Any wireless channel under IEEE 802.15.4 has a protocol function for valuating communication quality (Link Quality Indicator), whose action reduces to the determination of the power of the received signal P (dBm) [18]. Since under ideal conditions the power is inversely proportional to the square of the distance, the power logarithm is proportional to the distance with some coefficient, which is established empirically.

However, this method has a number of significant limitations, since the signal level is a highly volatile parameter due to the influence of the following factors:

- rapid and slow fading of signals on the circuit due to changes in the conditions of propagation of radio waves;
- multipath propagation due to reflections from various metallic objects;
- Distribution of output power of transmitters and sensitivity of receivers;
- The influence of the antenna orientation due to the irregularity of the directional pattern.

Due to the influence of these factors, the real power dependence on the distance turns out to be nonlinear and inconsistent in time, due to which the accuracy of measurements drops rapidly with increasing distance. The practical application of this method for IMS is limited to distances of up to 20 m, with the distance being measured with an accuracy of about 3 m.

6.2 ToF Method

Another approach is based on measuring the passage time (span) of the signal. A node sends a request to another node receives a response signal and determines its delay time. The complete delay consists of hardware delays in the processing of the received and the formation of response signals and the propagation time between the nodes. Because technical delays are known with good accuracy, they can be subtracted from the full value, and the remaining value will characterize the time of the flight of the signal there and back. Multiplying half the delay time by the speed of light, we get the distance between the nodes of the network. In this method, a linear connection between the distance and the measured value is provided. The accuracy of measurements is determined by a number of factors:

- frequency stability of the set generators;
- delayed fronts of pulses when receiving and emitting;
- signal/ noise ratio;
- errors due to multipath propagation.

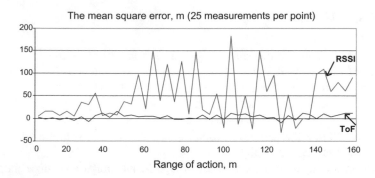

Fig. 9 Dependencies of distance measurement errors by RSSI and ToF methods on distance

To increase accuracy, repeated repeats of the measurement procedure are used. Filtration of abnormally large deviations caused by multipath propagation is used, which allows several times to increase the accuracy of measurements.

Figure 9 shows the experimental results of distance measurements using RSSI and ToF methods.

From the graphs, it is seen that the error in the use of the ToF method is significantly less than in the RSSI, which already at distances of more than 20 m, the measurement errors exceed the range itself. The ToF method is effective over the entire network range.

Distance measurement is not the only option for wireless networks. When using high-precision measurement algorithms, one can also determine the coordinates of the OESTM.

7 Calculation of the Range of Reliable Transmission of Messages for the Nodes of the Wireless Sensor Network

The range of wireless communication can be estimated by the well-known ratio

$$R_c \leq \sqrt{\frac{P_t \cdot G_t \cdot A_r \cdot K_{atten}}{4 \cdot \pi \cdot P_{thres}}}, \qquad (1)$$

where

R_c is communication range;
P_t is impulse power of the transmitter, W;
G_t is transmitter antenna gain;
A_r is effective receiver antenna area, m^2;

$K_{atten} = q_t \cdot q_r \cdot K_p$ is signal power attenuation ratio, $0 \leq K_{atten} \leq 1$;

$q_t(q_r)$ are relative gains of the transmitter (receiver) antennas towards each other $0 \leq q_t \leq 1, 0 \leq q_r \leq 1$;
$K_p = \prod_{i=0}^{k} K_{pi}$ is loss factor characterizing active energy losses associated with absorption in the propagation medium, etc., $0 \leq K_p \leq 1$;
P_{thres} is receiver threshold sensitivity, W.

An analysis of the technical characteristics of the radio modules of the unattended nodes of the sensory network showed that, for different variants for constructing topology of its loss in the radio channel due to mismatch maxima antenna patterns of the transmitter and the receiver, the loss due to absorption in the propagation medium can vary from 0 to 20 dB.

$$K_{atten} = -(0 \ldots 20)dB.$$

In Fig. 10 have been shown the dependence of the average power of a received signal on the communication range without taking into account the attenuation of the received signal and when it is attenuated in the radio channel by 10 and 20 dB.

Losses can be compensated by adjusting the gain of the transmitter power and the threshold sensitivity of the receiver depending on the conditions of the signal passing through the radio channel, the noise level, the range and the relative position of the nodes. As a source of data, you can use the results of assessing the quality of communication LQI (Link Quality Indicator) and measuring the range of communication using the methods RSSI or ToF. As a result of processing these data in the microcontroller, a command should be developed to control the automatic gain control (AGC) in the radio receiver and automatic power adjustment (APA) in the radio transmitter (Fig. 11).

The power at the transmitter output is determined in accordance with the condition: the energy of the received signal must exceed the threshold energy level

$$\frac{E_b(\theta, \varphi, R_c)}{E_{kr}} \geq 1,$$

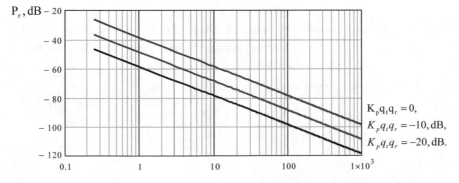

Fig. 10 The average power of the received signal depending on the communication range

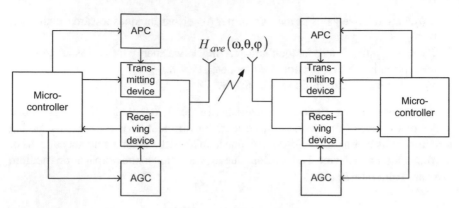

Fig. 11 Block diagram of the management of the energy characteristics of the transceiver, depending on radio conditions

where $E_b(\theta, \varphi, R_c)$ is dependence of the energy of one bit of information at the receiver output on the range and direction of signal reception.

The value of E_{kr} is equal to the energy of one bit of information at the receiver output, which ensures the specified reliability of message transmission.

Message reliability is the degree of consistency between the received and the committed message. When sending discrete messages, the accuracy is determined by the error rate

$$K_{er} = n_{er}/n_b \approx \Psi_b,$$

where

n_{er}	is the number of mistakenly received message elements;
n_b	is the total number of message elements;
$\Psi_b = F\left(\sqrt{2E_b/N_0}\right)$	is dependence of the bit error E_b probability on the signal-to-noise ratio—bit energy to the noise spectral density N_0;
$F(x) = \frac{1}{2\pi} \int_x^\infty \exp\left(-\frac{t^2}{2}\right)dt$	is Gaussian error integral, which is used in the description of probability with Gaussian distribution density

The signal-to-noise ratio can be represented as

$$\frac{E_b}{N_0} = \frac{S_b T_b}{N/\Delta\omega_c} = \frac{S_b/R_b}{N/\Delta\omega_c} = \frac{S_b}{N}\frac{\Delta\omega_c}{R_b},$$

where

$E_b = S_b T_b$ is energy, one bit of information;

N_0 is the spectral power density of white noise in the channel;
S_b is the average power of received bits;
N is the average power of received bits;
T_b is the bit duration;
R_b is the bit rate

Since the bit time and the bit rate are mutually inverse, it T_b can be replaced by $1/R_b$.

Dimensionless attitude E_b/N_0 this is a standard qualitative measure of the performance of digital communication systems [20]. Therefore, the necessary relation E_b/N_0 can be considered as a metric, allows you to compare the quality of different systems: the smaller the required ratio E_b/N_0, the more effective the detection process for a given probability of error.

When transmitting a digital signal with QPSK modulation format, the number of levels is defined as

$$L = \sqrt{M},$$

and the energy of the signal symbol is determined by the formula:

$$E_s = E_b \cdot \log_2 L.$$

When transmitting binary pulses $E_s = E_b$, and when transmitting pulses with QPSK modulation in the main band coinciding with the Nyquist band $\Delta\omega_N$

$$\Delta\omega_N = \frac{1}{2T_b},$$

received symbol power $S_s = \frac{E_b}{T_b} \cdot \log_2 L$ and noise power $N = N_0\left(\frac{1}{2T_b}\right)$.

Consequently

$$\frac{S_s}{N} = 2(\log_2 L)\frac{E_b}{N_0} = m\frac{E_b}{N_0},$$

where m is mapping ratio (number of bits per symbol information).

For QPSK, the difference between S_s/N and E_b/N_0 will be 3 dB.

If we take the IEEE 802.15.4 standard as a basis, then in this standard, to increase the base of the signal, we use a code sequence of pulses [18, 19], therefore, the signal-to-noise ratio increases in B_p time $(B_p = T_s\Delta\omega)$, where $\Delta\omega$ spreading spectrum width

$$\frac{S_s}{N} = mB_p\frac{ST_b\Delta\omega_N}{N}. \qquad (2)$$

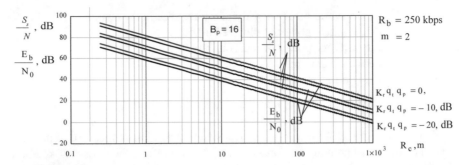

Fig. 12 Signal-to-noise ratio versus communication range (transmitter power 0 dBm; receiver sensitivity threshold −96 dBm)

Figure 12 shows the results of calculating the range of reliable messaging for WSN nodes. The operating frequencies of the IEEE 802.15.4 standard are from 2.405 to 2.455 GHz, the supported data transfer rate is 250 kbps.

8 Adaptation of the Wireless Sensor Network of the IMS Under the Conditions of Complex Electromagnetic Compatibility of Radio-Electronic Facilities of Landfill Facility

In the case of simultaneous operation of a wireless sensor network and other wireless telecommunication systems or radio-electronic means (REMs) centered on the landfill and using the same 2.4 GHz frequency resource it is required to solve the problem of electromagnetic compatibility (EMC), taking into account the limitations on the radio range for individual nodes IMS.

The spatial parameters of electromagnetic interference are characterized by the formation of a "conflict zone". Under the "conflict zone" we understand the area of space within which the level of energy and frequency spectrum of an electromagnetic field emitted by a technical tool (TT) does not allow simultaneous use of other TT without reducing the quality of their functioning. The dimensions of the "conflict zone" depending on the frequency band in which the field is generated, its energy level, as well as the method of its radiation and the surrounding propagation conditions. Accordingly, the objects should be placed on the landfill in such a way that their "conflict zone" do not overlap. To this end, you need to decide on the principles of adaptive control AGC of the automatic control system in radio receivers and APA workstations in radio transmitters of IMS nodes.

Fig. 13 Two-way radio link scheme

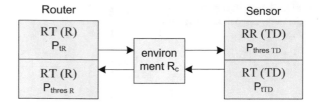

8.1 Adaptation Options for Energy Parameters

Analyze the work of the WSN IAT. It operates a lot of sensors, interconnected by two-way radio links. Consider, for the time being, one radio link that is formed: a terminal device (TD) and a router (R). Each device uses a radio transceiver (RT(TD)) and radio receiver (RR(TD)), RT(R) и RR(R). For all network sensors, receive threshold power P_{thres} radio receivers (RRs) and maximum radiation power P_t radio transceivers (RRs) are the same.

Figure 13 shows a block diagram of a two-way radio link, consisting of two network modules: RT(R)—RR(TD) and RT(TD)—RR(R). The diagram shows the parameters: P_{tR} is transmitter power Router (R); $P_{thres\,R}$ is receiver threshold sensitivity R; parameter R_c is radio range; P_{tTD} and $P_{thres\,TD}$ is relevant parameters TD.

Parameter R_c it takes the value of R_{min} is the smallest allowable range of approach of the receiver and transmitter devices TD and R to R_{max} is maximum range: $R_{min} \leq R_c \leq R_{max}$.

In a two-way radio link, energy balance must be observed.

$$P_{tR}P_{thresR} = P_{tTD}P_{thresTD}. \tag{3}$$

There is, therefore, an energy reserve that can be used for adaptation in the interest of better ensuring providing electromagnetic combination. With a range less than the maximum, it is permissible for the receiver to be crudity, i.e. increase $P_{thresTD}$ or P_{thresR}, or reduction of radiation power P_{tTD} or P_{tR}. This should be done in such a way that relation (3) is satisfied. Thus, when adopting, it is need to change at least two parameters.

8.2 Deterministic Radio Model

Suppose that in RR realized the ideal amplitude characteristic, i.e., $P_{out} = P_{o\,out} = $ const over the entire dynamic range of input signals $P_{por} \leq P_r \leq P_{max}$ (Fig. 14). Thus, RR is represented as a device with a threshold signal power P_{thres}, designed to receive signals not exceeding P_{max}.

Fig. 14 The dependence of
the output signal power RR
P_{out} (P_r)

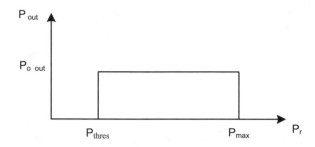

8.3 Statistical Estimate of the Threshold Power of the Radio

To assess the effectiveness of the automatic gain control system in the APA RR(TD), we use the network model with the star topology shown in Fig. 15.

In the center of the network is R, and the TD can be installed permanently or on a mobile object. The distance from the TD to the center R—random variable. Consequently, at the input of the control unit RR(TD), a random power signal will act:

$$P_r = P_t A R_c^{-2} K_{atten}, \qquad (4)$$

where P_t is transmitter power R; $A = G_t \cdot A_r / 4\pi$; K_{atten} is signal attenuation ratio between receiver and transmitter.

The random parameters in (4) are the distance R_c ($R_{min} \leq R_c \leq R_{max}$) and coefficient K_{atten}. Let one channel work in the WSN: RT(R)—RR(TD). We assume for simplicity AGC in RR(TD) has the ideal characteristic shown in Fig. 13. At the entrance RR(TD) power P_r will take random values ranging from $P_{thres} = P_t A R_{min}^{-2} K_{atten}$ to $P_{max} = P_t A R_{min}^{-2} K_{atten}$. Thanks to the ideal AGC, a constant power value is ensured throughout the entire range of distances $P_{0\,out}$.

Fig. 15 Wireless sensor
network IAT with star
topology

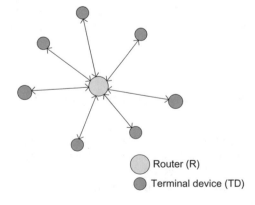

The location of the R within the WSN is considered equally likely throughout its area. Then the probability density of a random variable R_c:

$$\rho(R_c) = 2R_c/(R_{\max}^2 - R_{\min}^2), \ R_{\min} \leq R_c \leq R_{\max}. \tag{5}$$

In a rough terrain polygon, the random variable is also a signal attenuation coefficient K_{atten} in (4). So far, we can only give a rough estimate $\rho(K_{atten})$ or evaluate K_{atten} specific experimental conditions.

Formula (4) connects the signal power P_r at the receiver input and random parameters R_c and K_{atten}. This allows, by means of a functional transformation of the distribution (5) with regard to (4) and in the presence of a probability distribution $\rho(K_{atten})$ by which the average power of the signal is calculated $P_{r\,ave}$. With ideal, due to the AGC, amplitude response RR(TD) (Fig. 14), the value $P_{r\,ave}$ can be taken as the average receive threshold power.

Since we have no probability distribution $\rho(K_{atten})$, it can be calculated $\rho(P_r)$ at constant values K_{atten}, changing from 0 to 1 depending on the height of the location of the elements of the WSN, the terrain or buildings. In particular, for open space we take $K_{atten} = 1$. Then at a large dynamic range of signals when $P_{\max} \gg P_{por}$ we have

$$\rho(P_r) = P_{thres}/P_r^2, \ P_{thres} \leq P_r \leq P_{\max}.$$

Accordingly, the average value of the reception power threshold while the AGC is determined by the expression:

$$P_{r\,ave} = P_{thres} \ln(P_{\max}/P_{thres}). \tag{6}$$

Define for open space conditions ($K_{atten} = 1$) relative increase, with an ideal AGC, a threshold of the receiving power $P_{r\,ave}/P_{thres}$ of a radio receiver, if range R_c varies in range from $R_{min} = 10$ m to $R_{max} = 1$ km. The dynamic range of input signals

$$\frac{P_{\max}}{P_{thres}} = \frac{P_t A R_{\min}^{-2}}{P_t A R_{\max}^{-2}} = \frac{R_{\max}^2}{R_{\min}^2} = 10^4.$$

Using (6), we obtain

$$\frac{P_{r\,ave}}{P_{thres}} = \ln 10^4 \approx 9, 2.$$

8.4 Radio Interference

The aggregate of interference at the input RR of the control switchgear will be divided into two groups: interference, depending on the topology and the spatial arrangement of the WSN elements, and interference independent of this. At the input RR of the switchgear will act interference, internal and external, the total power of which is equal to P_{noise}. In addition, in the case of simultaneous operation of the WSN and other radio electronic systems (RESs) concentrated on the same territory, an additional specific interference is formed as a result of repeated use of radio channels coinciding in tuning. Let's call it a network radio interference whose power P_{NRI}. Thus, at the input of the control switchgear RR (TD), there will be an interference

$$P_{int} = P_{noise} + P_{NRI}, \tag{7}$$

where

P_{int} is interference power;
P_{noise} is noise power;
P_{NRI} is radio interference power.

As shown above for each WSN element carried by range selection. Set the minimum distance D between TD WSN and other radio electronic systems:

$$D = R_{max}\sqrt{2}.$$

As the range between the TD and other RESs increases, the propagation conditions of radio waves deteriorate sharply and if the range R exceeds the limit value D_{rl}, interference can be neglected.

Interference power can be calculated by the formula (4). The power $P_{NRI}(D_{rl})$ received by the antenna RR in the free space model is determined by the expression:

$$P_{NRI}(D_{rl}) = P_t A D_{rl}^{-2} K_{atten},$$

where D_{rl} is range limit.

We rewrite this equation in the form

$$P_{NRI}(D_{rl}) = P_t A D^{-2} K_{atten}\left(\frac{D}{D_{rl}}\right)^2.$$

Losses in the radio channel, $\overline{PL(D_{rl})}$ means that this signal attenuation between the receiving antenna of the TD WSN and the transmitting antenna of the interfering source is defined as

$$\overline{PL(D_{rl})} = \frac{P_t}{NRI(D_{rl})} = \frac{D^2}{AK_{atten}}\left(\frac{D_{rl}}{D}\right)^2.$$

In decibels, the above expression can be written like this

$$\overline{PL(D_{rl})} = PL(D)_{dB} + 20\log_{10}\left(\frac{D_{rl}}{D}\right).$$

The change P_{int} in the level at large distances due to rough terrain and the presence of large objects in the environment is the reason that two different receivers, equidistant from the transmitter, receive the transmitted signal with different power. The shading can be represented as an additional random component added to the path loss, and thus the effective path loss at a given distance from the transmitter is random and obeys the log-normal distribution with the mathematical expectation $PL(D_{rl})$:

$$PL(D_{rl})_{dB} = \overline{PL(D_{rl})}_{dB} + X_\sigma, \tag{8}$$

where $X_\sigma \sim N(0, \sigma^2)$ are Gaussian random changes.

The WSN model for calculating network radio interference thus becomes two-threshold; Interference range varies from D to D_{rl}. A graphic representation of such a model is shown in Fig. 16. Let nodes WSN and other sources cause interference evenly spread in the area of a circle centered at the point O' and radius R_o. Center study WSN is point O, which is also the center of the circles with radii $D_1 = R_{max}+D$ and $D_2 = R_{max} + D_{rl}$.

The distribution of the probability density of the distances R from these sources to the RR (TD) will have the form of a linear function:

Fig. 16 Graphic representation of the WSN model for calculating network radio interference

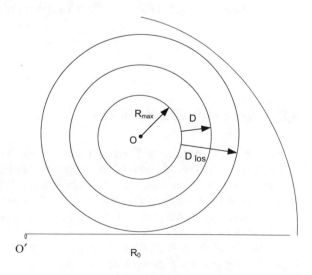

$$\rho(R) = \frac{2R}{D_{rl}^2 - D^2}, \ D \le R \le D_{rl}. \tag{9}$$

Having performed the functional transformation $\rho(R)$ taking into account dependence (4), which will be considered true for all distances $R \le D_{rl}$, we obtain

$$\rho(P_{NRI}) = 2 / \left\lfloor n \left(P_{in\,min}^{-2/n} - P_{inc\,max}^{-2/n} \right) P_r^{(2/n)+1} \right\rfloor$$

$$P_{inc\,min} \le P_r \le P_{inc\,max} \tag{10}$$

where $P_{incmin} = AP_t D_{rl}^{-2} K_{atten}$; $P_{incmax} = AP_t D^{-2} K_{atten}$; $n = \log_{Drl}$; $D_{rl}^2 K_{atten}$.

With the help of (10) you can find the average power of the network radio interference:

$$P_{ave}^{(n)} = \frac{2 \left(P_{inc\,max}^{(-2/n)+1} - P_{inc\,min}^{(-2/n)+1} \right)}{(n-2) \left(P_{inc\,min}^{(-2/n)} - P_{inc\,max}^{(-2/n)} \right)}, \tag{11}$$

$$(n \ne 2),$$

$$P_{ave}^{(2)} = \ln(P_{inc\,max} / P_{inc\,min}) / \left(P_{inc\,min}^{-1} / P_{inc\,max}^{-1} \right), \tag{12}$$

$$(n = 2).$$

The maximum interference power $P_{int\,max}$, which coincides with the operating frequency RR(TD) of the switch, can be calculated by the formula:

$$P_{int\,max} = N_{NRI} P_t A D^{-2} K_{atten},$$

where N_{NRI} is the number of sources of interference.

Under real conditions, the probability that all RR (TD) will operate on the transmission simultaneously, less than unity so the additional factor is introduced ($b < 1$) the use of radio transmitting devices. Then the power of radio interference

$$P_{NRI} = b P_{int\,max}. \tag{13}$$

8.5 Threshold Filtering in WSN

According to the plan, in the WSN, in order to improve the IMS, in addition to all other types of selection, the separation of signals by radio wave propagation distance is carried out. The main electronic circuit of this separation is the threshold device. Its action can be characterized as a threshold effect, which reduces to the fact that only signals (interference) with an input power are received RR $P_r > P_{thres}$, where P_{thres} is receive power threshold.

Some idea of the power ratios in the WSN at the input of the power supply unit is given in Fig. 17. Reception power threshold P_{thres}

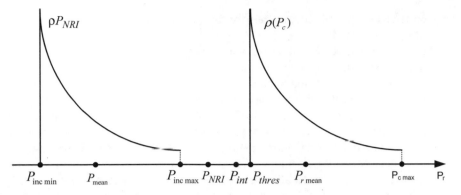

Fig. 17 The ratio of signal power and interference in the WSN at the input of the RR

$$P_{thres} = \gamma(P_{noise} + P_{NRI}) = \gamma\, P_{int}, \tag{14}$$

where γ is a factor required excess over the threshold signal interference.

With the help of the threshold separation of signals and interference, both the amplitude and the propagation distance of radio waves are provided.

The distribution $\rho(P_{NRI})$ is constructed in accordance with formula (10) for sources of network interference. For the normal operation of the WSN, the entire power range of the network interference should be in the subthreshold region. Moreover, in the same area, the total interference power P_{int} (7) and the maximum value of the power P_{NRI} (13) ultimately, at a minimum, must satisfy the relation (14).

In the above-threshold region when $P_r \geq P_{thres}$ should be the power of useful signals. Shown in Fig. 10 chart $\rho(P_c)$ the probability distribution of the power of the useful signal is also constructed in accordance with formula (10), but with different limits. These graphical dependencies reflect an idealized mechanism for separating signals and interference. In practice, interference in the RR is characterized by difficulty to describe statistical properties, since their probability distributions are compositions of many components. Often, an energy description of network radio interference is given by postulating any known power distribution laws, for example, lognormal. Their use gives some certainty for calculating the average value of interference P_{ave} by formulas (11), (12) and total power NRI according to the formula (13).

9 Conclusions

Thus, in the work done, the main directions were identified, on the construction of an automated optical IMS, which provides tracking for test objects (managed and unmanaged aircraft weapons, anti-aircraft guided missiles, artillery, and rocket launchers).

Main results:

- substantiated general requirements for the IMS, as well as the requirements for individual OESTM;
- has been further developed, information-analytical technology of detection and high-precision tracking of air objects;
- A new method for determining the location of each OESTM at the landfill using the technology of the wireless sensor network is proposed. The accuracy of positioning and orientation for each IMS should ensure that the purpose is clearly uncovered by reference to other targets. For this purpose, the mean-square error of the orientation determination should not exceed 1/10 of the minimum angle of the search camera. With a minimum angle of view of the search camera 2.3°, the mean-square error of orientation determination should not exceed 0.23°. Accordingly, with a maximum detection range of targets 5 km, the mean square error determination OESTM position should not exceed 20 m. A comparison of RSSI and ToF measurement methods showed that the RSSI method was unsuitable for measuring distances due to large errors. ToF method has the advantage of distance measurement accuracy compared with the method RSSI. In the conditions of the simultaneous operation of the wireless sensor network of the IMS and other wireless telecommunication systems or radio electronic equipment concentrated on the territory of the landfill and using the same frequency resource of 2.4 GHz, proposed to use radio communication distance limitation for individual nodes IAT.

Depending on the location of each OESTM on the territory of the polygon and the network topology, which combines them, there is an energy reserve that can be used to adapt the receiving and transmitting modules in order to better ensure electromagnetic compatibility. It is proposed to use methods of adaptive control of automatic gain control (AGC) in radio receivers and automatic power adjustment (APA) in radio transmitters. At a range lower than the maximum, it is permissible to boost the receiver or reduce the radiation power of the transmitter. As a result of the effect of AGC in APA, the probability of EMC increases due to a decrease in the average number of interfering signals entering the radio output. If APA only leads to the to boost the receiver, and the electromagnetic compatibility (EMC) does not change significantly, then the application of AGC in radio transmitters should reduce the intensity of electromagnetic fields. Both of them contribute to the better use of EMC.

References

1. Neofitnyj MV, Tevjashev AD, Shostko IS, Koljadin AV (2017) Informacionno-analiticheskaja tehnologija obnaruzhenija i soprovozhdenija podvizhnyh obektov v videopotoke(Information-analytical technology for detection and tracking of mobile objects in the video stream) in Materily 6-j mezhdunarodnoj nauchno tehnicheskoj konferencii «Informacionnye tehnologi» IST-2017 (Paper presented at the 6th international scientific and technical conference ICT-2017) pp 122–124. (In Russian)

2. Kopetz H, Ochsenreiter W (1987) Clock synchronization in distributed real-time systems. Comput IEEE Trans 100(8):933–940
3. Chaudhuri SP, Saha AK, Johnson DB (2004) Adaptive clock synchronization in sensor networks. In: Proceedings of the 3rd international symposium on Information processing in sensor networks, ACM, pp 340–348
4. Panfilo G, Tavella P (2008) Atomic clock prediction based on stochastic differential equations. Metrologia 45(6):108
5. Ping S (2003) Delay measurement time synchronization for wireless sensor networks. Intel Research Berkeley Lab, vol 6
6. Ren F, Lin C, Liu F (2008) Self-correcting time synchronization using reference broadcast in wireless sensor network. Wircl Commun IEEE 15(4):79–85
7. Rhee IK, Lee J, Kim J, Serpedin E, Wu YC (2009) Clock synchronization in wireless sensor networks: an overview. Sensors 9(1):56–85
8. Sundararaman B, Buy U, Kshemkalyani AD (2005) Clock synchronization for wireless sensor networks: a survey. Ad Hoc Netw 3(3):281–323
9. Paek J, Govindan R (2007) RCRT: rate controlled reliable transport for wireless sensor network. In: Proceedings of the ACM conference on embedded networked sensor systems. Sydney, Australia, pp 305–319
10. Tao S, Chan MC, Muravyov SV, Tarakanov EV (2010) A prioritized converge cast scheme using consensus ranking in wireless sensor networks. In: Proceedings of SAS 2010. Limerick, Ireland. pp 251–256
11. Li Z, Guo Z, Hong F, Hong L. E2dts: an energy efficiency distributed time synchronization algorithm for underwater acoustic mobile sensor networks. Ad Hoc Netw 11(4):1372–1380
12. Ageyev DV, Salah MT (2016) Parametric synthesis of overlay networks with self-similar traffic. Telecommun Radio Eng (Engl Transl Elektrosvyaz Radiotekhnika) 75(14):1231–1241
13. Ageyev D et al (2018) Classification of existing virtualization methods used in telecommunication networks. In: Proceedings of the 2018 IEEE 9th international conference on dependable systems, services and technologies (DESSERT), pp 83–86
14. Radivilova T, Kirichenko L, Ageiev D, Bulakh V (2020) The methods to improve quality of service by accounting secure parameters. In: Hu Z, Petoukhov S, Dychka I, He M (eds) Advances in computer science for engineering and education II. ICCSEEA 2019. Advances in intelligent systems and computing, vol 938. Springer, Cham
15. Kryvinska N, Zinterhof P, van Thanh D (2007) New-emerging service-support model for converged multi-service network and its practical validation. In: 1st international conference on complex, intelligent and software intensive systems (CISIS'07). IEEE, pp 100–110. https://doi.org/10.1109/cisis.2007.40
16. Kirichenko L, Radivilova T, Tkachenko A (2019) Comparative analysis of noisy time series clusterin. In: CEUR workshop proceedings vol 2362, pp 184–196. http://ceur-ws.org/Vol-2362/paper17.pdf
17. Shostko I, Tevyashev A, Neofitnyi M, Ageyev D, Gulak S (2018) Information and measurement system based on wireless sensory infocommunication network for polygon testing of guided and unguided rockets and missiles. In: Paper presented at the international scientific-practical conference problems of infocommunications science and technology, Kharkiv, Ukraine, 9–12 Oct 2018, pp 705–710
18. IEEE 802.15.4d–2009. IEEE standard for information technology—local and metropolitan area networks—specific requirements. Part 15.4 [Electronic recourse]/Institute of electrical and electronics engineers. Apr 17, 2009. Mode of access: www. http://standards.ieee.org/getieee802/download/802.15.4d-2009.pdf. 10 Oct 2012. Title from the screen
19. ZigBee Alliance. Open, Global Standards [electronic recourse]. Mode of access: www. http://www.zigbee.org. 17 July 2009. Title from the screen
20. Buratti C, Conti A, Dardari D, Verdone R (2009) An overview on wireless sensor networks technology and evolution. Sensors 6869–6896. https://doi.org/10.3390/s90906869
21. Blumrosen G, Hod B, Anker T, Dolev D, Rubinsky B (2013) Enhancing RSSI-based tracking accuracy in wireless sensor networks. ACM Trans Sen Netw 9(3):1–29. https://doi.org/10.1145/2480730.2480732

22. Yim J (2012) Comparison between RSSI-based and TOF-based indoor positioning meth-
    ods. Int J Multimed Ubiquitous Eng. Available via DIALOG: https://www.researchgate.
    net/publication/268269983_Comparison_between_RSSIbased_and_TOF-based_Indoor_
    Positioning_Methods. Accessed 1 Mar 2019

Development and Operation Analysis of Spectrum Monitoring Subsystem 2.4–2.5 GHz Range

Zhengbing Hu⬤, Volodymyr Buriachok⬤, Ivan Bogachuk⬤,
Volodymyr Sokolov⬤, and Dmytro Ageyev⬤

Abstract The paper presents a substantiation of the effectiveness of IEEE 802.11 wireless network analysis subsystem implementation using miniature spectrum analyzers. Also it was given an overview of firmware work scheme, development process of trial versions, monitoring system development approaches, current development stage, infrastructure for research system, reliability and scan check, our system design and hardware implementation, future work, etc. Paper also provides technical solutions on automation, optimal algorithms searching, errors correcting, organizing software according to the Model-View-Controller scheme, harmonizing data exchange protocols, storing and presenting the obtained results.

Keywords Dynamic channel allocation · Access point · Integrity · Availability · Spectrum analyzer

1 Introduction

To ensure the safety of roaming wireless infrastructure, namely, the integrity and availability at the same time. The development of backup and monitoring subsystems

Z. Hu
Central China Normal University, Wuhan, China
e-mail: hzb@mail.ccnu.edu.cn

V. Buriachok · I. Bogachuk · V. Sokolov (✉)
Borys Grinchenko Kyiv University, Kiev, Ukraine
e-mail: v.sokolov@kubg.edu.ua

V. Buriachok
e-mail: v.buriachok@kubg.edu.ua

I. Bogachuk
e-mail: v.bogachuk@kubg.edu.ua

D. Ageyev
Kharkiv National University of Radio Electronics, Kharkiv, Ukraine
e-mail: dmytro.aheiev@nure.ua

© The Editor(s) (if applicable) and The Author(s), under exclusive license to Springer Nature Switzerland AG 2021
T. Radivilova et al. (eds.), *Data-Centric Business and Applications*, Lecture Notes on Data Engineering and Communications Technologies 48,
https://doi.org/10.1007/978-3-030-43070-2_29

allows solving problems related to the availability and integrity of the transmitted data, thus increasing the security of the entire wireless system as a whole [1–5].

At the present stage of human activities development, its comprehensive for monitoring to be important. It's essence is to collect the necessary information and thoroughly analyze it. Modern studies of the processes occurring in technical systems have revealed their self-similar nature. So in works [6–8] a study of self-similar processes is carried out, both of signals of various nature and the properties of traffic in networks. Regular monitoring ensures timely detection of errors and corrections in the shortest possible time. Existing monitoring systems can conditionally divide us into systems that implement active and passive monitoring. Passive monitoring means receiving data in a read-out mode, for example, data collection systems for temperature, processor download, memory usage [9]. The term of active monitoring should be understood as environmental impact as operating system, applications and hardware elements monitoring. For example, a system that performs a corrective action under certain external conditions or parameter values. Monitoring systems are based on the client-server architecture. The interaction between the client and the server is carried out using standard or proprietary protocols, and the data is transmitted through the data transmission networks [7, 8]. The server stores, uses and modifies the current configuration for monitoring [9, 10]. Actually, the server conducts research of the system, gives notice, also on crashes, saves the research results in its configuration for their subsequent withdrawal in graphical form. The server itself cannot graphically display the network schema. To get a graphic image, as well as some types of failure alerts in the system, the client is used. The client does not store any information other than those located in RAM that is intended to be displayed [11]. The main purpose of the client—to draw a picture and in case of crashes or other issues, display it in the client console or other software, for easy of perception. The second client destination is a graphical user interface configured for the server. In order to determine the exact type of monitoring to be applied in one or another case, it is necessary to consider in detail the scope of both passive and active monitoring use [12].

The main problem of monitoring systems is that in such systems, there is usually a central server that performs all computational work and has the greatest load, so if it fails, there is a fatal performance of the entire system. A way of solving this problem may be to create a system that independently performs analysis, monitors the state of its nodes and, if necessary, redistributes the task of the failed node [13–16].

The problem of accessibility and integrity is solved in different ways: by modifying protocols [17], antenna solutions [18] or by changing the architecture [19]. Based on the scheme and algorithm given in [20, 21], the data analysis sub-system for the wireless system of the IEEE 802.11 standard is implemented in this paper (Fig. 1).

Wireless systems can be in three states:

- Regular work mode.
- Critical mode.
- Denial of service.

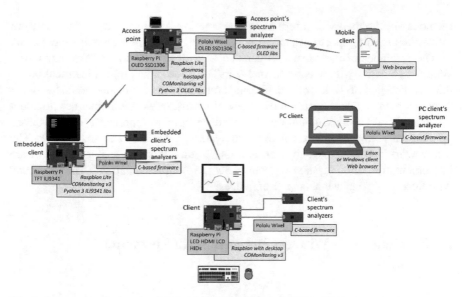

Fig. 1 Spectrum analysis subsystem with different types of wireless clients

Introduction to the wireless system of the subsystem with spectrum analyzers leads to the fact that the onset of a critical mode of operation is less likely, since subscribers on overloaded access points are redirected to adjacent access points not only at the maximum signal level but load distribution algorithms can be much more complicated.

In general, the system should be designed taking into account the following requirements:

- Cross-platform.
- Insignificant use of hardware resources to run on compact microcomputers.
- Infrastructure reliability.
- Data integrity in the network.
- User friendly interface.

At this moment, there are three versions of monitoring software implementation.

Each of them has a different architecture and working principles. Before the development of the second and third versions, the shortcomings of the previous ones were taken into account, a deep analysis was carried out, and more effective and optimal solutions were developed. The latest version is the final version of the implementation, which can be used for research purposes, industrial development, or as part of a comprehensive software solution. The implementation envisages and provides for further scale and improvement.

Object of research—*wireless technologies and systems*. The subject of research is the *integrity* and *accessibility* of information in sensory networks. The purpose of the work is the *development and research of a secure wireless* ad hoc *network*. Research methods are the theory of information, the creation of an ad hoc wireless network infrastructure using spectrum analyzers, and a series of tests of the monitoring system.

The reminder of the paper is organized as follows. Section 2 "Efficiency of a Wireless Network with Spectrum Analyzers" provides a theoretical justification for the effectiveness of the use of portable and embedded spectrum analyzers. Section 3 "Development Toolkit" gives an overview of firmware work scheme, development process of trial versions, monitoring system development approaches, current development stage, infrastructure for research system, reliability and scan check. The future work and plans are given in Sect. 4 "Future Work." Design and hardware implementation of our system is given in Sect. 5 "Implementation." The paper end with Sects. 6 "Conclusion" and "References."

2 Efficiency of a Wireless Network with Spectrum Analyzers

Determine the efficiency factor for the wireless system:

$$K = {}^E\!/_C \tag{1}$$

where E is the efficiency, and C is the cost.
Determine the cost of the usual wireless system:

$$C = P_i + P_s \tag{2}$$

where P_i is the cost of infrastructure; P_s is the cost of service.
The cost of a system with miniature spectrum analyzers, taking into account (1):

$$C_{sa} = P_i + P_s + P_{sa} = C + P_{sa} \tag{3}$$

where P_{sa} is the cost of the subsystem with spectrum analyzers. The cost of service is formed in both cases with the constant salary of the attendants, and therefore remains unchanged.

Wireless subscribers can be divided into three categories:

- Mobile (cell phones, tablets, etc.).
- Conditionally moving (laptops).
- Fixed (fixed PCs).

Determine the number of movable as N_m, conditionally moving is N_{cm} and fixed is N_f. The number of movable roughly equals the total number of conditionally moving and stationary, since virtually every worker has a personal mobile terminal (if the rules of the internal order do not provide for restrictions on the use of the corporate network):

$$N_m \approx N_{cm} + N_f \tag{4}$$

The cost of the subsystem with spectrum analyzers is:

$$P_{sa} = \left(N_{cm} + N_f\right)P_{sa}^* + P_c \tag{5}$$

where P_{sa}^* is the cost of one spectrum analyzer, P_c is the cost of the controller responsible for collecting, analyzing data from all spectrum analyzers, and generating tips for the main wireless controller. The number of mobile subscribers is not taken into account, since it is difficult to install additional equipment on them [22].

The cost of infrastructure depends on the number of subscribers:

$$P_i \sim N_m + N_{cm} + N_f + N_g \tag{6}$$

where N_g is number of guest-subscribers.

We introduce the openness parameter σ of the wireless system:

$$N_g = \sigma N_m. \tag{7}$$

Then from (4), (6) and (7) we have a full number of subscribers:

$$N \approx (2 + \sigma)\left(N_{cm} + N_f\right). \tag{8}$$

The efficiency of an ordinary wireless network is directly proportional to the subscriber's minimum access time to access point resources:

$$E \sim T_a^{min} = \frac{\Delta T}{N^{max}}, \tag{9}$$

where ΔT is the size of the time window, N_{max} is the maximum number of subscribers per access point (depending on the manufacturer is 30–50).

Wireless network performance with spectrum analyzers:

$$E_{sa} \sim \frac{N_{AP}}{N} \Delta T, \tag{10}$$

where N_{AP} is the number of access points.

Estimate the quality of work can be based on the ratio of efficiency coefficients:

$$\frac{K}{K_{sa}} = \frac{E \cdot C_{sa}}{E_{sa} \cdot C} = \frac{N}{N_{AP}} \cdot \frac{1}{N^{max}} \cdot \left(1 + \frac{P_{sa}}{P_i + P_s}\right). \tag{11}$$

3 Development Toolkit

The common monitoring system schema that was described above requires that the application embody such main features:

- Decentralization.
- Cross-platform.
- Ability to launch by the single-board computer (e.g. Raspberry Pi v3 or higher).

The Python is very suitable tool to satisfy our task's requirements. Beside of that Python has powerful standard library, it also has big repository with useful packages and modules for different development directions. For example, you can use *pyserial* package to works with Pololu Wixel modules [23]. This is great and simple tool for managing COM ports. As our system have to be cross-platform, we choose such group of used operating systems: *Raspbian*, *Ubuntu*, and *Windows*. The first two are Linux distributions that supplied with the Python interpreter installed by default. Other good points are:

- Great friendly community.
- Good documentation for modules and packages.
- Development speed.

Also a number of Linux distributions have Python 3.5 interpreter by default, it was decided to use Python of 3.6.6 stable version (Pyhton 3.7 stable is available now) [24], because it has important improvements for our app, in direction of:

- Security.
- CPython implementation.

3.1 Firmware Work Scheme

Before implementing the modules of Pololu Wixel in monitoring system, they are needed to be flashed with the necessary firmware. The spectrum analyzer firmware written by James Remington on C [25]. The wake-up receiver reports the signal level on all 256 channels. The signal level is measured by a number in the range from − 105 to −30 dBmW. Data transmitted by the module has the form:

```
A1-3D-DF-2F #57,
0x00007071 ms [ -102 -99 -101 <...> -84 -88] \r\n\n
```

where three points show a reduction, in arrays 256 values of the signal level. The data package carries information about:

- Unique identifier of microcontroller—"A1-3D-DF-2F". With this information, we can accurately identify the device and conduct a signal level analysis specifically for each module.

- Serial number of the package—"#57". This option allows to analyze the relationship between the number of successful packets and the total number of packets sent. Based on these data, we immediately detect possible problems during the work of the modules.
- The module operating time (from the moment of connection to a computer), which is presented in milliseconds in the sixteen-year system—"0x00007071 ms".
- An array of values of the signal level is "[−102 −99 −101 ... −84 −88]". For each of the programmed 256 channels, we get the signal level value, approximately every 0.5 s. These data are the main information load in the system, on the basis of which it is possible to conduct a statistical analysis, to monitor the state of the network, to exert a harmful external influence.

3.2 Development Process of Trial Versions

The first version of the monitoring system is based on client-server architecture in the form of two separate Python scripts for the client and server parts respectively [26]. The version was named *COMonitoring* from the COM component (the slang name of the serial port COM port) and the word "monitoring". Several client nodes and only one server can work in the system (Fig. 2). It is worth noting that the client server can also run a client script on the server node. The core work of the central node is to listen to the connections and retransmission of signal strength data to all those connected to it through a web browser. The client application must find a

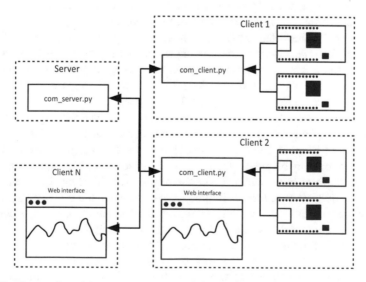

Fig. 2 The first version of the monitoring system infrastructure

functioning server in the monitoring system, initialize the Wixel microcontrollers, and start transferring the received data to the server.

Server and client scripts should be run in the console. First, the server script is started. The work it performs can be divided into two separate functions: the establishment of connections for the sending of static content (HTML, CSS, JavaScript), listening to the port to obtain data from client scripts and then forward them to the operator in a web browser. In order to perform this work simultaneously, you need to parallelize the work of the server script. Here you should look more closely at the asynchronous execution of the code in Python.

To accomplish several tasks simultaneously, we can use threads in the threaded module in the standard library It's surely worth to pay attention on the Python program launch, it already works in the stream, and it is called the main one. Flow is a unit of execution of instructions that can function with other threads within a single process, sharing resources. There may be a problem here when multiple threads access shared memory blocks at the same time.

Solves this question a tool called Global Interpreter Lock (GIL). GIL guarantees that only one thread can operate at any one time, working on the mutex principle, that is, one thread can capture GIL and only it can give back control. It turns out that the threads work in a certain sequence one after the other, and the interpreter is responsible for controlling this process.

Switching between threads can occur if the active thread is waiting for the I/O operation, a certain number of instructions is worked out, and the time for processing (time-out) has expired. It is worth noting that the switching process also takes some time and resources, but this operation takes place so fast that it seems like the threads work in parallel. In spite of this, you need to have a good idea of where the use of streams can be beneficial. In general, operations can be divided into two groups: those that are connected to the CPU bound, I/O bound.

The first group includes operations whose execution time depends on the performance of the processor, for example matrix processing or image conversion. To the second, the operations whose execution time depends on the performance of the transfer or receipt of data, for example, the creation of a record in the database, processing GET requests or sending an HTTP response.

Consequently, flows are not quite suitable for the operations dependent on the CPU, because during switching between flows, additional time and resources are spent, in which case the threads will switch over the timeout, as a result, the time for the operation may be greater than that of the synchronous the approach. On the other hand, for threading operations, the use of streams will be much more effective. One can conclude that the threads imitate the parallelism, although they are actually performed sequentially. If you really want to parallelize the work of the program, we will need a module of the standard library—multiprocessing.

It has a similar threading module interface, but in this case separate processes are created with it's interpreter and resources. To solve our problem, we have enough functional flows. The threading module will be actively used in all subsequent versions of the monitoring system.

Consequently, the server's work is based on two threads operating in an infinite loop. The server itself, which processes requests from client web browsers, works in the main stream. To implement the server, in the first version of the monitoring system, we chose the module *flask-socketio*—an extension of the microframe flask, which allows you to use the functionality of the library *SocketIO* on the client. *SocketIO* is a library that supports bidirectional data transfer in real time (WebSocket protocol).

Communication between the server and the client is based on the events principle, which act as a handler in the received data. In practice, we can register an event under a certain name and use it during data transfer. At that time, on the other side, when the data is received, an event whose name was specified during the transmission is executed. Another thread is used to listen to the port, when connected to which client applications send data from Wixel modules or other system information.

The client application also runs in an infinite loop, but the client runs several operations immediately after launch. From the beginning, the client starts an additional thread with a function that listens to the port to receive commands from the server.

The fact is that the debugging and monitoring of the monitoring system is due to commands that can be passed through the client's web browser to the server, and this in turn to the end client application.

However, in order to ensure better reliability of the system, the feedback scheme is working there, so confirmation of some changes requires confirmation from the end customer. That is why we need to accept the connection from the server on the client side, and handle the received commands. After that, the client application launches the automatic search for the server host to connect to the monitoring system. It has a rather primitive implementation, but as part of an experiment at that time, this solution was enough. The essence of the method is to overcome 256 IP addresses over the network mask 255.255.255.0, and attempt to connect. The port to which the clients are referred is 10,000. It is programmed in the code of the script itself. Since the monitoring system operates within a common access point, each client, knowing its IP address, conducts this transaction. In case if the connection failed, for example, waiting for the response time, the console displays the field for entering the IP address of the server manually.

Then runs an infinite loop in which every check iteration checks the command from the server and then a certain work depending on the team. There are two commands to work and pause. After the start, the client application works in regular mode, that is, it tries to get signal strength data—it is a team of work. The process of obtaining data from modules takes place in two stages: initialization of connected modules and data acquisition. To find the Wixel modules in the system, we use the functionality of the *pyserial* package. To initialize, you need to create the *Serial()* object and specify the port on which the device is mounted on the operating system. Depending on the platform (Linux, Windows), the port name may be different, for example, for Windows it will be ports COMN and for Linux/dev/ttyACMN. The module search function takes into account this feature to meet the requirements of the monitoring system. Realization of modules initialization function:

```
def serial_ports():
```

```
if platform.startswith('win'):
port = ['COM% s'% (i + 1) for i in range (256)]
elif platform.startswith('linux') or platform.startswith('cygwin'):
port = glob('/dev/tty[A-Za-z]*')
elif platform.startswith ('darwin'):
port = glob('/dev/tty.*')
else:
raise environmentError('unsupported platform…')
result = []
for port in ports:
try:
s = serial.serial(port)
s.close()
result.append(port)
except (OSError, serial.SerialException):
passport
return result
```

First, we check the platform on which the client application runs, to select the desired naming template and port range. On the basis of these data using a loop, at each iteration, an attempt is made to initialize the module. If the operation was successful, the port name is added to the array, which returns when the function is finished. To get data from the modules, a *Serial()* object is created again for each of them. Then the port opens with the method *open()* and *readline()* data is read. The *readline()* method reads the bytes until it receives the sequence "\n"—the next line. In the end, you must close the port using the *close()* method. Failure to do so may cause problems in the future operation of the system with the module.

```
for port in connected_ports:
ser = serial.Serial()
ser.port = port
try:
ser.open()
port_data = ser.readline().degode('utf-8')
ser.close ()
except:
passport
```

The resulting data is processed and added to an array, which is then transmitted to the server. Such an algorithm was used through some nuances:

- It was decided to work with the modules in a consistent way, rather than creating a separate stream for each of them. There was a problem of synchronization between module initialization function and data acquisition function. They both use the *Serial()* object, so if they work in different streams, there might be a situation where the open port will try to initialize again, resulting in an error. This problem will be solved in the third version of the monitoring system.
- The data extraction scheme is quite simple, so it is reliable and supports multiple modules at the same time.

Despite the fact that the scheme is really simple, we were expecting some problems using the *readline()* method:

- The method created a significant load on the central processor.
- Non-stable work on the Linux platform. Quite often, we received damaged packets that lacked approximately 40% bytes.

There were also cases when the module was not available for further initialization upon receiving the damaged package, and it was necessary to restart the client application. A serious issue was solving the problem with the dynamic switching off the modules. The fact is that under the current scheme of work, if during the data loop to disconnect the module, there was a big chance that an error would occur and the client application would stop working. In the future, the reason for such behavior was found, the problem was not completely correct exception handling—block try/except. However, at that moment, it was decided to use a paused scheme. Pause is the second pause command handled by the client application. To unlink the module, you need to send a pause command from the web browser, in turn, on the side of the client application, receiving the command, the script stops the cycle of work with the modules until it again receives the command. Therefore, we could safely disconnect the modules from the monitoring system. After collecting data from the modules, a package was passed to the server, which in turn handed it to all connected through the browser to clients.

The monitoring system of the first version had a very informative and convenient web interface as a one-page web application (Fig. 3). The interface was divided into two parts. In the left, there was a large graph of the signal level, and in the right—the control panel.

The JavaScript library for *Flot* was used to render graphs. It has a large number of adjustments, which allows you to flexibly adjust the graphics settings. On the right side, we can see all connected client applications and their modules. To visually identify the modules in the control panel and in the graph, each of them fixed a

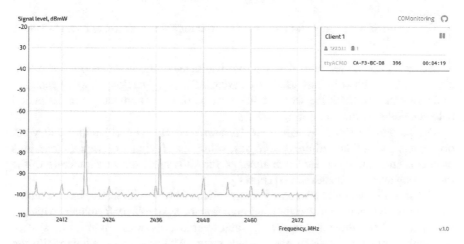

Fig. 3 Web interface view of the first version of the monitoring system

certain color that was specified in advance. From the functional part, we have access to:

- Sending a command to a client application.
- Change the client application name. Through the interface, we can provide different identifiers for client applications.

As a result, the first version demonstrated that we have quite successfully picked up the hardware and software for the development of a monitoring system. The entire system operates in the form of two scripts, which in total have no more than 300 lines of code (not taking into account HTML, CSS files and JavaScript scripts). The web interface was very convenient and agile solution. Despite the simple implementation, this solution provides the ability to access the system from a variety of devices, such as a smartphone or tablet. The use of the Wixel microcontrollers as a spectrum analyzer for the monitoring system is a rather attractive and promising idea. Among the main drawbacks of the first version are:

- The system has a client-server architecture and therefore rather vulnerable, has low fault tolerance.
- It is quite difficult to implement a functional for encrypting data during their transmission.
- Complicated and rather primitive scheme of system start-up.
- Low reliability of client application. Unfortunately, the used scheme of work with modules is not a good solution. The system is not stable; it needs to be restarted frequently. In addition, due to the algorithm used to collect data, we have a significant load on the central processor.
- Badly thought out the logic of working with modules. Since we process them one by one, it makes no sense to use multiple modules on the same machine. With the increase in the number of working modules, we do not get the best quality signal strength data.
- The system is not suitable for launch on microcomputers because of its architecture and workflow.

We reviewed and researched all the disadvantages of the first version, made conclusions about the solutions used. To develop the next version, it was necessary to analyze and re-think the whole scheme of work with modules, communication between nodes in the monitoring system.

The second version was called *COMonitoring2*. By its architecture and work principle, it is partially different from past solutions. Having reviewed the previous problems and conducted a deep analysis of the software, a new architecture of the monitoring system was developed (Fig. 4).

The second version is also based on the client server architecture, but we added an automation functionality to start and maintain the system's performance, which in theory should significantly improve its fault tolerance. Now the monitoring system is a single application in the form of a single script. All participants in the monitoring system are called nodes. The server part is responsible for the same functions as the previous version, but presented as a separate thread.

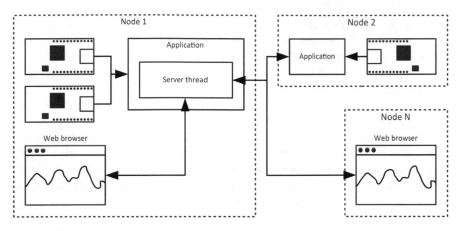

Fig. 4 Web interface schema of the first version of the monitoring system

When the monitoring system application is launched, the server is first searched for on the network. After analyzing the implementation of this feature in the previous version, a more thoughtful solution is used here. The application launches the function that initially initializes the TCP socket, and then sends a broadcast UDP packet that carries information about the node on which the application is running, including the IP address. In turn, there is a separate stream on the server that is listening to such broadcasts.

When a UDP packet is received from the application, the server analyzes the data in the package and establishes a TCP connection to it, knowing its host from the data received. Thus, the server sends its IP address which the application can use to further transmit signal level data. The application tries to contact the server tree times. If during this time the server did not respond, it is concluded that there is no working server in the system. The application itself starts the stream with the server and the following nodes will be able to connect to the monitoring system by a similar algorithm. This scheme is quite complicated and does not always work stable. Nevertheless, by such a solution we were able to automate the launch of the monitoring system. It is worth noting when the server receives the UDP packet, it records the IP address of the node from the packet into a global array. This array is transmitted to the application in response. That is, at each node there is information about the neighboring nodes that connect to the monitoring system. Information about operating nodes is updated with each new connection to the system. This array is used in case of a sudden shutdown of the server. If the server stops working, all nodes in the system will not be able to transmit signal strength data.

Accordingly, the application responds to such behavior—all nodes agree that if the server becomes inaccessible, it is necessary to transfer data to the first node stored in the global array. The first node in the array independently runs the thread with the server and the system continues to work.

In theory, such a scheme of interaction provides a system of significant failure-resistance. However, due to a complex implementation, we did not have enough time for quality testing of this method. As a result, there are rarely cases when there are errors in the system and a complete restart of the infrastructure is required. Otherwise, the functionality of the server remained similar to the previous version. The implementation of the server, as in the previous version, uses the flask *socketio* package.

In the previous version, the system of commands worked, which allowed the system to be suspended. In the second version, we rejected this solution, because the problem with the dynamic unplugging of the modules was solved. In this case, the detachment of the module during data readout by the system, no longer threatens the creation of errors, the implementation correctly works out the possible exceptions.

In addition, as in the previous version, we left the principle of processing modules in turn. Such a decision was made due to the fact that a significant part of the research resource was spent on another functionality of the monitoring system, and the previous scheme allowed to successfully receive data from the modules, although it was not effective enough. However, in the signal processing algorithm, one important implementation was implemented. The previous version of the monitoring system allowed to analyze data only in real time. That is, the web browser only received the latest signal level data packets. The problem is that under such a conditions, it will be complicated for the such system operator to assess the overall state of the network. In order to avoid this issue, it is recommended to store a certain amount of signal level data and find the mean values for the analysis. We immediately abandoned the use of databases, as this decision entails additional load and complexity of the system. At that time, it was more important to check the efficiency of the idea. We decided to store for each channel of the module the last 100 values of the signal level during the application work, that is, in RAM. This number seemed optimal in terms of the load on the system and the possibility of a qualitative assessment of the network. As a result, the integrity of the data has significantly increased. On the other hand, the load on hardware resources has increased, since for each module, for each of 256 channels, it is necessary to calculate the average value approximately two times per second. When considering the third version of the monitoring system, methods for storing and calculating averages will be more detailed.

In the first version of the monitoring system, we used the primitive scheme of visual identification of modules by color. The problem was that the colors were static and orderly for each node in the system, for example, the first connected module to the node had a blue color, the second one was red, and so on each node. In general, this could cause some embarrassment, since it would be more difficult for the operator to target the nodes if the monitoring system has several dozen participants. It was necessary to develop a more versatile and thought-out solution. The task was to create a unique color for each module. Three formats for color generation were considered: RGB, HEX, and HSL. RGB is an adaptive color model that describes a method of color synthesis, in which red, green and blue light overlap, mixing in a variety of colors. In this model, the color is encoded by the gradation of the constituent channels, therefore, with increasing the gradation of one of the colors, its intensity increases. In practice, the color can be written as an array of three integers in the range from 0 to 255 (1 byte). This range shows the intensity of each of the component colors, for example, the red color will look like {255, 0, 0}.

The HEX format essentially represents the same schema as the RGB, but has another form of recording. Instead of three integers, the color is written in the form of three bytes in a 16-year system. In addition, the symbol "#" is placed before the entry. For example, the red color in the HEX format will look like "#ff0000". HEX colors are widely used when creating styles for web pages.

HSL is a color model in which the color is given by three parameters: color tone (red, yellow, blue), saturation, and illumination. The first parameter is given by an integer in the range 0–360, the second and third—fractional number in the range 0–1.

To generate a unique color, you need to take into account the colors of the neighboring connected modules. Despite the fact that the colors may differ in value, while visually they may look almost the same. In addition, depending on the color format, there were additional issues. The approximate scheme of forming the RGB color was to generate random three numbers to indicate the intensity of the three components of colors. If some modules already work on the node, it was necessary to take into account the meaning of their colors. For example, if a module with color is running on a node {146, 154, 129}, and an array {134, 142, 156} is formed for the new module, then these colors will be visually similar. The algorithm took into account such a feature and established the rule that the difference in values for each component of colors on all modules should be not less than 20. Another problem was the very generation of random numbers. The fact is that generators of pseudorandom numbers generator (PRNG) are widely used in computer science. The result that generates PRNG is based on the initial random value, that is, all subsequent sequences will depend on the initial "grain". Indeed, random numbers can be obtained from an entropy source. Information entropy is a measure of the chaos of information, the disorderly appearance of any symbol of the source code. The source of entropy can be, for example, heat noise or a source of radioactive decay. If PRNG the sequence of each iteration is unique, does not depend on the previous results. The cost, the slow generation of the sequence and the complexity of the creation of PRNG are the reasons for the fact that in practice often PRNG is used. To improve the quality of PRNG sequences, sometimes a really random number is created for the original value.

To generate pseudorandom numbers in the standard Python library, there is a random module. To create a pseudorandom integer, it has the *randint(a, b)* method, where a and b are integers that specify a range for a pseudorandom number. After conducting an experimental study, the following results were obtained: on average, 10 formed packs of 10 random colors, in 4–5 packages there were 3–4 similar colors in tone. Such results did not suit us; the system needed a better solution. All the disadvantages of the formation of colors in the RGB format were inherent in the HEX format.

The HSL format allows to use another algorithm. Depending on the number of working modules on a node, we can evenly distribute the tone value (360) between all nodes, but the interval should be at least 20 if the same values of saturation and illumination are used. Otherwise, you can dynamically change the saturation or illumination settings. However, this decision has one important drawback—the relationship between the times spent on developing and testing the algorithm to the importance of the functional. As a result, this implementation was also rejected.

After the experiments and the solutions reviewed, we have developed a new algorithm for the formation of colors. Each module has its own unique MAC address—a unique identifier assigned to each unit of the active equipment or their interfaces. The essence of the solution is to use this identifier as data for creating a hash using the SHA256 algorithm from which HEX colors can be obtained. Since the hash SHA256 is a sequence of 64 characters long in a 16-year system, taking 60 of them, we can create 10 colors for each of the modules. The benefits of this solution are simplicity and guarantee that each module will have the same set of colors, regardless of external factors. Therefore, there is also partial inconvenience with the color of the set similarity, as with the generation of RGB format, but the uniqueness of the formed colors specifically for one module and the ease of implementation of the method overlays this drawback. Code of the color formation function:

```
def hash_color (date):
    return ['# {}'. format (hash (date) [2: -2] [c: c + 6])
    for c in range (0, len (hash (date) [2: -2]), 6)]
```

where the *hash()* function returns the hash SHA256 from the received data input.

Additional functionality in the form of color generation and calculation of averages required a solution for dynamically adjusting these parameters during the system monitoring. It was decided to save the results of the settings on the side of the application so that they were saved while re-connecting the modules. The very mechanism of setting new parameters we implemented in the form of tasks. This solution is similar to the team method that worked on the first version of the monitoring system, however, in this case from the web browser the application received a data packet containing the task identifier and the parameters that were to be installed. As a result, in each main loop iteration, we checked that the tasks did not arrive at the site, if so, they performed, and otherwise they collected the data from the modules and sent it to the server (Fig. 5).

The monitoring system operates with the following set of tasks:

- *set_color*—the color setup task for a specific module, at a specific node. Requires data about the IP address of the host, MAC module address and color values in HEX format.
- *set_chart_type*—the task of setting the type of the graph. To calculate the mean values, it was decided to process the following signal level data: the value from the last packet, the average values for the last ten packets, the average values for the last 100 packets, the average values or all packets received during the module work. Accordingly, each type was given its identifier: 0 for the last packet, 10 for the last ten, 100 for the last one and 1 for all packets. The task requires data about the IP address of the host, the MAC address of the module, and the value of the type of data signal level.

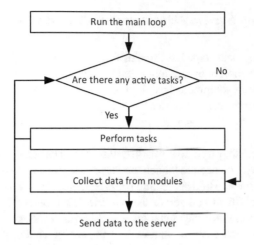

Fig. 5 Processing tasks algorithm

- *set_name*—the task of setting the identifier for the node. Requires IP address of the host and new text ID.
- *set_clients*—the task of installing a new array with nodes connected to the system. The task is performed by all participants in the monitoring system when a new node is connected to it. In this way all participants know about functioning of each other.

The new scheme of work with tasks required changes in the main web-interface of the monitoring system. Past solutions were reviewed and a new interface was developed. In general, the scheme was similar to the previous version, the software remained the same, but the control panel has undergone significant changes (Fig. 6).

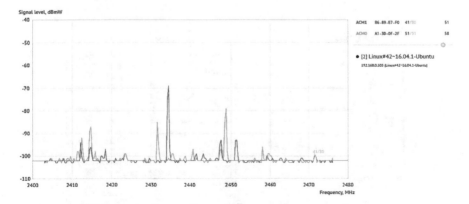

Fig. 6 Web interface view of the second version of the monitoring system

3.3 Monitoring System Development Approaches

The current development stage has three different app versions. The last app version is named *cmtg* (the short abbreviation of the first version name *COMonitoring*). The monitoring system development has significant path that associated with different solutions, issues, improvements and optimization methods.

The first version has client-server architecture with client and server Python script separated (Fig. 7).

To run the system they have to be on different machines that connected to the same access point. Firstly, you need to launch the server's script and then the client's script. The client's script have to find server and starts to handle Wixel modules data to send them to the server. Client and server use WebSocket protocol for communication. After that, you can monitor the signal level values via browser.

In general, this version shows that idea of the monitoring system on the Python works and browser interface is very convenient solution for data analyzation. However, there are some important disadvantages:

- This is not decentralized monitoring system, so it has low level of fault tolerance.
- Complicated launch scheme.
- Low server reliable.
- Bad thoughtful of app logic.

These disadvantages were the reason to make conclusions and create the system from scratch. The second app version has new functioning scheme and several important improvements. The key point is that we have single application that can do both of client's and server's jobs. We used the threading Python module to run the server's job on the separate thread. To provide this functional we came up with such algorithm:

- The app runs.

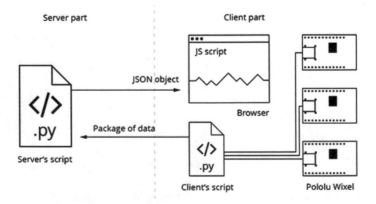

Fig. 7 COMonitoring functioning scheme

- It tries to find the working server in the current local network. The application sends the UDP package to signalize to the server, that new client was appeared in the network.
- If the worked server is in the current network and retrieved this package, it sends response to the client with self-address (URL) via TCP connection. After that, the client can communicate to the server. Otherwise if the server is absent, client runs the own server thread. In the future, other clients can connect to the current client's server, using the same steps.

In addition, there is the method to save the all connected clients at the current network. It means that all clients know who is connected to the network and who the server is. This list is used in the case, when the server goes down and all clients have to be able to reconnect to the new server. The selection of new server depends on the list orderliness, specific client have to run the server thread. This functional implements the system decentralization, but works very unstable and unpredictable. The reason is that we had time lack for testing stage. However, there are some other convenient features:

- Added new algorithm to store and handle signal level values. We save some part of values for each Wixel module's channel in the queue data structure (FILO) and can calculate the average values for specific range. This method allows to us to analyze the network state a much better.
- Significantly changed web-interface structure. There were added the unique color creation method for visual module identification. In addition, the interface functional provides the chart control. You can select the range for average signal values calculation and disable chart's rendering for specific modules.
- Some security improvement. The digital signature to check data integrity of incoming packages from client was added.
- Added simple console interface to output the information about the app working state.

Beside of this features, in the result it turned out that the system is unreliable. It has the same problem with app logic as the first version, but in this case, the system is very complicated and bulky. The most of entities in the app depend between themselves that creates additional problems with flexibility and scalability.

Thus, the third (current) version is combine from the both previous versions and new significant changes. It was developed the new functioning schema to solve all issues that previous versions had (Fig. 8).

We have the single application as in the second version. Each client in the system is named "node". The app structure combines two ideas: using threads (as the second version) and the micro-service architecture. You can see that each app includes three main entities: controller, node manager and device manager. These entities are analogue of the service in the micro-service architecture. They are fully separated and work in different threads.

The device manager provides functional for working with modules. First two versions operate with a single thread for every module. The device manager creates

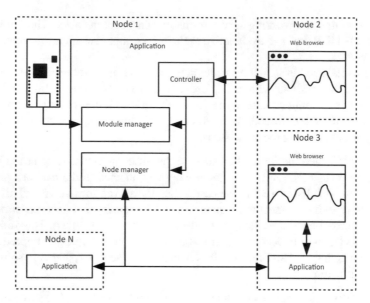

Fig. 8 Cmtg functioning scheme

new thread for each module to cooperate with it in the current version. The main task of device manager is to retrieves the data from module and store it in the special data structure.

The node manager is responding for nodes communication in the system. It binds the listening UDP socket that accept broadcast packages, in the one thread. With another thread, it sends broadcast UDP packages. This package contains data about the current node, such as, machine name, operating system name, MAC address, host, port. Therefor every node is listening and sending packages, so they all know about who is connected to the current network. Every node stores all this information in the system.

The controller entity handles client requests and communicates with node and device managers. This is the app server. The node manager and the device manager doesn't contribute, unlike the controller. For example, when the client sends request to get data about nodes that are connected now, controller takes specific data from the node manager storage and response to the client. It makes the same steps, when client want to get data about devices and signal level values, but uses the device manager.

The most interesting part is how web-interface allows us to communicate with all system nodes, because the centralized node that handle all connection does not exist, every node has its controller entity and server. To understand this scheme better, it would be good to divide the app lifecycle by steps:

1. The user run the app in the terminal (console). It can see the console app interface, that shows the system functioning state. The second version has the same functional.

2. The node manager and device manager start working in their own threads.
3. After that, user can see if the network has other connected nodes. The console interface shows node's URL address that allows us to communicate with this node via browser.
4. When we open browser and connect to the node, the node's server response the html page with JS script. The JS script makes three main functions:

- Store current URL.
- Start the infinity loop that always requests data about all connected nodes.
- Start the infinity loop that always requests data about devices of the current node.

Both of the loop is associated with the stored URL. Therefore, if we want to request data from another node, we just need to change this URL to the specific node's URL. This action can occur CORS errors in browser, so we use the *flask-cors* package to solve this issue.

In the third version, we have achieved that all previous issues are solved. The current app scheme works better. Besides of that it is decentralized system, it also uses separated threads, so system has high fault tolerance level. This is very important for our field. The monitoring system has powerful foundation for future scaling.

The node and module manager work in separate threads, the controller is in the main stream. Let's consider in detail the work of each of the components of the system.

In the third version, client-server architecture was removed in the standard sense of this term. We no longer have a central node to which we send data from modules and other system information. Instead, each node works locally without trying to connect to someone. The connection between the nodes of the system is carried out using such an algorithm. The nodes manager in one thread constantly listens to the broadcast UDP port, and in the other sends a broadcast UDP packet. The package contains information about the node from which the package was sent, in particular:

- *name*—host ID.
- *addr_inet*—the IP address of the host in the local wireless network.
- *port*—the port on which the server (controller) operates.
- *addr_hw*—hostname MAC.
- *system*—the operating system of the node.
- *access_point*—information about the access point on which the node operates.

There are a few changes here. We have removed the ability to change the host ID. Instead, the name used as the identifier is provided to the computer when it is connected to the network (hostname). We consider this solution to be more logical. In addition, nodes running on the Linux platform can transmit additional information about the access point they use. This information is very important if we have a large network infrastructure, where the nodes of one system can work on different access points.

Fig. 9 Algorithm for
module manager functioning

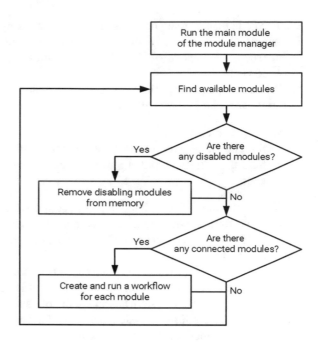

Work with the modules has also undergone significant changes, now the module manager is responsible for this process (Fig. 9), and work with each of the microcontrollers takes place in a separate thread. The duties of a manager include:

- Search for available ports.
- Monitoring the activity of modules, processing cases of dynamic disconnection, joining.
- Collection and storage of data from the module.
- Calculation of average values.

The first and second versions of the monitoring system used the same method to find the ports of available modules. It was implemented thanks to Python's *pyserial* library. In the current version, we used the operating system functional. Depending on the platform, its algorithm was developed. As a result, we resolved the issue with synchronizing two search and data operations. Let's take a closer look at the algorithm of module manager work. In the infinite loop, the port search function is initially executed. The result of its work is data containing the port name and the MAC address of the module. The manager temporarily stores this data to compare them with the next call to the search function. With temporarily stored data about available ports, we can make a comparison for each iteration of the loop and determine which ones were disconnected and which ones are connected. If the data did not change, compared to the temporary, all modules remained in place. In the event that new modules appear in the system, they need to be initialized and stored in a special global repository. If, in the system, disconnected modules are detected, they are simply removed from the global site repository. To save the functioning modules, we use the basic data type *dict*—associative array data structure (hash table).

Module parameters:

- *port* is the port name. Depending on the operating system, it may have a different name, for example, COM1 for Windows, /dev/ttyACM0 for Linux.
- *addr_hw*—module MAC.
- *accepted*—the number of accepted packages. As in the second version of the system is used to display the ratio between the number of received to the total number of sent packages.
- *connected*—stores the object of the class *UpdateTimer()*. Here's a similar mechanism for tracking activity, as in the nodes manager. There may be a situation where the module suddenly stops working and it is not available through the module search function. In this case, in order to complete the flow with the module cycle, it is necessary to determine if the module is still working. This is precisely what the timer uses to track activity.
- *serial*—object of class *Serial()*, with it we read data from modules.
- *counter*—*SignalCounter()* element, used to effectively store signal level data and to calculate averages.
- *firmware*—data sent by the firmware of the module.
- *packet_num*—total sent package number.
- *work_time*—the module's working time as an object of the class *datetime.timedelta()*.

It is worth considering the *SerialCounter()* class in more detail. Objects of this class store each module channel in a separate object of the class *Channel()*:

```
class Channel:
    __slots__ = ['freq', 'values', 'total']
    def __init__(self, freq):
    self.freq = freq
    self.values = deque(maxlen = 100)
    self.total = 0
    def add_value(self, value):
    self.values.append(value)
    self.total += value
```

Since for each module you want to create 256 objects of this class, for efficient use of memory, we use the special *__slots__* attribute, which lists the variables that will be used in the process of working with the module. By default, when Python is initialized, an *__dict__* attribute is created that stores the user attributes and internal attributes to interact with the interpreter; it also allows you to dynamically install new attributes when working with an object. Accordingly, in order to support such a functionality, we sacrifice a significant part of memory. The *_slots_* attribute allows you to reserve the required amount of memory for the specific attributes to be used, so the *_dict_* attribute is not created. Attributes of the class object *Channel()*:

- *freq*—stores the fractional value of the channel.

- *values*—stores the last 100 signal level values in the special *deque* data structure from the standard collections library. *deque* implements an abstract data type *dek*—a list of elements in which the addition and removal of elements can be performed at both ends. It also has a more effective implementation in terms of memory usage and the removal efficiency of elements than the standard list.
- *total*—stores the sum of all signal level values for the channel. This parameter is used when calculating averages for all packets accepted.

Class-object *Classes()* have their own interface to add new values of the signal level in the form of the *add_value()* method. This method sends the value of the signal level that is added to the queue values and to the total sum. The *SerialCounter()* class object operates with such a set of channels for storing data, adding signal level values, and calculating averages.

An important innovation is a new algorithm for reading data from modules. In previous versions, we just used the *readline()* method to get bytes. Among the disadvantages of this solution were a significant load on the central processor and the instability of work on Linux platforms. Now, the following sequence of commands is used to collect data:

```
first_byte = ser.read()
tools.sleep(0.5)
other_bytes = ser.read(ser.in_waiting)
data = b".join([first, other]).decode('utf-8')
```

Where *ser* is a *Serial()* object. First, we read the first byte. Then we wait 500 ms. During this time, the buffer will be filled with the rest of the bytes of the package. The number of bytes in the buffer is stored in the *in_waiting* attribute. Using it, we specify the exact number of bytes that you want to read. Next, we combine the received bytes and decode them by encoding UTF-8. As a result, we use resources very efficiently, which allowed us to reduce the load on the processor to the maximum. More about the study of the load will be discussed further in the work.

Component *Controller* acts as a server on the host. The third version of the monitoring system uses HTTP. Customers now have to generate queries themselves, since the server no longer has the ability to send data in real time. We refused to use the WebSocket protocol to simplify the system. The controller has access to global repositories of nodes and modules in the respective managers, so it can generate the necessary content for customers or operators of the monitoring system. The controller handles two main queries:

- Request to receive data about active nodes in the system. The controller refers to the global repository, which stores information about all connected nodes and generates the required response. These data are obtained during the work of the site manager.
- Request to receive data about the operating modules at the node. The algorithm is similar to the previous, but during this request, the average values for each module are calculated.

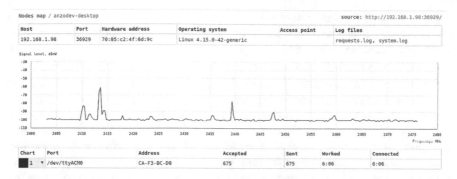

Fig. 10 Web interface view of the third version of the monitoring system

To interact with the controller, the web interface also required changes to the architecture and workflow (Fig. 10). Instead of installing WebSocket connections, client-side JavaScript code generated AJAX queries to the host server at a certain interval. AJAX is a technology for dynamically invoking the server in the background, that is, without reloading the page. Therefore, we simulate the scheme using WebSocket, so in the general mechanism of data processing from the server and the creation of elements on the page remained the same.

Earlier in the paper, it was considered that in the third version there is no central node and all participants work locally without establishing connections between themselves. Here you may wonder how to get data from modules of different nodes in this case, referring only to one. This question helps to resolve the controller. We came up with a rather interesting scheme of interaction between nodes, without the actual connection between them. When we open the webpage of the monitoring system application, the JavaScript script has access to the IP address and port of the current host. Using these data, we can contact the controller of the current node, which forms the necessary content for us. So, the problem can be formulated so—to get data from the modules of another node, we need access to the IP address and port of this node. Since the current controller has access to the node manager, it returns us the list of all the working nodes, so we can turn to the bud of one of them, first connecting to only one. Therefore, every node in the system works. However, here we are faced with the same origin policy. In programming, this is a very important concept in terms of security. The policy allows client scripts (such as JavaScript) that are generated on the same site to work with each other without restrictions, however, prohibits access to other sites. To address this, there is CORS technology—allowing web pages to retrieve data from other resources. To do this, the server from which the web page is downloaded must add the HTTP header Access-Control-Allow-Origin, which lists the domains to which the web page may access. We use the ready solution in the form of the Python module *flask-cors*, which automatically performs this work for us.

Significant changes were implemented in the method of creating colors for the module. The previous decisions were completely rejected, since they emphasized a lot of attention from the point of view of expediency. Colors are now stored only on the client side, in the web browser. In order to form a unique color for the module, the finished JavaScript-based color-hash library is used, which is intended for such an operation. As a data source, the MAC address of the module is also used. This solution is simple and logical. Each module has a surely unique color. In this case, the interface provides the possibility that several modules may have similar colors. To solve the problem, on the web page for each module there is an HTML5 input form field with a color type. When you click on a field of this type, depending on the platform, there is a separate window with a palette where we can choose any color.

For a better understanding on how the system works and for more efficient debugging, the file logging function was added. The system operates with two log files:

- *requests.log*—data about the requests received by the node is recorded here.
- *system.log*—there are records of important processes occurring in the system as a whole, for example, unplugging and connecting modules, disabling the node from the system.

Accordingly, the web interface provides an opportunity to analyze the content of log files in the web browser itself. This helps to actively monitor of the system status, identify the cause of possible problems, errors in the implementation code.

It can be seen that the monitoring system has undergone a thorny path of development, testing and improvement. Each version provided an opportunity to analyze the status of the network on its own, with its advantages and disadvantages. Gaining experience, with each new stage we solved important stumbling blocks of first-hand solutions and introduced a new functional. As a result, we have developed a final version that fully meets the requirements. It has all the features of a safe, reliable, productive monitoring system that is ready to work in the industrial environment of information protection. In addition, the third version provides improvements that are more convenient to be implemented, since the system has an agile architecture. Possible use of the final version can be considered as an independent product or an element of integrated software.

3.4   Current Development Stage

Today we work to implement new functional for such directions:

- Security.
- Logging system.
- Web-interface improvements.
- CPU load and memory usage optimization.

To make the app more safety, there is idea to use the encryption algorithm for packages that are transmitted between nodes in the monitoring system.

The user of the monitoring system needs to be able to trace the system working process in convenient form via browser. It is necessary feature to provide the monitoring system condition assessment. Based on the previous point, we need to change web-interface for logging functional supporting.

Most success we have achieved in optimization direction. It was decided to dive in through studying of used Python modules and built-in functionality. As a result, we were able to reduce CPU load by 80% and memory usage by 50%, but this is not the result. In addition, it is very important to provide good testing to get the output of reliable, stable and high-quality system.

3.5 Infrastructure for Research System

To conduct a full-fledged qualitative research, it is necessary to create an environment using the monitoring system, most suitable to the real conditions. The main tests are based on the third version of the monitoring system, but some solutions are considered in the context of the entire system branch. Another important factor in conducting research is the use of various configurations of nodes that will participate. The following infrastructure was approved (Fig. 1):

- Access point on Raspberry Pi 3. Application works with one Wixel module.
- Raspberry Pi node with a 3.2 in. screen on the TFT ILI9341 controller. Launches application, works with two Wixel modules.
- Raspberry Pi node with LED HDMI LCD monitor and peripherals. Works with two Wixel modules.
- Windows platform node. Launches application, works with one Wixel module.
- Client on a mobile platform. Connects through a web browser.

Since all microcomputers in our system are running on Linux distributions, we use packets to set up an access point:

- *dnsmasq* is a small, fast, non-resourceful DNS, DHCP and TFTP server designed to build small-network infrastructure. Can name computers that do not have global DNS records.
- *hostapd*—software that allows you to create a Wi-Fi access point using hardware resources.

To use the 3.2-in. screen with a microcomputer, we downloaded and installed the required driver.

By default, *Raspbian* has Python version 3.5. To install version 3.6 we can compile the interpreter from sources, but this operation may take long. Instead, we can use the *Berryconda3* package manager—the *Conda* batch manager for Raspberry microcomputers. Batch Manager is a set of software that allows you to control the process of installing, removing, debugging and updating various software components. The problem of using a separate version for microcomputers lies in the architecture of the processor.

At the nodes, the monitoring system application was downloaded and teams created for its automatic launch. Thus, we fully automated the system startup process. Simply turn on microcomputers to fully deploy.

3.6   Reliability Check

The reliability of the monitoring system was measured according to the following criteria:

- Number of critical errors during the continuous operation of the system, within 10 h.
- Number of unsuccessful dynamic switch-offs of modules during 30 attempts.
- Number of unsuccessful dynamic connections of modules during 30 attempts.

In the essence, the monitoring system can be attributed to the sensor wireless network. Such systems should continuously work a significant amount of time, which is measured in years. That is why we were supposed to provide reliable system operation for a large number of hours. It was decided to rely on values in 10 h, in our opinion, this is the optimal time in active development. As a result, all three versions of the system have successfully passed the testing stage. During all this time, we did not find any crucial error related to the system processes. There was only one nuance in the second version, sometimes the system suspended working with Wixel modules for a while, after which it continued again. We could not finally identify the source of this behavior, since any errors were missing. There is an assumption that this situation could cause the wrong closed port of the module during data readout, as a result, the program could wait a long time to reconnect with this module, because in the second version, we did not use the time out for standby during the connection.

With other tests, the first version of the monitoring system also managed well. Due to its functioning scheme of stopping the work with the modules, we could at any time, securely connect and disconnect the module from the system. At that, several modules at the same time. We did not get any errors for each of the operations. It should be noted that this is not a completely correct test for the first version, because it did not provide the ability to dynamically disconnect and connect modules.

The second version failed a lot during subsequent tests. Approximately 90% of the attempts to dynamically disconnect the module, ended with a critical error with the subsequent stopping of the site. A bit better was the result for a reverse operation, in about 25% of attempts, there were the same critical errors. However, both results were unacceptable for a reliable monitoring system. As mentioned earlier, we found the source of the problem. This behavior was caused by the incorrect handling of exceptions in the code.

Like the first version, the software implementation of the third version coped with the following tests on excellent. Because we use threads to work with each module, the errors that arise in them do not threaten the functioning of the system as a whole.

In spite of this, as in the first version, we did not get any errors during 30 attempts for each of the disassembly and plug-in operations.

In addition, a test was conducted for the third version of the monitoring system. Since, this implementation met the requirements at the initial setting of the task; we had to test how far the system is fault-tolerant, to check the decentralized architecture. The essence of the test was simple enough—make sure that the node's shutdown of the system is handled correctly, there are no errors. The system successfully performs this operation, the response time is approximately 1–4 s.

With these tests, we found that the first version, despite the unsuccessful architecture and logic of work, still coped with the task at a decent level. The second version can run stably for a long time without dynamically connecting and unplugging modules. Such behavior did not suit us, the system required significant changes and improvements. The third version managed the tasks best, works stably, reliably and anticipated. The decentralized architecture is implemented in a very simple scheme, which also increases the reliability of the system, high-level fault tolerance.

3.7 Scan Check

During research, the system should be ready to maintain a significant number of both nodes and modules on each of them. Depending on the area of the monitoring object and the complexity of the network, different configurations of the monitoring system infrastructure can be used. Here the main thing is to provide the flexibility of the system, to allow you to easily modify the required functionality or quickly adjust the system parameters for different conditions. Another important parameter is the load on the hardware resources. Unfortunately, during the development of each of the versions of the systems, it conducted its set of tests using the available hardware. That is, during the development of the first and second version of the monitoring system, we were not able to use Raspberry microcomputers, testing was conducted on computers. Then, we were not able to conduct a universal testing of all versions, because the first two needed a significant refactoring to run on microcomputers, which did not fit into the period. Next, some tests will be considered in the direction of scaling the system for each version separately.

First version. Requirements were required to support up to five client applications and up to five modules at each of them simultaneously. Due to problems in architecture and poorly thought out logic, out of 10 attempts to connect a client application, 7–9 of them ended with a critical mistake, along with a server stop. However, thanks to the simple sequential scheme of work with the modules, we could easily support five or more modules on the same client; however, it did not carry a special meaning. Regarding the load, the Python interpreter takes about 9 MB of RAM. Together with the micro framework flask, this value increases to 20. Thus, the server application used approximately 25 MB of RAM in its work, while processing requests could use up to 3–5 MB. The client application does not exceed the mark of 15 MB. Load testing was performed on the Intel Core i5 3317U processor. This is a mobile processor built with 22 nm technology, with a base clock of 1.7 GHz. The server application, on average, created a load in the range of 2–6% when working with one client. The client application could load the processor 30–40%; sometimes the value reached 80%. Such a load caused an inefficient way to read data from modules. From the point of view of memory usage, the monitoring system has acceptable metrics. However, the system creates a significant load, even on a productive hardware part, as for such a level of software. Because of the testing, we conclude that the system is not scalable, and a significant load on the CPU will not allow the system to work comfortably on microcomputers.

Second version. The monitoring system has completely similar indicators for the same set of tests. However, on the nodes in which server flow worked, memory consumption was much higher. Units without a server stream occupied about 20 MB of RAM in the course of work. The application with the server after launch occupied 30–35 MB of RAM. This was also because the new functionality was working with the preservation of signal level values. During the long work, it was noticed that the memory consumption is increasing linearly. After 4–5 h of continuous operation, the application could use 45–50 MB of memory. This situation was rather weird, because Python has built-in function for garbage collection (automatic clearing of memory from the resources that are no longer used), that is, the leak of memory is extremely unlikely. Unfortunately, at that time we could not find the answer to this question, but the fact that the system had problems with memory use actually made the second version of the system less effective than the first, in spite of a more primitive implementation. All this contributed to the creation of a new system that would be deprived of past problems.

Third version. During the development of the third version we could use Raspberry microcomputers, so testing was done on them. At this time, we managed to create an infrastructure that successfully supported simultaneous work of more than 4 nodes. Now, further scaling depended on the performance of the access point and network equipment. With this task, the system coped; it can function with a huge number of nodes at a time. Regarding the use of RAM, during the long run, the application saves 22–26 MB in a stable way. This is valid even for resourceful systems. Thanks to the new method for collecting data from modules, we have achieved that the load on the central processor has been reduced to 0–4%. Using more powerful systems, such as stationary computers or laptops, this figure does not exceed 2%. It should be noted that such a load is stored for any number of simultaneously connected modules to the node. From the point of view of horizontal scaling, the system uses hardware resources extremely efficiently without adversely affecting the reliability of the system. The load may increase if the host receives requests from web browsers. For example, if tree different client web browsers start sending requests for signal strength data at one and the same time, the processor load does not exceed 8–10%. In our opinion, this is a great result, given that the system works on microcomputers. Probably using WebSocket protocol can improve this result. Summing up, in the third version, we have achieved the most effective performance in terms of resource utilization. The system is scalable, which increases the security and power of the infrastructure at times. With such results, the system still provides the opportunity for further improvement.

The monitoring system of the third version, thanks to its flexible implementation, provides good opportunities for further improvement. After research, we have developed several solutions for improving the system's performance in the following areas:

- Safety improvement [27, 28].
- Optimization (processing of modules in processes, database) [29].

It should be noted that the latest version of the system deliberately does not use any means to protect data on the network. This decision was made on the basis of several facts. The information transmitted between the nodes (data about the nodes themselves) is not critical for the monitoring system. The only possible weakness may be the transmission of signal strength data through an unsecured HTTP request between the web browser and the host. We believe that this is a rather serious field for developing a variety of solutions that will most likely depend on specific problems that need to be addressed. In our case, we have created a system with a bias to conduct experiments. However, based on our experience, we can recommend the use of such developments.

From the optimization purposes, we have made significant progress in the latest version of the system. Nevertheless, the system has inspirational prospects for further improvement in this direction:

- The use of separate processes for working with modules, can reduce the load on the hardware resources of the site. In contrast to the threads, processes are processed in all core of the CPU in parallel. Another advantage is the use of dedicated resources for the process. This greatly improves system security, and simplifies debugging.
- This recommendation comes partially from the previous one. To save signal level data, we can use databases. Among the advantages is that the database has its own distribution system of access rights, reducing the amount of memory in the nodes, respectively, and the load. With databases, you can create a distributed system using the micro-server architecture. That is, some applications will work exclusively with modules for storing data, others—to calculate average values.

4 Future Work

To ensure better security following guidelines can be used:

- Use HTTPS protocol while exchanging data between the node and the client web browser. The flask micro-frame provides several methods for using a secure connection. The server that uses the monitoring system is debugging, that is, it

(a)   (b)

Fig. 11 Pololu Wixel sensors (**a**) and its modification with OLED (**b**)

was created for testing. For our needs it was enough. When initializing a server, we can pass a certain parameter to use SSL.

- Introduce the blockade technology for storing signal level data. Due to the fact that we have a decentralized architecture, we can significantly increase the integrity of the data thanks to the blockade technology—a distributed database, presented in the form of a sequence of data blocks that are linearly interconnected. Such a database is protected from data damage from the side of the intruders.
- Develop a system for sharing rights of access. Depending on security policy, select individual groups for partial access to the functional monitoring system. In some cases, this can be a very important introduction, for example, for wireless networks with a complex structure on large monitoring objects.

5 Implantation

As a sensor you can use separate Pololu Wixel modules or together with OLED SSD1306 128 × 64 [25]. Examples of such sensors are shown in Fig. 11.

As a controller, Raspberry Pi versions higher than 3rd are used together with the liquid crystal TFT display or others, including electronic ink technology (see an example in Fig. 12).

6 Conclusion

The monitoring system can be used to provide safe and stable working of the network. Spectrum analyzers can help us to create such type of the system. For our system, we choose to use Pololu Wixel modules with specific C firmware on the board.

Fig. 12 General view of the controller

The first version demonstrated the promise of the idea of using spectrum analyzers for the construction of monitoring systems. The software fully complies with the requirements. In spite of significant disadvantages in the architecture and software implementation logic, in terms of reliability, the system showed good results during testing.

The second version had an effective method of storing the signal level data for the subsequent calculation of averages. Functional provided an opportunity to better analyze the state of the network, to conduct statistics. Also, in the second version were the first attempts to create a fault-tolerant monitoring system. The use of threads for the server to work with the client application in the future has turned into a decentralized architecture.

The third version of the system has gained the advantages and solved all the disadvantages of the two previous ones. A simple workflow with decentralized architecture has created a flexible implementation that is easy to maintain, debug, and improve. A great deal of work was done in terms of optimization. We managed to reduce the load to the minimum values, which allows you to run the system on microcomputers with low power.

After conducting the research, we checked the effectiveness of the monitoring system for the analysis of wireless networks. It turned out that the system can function successfully as an independent solution, as well as in the complex software. Based on research data, recommendations were developed for improving the system.

There were consider several methods of the monitoring system building that are based on programmable modules and Python programming language. Every version of the monitoring system is different by own key features. During the development process was created and tested many solutions, that formed the basis of the final system version. It was shown that the current realized application scheme is enough viable and can be used in commercial or industrial fields.

The development of monitoring systems will allow to increase the efficiency of frequency resource use, will enable to monitor regular narrowband interference and flexibly redistribute the load, which will increase the availability and integrity of the transmitted data.

Acknowledgements This scientific work was partially supported by RAMECS and self-determined research funds of CCNU from the colleges' primary research and operation of MOE (CCNU19TS022).

References

1. Johnson D, Ketel M (2019) IoT: application protocols and security. Int J Comput Netw Inf Secur 11(4):1–8. https://doi.org/10.5815/ijcnis.2019.04.01
2. Noman Riaz M, Buriro A, Mahboob A (2018) Classification of attacks on wireless sensor networks: a survey. Int J Wirel Microw Technol 8(6):15–39. https://doi.org/10.5815/ijwmt.2018.06.02
3. Radivilova T, Hassan HA (2017) Test for penetration in Wi-Fi network: attacks on WPA2-PSK and WPA2-enterprise. In: 2017 international conference on information and telecommunication technologies and radio electronics (UkrMiCo). IEEE, pp 1–4
4. Dobrynin I et al (2018) Use of approaches to the methodology of factor analysis of information risks for the quantitative assessment of information risks based on the formation of cause-and-effect links. In: Proceedings of the 2018 international scientific-practical conference problems of infocommunications. Science and technology (PIC S&T). IEEE, pp 229–232. https://doi.org/10.1109/infocommst.2018.8632022
5. Ageyev D et al (2018) Provision security in SDN/NFV. In: 2018 14th international conference on advanced trends in radioelecrtronics, telecommunications and computer engineering (TCSET). IEEE, pp 506–509. https://doi.org/10.1109/tcset.2018.8336252
6. Ageyev DV, Salah MT (2016) Parametric synthesis of overlay networks with self-similar traffic. Telecommun Radio Eng (English translation of Elektrosvyaz and Radiotekhnika) 75(14):1231–1241
7. Kirichenko L, Radivilova T, Bulakh V (2018) Classification of fractal time series using recurrence plots. In: 2018 international scientific-practical conference problems of infocommunications. Science and technology (PIC S&T). IEEE, pp 719–724. https://doi.org/10.1109/infocommst.2018.8632010
8. Daradkeh YI, Kirichenko L, Radivilova T (2018) Development of QoS methods in the information networks with fractal traffic. Int J Electron Telecommun 64(1):27–32
9. Noman Riaz M (2018) Clustering algorithms of wireless sensor networks: a survey. Int J Wirel Microw Technol 8(4):40–53. https://doi.org/10.5815/ijwmt.2018.04.03
10. Kasim Ibraheem I, Al-Hussainy AA-H (2018) A multi QoS genetic-based adaptive routing in wireless mesh networks with Pareto solutions. Int J Comput Netw Inf Secur 10(9):1–9. https://doi.org/10.5815/ijcnis.2018.09.01
11. Saini R, Khurana SS (2016) Deployment of coordinated multiple sensors to detect stealth man-in-the-middle attack in WLAN. Int J Inf Technol Comput Sci 8(6):44–51. https://doi.org/10.5815/ijitcs.2016.06.06
12. Abidoye AP (2018) Energy efficient routing protocol for maximum lifetime in wireless sensor networks. Int J Inf Technol Comput Sci 10(4):33–45. https://doi.org/10.5815/ijitcs.2018.04.04
13. Chandrappa S, Dharmanna L, Shyama Srivatsa Bhatta UV, Sudeeksha Chiploonkar M, Suraksha MN, Thrupthi S (2017) Design and development of IoT device to measure quality of water. Int J Modern Educ Comput Sci 9(4):50–56. https://doi.org/10.5815/ijmecs.2017.04.06
14. Chandrappa S, Dharmanna L, Poojary SV, Meghana NU (2017) Automatic control of railway gates and destination notification system using internet of things (IoT). Int J Educ Manag Eng 7(5):45–55. https://doi.org/10.5815/ijeme.2017.05.05

15. Alakbarov GR, Hashimov MA (2018) Application and security issues of internet of things in oil-gas industry. Int J Educ Manag Eng 8(6):24–36. https://doi.org/10.5815/ijeme.2018.06.03
16. Mahadevaswamy UB (2018) Automatic IoT based plant monitoring and watering system using Raspberry Pi. Int J Eng Manuf 8(6):55–67. https://doi.org/10.5815/ijem.2018.06.05
17. Karakaya A, Akleylek S (2018) A survey on security threats and authentication approaches in wireless sensor networks. In: 6th international symposium on digital forensic and security, Antalya, Mar 2018, pp 1–4. https://doi.org/10.1109/isdfs.2018.8355381
18. Astapenya VM, Sokolov VY (2017) Experimental evaluation of the shading effect of accelerating lens in azimuth plane. In: 11th international conference on antenna theory and techniques, Kyiv, May 2017, pp 388–390. https://doi.org/10.1109/icatt.2017.7972671
19. Chatfield B, Haddad RJ (2017) RSSI-based spoofing detection in smart grid IEEE 802.11 home area networks. In: IEEE power & energy society innovative smart grid technologies conference, Washington, 2017, pp 1–5. https://doi.org/10.1109/isgt.2017.8086064
20. Sokolov V, Carlsson A, Kuzminykh I (2017) Scheme for dynamic channel allocation with interference reduction in wireless sensor network. In: 4th international scientific-practical conference problems of infocommunications. Science and technology, Kharkiv, Oct 2017, pp 564–568. https://doi.org/10.1109/infocommst.2017.8246463
21. Kuzminykh I, Carlsson A, Yevdokymenko M, Sokolov V (2019) Investigation of the IoT device lifetime with secure data transmission. In: Galinina O, Andreev S, Balandin S, Koucheryavy Y (eds) Internet of things, smart spaces, and next generation networks and systems. NEW2AN 2019, ruSMART 2019. Lecture notes in computer science, vol 11660. Springer, Cham, pp 16–27. https://doi.org/10.1007/978-3-030-30859-9_2
22. Chakkor S, Cheikh EA, Baghouri M, Hajraoui A (2014) Efficiency evaluation metrics for wireless intelligent sensors applications. Int J Intell Syst Appl 6(10):1–10. https://doi.org/10.5815/ijisa.2014.10.01
23. Pololu Corporation (2015) Pololu Wixel user's guide. https://www.pololu.com/docs/pdf/0J46/wixcl.pdf. Accessed 15 Oct 2019
24. Python Software Foundation (2018) Python 3.6.6 Documentation. https://docs.python.org/3/download.html. Accessed 15 Oct 2019
25. Remington J (2011) Spectrum analyzer app code. https://forum.pololu.com/t/spectrum-analyzer-app-code/3394. Accessed 15 Oct 2019
26. Bogachuk I, Sokolov V, Buriachok V (2018) Monitoring subsystem for wireless systems based on miniature spectrum analyzers. In: 5th international scientific and practical conference problems of infocommunications. Science and technology, Kharkiv Oct 2018, pp 581–585. https://doi.org/10.1109/infocommst.2018.8632151
27. Kryvinska N (2008) An analytical approach for the modeling of real-time services over IP network. Math Comput Simul 79(4):980–990. https://doi.org/10.1016/j.matcom.2008.02.016
28. Babatope LO, Babatunde L, Ayobami I (2014) Strategic sensor placement for intrusion detection in network-based IDS. Int J Intell Syst Appl 6(2):61–68. https://doi.org/10.5815/ijisa.2014.02.08
29. Kryvinska N (2004) Intelligent network analysis by closed queuing models. Telecommun Syst 27:85–98. https://doi.org/10.1023/B:TELS.0000032945.92937.8f

Computer-Integrated Technologies for Fitomonitoring in the Greenhouse

Taras Lendiel⑩, Vitaliy Lysenko⑩, and Kateryna Nakonechna⑩

Abstract The phytomonitoring subsystem software and hardware in a modern greenhouse building is provided through the use of LabVIEW software and Arduino equipment tested directly at the production site. It is shown that when growing vegetables in greenhouses it is important to take into account not only the temperature of the atmosphere in the greenhouse, but also the temperature of the plant itself. The dependence of the plant temperature on the light in the greenhouse has been analyzed, an advanced mathematical model of the greenhouse has been obtained as an object of control, suitable for forming control effects taking into account the spatial variance of the object. Purpose: to develop a subsystem of phytomonitoring in the greenhouse, which will complement the traditionally existing microclimate control system (in our case tomatoes) and clarify the mathematical model of the greenhouse as a spatially dispersed control object.

Keywords Plant · Phytomonitoring · Greenhouse · Phytometric parameters · LabVIEW · Subsystem

1 Question Condition

According to the reference literature of biologists, it is noted that an important factor affecting plants in protected soil structures is the temperature provided by the microclimate system [1, 2]. Maintaining a set temperature in the space of a greenhouse is a difficult task, since the area of the greenhouses itself is significant, which contributes to heat loss, uneven heat dissipation from the heating system and the effects of external disturbances [1, 3–8]. This leads to an uneven distribution of temperature, which confirms the need to consider the greenhouse as an object with dispersed parameters [2, 9–11].

T. Lendiel (✉) · V. Lysenko · K. Nakonechna
National University of Life and Environmental Sciences of Ukraine, Kyiv, Ukraine
e-mail: taraslendel@gmail.com

© The Editor(s) (if applicable) and The Author(s), under exclusive license to Springer Nature Switzerland AG 2021
T. Radivilova et al. (eds.), *Data-Centric Business and Applications*, Lecture Notes on Data Engineering and Communications Technologies 48,
https://doi.org/10.1007/978-3-030-43070-2_30

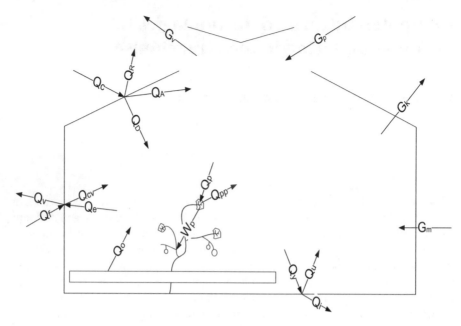

Fig. 1 Schematic representation of heat flows in a greenhouse

The thermal balance of a protected soil structure is the sum of all thermal flows entering and exiting from it (Fig. 1). For a static mode, that is, for a period when the temperature inside and outside the building is constant, the thermal balance is zero. In this case, there is a temperature equilibrium. In transition (dynamic) modes the relationship between the influx and loss of thermal energy changes and the temperature in the building will increase or decrease [12, 13]. When forming the thermal balance of greenhouses should take into account the participation of plants in this process (plants absorb or reflect a certain amount of energy).

Taking into account the above, there is a need to improve the algorithms of microclimate control systems functioning in greenhouses due to the lack of ability to monitor plant response to the effects of natural disturbances. Therefore, it is proposed to supplement the existing components of the greenhouse equipment for phyto-monitoring, which will allow to measure the phytoclimatic and phytometric parameters of plants in the greenhouse space.

In addition, microclimate control strategies are based on simplified mathematical models that do not take into account the dispersion of such an object. There is a need to refine the mathematical model of the greenhouse, taking into account its spatial dispersion.

where Q_o is the amount of heat from the heating system; Q_c is the amount of heat from solar radiation entering the greenhouse; Q_R is amount of heat reflected by the surface of the greenhouse; Q_A is amount of heat of solar energy absorbed by the surface of the greenhouse; Q_D is amount of heat of solar energy entering the greenhouse through the surface of the greenhouse; Q_f is amount of heat of all external perturbations; Q_e is amount of heat lost for heating the fence; Q_{cv} is amount of heat of external perturbations that arrived in the greenhouse; Q_R is quantity of heat received to the plant; Q_{RP} is amount of heat reflected by the plant; W is amount of energy absorbed by the plant; Q_u is amount of heat received by the soil in the greenhouse; Q_i is amount of heat absorbed by the soil of the greenhouse; Q_u is amount of heat reflected by the soil in the greenhouse; G_R is amount of air entering the greenhouse through the transom and the ventilation system; G_v is amount of air, with air exchange through the transom and ventilation system; G_m is air performance through the greenhouse enclosure; G_k is air productivity through the greenhouse surface.

When growing vegetables in greenhouses, technology should be followed to ensure the required yield and quality of produce. Maintenance of permissible limits of technological parameters is realized by traditional systems of microclimate. Such systems do not provide the opportunity to monitor and take into account the condition of plants as a reaction to the effect of natural disturbances. One of the most important conditions of the plant is its temperature.

The temperature of the plant depends on both the environmental impact (air temperature, radiation level, speed of its movement), and the properties and characteristics of the plants themselves (optical properties of the plant, size and location of the leaves, etc.) [14]. In combination with all these factors, the temperature of the plant, as a rule, differs from the ambient temperature. At the same time, the temperature of the plant may be relative to the environment:

- higher
- lower
- equal or very close to her.

The first is possible only with the heating of tomatoes, stems and leaves of sunlight, which can lead to overheating of plants [15]. Therefore, it is expedient to supplement the standard control systems of the microclimate parameters—devices that measure the temperature of the plant—the corresponding algorithms, when the control strategy is formed taking into account the temperature of the plant.

2 Materials and Methods

2.1 Realization the Phytomonitoring Subsystem

To monitor the change in the parameters controlled in each zone of the industrial greenhouse, a subsystem of fitomonitoring based on the Arduino Mega2560 and three DS18B20 temperature sensors, providing a moisture content of DHT11 and two light sensors. The connection and communication between the chip and the computer is via the Universal Serial Bus port [16–31]. The simplified structure of this subsystem is shown in Fig. 2.

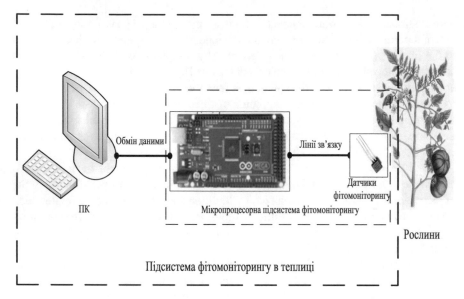

Fig. 2 The block diagram of the subsystem of phytomonitoring in the greenhouse

Power supply circuits via Universal Serial Bus. The subsystem frequency is equal to 16 MHz, which includes 54 digital input/output channels, 14 of which are capable of operating in Pulse-Width Modulation, 16 analog inputs, 4 universal asynchronous receiver/transmitter hardware serial ports for communication with an industrial computer and others devices. In the event of a malfunction, a "Reset" button is provided.

The block diagram of measurement and recording in the database of technological parameters from the sensors is shown in Fig. 3.

2.2 Software

The software for this subsystem is implemented in the LabVIEW environment, which made it possible:

- to read out the information from the sensors, which made it possible to estimate the change of the parameter controlled in the greenhouse space where the sensor is installed;
- saving measured values to a database for expert processing;
- develop an operator interface.

In the block diagram, all measured data is stored as Microsoft Office Excel spreadsheets, which in turn facilitates further analysis of the measurement results.

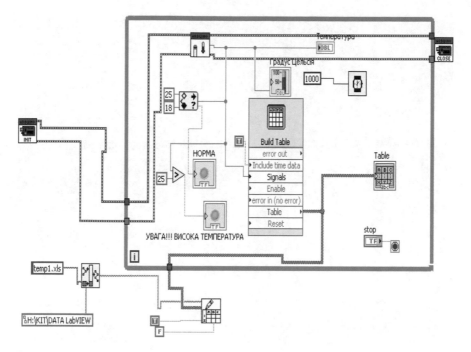

Fig. 3 The block diagram of measurement and recording in the database of technological parameters

To refine the mathematical model, the greenhouses used a method of balance of energy flows, which was different in that the greenhouse was divided into zones by design features, the mutual influence of which was taken into account. The adequacy of the mathematical model was tested with the Matlab Simulink software environment and the results of natural temperature measurements directly in the greenhouse.

3 Results and Discussion

Phytomonitoring was carried out in specific places of the greenhouse—air temperature in the plant's phytoclimatic environment, its temperature and humidity are measured (Fig. 4).

Schematically, the locations of phytomonitoring in the greenhouse area are shown in Fig. 5.

The described subsystem of phytomonitoring has been tested at PJSC "Combine" "Teplichny", which confirmed its workability.

Measurements of light and temperature of plants and air in the greenhouse were carried out in different rows of greenhouses.

Fig. 4 Sensors in the greenhouse space

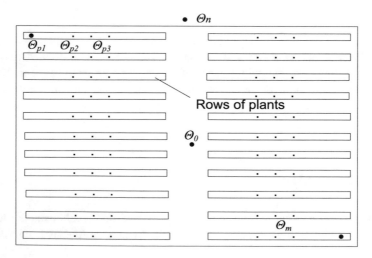

Fig. 5 Schematic arrangement of rows for fitomonitoring throughout the greenhouse area: Θ_p is the temperature of the plant in the i-th zone of the greenhouse; Θ_o is the air temperature in the i-th zone of the greenhouse; Θ_n is the temperature of the outside air, m is the serial number

According to the analysis of the obtained information, regression equations and root mean square deviation are obtained, namely:

- For the temperature of stem

$$y1(x) = 24.902 + 2.489 \times 10^{-5}x, \tag{1}$$

$$\delta1 = 0.829 \text{ °C};$$

- For the temperature of fruit

$$y2(x) = 24.443 + 2.707 \times 10^{-4}x, \tag{2}$$

$$\delta2 = 0.507 \text{ °C};$$

- For the temperature of leaf

$$y3(x) = 23.665 + 2.155 \times 10^{-5}x \tag{3}$$

$$\delta3 = 1.188 \text{ °C};$$

- For the temperature of air near the plant

$$y4(x) = 32.142 + 2.515 \times 10^{-4}x, \tag{4}$$

$$\delta4 = 1.37 \text{ °C},$$

where x is illumination.

The graphical dependences of the measurement results are shown in Fig. 6. Their analysis makes it possible to establish that during the day the temperature of the plant rises due to the influence of solar radiation, despite the fact that the air temperature in the greenhouse remains unchanged. As overheating of plants can adversely affect their tendency of development, it is necessary to monitor and take into account not only the temperature of the environment, but also the temperature of the plant itself, and use these to form control effects. This is what is provided by the developed phytomonitoring subsystem.

The carried out phytoconditioning measurements in the greenhouse revealed a significant temperature irregularity due to the large number of heat transfer surfaces, which contribute to heat loss, as well as irregular plant development [32]. This fact, namely that the greenhouse is an object with distributed parameters (Fig. 7), must be taken into account for the evaluation of plant conditions and the formation of control effects.

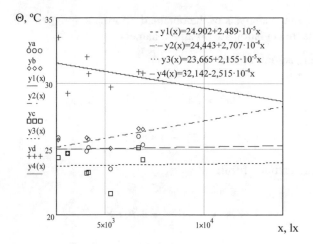

Fig. 6 Dependence of the temperature of the plant and air near the plant on the illumination: ya are results of measurements of the temperature of the stem, °C; yb are results of measurements of fetal temperature, °C; yc are results of measurements of leaf temperature, °C; yd are results of measurements of air temperature near the plant, °C; x are illumination, lx

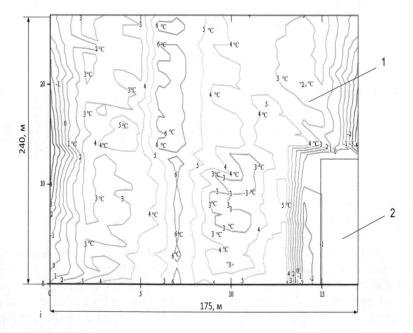

Fig. 7 Differences in the actual values of temperatures in the greenhouse and technological problems: 1—area for growing tomatoes; 2—area of the main and auxiliary control systems

To take into account the spatial dispersion of the greenhouse and its effect on the development of plants, a mathematical model that takes into account the spatial coordinate by width and created the conditions for calculating the air temperature in the greenhouse, depending on the influence of external perturbations and taking into account the mutual influence of the zones on each other, which are divided to the greenhouse. To do this, the space of the section of the greenhouse is divided into temperature zones along the width of the greenhouse, taking into account the structural features of the section. As a consequence, the number of such zones forming the greenhouse section will be 8 (Fig. 8).

It should be noted that the set of temperature zones will form sections which, in turn, form a block of greenhouses, and the lateral zones 1 and 8 are in contact with the lateral surfaces of the zones of other sections or with external air.

The parametric scheme of the greenhouse climate microclimate model is shown in Fig. 9. When designing the model, the following assumptions are made:

- for each zone, the corresponding climate parameter is considered to be homogeneous throughout its area;
- the heat exchange with the greenhouse of the greenhouse will not be taken into account, due to the specifics of the thermal protection of the soil from the action of the external environment;
- the concentration of carbon dioxide is the same for the whole section;
- the heat transmitted from the sun depends on the time period and the angle of the fall to the roof of the greenhouse (in zones 1–4, 5–8 this angle is different due to the peculiarities of the slope of the roof of the greenhouse).

It was assumed that each section affects the temperature balance of the greenhouse, and the amount of heat for the i-th zone Q_i will depend on the amount of heat Q_{i+1} and Q_{i-1} that is given or received from nearby zones $(i + 1, i - 1)$. Thus, we can write:

$$Q_i = Q_{t,i} + Q_{s,i} - Q_{sr,i} - Q_{v,i} + Q_{i+1} + Q_{i-1}, \tag{5}$$

Fig. 8 Dividing the volume of greenhouses into zones: **a** is soil surface of greenhouses; **b** is face surfaces; **c**, **d** are the roof of the greenhouse

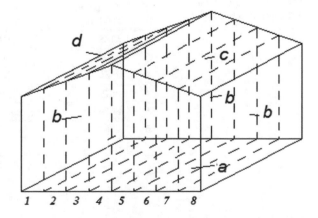

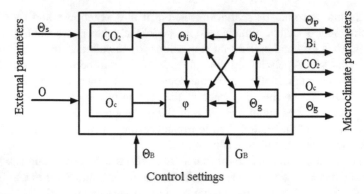

Fig. 9 Parametric scheme of the microclimatic model used in greenhouses: Θ_g—soil temperature; Θ_P—average air temperature; CO_2—concentration of carbon dioxide in the section; O_c—the intensity of solar radiation in the section; Θ_B—hot water temperature; Θ_c—ambient air; G_B—the cost of hot water; φ—humidity; O—the intensity of solar radiation; Θ_s—the temperature outside; Θ_i—the temperature in the section

where i = 1...8, Q_t is the amount of heat coming from the heating system; Q_s is amount of heat received from the sun, J; Q_v is loss of heat through the roof of the greenhouse and the end walls, J; Q_{i+1}, Q_{i-1} is amount of heat coming from nearby temperature zones, J; $Q_{sr,i}$ is amount of heat absorbed by plants in the i-th zone, J.

Taking into account the geometric dimensions of the temperature zones of the greenhouse (Table 1), its thermal characteristics, the thermal balance equation for the i-th the zone will look like:

$$6.443 \times 10^5 \cdot \Theta_i = 1.658 \times 10^3 \cdot (\Theta_{w,i} - \Theta_i) + 3 \cdot S_{k,i} + 4 \cdot (S_{b,i} - S_{k,i})(\Theta_i - 20)$$
$$+ 0.026 \cdot S_{i-1,i}(\Theta_{i-1} - \Theta_i) + 0.026 \cdot S_{i+1,i}(\Theta_{i+1} - \Theta_i)$$
$$- 3.3 \cdot S_{k,i}. \tag{6}$$

Table 1 Design dimensions of the greenhouse section

No.	The height of the zone (m)	Area of the lateral surface (m^2)	Volume (m^3)
1	5155	9155	457,735
2	6309	11,464	573,205
3	7464	13,774	688,675
4	8619	16,083	804,145
5	8619	16,083	804,145
6	7464	13,774	688,675
7	6309	11,464	573,205
8	5155	9155	457,735

where Θ_i is the air temperature in the i-th zone; $\Theta_{w,i}$ is the temperature of coolant in the i-th zone; $S_{b,i}$ is the area of the lateral surface in the i-th zone; $S_{k,i}$ is roof area in the i-th zone.

The mathematical model is developed for a greenhouse with an area of 4.2 ha., with dimensions: length of the block—$L_2 = 50$ m, section width—16 m, angle of the roof of the greenhouse of 30°. Due to the inter-row characteristics, the width of one zone is taken to be equal to $L_1 = 2$ m, the height of the final supporting wall $H_1 = 4$ m. In calculations, it was taken into account that the surfaces between the zones from the edge of the greenhouse to the center will increase due to the increase in the height of each zone (the angle of inclination of the roof is taken into account) (Fig. 10).

H_i is height of a separate zone; Sb_i is area of the lateral surface of the zone; 1...8—zones.

The composite model of each zone takes into account the influence of adjacent temperature zones, namely in the first zone and the eighth zone thermal exchange with the environment occurs precisely through the end surfaces. Taking into account the angle of inclination of the roof, solar energy entering the greenhouse is taken into account. Given the recorded, in general, the final balance for the greenhouse will be recorded as a system of equations:

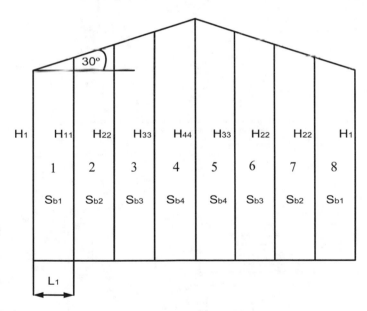

Fig. 10 Zones of the section of the greenhouse at its transverse section

$$\begin{cases} 6.443 \times 10^5 \cdot \Theta_i = 11 \cdot S_t(\Theta_{w,1} - \Theta_1) + 1.181 \times 10^3 \cdot (\Theta_1 - 20) \\ +8.19 \cdot (\Theta_2 - \Theta_1) - 25.8; \\ 6.443 \times 10^5 \cdot \Theta_i = 11 \cdot S_t(\Theta_{w,i} - \Theta_i) + 3 \cdot S_{k,i} + 4 \cdot S_{b,i} + S_{k,i})(\Theta_i - \Theta_3) \\ +0.026 \cdot S_{i-1,i}(\Theta_{i-1} - \Theta_i) + 0.026 \cdot S_{i+1,i}(\Theta_{i+1} - \Theta_i) - 3.3 \cdot S_{k,i}, \quad i = 2 \ldots 7; \\ 6.443 \times 10^5 \cdot \Theta_i = 11 \cdot S_t(\Theta_{w,8} - \Theta_8) + 3 \cdot S_{k,8} + 1.181 \times 10^3 \cdot (\Theta_8 - \Theta_{i+1}) \\ +8.19 \cdot (\Theta_7 - \Theta_8) - 25.8. \end{cases} \quad (7)$$

The mathematical model (7) created was studied in the MATLAB Simulink software environment (Fig. 11) for the following input parameters: $O_s = 300 \text{ W/m}^2$ and $\Theta_z = 20 \,^\circ\text{C}$.

The transient processes show that the air temperature in different zones stabilizes at the level of 23.5–26 °C, and the transition time—from 2000 to 6000 s (Fig. 12). This difference is due to the different volume of each zone and different the effect of external perturbations on the extreme zones, which leads to uneven heat exchange of the greenhouse environment with the heating system [3, 7]. A comparison of the results of the mathematical model calculation and the experimental temperature measurements is shown in the cross section (temperature relief) of the greenhouse cross section from 1 to 8 zones (Fig. 13). From this analysis, the mean square deviation

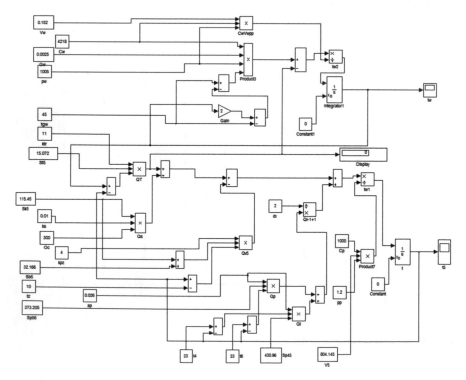

Fig. 11 Simulation model of one of the temperature zones of a greenhouse section using a software environment MATLAB Simulink

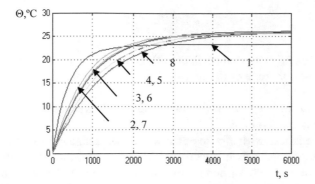

Fig. 12 Temperature change in the greenhouse: 1 ... 8 are zone sections

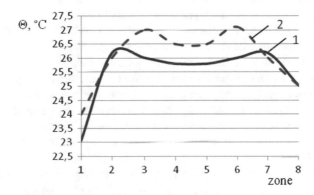

Fig. 13 Average temperature: 1—calculated temperature; 2—measured temperature value used for greenhouses (static mode)

of 1.05 °C was determined, which confirms the adequacy of the model and the possibility of its use for forming control effects.

To assess the adequacy of the developed mathematical model is used mean-square deviation calculated by the known algorithm:

$$\delta = \sqrt{\frac{\sum_{i=1}^{m} (t_i - t_{pi})^2}{m - 1}}, \tag{8}$$

where δ—calculating standard deviation of measurements; t_i—the result of the calculation of the model that investigated; t_{pi}—parameter value based on the results of experimental measurements; m—sample size.

The calculated mean-square deviation of the model results does not exceed 1.05 °C compared to the real values.

The reaction of the plant in the process of providing the temperature regime is taken into account by using the phytothermic criterion for assessing the state of the plant [3], which can be introduced into the mathematical model of the control object:

$$r = \frac{\Theta_p - \Theta_H}{\Theta_i - \Theta_H}, \tag{9}$$

where Θ_p is plant's temperature in the i-th zone of the greenhouse; Θ_i is air temperature in the i-th zone of the greenhouse; $\Theta_н$ is the temperature of the outside air.

To calculate the expression (9), the temperature is measured by the proposed subsystem phytomonitoring.

In the greenhouse where the research was conducted, the distribution of thermal energy in its space passes in accordance with the allocation of heating registers: heating pipes are laid along a number of plant placements, so heat comes to the plants separately for each row, but not evenly due to the influence of perturbing factors, the inertia of the heating system. In view of these circumstances, it is advisable to measure phytometric parameters for each row, replacing the temperature of the plant with the average value of the temperature of plants in a row.

The criterion r will be 1 in the case when the temperature of the plants will be equal to the temperature of the air in the greenhouse. However, this is difficult to achieve, due to the effect of disturbances, the inertia of the heating system.

Our long-term studies [3] have shown that the temperature of the plant during the day may be less than 2 °C and more than 7 °C from the atmosphere temperature in the greenhouse (Fig. 14). This allows to recommend to set the value of the criterion r from 0.9 to 1.1, and according to the technology of growing tomatoes, the temperature of the air in the greenhouse should be 21...25 °C during the day and—18...22 °C at night (Table 2).

Thus, the use of the phytothermic criterion for assessing the plant condition allows for the control of the temperature of the biological object to be taken into account during the control process.

The need for using the phytothermic criterion also confirms the fact of obtaining uneven yield from each row of greenhouses [33]. Using this criterion allows you to get more products, while using traditional control systems [6–8].

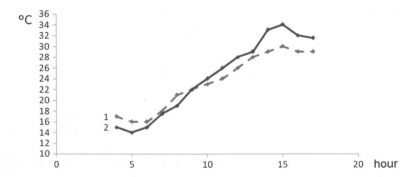

Fig. 14 Value of temperature of the tomato leaf and air temperature in the greenhouse for a day: 1—air temperature in the greenhouse; 2—the temperature of the tomato leaf

Table 2 The value of the phytothermic criterion for determining the temperature of the plant and its temperature environment

Hour of day	The temperature of the plant (°C)	Air temperature in the greenhouse (°C)	The ambient temperature (°C)	Phytotemperature (r_m)
2	15	17	8	0.77
3	14	16	8	0.75
4	15	16	8	0.875
5	17,5	18	8	0.95
6	19	21	8	0.84
7	22	22	9	1
8	24	23	10	1.08
9	26	24	12	1.16
10	28	26	13	1.15
11	29	28	14	1.07
12	33	29	14	1.26
13	34	30	15	1.26
14	32	29	15	1.21
15	31.5	29	15	1.17
16	28	27	14	1.08
17	23	25	14	0.81
18	23	25	14	0.81
19	22	24	14	0.8
20	22	24	13	0.81

From literary sources [34] it is known that the growth of the yield at sunlight is irradiated with 8 MJ (2.22 kWh) per day and the usual content in air CO_2 (0.03%), ranging from 40 to 150 g per bush per day and depends on the level of air temperature in the greenhouse, which is presented in Table 3.

The microclimate of the greenhouse is influenced by the phytothermic criterion level, which is achieved by supplementing the existing heating system with additional equipment, which will provide differential heat distribution with allowance for external perturbations.

From the data given, one can determine the hourly growth of the crop at the specified temperature regimes of air

$$m_{cp} = \frac{m_t}{24},$$ (10)

m_{cp} average weight gain per hour;
m_t weight gain at air temperature Θ.

Calculated data are shown in Table 4.

Table 3 Influence of air temperature on yield of tomatoes

Temperature (°C)	Tomato yield (g)
10	40
11.5	61
13	81
14.5	100
16	114
17.5	125
10	134
20.5	141
22	146
23.5	150
25	151
26.5	150
28	147

Table 4 Influence of temperature of air on increase of yield of tomatoes hourly

Temperature (°C)	Tomato yield (g)
10	1.6
11.5	2.5
13	3.4
14.5	4.1
16	4.7
17.5	5.2
19	5.6
20.5	5.8
22	6
23.5	6.2
25	6.3
26.5	6.2
28	6.1

The results of the analysis suggest that in the existing greenhouse with the traditional microclimate control system, the largest increase in production was from 8 to 9 h at temperatures 23 … 24 °C and was 6.2… 6.3 g/h. As a result, from one bush we get less than 151 g of gain per day, because under temperature conditions of 17…22 °C the plant receives insufficient energy for development and yield growth is 5.2 … 6 g/h, and at temperatures above 25 °C increase the harvest will be more than 6 g/h.

Using the phytothermic criterion when adjusting the temperature regime gives an opportunity to receive 150–151 g of the yield from one bush per day. Use in the traditional microclimate control system in the greenhouse of this criterion brings the plants closer to a comfortable temperature zone for them and promotes their higher productivity, and in turn leads to an increase in the company's profit.

4 Conclusion

1. It has been experimentally determined that solar radiation influences the temperature of a plant in a greenhouse, and therefore a significant difference between the temperature of the plant and the air temperature in the greenhouse is possible.
2. Developed by

   - tests of the phytomonitoring subsystem in the greenhouse directly at the production confirmed its efficiency;
   - software of the subsystem of phytomonitoring, which made it possible to create a convenient for the effective operation of the staff of the interface.

3. The mathematical model of the greenhouse as an object with dispersed parameters has been improved (the calculated mean-square deviation of the model results does not exceed 1.05 °C compared to the real values), which gives grounds to recommend it for use in the formation of control influences.
4. The established phytomonitoring subsystem in the greenhouse has shown its effectiveness and can complement the existing microclimate control systems in industrial greenhouses.

References

1. Zeeman I (1961) Greenhouses and ego climate regulation. Selhozizdat, Moscow
2. Zhang P, Xu L (2018) Unsupervised segmentation of greenhouse plant images based on statistical method. Sci Rep 8:4465. https://doi.org/10.1038/s41598-018-22568-3
3. Bodrov VI, Bodrov MV, Kuzin VY (2017) Ensuring the parameters of microclimate of hothouses during a warm season. ARPN J Eng Appl Sci 12(6):1864–1869
4. Fourati F (2014) Multiple neural control of a greenhouse. Neurocomputing 139:138–144
5. Vovna O, Laktionov I, Sukach S, Kabanets M, Cherevko E (2018) Method of adaptive control of effective energy lighting of greenhouses in the visible optical range. Bul J Agric Sci 24(2):335–340
6. Laktionov I, Vovna O, Zori A (2017) Copncept of low cost computerized measuring system for microclimate parameters of greenhouses. Bul J Agric Sci 23(4):668–673
7. Reshetiuk VM, Lendiel TI, Kuliak BV (2016) Vymiriuvalnyi elektrotekhnichnyi kompleks dlia monitorynhu parametriv biometrychnoho stanu roslyny ta mikroklimatu v teplytsi (Measuring range of electrical parameters for monitoring the state of biometric plants and microclimate in the greenhouse). Visnyk Kharkivskoho natsionalnoho tekhnichnoho universytetu silskoho hospodarstva imeni Petra Vasylenka 176:51–53
8. Trokhaniak V, Klendii O (2018) Numerical simulation of hydrodynamic and heat-mass exchange processes of a microclimate control system in an industrial greenhouse. Bull Transilvania Univ Brasov, Ser II: For Wood Ind Agric Food Eng 11(2):171–184

9. Korobiichuk I, Lysenko V, Reshetiuk V, Lendiel T, Kamiński M (2017) Energy-efficient electrotechnical complex of greenhouses with regard to quality of vegetable production. In: Szewczyk R, Kaliczyńska M (eds) Recent advances in systems, control and information technology. SCIT 2016. Advances in intelligent systems and computing, vol 543. Springer, Cham

10. Hart J, Hartova V (2018) Development of new elements to automatized greenhouses. Agron Res 16(3):717–722. https://doi.org/10.15159/AR.18.105

11. Revathi S, Radhakrishnan TK, Sivakumaran N (2017) Climate control in greenhouse using intelligent control algorithms. Paper presented at the proceedings of the American control conference, pp 887–892. https://doi.org/10.23919/acc.2017.7963065

12. Story D, Kacira M (2014) Automated machine vision guided plant monitoring system for greenhouse crop diagnostics. doi: 10.17660/ActaHortic.2014.1037.81

13. Abhfeeth KA, Ezhilarasi D (2013) Monitoring and control of agriculture parameters in a greenhouse through internet. Sens Transducers 150(3):106–111

14. Outanoute M, Lachhab A, Ed-Dahhak A, Selmani A, Guerbaoui M, Bouchikhi B (2015) A neural network dynamic model for temperature and relative humidity control under greenhouse. Paper presented at the proceedings—2015 3rd international workshop on RFID and adaptive wireless sensor networks, RAWSN 2015—in conjunction with the international conference on NETworked sYStems, NETYS 2015, pp 6–11. https://doi.org/10.1109/rawsn.2015.7173270

15. Lysenko V, Lendiel T, Komarchuk D (2018) Phytomonitoring in a greenhouse based on Arduino Hardware. In: 2018 International scientific-practical conference problems of info-communications. Science and technology (PIC S&T), IEEE, pp 365–368. https://doi.org/10.1109/infocommst.2018.8632030

16. Kryvinska N (2010) Converged network service architecture: a platform for integrated services delivery and interworking. Electronic business series, vol 2. International Academic Publishers, Peter Lang Publishing Group

17. Kryvinska N (2008) An analytical approach for the modeling of real-time services over IP network. Math Comput Simul 79(4):980–990. https://doi.org/10.1016/j.matcom.2008.02.016

18. Kryvinska N (2004) Intelligent network analysis by closed queuing models. Telecommun Syst 27:85–98. https://doi.org/10.1023/B:TELS.0000032945.92937.8f

19. Kryvinska N, Zinterhof P, van Thanh D (2007) An analytical approach to the efficient real-time events/services handling in converged network environment. In: Enokido T, Barolli L, Takizawa M (eds) Network-based information systems. NBiS 2007. Lecture notes in computer science, vol 4658. Springer, Berlin, Heidelberg

20. Ageyev DV, Salah MT (2016) Parametric synthesis of overlay networks with self-similar traffic. Telecommun Radio Eng (English translation of Elektrosvyaz and Radiotekhnika) 75(14):1231–1241

21. Ignatenko AA, Ageyev DV (2013) Structural and parametric synthesis of telecommunication systems with the usage of the multi-layer graph model. In: Proceedings of the 2013 23rd international crimean conference microwave and telecommunication technology (CriMiCo 2013), pp 498–499

22. Ageyev DV, Kopylev AN (2013) Modelling of multiservice streams at the decision of tasks of parametric synthesis. In: The 2013 23rd international crimean conference microwave and telecommunication technology (CriMiCo 2013). IEEE, pp 505–506

23. Ageyev DV, Ignatenko AA, Fouad W (2013) Design of information and telecommunication systems with the usage of the multi-layer graph model. In: Proceedings of the XIIth international conference the experience of designing and application of CAD systems in microelectronics (CADSM). Lviv-Polyana. Lviv Polytechnic National University, Ukraine, pp 1–4

24. Ageyev DV (2010) NGN network planning according to criterion of provider's maximum profit. In: 2010 international conference on modern problems of radio engineering, telecommunications and computer science (TCSET—2010). Lviv-Slavske, p 256

25. Radivilova T et al (2018) Decrypting SSL/TLS traffic for hidden threats detection. In: Proceedings of the 2018 IEEE 9th international conference on dependable systems, services and technologies (DESSERT). IEEE, pp 143–146. https://doi.org/10.1109/dessert.2018.8409116

26. Ageyev D et al (2018) Classification of existing virtualization methods used in telecommunication networks. In: Proceedings of the 2018 IEEE 9th international conference on dependable systems, services and technologies (DESSERT), pp 83–86

27. Radivilova T, Kirichenko L, Ageiev D, Bulakh V (2020) The methods to improve quality of service by accounting secure parameters. In: Hu Z, Petoukhov S, Dychka I, He M (eds) Advances in computer science for engineering and education II. ICCSEEA 2019. Advances in intelligent systems and computing, vol 938. Springer, Cham

28. Kirichenko L, Radivilova T, Tkachenko A (2019) Comparative analysis of noisy time series clusterin. CEUR workshop proceedings 2362:184–196. http://ceur-ws.org/Vol-2362/paper17.pdf

29. Radivilova T, Kirichenko L, Yeremenko O (2017) Calculation of routing value in MPLS network according to traffic fractal properties. In: 2017 2nd international conference on advanced information and communication technologies (AICT). IEEE, pp 250–253. https://doi.org/10.1109/aiact.2017.8020112

30. Ivanisenko I, Radivilova T (2015) The multifractal load balancing method. In: 2015 second international scientific-practical conference problems of info communications science and technology (PIC S&T). IEEE, pp 122–123. https://doi.org/10.1109/infocommst.2015.7357289

31. Kryvinska N, Zinterhof P, van DT (2007) New-emerging service-support model for converged multi-service network and its practical validation. In: First international conference on complex, intelligent and software intensive systems (CISIS'07). IEEE, pp 100–110. https://doi.org/10.1109/cisis.2007.40

32. Chen L, Zhang B, Yao F, Cui L (2016) Modeling and simulation of a solar greenhouse with natural ventilation based on error optimization using fuzzy controller. In: 2016 35th Chinese control conference (CCC), Chengdu. IEEE, pp 2097–2102. https://doi.org/10.1109/chicc.2016.7553676

33. Hill LS (2008) Modern technologies of vegetable closed and open ground. Part 1 indoor ground. L.S. Hill Ball

34. Lysenko V, Komarchuk D, Opryshko O, Pasichnyk N, Zaets N (2018) Determination of the not uniformity of illumination in process monitoring of wheat crops by UAVs. In: 2017 4th international scientific-practical conference problems of info communications science and technology, (PIC S&T). IEEE, pp 265–267. https://doi.org/10.1109/infocommst.2017.8246394

35. Tregub V, Korobiichuk I, Klymenko O, Byrchenko A, Rzeplińska-Rykała K (2020) Neural network control systems for objects of periodic action with non-linear time programs. In: Szewczyk R, Zieliński C, Kaliczyńska M (eds) Automation 2019. Advances in intelligent systems and computing, vol 920. Springer, Cham

Fusion the Coordinate Data of Airborne Objects in the Networks of Surveillance Radar Observation Systems

Ivan Obod⬡, Iryna Svyd⬡, Oleksandr Maltsev⬡, Ganna Zavolodko⬡, Daria Pavlova⬡, and Galyna Maistrenko⬡

Abstract In this paper, we provide a classification of surveillance radar surveillance systems of airspace, which are among the main information sources of the airspace control system and air traffic control. A brief description of the information processing process in survey radar systems for observing airspace is given and it is shown that the complexity of the processing system does not allow formalization and analysis of its robots as a whole; therefore, it is necessary to preliminarily divide the system into elements and study their functioning separately. The tasks of information processing at the stage of signal processing are considered, as well as a brief description of the primary, secondary and tertiary data processing. It is shown that the fusion of information from the same air objects can be carried out at all stages of data processing. It is shown that the transition to the assessment of the four-dimensional location (4D) of an airborne object changes the procedures for merging individual measurements carried out by various radar observation systems with different rates of data output. This is due to the fact, that from the output of the primary data processing by monitoring systems, an airborne object form is issued, which includes the time to estimate the coordinates of the airborne object with the necessary accuracy.

Keywords First keyword · Second keyword · Third key surveillance systems · Air object · Secondary surveillance radar · Data processing · Multi-radar data processing

I. Obod · I. Svyd (✉) · O. Maltsev · G. Maistrenko
Kharkiv National University of Radio Electronics, Kharkiv, Ukraine
e-mail: iryna.svyd@nure.ua

G. Zavolodko · D. Pavlova
National Technical University «KhPI», Kharkiv, Ukraine

T. Radivilova et al. (eds.), *Data-Centric Business and Applications*, Lecture Notes on Data
Engineering and Communications Technologies 48,
https://doi.org/10.1007/978-3-030-43070-2_31

1 Information Processing of Airspace Surveillance Systems

1.1 Classification of Radar Airspace Surveillance Systems

The main source of information about the air situation in the airspace control system [1–3] is survey radar surveillance systems (SS). These include primary [4, 5] and secondary [6–8] radar SS. The use of primary radars for monitoring the air situation does not require any additional equipment on board, that is, such an SS is completely independent. Using primary radars, two coordinates of an air object (AO) are determined: sloped range and azimuth. The accuracy and resolution are determined by the technical parameters of the radar. The main disadvantages of surveillance systems operating on the basis of primary surveillance radar systems are:

- low information content associated with the lack of the ability to obtain additional flight information;
- high energy consumption;
- high level of interference associated with reflections of signals from local objects;
- the limitations of the field of view due to a change in the antenna radiation pattern in the vertical plane and the need to fulfill the line of sight condition between the radar and the airborne object.

Secondary surveillance radar (SSR) solves the following problems [6–8]:

- determination of AO coordinates;
- receiving flight information from the AO;
- transfer on board AO information necessary for monitoring and control of flights and guidance of AO;
- dispatcher identification of AO;
- radar identification of AO state affiliation.

To solve these problems, the SSR has the following operating modes (A, B, C, D, S) and 1, 2, 3, 4, 5 [9].

Common advantages for secondary radar SS along with primary, regardless of the class and type of radars, are:

- increased, compared with primary radars, information ability that allows to automatically identify objects of observation and carry out ATC by four coordinates: sloped range, azimuth, altitude and time;
- large instrumental range with low energy costs;
- low interference caused by reflections of signals from local objects and meteorological formations;
- low levels of radiated power.

Common disadvantages of surveillance systems based on the use of secondary radars are:

- the need to equip all AOs with aircraft responders;

- the need for introducing into the equipment of requesters and responders the systems suppression by request and response of the side lobes signals in the antenna radiation pattern;
- high intra-system interference
- low interference immunity and interference protection when acting of intentional correlated interference to request and response channels.

1.2   Information Processing of Surveillance Radar Systems for Airspace Surveillance

The source of dynamic information about the air situation is the SS, including the cooperative [10–14], whose data processing is the basis for decision-making.

Data processing of surveillance systems—the process of bringing information obtained from the SS into a view that can be further transmitted to users.

At this time, SS data processing is not possible without the widespread use of IT, which enables the automatic collection, processing, storage, transmission and delivery of information to consumers, while enhancing practically all quality indicators.

The SS information processing system is directly joined to the signal sources and provides the following tasks:

- spatial processing of coherent signals;
- temporal processing of coherent signals;
- inter-period compensation for passive interference;
- detection of useful signals received from the AO and elimination of interferences;
- determination of the parameters of the received signals;
- air objects detection;
- measurement of coordinates and parameters of movement of air objects;
- obtaining flight information from the aircraft object;
- identification of the air object on the basis of "one-stranger";
- "ties" of detected AO in the trajectory, and determination of the parameters of these trajectories;
- calculation of smoothed and outstripped coordinates of aerial objects for some period of time;
- formation of a generalized air environment in the control area from several sources.

The solution of these problems leads to a variety of functions performed by the system associated with the step-by-step processing of large flows of information. At each stage of the processing, certain operations are performed on the input signals of individual devices of different complexity. The processing system can be represented as a set of elementary subsystems with complex relationships. The complexity of the processing system does not allow the formalization and analysis of its operation as a whole, so we have to pre-break the system into elements and study their functioning.

Therefore, it is advisable to elements of the processing system have a clear purpose, and that they can be described from mathematical quite common position. This approach allows the process of information processing of the airspace SS to be divided into the following functionally completed steps:

- signal processing of surveillance systems;
- primary processing of information;
- secondary processing of information;
- tertiary information processing.

Signal processing of surveillance systems In the signal processing stage, the first five airspace SS data processing tasks are solved. To solve these tasks, the signal processing in the SS can be divided into the following functionally completed operations:

1. Spatial processing of coherent signals by a multi-element antenna system located at one or more receiving points.
2. Temporal inter-period coherent signal processing including nonlinear processing (constraints, logarithms, etc.) and matched filtering. The filtering of the received high-frequency SS signals is carried out in the radio frequency path of the SS receiver. The optimization of the CC signals filtering is generally estimated by the criterion of the maximum ratio of the amplitude of the received signal to the mean square value of the interference voltage at the detector input and, therefore, at its output at the monotonic characteristic of the detector. It should be noted that filters designed to ensure optimal filtering of useful signals are not crucial devices because they make no decisions about the presence or absence of a signal. Their role is to make an assessment (decision) about the signal to the subsequent (after the detector) decision device.
3. Inter-period compensation of correlated noise caused by reflection from local objects, hydrometeors, and special reflectors (artificial passive noise).
4. Detection of useful signals (reflected or emitted by AO). The task of detecting useful signals is solved in the devices of post-detector signal processing and consists in making a unambiguous decision: either the signal is ($x_i = 1$), or no signal ($x_i = 0$). The optimality of the task solution of signals detection is accepted, as a rule, according to the Neumann-Pearson criterion, which reduces to the maximization of the probability of correct detection of signals D with restrictions on the probability of false detection $F = const$ [14, 15].
5. Measurement of parameters of detected (received) signals. Signal parameter estimation operations are generally optimized by the medium risk minimum criterion.

A brief description of the tasks of primary information processing At the stage of primary information processing (PIP), the following four problems of processing information of the airspace SS are solved. The essence of the first stage is to form an AO form that requires information from the primary and secondary SS. To solve these tasks, the act of information processing in a joint SS can be divided into the following functionally completed operations:

1. AO Detection. The task of detecting AO is solved in the devices of primary information processing and consists in making a unambiguous decision: either the AO is $(x_j = 1)$, or the AO is not $(x_j = 0)$. The optimality of the task solution of detection of AO is accepted, as a rule, according to the Neumann-Pearson criterion, which is reduced to maximize the probability of correct detection of AO D with restrictions on the probability of false detection $F = const$ at the output of PIP.

2. Determination (estimation) of the instantaneous position (coordinates) of the AO in space based on the results of one review of the primary and secondary SS. During this operation, the detection of AO by a bundle of reflected (emitted) signals, the statistical estimation of the delay time of the reflected (emitted) signals at the moment of sending the probing signals (the statistical estimation of the distance to the AO relative to the location of the SS). As well as the statistical evaluation of the angular coordinates AO by the angular position of the SS antenna at the moment of passing the maximum radiation pattern through the AO. The optimal parameter estimate is the maximum approximation of the estimate to the true value of the parameter being evaluated (ie, minimizing estimation errors). In the general case, the accuracy of coordinate estimation is characterized by the correlation matrix of the estimation error, and in the simplest case the variance of the estimation error σ_i^2.

3. Identification on the basis of "friend or foe".

4. Detection and decoding of flight (on-board) AO information received through the secondary SS response channel.

5. Combination of coordinate and flight information obtained via a secondary channel from the aircraft responder.

6. The combination of coordinate information obtained from primary and secondary SS.

7. Compilation of a complete AO form based on information from primary and secondary SSs by comparing the coordinate information of these SSs.

8. The coordinates of the observed objects, as a rule, are transformed from a natural SS polar coordinate system into a rectangular (Cartesian).

9. Generate a standard message for transmission to consumers using a data link (DL).

These functions are the main ones. Additional include:

- inter-overview processing of coordinate information obtained from primary and secondary SS;
- spatial and temporal (frequency) selection of signals, in particular registration of information in specified zones;
- processing, conversion and issuing in DL the direction to the source of interference;
- generation of messages on the status and modes of the radar;
- receiving radar mode commands coming from the DL;
- secondary SS code management.

Brief description of the secondary information processing tasks The purpose of secondary processing is to obtain trajectory information and AO motion parameters for the decision-making process. Secondary processing—a process that takes place in real time.

At this stage of processing solves the seventh and eighth tasks of processing information airspace SS, which involves performing the following functionally completed operations:

1. Detecting the AO trajectory by a set of estimates obtained in a series of consecutive SS reviews. During this operation, the estimation of the belonging of several estimates from different SS inspection periods to one AO, the decision on the presence or absence of AO, and the initial values of the trajectory parameters of the detected AO are calculated.
2. AO trajectory observation (trajectory tracking). In the process of observing the trajectory, in each review selects new estimates for the continuation of the trajectory, specifies the trajectory parameters taking into account the coordinates of the new estimates, and also smoothes and predicts (extrapolates) the coordinates.

These operations are performed on the basis of SS estimates obtained during the initial processing of information in the general case from several SSs, which are generally in the same positions. The inputs of the secondary processing device are labels (marks) that contain:

- vector of observed AO states;
- correlation matrix of errors of state vector estimation;
- the time to determine the coordinates of the AO that is needed to combine the information of several SS.

Due to the secondary processing of information SS can:

- improve the likelihood of correct decisions when detecting AO;
- to improve the accuracy of AO coordinate measurement by means of inter-review (discrete) filtering of the coordinates of the estimates;
- determine the full velocity vector AO;
- to determine extrapolated values of coordinates and parameters of AO motion.

Secondary information processing is carried out sequentially over time as new estimates are received from the primary signal processing device and includes the following processing steps:

- identification of accepted estimates with already tracked trajectories;
- identifying new trajectories;
- filtering of coordinates and parameters of trajectories;
- resetting trajectories.

In the case of multiple SSs, the task of information fusion from them arises in order to obtain a single AO form, which, by including coordinates in the AO form, provides the time to coordinate SS information with different rates of issuance at the AO tracking stage.

Brief description of the tasks tertiary information processing Tertiary processing of SS information is the fusion of information from different SSs by the AOs of the same name in order to improve the observation characteristics:

- detection characteristics;
- recognition characteristics;
- characteristics of measurement of coordinates and parameters of motion AO.

At this stage of processing, the ninth problem of processing airspace information from SS is solved, which involves performing the following functionally completed operations:

- aligning AO placemarks to a single coordinate system;
- bringing the AO placemarks to a single reference time;
- identification of trajectories obtained from multiple sources by the same AO;
- calculating the parameters of combined (averaged) trajectories.

Because the ordinary extrapolation time is short, linear extrapolation is used.

The problem of label identification is solved in two steps. Initially labels are grouped according to their hit in strobe permissible deviations determined by evaluation errors of coordinates. Then the labels are identified and merged.

In most known information processing systems, tertiary processing is performed at the stage of AO route integration. Indeed, in computing, it is preferable to first plot the AO trajectories independently, according to each source, and in the next step, the tertiary stage, to use them to improve the quality of information provided to users. The trajectories of each AO observed from different angles of several SS, when generalized, allow not only to more accurately determine and predict the location of the object, but also to quickly track the current error values of each measurement source

2 Fusion the Coordinate Data of Airborne Objects in the Networks of Survey Radar Systems

The fusion of coordinate data from same-name AOs can be carried out at all stages of data processing [16]. However, as we noted above, tertiary data processing sets as its task precisely data merging.

In papers [17–20] various aspects of the implementation of a four-measurement trajectory are considered. So, in [17], the issues of modeling external interference for an aircraft in flight to build reliable 4D trajectories are considered, in [18] a method for detecting and resolving conflicts with several aircraft for a four-measuremental flight trajectory is considered, in [19] a multi-purpose a multi-purpose multi-memetic algorithm is proposed is proposed for conflict-free planning of 4D flight paths across the entire network, in [20] air traffic control issues based on 4D trajectories: reliability analysis using the theory of multi-state systems, in [21] 4D trajectory optimization question reviewed.

2.1 Data Fusion at the Tertiary Data Processing Stage

As noted above, one of the tasks of tertiary data processing is the efficient use of detection or tracking data from various surveillance systems integrated into a network and controlling the same area of space. Therefore, tertiary data processing is a combination of data from various observation systems of the same airborne objects in order to improve the observation characteristics such as:

- detection characteristics;
- coordinate measurement characteristics;
- characteristics of the parameters of an air object movement.

Since the coordinates of the airborne object are measured in the coordinate system of the observation systems, therefore, when transmitting data to the tertiary data processing point, it is necessary to recalculate them to the standing of the data receiver point. In automated control systems, the transmission of coordinates of airborne objects is usually carried out in a rectangular coordinate system. The processing station also uses a rectangular system. Therefore, the task is to transform the rectangular coordinates of the airborne object relative to the point of standing of the source into rectangular coordinates relative to the point of the data processing station.

Common time is required to determine the position of the marks being processed at any one point in time. This operation makes it much easier to identify the air objects marks.

To reduce to a single time, a model of motion of air objects in a single coordinate system with constant velocity is used, ie

$$\vec{W}(i) = \vec{\Phi}(i, i-1)\vec{W}(i-1), \tag{1}$$

$\vec{\Phi}$—the transition matrix is equal to

$$\vec{\Phi}(i, i-1) = \left[\begin{array}{ccc|ccc} 1 & 0 & 0 & T & 0 & 0 \\ 0 & 1 & 0 & 0 & T & 0 \\ 0 & 0 & 1 & 0 & 0 & T \\ \hline 0 & 0 & 0 & 1 & 0 & 0 \\ 0 & 0 & 0 & 0 & 1 & 0 \\ 0 & 0 & 0 & 0 & 0 & 1 \end{array}\right] \tag{2}$$

where

$$T = t_i - t_{i-1}. \tag{3}$$

Aligning the state vector to a given time is performed accordingly

$$\overrightarrow{W}(s) = \Phi(s, i)\overrightarrow{W}(i) \tag{4}$$

for $T = t_s - t_i$.

The problems of tertiary data processing are currently solved by two main methods [16, 22–27]:

- mosaic processing;
- multiradar processing.

In the mosaic processing of each surveillance system is allocated its own viewing area, which does not intersect with the viewing areas of other surveillance systems. Only one surveillance system is used to compose a single air object form.

The disadvantages of this method include the task of tracking the trajectories of the air object when crossing the boundaries of the viewing areas, and not to take advantage of the overlap of the detection areas of neighboring surveillance systems.

Multiradar processing uses all available surveillance systems to form a single air object form.

Multiradar processing should provide stable AO tracking and picture formation of the air situation by analyzing information coming from multiple information systems. Typically, information systems have different characteristics, so that in specific circumstances may be more effective one or another surveillance system. Other things being equal, long-range, higher-power radar will be effective over long distances. In areas of high traffic density and increased maneuverability of aerial objects, a radar with a short review period is required. In addition, the effectiveness of the radar depends on its location in relation to the surrounding interference. Buildings, natural terrain, and other elements of the surrounding area can shield, reflect, or repel radiation, resulting in specific areas appears (or missing the real ones) numerous misprints. Therefore, to obtain the most informative picture, it is advisable to use information from several surveillance systems, taking into account the peculiarities of these radars and their capabilities with respect to specific areas of the airspace control system. The result of the processing is multi-radar trajectories calculated from real ones by special algorithms. Depending on the circumstances, the trajectory of only one radar or several radars can be used to form a multi-radar trajectory, the measurements of which are averaged with different weights.

The problem of label identification is solved in two steps.

If the labels from many surveillance systems belonging to multiple aerial objects getting into strobe, then the grouping problem is solved as follows:

- all possible variants of grouping are drawn up;
- calculates the differences of coordinates in each group;
- the correlation error matrix $\overrightarrow{C}_i^{-1}$, is calculated as the sum of the correlation error matrices of the coordinate groups;

- for each variant of grouping, a quadratic form $\vec{Q}_i = \vec{Z}_i \vec{C}_i^{-1} \vec{Z}_i$ is formulated and a variant of grouping for which the value $\vec{Q}_i$, is minimal is accepted.

For example, let us say that two labels with state vectors $\vec{W}_{1,1}$ and $\vec{W}_{1,2}$ obtained from the first observation system, and one label with state vectors $\vec{W}_2$ from the second observation system are included in the tolerance gate. Correlation matrix of errors respectively $\vec{C}_1^{-1}$ and $\vec{C}_2^{-1}$.

There are two options for grouping:first

$$\vec{Z}_1 = \left\| \begin{array}{c} \vec{Q}_{1,2} \to \vec{Q}_2 \\ \vec{Q}_{1,2} \end{array} \right\|, \tag{5}$$

second

$$\vec{Z}_2 = \left\| \begin{array}{c} \vec{Q}_{1,1} \\ \vec{Q}_{1,2} \to \vec{Q}_2 \end{array} \right\|. \tag{6}$$

The correlation error matrices for the first and second grouping variants are the same and are defined as

$$\vec{C}_0^{-1} = \vec{C}_1^{-1} + \vec{C}_2^{-1}. \tag{7}$$

Competing quadratic forms can be written as:

$$\vec{Q}_1 = \vec{Z}_1^T \vec{C}_0^{-1} \vec{Z}_1; \quad \vec{Q}_2 = \vec{Z}_2^T \vec{C}_0^{-1} \vec{Z}_2. \tag{8}$$

If $\vec{Q}_1 < \vec{Q}_2$ the first option is selected and vice versa.

For m SS and n AO the number of grouping variants can be defined as

$$k = (m - 1)! n!. \tag{9}$$

This formula shows that it increases sharply with m and n.

The problem of forming single measurements is solved by averaging the coordinates of the AO with the weights, inversely proportional to the variance of the errors of the individual measurements of each observation system.

The above shows the complexity of the data merge process in an non-coherent network of surveillance systems, which leads to a decrease in the quality of information support. Indeed, data fusion by air objects of the same name in an non-coherent network, does not allow for the benefit of improving the observation characteristics of both the additional energy and correlation of the received signals, as well as the spatial similarity of the primary data about one object from different systems observation.

2.2   Fusion of Coordinate Data in 4D Measurements of Airspace Surveillance Radar Systems

The transition to an estimate of the four-measuremental location (4D) of an air object [17–21, 28–33] changes the procedures for merging individual measurements carried out by different observation systems with different data output rates. Indeed, from the output of the primary data processing, the observation system will issue a form of the air object, which includes time T_i of obtaining the coordinates of the air object with the required accuracy. This can be used to merge data from different surveillance systems during the secondary data processing stage [34]. Show it.

It should be noted that the concept of synchronicity is closely related to the concept of simultaneity. Indeed, in the locations of the surveillance systems should be simultaneously produced the same time signals. The concept of simultaneity in the general theory of relativity is not monosemantic. However, it can be argued that the only self-consistent definition of simultaneity is the following definition. To analyze any explicit data in the framework of the general theory of relativity, one can introduce a certain four-dimensional coordinate system (coordinate system), which can contain one time coordinate. Two events fixed in some coordinate system by values (t_1, x_1, y_1, z_1) and (t_2, x_2, y_2, z_2), are considered simultaneous with respect to this coordinate system, if the corresponding values of the time coordinate coincide: $t_1 = t_2$. In the following, such a definition of simultaneity (and the corresponding definition of clock synchronization) will be called coordinate. This definition allows one to introduce, within the framework of the general theory of relativity, a self-consistent unified time scale in the most diverse areas of space-time and with any reasonable accuracy. The fact that the choice of the coordinate system according to the coordinate time of which synchronization is performed is arbitrary should not cause concern: it is easy to move from synchronization on coordinate time of one coordinate system to synchronization on coordinate time any other coordinate system.

The model of discrete change in the parameters of an air object contains non-random transformations of the previous value of the state vector $\vec{W}_k$, what is generally non-stationary and non-linear, and a sensor of random multimeasurement variables distributed according to Gauss's law with zero expectation is needed to take into account the possible current maneuver of an air object

$$\vec{W}_{k+1} = \vec{b}\left(\vec{W}_k\right) + \vec{\mu}_k. \tag{10}$$

Values $\vec{\mu}_k$ for different k statistically independent. Consequently, each value $\vec{W}_{k+1}$ depends only on the previous value $\vec{W}_k$.

Resulted errors $\vec{\varepsilon} = \vec{W}_k - \widehat{\vec{W}}_k$ on the measurement k-step significantly affects the magnitude of the random prediction error. Error $\vec{\varepsilon}_k$ is not significant, which allows the use of no more than two members of the expansion function $\vec{b}\left(\vec{W}_k\right)$

Taylor series near evaluation $\widehat{\overrightarrow{W}}_k$. This produces a prior linearization function $\vec{b}(\overrightarrow{W})$ in the vicinity of the assessment $\widehat{\overrightarrow{W}}$:

$$\vec{b}(\overrightarrow{W}) = \vec{b}(\widehat{\overrightarrow{W}}) + \overrightarrow{B}(\overrightarrow{W} - \widehat{\overrightarrow{W}}),\tag{11}$$

where $\overrightarrow{B}$ —forecast matrix

$$\overrightarrow{B} = \left\| db^{(i)}/dW^{(i)} \right\|_{\overrightarrow{W}=\widehat{\overrightarrow{W}}},\tag{12}$$

includes partial derivatives of vector components $\vec{b}(\overrightarrow{W})$ behind the vector components $\overrightarrow{W}$. In this case, the linearized version of the model it is possible to write as

$$\overrightarrow{W}_{k+1} = \vec{b}\left(\widehat{\overrightarrow{W}}_k\right) + \overrightarrow{B}_k \vec{\varepsilon}_k + \vec{\mu}_k.\tag{13}$$

Projected estimate $\widehat{\overrightarrow{W}}_{0(k+1)}$ find as the conditional expectation of the vector (13) after k measurement steps in the form

$$\widehat{\overrightarrow{W}}_{0(k+1)} = M\left(\overrightarrow{W}_{k+1}\right) = \vec{b}_k\left(\widehat{\overrightarrow{W}}_k\right).\tag{14}$$

Random prediction error on $(k + 1)$-step can be found as the difference (13) and (14):

$$\vec{\varepsilon}_{0(k+1)} = \overrightarrow{B}_k(\vec{\varepsilon}_k) + \vec{\mu}_k.\tag{15}$$

It consists of the listed error of the previous measurement $\overrightarrow{B}_k(\vec{\varepsilon}_k)$ and the consequent random maneuver of an air object $\vec{\mu}_k$. Since these values are independent, the correlation prediction error matrix is reduced to the sum of the correlation matrix of the discrete maneuver of the air object $\overrightarrow{Q}_k = M(\vec{\mu}_k \vec{\mu}_k^T)$ and the listed correlation matrix of previous measurement errors

$$M\left[\left(\overrightarrow{B}_k \vec{\varepsilon}_k\right)\left(\overrightarrow{B}_k \vec{\varepsilon}_k\right)^T\right] = \overrightarrow{B}_k \vec{C}_k^{-1} \overrightarrow{B}_k^T,\tag{16}$$

where $\vec{C}_k^{-1} = M(\vec{\varepsilon}_k \vec{\varepsilon}_k^T)$ is the resulting correlation measurement error on k-th step. The correlation matrix of forecast errors is

$$\vec{C}_{0(k+1)}^{-1} = \overrightarrow{B}_k \vec{C}_k^{-1} \overrightarrow{B} + \overrightarrow{Q}_k.\tag{17}$$

The distribution law of the random vector (13), as well as its components, can be considered normal. Expressions (14) and (16) define the parameters of this law.

The expression of the resulting accuracy matrix and the resulting optimal estimate taking into account the data of the current measurement are:

$$\vec{C}_{k+1}^{-1} = \left(\vec{B}_k \vec{C}_k^{-1} \vec{B} + \vec{Q}_k \right)^{-1} + \vec{C}_{y(k+1)}, \tag{18}$$

$$\widehat{\vec{W}}_{k+1} = \vec{b}_k \left(\widehat{\vec{W}}_k \right) + \vec{C}_{k+1}^{-1} \vec{C}_{y(k+1)} \left[\widehat{\vec{W}}_{y(k+1)1} - \vec{b}_k \left(\widehat{\vec{W}}_k \right) \right]. \tag{19}$$

Expression (19) is the main linearized equation for quasilinear recurrent filtering of vector estimates $\widehat{\vec{W}}$.

The block diagram of a discrete system that implements the filtering of estimates (19) is shown in Fig. 1. Filtered estimates are stored using the delay element and are used to predict the next step. The forecast is made using the function $\vec{b}_k \left(\widehat{\vec{W}}_k \right)$, which is adjusted generally at every step. The inconsistency of the current estimate and forecast data with the established matrix weight is attached to the predicted estimate, and gives the resulting estimate.

Equation (18) determines the accuracy matrix of the resultant measurement at each step $\vec{C}_{k+1}$, and also, the matrix coefficient accounting inconsistencies $\vec{C}_{k+1}^{-1} \vec{C}_{y(k+1)}$, that change.

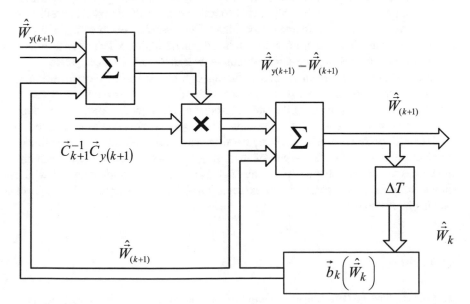

Fig. 1 Filtering the trajectory of airborne objects in a network of surveillance systems with 4D measurement

After any filtering step, you can proceed to the evaluation predicted $\widehat{\overrightarrow{W}}_0$ at any time in advance, and to the prediction error correlation matrix $\left|\overrightarrow{C}_0^{-1}\right|$, using for this relationship (14) and (16).

The structure allows filtering the estimates of the location of an air object for individual observation systems, in which the known rate of receipt of current estimates of the air object states, that is, the known delay time ΔT.

Upon transition to 4D measurement in the form of the air object is provided when the mark arrives T_i, which allows to always get the delay time from the previous mark as

$$\Delta T = T_{i+1} - T_i. \tag{20}$$

Thus, using (20), it is possible to filter estimates of the state vector of the same air object in accordance with Fig. 1 according to data from various airspace surveillance systems. This feature of information processing eliminates the need for tertiary information processing.

3 Conclusion

The article proposes a model for using a four-dimensional coordinate system for merging data from disparate airspace surveillance systems. This will enable the transition to a synchronous or partially synchronous network of airspace surveillance systems. As a result, it is possible to coordinate the processes of obtaining and processing data from disparate information tools, which, basically, will resolve technical inconsistencies in existing non-synchronous networks of surveillance systems.

Based on the research results, a block diagram of filtering the trajectory of airborne objects in a network of observation systems with 4D measurements is proposed. It allows you to filter estimates of the state vector of an air object according to various according to airspace surveillance systems. This makes it possible to eliminate the need for tertiary data processing. The above allows us to conclude that the transition to 4D measurements of the state vector of an air object in airspace surveillance systems allows you to consistently go to a partially synchronous information network, which allows the data to be merged from existing surveillance systems at the stage of tracking an air object.

References

1. Farina A, Studer F (1993) Digital processing of radar information. Radio i svyaz, Moscow
2. Ueda T, Shiomi K, Ino M, Imamiya K (1998) Passive secondary surveillance radar system for satellite airports and local ATC facilities. 43rd annual air traffic control association. Atlantic City, Air Traffic Control Association, pp 20–24

3. Skolnik M (2008) Radar handbook, 3rd edn. McGraw-Hill, New York
4. Chernyak V (1998) Fundamentals of multisite radar systems, 1st edn. Gordon & Breach, Amsterdam, p 492
5. Lynn P (1987) Radar systems. Springer, New York, Boston, p 132
6. Stevens M (1988) Secondary surveillance radar. Artech House, Norwood
7. Kim E, Sivits K (2015) Blended secondary surveillance radar solutions to improve air traffic surveillance. Aerosp Sci Technol 45:203–208
8. Bloisi D, Iocchi L, Nardi D, Fiorini M, Graziano G (2012) Ground traffic surveillance system for air traffic control. In: 2012 12th international conference on ITS telecommunications, pp 160–164. https://doi.org/10.1109/ITST.2012.6425151
9. Ahmadi Y, Mohamedpour K, Ahmadi M (2010) Deinterleaving of interfering radars signals in identification friend or Foe Systems. In: 18th telecommunications forum Telfor, Telecommunications Society. ETF School of EE, University in Belgrade, IEEE Serbia & Montenegro COM CHAPTER, Belgrade, pp 729–733
10. Jackson D (2016) Ensuring honest behaviour in cooperative surveillance systems. The Centre for Doctoral Training in Cyber Security, Oxford
11. Siergiejczyk M, Krzykowska K, Rosiński A (2014) Reliability assessment of cooperation and replacement of surveillance systems in air traffic. In: Proceedings of the ninth international conference on dependability and complex systems DepCoS-RELCOMEX, June 30–July 4, 2014, Brunów, Poland, pp 403–411. https://doi.org/10.1007/978-3-319-07013-1_39
12. Ramasamy S, Sabatini R, Gardi A (2016) Cooperative and non-cooperative sense-and-avoid in the CNS+A context: a unified methodology. In: 2016 international conference on unmanned aircraft systems (ICUAS), pp 531–539. https://doi.org/10.1109/ICUAS.2016.7502676
13. Svyd I, Obod I, Maltsev O, Shtykh I, Maistrenko G, Zavolodko G (2019) Comparative quality analysis of the air objects detection by the secondary surveillance radar. In: 2019 IEEE 39th international conference on electronics and nanotechnology (ELNANO). https://doi.org/10.1109/ELNANO.2019.8783539
14. Obod I, Svyd I, Maltsev O, Maistrenko G, Zubkov O, Zavolodko G (2019) Bandwidth assessment of cooperative surveillance systems. In: 2019 3rd international conference on advanced information and communications technologies (AICT). https://doi.org/10.1109/AIACT.2019.8847742
15. Obod I, Svyd I, Maltsev O, Vorgul O, Maistrenko G, Zavolodko G (2018) Optimization of data transfer in cooperative surveillance systems. In: 2018 international scientific-practical conference problems of infocommunications. Science and Technology (PIC S&T). https://doi.org/10.1109/INFOCOMMST.2018.8632134
16. Olivier B, Pierre G, Nicolas H, Loc O, Olivier T, Philippe T (2009) Multi sensor data fusion architectures for air traffic control applications. Sensor and data fusion, pp 103–122
17. Pabst T, Kunze T, Schultz M, Fricke H (2013) Aircraft trajectory simulator using a three degrees of freedom aircraft point mass model. In: 3rd international conference on application and theory of automation in command and control systems (ATACCS'2013). ATACCS'2013, Naplesm pp 132–135
18. Hao S, Cheng S, Zhang Y (2018) A multi-aircraft conflict detection and resolution method for 4-dimensional trajectory-based operation. Chin J Aeronaut 31(7):1579–1593. https://doi.org/10.1016/j.cja.2018.04.017
19. Yan S, Cai K (2017) A multi-objective multi-memetic algorithm for network-wide conflict-free 4D flight trajectories planning. Chin J Aeronaut 30(3):1161–1173. https://doi.org/10.1016/j.cja.2017.03.008
20. Ruiz S, Piera M, Nosedal J, Ranieri A (2014) Strategic de-confliction in the presence of a large number of 4D trajectories using a causal modeling approach. Transp Res Part C: Emerg Technol 39:129–147. https://doi.org/10.1016/j.trc.2013.12.002
21. Alejo D, Cobano J, Heredia G, Ollero A (2013) Particle swarm optimization for collision-free 4D trajectory planning in unmanned aerial vehicles. In: 2013 international conference on unmanned aircraft systems (ICUAS), pp 298–307. https://doi.org/10.1109/ICUAS.2013.6564702

22. Pjatko S, Krasnova A (2004) Automated air traffic control systems: new information technologies in aviation. Politehnika, Saint-Petersburg, p 446
23. Dang K, Le T, Fung K (2017) The trajectory of the observed objects in the problem of tertiary processing of radar information. Nat Tech Sci 10(112):97–105
24. Mazimpaka J, Timpf S (2016) Trajectory data mining: a review of methods and applications. J Spatial Inf Sci 13:61–99. https://doi.org/10.5311/josis.2016.13.263
25. Yepes J, Hwang I, Rotea M (2007) New algorithms for aircraft intent inference and trajectory prediction. J Guid Control Dyn 30(2):370–382. https://doi.org/10.2514/1.26750
26. Chen X, Landry S, Nof S (2011) A framework of enroute air traffic conflict detection and resolution through complex network analysis. Comput Ind 62(8–9):787–794. https://doi.org/10.1016/j.compind.2011.05.006
27. Paielli R, Erzberger H, Chiu D, Heere K (2009) Tactical conflict alerting aid for air traffic controllers. J Guid Control Dyn 32(1):184–193
28. Svyd I, Obod I, Maltsev O, Maistrenko G, Zavolodko G, Pavlova D (2019) Fusion of airspace surveillance systems data. In: 2019 3rd international conference on advanced information and communications technologies (AICT). https://doi.org/10.1109/AIACT.2019.8847916
29. Pavlova D, Zavolodko G, Obod I, Svyd I, Maltsev O, Saikivska L (2019) Optimizing data processing in information networks of airspace surveillance systems. In: 2019 10th international conference on dependable systems, services and technologies (DESSERT). https://doi.org/10.1109/DESSERT.2019.8770022
30. Svyd I, Obod I, Maltsev O, Vorgul O, Zavolodko G, Goriushkina A (2018) Noise immunity of data transfer channels in cooperative observation systems: comparative analysis. In: 2018 international scientific-practical conference problems of infocommunications. Science and Technology (PIC S&T). https://doi.org/10.1109/INFOCOMMST.2018.8632019
31. Pavlova D, Zavolodko G, Obod I, Svyd I, Maltsev O, Saikivska L (2019) Comparative analysis of data consolidation in surveillance networks. In: 2019 10th international conference on dependable systems, services and technologies (DESSERT). https://doi.org/10.1109/DESSERT.2019.8770008
32. Rodríguez-Sanz Á, Álvarez D, Comendador F, Valdés R, Pérez-Castán J, Godoy M (2018) Air traffic management based on 4D trajectories: a reliability analysis using multi-state systems theory. Transp Res Procedia 33:355–362. https://doi.org/10.1016/j.trpro.2018.11.001
33. Svyd I, Obod I, Maltsev O, Shtykh I, Zavolodko G, Maistrenko G (2019) Model and method for request signals processing of secondary surveillance radar. In: 2019 IEEE 15th international conference on the experience of designing and application of CAD systems (CADSM). https://doi.org/10.1109/CADSM.2019.8779347
34. Obod I, Svyd I, Maltsev O, Zavolodko G, Pavlova D, Maistrenko G (2019) Fusion of discrete evaluation of the state vector of air objects based on 4D measurement. In: 2019 international scientific-practical conference problems of infocommunications. Science and Technology (PIC S&T). https://doi.org/10.1109/PICST47496.2019.9061562

Diakoptical Method of Inter-area Routing with Load Balancing in a Telecommunication Network

Oleksandr Lemeshko⑩, Tetiana Kovalenko⑩, Olena Nevzorova⑩, and Andriy Ilyashenko⑩

Abstract In this paper, the diakoptical method of inter-area routing with load balancing in a telecommunication network was proposed. The method allows to increase the scalability of routing solutions in comparison with the centralized approach without reducing the efficiency of the network, estimated by the maximum value of link load threshold. The method involves the decomposition of the general routing problem in a multi-area network into several routing subtasks of smaller size that can be solved for each individual area followed by combining the solutions obtained for the whole telecommunication network. The foundation of the method is a flow-based routing model based on the implementation of the concept of Traffic Engineering and focused on minimizing the maximum value of link load threshold. The results of the analysis confirmed the operability of the method on a variety of numerical examples and demonstrated the full correspondence of the efficiency of the obtained diakoptical routing solutions to the centralized approach. The advantage of the proposed method is also the absence of the need to coordinate routing solutions received on subnetworks, which positively affects both the time of solving the set task and the amount of service traffic circulated in the network associated with the transfer of data on the state of network areas and coordinating information.

Keywords Inter-area routing · Traffic engineering · Load balancing · Diakoptical method

O. Lemeshko (✉) · T. Kovalenko · O. Nevzorova · A. Ilyashenko
Kharkiv National University of Radio Electronics, Kharkiv 61000, Ukraine
e-mail: oleksandr.lemeshko.ua@ieee.org

T. Kovalenko
e-mail: tetiana.kovalenko@nure.ua

O. Nevzorova
e-mail: olena.nevzorova.ua@ieee.org

A. Ilyashenko
e-mail: andy.ilyashenko@gmail.com

T. Radivilova et al. (eds.), *Data-Centric Business and Applications*, Lecture Notes on Data Engineering and Communications Technologies 48,
https://doi.org/10.1007/978-3-030-43070-2_32

747

1 Introduction

Modern telecommunication networks (TCN) are characterized by a constant increase in territorial distribution, the number of switching and terminal devices, the expansion of the number of provided services, that complicates their structure and functioning algorithms significantly and requires new approaches in the development of models and methods for their analysis and synthesis. Therefore, one of the main problems arising in traffic control in general and routing is the scalability of the proposed solutions to ensure the required level of Quality of Service (QoS) [1, 2]. In this regard, the majority of supported technological solutions and modern routing protocols used in the transport technologies such as IP (Internet Protocol), ATM (Asynchronous Transfer Mode) and MPLS (Multiprotocol Label Switching) are characterized by a decompositional, hierarchical approach [3–5]. The main aim of such approach is to reduce the size of routing tables, the volume of service traffic circulating in the network, as well as the time for solving routing tasks. At the same time, the effectiveness of the protocol solution of hierarchical routing problems is to a great extent determined by the level of adequacy of the mathematical models and methods used in this process, within which it is important to ensure that the multi-area and hierarchical nature of the modern TCN design is considered. Thus, increasing the scalability of routing solutions based on the improvement of the corresponding mathematical models and methods is one of the urgent tasks in modern telecommunication networks.

2 Overview of Known Hierarchical Routing Solutions

As a result of the analysis of mathematical routing models, it was established that there are two main types of them: graph and flow-based models [6–12]. In the framework of graph models, the structural features of the simulated TCN are taken into account first of all, and the routing problem is reduced to finding the shortest path on the graph—for one-way routing or finding the optimal multi-path—for multi-path routing [6–8]. Modern routing protocols are based on graph models in which two basic algorithms are used: the Dijkstra algorithm is used, for example, in the OSPF, IS-IS, PNNI protocols, and Bellman-Ford algorithm is used in such protocols as RIP, IGRP and BGP. In flow-based models [9–12], the main attention is paid to the description of the functional properties of the simulated process, and besides the network structure, the parameters of communication links and transmitted traffic are considered. Since modern network traffic is predominantly multi-media and has a flow nature, the use of flow-based models makes it possible to obtain the most efficient solutions for routing problems.

In addition, as the analysis showed [10–22], there exist a great number of flow-based models of multi-path routing, which are based on the use of queuing networks, tensor representation, algebraic, integral-differential and difference state equations of TCN, and these models are actively used. However, according to many scientists that work in the field of network technologies, a compromise option combining the adequacy of the description and the acceptable complexity of the calculations are the routing solutions with the load balancing according to the principles of Traffic Engineering (TE) technology, proposed in [20–22]. Unfortunately, this solution is focused on the centralized calculation of routes, representing more theoretical interest than applied value. Therefore, in this article, the solution proposed in [20–22] will be adapted to the implementation of hierarchical routing in a multi-area network.

3 Routing Model with Load Balancing Based on Traffic Engineering Concept

Within the framework of the TE model proposed in [20–22], the structure of TCN is described using a directed graph $G = (R, E)$, where R is the set of vertices of the graph that models the network routers, and E is the set of graph arcs that represents communication links. For each communication link $E_{i,j} \in E$, its bandwidth $c_{i,j}$ is known. We denote by K the set of flows circulating in the network, then $|K| = \tilde{K}$ is the power of the set K, which quantitatively characterizes the total number of flows in the TCN. For each kth flow ($k \in K$), the source and destination nodes (s_k and t_k respectively) are known, as well as its average intensity (transmission rate) λ^k, which is measured in packets per second (1/s).

Let us assume that the value $x_{i,j}^k$ is a route variable that characterizes the fraction of the kth flow transmitted in the communication link $E_{i,j} \in E$. Let also the value α determine the dynamically controlled threshold of the maximum utilization of the TCN links. Then the routing problem with the support of Traffic Engineering technology can be formulated as a linear programming task to minimize the threshold α

$$\min_{x} \alpha \tag{1}$$

subject to the following constraints:

$$\sum_{E_{i,j} \in E} x_{i,j}^k - \sum_{E_{j,i} \in E} x_{j,i}^k = 0 \quad \text{if } k \in K, R_i \neq s_k, t_k; \tag{2}$$

$$\sum_{E_{i,j} \in E} x_{i,j}^k - \sum_{E_{j,i} \in E} x_{j,i}^k = 1 \quad \text{if } k \in K, R_i = s_k; \tag{3}$$

$$\sum_{E_{i,j} \in E} x_{i,j}^k - \sum_{E_{j,i} \in E} x_{j,i}^k = -1 \quad \text{if } k \in K, R_i = t_k; \tag{4}$$

$$\sum_{k \in K} \lambda_k x_{i,j}^k \leq c_{i,j}\alpha; \quad E_{i,j} \in E. \tag{5}$$

Constraints (2)–(4) are associated with the need to fulfill the conditions for the flow conservation, that is, they are responsible for the absence of packet loss on routers and in the whole network. Besides that, conditions (5) are responsible for preventing network overload. In addition, based on the physical meaning of the variables α and $x_{i,j}^k$, additional constraints on them must be taken into account, these are related to the implementation of the multi-path routing strategy:

$$0 \leq x_{i,j}^k \leq 1, \tag{6}$$

$$0 \leq \alpha \leq 1. \tag{7}$$

The advantages of the model (1)–(7) is that Traffic Engineering is consistent with multi-path routing, and the threshold (1) grows linearly when network load increases [23], which ensures that there are no fluctuations in the numerical values of the main Quality of Service indicators. However, despite these advantages of the described model, it also has several drawbacks, some of which were considered in [24–26]. The authors of the article proposed options for minimizing the shortcomings of the described model associated with the inadequacy of its use in networks with half-duplex and/or duplex communication links, leading to packet looping, as well as a significant decrease in the quality of balancing for networks with a heterogeneous topology when the network connectivity was not constant, but varied within certain limits. It is important to consider the problem of scalability as another significant drawback of the model (1)–(7). An increase in the number of nodes and/or communication links in the considered TCN will lead to a significant increase in the number of variables and restrictions in the model, that affects the complexity, required time for solving the routing problem and the volumes of the resulting routing tables.

An effective direction for increasing the scalability of network solutions is the application of hierarchical (multi-level) routing based on the decomposition representation of the flow-based model (1)–(7), considering the multi-area nature of modern TCN. The principles of decomposition and hierarchy are widely used now in the most popular transport technologies. For example, in IP technology, routing is implemented on the basis of dividing the initial network into many subnets named Autonomous Systems (AS), each of which uses its own Interior Gateway Protocol (IGP), and Exterior Gateway Protocol (EGP), such as BGP (Border Gateway Protocol) is used for routing between AS [3–7]. At the same time, some IGP protocols, for example, OSPF (Open Shortest Path First) and Integrated IS-IS, require further decomposition of AS into subnets called Domains or Areas. In ATM technology, hierarchical routing functions are implemented using the PNNI protocol (Private Network-to-Network Interface), which involves dividing a network into multiple peer groups.

In [27–32], solutions to the hierarchical routing problem in multi-area TCNs were proposed based on the introduction of a two-level hierarchy. Moreover, the nature of the optimality criteria used in [27–32] is oriented toward considering the requirements of the Traffic Engineering concept. As usual, at the lower level, routing problems are solved in each of the areas, and this way the dimension and complexity of the optimization problem being solved is significantly reduced. The upper level of the calculation hierarchy is functionally implemented on the route server and is responsible for coordinating the distributed work of individual network areas to ensure inter-area routes connectivity in order to approximate the effectiveness of distributed routing solutions to the quality of centralized calculations. A separate problem in this case was to ensure the fast convergence of solutions to their optimal values within the minimum number of iterations of the coordinating procedure, since the number of iterations directly affects both the time it takes to solve routing problems and the amount of service information transmitted in the network.

In this regard, in this work, we propose a diakoptical method of hierarchical routing in a multi-area TCN, which is based on the calculation of the routing order in each of the areas separately, followed by their non-iterative combination (generalization) for the whole network. At the same time, the advantage of the obtained routing solutions is that their effectiveness according to criterion (1) is fully consistent with the results obtained during centralized routing. The absence of the need for iterative coordination of the obtained multilevel routing solutions has a positive effect on the efficiency of solving routing problems in a network with a multi-area architecture.

4 Diakoptical Method of Hierarchical Routing with Load Balancing in Multi-area TCN

In the framework of the proposed diakoptical method of hierarchical routing, we are supposed to divide the general optimization problem (1) into N subproblems (by the number of areas) of lower dimension, followed by a generalized synthesis of the solution of the routing problem for the whole network [33]. The borders of each area are defined by the communication links, which corresponds to the principles of network decomposition applied in the IS-IS protocol. Then let each separate pth area in the TCN be described using a subgraph $G^p = (R^p, E^p)$ of a graph G, where $R^p = \left\{R_i^p; i = \overline{1, m_p}\right\}$ is the set of routers of the p th area, and m_p is their total number in the area; $E^p = \left\{E_{i,j}^p; i, j = \overline{1, m_p}, i \neq j\right\}$ is a set of intra-area links connecting routers within the pth area. For each intra-area communication link of the pth area, we denote its bandwidth $c_{i,j}^p$, measured in packets per second (1/s). We denote the set of links connecting routers of the areas p and q(inter-area links) by $E^{p,q} = \left\{E_{i,j}^{p,q}; i = \overline{1, m_p}, j = \overline{1, m_q}; p, q = \overline{1, N}, p \neq q\right\}$. The bandwidths of inter-area links between routers of the areas p and q are denoted by $c_{i,j}^{p,q}$.

For clarity, let's consider the description and analysis of the proposed diakoptical method of hierarchical routing using the example of the network structure shown in Fig. 1. This network, consisting of 13 routers and 19 communication links (CL), is divided into 4 network areas, i.e. $N = 4$. For example, for a subgraph $G^1 = (R^1, E^1)$ of the first network area shown in Fig. 1, we denote the set of routers as

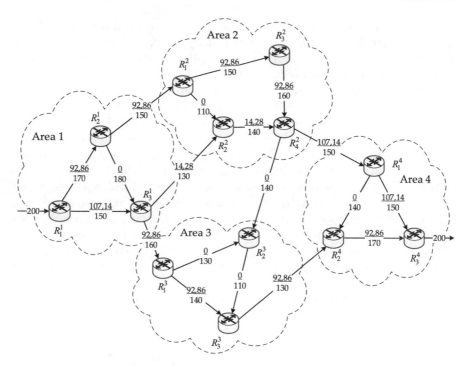

Fig. 1 The procedure for centralized routing in a multi-area TCN

$R^1 = \{R_1^1, R_2^1, R_3^1\}$ ($m_1 = 3$), and the set of intra-area links connecting routers within the area is $E^1 = \{E_{1,2}^1, E_{1,3}^1, E_{2,3}^1\}$. The set of inter-area links connecting routers of the 1st and 2nd areas is $E^{1,2} = \{E_{2,1}^{1,2}, E_{3,2}^{1,2}\}$. Suppose that it is necessary to transmit packets with an intensity λ that varies in the range from 1 to 200 1/s between the first router of the first area (R_1^1) and the third router of the fourth area (R_3^4).

For the further analysis of the effectiveness of the obtained solutions when the model (1)–(7) is used, the order of centralized routing of the flow for intensity $\lambda = 200$ 1/s was obtained in the considered network example (Fig. 1). A fraction indicated in the breaks in the communication links shows the intensity of the flow that is transmitted in this link in the numerator, and the bandwidth of the link (1/s) in the denominator. For the obtained solution (Fig. 1), the numerical value of the threshold of the maximum utilization of the TCN links (1) is 0.7143.

The studies showed that in the centralized calculation of the route variables (6), when the flow intensity increases, the threshold of the network links utilization (1) grows linearly (Fig. 2). Moreover, the results of the centralized TE routing (Fig. 2) were used in further analysis as a reference when comparing with the results obtained by using the proposed diakoptical routing method.

At the first step of the proposed method, it is necessary to determine the generalized throughput of each area of the given TCN. At the same time, each subgraph G^p

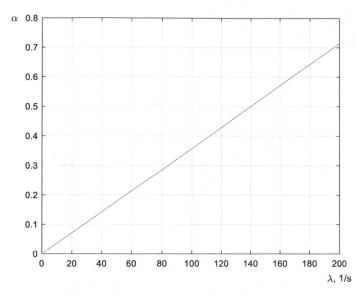

Fig. 2 The threshold of maximum utilization of network links subject to the intensity of the input flow for a centralized solution of the routing problem

describing the structure of the pth area must be transformed into a subgraph G^p as follows: the bandwidth values of all of the inter-area links connecting the routers of this area with the routers of any other area must be assumed to be infinity:

$$c_{i,j}^{p,q} = \infty, \quad \forall E_{i,j}^{p,q} \in E^{p,q}, \; E_{i,j}^{q,p} \in E^{q,p}, \quad q = \overline{1,N}, q \neq p. \tag{8}$$

This is done because the values of the criterion (1), calculated in each separate pth area, should not be affected by the values of utilization of inter-area links. If the number of inter-area links incoming to the pth area ($E_{i,j}^{q,p}$) or outgoing from the pth area ($E_{i,j}^{p,q}$) is more than one, then an equivalent (fictitious) source node R_s^p and/or a fictitious destination node R_t^p are introduced into the subgraph G^p, such that:

$$R_i^p = R_s^p \forall E_{i,j}^{q,p} \in E^{q,p}, \; q = \overline{1,N}, \; q \neq p; \tag{9}$$

$$R_j^p = R_t^p \forall E_{i,j}^{p,q} \in E^{p,q}, \; q = \overline{1,N}, \; q \neq p. \tag{10}$$

Thus, the modified subgraph $G'^p = (R'^p, E'^p)$ contains a set vertices $R'^p = R^p \bigcup \{R_s^p, R_t^p\}$ and arcs $E'^p = E^p \bigcup \{E_{s,j}^p, E_{i,t}^p\}$.

An example of converting a subgraph G^1 into a subgraph G'^1 in accordance to the described procedures is presented in Fig. 3. The subgraph under consideration G^1 has two outgoing inter-area links, therefore, an equivalent (fictitious) destination

Fig. 3 An example of
introducing a fictitious
destination node in the first
area with an indication of the
routing order of a flow with
intensity 200 1/s

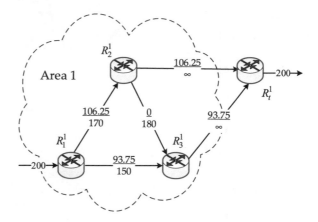

node R_t^1 was introduced into it, while $E_{2,t}^1 = E_{2,1}^{1,2}$, $E_{3,t}^1 = E_{3,2}^{1,2}$, $c_{2,t}^1 = c_{3,t}^1 = \infty$, in accordance with the conversion procedures (8)–(10) described above.

Thus, the modified subgraph $G'^p = (R'^p, E'^p)$ has the set of vertices $R'^p = R^p \bigcup \{R_s^p, R_t^p\}$ and arcs $E'^p = E^p \bigcup \{E_{s,j}^p, E_{i,t}^p\}$.

The conversion order of the subgraphs of the second, third, and fourth areas of the network under study, as well as the results of calculating their throughput, are shown in Figs. 4, 5 and 6 respectively. The subgraph G^2 of the second area has two outgoing and two incoming inter-area links; thus, additional fictitious source R_s^2 and destination R_t^2 nodes were introduced into it, while in accordance with the conversion rules (8)–(10) $E_{2,1}^{1,2} = E_{s,1}^{1,2}$, $E_{3,2}^{1,2} = E_{s,2}^{1,2}$, $E_{4,1}^{2,4} = E_{4,t}^{2,4}$, $E_{4,2}^{2,3} = E_{4,t}^{2,3}$, $c_{s,1}^{1,2} = c_{s,2}^{1,2} = c_{4,t}^{2,4} = c_{4,t}^{2,3} = \infty$ (Fig. 4).

The subgraphs of the third and fourth areas have two incoming inter-area links, so the additional fictitious source nodes R_s^3 and R_s^4 were introduced and, in addition, in accordance with the conversion rules (8)–(10) $E_{3,1}^{1,3} = E_{s,1}^{1,3}$, $E_{4,2}^{2,3} = E_{s,2}^{2,3}$, $E_{4,1}^{2,4} = E_{s,1}^{2,4}$, $E_{3,2}^{3,4} = E_{s,2}^{3,4}$, $c_{s,1}^{1,3} = c_{s,2}^{2,3} = \infty$, $c_{s,1}^{2,4} = c_{s,2}^{3,4} = \infty$ (Figs. 5 and 6).

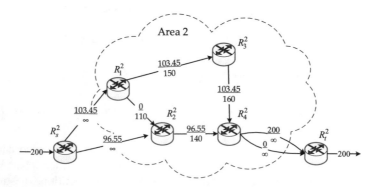

Fig. 4 An example of the introduction of fictitious source and destination nodes in the second area with an indication of the routing order of the flow with intensity 200 1/s

Fig. 5 An example of the introduction of a fictitious source node in the third area with an indication of the routing order of the flow with intensity 200 1/s

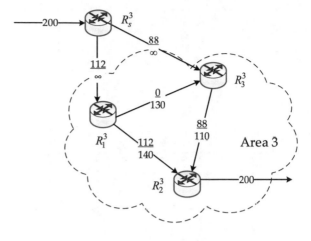

Fig. 6 An example of the introduction of a fictitious source node in the fourth area with an indication of the routing order of the flow with intensity 200 1/s

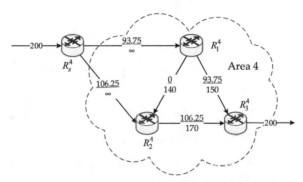

Next, for each of the areas represented by subgraphs G'^p, we define the routing order and load balancing using the model (1)–(7), and we denote the dynamically controlled threshold for the maximum utilization of the links in the pth area, which must be minimized, by α^p.

During this research the input flow intensity λ for each area, as well as for the whole network, was varied in the range from 1 to 200 1/s. For example, Figs. 3, 4, 5 and 6 show the routing order of a flow with intensity 200 1/s in each of the network areas. In the breaks of the communication links, the values of their bandwidth (denominator of the fraction) and the intensity of the packet flows being transmitted in this link (numerator of the fraction) are indicated. The obtained calculation results showed that when the input flow intensity increases, the threshold of the maximum utilization of a link α^p in each pth area of the network grows linearly (Fig. 7).

Due to the linearity of the obtained dependencies (Fig. 7), we can conclude that the throughput of the pth area is constant when transmitting packets of the flow in the selected direction and this throughput can be defined as

$$c^p = \alpha^p \cdot \lambda. \tag{11}$$

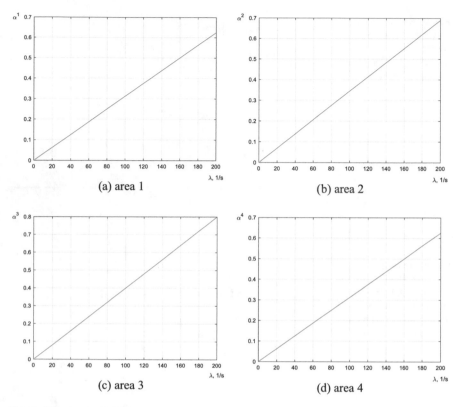

Fig. 7 The threshold of the maximum utilization of the links of the pth area α^p subject to the intensity of the input flow λ

So, for the first TCN area when the value of the intensity is, for example, $\lambda = 200$ 1/s (Figs. 3 and 7a), the calculated value of the threshold of the maximum link utilization is $\alpha^1 = 0.625$. Therefore, the throughput of the first area (11) is $c^1 = 320$ 1/s. The same value of the throughput of the first area can also be obtained by considering any other point on the graph shown in Fig. 7a. Based on the results obtained (Figs. 4, 5, 6 and 7), in accordance with expression (11), the throughput of the second, third and fourth areas were also calculated in a similar way: $c^2 = 290$ 1/s, $c^3 = 250$ 1/s and $c^4 = 320$ 1/s.

At the second step of the proposed diakoptical method, a model of the aggregated network structure is constructed, in which, based on the results of the calculations of the first step, each network area is represented by an equivalent communication link $\overline{E}^p$ with bandwidth c^p that connects the routers R_s^p and R_t^p. The aggregated structure of the TCN obtained as a result of such a transformation can be represented in the form of a graph $G^A = (R^A, E^A \bigcup E^{p,q})$, where $R^A = \left\{ R_s^p, R_t^p; p = \overline{1,N} \right\}$ is the set of equivalent routers; $E^A = \left\{ \overline{E}^p; p = \overline{1,N} \right\}$ is the set of equivalent communication links; $E^{p,q}$ is the set of inter-area communication links. The aggregated structure of

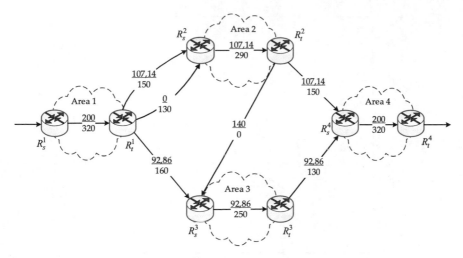

Fig. 8 The aggregated network structure and the routing procedure obtained for it

the studied network from Fig. 1 obtained in the second step of the proposed method is shown in Fig. 8.

At the third step of the method based on the models (1)–(7), the routing order is calculated for the obtained aggregated network structure. As a result of this calculation, the intensities of the flows are determined both in the inter-area communication links and of the flows that arrive to and leave from each pth area. The solution obtained at the third step is called a macro-solution of the routing problem in a given network. The macro-solution of the routing problem for the considered example of the network structure (Fig. 1) with an input flow intensity 200 1/s is also shown in Fig. 8. For the obtained solution, the numerical value of the threshold of the maximum utilization of TCN links (1) is 0.7143, same as for the case of the centralized calculation (Fig. 1)

Figure 9 shows how the threshold value of the maximum utilization of TCN links depends on the intensity of the input flow for the aggregated structure of the network under study. As can be seen from this graph, this dependency is also linear.

At the fourth step of the proposed method, we solve the routing problem in each of the areas separately, based on the obtained values of the flow intensity entering each of the areas and applying the mathematical model (1)–(7) to subgraphs G^p of each area. This way, a micro-solution of the routing problem is obtained, clarifying the final order of the distribution of flows transmitted in the communication links of each individual area. The micro solutions obtained at the fourth step for each area together with the macro solution for inter-area links determine the diakoptical solution of the routing problem for the network under study, which is shown in Fig. 10.

It should be noted that the routing order obtained using the proposed diakoptical method (Fig. 10) in general case does not correspond to the distribution of flows obtained during centralized routing (Fig. 1). But the effectiveness of the obtained route solutions evaluated by criterion (1) is completely identical. One can see that

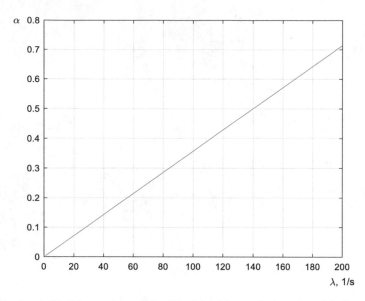

Fig. 9 The threshold of the maximum link utilization subject to the intensity of the input flow for the macro-solution of the routing problem in the aggregated structure of the TCN

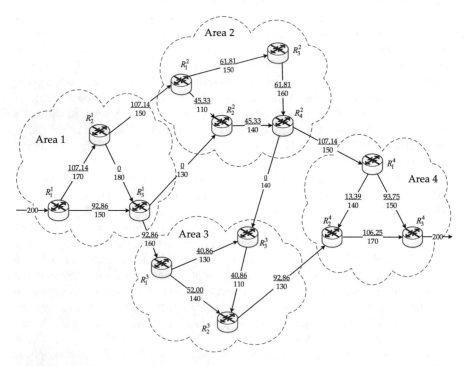

Fig. 10 Diakoptical solution of the routing problem in the TCN

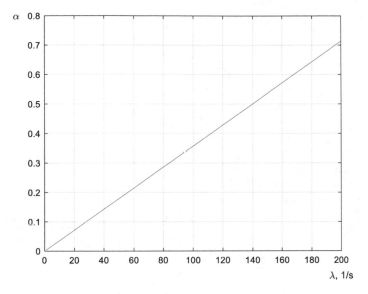

Fig. 11 The threshold of the maximum link utilization subject to the intensity of the input flow for the diakoptical solution of the routing problem in the TCN

for the above solution (Fig. 10), the numerical value of the threshold of the maximum utilization of TCN links (1) is 0.7143, coinciding with the corresponding value obtained in the case of the centralized calculation (Fig. 1).

5 Analysis of the Diakoptical Method of Inter-area Routing with Load Balancing in a Telecommunication Network

For the purpose of an extended analysis of the proposed diakoptical method of inter-area routing, we consider the second example of the network, in which, unlike the previous example (Fig. 1), the bottleneck in terms of the bandwidth of communication links is not inside the area, but in one of the inter-area links. The structure of the network under study coincides with the one that was considered earlier (Fig. 1), except the link bandwidth between the areas 1 and 3, the areas 2 and 3, as well as the areas 2 and 4, which were reduced from 160 1/s, 140 1/s and 150 1/s to the values of 100 1/s, 120 1/s and 120 1/s, respectively.

An example of a centralized solution to the routing problem of a flow with intensity 200 1/s being transmitted in a given network between routers R_1^1 and R_3^4 is shown in Fig. 12. In the breaks of the communication links, the intensity of the flow transmitted by this link (in the fraction numerator) and the link bandwidth (in the fraction denominator) are given. For the obtained solution (Fig. 12), the numerical value of the threshold of the maximum utilization of TCN links (1) is 0.8.

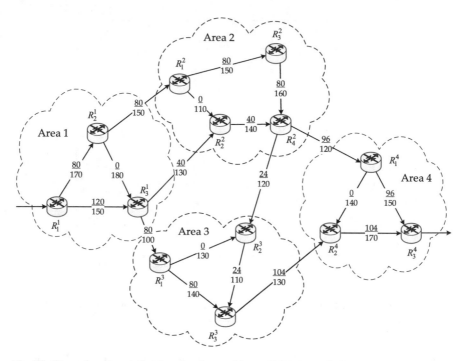

Fig. 12 The order of centralized routing in a multi-area TCN (example 2)

The results of the analysis of the centralized solution of the routing problem showed that with an increase in the input flow intensity from 1 to 200 1/s, the threshold value of the maximum utilization of network links, also grows linearly (Fig. 13) as it was observed for the first example (Fig. 2),

In accordance to the proposed diakoptical method, the throughputs of each area of the given TCN are determined. But since the structures of all areas and the bandwidth of all intra-area links remained the same, we can use the calculation results obtained in Figs. 3, 4, 5, 6 and 7, i.e. $c^1 = 320$ 1/s, $c^2 = 290$ 1/s, $c^3 = 250$ 1/s and $c^4 = 320$ 1/s.

The structure of the aggregated network with the introduction of equivalent communication links corresponding to each individual area is shown in Fig. 14. The same figure shows a macro-solution of the routing problem for the flow with intensity 200 1/s, where in the breaks of the communication links the bandwidth of this link is indicated in the denominator of the fraction, and the numerator of the fraction shows the intensity of the flow transmitted in this link. For the obtained solution (Fig. 14), the threshold value of the maximum utilization of TCN links (1) is also 0.8.

Figure 15 illustrates the dependence of the threshold value of the maximum utilization of network links on the intensity of the input flow for the aggregated structure of the second example of the studied network. Comparing the obtained solutions shown in Figs. 12 and 14, it can be concluded that the distribution of flows in inter-area links,

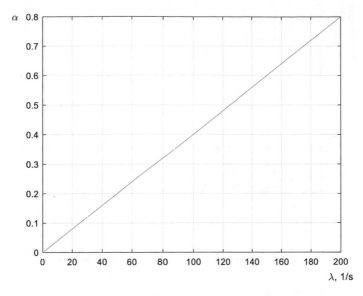

Fig. 13 The threshold of the maximum utilization of network links subject to the intensity of the input flow for a centralized solution of the routing problem (example 2)

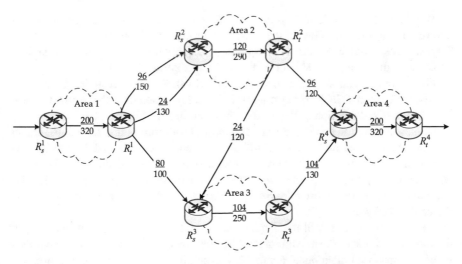

Fig. 14 Aggregated network structure and the routing solution obtained for it (example 2)

calculated as a result of solving the routing problem for an aggregated network, does not coincide with the centralized solution. However, the threshold values of the maximum utilization of network links presented in Figs. 13 and 15 match each other completely.

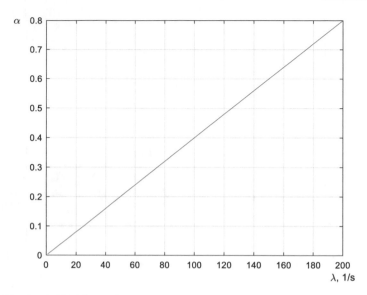

Fig. 15 The threshold of the maximum utilization of links subject to the intensity of the input flow for the macro-solution of the routing problem in the aggregated structure of the TCN (example 2)

Within the framework of the proposed method, the solutions presented in Fig. 14 and relating to the intensities of the flows in the inter-area links and the areas themselves, act as initial data for calculating the routing order in each of the areas and the subsequent diakoptical combination into a single solution (Fig. 16).

For the final solution (Fig. 16), the numerical value of the threshold of the maximum utilization of TCN links (1) is also 0.8, i.e. fully coincides with the effectiveness of the centralized solution of the routing problem in a multi-area TCN.

Comparison of the centralized and diakoptical solutions (Figs. 12 and 16) shows that, same as for the first example, different options were obtained for routing and balancing of the flows in the network links, both inside and between the areas, however, the threshold of the maximum utilization of the network links coincide, i.e. the diakoptical approach provides a solution of the routing problem that is similar to the centralized one.

To prove the similarity of the obtained solution Fig. 17 shows the dependency of the threshold value of the maximum link utilization on the intensity of the input flow when the proposed diakoptical approach was applied to solve the routing problem for the second example of the network structure.

Same as for the first example, this dependency is linear in the case of a centralized solution of the routing problem, as well as for solutions obtained at all steps of the proposed diakoptical method: for each separate network area, for the macro solutions of the routing problem in the aggregated network structure and for the final diakoptical solution, obtained by combining the solutions calculated at the macro and micro level (area level).

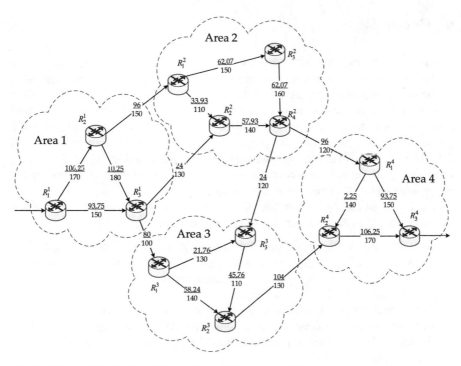

Fig. 16 Diakoptical solution of the routing problem (example 2)

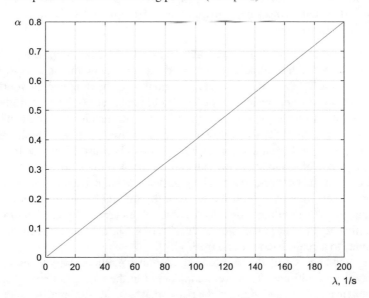

Fig. 17 The threshold of the maximum link utilization subject to the input flow intensity for the diakoptical solution of the routing problem in the TCN (example 2)

6   Conclusions

The article proposes a diakoptical method of inter-area routing with load balancing in a telecommunication network, which rejects the centralized calculations and based on the idea of dividing the general task of routing in a multi-area telecommunication network into several subtasks of lower complexity, solved for each individual area separately, with their subsequent combination into a generalized solution for the whole TCN. At various steps of the proposed method, a flow-balanced routing model (1)–(7) with load balancing was used that meets the requirements of the Traffic Engineering concept and is associated with minimizing the threshold value of the maximum utilization of communication links.

The proposed diakoptical method of inter-area routing with load balancing in a telecommunication network consists of the following steps:

1. Determination of the generalized throughputs of each area of the analyzed TCN.
2. Construction of an aggregated network structure model in which each network area is represented by an equivalent communication link with the already known (determined at the first step) bandwidth.
3. Based on the obtained model of the aggregated network structure, first, the routing order of flows in the inter-area communication links is calculated, and second, the intensities of the flows that arrive to and leave from each pth area of the TCN are determined.
4. The final order of flow routing with load balancing in each of the TCN areas is specified, which is consistent with the obtained routing order of flows in inter-area communication links.

It is important to note that the calculations of the first two steps are performed only once, since the throughput of the areas does not depend on the intensity of the flow entering the TCN. Thus, the basis of the method is the third and the fourth step, which are involved when the intensity of the input flow changes. The main advantage of the diakoptical method is the rejection of the centralization of the calculations, which positively affects the scalability of the final routing solutions and the required processing power of the routers (route server). This is because within the framework of the fourth step of the method, the calculations related to determining the routing order in each of the network area can be parallelized and carried out autonomously from each other. Moreover, with the increase in the number of areas in the TCN, the dimension of the routing problems that are solved separately in each of the area will significantly decrease in relation to the size of the initial routing problem, which corresponds to a centralized routing method.

During the analysis of the proposed diakoptical method, two examples of the structure of the multi-area TCN were considered in the work: in the first case, the bottleneck (in terms of bandwidth) in the network was inside the area, in the second—at the level of inter-area communication links.

References

1. Vegesna S (2001) IP quality of service. Cisco Press
2. Barreiros M, Lundqvist P (2016) QOS-enabled networks: tools and foundations, 2nd ed. Wiley Series on Communications Networking & Distributed Systems. Wiley (2016)
3. Halabi S (2000) Internet routing architectures, 2nd ed. Cisco Press
4. Medhi D, Ramasamy K (2018) Network routing: algorithms, protocols, and architectures, 2nd ed. The Morgan Kaufmann Series in Networking. Elsevier Inc., Cambridge
5. Misra S, Goswami S (2017) Network routing: fundamentals, applications, and emerging technologies, 1st ed. Wiley (2017)
6. Wójcik R, Domżał J, Duliński Z, Rzym G, Kamisiński A, Gawłowicz P, Jurkiewicz P, Rząsa J, Stankiewicz R, Wajda K (2016) A survey on methods to provide interdomain multipath transmissions. Comput Netw 108:233–259. https://doi.org/10.1016/j.comnet.2016.08.028
7. Cisco Networking Academy (2014) Routing protocols companion guide. Pearson Education, Cisco press
8. Black UD (2000) IP routing protocols: RIP, OSPF, BGP, PNNI and Cisco routing protocols. Prentice Hall PTR
9. Lee GM, Choi JS (2005) A survey of multipath routing for traffic engineering. Springer, pp 635–661
10. Segall A (1977) The modeling of adaptive routing in data communication networks. IEEE Trans Commun 25:85–95. https://doi.org/10.1109/TCOM.1977.1093715
11. Rak J (2015) Resilient routing in communication networks (Computer communications and networks), 1st ed. Springer
12. Lemeshko OV, Yeremenko OS (2016) Dynamics analysis of multipath QoS-routing tensor model with support of different flows classes. In: 2016 International conference on smart systems and technologies (SST). Josip Juraj Strossmayer University of Osijek, Croatia, pp 225–230. https://doi.org/10.1109/sst.2016.7765664
13. Lemeshko O, Yeremenko O (2016) Dynamic presentation of tensor model for multipath QoS routing. In: 2016 13th international conference on modern problems of radio engineering, telecommunications and computer science (TCSET). Lviv, Ukraine, pp 601–604. https://doi.org/10.1109/tcset.2016.7452128
14. Mersni A, Ilyashenko A, Vavenko T (2017) Model of multicast routing with support of shared explicit reservation of link resources. In: 2017 IEEE first Ukraine conference on electrical and computer engineering (UKRCON). Kyiv, Ukraine, pp 1145–1148. https://doi.org/10.1109/ukrcon.2017.8100429
15. Lemeshko O, Drobot O (2006) A mathematical model of multipath QoS-based routing in multiservice networks. In: Proceedings of international conference modern problems of radio engineering, telecommunications and computer science (TCSET). Lviv-Slavsko, pp 72–74. https://doi.org/10.1109/tcset.2006.4404448
16. Lin SC, Akyildiz IF, Wang P, Luo M (2016) QoS-aware adaptive routing in multi-layer hierarchical software defined networks: a reinforcement learning approach. In: 2016 IEEE international conference on services computing (SCC). San Francisco, CA, USA, pp 25–33. https://doi.org/10.1109/scc.2016.12
17. Nevzorova O, Arous K, Hailan A (2015) Flow-based model of hierarchical multicast routing. In: 2015 2nd international scientific-practical conference problems of infocommunications science and technology (PIC S and T'2015), pp 50–53. https://doi.org/10.1109/infocommst.2015.7357266
18. Wright B (2003) Inter-area routing, path selection and traffic engineering. White paper. Data Connection Limited
19. Vutucury S (2001) Multipath routing mechanisms for traffic engineering and quality of service in the Internet. Ph.D. dissertation, University of California, USA
20. Vutukury S, Garcia-Luna-Aceves JJ (2000) A traffic engineering approach based on minimum-delay routing. In: 2000 IEEE ninth international conference on computer communications and networks. Las Vegas, pp 42–47. https://doi.org/10.1109/icccn.2000.885468

21. Seok Y, Lee Y, Choi Y, Kim C (2002) Dynamic constrained traffic engineering for multi-cast routing. Inf Netw Wired Commun Manage 2343:278–288. https://doi.org/10.1007/3-540-45803-4_26
22. Wang Y, Wang Z (1999) Explicit routing algorithms for internet traffic engineering. In: Proceedings eight international conference computer communications and networks. Boston, USA, pp 582–588. https://doi.org/10.1109/icccn.1999.805577
23. Lemeshko O, Vavenko T, Ovchinnikov K (2013) Design of multipath routing scheme with load balancing in MPLS-network. In: 2013 IEEE XIIth international conference the experience of designing and application of CAD systems in microelectronics. Polyana-Svalyava-(Zakarpattya), UKRAINE, pp 211–213
24. Mersni A, Ilyashenko A, Vavenko T (2017) Complex optimality criterion for load balancing with multipath routing in telecommunications networks of nonuniform topology. In: 14th international conference the experience of designing and application of CAD system in microelectronic (CADSM). Polyana-Svalyava (Zakarpattya), Ukraine, pp 100–104. https://doi.org/10.1109/cadsm.2017.7916095
25. Mersni A, Ilyashenko AE (2017) Complex criterion of load balance optimality for multipath routing in telecommunication networks of nonuniform topology. Telecommun Radio Eng 76(7):579–590. https://doi.org/10.1615/TelecomRadEng.v76.i7.20
26. Lemeshko O, Yeremenko O, Nevzorova O (2017) Hierarchical method of inter-area fast rerouting. Transp Telecommun J 18(2):155–167. https://doi.org/10.1515/ttj-2017-0015
27. Lemeshko O, Yeremenko O (2017) Enhanced method of fast re-routing with load balancing in software-defined networks. J Electr Eng 68(6):444–454. https://doi.org/10.1515/jee-2017-0079
28. Yeremenko O, Lemeshko O, Nevzorova O, Hailan AM (2017) Method of hierarchical QoS routing based on network resource reservation. In: 2017 IEEE first Ukraine conference on electrical and computer engineering (UKRCON). Kyiv, Ukraine, pp 971–976. https://doi.org/10.1109/ukrcon.2017.8100393
29. Lemeshko O, Nevzorova O, Vavenko T (2016) Hierarchical coordination method of inter-area routing in telecommunication network. In: 2016 IEEE international scientific conference "radio electronics and info communications", (UkrMiCo 2016). Kyiv, Ukraine, pp 1–4. https://doi.org/10.1109/ukrmico.2016.7739626
30. Nevzorova YS, Arous KM, Salakh MTR (2016) Method for hierarchical coordinated multicast routing in a telecommunication network. Telecommun Radio Eng 75:1137–1151. https://doi.org/10.1615/telecomradeng.v75.i13.10
31. Yeremenko AS (2018) A two-level method of hierarchical-coordination QoS-routing on the basis of resource reservation. Telecommun Radio Eng 77(14):1231–1247. https://doi.org/10.1615/TelecomRadEng.v77.i14.20
32. Yevsyeyeva O (2010) Diakoptical approach in telecommunication engineering. In: 2010 IEEE modern problems of radio engineering, telecommunication and computer science (TCSET'2010). Lviv, Ukraine, p 29
33. Kron G (1963) Diakoptics; the piecewise solution of large-scale system. MacDonald, London

Analysis of Influence of Network Architecture Nonuniformity and Traffic Self-similarity Properties to Load Balancing and Average End-to-End Delay

Oleksandr Lemeshko⬤, Amal Mersni⬤, and Olena Nevzorova⬤

Abstract The analysis of influence of network architecture nonuniformity and traffic self-similarity properties to load balancing and average end-to-end delay is presented. Therefore, by non-uniformity of network architecture was implied that its structure could be represented by a separable graph or the one close to it. This means that the telecommunication network contained routers and links, which were simulated by articulation points and bridges, respectively. And, by non-uniformity may be implied the fact that the network could have a minimum cut, the rate of which was much less than the bandwidth of other cuts of the network. And for the analysis of influence of network architecture nonuniformity and traffic self-similarity properties to load balancing and average end-to-end delay the mathematical model of load balancing in the telecommunication network was used, within which not only the upper threshold of traffic load of the network links in general, but also certain coefficients of link utilization are minimized for maximally satisfaction of the requirements of the concept of Traffic Engineering. This made it possible to organize the load balancing process in the network more effectively and provide the best value of such an important quality of service indicator as the average end-to-end packet delay in the network.

Keywords Network · Traffic · Self-similar · Nonuniformity · Load balancing

O. Lemeshko (✉) · A. Mersni · O. Nevzorova
Kharkiv National University of Radio Electronics, Nauky Ave. 14, Kharkiv 61166, Ukraine
e-mail: oleksandr.lemeshko.ua@ieee.org

A. Mersni
e-mail: amal.mersni.ua@ieee.org

O. Nevzorova
e-mail: olena.nevzorova.ua@ieee.org

T. Radivilova et al. (eds.), *Data-Centric Business and Applications*, Lecture Notes on Data Engineering and Communications Technologies 48,
https://doi.org/10.1007/978-3-030-43070-2_33

767

1   Introduction

In accordance with the requirements of the Traffic Engineering (TE) concept while ensuring quality of service (QoS), the telecommunications network resource (TCN) should be used in a balanced manner [1–4], without causing overload of individual network elements (routers and links) when underload of others. In this case, the network resource is classified into a buffer and link. The buffer resource is a packet queue organized on the TCN routers, and the link resource, in turn, is characterized by bandwidth of links. Buffer load queues and communication channels significantly affect the numerical values of key QoS indicators (average packet delay, jitter, packet loss rate) and TCN performance. The Traffic Engineering concept is being actively implemented in modern TCN at various functional levels of the QoS support architecture: queue management, routing, resource reservation, traffic profiling, etc.

From the point of view of the solution for routing problems, the mathematical model of load balancing described in [5–7] is used as a reference one and focused on minimizing the upper threshold value of the TCN communication links. This, as it is known [8–14], contributes to the improvement of QoS-indicators.

However, as shown in [9], the approach based on minimizing the high level of network traffic provides relatively low efficiency of multipath routing tasks in TCNs that have a non-uniform architecture. At the same time, architectural non-uniformity can be caused both by the structural features of the construction of the network [9] and by the functional asymmetry. Such asymmetry is manifested, for example, in such an allocation of bandwidth of communication links that in the end may cause the formation of "bottlenecks" in the TCN, which will determine the maximum values of the load on the network elements. This negatively affects the optimization of the use of communication links that are not a part of the "bottleneck" network. However, load of the communication link also largely depends on the value of end-to-end QoS indicators, including the average end-to-end packet delay. Therefore, the proposed article is devoted to the improvement of load balancing processes when solving the problems of multipath routing in TCN with non-uniform architecture.

2   Description of TE Model of Load Balancing When Solving Problems of Single Path Routing in Telecommunication Network

As shown in works [5, 6] devoted to the simulation of Traffic Engineering, the following output data are known for the description of multipath routing tasks for each kth flow:

r_k is average packet rate (intensity), which is measured in packets per second (1/s);

s_k and d_k are the source router and the destination router for packets of the kth flow, respectively.

Then, the order of unicast routing and load balancing in TCN is determined by the routing variables $x^k_{(i,j)}$. Each of the variables characterizes the portion (part) of the kth flow occurring in the communication link between the ith and jth routers. Based on the physical content of the introduced routing variables, the conditions of the following form are imposed to them

$$0 \leq x^k_{i,j} \leq 1. \tag{1}$$

The introduction of constraints (1) is responsible for the implementation of the multipath routing strategy in TCN. However, this thereby prohibits the use of single path solutions, in which variables $x^k_{(i,j)}$ can take extreme of their possible values of zero or one (1). In addition, during the calculation of routing variables, the conditions for conservation of the flow on TCN routers must be fulfilled [5, 6]:

$$\begin{cases} \sum\limits_{j:(i,j)\in E} x^k_{(i,j)} - \sum\limits_{j:(j,i)\in E} x^k_{(j,i)} = 1, k \in K, i = s_k, \\ \sum\limits_{j:(i,j)\in E} x^k_{(i,j)} - \sum\limits_{j:(j,i)\in E} x^k_{(j,i)} = 0, k \in K, i \neq s_k, d_k, \\ \sum\limits_{j:(i,j)\in E} x^k_{(i,j)} - \sum\limits_{j:(j,i)\in E} x^k_{(j,i)} = -1, k \in K, i = d_k. \end{cases} \tag{2}$$

During the fulfillment of the conditions (2), the absence of packet loss on each router and on the whole network is ensured, and the connectivity of the calculated routes between the source and the destination of the packet of the kth flow is provided.

To prevent the overload of TCN communication links, the following conditions must be satisfied [5, 6]:

$$\sum_{k \in K} r_k \cdot x^k_{(i,j)} \leq \alpha \cdot \varphi_{(i,j)}, (i, j) \in E, \tag{3}$$

the number of which corresponds to the number of communication links in the TCN.

In conditions (3), the unknown parameter α is an additionally introduced control variable, to which according to its physical content the following constraint is imposed:

$$0 \leq \alpha \leq 1. \tag{4}$$

If we write the conditions (3) in the form

$$\frac{\lambda_{(i,j)}}{\varphi_{(i,j)}} \leq \alpha, (i, j) \in E$$

under $\lambda_{(i,j)} = \sum\limits_{k \in K} r_k \cdot x^k_{(i,j)}$, then with consideration of the equality

$$\rho_{(i,j)} = \frac{\lambda_{(i,j)}}{\varphi_{(i,j)}}$$

it can be stated that the variable α describes the maximum value of the load threshold (utilization coefficient) of the network communication links.

To improve the quality of service during the implemented load balancing in the TCN, this variable needs to be minimized. In such a way we define the type of optimality solution for routing and balancing load in the network [5, 6]

$$\alpha \rightarrow \textbf{min}. \tag{5}$$

The optimization problem (1)–(5) relates to the class of linear programming problems, since the objective function (5) to be minimized is linear. Constraints (4), (3) are also linear and they are introduced for routing variables $x^k_{(i,j)}$ (1)–(3) and the control variable α. It can be effectively solved using appropriate methods, for example, the simplex method, "branch and bounds," dynamic programming, etc. In this case, the dimension of this optimization problem corresponds to the number of variables that are calculated and is $|K| \cdot |E| + 1$.

When solving the optimization problem formulated, the resulting minimum value of the variable α in accordance with the conditions (3) and (4) corresponds quantitatively to the utilization coefficient of the most loaded link. As shown in [14, 15], in addition to the linearity, there is an undeniable advantage of this model (1)–(5). When load on the network increases, the threshold value of the variable α increases linearly. Therefore, it contributes to the predicted (without sharp fluctuations) change in the main indicators of the quality of service—the average delay, jitter and packet loss probability.

2.1 Analysis of Directions in Improvement of the Basic TE Model for Load Balancing in Network

For some specific cases, which are limited to the representation of the network in the form of a separable graph, in [9], it is proposed to carry out the balancing of the load "in parts". This means that is done separately in each connected component of the graph joined by the bridge through the articulation points. In this case, the maximum bandwidth of the network communication links represented by the value of the variable α will correspond to the coefficient of link utilization simulated by the bridge of the graph. Optimization of balancing in connected network components joined by a bridge allows improving the final solutions in terms of improving the quality of service [9].

However, such an approach can adequately be applied only in those cases where the physical or logical architecture of the network can be represented by a separable graph with several articulation points and a bridge. If the low-productive area (bottleneck) of the network on the graph model of TCN cannot be described by a single bridge or the network structure does not contain articulation points, then the approach based on a separate solution to the balancing problems for individual network fragments cannot be applied. This is since there is ambiguity of the resulting solutions in the connected fragments of the network and in communication links that form a "bottleneck" of the TCN accompanied by a violation of the connectivity between end-to-end routes or the order of balancing the load in them. Therefore, in [7], the improvement of the basic TE model (1)–(5) relates not to revising the order of calculating routing variables, but to replacing the very type of optimality criterion used and the objective function laid down in its basis.

As a rule, to increase the sensitivity of the optimality criterion to the load of not only the "bottleneck" of the network, but also other communication links, the components that are numerically related to the utilization coefficient of all TCN links are introduced into the criterion. In this case, by analogy with the approach given in [7], the objective function to be minimized can maintain its linear form

$$J_l = \vec{f}^t \cdot \vec{x} + g \cdot \alpha \rightarrow \mathbf{min}, \qquad (6)$$

where $\vec{x}$ is the vector of routing variables $x^k_{(i,j)}$; $\vec{f}$ is the vector of routing metrics that orients towards inclusion communication links with the highest bandwidth into the optimal path (multipath). Its coordinates are represented by the variables $1/\varphi_{(i,j)}$; g is the additional weight coefficient, the value of which determines the importance of the second summand in the criterion (6); $[\cdot]^t$ is the transposition operation of matrix (vector).

To minimize, first, the upper load threshold of the TCN links, i.e., the second summand in the expression (6), the value of the coefficient g must be sufficiently large.

In addition, the optimality criterion is proposed in the work, which modifies the expression (6) and takes the form:

$$J_{lq} = \vec{x}^t \cdot H_x \cdot \vec{x} + g \cdot \alpha \rightarrow \mathbf{min}. \qquad (7)$$

where H_x is the diagonal matrix of the size $|E| \cdot |K| \times |E| \cdot |K|$, the coordinates of which are route metrics (values $1/\varphi_{(i,j)}$).

Thus, the criterion (7) has a linear quadratic character, since the first summand has a quadratic form from the vector of routing variables. The second summand is a linear function from the threshold value of the TCN communication links. The introduction of a quadratic summand into the criterion (7) pursued two main goals:

(1) providing more detailed consideration of the load of each network link separately. This facilitates the selection of routes with maximum bandwidth and minimum number of hops;

(2) implementation of the multipath routing strategy.

To maintain the basic functionality of the model (1)–(5) from the point of view of compliance with the requirements of the Traffic Engineering concept, the second summand in the criterion (7) should be decisive. This is achieved by choosing the weight coefficient according to the condition $g \gg 1$.

Therefore, in the given work, a study was conducted that consists in the comparative analysis of solutions for the load balancing problem in TCN with non-uniform architecture obtained based on using different optimality criteria (5), (6) and (7). An indicator of the efficiency of the resulting solutions was presented by the values of the end-to-end packet delay [16].

To calculate the average end-to-end delay of the packets of the kth flow, the expression [17–19] was used:

$$\tau^k = \sum_{p \in P} x_p^k \tau_p^k, \tag{8}$$

where P is a set of routes between the pair of «source» and «destination» routers; x_p^k is a portion of the kth flow transmitted over the pth path;

$$\tau_p^k = \sum_{(i,j) \in p} \tau_{(i,j)}^k \tag{9}$$

is the average packet delay of the k flow along the pth path; $\tau_{(i,j)}^k$ is the average packet delay of the kth flow in the link represented by the arc (i, j).

Without losing the universality of the results, it was assumed that the expression was used to calculate the average packet delay in the link [17–19]

$$\tau_{(i,j)} = \frac{1}{\varphi_{(i,j)}} + \rho_{(i,j)} \frac{\rho_{(i,j)}^{1/2(1-H)}}{(1 - \rho_{(i,j)})^{H/(1-H)} \sum_{k \in K} r_k \cdot x_{(i,j)}^k}, \tag{10}$$

where

$$\rho_{(i,j)} = \frac{\sum_{k \in K} r_k \cdot x_{(i,j)}^k}{\varphi_{(i,j)}}$$

is the coefficient of link utilization represented by the arc (i, j); $0.5 \le H \le 1$ is the self-similarity parameter (Hurst parameter). Its value depends on the type of traffic transmitted on the network (Table 1) [20].

Such a delay (10) will be experienced by all packets of flows that claim the bandwidth of a link specified by the value $\varphi_{(i,j)}$. Therefore, in the Formula (7) the index k is omitted in the value $\tau_{(i,j)}$.

Table 1 Hurst parameter value for modeling different types of traffic (according to ITU-T Q.3925 recommendations)

Traffic type		Flow type	Range of Hurst parameter values (H)
WWW-traffic		Self-similar	$H = 0.7...0.9$
Transmission of data (files)			$H = 0.85...0.95$
E-mail			$H = 0.75$
P2P traffic			$H = 0.55...0.6$
IPTV traffic	Unicast		$H = 0.75...0.8$
	Multicast		$H = 0.55...0.6$
Telemetry traffic in sensor networks			$H = 0.67...0.69$

2.2 Comparative Analysis of Solutions for the Problem of Load Balancing in Telecommunication Network with Non-uniform Architecture

A comparative analysis of the solutions for the load balancing problem in TCN with non-uniform architecture obtained using different optimality criteria (5), (6) and (7) was carried out on several network architectures that differ in size, degree of non-uniformity and productivity. For clarity, the results of the comparative analysis will be presented as an example of four network configurations.

Therefore, Fig. 1 presents a TCN. Its structural non-uniformity is expressed through the fact that the network can be described by a separable graph where the articulation point is the vertex that simulates the third router. Under the router failure, the output TCN splits into two connected components. The first of them includes the first, second and fourth routers. The second one includes routers with numbers of five, six, seven and eight. In Fig. 1, the gaps of communication links present their bandwidths. The first router acts as a source, and the eighth was the destination of packets transmitted, for example, with the rate of 400 1/s. Functional non-uniformity

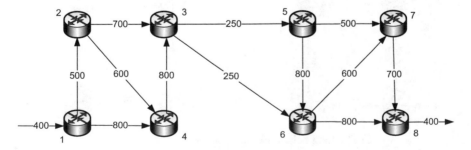

Fig. 1 An example of the first studied network architecture

of the presented in Fig. 1 TCN architecture is determined by the fact that when packets are transferred from the first router to the eighth in the network, a network cut is formed. In addition, such a cut has very limited performance—a "bottleneck"—in comparison with other cuts.

This cut includes the links connecting the third router with the sixth and fifth, and has a bandwidth of 500 1/s.

In Fig. 2, the examples of optimal solutions for the load balancing problem in the TCN are presented based on model (1)–(4) utilization with different optimality

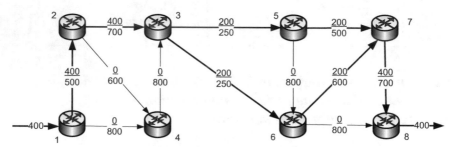

(a) The load balancing order using the criterion (5)

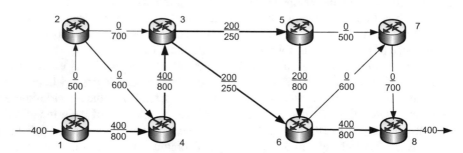

(b) The load balancing order using the criterion (6)

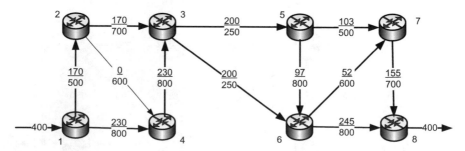

(c) The load balancing order using the criterion (7)

Fig. 2 An example of solving the load balancing problem using the model (1)–(4) and various optimality criteria for the first network architecture (Fig. 1)

criteria: for the criterion (5)—Fig. 2a, for the criterion (6)—Fig. 2b, for criterion (7)—Fig. 2c. In the gaps of each communication link in Fig. 2, there is a fraction in which the numerator shows the intensity of the transmitted packet flow and in the denominator shows the bandwidth of this communication link.

Each of the solutions shown in Fig. 2 is focused on the implementation of the multipath routing strategy and providing the same parameter value $\alpha = 0.8$, which is determined for communication links $(3, 5)$ and $(3, 6)$ of the minimum network cut. However, the order of routing and balancing the load on the communication links not included in this cut when using different criteria of optimality is quite different. Table 2 gives a detailed description of the multipaths and paths that they form in terms of the rate of the flows transmitted over them (Fig. 2) and the resulting values of the average end-to-end packet delay (8). At the same time in Table 2 the results of the calculation of the average delay (8), (9), (10), for example, are detailed for the Hurst parameter value $H = 0.5$.

Figure 3 shows the results of the comparison of the average end-to-end packet delay obtained using the optimality criteria (5), (6) and (7) for various values of the Hurst parameter. The results are referred to the TCN architecture shown in Fig. 1. The application of the proposed criterion (7) allows improving the numerical values of the average end-to-end delay of packets by 31–34% compared with the use of the criterion (5) and by 3–4.5% compared with the use of the criterion (6).

The second version of the network architecture, which has been analyzed, is given in Fig. 4 and differs from the first architecture by the presence of an additional input link between fourth and sixth routers. In this Figure, the gaps in communication links indicate their bandwidth. As in the previous case (Fig. 1), the first router acts as a source, and the eighth receives packets transmitted, for example, at a rate of 600 1/s.

Table 2 Characteristics of the solution for the load balancing problem using different optimality criteria for the first network architecture

Criterion used	Routes included into the multipath	Portion of the flow passing along the path	Packet transmission rate (1/s)	Average delay along the route (ms)	Average end-to-end packet delay (ms)
(5)	$1 \to 2{\to}3 \to 5{\to}7 \to 8$	0.5	200	40	39.6
	$1 \to 2{\to}3 \to 6{\to}7 \to 8$	0.5	200	39.2	
(6)	$1 \to 4{\to}3 \to 5{\to}6 \to 8$	0.5	200	29.2	28.3
	$1 \to 4{\to}3 \to 6{\to}8$	0.5	200	27.5	
(7)	$1 \to 4{\to}3 \to 6{\to}8$	0.5	200	25.3	27.1
	$1 \to 2{\to}3 \to 5{\to}7 \to 8$	0.2575	103	29.3	
	$1 \to 2{\to}3 \to 5{\to}6 \to 7{\to}8$	0.13	52	30	
	$1 \to 2{\to}3 \to 5{\to}6 \to 8$	0.035	14	28.1	
	$1 \to 4{\to}3 \to 5{\to}6 \to 8$	0.0775	31	26.7	

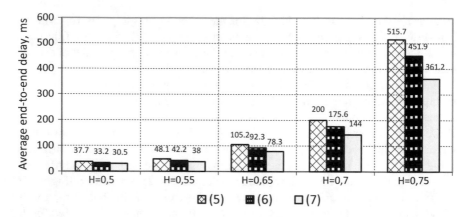

Fig. 3 The results of the comparison of the average end-to-end packet delay obtained using the optimality criteria (5), (6), (7) for various values of the Hurst parameter (TCN architecture is shown in Fig. 1)

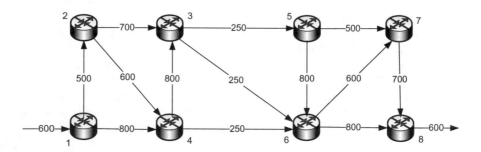

Fig. 4 An example of the second studied network architecture

The TCN presented in Fig. 4 does not possess structural non-uniformity, because it cannot be represented by a separable graph. This means it does not contain any articulation points or bridges. However, this network architecture has functional non-uniformity, since the cut represented by links (3, 5), (3, 6), (4, 6) is minimal and it has a bandwidth of 750 1/s, which is approximately two times less than performance of any other cut of the network.

Figure 5 presents the examples of solutions for the load balancing problem in TCN given in Fig. 4, based on the use of the model (1)–(4) with different optimality criteria: for the criterion (5)–Fig. 5a, for criterion (6)—Fig. 5b, for the criterion (7)—Fig. 5c. By analogy with Fig. 2 the gaps of each communication link in Fig. 5 indicate the fraction, in which the numerator shows the intensity of the packet flow, and in the denominator denotes the bandwidth of the given communication link. As for the previously obtained results (Fig. 2), each of the presented in Fig. 5 solutions has two general features. First, the multipath routing strategy is used. Second, the same values of the maximum bandwidth of the TCN communication links ($\alpha = 0.8$) are

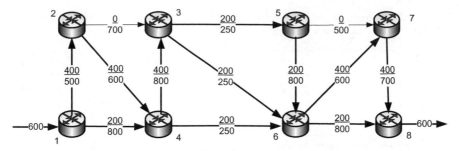

(a) The load balancing order using the criterion (5)

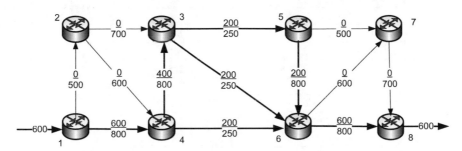

(b) The load balancing order using the criterion (6)

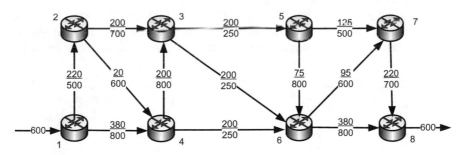

(c) The load balancing order using the criterion (7)

Fig. 5 An example of solving the load balancing problem using the model (1)–(4) and various optimality criteria for the second network architecture (Fig. 7)

ensured due to the presence of the minimum cut presented by the links (3, 5), (3, 6), (4, 6). The use of different forms of optimality criteria also determines the different order of routing and balancing the load in the TCN, which is especially specific for communication links that are not a part of the minimum cut. By analogy with Tables 2 and 3 shows the characteristics of the routes involved (Fig. 5) and the resulting values of the average end-to-end delay of packets (8) for the Hurst parameter value $H = 0.5$.

Table 3 Characteristics of the solution for the load balancing problem using different optimality criteria for the second network architecture

Used criterion	Routes included into the multipath	Portion of the flow passing along the path	Packet transmission rate (1/s)	Average delay along the route (ms)	Average end-to-end packet delay (ms)
(5)	$1 \rightarrow 2 \rightarrow 4 \rightarrow 3 \rightarrow 5 \rightarrow 6 \rightarrow 7 \rightarrow 8$	1/3	200	47.5	38.9
	$1 \rightarrow 2 \rightarrow 4 \rightarrow 3 \rightarrow 6 \rightarrow 7 \rightarrow 8$	1/3	200	45.8	
	$1 \rightarrow 4 \rightarrow 6 \rightarrow 8$	1/3	200	23.3	
(6)	$1 \rightarrow 4 \rightarrow 3 \rightarrow 5 \rightarrow 6 \rightarrow 8$	1/3	200	34.2	31.97
	$1 \rightarrow 4 \rightarrow 6 \rightarrow 8$	1/3	200	30	
	$1 \rightarrow 4 \rightarrow 3 \rightarrow 6 \rightarrow 8$	1/3	200	31.7	
(7)	$1 \rightarrow 2 \rightarrow 3 \rightarrow 5 \rightarrow 7 \rightarrow 8$	0,21	125	30.3	27.4
	$1 \rightarrow 4 \rightarrow 6 \rightarrow 8$	1/3	200	24.8	
	$1 \rightarrow 2 \rightarrow 3 \rightarrow 5 \rightarrow 6 \rightarrow 7 \rightarrow 8$	0.125	75	31	
	$1 \rightarrow 2 \rightarrow 4 \rightarrow 3 \rightarrow 6 \rightarrow 7 \rightarrow 8$	0.033	20	31	
	$1 \rightarrow 4 \rightarrow 3 \rightarrow 6 \rightarrow 8$	0.3	180	26.4	

Figure 6 shows the results of the comparison of the average end-to-end packet delay obtained using the optimality criteria (5), (6) and (7) for different values of the Hurst parameter. The results are referred to the TCN architecture shown in Fig. 4. The application of the proposed criterion (7) allows improving the numerical average

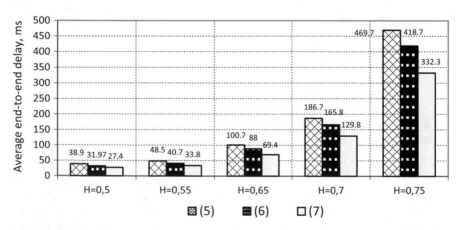

Fig. 6 The results of the comparison of the average end-to-end packet delay obtained using the optimality criteria (5), (6), (7) for various values of the Hurst parameter (the TCN architecture is shown in Fig. 4)

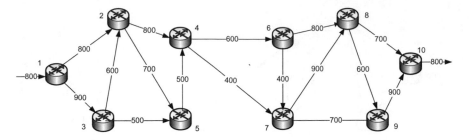

Fig. 7 An example of the third studied network architecture

end-to-end packet delay by 29–31% compared with the use of criterion (5) and by 15–21% compared with the use of the criterion (6).

The third example of the analyzed network architecture is shown in Fig. 7. As in the first case (Fig. 1), the TCN structure can be represented by a separable graph, in which the vertex that simulates the fourth router is the articulation point. It is precisely when the router fails, the original TCN splits into two connected components. The first component includes the first, second, third, and fifth routers. The second component includes the routers with numbers from six to ten. This determines the existence of structural non-uniformity in the architecture under consideration. However, unlike the first version of the network architecture (Fig. 1), these components include more routers.

In Fig. 7, in the communication links gaps, their bandwidths are again set. The first router acts as a source. The tenth receives packets transmitted, for example, with the rate of 800 1/s. Functional non-uniformity, as in the previous cases, is determined by the fact that in the network architecture there is a minimal cut represented by links (4, 6) and (4, 7), which has, in comparison with other cuts, a rather low total bandwidth of 1000 1/s.

Figure 8 shows the results of solving the load balancing problem in the TCN. Its architecture is presented in Fig. 8 based on the use of the model (1)–(4) with different optimality criteria: for the criterion (5)—Fig. 8a, for the criterion (6)—Fig. 8b, for the criterion (7)—Fig. 8c.

By analogy with Figs. 2 and 5, the gaps of each communication link in Fig. 8 show the fraction. The numerator shows the intensity of the packet flow transmitted. The denominator shows the bandwidth of the given communication link. For the obtained solutions, the problem of balancing the load in the TCN is presented in Fig. 8. Table 4 shows the final values of the average end-to-end packet delay (8) and the characteristics of the paths used ($H = 0.5$). All three solutions provide the same threshold values for the communication links ($\alpha = 0.8$), primarily due to the minimum cut presented by links (4, 6), (4, 7) in the network structure.

Figure 9 shows the results of comparing the average end-to-end packet delay obtained using the optimality criteria (5), (6) and (7) for different values of the Hurst parameter. The results are referred to the TCN architecture shown in Fig. 7.

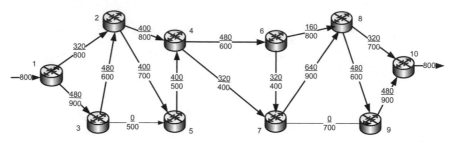

(a) The load balancing order using the criterion (5)

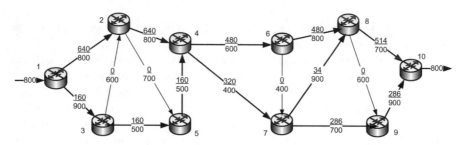

(b) The load balancing order using the criterion (6)

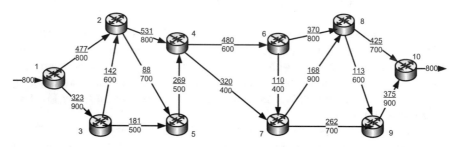

(c) The load balancing order using the criterion (7)

Fig. 8 An example of solving the load balancing problem using the model (1)–(4) and various optimality criteria for the third network architecture (Fig. 10)

For this network architecture, the application of the proposed criterion (7) allows improving the numerical values of the average end-to-end delay of the packets by 43–63% compared with the use of the criterion (5) and by 20–45% compared with the use of the criterion (6).

The final example of the analysis for the fourth network architecture is shown in Fig. 10. This architecture cannot be represented by a separable graph, which means it has no structural non-uniformity. However, it includes a minimal cut. It is represented by links (4, 6), (4, 7) and (5, 7) and has a bandwidth of 1500 1/s, which determines the functional non-uniformity. Figure 10 shows the links bandwidths in the gaps

Table 4 Characteristics of the solution for the load balancing problem using different optimality criteria for the third network architecture

Used criterion	Routes included into the multipath	Portion of the flow passing along the path	Packet transmission rate (1/s)	Average delay along the route (ms)	Average end-to-end packet delay (ms)
(5)	$1 \to 3 \to 2 \to 5 \to 4 \to 6 \to 7 \to 8 \to 9 \to 10$	0.4	320	59.4	41
	$1 \to 3 \to 2 \to 4 \to 7 \to 8 \to 9 \to 10$	0.2	160	40.3	
	$1 \to 2 \to 5 \to 4 \to 7 \to 8 \to 10$	0.1	80	34.4	
	$1 \to 2 \to 4 \to 7 \to 8 \to 10$	0.1	80	23.6	
	$1 \to 2 \to 4 \to 6 \to 8 \to 10$	0.2	160	17.1	
(6)	$1 \to 2 \to 4 \to 6 \to 8 \to 10$	0.6	480	29.3	28.3
	$1 \to 2 \to 4 \to 7 \to 9 \to 10$	0.2	160	29	
	$1 \to 3 \to 5 \to 4 \to 7 \to 9 \to 10$	0.1575	126	24	
	$1 \to 3 \to 5 \to 4 \to 7 \to 8 \to 10$	0.0425	34	26.5	
(7)	$1 \to 2 \to 4 \to 7 \to 9 \to 10$	0.3275	262	23.5	23.4
	$1 \to 2 \to 4 \to 7 \to 8 \to 9 \to 10$	0.0725	58	24.6	
	$1 \to 3 \to 5 \to 4 \to 6 \to 8 \to 10$	0.2263	181	23.5	
	$1 \to 2 \to 4 \to 6 \to 8 \to 10$	0.1962	157	21.1	
	$1 \to 3 \to 2 \to 4 \to 6 \to 7 \to 8 \to 10$	0.0675	54	24.4	
	$1 \to 3 \to 2 \to 5 \to 4 \to 6 \to 8 \to 10$	0.04	32	24.2	
	$1 \to 3 \to 2 \to 5 \to 4 \to 6 \to 7 \to 8 \to 9 \to 10$	0.07	56	27	

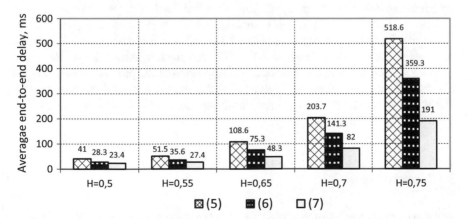

Fig. 9 The results of the comparison of the average end-to-end packet delay obtained using optimality criteria (5), (6), (7) for various values of the Hurst parameter (TCN architecture is shown in Fig. 7)

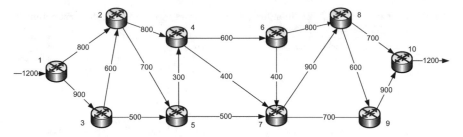

Fig. 10 An example of the fourth studied network architecture

of communication links. The first router acts as a source. The tenth router is the destination for packets transmitted, for example, at 1200 1/s.

The results of solving the task of load balancing in TCN, the architecture of which is presented in Fig. 10, using the model (1)–(4) with different optimality criteria are shown in Fig. 11. The use of the criterion (5) corresponds to Fig. 11a. The criterion (6) corresponds to Fig. 11b. The criterion (7) corresponds to Fig. 11c. By analogy with Fig. 7 the gaps of each communication link in Fig. 11 indicate the fraction. The numerator shows the intensity of the packet flow transmitted. The denominator presents the bandwidth of the given communication link. Characteristics of the calculated paths and the resulting values of the average end-to-end packet delay (6) for the value of the Hurst parameter $H = 0.5$, obtained when solving the load balancing problem in the TCN (Fig. 8), are presented in Table 5. As before, all three solutions provide the same threshold values for link loading ($\alpha = 0.8$).

Figure 12 shows the results of comparing the average end-to-end packet delay obtained using the optimality criteria (5), (6) and (7) for different values of the Hurst parameter. The results are referred to the TCN architecture shown in Fig. 10.

For the analyzed network architecture, application of the proposed criterion (10) allows improving the numerical values of the average end-to-end packet delay by 20–30% compared with the use of the criterion (5) and by 9–20% compared with the use of the criterion (9).

3 Conclusions

In this paper the research of the flow-based load balancing model in the telecommunication network with non-uniform architecture has been conducted. At the same time, non-uniformity of network architecture was divided into structural and functional. The structural non-uniformity of TCN implied that its structure could be represented by a separable graph or the one close to it. This means that the telecommunication network contained routers and links, which were simulated by articulation points and bridges, respectively. The functional non-uniformity of TCN was manifested particularly in the fact that the network could have a minimum cut, the rate of which was much less than the bandwidth of other cuts of the network. As shown by the research,

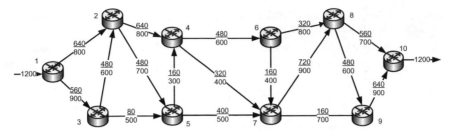

(a) The load balancing order using the criterion (5)

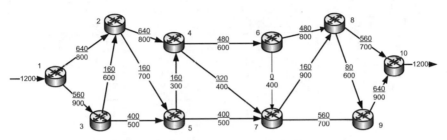

(b) The load balancing order using the criterion (6)

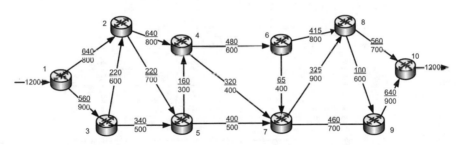

(c) The load balancing order using the criterion (7)

Fig. 11 An example of solving the load balancing problem using the model (1)–(4) and various optimality criteria for the fourth network architecture (Fig. 10)

the presence of both structural and functional non-uniformity of the TCN leads to the formation of the so-called "bottlenecks" in the network. This negatively affects the efficiency of load balancing in terms of providing extreme values of service quality indicators, for example, the average end-to-end delay of packets.

In this regard, the paper proposes the improvement for the mathematical model of load balancing in the TCN [5–7]. Such improvement maximally satisfies the requirements of the concept of Traffic Engineering. The improvement concerned the review of the optimality criterion of the route solutions used. It is proposed to proceed to the linear quadratic criterion (7), within which not only the upper threshold of traffic load of the network communication links in general, but also certain coefficients of link utilization are minimized. This made it possible to organize the load balancing

Table 5 Characteristics of the solution for the load balancing problem using different optimality criteria for the fourth network architecture

Used criterion	Routes included into the multipath	Portion of the flow passing along the path	Packet transmission rate (1/s)	Average delay along the route (ms)	Average end-to-end packet delay (ms)
(5)	$1 \rightarrow 2 \rightarrow 4 \rightarrow 6 \rightarrow 8 \rightarrow 10$	0.27	320	30.1	37.7
	$1 \rightarrow 2 \rightarrow 4 \rightarrow 6 \rightarrow 7 \rightarrow 9 \rightarrow 10$	0.13	160	30.7	
	$1 \rightarrow 3 \rightarrow 2 \rightarrow 5 \rightarrow 7 \rightarrow 8 \rightarrow 9 \rightarrow 10$	0.33	400	43.6	
	$1 \rightarrow 3 \rightarrow 2 \rightarrow 5 \rightarrow 4 \rightarrow 7 \rightarrow 8 \rightarrow 9 \rightarrow 10$	0.07	80	53.2	
	$1 \rightarrow 3 \rightarrow 5 \rightarrow 4 \rightarrow 8 \rightarrow 10$	0.07	80	37.7	
	$1 \rightarrow 2 \rightarrow 4 \rightarrow 7 \rightarrow 8 \rightarrow 10$	0.13	160	37.7	
(6)	$1 \rightarrow 3 \rightarrow 5 \rightarrow 7 \rightarrow 9 \rightarrow 10$	0.33	400	33.9	33.2
	$1 \rightarrow 2 \rightarrow 4 \rightarrow 6 \rightarrow 8 \rightarrow 10$	0.4	480	31	
	$1 \rightarrow 3 \rightarrow 2 \rightarrow 5 \rightarrow 4 \rightarrow 7 \rightarrow 9 \rightarrow 10$	0.13	160	38.1	
	$1 \rightarrow 2 \rightarrow 4 \rightarrow 7 \rightarrow 8 \rightarrow 9 \rightarrow 10$	0.07	80	32.1	
	$1 \rightarrow 2 \rightarrow 4 \rightarrow 7 \rightarrow 8 \rightarrow 10$	0.07	80	33.4	
(7)	$1 \rightarrow 3 \rightarrow 5 \rightarrow 7 \rightarrow 9 \rightarrow 10$	0.283	340	27.2	30.5
	$1 \rightarrow 3 \rightarrow 2 \rightarrow 5 \rightarrow 4 \rightarrow 6 \rightarrow 8 \rightarrow 10$	0.133	160	32.9	
	$1 \rightarrow 3 \rightarrow 2 \rightarrow 5 \rightarrow 7 \rightarrow 8 \rightarrow 9 \rightarrow 10$	0.05	60	25.6	
	$1 \rightarrow 2 \rightarrow 4 \rightarrow 6 \rightarrow 8 \rightarrow 10$	0.213	255	30.6	
	$1 \rightarrow 2 \rightarrow 4 \rightarrow 6 \rightarrow 7 \rightarrow 8 \rightarrow 9 \rightarrow 10$	0.054	65	31.8	
	$1 \rightarrow 2 \rightarrow 4 \rightarrow 7 \rightarrow 9 \rightarrow 10$	0.1	120	33	
	$1 \rightarrow 2 \rightarrow 4 \rightarrow 7 \rightarrow 8 \rightarrow 10$	0.121	145	33.9	
	$1 \rightarrow 2 \rightarrow 4 \rightarrow 7 \rightarrow 8 \rightarrow 9 \rightarrow 10$	0.046	55	33	

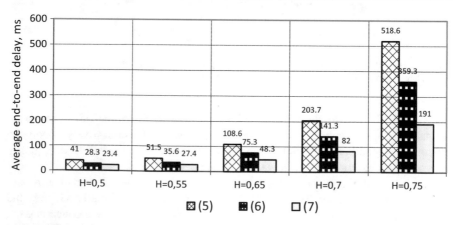

Fig. 12 The results of the comparison of the average end-to-end packet value obtained using the optimality criteria (5), (6), (7) for various values of the Hurst parameter (the TCN architecture is shown in Fig. 10)

process in the TCN more effectively and provide the best value of such an important QoS indicator as the average end-to-end packet delay in the network (8).

During the quantitative analysis of the improvement advantages, the obtained results were compared with the calculations based on other optimality criteria. It was done according to criterion (5), which corresponded to the model of Traffic Engineering [5–7], and by criterion (6), which was a linear function of utilized network communication links. At the same time, calculations were made for a set of network architectures that differed in their dimension and degree of non-uniformity. The numerical values of the average end-to-end packet delay (8), (9) and (10) were evaluated for various values of the Hurst parameter (Table 1), which corresponded to a certain type of network traffic. During the research it was established that the use of the proposed criterion (7) in organizing the load balancing in the TCN with non-uniform architecture allowed reducing the average end-to-end delay of packets in comparison with solutions based on the use of criterion (5) on average from 20–30% to 40–60%. In comparison to solutions based on the use of the criterion (6) it allowed reducing on average from 5–9% to 20–40% (Figs. 3, 6, 9 and 12).

As shown by the results of calculations, the use of the proposed load balancing criterion (7) is most effective in the conditions of high non-uniformity of network architecture. Such conditions include in cases when the structure of the network was modeled by a separable graph (Fig. 7), and the bandwidth of the minimum cut significantly differed (to the lesser side) from the bandwidth of other cuts of the network. In addition, the effectiveness of the proposed solution increased with traffic routing with high values of the Hurst parameter.

In addition, it was found out that the use of the linear criterion (6) focused mainly on the implementation of single path routing (Figs. 2b and 5b). This means there is no support of the load balancing in the connected components of the network, into which the network had been split by minimum cut. With the availability of several roughly equivalent routes, the use of the criterion (7) always provided the maximum gain compared to the criterion (6), as shown in Figs. 6 and 9.

It has been established that the more different the minimum cut and other cuts of the network were in their bandwidths, the higher was the gain due to the average end-to-end delay, which provided the use of the optimality criterion (7). For example, for the fourth TCN architecture presented in Fig. 10, the minimum cut had the bandwidth that was only 12–15% less than the bandwidth of other network cuts. This was accompanied by an improvement in the average end-to-end delay of 20–30% compared with the use of the criterion (5) and 9–20% compared with the use of the criterion (6) (Fig. 12). If, however, the bandwidth of the minimum cut was on average 42–50% less than the bandwidth of other network cuts, which is typical for the third TCN architecture (Fig. 10), then the gain over the average end-to-end delay was already 43–63% and 20–45% respectively (Fig. 9).

The proposed improvement will not lead to a significant complication of the algorithmic support and software of modern routers due to the mainly linear nature of the model (1)–(4). If it is necessary to consider explicitly the conditions for ensuring the quality of service by a set of QoS indicators, it is needed to switch to more complex routing models from the computational point of view, for example, tensor ones [21–23]. In case of considering the dynamics of the state of the TCN, the proposed criterion (10) can be used as the basis of optimization functional using dynamic routing models [24] represented by integral or difference-differential equations of the network state.

References

1. White R, Banks E (2018) Computer networking problems and solutions: an innovative approach to building resilient, modern networks, 1 ed. Addison-Wesley Professional
2. Marsic I (2013) Computer networks: performance and quality of service. Rutgers University
3. Cisco Networking Academy (ed.) Routing protocols companion guide. Pearson Education
4. Simha A, Osborne E (2002) Traffic engineering with MPLS. Cisco Press
5. Wang Y, Wang Z () Explicit routing algorithms for internet traffic engineering. In: 8th international conference on computer communications and networks. Paris, pp 582–588. https://doi.org/10.1109/icccn.1999.805577
6. Seok Y, Lee Y, Kim C, Choi Y (2001) Dynamic constrained multipath routing for MPLS networks. In: IEEE international conference on computer communications and networks. Scottsdale, AZ, USA, USA, pp 348–353. https://doi.org/10.1109/icccn.2001.956289
7. Seok Y, Lee Y, Choi Y, Kim C (2002) Dynamic constrained traffic engineering for multicast routing. Proc Wired Commun Manag 2343:278–288. https://doi.org/10.1007/3-540-45803-4_26
8. Lemeshko OV, Garkusha SV, Yeremenko OS, Hailan AM (2015) Policy-based QoS management model for multiservice networks. In: International Siberian conference on control and communications (SIBCON), 21–23 May 2015. IEEE, Omsk, Russia, pp 1–4
9. Nevzorova YS, Arous KM, Salakh MTR (2016) Method for hierarchical coordinated multicast routing in a telecommunication network. Telecommun Radio Eng (English translation of Elektrosvyaz and Radiotekhnika) 75(13):1137–1151. https://doi.org/10.1615/telecomradeng.v75.i13.10
10. Pioro M, Medhi D (2000) Routing, flow, and capacity design in communication and computer networks. Morgan Kaufmann Series in Networking. Elsevier Digital Press
11. Yeremenko OS, Lemeshko OV, Nevzorova OS, Hailan AM (2017) Method of hierarchical QoS routing based on network resource reservation. In: 2017 IEEE 1st Ukraine conference on electrical and computer engineering, UKRCON. Kiev, Ukraine, pp 971–976. https://doi.org/10.1109/ukrcon.2017.8100393
12. Wójcik R, Domżał J, Duliński Z, Rzym G, Kamisiński A, Gawłowicz P, Jurkiewicz P, Rząsa J, Stankiewicz R, Wajda K (2016) A survey on methods to provide interdomain multipath transmissions. Comput Netw 108:233–259. https://doi.org/10.1016/j.comnet.2016.08.028
13. Lemeshko O, Yeremenko O (2017) Enhanced method of fast re-routing with load balancing in software-defined networks. J Electr Eng Open Access 68(06):444–454. https://doi.org/10.1515/jee-2017-0079
14. Osunade O (2012) A packet routing model for computer networks. I J Comput Netw Inf Secur 4:13–20. https://doi.org/10.5815/ijcnis.2012.04.02
15. Lee GM (2005) A survey of multipath routing for traffic engineering. In: Proceedings of LNCS 3391, vol 4. Springer, pp 635–661
16. Vutukury S (2001) Multipath routing mechanisms for traffic engineering and quality of service in the Internet. PhD Dissertation. University of California
17. Mersni A, Ilyashenko AE (2017) Complex criterion of load balance optimality for multipath routing in telecommunication networks of nonuniform topology. Telecommun Radio Eng 76(7):579–590. https://doi.org/10.1615/TelecomRadEng.v76.i7.20

18. Kirichenko L, Radivilova T (2017) Analyzes of the distributed system load with multifractal input data flows. In: 14th international conference the experience of designing and application of CAD systems in microelectronics (CADSM), Lviv, pp 260–264. https://doi.org/10.1109/cadsm.2017.7916130

19. Zaman RU, Shehnaz Begum S, Ur Rahman Khan K, Venugopal Reddy A (2017) Efficient adaptive path load balanced gateway management strategies for integrating MANET and the Internet. Int J Wirel Microwave Technol (IJWMT) 7(2):57–75. https://doi.org/10.5815/ijwmt.2017.02.06

20. Network performance objectives for IP-based services, ITU-T Recommendation Y.1541. ITU-T (2006)

21. Lemeshko O, Yeremenko O, Yevdokymenko M (2018) Tensor model of fault-tolerant QoS routing with support of bandwidth and delay protection. In: 2018 IEEE 13th international scientific and technical conference on computer sciences and information technologies (CSIT). Lviv, Ukraine, pp 135–138. https://doi.org/10.1109/stc-csit.2018.8526707

22. Lemeshko O, Yevsieieva O, Yevdokymenko M (2018) Tensor flow-based model of quality of experience routing. In: 2018 14th international conference on advanced trends in radioelecrtronics, telecommunications and computer engineering (TCSET). Lviv-Slavske, Ukraine, pp 1005–1008. https://doi.org/10.1109/tcset.2018.8336364

23. Oleksandr L, Olena N, Tetiana V (2016) Hierarchical coordination method of inter-area routing in telecommunication network. In: 2016 IEEE international scientific conference "radio electronics and info communications". (UkrMiCo 2016). Kyiv, Ukraine, pp 1–4. https://doi.org/10.1109/ukrmico.2016.7739626

24. Lemeshko O, Yeremenko O (2016) Dynamic presentation of tensor model for multipath QoS routing. In: The international conference modern problems of radio engineering, telecommunications and computer science. TCSET'2016. Lviv-Slavske, Ukraine, 23–26 Feb, Publishing House of Lviv Polytechnic, pp 601–604. https://doi.org/10.1109/tcset.2016.7452128

Printed in the United States
By Bookmasters